HYDROLOGY
An Introduction to
Hydrologic Science

HYDROLOGY
An Introduction to Hydrologic Science

Rafael L. Bras

Massachusetts Institute of Technology

ADDISON-WESLEY PUBLISHING COMPANY

Reading, Massachusetts • Menlo Park, California • New York
Don Mills, Ontario • Wokingham, England • Amsterdam • Bonn
Sydney • Singapore • Tokyo • Madrid • San Juan

This book is in the **Addison-Wesley Series in Civil Engineering**

Library of Congress Cataloging-in-Publication Data

Bras, Rafael L.
Hydrology: an introduction to hydrologic science.
Bibliography: p.
Includes index.
1. Hydrology. I. Title.
GB661.2.B7 1990 551.48 88-16793
ISBN 0-201-05922-3

Cover Illustration: Used with permission of Eric F. Wood and Dominique Thongs, Water Resources Program, Princeton University. Produced at the Interactive Computer Graphics Laboratory, Princeton University.

ABCDEFGHIJ-HA-89

To Rafael E. and Alejandro L.:

Build me a son, O Lord, who will be strong enough to know when he is weak, and brave enough to face himself when he is afraid; one who will be proud and unbending in honest defeat, and humble and gentle in victory.

Build me a son whose heart will be clear, whose goal will be high; a son who will master himself before he seeks to master other men; one who will reach into the future, yet never forget the past.

And after all these things are his, add, I pray, enough of a sense of humor, so that he may always be serious, yet never take himself too seriously. Give him humility, so that he may always remember the simplicity of true greatness, the open mind of true wisdom, and the meekness of true strength.

Then I, his father, will dare to whisper, 'I have not lived in vain.'

Excerpts from "A Father's Prayer" by Douglas MacArthur, written for his son Arthur.

Preface

> The remarkable levels of public health and safety enjoyed by the urban population of the developed world are due in considerable part to investments in hydrology over the past century. While we have spent lavishly to cope with the scarcity and excesses of water, we have invested little in the basic science underlying water's role in shaping and reshaping our planet. Hydrology, the science of water, has a natural place alongside oceanography, meteorology, geology, and others as one of the geosciences; yet in the modern science establishment, this niche is vacant. Why is this?*

The answer to the above question lies in the fact that "the elaboration of the field [of hydrology], the education of its practitioners, and the creation of its research culture" have been problem driven. There is nothing wrong with knowledge responding to problems. But when knowledge *only* follows problems, the human mind and condition stagnate, and in a self-fulfilling prophesy we become unable to predict the new challenges our water environment will bring.

It is a firm belief in hydrology as an earth science, as a useful earth science, that has driven my career and the writing of this book. I have been inspired by many with the desire to help inform people of what hydrology is. That challenge has come from a variety of places, some strange indeed. I will never forget arriving in New York in 1983 from a year's sabbatical abroad, including a three-week trip to the People's Republic of China. My family was

*From "Hydrology, the Forgotten Earth Science," by R. L. Bras and P. S. Eagleson, editorial in *EOS,* Vol. 68, No. 16, April 21, 1987.

received by your "friendly" immigration official. After responding the best I could to her question "What is hydrology?," I had to restrain my anger when she asked why the Chinese, or anyone else, would care about that. My dedication to the education of hydrologists was reaffirmed.

Although it has changed and developed considerably, the core of this book was ready nine years ago. At that time, hydrology in the United States was dominated by only two introductory textbooks, both of which had been on the market for a considerable amount of time. Many of us felt the need for a different approach, somewhere between the "engineering hydrology" of the time and Pete Eagleson's *Dynamic Hydrology,* which ahead of its time espoused an advanced and more scientific, but quantitative, view of hydrology. Circumstances conspired to delay my project. To the chagrin of my editor (sorry, Tom) in the past two years alone, more introductory hydrology books have been published than I care to remember. I should be upset at myself for losing the edge, for fumbling the opportunity to be the first of a new generation. But I am not. In whatever small way, I believe that my activities (yes, including drafting and redrafting this book and making my students suffer through the changes) have helped to spur renewed interest in hydrology. It is an exciting time for hydrologists, a time of major changes and opportunities. The field is beginning to carve its niche in the science establishment. We are moving, and it is fun to be part of it!

Like all efforts of this nature, this book has an ideal as a soul and compromise as its body. It is not all I wanted, nor will it be for anybody else. I intended it to be a textbook for advanced undergraduates or first-year graduate students. As a prerequisite it requires some Newtonian physics and some thermodynamic concepts. Elementary fluid mechanics is also needed. Introductory knowledge of probability and statistics would be helpful to the understanding of some ideas, in particular those of Chapter 11, but is not required. Metric units are favored, but in response to the reality of practice in the United States, other commonly used units are intentionally used.

The book emphasizes physical hydrology. The processes inherent to the hydrologic cycle are discussed. The origins and, to the level possible, the scientific foundations of the hydrologic processes are presented in a descriptive and quantitative way. The book is unique in keeping a reasonable balance in the discussion of the occurrence and movement of water in the atmosphere and land masses. More than the usual attention is given to a unified view of energy, climate, meteorology, and land–atmospheric fluxes like precipitation and evaporation (Chapters 2–6). Chapters 7 and 8 deal with subsurface waters, saturated and unsaturated. This knowledge is used to develop the concepts of infiltration and exfiltration as well as percolation and capillary rise. Chapter 9 conceptualizes the river basin response, leading to the definition of the discharge hydrograph. Chapter 10 discusses the flow of water in channels, which is necessary for streamflow routing and flood forecasting.

The interpretation of hydrologic variables as random is inherent to modern hydrologic practice. Chapter 11 introduces and presents ideas for the analysis of random variables, particularly oriented to the definition of the return period or recurrence of a hydrologic event. The river basin is the product of climate, geology, and hydrology. It is the reflection of nature's balance and dynamics. Chapter 12 ends with a discussion of hydrology and fluvial geomorphology, one of the many exciting frontiers in hydrology.

Although all material is intended to emphasize basic understanding, it remains very quantitive and useful. The student should finish this book with a practical working knowledge of hydrology that goes beyond the cookbook approach that has plagued us for many years.

Many hydrologic subjects are missing from this book, like any other. A balance had to be achieved between the book's goals, the nature and level of the subject matter, space, novelty, and the intended audience. In my opinion, the biggest gap is in the description of sediment transport and erosion. This is the result of a tradeoff in which increased attention to meteorology, snow, and groundwater carried the day.

Others will point out that there is little or no reference to models, the integration of concepts into the tools that make up the day-to-day life of the practitioner. This was to have been Part II of the book, but it lost out in the competition for space. Given my objectives and my view of what a solid introduction to hydrology is, I see this as an unfortunate but correct definition of priorities.

Drafts of this book have been used at MIT for many years (it has also found its way, via former students, to other universities). There is an extraordinary amount of material to cover in one semester. Generally, I go over the great majority of the subjects but not all of them. I have starred sections that are more expendable, either because they are elaborations of a subject or because they are secondary to the main concepts. Depending on the audience and geography, Chapter 6, Snowpack and Snowmelt, can be given variable emphasis; I nevertheless suggest some study of this chapter, however brief it may be. Chapter 12, Concepts of Fluvial Geomorphology, is not process-oriented and hence is not part of the mainstream. The temptation to treat Chapter 11, Concepts of Probability in Hydrology, in a similar way should be avoided. Probability is crucial to hydrology and should be the next level of sophistication to the student.

The book has a respectable collection of end-of-chapter problems. Commonly, these are used as didactic tools to either expand or introduce new concepts. The teacher should carefully select homework to achieve the goal of teaching by doing. Introductions and summaries to chapters should help in continuity of thought and link the variety of concepts presented.

Albert Einstein once said:

> A hundred times every day I remind myself that my inner and outer life are based on the labors of other men, living and dead, and that I must exert myself in order to give in the same measure as I have received and am still receiving.

These words are framed in the Arthur T. Ippen Conference Room at the Ralph M. Parsons Laboratory for Water Resources and Hydrodynamics at MIT. I cannot think of a better way to express my appreciation for the people that make up that laboratory and for the many other friends and colleagues I have. Sharing the dreams, the excitement, the opportunities, and the collegiality of the Ralph M. Parsons Lab has been the reward of a career. Having the honor of serving as its director fulfilled a dream. I want to particularly thank the students, both undergraduate and graduate, who for many years suffered the development of this book and the maturing of my thoughts. I have been blessed with extraordinary students. Seeing them succeed is my greatest enjoyment and satisfaction. Some have contributed particular problems and/or portions of text. They are acknowledged where the material appears. Dr. Carlos Puente was also very helpful in the development of some end-of-chapter problems.

Besides challenges and opportunities, hydrology has provided me with extraordinary friends and mentors. Even in competition, I can't think of a better set of colleagues. There are two who deserve special attention; I would like to mention Professor Peter S. Eagleson, a friend, example, and mentor, and Professor I. Rodriguez-Iturbe, my closest collaborator.

The career of an academician is impossible without the support of many agencies that sponsor research and facilitate our thinking. Many have provided me with financial support. The Office of Hydrology of the National Weather Service and the National Science Foundation (Engineering and Science Directorates) have always been there when I needed them.

For many years now, Elaine Healy has served as my personal secretary and friend. She went through three word processors and uncountable drafts of this book (not to mention my many other wild projects). Believe it or not, she actually likes it! If I ever write another book (two may be enough for me!) I hope to count on her excellent support again. Thanks, Elaine.

Tom Robbins, Sponsoring Editor, and Sherry Berg, Production Coordinator, have worked on my two books. I thank them for their efforts and confidence.

Versions of this book have been reviewed by many. Drs. Stephen J. Burges, University of Washington; Daniel D. Evans and Soroosh Sorooshian, University of Arizona; Konstantine P. Georgakakos, University of Iowa, Iowa Institute of Hydraulic Research; and Antonis D. Koussis, Vanderbilt University, provided formal reviews and criticisms that were very helpful in preparing the final manuscript.

Finally, I want to acknowledge the love of my wife, Pat, and my children, Rafael E. and Alejandro L., and my parents, Amalia and Rafael. Without them any project would be senseless.

Cambridge, Massachusetts R.L.B.

Contents

1 INTRODUCTION 1

1.1 Introduction 1
1.2 The Hydrologic Cycle and the Hydrologic Budget Equation 3
1.3 Scope of Work 12
1.4 Summary 13
References 13
Problems 14

2 SOLAR RADIATION AND THE EARTH'S ENERGY BALANCE 19

2.1 Introduction 19
2.2 Planetary Motions and the Distribution of Radiation 21
2.3 Radiation Physics 31
2.3.1 Shortwave Radiation 34
2.3.2 Longwave Radiation 42
2.4 Summary 47
References 47
Problems 49

3 PRINCIPLES OF METEOROLOGY: THE EARTH–ATMOSPHERE SYSTEM 53

3.1 Introduction 53
3.2 Composition and General Characteristics of the Atmosphere 54
3.3 Transport Processes 56
3.4 Temperature Distribution 58
3.4.1 Temporal Distribution 58
3.4.2 Horizontal Distribution 60
3.4.3 Vertical Temperature Distribution in the Lower Atmosphere 63
3.4.4 Temperature Measurements 66
3.5 Pressure Distribution 68
3.6 Advection by Winds and Ocean Currents 73
3.6.1 Atmospheric Circulation 73
3.6.2 Circulation in the Oceans 80
3.7 Atmospheric Humidity 82
3.7.1 The Phases of Water 82
3.7.2 Vapor Pressure and Humidity 84
3.7.3 Measurement and Estimation of Humidity and Vapor Pressure 86
3.7.4 Distribution of Atmospheric Moisture 88
3.8 Atmospheric Stability and Condensation 92
3.8.1 Adiabatic Cooling 93
3.8.2 Condensation by Pseudo-adiabatic Cooling 95
3.8.3 Further Comments on Thermal Convection and Stability 100
3.9 Summary 102
References 103
Problems 104

4 PRECIPITATION OCCURRENCE AND MEASUREMENT 109

4.1 Introduction 109
4.2 Cooling and Lifting Processes 110
4.3 An Introduction to Cloud Physics 116
4.3.1 Nucleation 116
4.3.2 Growth and Distribution of Precipitation Particles 117
4.3.3 Terminal Velocities of Hydrometeors 120
4.3.4 Evaporation of Precipitating Hydrometeors 120
4.4 Forms of Precipitation 126

4.5 Storm Structure 127
4.6 Measurement of Precipitation 132
4.6.1 Gages 132
4.6.2 Radar 134
4.6.3 Satellites 140
4.7 Precipitation Data Analysis 146
4.7.1 Estimation of Missing Data 149
4.7.2 Consistency Checks 153
4.7.3 Mean Areal Precipitation 154
4.7.4 Frequency Analysis 157
4.7.5 Network Design 158
4.8 Summary 170
References 171
Problems 175

5 EVAPORATION, TRANSPIRATION, INTERCEPTION, AND DEPRESSION STORAGE 183

5.1 Introduction 183
5.2 Evaporation from Free Water Surfaces 188
5.2.1 Water Balance Method 189
5.2.2 Energy Balance Method 190
5.2.3 Mass-Transfer Methods: The Dalton Law Analogy 197
5.2.4 Combined Mass-Transfer and Energy Methods: The Penman Equation 201
5.2.5 Empirical Equations 203
5.2.6 Direct Measurement of Evaporation 210
5.3 Transpiration and Evapotranspiration 219
5.4 Evaporation from Snow 231
5.5 Interception 232
5.6 Depression Storage 234
5.7 Summary 236
References 237
Problems 241

6 SNOWPACK AND SNOWMELT 247

6.1 Introduction 247
6.2 Snow Accumulation and Measurement 248
6.3 Snowpack 256
6.3.1 Density 257
6.3.2 Cold Content 259

6.3.3 Thermal Quality 260
6.3.4 Liquid-Water Content 261
6.3.5 Albedo 262
6.4 Energy Budget and Snowmelt 264
6.4.1 Net Radiation 265
6.4.2 Advected Heat in Precipitation 265
6.4.3 Energy Consumed in Evaporation, Condensation, and Sensible-Heat Transfers 266
6.4.4 Heat of Conduction from the Soil 267
6.4.5 Energy Released by Freezing of Liquid-Water Content 268
6.5 Air Temperature as an Index of Snowmelt 268
6.6 Routing of Melt through Snowpack 275
6.7 Summary 278
References 278
Problems 280

7 GROUNDWATER FLOW IN SATURATED POROUS MEDIA 283

7.1 Introduction 283
7.2 The Soil–Rock Profile and Subsurface Waters 284
7.3 Darcy's Law 290
7.4 Mass Balance Equations—Flow in Saturated Porous Media 294
7.4.1 Confined Aquifers 294
7.4.2 Unconfined Aquifer 296
7.4.3 Horizontal-Plane Flow and the Dupuit Approximation 296
7.4.4 Initial and Boundary Conditions 300
7.4.5 Linearity and the Superposition Principle 300
7.5 Hydraulics of Wells 313
7.5.1 Steady-State Solution of a Fully Penetrating Well in a Confined Aquifer 313
7.5.2 Steady-State Solution of a Fully Penetrating Well in an Unconfined Aquifer 316
7.5.3 Unsteady Flow in Wells 317
7.5.4 Wells in Leaky Aquifers 323
7.5.5 Superposition of Wells and the Method of Images 323
7.5.6 Aquifer Tests 325
7.6 Summary 336
References 337
Problems 339

8 FLOW IN UNSATURATED POROUS MEDIA AND INFILTRATION 349

8.1 Introduction 349
8.2 Flow in Unsaturated Porous Media 350
8.2.1 Conservation of Mass in Unsaturated Porous Media 352
8.3 Infiltration and Exfiltration 355
8.3.1 Empirical Infiltration Equations 362
8.3.2 Storm Runoff 368
8.3.3 Actual Evaporation 377
8.4 Percolation and Capillary Rise 382
8.5 Summary 384
References 385
Problems 388

9 THE HYDROGRAPH AND SIMPLE RAINFALL–DISCHARGE RELATIONSHIPS 395

9.1 The Hydrograph 395
9.2 Hydrograph Separation 399
9.3 Streamflow Measurements 401
9.4 Rainfall–Discharge Relationships 404
9.4.1 Peak Discharge Formulas—The Rational Formula 405
9.4.2 The Unit Hydrograph 409
9.4.3 Synthetic Unit Hydrographs 419
9.5 The Instantaneous Unit Hydrograph 430
*9.5.1 Fourier Series 432
*9.5.2 Fourier and Laplace Transforms 436
9.5.3 Moments and Cumulants 437
9.6 Conceptual Instantaneous Unit Hydrographs 441
9.7 Summary 449
References 450
Problems 453

10 FLOOD ROUTING 465

10.1 Routing 465
10.2 Conceptual Models 466
10.2.1 Channel Routing: The Muskingum Method 466

10.2.2 Reservoir Routing 475
10.3 Hydraulic Routing: The St. Venant Equations 478
10.3.1 Solutions to St. Venant Equations 482
10.3.2 Numerical Solutions 486
10.4 Black-Box Models 491
10.5 The Diffusion Analogy 492
10.6 Summary 494
References 494
Problems 496

11 CONCEPTS OF PROBABILITY IN HYDROLOGY 505

11.1 Introduction 505
11.2 Review of Probability 506
11.3 Models of Probability 515
11.3.1 Models of Discrete Random Variables 515
11.3.2 Models of Continuous Random Variables 521
11.4 Nonparametric Estimates of Exceedance Probability 543
*11.5 Novel Approaches and Future Directions 545
11.5.1 Derived Distributions 545
11.5.2 Regional Analysis 548
11.5.3 Paleohydrology and the Value of Historical Information 550
11.6 Summary 554
References 555
Problems 557

12 CONCEPTS OF FLUVIAL GEOMORPHOLOGY 567

12.1 Introduction 567
12.2 Descriptions of Drainage Basin Composition 568
12.2.1 Two-Dimensional Planar Descriptors 568
12.2.2 Descriptors of Relief 582
12.2.3 Stream Channel Geometry 587
*12.3 Fluvial Geomorphology and Hydrology 589
12.3.1 Geomorphologic Instantaneous Unit Hydrograph 590

12.3.2 Geomorphoclimatic Instantaneous Unit Hydrograph 597

12.3.3 Comments and Further Developments of the Geomorphologic Instantaneous Unit Hydrograph 599

12.3.4 Link-Based Derivations of the Geomorphologic Instantaneous Unit Hydrograph 602

12.4 Summary 605

References 605

Problems 610

APPENDIX A TABLES OF WATER PROPERTIES 619

APPENDIX B DEVELOPMENT OF UNSTEADY FLOW EQUATIONS FOR SATURATED MEDIA 629

INDEX 633

HYDROLOGY
An Introduction to Hydrologic Science

Chapter 1

Introduction

1.1 INTRODUCTION

Hydrology is the study of water in all its forms and from all its origins to all its destinations on the earth. Although the hydrologic umbrella would include water-quality issues, this work will essentially concentrate on the questions of water quantity.

As a requirement for life as we know it, water has been a source of continuous preoccupation for humans since the beginning of mankind. The same questions and issues of the past are prevalent today. How much water is there? Where is the water coming from? Where is it going? What is the quality of the water and how can we control it? What should we do when we have too much or too little of it?

Since our elementary-science-course days, we all have a feeling for the movement and location of water. The ocean is clearly recognized as the biggest source of water. Rainfall is always associated with streamflow as well as with climatic and meteorologic phenomena. Water losses due to evaporation and infiltration are not abstract concepts. But these seemingly trivial ideas were not always so clearly understood; our predecessors in the study of hydrology did not reach this level of knowledge until relatively recent times.

The history and development of hydrology are fascinating subjects. The reader is referred to an excellent book by Biswas [1972] on the history of hydrology. A few remarks are useful to create the correct historical perspective.

Early thinkers and philosophers did not understand three basic hydrologic principles (Eagleson [1970]):

1. conservation of mass,
2. evaporation and condensation, and
3. infiltration.

They were worried about how water gets up to the mountains, flows down to the sea, and fails to raise the level of the latter. Because of what may be called limited spatial awareness, they could not see rainfall as a sufficient source of streamflow. To account for observed water behavior, underground reservoirs (beneath mountains) were hypothesized. Water was believed to be pushed up the mountains by vacuum forces, capillary action, or "rock pressure" and surfaced as streamflow. The underground reservoirs were replenished by the sea.

Vitruvius, during the first century B.C., stated that the mountains received precipitation that then gave rise to streamflow. A filtration process by which water percolated into soil was also acknowledged by Vitruvius and later by da Vinci.

It was in the seventeenth century that Perrault proved by measurement that precipitation could account for streamflow in the Seine River, France. Similar quantitative studies were made by Mariotte and Halley during the same historical period. At this stage, the mass balance concept was pretty well established, although questioning of it continued well into the twentieth century.

The eighteenth century saw advances in hydraulics and the mechanics of water movement by Bernoulli, Chezy, and many others. The nineteenth century saw experimental work on water flow by people like Darcy and Manning. The above names are familiar to students of groundwater and surface-water movement.

Until the 1930s hydrology remained a science filled with empiricism, qualitative descriptions, and little overall understanding of ongoing processes. At that time, people such as Sherman [1932a] and Horton [1940] initiated a more theoretical, quantitative, approach. Sherman's unit hydrograph concept still remains with us as the most successful (but not necessarily the best) and most well-known explanation of river-basin behavior. Horton's ideas on infiltration, soil-moisture accounting, and runoff are still recognized by present-day hydrologists.

All these centuries of experience and study have converged to form the concept of the hydrologic cycle. The concept is simply that water changes state and is transported in a closed system: the earth and its atmosphere. The cycle is closed only earthwide, each drop of water following a path from the ocean to the atmosphere to the earth (through surface or underground movement). Energy to keep this cycle going is provided by the sun. Processes involved are evaporation, condensation, precipitation, infiltration, and runoff.

The practicing hydrologist is usually concerned with local conditions and is therefore facing an open system. Nevertheless, unless the basic hydrologic cycle is recognized, he or she will again wonder how the water gets up the mountain.

1.2 THE HYDROLOGIC CYCLE AND THE HYDROLOGIC BUDGET EQUATION

A schematic view of the global hydrologic cycle is shown in Figure 1.1. This diagram shows the interactions and mass transfers (water in different states) that occur between the atmosphere, land surfaces, and the oceans. Note that water appears in liquid, solid, and gaseous states. Emphasis is given to processes on or within the land surface with no detail of water-transport mecha-

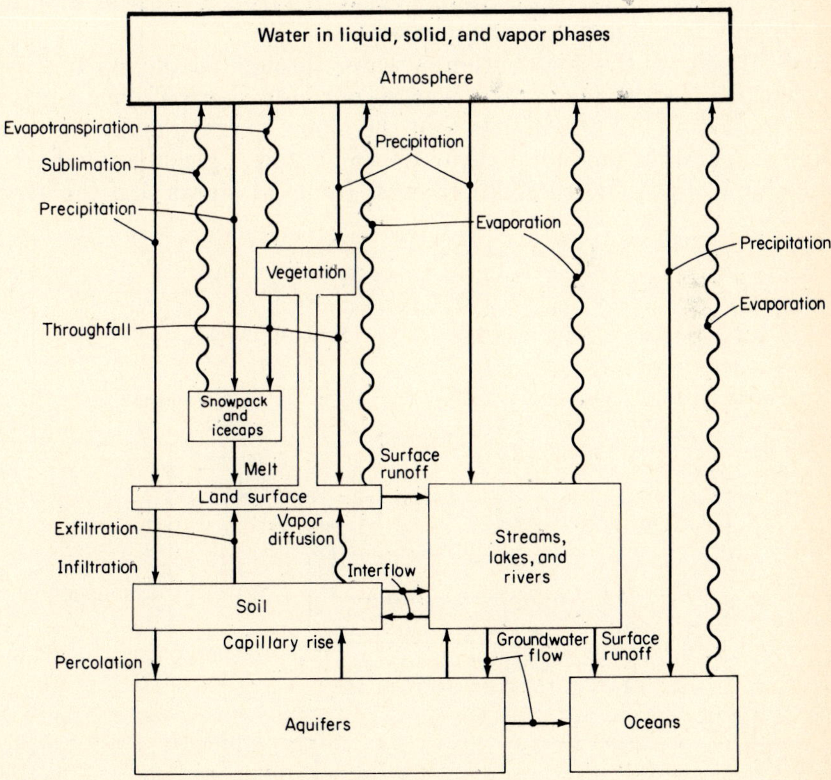

FIGURE 1.1 A schematic view of the hydrologic cycle. Transport of water as vapor is indicated by wavy lines. Source: Eagleson [1970].

nisms operating within the atmosphere and oceans. These details will be discussed in later chapters and are extensively covered in the associated fields of meteorology and oceanography.

A few possible new terms need initial definition. *Sublimation* is the change of ice to vapor. *Throughfall* is the water not intercepted by vegetation. *Evapotranspiration* is the combined consumptive–evaporative process by which water is released to the atmosphere through vegetation and soil. *Exfiltration* is the rising of soil moisture due mostly to tension and capillary forces. *Interflow* is the water flow at shallow depths within the soil structure. *Infiltration* is water absorption by the soil surface. *Percolation* refers to water movement into deep groundwater reservoirs called aquifers.

Figure 1.2 indicates the magnitudes and distribution of global annual average precipitation. In the figure water quantities are measured in volume per unit area of land or ocean. For example, 31 cm of runoff over land is equivalent to 13 cm over the oceans, given the much larger ocean area. Note that more water falls directly into the oceans than over land (30% of total surface area). Fifty-seven percent (41/72) of precipitation falling on land never reaches the ocean. More water evaporates from the ocean than it receives directly in the form of precipitation.

Table 1.1 shows the distribution of water throughout the earth. Clearly, the oceans and icecaps dominate as sources of water. Using Table 1.1 and Figure 1.2, Eagleson [1970] computes that the global average annual precipitation and evaporation each total about 100 cm, giving a global annual precipitation volume of 511,000 cubic kilometers. Table 1.1 shows that the average

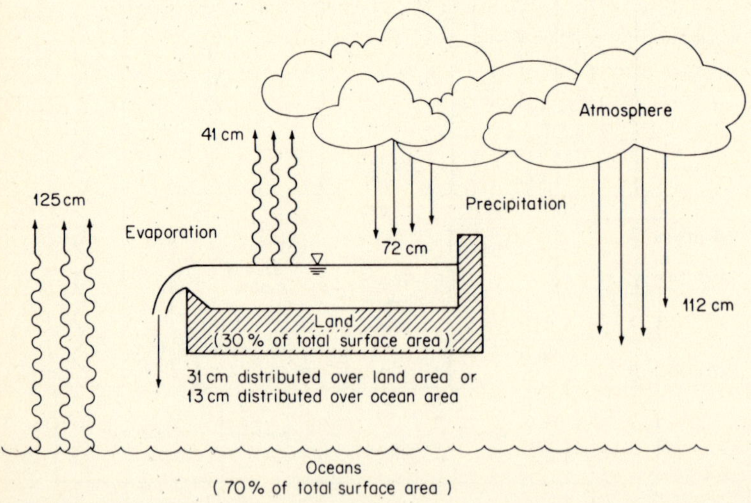

FIGURE 1.2 Disposition of global annual average precipitation. Source: Eagleson [1970]. Data from Budyko et al. [1962].

TABLE 1.1 Distribution of World's Estimated Water Resources

LOCATION	SURFACE AREA (km^2)	WATER VOLUME (km^3)	PERCENTAGE OF TOTAL WATER
Surface water			
Freshwater lakes	855,100	125,100	0.009
Saline lakes	699,700	104,300	0.008
Stream channels	....	1,300	0.000
Subsurface water			
Groundwater (less than ½ mi deep)	129,565,000	4,171,400	0.307
Groundwater (more than ½ mi deep)	129,565,000	4,171,400	0.307
Soil moisture, etc.	129,565,000	66,700	0.005
Icecaps and glaciers	17,880,000	29,199,700	2.147
Atmosphere (at sea level)	510,486,000	12,900	0.000
Oceans	361,486,000	1,322,330,600	97.217
Approximate totals		1,360,183,400	100.000

Source: Adapted from R. L. Nace, "Water of the World," *Natur. Hist.*, Vol. 73, No. 1, January 1964.

atmospheric moisture content is 12,900 km^3. So atmospheric moisture must be replaced on the average 40 times a year, which implies a moisture residence time of nine days. Quite an active cycle indeed!

The hydrologist usually faces an open system of the type illustrated in Figure 1.3. The quantification of the hydrologic cycle in such a system becomes a simple mass balance equation, where inputs are equal to the change in storage. The general water budget equation is then

$$\frac{dS}{dt} = I - Q \tag{1.1}$$

where I is inputs and Q represents outputs.

The hydrologist must be careful in defining the region or control volume over which Eq. (1.1) is to be written. Only then can the terms comprising the inputs I and outputs Q be defined.

Generally, the mass balance can be written over surface and underground water systems. Using Figure 1.3 as a control volume, the surface budget equation is

$$P + Q_{in} - Q_{out} + Q_g - E_s - T_s - I = \Delta S_s , \tag{1.2}$$

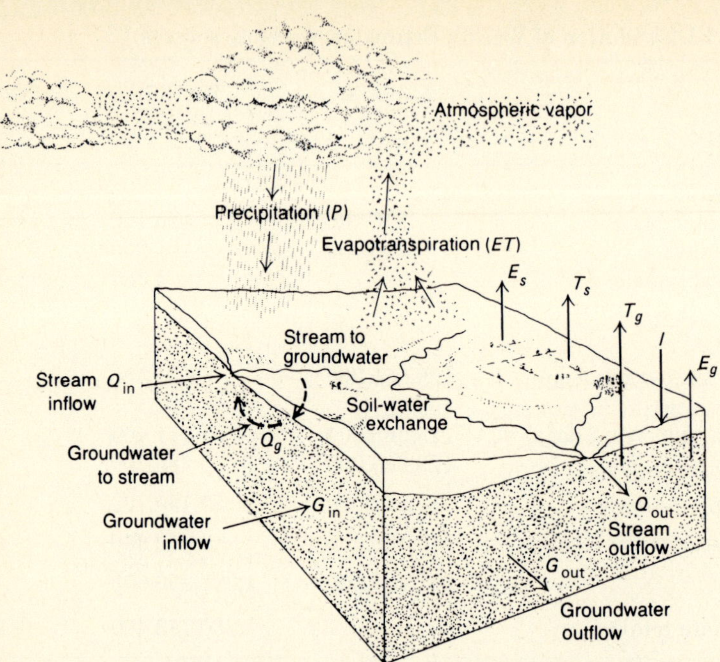

FIGURE 1.3 Components of hydrologic cycle in an open system: the major inflows and outflows of water from a parcel of land. Source: W. M. Marsh and J. Dozier, *Landscape: An Introduction to Physical Geography*. Copyright © 1986 by Wiley. Reprinted by permission of John Wiley & Sons, Inc.

where P is precipitation over the period of interest, Q_{in} and Q_{out} are surface-water flows into and out of the control volume over the period of interest, Q_g is groundwater rate of flow into surface streams, E_s is surface evaporation rate, T_s is plant transpiration rate of surface moisture, I is infiltration rate, and ΔS_s is change in water storage on the surface over the period of interest.

Similarly, the groundwater budget equation is

$$I + G_{in} - G_{out} - Q_g - E_g - T_g = \Delta S_g \, , \tag{1.3}$$

where G_{in} and G_{out} are groundwater flow rates in and out of the control volume. All other variables have definitions equivalent to the surface counterparts but refer to water in and out of the ground.

The total mass balance, obtained by adding both equations, is

$$P - (Q_{out} - Q_{in}) - (E_s + E_g) - (T_s + T_g) - (G_{out} - G_{in}) = \Delta(S_s + S_g) \, . \tag{1.4}$$

The units of the above equations are volume per unit time. Using net mass exchanges, Eq. (1.4) can be summarized as

$$P - Q - G - E - T = \Delta S,$$

or

$$P - Q - G - ET = \Delta S, \tag{1.5}$$

where ET is a combined evapotranspiration term.

Generally, more than one of the elements of Eq. (1.5) are unknown, making the solution of the algebraic equation impossible. Only in cases where reasonable approximations of the majority of the values can be made or where one or more terms can be related is it possible to solve Eq. (1.5).

The simple hydrologic equation can, nevertheless, lead to highly significant and useful results. An example is the study of the potential water yield of swampy regions in the upstream reaches of the Nile River Basin. The point of complete utilization of the Nile River streamflows has nearly been reached. Therefore Egyptian and Sudanese water planners are seriously studying the channelization and drainage (one such project, the Jonglei Canal, was initiated) of some of these swampy regions in order to decrease water losses and increase annual water yield into Lake Nasser (and the Aswan High Dam System) in Egypt, which is the main water-storage facility for both countries. The reservoir holds several years of Nile inflows.

Figure 1.4 gives the general location of the Machar marshes near the Sobat and Baro rivers; the swamps along the Bahr El Jebel and Bahr El Zeraf tributaries; and the Bahr El Ghazal region, covered by swamps near its outlet to Lake No, where it joins the main White Nile channel. In order to assess the potential of these regions as water sources, detailed water-balance studies are required.

Chan and Eagleson [1980] used published literature to make an initial water balance in the permanent swamp regions of the three areas involved. Their summary appears in Table 1.2. They used average annual quantities; they assumed that over the year, on the average, the change of storage in the system is zero. This is a reasonable assumption over many years in a hydrologic system dominated by an annual meteorologic and climatic cycle. Therefore ΔS in Eq. (1.5) is taken as zero. They had no information on lateral groundwater movement. It is then assumed that either the groundwater systems of these hydrologic units are fully enclosed within the chosen unit volume of study (which is not really the case in this particular situation) or more reasonably that the groundwater system is in reasonable steady state and dominated by the other inputs and outputs to the system. These arguments lead to $G = 0$ in Eq. (1.5). Surface inputs to the system are of two types: gaged inflow through channels and uncontrolled over-bank spillage from

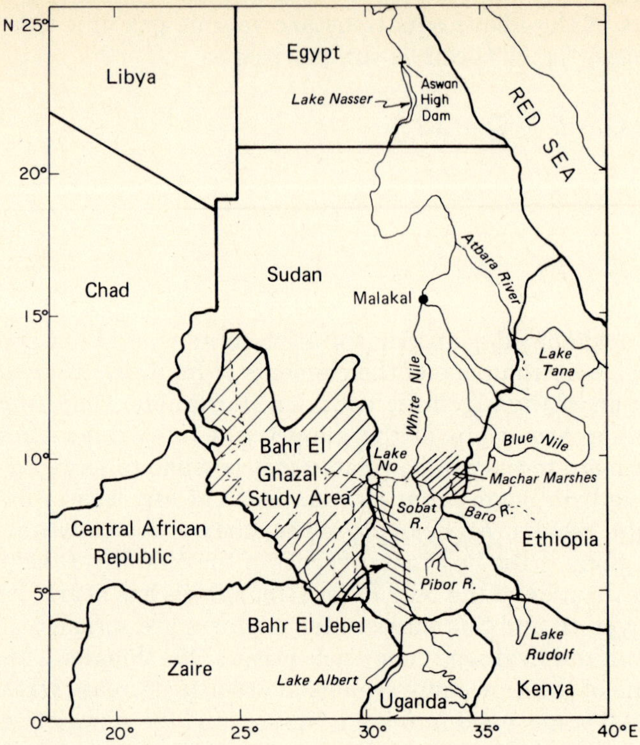

FIGURE 1.4 The Nile River in the Sudan, showing location of marsh areas. Source: Chan and Eagleson [1980].

channels into the swamps. The hydrologic equation for all three systems then becomes (in units of 10^9 cubic meters or milliards [md]):

	$P - Q$	ET
Machar	7.3 − (0.1 − 3.5 − 2.0)	12.7
Jebel-Zeraf	7.5 − (14.3 + 6.0 − 27.0)	14.2
Ghazal	15.0 − (0.6 − 6.0 − 12.7)	33.1

In units of depth per unit area of swamp, Table 1.2 indicates annual evapotranspired water depth of 1.5 m for Machar, 1.7 m for Jebel-Zeraf, and 2.0 m for Ghazal.

The above analysis ignores potentially important terms such as ΔS and G. Nevertheless, Chan and Eagleson [1980] point out that these permanent swamp regions are covered by papyrus plants and tall grasslands. Independent assessment of the evapotranspiration from these plants yields values of 2.2 m for papyrus and 1.9 m for a composite of papyrus and grasslands weighted by area. These numbers agree with estimated water losses and serve as checks on the reasonableness of the assumptions made.

TABLE 1.2 Apparent Water Losses from Major Nile Swamps

LOCATION	AREA OF PERMANENT SWAMP	PRECIPITATION	GAGED INFLOW	ESTIMATED SPILLAGE	OUTPUT GAGED	ESTIMATED LOSSES		LOSSES TO WHITE NILE FLOW
	km^2	md*	md	md	md	md	m	
Machar	8,700	7.3	2.0	3.5	0.1	12.7	1.5	0.5
Jebel-Zeraf	8,300	7.5	27.0	−6.0	14.3	14.2	1.7	0.5
Ghazal	16,600	15.0	12.7	6.0	0.6	33.1	2.0	1.2

*1 milliard (md) = 10^9. Here it will always refer to the number of cubic meters per year.

Source: Adapted from Chan and Eagleson [1980].

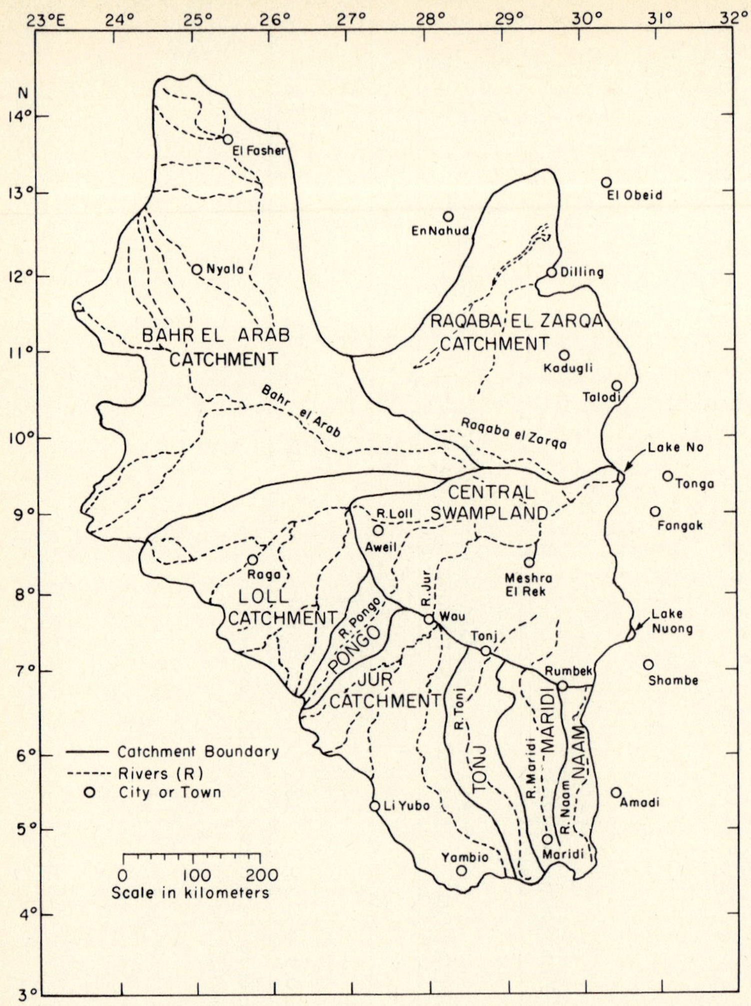

FIGURE 1.5 The Bahr El Ghazal Basin and demarcation of boundaries of subcatchments. Source: Chan and Eagleson [1980].

Chapter 5 will discuss methods to estimate evapotranspiration from various surfaces and vegetation. Most impressive to the new student of hydrology should be the proportion of annual water flow in the White Nile that is lost, mainly through evapotranspiration. The last column of Table 1.2 gives the ratio of losses to White Nile flow before its junction with the Blue Nile (see Fig. 1.4). Note that in the Bahr El Ghazal region losses are more than the gaged average annual streamflow of the Nile River at Malakal (see Fig. 1.4) and over 30% of the annual average flow of the Nile at Aswan (84×10^9 m^3). The potential water gains from reduced evapotranspiration are immense!

In order to further define how some of this evapotranspired water could be recovered, Chan and Eagleson [1980] studied the Bahr El Ghazal region as a whole, taking into account the central swampland and the feeding tributary catchments, as shown in Figure 1.5. Individual water balances were performed for each tributary basin. Still taking ΔS and G as initially zero in Eq. (1.5), the rest of the terms—P, Q, and ET—were independently assessed, to the extent possible, using available data on precipitation, streamflow, evapotranspiration, and distribution of vegetation. The result of this exercise is given in Figure 1.6, where P, ET, and Q denote annual average precipitation, evapotranspiration, and surface runoff, respectively. Water spilled into the Bahr El Ghazal system from the nearby Jebel system is given by D_O. Groundwater seepage or unattributed inflows into the central swampland is represented by G_L. Gaged surface input into the central swampland is represented by Y_L. Subscripts L, O, and B indicate inputs or outputs into tributary systems, the central swampland, and the basin as a whole, respectively. Numbers in parentheses are water volumes per unit area (in millimeters) over tributary, swampland, or total basin areas as indicated by the

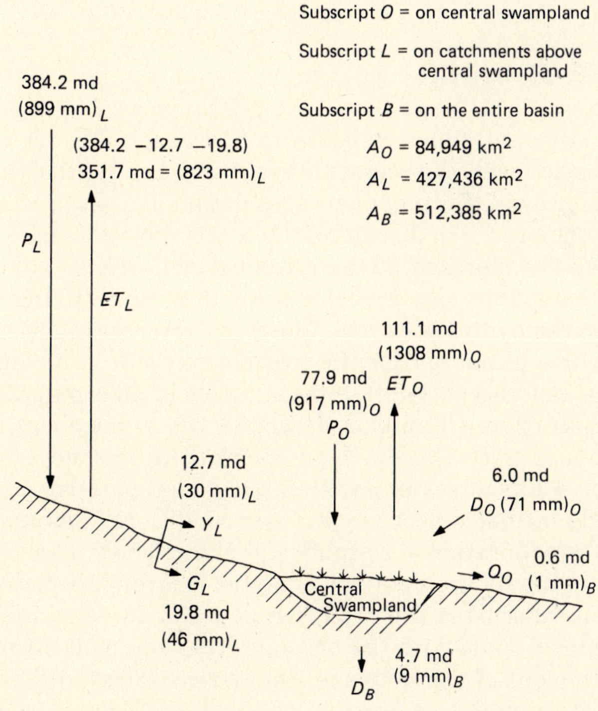

FIGURE 1.6 Mean annual water balance of the Bahr El Ghazal Basin. Source: Chan and Eagleson [1980].

subscripts. The hydrologic budget equation as applied to the swampland becomes,

$$P_O - ET_O + D_O - Q_O - D_B + Y_L = -G_L .$$

All the above quantities were independently assessed except G_L, which is groundwater seepage into the central swampland or ungaged surface flows. Substitution of the quantities yields $G_L = 19.8$ md m^3. Evapotranspiration in the tributaries, ET_L, was then obtained by solving the hydrologic budget equation for the tributary systems

$$ET_L = P_L - Y_L - G_L$$

leading to $ET_L = 351.7$ md m^3.

In summary, the maximum amount of water (annual average) that could possibly be recovered by intercepting all flows (surface and groundwater) into the central swampland of the Bahr El Ghazal is on the order of 32.5×10^9 m^3 $(19.8 \times 10^9 + 12.7 \times 10^9)$.

1.3 SCOPE OF WORK

In contrast to other introductory books in hydrology, this one intends to emphasize, at a basic level, the scientific reasoning behind the various subjects in hydrology. However, this emphasis does not imply that practical and useful procedures are ignored. Furthermore, there are subjects where empiricism is the only reasonable approach at the introductory level.

The first ten chapters cover basic principles, quantifying the parts and the processes of the hydrologic cycle. Chapter 2 discusses solar radiation. This is the main source of energy for the hydrologic cycle. It dictates the earth's energy balance and also plays an important role in determination of evaporation and transpiration. Chapter 3 describes the atmosphere as an equally important partner with the earth in controlling the movement of water. Emphasis is on a global scale, pointing out global patterns and phenomena that can and do influence local hydrologic conditions. Chapter 4 covers the main forms of precipitation—rainfall and snow. Their genesis, characteristics and measurements are detailed. Chapter 5 covers evaporation and transpiration. It also discusses the retention of water in vegetation and surface ponding. Chapter 6 deals with the accumulation and melt of snow. Chapter 7 covers the movement of water in the soil system. It describes the occurrence and dynamics of groundwater flow and discusses groundwater flow under saturated conditions and the hydraulics of wells. Unsaturated flow equations and infiltration are studied in Chapter 8. Chapter 9 quantifies surface runoff from precipitation and snowmelt. Chapter 10 deals with the mechanics of flow in channels and flood routing. Emphasis is on theory and methodologies. In summary, the first ten chapters provide insight and tools for independent evaluation of the various terms in the water balance (Eq. 1.5). Chapter 11 introduces the concepts of probability most commonly used in the analysis of the frequency of occurrence of hydrologic events. Chapter 12 talks about river basin geometry and fluvial geomorphology and its implications in hydrology.

Not all the material presented here is necessary for an introductory study of hydrology. Advanced topics are indicated throughout the book with an asterisk. The Preface will help guide the teacher, student, and general reader in the selection of topics.

1.4 SUMMARY

Hydrology is the study of water in all its forms, from all its origins, to all its destinations on the earth. Traditionally, hydrologic science has concentrated on the fluxes between land surfaces and the atmosphere and oceans. Many of the details of the study of the processes within the atmosphere and oceans must be left to the sister disciplines of meteorology and oceanography, respectively. Nevertheless, the expediency of this separation should not be interpreted as encouragement for sharp disciplinary boundaries. The earth processes do not recognize such artificial groupings.

This chapter introduces the hydrologic cycle and its elements. The concept of conservation of mass within this truly active cycle is presented and illustrated with an example. From the example the reader should come to the realization that it is generally hard to quantify some of the elements of the hydrologic cycle. The chapter also intends to provide a feeling of the magnitude, at global and local scales, of some of the elements of the cycle.

Most of the rest of the book will be spent in detailing and quantifying the processes that control the hydrologic cycle. We will begin in Chapter 2 by discussing the fuel of this cycle — solar radiation — and the related energy balance on global and local scales.

REFERENCES

Biswas, A. K. [1972]. *History of Hydrology.* Amsterdam: North-Holland.

Branson, F. A., G. F. Gifford, K. G. Renard, and R. F. Handley [1981]. *Rangeland Hydrology.* Dubuque, Iowa: Kendall/Hunt.

Budyko, M. I., N. A. Efimova, L. I. Zubenok, and L. A. Strokina [1962]. "The Heat Balance of the Earth's Surface." *Akad. Navk. USSR, I-Z v. Ser. Georgr. No. 1.*

Chan, S.-O., and P. S. Eagleson [1980]. "Water Balance Studies of the Bahr El Ghazal Swamp." Cambridge, Mass.: MIT Department of Civil Engineering, Ralph M. Parsons Laboratory. (Technical report no. 261.)

Chow, V. T., ed. [1964]. *Handbook of Applied Hydrology*. New York: McGraw-Hill.

Eagleson, P. S. [1970]. *Dynamic Hydrology*. New York: McGraw-Hill.

Eagleson, P. S., and S.-O. Chan [1979]. "Water Balance Estimates of a Sudd Tributary." *Proc. Conf. Water Resources Plan. Egypt*. Cairo: CU/MIT Technological Planning Program. June 25–27, 1979, p. 538.

Gray, D. M., ed. [1973]. *Handbook on the Principles of Hydrology*. Port Washington, N.Y.: National Research Council of Canada. (Reprinted by Water Information Center, Inc., Port Washington, N.Y.)

Haan, C. T., H. P. Johnson, and D. L. Brakensiek, eds. [1982]. *Hydrologic Modelling of Small Watersheds*. St. Joseph, Mich.: American Society of Agricultural Engineers. (ASAE monograph no. 5.)

Horton, R. E. [1935]. "Surface Runoff Phenomena: Part I. Analysis of the Hydrograph." Ann Arbor, Mich.: Edwards Brothers, Inc. (Horton Hydrology Laboratory, series no. 101.)

Idem. [1938]. "The Interpretation and Application of Runoff Plot Experiments with Reference to Soil Erosion Problems." *Soil Sci. Soc. Am. Proc*. 3:340–349.

Idem. [1940]. "An Approach Toward a Physical Interpretation of Infiltration Capacity." *Soil Sci. Am. Proc*. 5:399–417.

Marsh, W. M., and J. Dozier [1986]. *Landscape: An Introduction to Physical Geography*. New York: Wiley.

Nace, R. L. [1964]. "Water of the World." *Natur. Hist*. 73(1).

Linsley, R. K., Jr., M. A. Kohler, and J. L. H. Paulhus [1982]. *Hydrology for Engineers*. 3rd ed. New York: McGraw-Hill.

Raudkivi, A. J. [1979]. *Hydrology—An Advanced Introduction to Hydrological Processes and Modelling*. Oxford: Pergamon.

Sherman, L. K. [1932a]. "Stream Flow from Rainfall by the Unit-Graph Method." *Engin. News. Rec*. 108:501–505.

Idem. [1932b]. "The Relation of Hydrographs of Runoff to Size and Character of Drainage Basins." *Am. Geophys. Union Trans*. 13:332–339.

Viessman, W., Jr., J. W. Knapp, G. L. Lewis, and T. E. Harbaugh [1972]. *Introduction to Hydrology*. 2nd ed. New York: Harper & Row.

PROBLEMS

1. Unfortunately, hydrologists throughout the world use a variety of units. Although S.I. and metric systems are fairly universal, it is useful to be familiar with the most common unit conventions. Some exercises to help achieve that familiarity follow.

a) Volume per unit time is commonly measured in cubic feet per second (cfs). What is the equivalent, in cubic feet per second, of 100 cubic meters per second ($m^3 s^{-1}$)?

b) The concept of volume is commonly expressed in terms of a volume per unit area, or a depth. That is generally the case in measuring rainfall over a known area, like a river basin. A fairly wet region may receive about 1700 mm of rainfall per year. How many inches of rain fall in the area?

c) Boston gets about 40 in. of rainfall per year. How many centimeters is that?

d) If 40 in. of rain fall per year over a river basin that is 1000 km^2 in area, what is the volume of water received over a year in cubic meters?

e) An acre is about 4000 m^2 (4047, to be exact) and a hectare (ha) is 10,000 m^2 or about 2.5 acres. In the United States, a commonly used unit of volume is the acre-ft or the volume of water required to cover one acre of land with water 1 ft deep. How many cubic feet in an acre-ft? How many cubic meters?

f) Another commonly used unit of volume is the U.S. gallon. The city of Boston water-supply system can safely handle about 300 million gallons a day (mgd) of demand. How many cubic meters per second is that?

g) The mile (1 mi = 5280 ft) is another common unit of length. How many acres to a square mile? How many square kilometers to a square mile?

2. 1.25×10^4 m^3 of runoff occurs from a 100-ha plot of uniformly sloping land in a certain half-hour period during which the rainfall averages 10 cm hr^{-1}. Compute the magnitude of change in storage in cubic meters and tell the probable forms of storage into which it goes during the half hour.

3. A lagoon has a surface area of 350.5×10^6 m^2. The average annual rainfall and evaporation are obtained as 1850.4 mm and 1142.7 mm, respectively. If the increase in storage is 247.8×10^6 $m^3 yr^{-1}$, obtain the net annual inflow into the lagoon. What are the hydrologic components included in the net inflow?

4. Determine the volume of water lost through evapotranspiration during a year from the surface of a 1500-ha lake located in a region where the annual rainfall is 135 cm. The increase in the depth of the lake over the year is 10 cm. Neglect the effect of groundwater flow.

5. A city is supplied by water from a 1250-ha catchment area. The average water consumption of the community is 50,000 $m^3 day^{-1}$. The annual precipitation in the region is 412 cm. A river with an average annual flow of 0.35 $m^3 s^{-1}$ originates in and flows out of the catchment area. If the net annual groundwater outflow from the area is equivalent to a 16-cm depth of water, what is the evapotranspiration loss in cubic meters per year, which, if exceeded, would cause a shortage of the water supply to the community? Assume that the storage of water in the area at the beginning and at the end of the year are equal.

6. The following information was either measured or estimated for the Great Salt Lake of Utah:

YEAR	LAKE LEVEL AT THE END OF THE YEAR (FEET ABOVE SEA LEVEL)	TOTAL ANNUAL LAKE PRECIPITATION (INCHES)	TOTAL ANNUAL LAKE STREAMFLOW INPUTS (ACRE-FEET)	ANNUAL LAKE EVAPORATION (INCHES)
1980	4198.6			
1981	4197.7	9.46	1448900.0	43.3
1982	4199.4	16.78	2443000.0	41.4
1983	4203.4	17.43	5113390.0	40.9
1984	4207.7	28.00	6359170.0	39.7

Using the following elevation–area–volume table for the lake, compute the implied unaccounted lake inputs or losses, in millimeters, during the years 1981 to 1984. The lake is terminal, i.e., has no outflows. State any assumptions you make.

Great Salt Lake Elevation–Area–Volume

ELEVATION (FEET)	AREA (ACRES)	VOLUME (ACRE-FEET)
4197.0	839809.0	12556430.0
4198.0	890047.0	13421890.0
4199.0	969949.0	14350140.0
4200.0	1079259.0	15370180.0
4201.0	1140000.0	16481450.0
4202.0	1175000.0	17640700.0
4203.0	1201000.0	18828700.0
4204.0	1223000.0	20040700.0
4205.0	1250468.0	21275600.0
4206.0	1330000.0	22541900.0
4207.0	1375000.0	23808300.0
4208.0	1410000.0	25074700.0
4209.0	1450000.0	26341000.0
4210.0	1490000.0	27607300.0
4212.0	1572000.0	30669000.0
4216.0	2228000.0	38671000.0
4218.0	2519000.0	43417000.0

7. In 1985 researchers predicted that the total streamflow inputs to the Great Salt Lake for the following three years would be bounded as follows:

YEAR	MINIMUM INFLOW (ACRE-FEET)	MAXIMUM INFLOW (ACRE-FEET)
1985	4000000.0	4950000.0
1986	3500000.0	6500000.0
1987	3000000.0	4500000.0

If rainfall and evaporation conditions for these years are 20% more than the averages for years 1981 to 1984, find minimum and maximum lake elevation profiles, in feet, for the years 1985 to 1987.

8. A typical monthly precipitation over a 78-mi^2 area is 5 in. A river goes through the area and brings (input) 19188.8 acre-ft per month. River outflow is 29321.6 acre-ft per month. The area is much larger than any existing aquifer and there is no significant seasonality in the hydrology. Estimate the monthly evapotranspiration in the area. Explain and justify assumptions.

9. A river basin discharges water at a rate linearly proportional to the amount it has in storage,

$$Q = K_1 S .$$

The only input into the basin is rainfall. Any rainfall will infiltrate. (Assume no evaporation during storm.) The rate of infiltration f is linearly proportional to the rainfall

$$f = K_2 I ,$$

where $K_2 < 1$. If a rainfall of constant intensity I and duration t_d occurs at a time when the storage is S_o, write an equation for the change in storage in the basin. Solve that equation and obtain the discharge as a function of time resulting from the described rainfall input.

10. Go to your library and find records of precipitation and streamflow for a nearby river basin. In the United States, precipitation records are published by the National Weather Service. Streamflow records are published mainly by the U.S. Geological Survey.

11. Figure 1.2 showed a global water balance. Rank the elements of the balance from least to most uncertain. Give your reasons.

12. Given that groundwater is so abundant, why is it not the main source of water in the world?

13. It is accepted that if the icecaps were to melt, large portions of the earth land surface would flood. What does this tell you, in general, of the land masses? How much would the ocean surface rise if its area did not change after the icecaps melt?

14. Given the amount of fresh water on earth, why are there droughts?

Chapter 2

Solar Radiation and the Earth's Energy Balance

2.1 INTRODUCTION

The hydrologic cycle is like an engine principally fueled by radiant energy from the sun. In fact, all of earth's processes are dependent mostly on solar radiation, with a secondary role played by heat sources from within the earth. In hydrology, we are interested in events within the uppermost layer of the earth's surface and in the lower layers of the atmosphere. The dominance of solar radiation at those levels is unquestionable. Figure 2.1 summarizes the energy flow of the earth.

The distribution of radiation over the globe and surrounding atmosphere leads to heat imbalances that drive most hydrologic and meteorologic events. In this chapter we will study how planetary geometry and motions affect radiation incidence, leading to an uneven distribution of energy over the globe, and how to quantify the amount of radiation received and emitted by the earth and its atmosphere.

The main fact to remember throughout this chapter is that radiation is absolute temperature-dependent. All objects with a temperature emit radiation. The details of this dependence will be seen soon. Nevertheless, at this point it is important to state that the sun at a temperature of 6000 °K (degrees Kelvin) is the main radiation source of the earth. Hotter bodies, such as other stars, are too far away to play a role in earth's radiation balance. Given that the earth and atmosphere also have a temperature ($\approx$287 °K), they will also emit radiation. The net exchange of incoming solar radiation

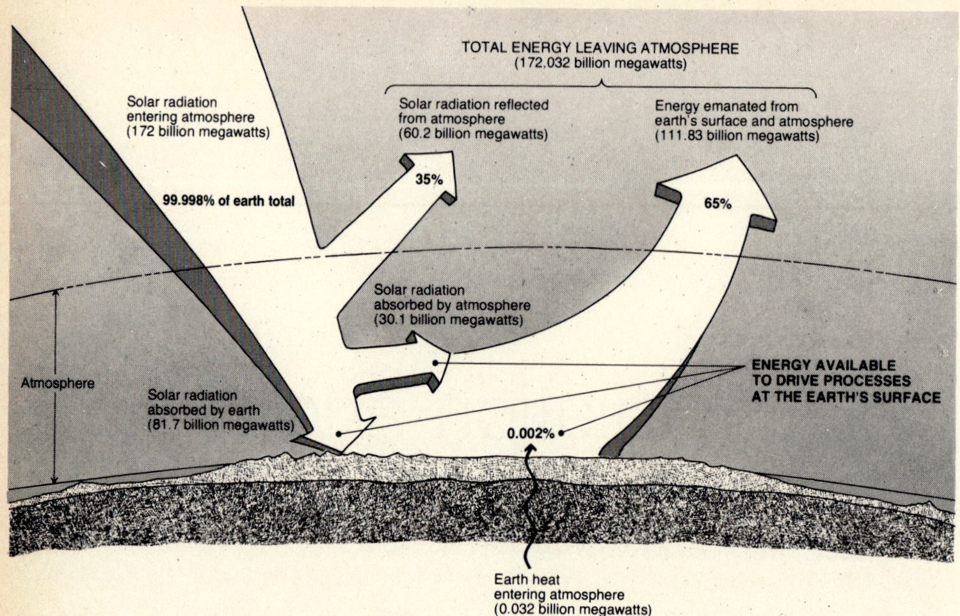

FIGURE 2.1 Energy-flow data for the earth as a whole in megawatts of power per year. (Illustration by William M. Marsh.) Source: W. M. Marsh, *Earthscape: A Physical Geography.* Copyright © 1987 by Wiley. Reprinted by permission of John Wiley & Sons, Inc.

and outgoing terrestrial radiation is critical in hydrology. Figure 2.1 advances some of the nature and magnitude of this exchange.

Before explaining the details we must establish our units of measurement. Temperature will always be given in degrees Kelvin (°K). Degrees Kelvin are obtained by adding 273.15 to temperature in degrees Centigrade. The relationship between degrees Kelvin, Centigrade, and Fahrenheit is summarized in Table 2.1. In converting from degrees Centigrade to degrees Fahrenheit and vice versa, remember the following equations:

$$°C = (°F - 32) \times 5/9 \qquad (2.1)$$

$$°F = (9/5 \times °C) + 32 \qquad (2.2)$$

Radiation rate is measured in units of energy, usually per unit area and time. The Système Internationale (S.I.) for units recommends joules for energy (J), square meters (m^2) for area, and seconds (s) for time. Therefore, radiation is given in $J\,m^{-2}\,s^{-1}$. A joule is equal to one unit of force, a newton (N), applied over 1 meter. In turn, a newton is equal to 1 kilogram of mass multiplied by an acceleration of 1 meter per second squared. Power (energy per time) is also given in watts (W), defined as one joule per second.

TABLE 2.1 Key Temperatures on the Centigrade, Fahrenheit, and Kelvin Scales

	°C	°F	°K
Absolute zero	−273.15	−459.67	0
Normal freezing point of H_2O*	0	32	273.15
Normal boiling point of H_2O*	100	212	373.15

*At sea level.

Source: W. M. Marsh and J. Dozier, *Landscape: An Introduction to Physical Geography*. Copyright © 1986 by Wiley. Reprinted by permission of John Wiley & Sons, Inc

A large body of literature, scientific and engineering, still prefers to use the concept of a calorie (cal) to represent energy. A calorie is 4.186 joules. (Vice versa, a joule is 0.239 calories). Radiation per unit area is commonly given in terms of calories per square centimeter (cal cm^{-2}). One calorie per square centimeter is called a langley (ly). The rate of radiation incidence then appears as langleys per second (ly s^{-1}), langleys per minute (ly min^{-1}) or langleys per day (ly day^{-1}).

2.2 PLANETARY MOTIONS AND THE DISTRIBUTION OF RADIATION

The earth–atmosphere system receives a small portion of the sun's total energy (radiation) output, a fraction equivalent to about 2×10^{-5} of the emitted solar radiation per unit area. Basically, the earth is a speck in the part of the universe influenced by the sun. The portion of solar radiation output that we receive is of reasonably constant intensity. This intensity, at a plane on the upper atmosphere perpendicular to incoming radiation, is about $1353 \text{ J m}^{-2} \text{s}^{-1}$, or $1.94 \text{ cal cm}^{-2} \text{min}^{-1}$. This number is called the *solar constant* and is commonly approximated as $2.0 \text{ cal cm}^{-2} \text{min}^{-1}$ (2.0 ly min^{-1}). But since the earth is approximately a sphere, which rotates on a tilted axis while revolving around the sun, the intensity of radiation received at a plane tangent to the top of the atmosphere varies in time and from one location to another.

At any given time, at some point on earth the sun will be directly overhead, so that a line connecting the centers of earth and sun will be perpendicular to a plane tangent to the earth–atmosphere surface at the point of interest. The latitude at which this occurs at any one time is called the *declination of the sun* δ. Figure 2.2 is a brief review of latitudes (parallels) and longitudes (meridians) that constitute the earth's coordinate system. Moving

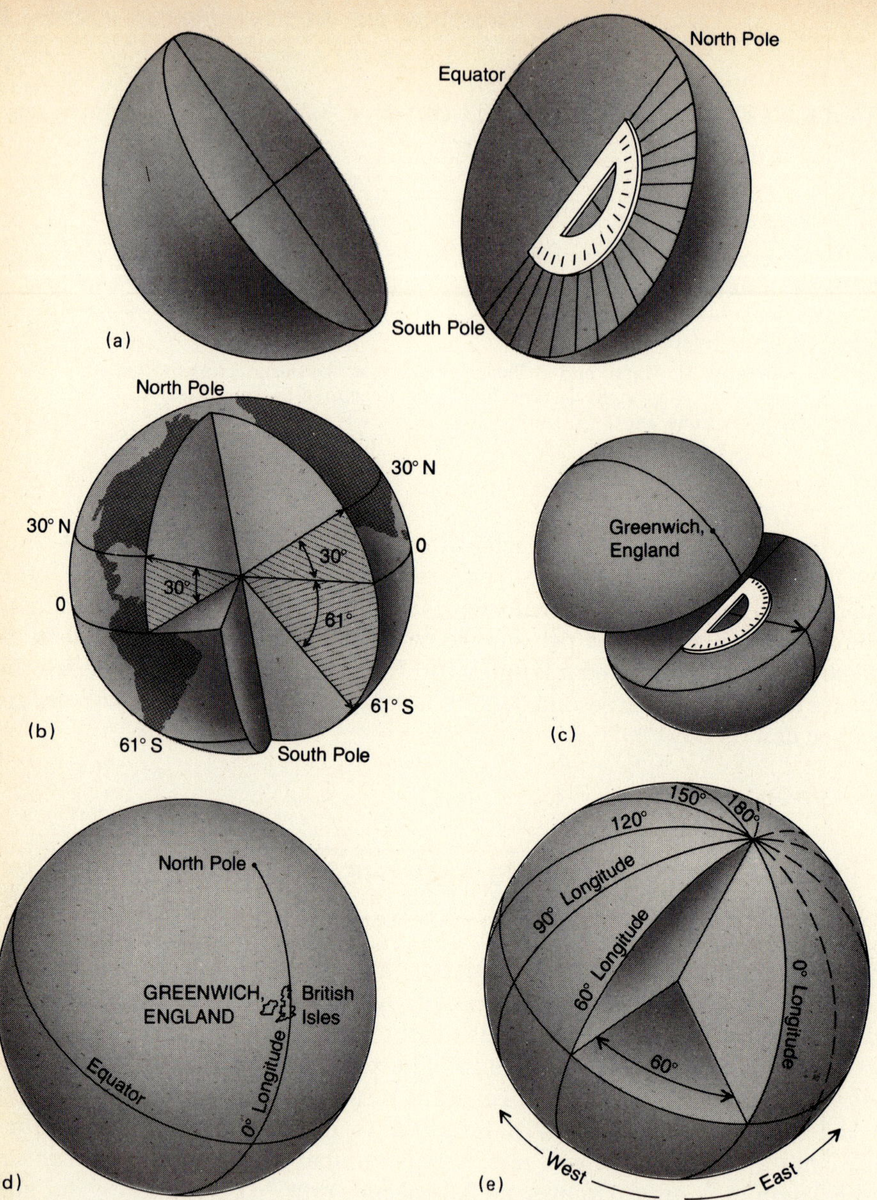

FIGURE 2.2 Constructing lines of latitude, longitude, and meridians. (a) If the earth is bisected along the polar axis, angles can be measured northward and southward from the equator with the aid of a protractor. (b) The latitude of any location represents an angle constructed through the equator, the center of the earth, and the location on the earth's surface. (c) Meridians are constructed on the perimeter of the equatorial plane, with the protractor fixed on the earth's center. (d) Since there is no geometrically convenient place to begin, the 0 meridian is fixed on the Royal Observatory, Greenwich, England, and is called the Greenwich Meridian, or the prime meridian. (e) Every meridian is half of a great circle. Longitude is given as degrees east or west of the Greenwich Meridian, which is 0° longitude. Source: W. M. Marsh, *Earthscape: A Physical Geography.* Copyright © 1987 by Wiley. Reprinted by permission of John Wiley & Sons, Inc.

away from the declination latitude implies that the earth–atmosphere surface is then at an angle relative to the plane perpendicular to solar radiation. When projected to the surface of the earth the solar radiation (solar constant) is spread over a larger area leading to less radiation per unit area at the surface. This is illustrated in Figure 2.3. If we denote the angle of the incoming radiation with a tangent plane at some point on the earth–atmosphere surface as α, then the radiation per unit area per unit time, or the intensity, will be given by

$$I_\circ = W_\circ \sin \alpha \,, \tag{2.3}$$

where $W_\circ$ is the solar constant, $I_\circ$ is the effective radiation intensity at the point of interest, and $\sin \alpha$ is called the solar altitude. The radiation intensity $I_\circ$ is also called insolation at the top of the atmosphere.

If the earth's axis of rotation were perpendicular to its plane of revolution (a plane across the center of the earth and the sun at all times during the 365.242 days of revolution) then α, at local noontime, would be $90° - \Phi$, where Φ is the latitude of the point of interest (see Fig. 2.3). Furthermore, there would be no seasons, since all points on earth (given the 24-hour rotation) would be illuminated by the sun an equal proportion of the time throughout the year. On such an earth, it would be colder (less radiation intensity) as we move toward the poles, but, ignoring all other factors, the temperature would tend to be uniform throughout the year. The earth's axis of rotation, though, is inclined relative to the plane of revolution. The inclination is 23°27′ off the (vertical) line perpendicular to the plane of revolution.

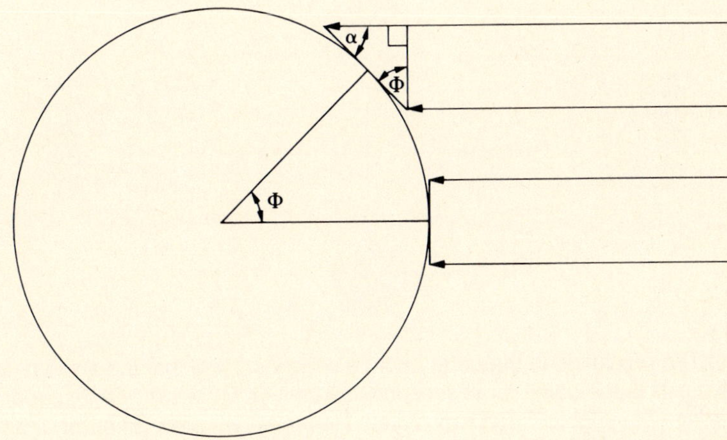

FIGURE 2.3 Projection of solar radiation on a plane tangent to the top of the atmosphere.

This is illustrated in Figure 2.4. During the winter solstice for the Northern Hemisphere (around December 22) the declination of the sun is 23°27′ south, its southernmost position. At the opposite end of the revolution, the declination is 23°27′ north (in the Northern Hemisphere). This is called the summer solstice, which occurs around June 22. At the autumnal (September 22) and vernal (March 21) equinoxes, the sun is directly over the equator, a 0° declination. Therefore, the sun is moving down toward the equator between the summer solstice and the autumnal equinox. It continues going down until the winter solstice, after which it starts the trip upward toward the north. The most northern and southern declinations are called the Tropic of Cancer and the Tropic of Capricorn, respectively.

The variation of declination with time implies that

1. The projection of a plane perpendicular to radiation on a plane tangent to the surface is varying in time since the solar angle, the angle between the noon sun and the horizon, is changing with the seasons.
2. The duration of daylight is also varying, implying that different locations receive different total amounts of radiation, resulting in seasonal climate patterns.

The second point is illustrated in Figure 2.5. At the equinox with a declination of 0° the hemispheres have exactly 12 hours of daylight and night. At the winter solstice the North Pole and Arctic regions never see daylight, while the South Pole is constantly illuminated. Points near the North Pole have very short days, since they are exposed to the sun for limited segments of their rotational circles. During the summer solstice the opposite effects are observed.

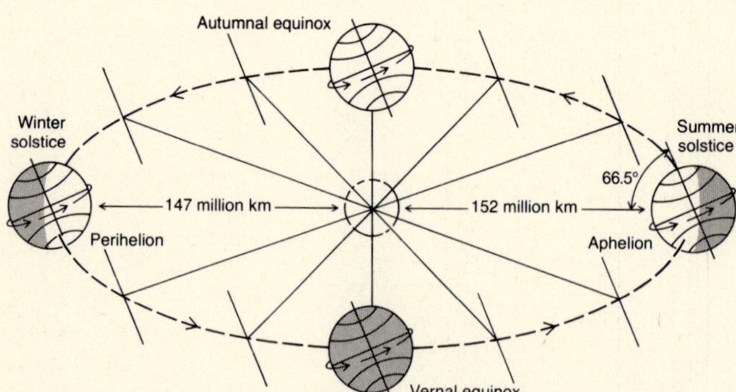

FIGURE 2.4 The revolution of the earth and the seasons. Note that the angle of inclination of the earth's axis is the same in all seasons. Source: W. M. Marsh, *Earthscape: A Physical Geography.* Copyright © 1987 by Wiley. Reprinted by permission of John Wiley & Sons, Inc. Adapted from Byers [1974].

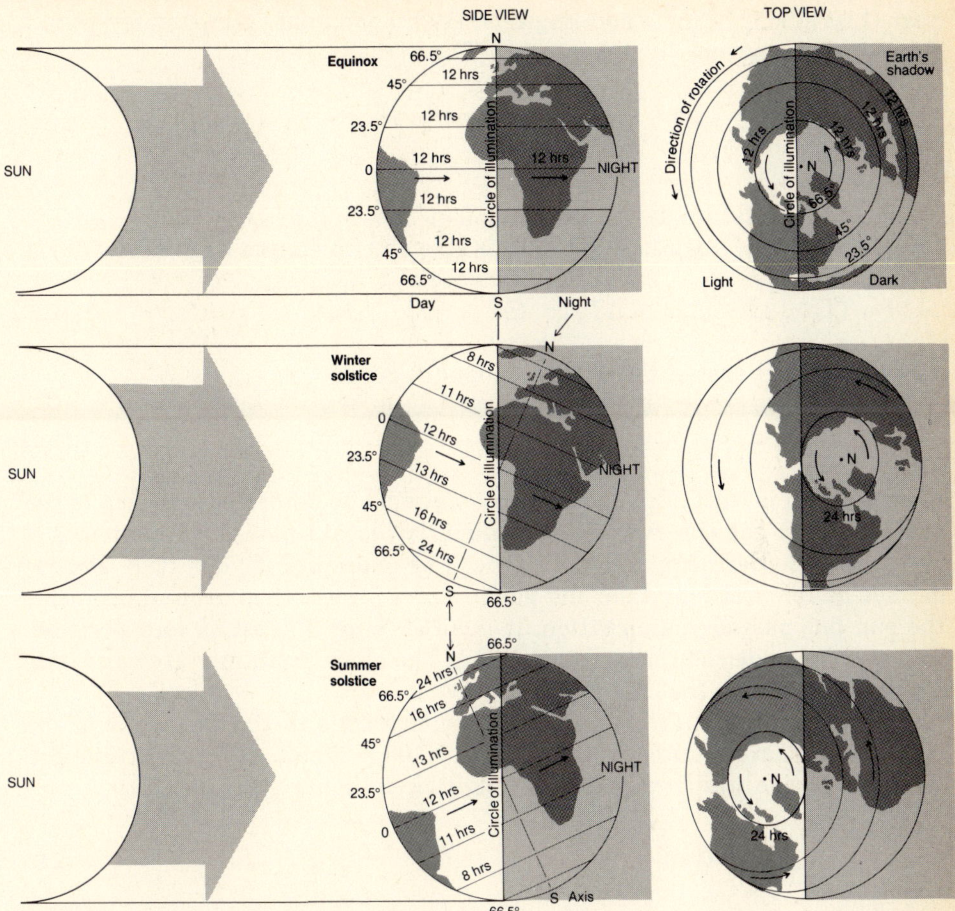

FIGURE 2.5 Portions of the earth illuminated by the sun at the solstices and equinoxes. Source: W. M. Marsh, *Earthscape: A Physical Geography.* Copyright © 1987 by Wiley. Reprinted by permission of John Wiley & Sons, Inc.

Radiation at the top of the atmosphere is finally dependent on the hour angle of the sun; that is, the position relative to solar noon. Using planetary geometry (Eagleson [1970]), it is then possible to deduce the following expression for solar altitude, in Eq. (2.3)

$$\sin \alpha = \sin \delta \sin \Phi + \cos \delta \cos \Phi \cos \tau, \tag{2.4}$$

where δ is the declination of the sun, Φ is local latitude, and τ is the hour angle of the sun. The sun's declination is published in tabular form for every year (List [1963]). Curtis and Eagleson [1982] quote an approximate formula from the Tennessee Valley Authority (TVA) [1972].

$$\delta = \frac{23.45\pi}{180} \cos\left[\frac{2\pi}{365}(172 - D)\right], \tag{2.5}$$

where D is the Julian day ($1 \leq D \leq 365$ or 366) and δ is in radians.

The local hour angle, $0 \leq \tau \leq 360$, is given by Curtis and Eagleson [1982] as

$$\tau = (T_S + 12 - \Delta T_1 + \Delta T_2) \times 15 \tag{2.6}$$

when the sun is east of the observer's longitude, or

$$\tau = (T_S - 12 - \Delta T_1 + \Delta T_2) \times 15 \tag{2.7}$$

when the sun is west of the observer's longitude. To determine position of the sun relative to the observer, keep in mind that local time is given in terms of the longitude defining the time zone. For example, if your time zone is defined at 75° west longitude and you are at 72° west, then at local noontime the sun has passed your position; it is at 75°W or 3° west of your position. This will be made clearer in Example 2.1. The above equations are valid only for values of τ such that $\cos\tau$ is positive.

In Eqs. (2.6) and (2.7), T_S is the standard time in the time zone of the observer in hours counted from midnight (0.00 − 23.59). ΔT_1 is the time difference between standard and local longitude in hours given as

$$\Delta T_1 = \frac{i}{15}(\theta_S - \theta_L), \tag{2.8}$$

where $i = -1$ for west longitude and $i = 1$ for east longitude, relative to Greenwich. θ_S is the longitude of the standard meridian (meridian where the observer's time zone is centered), and θ_L is the longitude of the observer meridian. Finally, ΔT_2 is the difference between true solar time and mean solar time in hours, which is usually neglected.

TVA [1972] suggests a modification of Eq. (2.3) to account for the elliptical nature of the earth's orbit around the sun, which leads to variable distance from the sun. They suggest

$$I_\circ = \frac{W_\circ}{r^2} \sin\alpha, \tag{2.9}$$

where r is the ratio of actual earth–sun distance to mean earth–sun distance, given by

$$r = 1.0 + 0.017 \cos\left[\frac{2\pi}{365}(186 - D)\right]. \tag{2.10}$$

In order to compute incident solar radiation over a given finite period $\Delta t = t_2 - t_1$, Eq. (2.9) can be integrated, keeping δ and Φ constant over Δt. The result (TVA [1972]) is

$$I_{\Delta t} = \frac{W_\circ}{r^2}\{(t_2 - t_1) \sin \delta \sin \Phi + \frac{12}{\pi} \cos \delta \cos \Phi[\sin(\tau_2) - \sin(\tau_1)]\}. \tag{2.11}$$

The net result of planetary geometry is the pattern of total daily radiation at the outer edge of the atmosphere shown in Figure 2.6. The most significant features are the relatively constant radiation near the equator and the increasing seasonal variation as we move to the poles.

Radiation at the outer edge of the atmosphere is significantly altered in its journey to the land–sea surface. It is reduced by reflection, refraction, and absorption by atmospheric constituents. The atmospheric moisture of clouds is particularly effective in absorbing solar radiation. The following section will discuss how to quantify all these effects.

Average annual radiation received at the earth surface is shown in Figure 2.7. Note that radiation at the equator is lower than at the subtropics. This is a reflection of the persistent cloudiness of the region, which can effectively reduce radiation by as much as 50%.

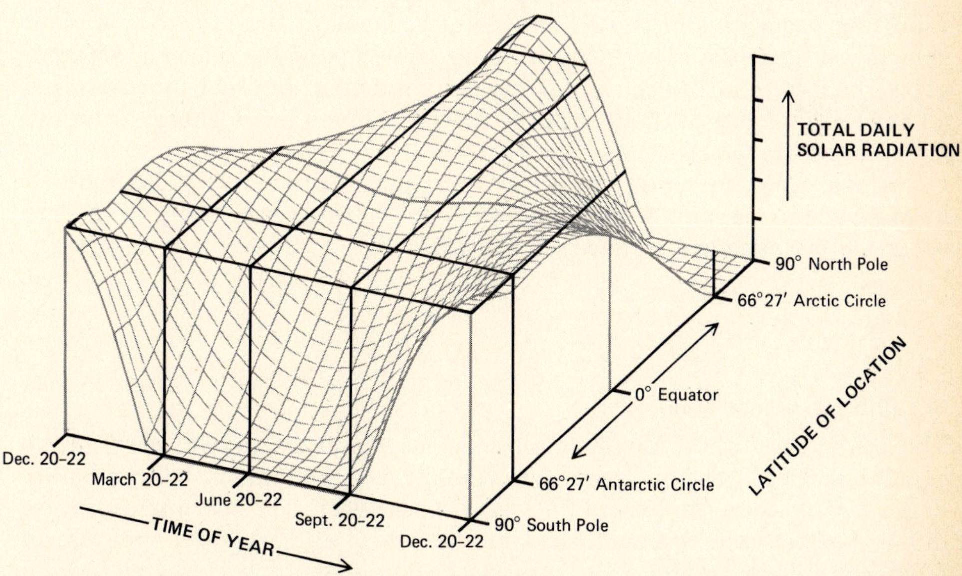

FIGURE 2.6 The variation in total daily solar radiation at the outer edge of the atmosphere. (Illustration by Jeff Dozier.) Source: W. M. Marsh, *Earthscape: A Physical Geography.* Copyright © 1987 by Wiley. Reprinted by permission of John Wiley & Sons, Inc.

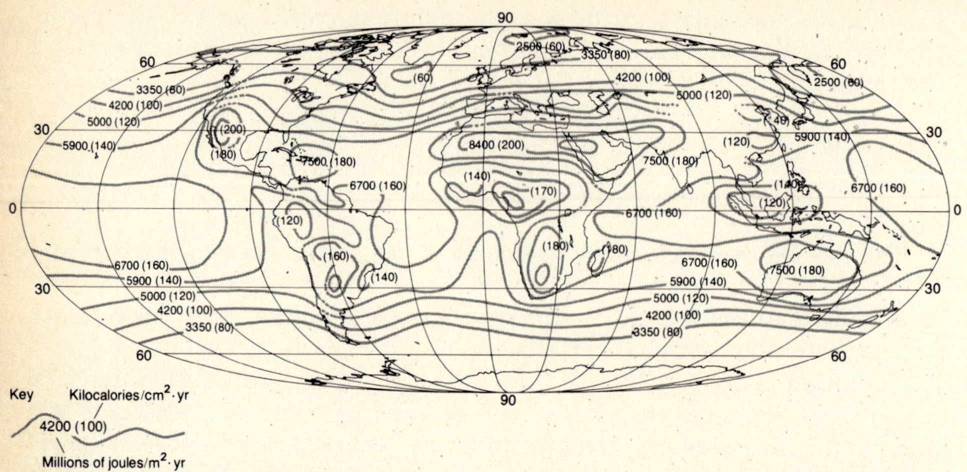

FIGURE 2.7 The worldwide distribution of solar radiation in millions of joules per square meter per year and kilocalories per square centimeter per year. Source: W. M. Marsh, *Earthscape: A Physical Geography.* Copyright © 1987 by Wiley. Reprinted by permission of John Wiley & Sons, Inc. Map projection by Waldo Tobler. Data from Budyko [1974].

The unevenness of incoming solar radiation, coupled with nonuniform radiating properties of the earth's mantle, leads to the radiation budget shown in Figure 2.8. The latitudes near the equator have a net radiation gain. The higher latitudes exhibit a radiation deficit. In order to maintain a global energy balance there must be mechanisms to move energy from the lower to the higher latitudes.

Winds, ocean currents, and water vapor play these carrier roles and will be discussed in Chapter 3. The magnitude and nature of the energy transfers occurring by latitude are shown in Table 2.2.

EXAMPLE 2.1

Computation of Insolation

The Commonwealth of Massachusetts is centered at around 42° north of the equator and 72° west of Greenwich. Nassau, in the Bahama Islands, is 25° north of the equator and about 78° west of Greenwich. The sharp temperature differences between the two places are known to all those who snow ski in Massachusetts and water ski in Nassau. Local temperature is related to insolation, so let us find its value at local noon on a typical January 1 in both places.

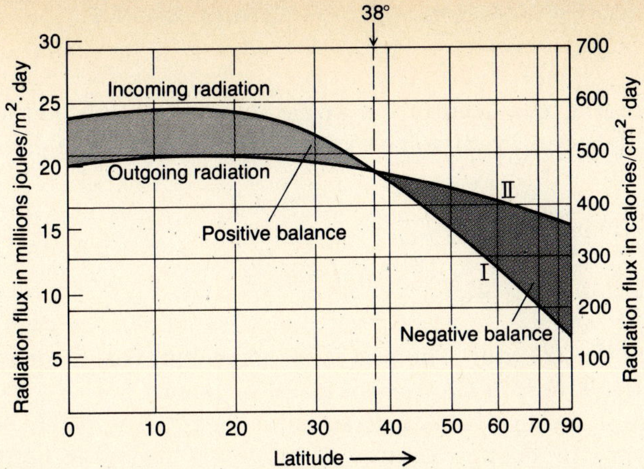

FIGURE 2.8 Distribution with latitude of absorbed and outgoing radiation. Curve I shows the solar radiation absorbed by the earth and atmosphere; curve II, shortwave and longwave radiation leaving the atmosphere. The scaling of the latitude axis is representative of the areas on the earth's surface. Source: W. M. Marsh, *Earthscape: A Physical Geography.* Copyright © 1987 by Wiley. Reprinted by permission of John Wiley & Sons, Inc.

TABLE 2.2 Components of the Energy Balance at the Earth's Surface (Mean Values in kcal per cm² per Year)

	OCEANS				CONTINENTS			EARTH AS A WHOLE			
LATITUDE	*R*	*E*	*T*	*S*	*R*	*E*	*T*	*R*	*E*	*T*	*S*
70–60°N	23	33	16	−26	20	14	6	21	20	9	−8
60–50	29	39	16	−26	30	19	11	30	28	13	−11
50–40	51	53	14	−16	45	24	21	48	38	17	−7
40–30	83	86	13	−16	60	23	37	73	59	23	−9
30–20	113	105	9	−1	69	20	49	96	73	24	−1
20–10	119	99	6	14	71	29	42	106	81	15	10
10–0	115	80	4	31	72	48	24	105	72	9	24
0–10°S	115	84	4	27	72	50	22	105	76	8	21
10–20	113	104	5	4	73	41	32	104	90	11	3
20–30	101	100	7	−6	70	28	42	94	83	15	−4
30–40	82	80	9	−7	62	28	34	80	74	12	−6
40–50	57	55	9	−7	41	21	20	56	53	9	−6
50–60	28	31	8	−11	31	20	11	28	31	8	−11
Total	82	74	8	0	49	25	24	72	60	12	0

R = net radiation balance.
E = loss of heat by evaporation.
T = turbulent heat transfer.
S = redistribution of heat by ocean currents.

Source: M. I. Budyko, *Climatic Changes,* 1977, p. 33. Copyright by the American Geophysical Union.

Begin by finding the declination angle, using Eq. (2.5), which is only a function of the date (and year to be exact). The Julian day of January 1 is 1.

$$\delta = \frac{23.45\pi}{180}\cos\left[\frac{2\pi}{365}(172 - 1)\right]$$
$$= -0.4 \text{ radians} = -22.9°$$

or 22.9° south of the equator. The sun is, as expected, very near its southernmost position, the Tropic of Capricorn, at 23°27′ south.

Global time is generally measured relative to Greenwich, England, at meridian 0°. Every 15° east of Greenwich implies a lead of an hour relative to Greenwich. There are local variations to this rule due to constraints imposed by political, geographic, and other boundaries. Similarly, there are local alterations, such as daylight savings time (Strahler and Strahler [1983]).

During January, both Boston and Nassau are within the time zone centered at 75° west of Greenwich; this puts them five hours behind Greenwich time. To find the hour angle, we then use Eqs. (2.6) through (2.8). Using Eq. (2.8), we find the adjustment necessary to find true solar time at the location of interest.

Massachusetts:

$$\Delta T_1 = \frac{-1}{15}(75 - 72) = -0.2 \text{ hr}$$

Nassau:

$$\Delta T_1 = \frac{-1}{15}(75 - 78) = 0.2 \text{ hr}$$

From Eq. (2.6), ignoring ΔT_2, the hour angle is then:

Massachusetts (at 12 noon at the 75°W meridian, the sun is west of the 72°W meridian):

$$\tau = [12 - 12 - (-0.2)]15 = 3°$$

Nassau (at 12 noon at the 75°W meridian, the sun is east of the 78°W meridian):

$$\tau = [12 + 12 - (0.2)]15 = 357°$$

The radius adjustment r is obtained from Eq. (2.10):

$$r = 1.0 + 0.017\cos\left[\frac{2\pi}{365}(186 - 1)\right]$$
$$= 0.983\,.$$

The solar altitude for Massachusetts (Eq. 2.4) is then

$$\begin{aligned}\sin\alpha &= \sin(-22.9°)\sin(42°) + \cos(-22.9°)\cos(42°)\cos(3°)\\ &= 0.42\,.\end{aligned}$$

For Nassau, the solar altitude is

$$\begin{aligned}\sin\alpha &= \sin(-22.9°)\sin(25°) + \cos(-22.9°)\cos(25°)\cos(357°)\\ &= 0.67\,.\end{aligned}$$

Using Eq. (2.9), the insolation in both places is

Massachusetts:

$$I_\circ = \frac{2}{(0.983)^2}0.42 = 0.87\ \text{ly}\,\text{min}^{-1}$$

Nassau:

$$I_\circ = \frac{2}{(0.983)^2}0.67 = 1.39\ \text{ly}\,\text{min}^{-1}$$ ◆

2.3 RADIATION PHYSICS

There are several basic laws that must be understood in radiation studies. The first one is Kirchhoff's law, which says that at thermal equilibrium the ratio of radiation intensity (radiation per unit area per unit time, emissive power, emittance, or radiant flux density) W_i to absorptivity a_i is equal for each body in equilibrium. Two bodies in an insulated box, once in thermal equilibrium, will satisfy

$$W_B A_1 a_1 = W_1 A_1; \qquad W_B A_2 a_2 = W_2 A_2\,, \tag{2.12}$$

where W_B is the box radiation intensity and A_i is the surface area of body i. Equation (2.12) leads to the Kirchhoff law statement, in thermal equilibrium,

$$\frac{W_1}{a_1} = \frac{W_2}{a_2} = W_B\,. \tag{2.13}$$

A body with absorptivity of 1 is called a *black body*. The ratio of the emissive power of a body to that of a black body is called the *emissivity, E*. Emissivity and absorptivity are equal under thermal equilibrium.

Planck's law gives the distribution of radiation energy from a black body at different wavelengths. All radiation, as commonly known, has varied wavelengths. Define $W_{B_\lambda} d\lambda$ as the radiation intensity of a black body at wavelength band $d\lambda$, such that

$$\int_0^\infty W_{B_\lambda} d\lambda = W_B \,, \tag{2.14}$$

where W_B is *black-body radiation intensity*.

Planck's law states

$$\frac{W_{B_\lambda}}{T^5} = \frac{2\pi hc^2\lambda^{-5}T^{-5}}{e^{ch/k\lambda T} - 1} \,, \tag{2.15}$$

where c is the velocity of light (2.998×10^{10} cm s^{-1}), h is Planck's constant (6.625×10^{-34} J s), k is the Boltzmann constant (1.38×10^{-23} J °K^{-1}), and λ is wavelength in centimeters. Temperature T is in degrees Kelvin and W_{B_λ}, to be consistent, must be in joules per cubic centimeter per second (J cm^{-3} s^{-1}) or radiation intensity per wavelength.

Figure 2.9 shows the spectrum of thermal radiation from a black body. The maximum of the spectral distribution of radiation intensity occurs at the point

$$\lambda T = 0.2898 \text{ cm °K} = 2898 \text{ micron °K} \,. \tag{2.16}$$

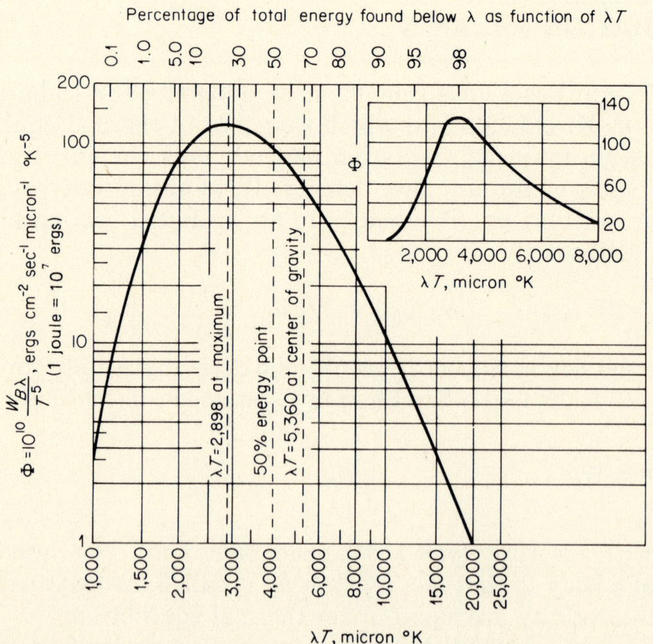

FIGURE 2.9 Spectrum of thermal radiation from a black body. Source: W. H. McAdams, *Heat Transmission,* 3rd ed., McGraw-Hill, 1954. Reproduced by permission.

The above shows that the wavelength of maximum intensity is temperature-dependent, since as T varies λ must change to keep their product constant. A *gray body* radiates a fixed proportion of a black-body radiation in all wavelengths for a given temperature.

The Stefan–Boltzmann law is probably the most important for applications. It states

$$W_B T^{-4} = \int_0^\infty \frac{W_{B_\lambda}}{T^5} d(\lambda T) = \sigma, \tag{2.17}$$

where σ is a constant 0.826×10^{-10} cal cm^{-2} min^{-1} °K^{-4} (5.67×10^{-8} J m^{-2} s^{-1} °K^{-4}). In other words, the radiation of a black body is proportional to the fourth power of the absolute temperature (in degrees Kelvin) of the body.

Based on the above, the total energy emitted by the sun is on the order of 100,000 ly min^{-1}. The solar constant, taken as 2 ly min^{-1}, is but a small fraction of the total energy emitted.

As implied by Eq. (2.16), the higher the temperature, the shorter are the wavelengths that dominate radiation intensity. Radiation is then commonly divided in shortwaves and longwaves. The sun emits shortwave radiation with the wavelength spectrum shown in Figure 2.10. The spectrum of longwave emissions from the earth is also shown in the figure.

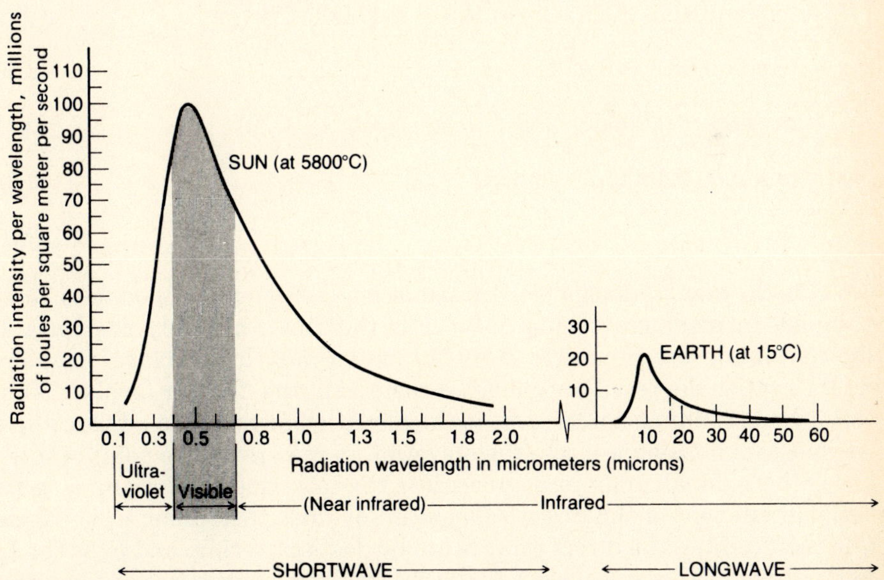

FIGURE 2.10 The distribution of intensities of radiation produced by the sun and the earth. The vertical axis represents radiation intensity of output; the horizontal axis, in micrometers, represents wavelength. Solar radiation is concentrated around 0.5 μm between the ultraviolet and infrared wavelengths, whereas earth radiation is entirely infrared. Source: W. M. Marsh and J. Dozier, *Landscape: An Introduction to Physical Geography.* Copyright © 1986 by Wiley. Reprinted by permission of John Wiley & Sons, Inc.

As seen in Section 2.2 (Eq. 2.9), the radiation intensity received at the surface of the outer atmosphere (insolation) is less than the solar constant due to the angle between the surface and the incident radiation. At the land–sea surface the net radiation is even less. The incoming insolation is reflected, refracted, and absorbed. Furthermore, the earth's surface reradiates some energy. The following paragraphs will present methods for quantifying this radiation exchange.

2.3.1 Shortwave Radiation

Shortwave radiation, mostly resulting from the high solar temperatures, is scattered and absorbed when passing through the atmosphere. The most commonly used formulas for quantifying these effects are mostly empirical in nature. As such, many alternative expressions can be found in the literature. Here we give approaches compiled by Eagleson [1970] and Curtis and Eagleson [1982]. The following paragraph is as given by Curtis and Eagleson [1982].

Klein [1948] and TVA [1972] obtain the *clear sky shortwave radiation*, I_c, after accounting for atmospheric effects, as

$$\frac{I_c}{I_o} = \frac{a' + 0.5(1 - a' - d) - 0.5d_a}{1 - 0.5R_g(1 - a' + d_s)} \tag{2.18}$$

$$a' = \exp[-(0.465 + 0.134w)(0.129 + 0.171e^{-0.880m_p})]m_p \tag{2.19}$$

$$w = \exp[(-0.981 + 0.0341T_d)] \tag{2.20}$$

$$m_p = m[(288 - 0.0065z)/288^{5.256}] \tag{2.21}$$

$$m = [\sin\alpha + 0.1500(\alpha + 3.885)^{-1.253}]^{-1} \tag{2.22}$$

$$d = d_s + d_a \tag{2.23}$$

where a' is the mean atmospheric transmission coefficient for cloudless, dust-free, moist air after scattering only; w is the mean monthly precipitable water content in centimeters; T_d is the mean monthly surface dewpoint [temperature], in degrees Fahrenheit, measured at the 2-meter level; m is the optical air mass, dimensionless; m_p is the elevation or pressure adjusted optical air mass, dimensionless; z is the elevation in meters [of the point of interest]; α is the solar altitude in degrees; d is the total dust depletion; d_a is the depletion coefficient of the direct solar beam by dust absorption; d_s is the depletion coefficient of the direct solar beam by dust scattering; and R_g is the total reflectivity of the ground. TVA [1972] provides a summary of total dust depletion coefficient at different locations and seasons as a function of optical air mass, m (or relative thickness of air mass). Table 2.3 gives typical values. Other references are Kimball [1927], [1928], and [1930]; Fritz [1949]; Bolsenga [1964]; and Reitan [1960].

TABLE 2.3 Total Dust Depletion Coefficient d

	WASHINGTON, D.C.		MADISON, WIS.		LINCOLN, NEB.	
SEASON	$m = 1$	$m = 2$	$m = 1$	$m = 2$	$m = 1$	$m = 2$
Winter	—	0.13	—	0.08	—	0.06
Spring	0.09	0.13	0.06	0.10	0.05	0.08
Summer	0.08	0.10	0.05	0.07	0.03	0.04
Fall	0.06	0.11	0.07	0.08	0.04	0.06

Sources: Adapted from TVA [1972], Walter O. Wunderlich, Tennessee Valley Authority Engineering Laboratory. Summarized by Bolsenga [1964] and based on data by Kimball [1927], [1928], and [1980].

An alternative to obtain clear sky shortwave radiation is given by Eagleson [1970]. Atmospheric absorption and scattering is approximated by

$$\frac{I_c}{I_\circ} = \exp[-(a_1 + a_2 + a_3)m] = \exp(-a_t m), \tag{2.24}$$

where m is the optical air mass given by Eq. (2.22) or approximated by the cosecant (l/sin α) of the solar altitude α under 1 standard atmosphere. Coefficients a_1, a_2, and a_3 correspond to molecular scattering, absorption, and particulate scattering, respectively.

Equation (2.24) is also expressed as

$$\frac{I_c}{I_\circ} = \exp(-na_1 m), \tag{2.25}$$

where n is (a_t/a_1), a turbidity factor of air that varies from about 2.0 for clear mountain air to 4 or 5 for smoggy urban areas.

The molecular scattering coefficient a_1 is defined as a function of the effective thickness of the atmosphere,

$$a_1 = 0.128 - 0.054 \log_{10} m\,. \tag{2.26}$$

Curtis and Eagleson [1982] point out that the form of Eq. (2.25) should only be valid for radiation at a single wavelength (TVA [1972]). Nevertheless they argue for the simplicity of its use.

Cloudy skies further reduce total net radiation at the earth's surface. Absorption and scattering by clouds can be significant, depending on cloud type, thickness, and elevation.

Eagleson [1970] quotes from the U.S. Army Corps of Engineers [1956] the following relationship to obtain net radiation I_s' after accounting for clouds,

$$\frac{I_s'}{I_c} = 1 - (1 - K)N, \tag{2.27}$$

where K is the fraction of cloudless sky insolation received on a day with overcast skies and N is the fraction of sky covered by clouds, which takes a value of 1 for completely overcast skies. The factor K is obtained as a function of the cloud-base altitude according to the formula,

$$K = 0.18 + 0.0853z\,, \tag{2.28}$$

where z is the cloud-base altitude in kilometers. The above is a close approximation to the English-unit equation given by Eagleson [1970].

Curtis and Eagleson [1982] point out that the difficulty in using Eq. (2.27) lies in the nature of the available data on cloud-base elevation. Cloud-base altitude is only reported when $N > 0.50$ and for lesser values it is given as "unlimited ceiling." A simpler relation, not dependent on z is given by the TVA [1972]:

$$\frac{I'_s}{I_c} = 1 - 0.65N^2. \tag{2.29}$$

The above equation may overestimate attenuation as N approaches 1 for high and thin clouds. To avoid this problem it is recommended to use total opaque cloud cover instead of cloud cover. This parameter is usually reported in stations that measure cloud cover.

Vegetation also reflects and absorbs incident net radiation. If I'_s is the radiation above the vegetative cover, the net radiation on the ground level can be obtained using

$$I'_{sg} = K_t I'_s\,, \tag{2.30}$$

where K_t is a transmission coefficient function of density, type, and condition of vegetation. Table 2.4 and Figure 2.11 give some values of K_t for grass and a forest canopy.

TABLE 2.4 Typical Extinction of Insolation by Grass

HEIGHT OF GRASS	$K_t = I'_{sg}/I'_s$
1 m	0.18
50 cm	0.18
10 cm	0.68

Source: Data from O. G. Sutton, *Micrometeorology*, McGraw-Hill, 1953. Adapted from Eagleson [1970].

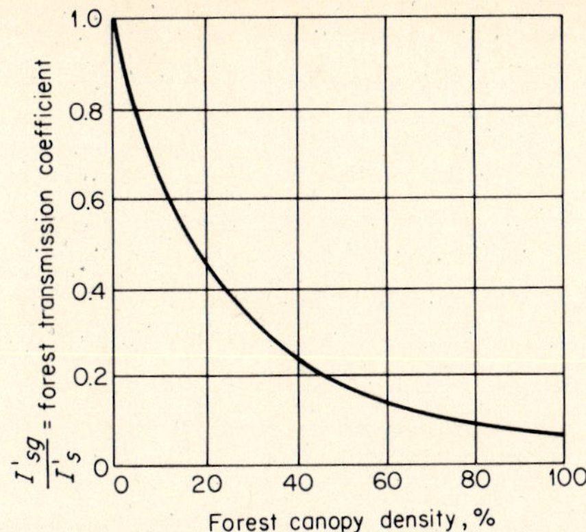

FIGURE 2.11 Transmission of insolation by coniferous forest canopy. Canopy density is the percentage of forested area that is covered by a horizontal projection of the vegetation canopy. Sources: U.S. Army Corps of Engineers [1956] and Eagleson [1970].

Albedo is the ratio of reflected radiation to incident radiation,

$$A = \frac{I''_s}{I'_s}. \tag{2.31}$$

The average albedo for the earth is 0.34, although each individual material and surface shows a different value (Table 2.5). Albedo of water surfaces

TABLE 2.5 Albedo of Natural Surfaces

SURFACE	ALBEDO A	SURFACE	ALBEDO A
Water (see Fig. 2.12)	0.03–0.40	Spring wheat	0.10–0.25
Black, dry soil	0.14	Winter wheat	0.16–0.23
Black, moist soil	0.08	Winter rye	0.18–0.23
Gray, dry soil	0.25–0.30	High, dense grass	0.18–0.20
Gray, moist soil	0.10–0.12	Green grass	0.26
Blue, dry loam	0.23	Grass dried in sun	0.19
Blue, moist loam	0.16	Tops of oak	0.18
Desert loam	0.29–0.31	Tops of pine	0.14
Yellow sand	0.35	Tops of fir	0.10
White sand	0.34–0.40	Cotton	0.20–0.22
River sand	0.43	Rice field	0.12
Bright, fine sand	0.37	Lettuce	0.22
Rock	0.12–0.15	Beets	0.18
Densely urbanized areas	0.15–0.25	Potatoes	0.19
Snow (see Figs. 2.13 and 2.14)	0.40–0.85	Heather	0.10
Sea ice	0.36–0.50		

Source: Eagleson [1970].

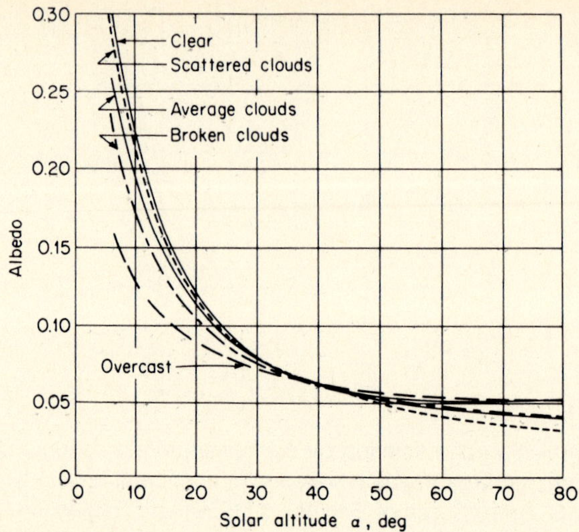

FIGURE 2.12 Effect of solar altitude and cloud cover on albedo for a horizontal water surface. Sources: Reprinted from *J. Power Div.*, ASCE Paper 3200 (Raphael), July 1962, with permission; and Eagleson [1970].

varies with cloud cover and solar altitude (Fig. 2.12). Over snow, it decreases with age or cumulative temperature index. Table 2.5 and Figures 2.13 and 2.14 show these effects.

The effective incoming shortwave radiation is then given by

$$I_s^* = I_s' - I_s'' = I_s'(1 - A). \tag{2.32}$$

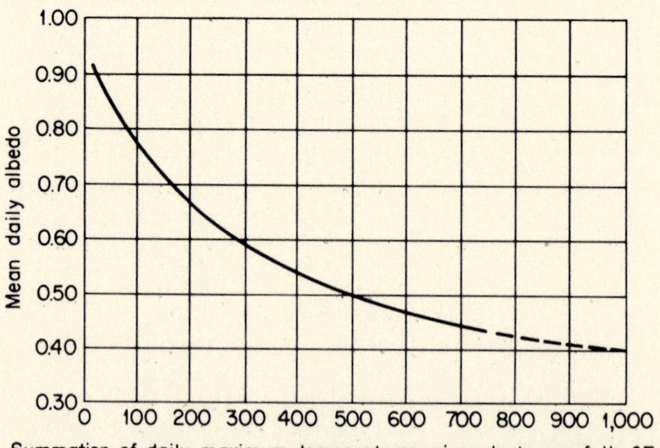

FIGURE 2.13 Variation of albedo of a snow surface with accumulated temperature index. Sources: U.S. Army Corps of Engineers [1956] and Eagleson [1970].

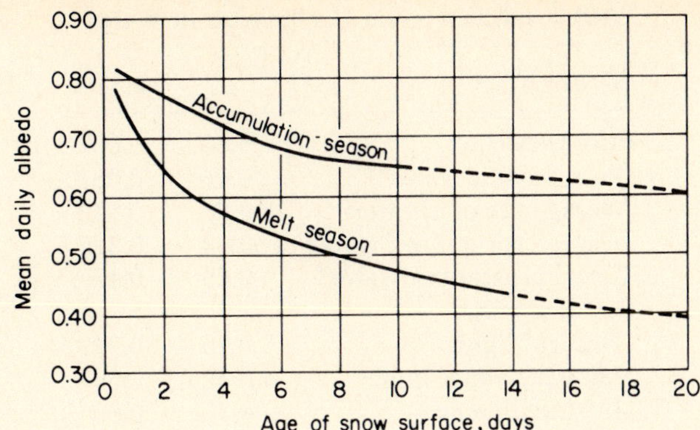

FIGURE 2.14 Time variation in albedo of a snow surface. Sources: U.S. Army Corps of Engineers [1956] and Eagleson [1970].

Effective incoming radiation incident on a water or snow surface suffers an extinction effect with penetration depth. For water, this can be represented by

$$\frac{I'_{sz}}{I_s^*} = (1 - \beta)e^{-kz}, \tag{2.33}$$

where I_s^* is the effective incoming radiation; I'_{sz} is the effective radiation at location z beneath the surface; and β and k are parameters given by Table 2.6.

For snow, the extinction formula is

$$\frac{I'_{sz}}{I_s^*} = e^{-kz}, \tag{2.34}$$

where the coefficient k is given in Table 2.7 as a function of pack density.

The possible paths and disposition of incident shortwave radiation are summarized in Figure 2.15.

EXAMPLE 2.2

Computation of Shortwave Radiation

Boston, Massachusetts, can suffer days of considerable smog and cloudiness. Let us compare the net shortwave radiation reaching the earth's surface in Boston under smoggy overcast conditions (cloud base at 1500 m) and under clear skies. Let us assume, for simplicity, the location, day, and hour given for central Massachusetts in Example 2.1.

TABLE 2.6 Extinction Coefficients in Water

CONDITIONS	β	k, m^{-1}
Pure water*	0.63	0.052
Clear oceanic water*	0.64	0.081
Average oceanic water†	0.68	0.122
Average coastal sea water	0.69	0.325
Turbid coastal sea water*	0.69	0.425
Distilled water		
Natural light	0.75	0.029
Mercury-vapor lamps	0.62	0.60
Infrared lamps	0.75	0.60
Natural lake		
Clear (Lake Tahoe)	0.40	0.05
Turbid (Lake Castle)	0.40	0.27
Lake Mendota‡	0.58	0.720
Trout Lake‡	0.50	1.400
Big Ridge Lake§	0.24	1.110
Fontana Lake§	0.24	0.930

*Equation valid after 2 meters.

†Equation valid after 3 meters.

‡Equation valid after 1 meter.

§Equation valid after 0.5 meter.

Source: Adapted from Dake and Harleman [1966]; Eagleson [1970]; and TVA [1972], Walter O. Wunderlich, Tennessee Valley Authority Engineering Laboratory.

One alternative would be to use Eq. (2.18), but not enough information is given. On the other hand, there is enough information to use Eq. (2.25),

$$\frac{I_c}{I_o} = \exp(-na_1 m).$$

TABLE 2.7 Extinction Coefficients in a Snowpack

SNOWPACK DENSITY (%)	EXTINCTION COEFFICIENT (k, cm^{-1})
26.1	0.280
32.2	0.184
39.7	0.106
44.8	0.106

Source: U.S. Army Corps of Engineers, [1956] and adapted from Eagleson [1970].

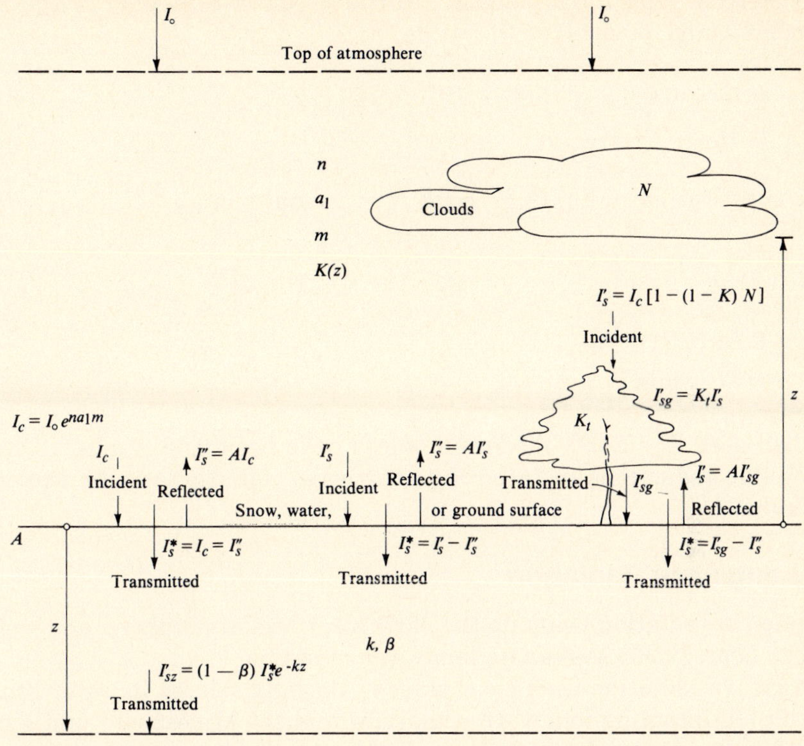

FIGURE 2.15 Summary disposition of shortwave radiation. Source: Eagleson [1970].

The relative thickness of the atmosphere m is defined by the cosecant of α. From Example 2.1, sin α is 0.42. The cosecant $(1/\sin \alpha)$ of α is $m = 2.4$.

Assuming that $n = 2.0$ for clear skies in Boston and using $n = 5.0$ for smoggy conditions, the molecular scattering coefficient is

$$\begin{aligned} a_1 &= 0.128 - 0.054 \log_{10} 2.4 \\ &= 0.107\,. \end{aligned}$$

Therefore, the net radiation under clear skies is

$$\begin{aligned} I_c &= I_\circ \exp(-na_1m) \\ &= 0.87 \exp(-2 \times 0.107 \times 2.4) \\ &= 0.52 \text{ ly min}^{-1}. \end{aligned}$$

For smoggy conditions,

$$\begin{aligned} I_c &= 0.87 \exp(-5 \times 0.107 \times 2.4) \\ &= 0.24 \text{ ly min}^{-1}. \end{aligned}$$

Under overcast skies the last value is further reduced according to Eq. (2.27). The reduction factor K takes the value,

$$
\begin{aligned}
K &= 0.18 + 0.0853 \times 1500 \times 10^{-3} \\
&= 0.31 ,
\end{aligned}
$$

which, substituting in Eq. (2.27) with $N = 1$ results in

$$
\begin{aligned}
I_s' &= I_c[1 - (1 - K)N] \\
&= I_c \cdot K \\
&= 0.24\ (0.31) \\
&= 0.07 \text{ ly min}^{-1}.
\end{aligned}
$$

If we had used Eq. (2.29), the result would have been 0.08 ly min^{-1}. Clearly, the possible reductions in incoming shortwave radiation can be extremely significant. ◆

2.3.2 Longwave Radiation

Absorption of radiation heats up the earth's surface, atmosphere, and vegetation. The heated bodies emanate longwave radiation.

Longwave radiation from clear skies is strongly related to water content (Idso [1981]). Incoming longwave radiation from the atmosphere under clear sky conditions is given by

$$
\frac{I_\ell'}{W_B} = E_a = 0.740 + 0.0049e, \tag{2.35}
$$

where E_a is *atmospheric emissivity;* W_B is black-body emissive power, a function of air temperature T_a at 2-m elevation; and e is vapor pressure in millibars.

TVA [1972] and Curtis and Eagleson [1982] cite several other alternatives to Eq. (2.35); some of them depend only on temperature:

Brunt [1932]:

$$
E_a = a + b(e)^{1/2} \tag{2.36}
$$

Angström [1915, 1936]:

$$
E_a = a - be^{-\gamma e} \tag{2.37}
$$

Swinbank [1963]:

$$
E_a = a \times 10^{-5} T_a^2 \tag{2.38}
$$

TABLE 2.8 Coefficients of Atmospheric Emissivity Formulas

INVESTIGATOR/SITE			
BRUNT'S FORMULA		*a*	*b* (in $mb^{-1/2}$)
Kimball (Washington, D.C.)		0.44	0.061
Angström (California)		0.50	0.032
Anderson (Oklahoma)		0.68	0.036
Eckel (Austria)		0.47	0.063
Goss and Brooks (Davis, Calif.)		0.66	0.039
ANGSTRÖM FORMULA	*a*	*b*	γ (in mb^{-1})
Angström (Sweden)	0.806	0.236	0.115
Kimball (Virginia)	0.800	0.326	0.154
Eckel (Austria)	0.710	0.240	0.163
Anderson (Oklahoma)	1.107	0.405	0.022
Linke's Meteorol. Taschenbuch, Vol. 2, Geest and Portig, Leipzig, 1953	0.790	0.174	0.041
SWINBANK FORMULA	*a*		
Australia/Indian Ocean	0.937		
IDSO AND JACKSON [1969]	*c*	*d*	
Phoenix, Ariz.	0.261	7.77×10^{-4}°C^{-2}	

Source: Adapted from TVA [1972], Walter O. Wunderlich, Tennessee Valley Authority Engineering Laboratory.

Idso and Jackson [1969]:

$$E_a = 1 - c \exp[-d(273 - T_a)^2] \tag{2.39}$$

Idso [1981]:

$$E_a = 0.70 + 5.95(10^{-5})e \cdot \exp(1500/T_a) \tag{2.40}$$

for 245°K $\leq T_a \leq$ 325°K, 3 mb $\leq e \leq$ 28 mb.

In all the above e is air vapor pressure in millibars at the 2-m elevation; T_a is air temperature in degrees Kelvin at the same level; and a, b, and c are empirical constants. Sample values of the constants are given in Table 2.8.

Over snow, the clear sky longwave atmospheric emissivity is taken as a constant,

$$\frac{I'_\ell}{W_B} = E_a = 0.757\,. \tag{2.41}$$

The presence of clouds increases longwave radiation. TVA [1972] suggests the following correction factor for a variety of conditions.

$$I'_\ell = KE_a\sigma T_a^4, \tag{2.42}$$

where coefficient K is related to cloud cover by

$$K = (1 + 0.17N^2). \tag{2.43}$$

Water vapor in the atmosphere is an effective radiation absorber. Because of this, clouds radiate very much like black bodies. This is the assumption made under overcast conditions for atmospheric longwave radiation. Hence, under overcast conditions radiation may also be estimated as

$$I'_\ell = \sigma T_C^4, \tag{2.44}$$

where T_C is absolute temperature of air at the cloud base.

Longwave albedo for water is about 0.03 and is essentially zero for all other granular surfaces. Net incoming longwave radiation is then given by

$$I^*_\ell = I'_\ell - I''_\ell = (1 - A)E_aW_B. \tag{2.45}$$

For water, Eq. (2.45) becomes

$$I^*_\ell = 0.97\sigma E_aT_a^4. \tag{2.46}$$

For other surfaces the net longwave radiation is simply,

$$I^*_\ell = \sigma E_aT_a^4, \tag{2.47}$$

where T_a is surface air temperature and E_a is defined according to Eqs. (2.35) through (2.40), depending on atmospheric conditions.

The effective longwave back-radiation or the net exchange of longwave back-radiation between the atmosphere and the earth's surface is given by

$$R_\ell = I^*_\ell - I_\ell, \tag{2.48}$$

where I^*_ℓ is atmospheric radiation and I_ℓ is surface back-radiation. Water emissivity is 0.97. All other earth surfaces are assumed to radiate as black bodies, according to the Stefan–Boltzmann law with temperature T of the surface. Summarizing, *clear sky net longwave radiation* is

$$R_\ell = (1 - A)\sigma E_aT_a^4 - \sigma ET^4, \tag{2.49}$$

where T_a is air temperature and T is surface temperature.

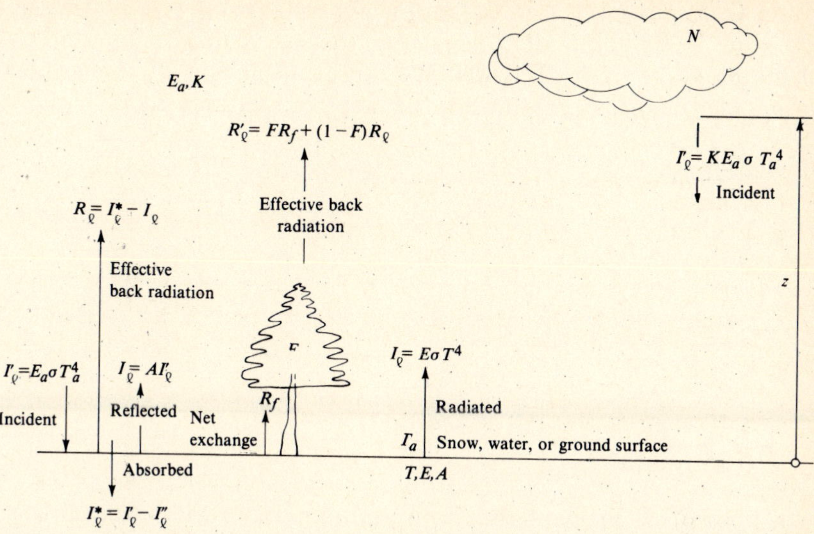

FIGURE 2.16 Summary disposition of longwave radiation. Source: Eagleson [1970].

The effect of forest cover on longwave radiation at the ground surface is considered like that of the overcast sky situation, black-body radiation at the temperature of the bottom of the canopy, T_f. Therefore,

$$R_f = \sigma(T_f^4 - ET^4), \tag{2.50}$$

where T_f is ambient air temperature and R_f is net radiation exchange between solid canopy and surface. The total exchange is then proportioned according to canopy density, $F(0 \leq F \leq 1)$

$$R'_\ell = FR_f + (1 - F)R_\ell. \tag{2.51}$$

Figure 2.16 gives the disposition of longwave radiation.

EXAMPLE 2.3

Net Radiation—The Greenhouse Effect

Given the radiation data seen in this chapter, you decide to go scuba diving off the coast of Nassau on a January 1 at noontime. There are, as expected, clear skies and clear water. You dive to a 10-m depth. Assume that the parcel of water you are diving in is in thermal equilibrium with surroundings and effectively re-radiates all shortwave radiation it receives. What would the water temperature be?

From Example 2.1, the radiation incident at the top of the atmosphere is 1.39 ly min^{-1}. The solar altitude, sin α, was previously computed as 0.67, which implies $\alpha = 42.1°$. To calculate the atmospheric effect on shortwave radiation, use Eq. (2.25). Following the steps of Example 2.2,

$$\begin{aligned} m &= \text{cosecant } \alpha \\ &= 1/\sin \alpha \\ &= 1.49 \end{aligned}$$

$$n = 2$$

$$\begin{aligned} a_1 &= 0.128 - 0.054 \log_{10} 1.49 \\ &= 0.119 \end{aligned}$$

$$\begin{aligned} I_c &= 1.39 \exp(-2 \times 0.119 \times 1.49) \\ &= 0.98 \text{ ly min}^{-1}. \end{aligned}$$

The above I_c is incident radiation and is partly reflected from the water surface. From Figure 2.12 with $\alpha = 42.1$, the albedo is approximately 0.065. Therefore, the net radiation at the water surface is

$$\begin{aligned} I_s^* &= 0.98(1 - 0.065) \\ &= 0.92 \text{ ly min}^{-1}. \end{aligned}$$

In going through 10 m of water, the above radiation is further reduced according to Eq. (2.33). From Table 2.6, we may use the extinction coefficients of clear oceanic water. This leads to $\beta = 0.64$ and $k = 0.081 \text{ m}^{-1}$. Substituting in Eq. (2.33),

$$\begin{aligned} I''_{sz} &= 0.92(1 - 0.64) \exp(-0.081 \times 10) \\ &= 0.15 \text{ ly min}^{-1}. \end{aligned}$$

To approximate the water temperature, assume that it is radiating as a black body everything it receives. Black-body radiation is given by Eq. (2.17), which can be equated to 0.15 ly min^{-1}.

$$T = [(0.15)/\sigma]^{1/4}.$$

Using σ of 0.826×10^{-10} ly $\text{min}^{-1}\,°\text{K}^{-4}$ results in,

$$T = 206.4°\text{K}$$

or about −66.7°C. This is an impossible answer! Where is the error? The incorrect assumptions are many; two are as follows:

1. The layer of water at 10 m is not isolated from its environment. There is heat transport from other layers by turbulent mixing, conduction, and advection, as will be discussed in Chapter 3.

2. The upper layers of water serve as insulators of the lower layer. The 10-m layer simply cannot radiate everything it absorbs, since there is a net exchange of radiation and temperature with upper layers that effectively establishes a higher equilibrium temperature. This *"greenhouse" effect* occurs because the medium is selective in the type of radiation that goes through. Shortwave radiation can go through water more easily than longwave radiation can go out. This effect is also the one that keeps our planet warm, with the atmosphere playing the role of insulator, or the glass in the greenhouse. Without atmosphere our planet would be frigid. ◆

2.4 SUMMARY

The study of radiation is necessary to understand the energy balance of the earth. As we have seen, planetary geometry conspires to create areas of surplus and deficits of energy. The established gradients drive global scale energy transfer processes like winds and ocean currents, as we will see in Chapter 3.

As seen in Table 2.2, large quantities of energy are also carried by water vapor, energy absorbed during the change of phase from water to vapor. As we will study in Chapter 5, radiation is again the energy source leading to evaporation. At that point the radiation physics techniques seen in Section 2.3 will become extremely useful, even at the level of a specific location.

REFERENCES

Angström, A. [1915]. "A Study of the Radiation of the Atmosphere." *Smithsonian Misc. Collec.* 65(3):1–159.

Idem. [1936]. "A Study of the Radiation During the Second International Polar Year." Stockholm: Meddelanden Fran Statens Meteorological Hydrology Anstalt. 6(8).

Bolsenga, S. J. [1964]. *Daily Sums of Global Radiation for Cloudless Skies.* Hanover, N.H.: U.S. Army Material Command, Cold Regions Research and Engineering Laboratory. (Technical report no. 160.)

Brunt, D. [1932]. "Notes on Radiation in the Atmosphere." *Q. J. R. Meteorol. Soc.* 58:389–418.

Budyko, M. I. [1974]. *Climate and Life.* New York: Academic Press.

Idem. [1977]. *Climatic Changes.* Washington, D.C.: American Geophysical Union.

Budyko, M. I., et al. [1962]. "The Heat Balance of the Surface of the Earth." *Soviet Geogr. Rev. Transl.* 3(5):3–16.

Byers, H. R. [1974]. *General Meteorology.* 4th ed. New York: McGraw-Hill.

Chan, S.-O., and P. S. Eagleson [1980]. "Water Balance Studies of the Bahr El Ghazal Swamp." Cambridge, Mass.: MIT Department of Civil Engineering, Ralph M. Parsons Laboratory. (Technical report no. 261.)

Charney, J. G. [1977]. "A Comparative Study of the Effects of Albedo Change in Drought in Semi-Arid Regions." *J. Atmos. Sci.* 34(9):1366–1385.

Curtis, D. C., and P. S. Eagleson [1982]. "Constrained Stochastic Climate Simulation." Cambridge, Mass.: MIT Department of Civil Engineering, Ralph M. Parsons Laboratory. (Technical report no. 274.)

Dake, J. M. K., and D. R. F. Harleman [1966]. "An Analytical and Experimental Investigation of Thermal Stratification in Lakes and Ponds." Cambridge, Mass.: MIT Department of Civil Engineering. (Hydrodynamics Laboratory report no. 99.)

Eagleson, P. S., and S.-O. Chan [1979]. "Water Balance Estimates of a Sudd Tributary." Proceedings of the Conference on Water Resources Planning. CU/MIT Technological Planning Program. Cairo, Egypt, June 25–27, 1979.

Eagleson, P. S. [1970]. *Dynamic Hydrology.* New York: McGraw-Hill.

Fritz, S. [1949]. "Solar Radiation During Cloudless Days." *Heat. Ventil.* 46:69–74.

Idso, S. B. [1981]. "A Set of Equations for Full Spectrum and 8- to 14-μm and 10.5- to 12.5-μm, Thermal Radiation from Cloudless Skies." *Water Resources Res.* 17(2):295–304.

Idso, S. B., and B. L. Blad [1971]. "The Effect of Air Temperature Upon Net and Solar Radiation Relations." *J. Appl. Meteorol.* 10:604–605.

Idso, S. B., and R. D. Jackson [1969]. "Thermal Radiation from the Atmosphere." *J. Geophys. Res.* 74:5397–5403.

Kimball, H. H. [1927]. "Measurements of Solar Radiation Intensity and Determination of Its Depletion by the Atmosphere." *Monthly Weather Rev.* 55:155–169.

Idem. [1928]. "Amount of Solar Radiation That Reaches the Surface of the Earth on the Land and on the Sea and Methods by Which It Is Measured." *Monthly Weather Rev.* 56:393–398.

Idem. [1930]. "Measurement of Solar Radiation Intensity and Determinations of Its Depletion by the Atmosphere." *Monthly Weather Rev.* 58:43–52.

Klein, W. H. [1948]. "Calculations of Solar Radiation and Solar Heat Load on Man." *J. Meteorol.* 5(4):119–129.

List, R. T. [1963]. *Smithsonian Meteorological Tables.* 6th rev. ed. Washington, D.C.: Smithsonian Institution.

Marsh, W. M. [1987]. *Earthscape: A Physical Geography.* New York: Wiley.

Marsh, W. M., and J. Dozier [1986]. *Landscape: An Introduction to Physical Geography.* New York: Wiley.

McAdams, W. H. [1954]. *Heat Transmission.* 3rd ed. New York: McGraw-Hill.

Raphael, J. M. [1962]. "Prediction of Temperature in Rivers and Reservoirs." *Proc. Am. Soc. Civil Eng., J. Power Div.* No. PO2, Paper 3200.

Reitan, C. H. [1960]. "Distribution of Precipitable Water Vapor Over the Continental United States." *Bull. Am. Meteorol. Soc.* 41(2):79–87.

Strahler, A. N., and A. H. Strahler [1983]. *Modern Physical Geography.* 2nd ed. New York: Wiley.

Sutton, O. G. [1953]. *Micrometeorology.* New York: McGraw-Hill.

Swinbank, W. C. [1963]. "Long-Wave Radiation from Clear Skies." *Q. J. R. Meteorol. Soc. Lond.* 89:339–348.

Tennessee Valley Authority [1972]. "Heat and Mass Transfer Between a Water Surface and the Atmosphere." Norris, Tenn.: Tennessee Valley Authority. (Laboratory report no. 14; Water resources research report no. 0-6803.)

Trewartha, G. T. [1968]. *An Introduction to Climate.* New York: McGraw-Hill.
Trewartha, G. T., and L. H. Horn [1980]. *An Introduction to Climate.* 5th ed. New York: McGraw-Hill.
U.S. Army Corps of Engineers [1956]. *Snow Hydrology.* Portland, Oreg.: U.S. Army Corps of Engineers, North Pacific Division.

PROBLEMS

1. Find the effective incoming shortwave radiation I_s^* in langleys per minute under the following set of conditions: latitude, 15°N; date, July 10; hour, noon; clouds, overcast at 3000 ft; air turbidity, smoggy; surface, grass-covered ground. (From P. S. Eagleson, *Dynamic Hydrology,* McGraw-Hill, 1970.)

2. Vapor pressure is 36.1 mb and air temperature is 90°F. Calculate the clear sky net longwave radiation exchange between the atmosphere and a water surface, where the temperature of the latter is 85°F.

3. Suppose the albedo of the planet earth is 0.34 with respect to solar radiation. Assume that the earth radiates as a black body, neglect atmospheric absorption of the terrestrial radiation, and calculate the earth's black-body temperature. Compare this with the actual mean of 287°K and comment on any difference observed. (From P. S. Eagleson, *Dynamic Hydrology,* McGraw-Hill, 1970.)

4. Large clouds of dust and smoke are expected to occur in the event of a major nuclear exchange. In such an event, the total air's absorption–scattering coefficients a_t would be 2.3 and 13 for smoke and dust clouds. The net radiation I_s' will be 10% of the incoming "clear" sky shortwave radiation I_c.

For the location and general conditions of Problem 1 compute the incoming shortwave radiation in the ground for both dust and smoke clouds. Assume that the atmosphere's ozone layer does not suffer enough to change the earth's solar constant. What are the turbidities of dust and smoke clouds? Comment on possible climatic effects caused by a major nuclear war.

5. Find the net longwave and shortwave radiation on a lake using the following information: latitude, 15°N; date, July 9; hour, noon; overcast (cloudy) with 3000-ft cloud-base elevation; smoggy air conditions; temperature of water, 18.33°C; emissivity of water, 0.8; cloud-base temperature, 20°C; albedo of water, 0.1; temperature of air, 26.67°C; relative humidity of air, 0.85.

6. Assume that the earth is a rotating sphere with an axis perpendicular to a line from the center of the sun to the center of the earth. What would be the annual total radiation at latitudes 0, 20, 40, 60, 80, and 90?

7. Discuss what may happen to the earth's energy balance if all of the Amazonian jungle were cut and razed to bare soil. How would the global water balance be affected? Are there relations between the energy and water balances?

8. If you had a photoelectric cell that converted solar radiation to electricity with 5% efficiency, what area would be required to produce 1000 megawatts of power—the equivalent of a good-sized fossil or a nuclear power plant?

9. Discuss and quantify to the extent possible what may happen if the earth's axis tilt were to become 30° rather than 23°37′.

10. The hour angle of the sun τ is given (in degrees) by the simplified expressions

$$\tau = (T_S + 12 - \Delta T_1) \times 15 \qquad \text{east of observer's meridian,}$$

$$\tau = (T_S - 12 - \Delta T_1) \times 15 \qquad \text{west of observer's meridian,}$$

where T_S = standard time at observer's site counted from midnight in hours

$(0 \leqq T_S < 24)$

$$\Delta T_1 = \frac{i}{15}(\theta_S - \theta_L);$$

$i = -1$ for west longitude (relative to Greenwich)

$+1$ for east longitude;

θ_S = longitude of standard meridian, where the observer's time zone is centered (every 15° east of Greenwich adds an hour relative to Greenwich); and

θ_L = longitude of observer's meridian.

a) Compute analytically the incident solar radiation over a period $\Delta t = t_2 - t_1$ (in hours) at a specific site and day of the year.

b) Obtain an estimate of the daily insolation outside the earth's atmosphere at latitude $\Phi = 20°$N on the Greenwich meridian and on May 31.

(Contributed by Dr. Angelos Protopapas, based on the work of Curtis and Eagleson [1982].)

11. Assume that some chemicals are injected in the upper atmosphere. The chemicals are such that they allow the incoming solar radiation to penetrate fully, but do not allow reflection from the soil surface to escape back to space. Assume that the effective transmissivity of the atmosphere and chemicals is f. That is, the radiation reaching the ground is $R \times f$, where R is the radiation perpendicular to the top of the chemical layer. The albedo of the earth surface is a_s and of the chemical layer is a_ℓ. Think of a simplified conceptualization like the figure below.

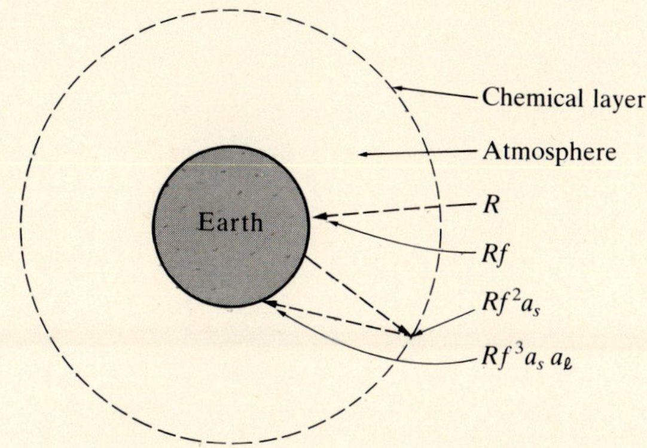

a) Compute the total radiation flux absorbed by the earth.

b) Sketch the total absorbed radiation flux R_a as we change a_ℓ from 0 to 1. (*Hint:*

$$\sum_{i=1}^{n} 1 + a + a^2 + \cdots + a^n = \frac{a^n - 1}{a - 1} \to \frac{1}{1 - a}$$

$$n \to \infty$$

$$|a| < 1).$$

(Contributed by Dr. Angelos Protopapas.)

Chapter 3

Principles of Meteorology: The Earth–Atmosphere System

3.1 INTRODUCTION

In the previous chapter we studied solar radiation, the ultimate fuel of the hydrologic cycle. The uneven temporal and spatial distribution of radiation leads to an uneven distribution of temperature throughout the earth–atmosphere system and, effectively, a lack of balance of energy. All active meteorologic and hydrologic processes originate in order to redistribute energy throughout the system. The earth and the atmosphere are the media through which the energy transport commonly occurs, involving large transfers of mass. These transport activities dictate climate and weather as we know it.

The extent to which the atmosphere or the earth (land masses and oceans) dominates climate is unclear. The land masses are mostly moisture sinks and play a somewhat passive, but important, part in climate through topography, temperature, and surface albedo characteristics. Vegetation on land is also important in moisture recycling through evapotranspiration, which will be studied in Chapter 5. The oceans and the atmosphere are more active participants in redistributing mass and energy. This chapter will concentrate on the atmospheric and oceanic activities—subjects of meteorology and oceanography. The rest of the book will emphasize activities over, on, or under land, which are more commonly related to hydrology.

This chapter will cover the dominating atmospheric characteristics such as temperature, pressure, winds, and water content of the atmosphere. Ocean currents will also be discussed. It is intended to provide adequate background for the hydrologic subjects to follow.

3.2 COMPOSITION AND GENERAL CHARACTERISTICS OF THE ATMOSPHERE

The gases surrounding the earth are usually divided into upper and lower atmospheres, the demarcation being at an elevation of about 50 km. The upper atmosphere plays, as far as is known, a secondary role to climate determination.

The lower atmosphere is where most of the critical mass and energy transfers occur. It is usually divided in two parts: the stratosphere and troposphere. The atmospheric layers are defined in terms of the temperature distribution. A typical temperature profile is shown in Figure 3.1.

The troposphere is the atmospheric layer in direct contact with the earth's surface. It is the layer where most of the energy, momentum, and mass transfer occur. As such, it is where we observe the development of winds and precipitation on the earth's surface.

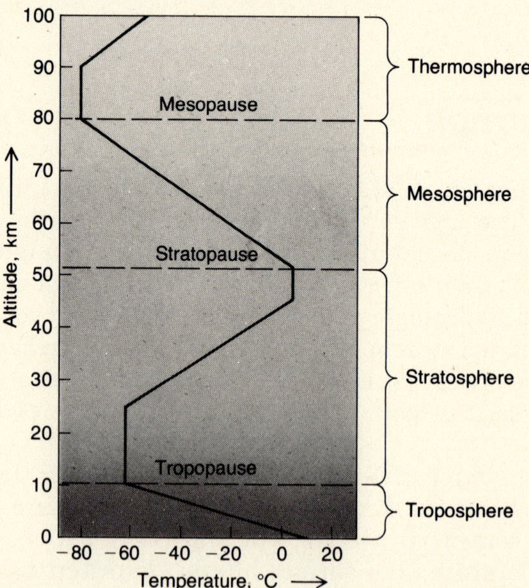

FIGURE 3.1 Temperature profile of the atmosphere, with subdivisions. This change in temperature with altitude is considered to be typical in the middle latitudes and is called the U.S. Standard Atmosphere. The boundaries between the subdivisions are termed pauses, meaning change. Source: W. M. Marsh and J. Dozier, *Landscape: An Introduction to Physical Geography.* Copyright © 1986 by Wiley. Reprinted by permission of John Wiley & Sons, Inc.

A few tropospheric characteristics are

1. Variable thickness, 8 km at the poles and 16 km at the equator;
2. Decreasing temperature with elevation;
3. Well-defined pressure gradients;
4. Generally well-defined distribution of moisture and suspended particles; and
5. Sharp velocity gradient, starting at 0 at the earth's surface and developing over a 2700-m-thick boundary layer that seriously influences microclimatic conditions.

The most common gases in the atmosphere are nitrogen and oxygen; between them they compose about 99% of the atmosphere by volume. Table 3.1

TABLE 3.1 Composition of the Atmosphere

GAS	% BY VOLUME IN TROPOSPHERE	NOTES
Nitrogen (N_2)	78.084	
Oxygen (O_2)	20.946	Has developed with the evolution of plant life in past billion years.
Argon (A)	0.934	
Carbon dioxide (CO_2)	0.033	Only 0.029 in nineteenth century; absorbs long-wave radiation in the 1–5 and 12–14-μm range.
Neon (Ne)	0.00182	
Helium (He)	0.000524	
Methane (CH_4)	0.00016	
Krypton (Kr)	0.00014	
Hydrogen (H_2)	0.00005	
Nitrous oxide (N_2O)	0.000035	Absorbs radiation above 1 μm.
Important Variable Gases		
Water vapor (H_2O)	0–4	Absorbs radiation in the 0.85–6.5-μm range and the range longer than 18 μm.
Ozone (O_3)	0–0.000007 at ground level (0.00001–0.00002 in stratosphere and mesosphere)	Absorbs ultraviolet radiation in upper atmosphere.

Source: W. M. Marsh and J. Dozier, *Landscape: An Introduction to Physical Geography*. Copyright © 1986 by Wiley. Reprinted by permission of John Wiley & Sons, Inc.

gives a complete accounting of gaseous components. The so-called permanent constituents do not show considerable temporal and spatial variations. The variable components usually decrease with increasing elevation and vary widely in time and space.

3.3 TRANSPORT PROCESSES

The inequalities in heat balance seen in Chapter 2 drive a series of transport processes that spread the available energy, as well as existing mass and momentum, over the earth–atmosphere system. This exchange and transport occurs between and within the oceans, land masses, and atmosphere.

The main transport processes are conduction, convection, and advection. Conduction is the interchange of molecules between adjacent substances due to molecular movement. Momentum, energy, and mass can be transported in this manner. Conduction is characterized by a transport rate proportional and in the direction of the decreasing gradient of temperature, concentration, or velocity. The proportionality constant is usually only a function of the substance or medium.

Heat transport by conduction is given by

$$q = -k\, dT/dx, \tag{3.1}$$

where q is the rate heat transfer per unit area (heat flux) in the x direction; dT/dx is the temperature gradient; and k is the thermal conductivity in calories per degree per centimeter per second.

Momentum transport in laminar flow is given by

$$\tau_{zx} = -\mu\, du/dz, \tag{3.2}$$

where u is local velocity (meters per second); μ is the coefficient of dynamic viscosity, a function of the fluid (newtons per second per square meter or kilograms per meter per second); and τ_{zx} is stress (momentum flux) on a plane normal to z (newtons per square meter).

Mass transport is given by Fick's law of diffusion,

$$J_{AB} = -D_{AB}\, dC_A/dx, \tag{3.3}$$

where J_{AB} is the mass flux of diffusant in x direction in grams per square meter per second; C_A is the local concentration of diffusant; and D_{AB} is molecular diffusivity, a function of the medium and diffusant in grams per meter per second.

Vapor transport in air is partly accounted for by molecular diffusion. Molecular diffusion also plays a part in heat transport processes.

EXAMPLE 3.1

Shear Stress in Laminar Flow

To illustrate transport by conduction, assume you have water at 20°C, with a dynamic viscosity of 0.01 $\mathrm{g\,cm^{-1}\,s^{-1}}$. The density of water at 20°C is 0.99821 $\mathrm{g\,cm^{-3}}$. Assume the water is flowing in an open channel with a velocity profile of the following form:

$$u = 100 - 100e^{-0.0054z},$$

where u is in centimeters per second and z is the depth of the water measured from the bottom in centimeters. The gradient of the above velocity profile is

$$\frac{du}{dz} = 100(0.0054)e^{-0.0054z} = 0.54e^{-0.0054z}.$$

At $z = 100$ cm, that would give,

$$\frac{du}{dz} = 0.315\ \mathrm{s^{-1}}.$$

If the flow is laminar, then shear or momentum transfer at $z = 100$ cm will be (using Eq. 3.2),

$$\begin{aligned}\tau_{zx} &= -0.01(\mathrm{g\,cm^{-1}\,s^{-1}})0.315\ \mathrm{s^{-1}}\\ &= -0.00315\ \mathrm{g\,cm^{-1}\,s^{-2}}.\ \blacklozenge\end{aligned}$$

Convection may be of a turbulent or a thermal nature. Turbulent convection results from the random movements of eddies of various sizes and intensities. These rotating vortices move about haphazardly, carrying and distributing their cargoes of energy, mass, and momentum. Turbulent convection accounts for considerably more transport activities than diffusion, which is usually limited to stationary conditions or laminar flow in fluids.

By analogy with diffusion, and resulting from a probabilistic analysis where turbulence is seen as a random, statistically stationary, component superimposed on a constant mean behavior, turbulent convection takes the following forms:

For evaporation,

$$E = -\rho k_w \frac{\partial \overline{q}_h}{\partial z}, \tag{3.4}$$

where E is the vertical flux of water vapor in kilograms per square meter per second; $\overline{q}_h$ is the specific humidity, temporal mean mass of water per unit mass of moist air; k_w is the vapor eddy diffusivity in square meters per second;

and ρ is the density of water vapor in kilograms per cubic meter. The above equation will be discussed extensively in Chapter 5.

For momentum transfer,

$$\tau_{zx} = -\rho k_m \frac{\partial \bar{v}}{\partial z}, \tag{3.5}$$

where τ_{zx} is vertical flux of momentum per unit area in newtons per square meter (also shear on a horizontal plane); $\bar{v}$ is the time averaged fluid velocity in meters per second; and k_m is the kinematic eddy viscosity in square meters per second. Proportionality factors in turbulent diffusion are not only a function of medium but also of the system state.

The rate of turbulent convection is usually several orders of magnitude larger than that of molecular diffusion. Turbulent transport plays an important role in the redistribution of energy mass and momentum over the earth. Turbulent evaporation leads to moisture recycling as well as tremendous latent heat transfers (see Section 3.7.1). Turbulent momentum transfer seriously influences winds and ocean currents, which are the major advective mechanisms discussed in Section 3.6.

The following sections will describe the spatial and temporal distributions of properties that drive the transport processes.

3.4 TEMPERATURE DISTRIBUTION

The nonuniformity of radiation and heat budgets is reflected in the variability of temperatures throughout the earth and the atmosphere. This temperature distribution has a great influence on climate as we know it and affects everything from our daily routines to the genesis of precipitation and ocean currents.

3.4.1 Temporal Distribution

Usually, air temperature rises during the day and falls at night, the peak temperature lags behind the peak daily incoming radiation by several hours. This lag is due to the lag in maximum radiation leaving the earth relative to the maximum daily incoming radiation, as illustrated in Figure 3.2. A typical daily fluctuation of temperature is shown in Figure 3.3 for clear and cloudy days. The implication is that cloud cover buffers the maxima of incoming and outgoing radiation. In fact, on any given day the local meteorologic conditions may destroy or even reverse the expected daily temperature fluctuation. The daily temperature fluctuations are, on the average, of larger magnitude over continental land masses than over oceans. An ocean distributes heat more efficiently throughout its fluid mass, thereby preventing large and quick temperature fluctuations.

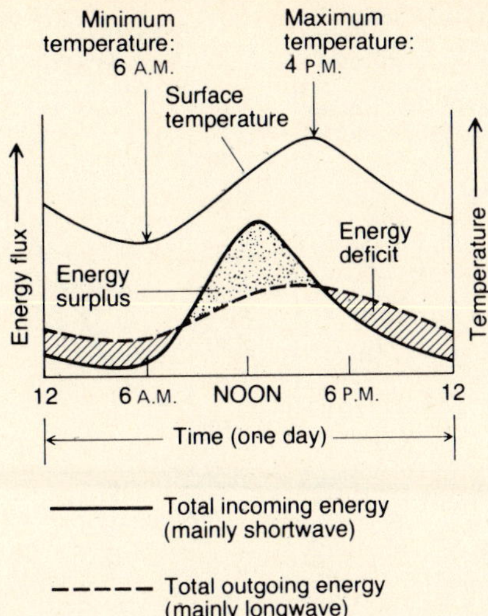

FIGURE 3.2 Peak period of daily outgoing longwave radiation. Source: After Oke [1978]. W. M. Marsh and J. Dozier, *Landscape: An Introduction to Physical Geography.* Copyright © 1986 by Wiley. Reprinted by permission of John Wiley & Sons, Inc.

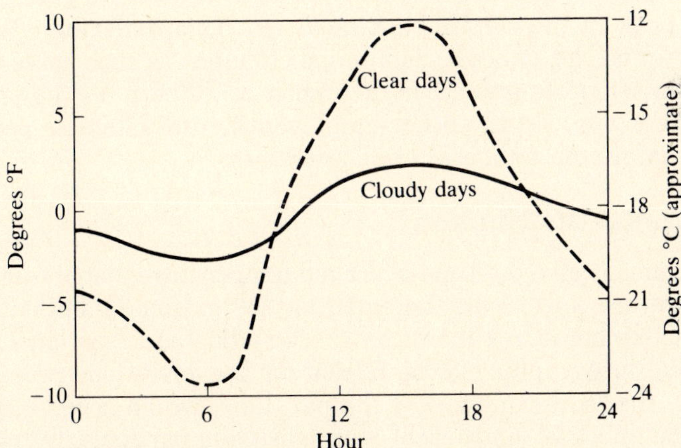

FIGURE 3.3 Daily movement of air temperature at Washington, D.C., on clear and cloudy days. The data represent deviations from the 24-hour mean. Each curve is the mean of 10 days at about the time of the autumn equinox, when days and nights are nearly equal. The selected days had a minimum of advection. Source: After Landsberg [1958]. G. T. Trewartha, *An Introduction to Climate,* 4th ed., McGraw-Hill, 1968. Reproduced with permission.

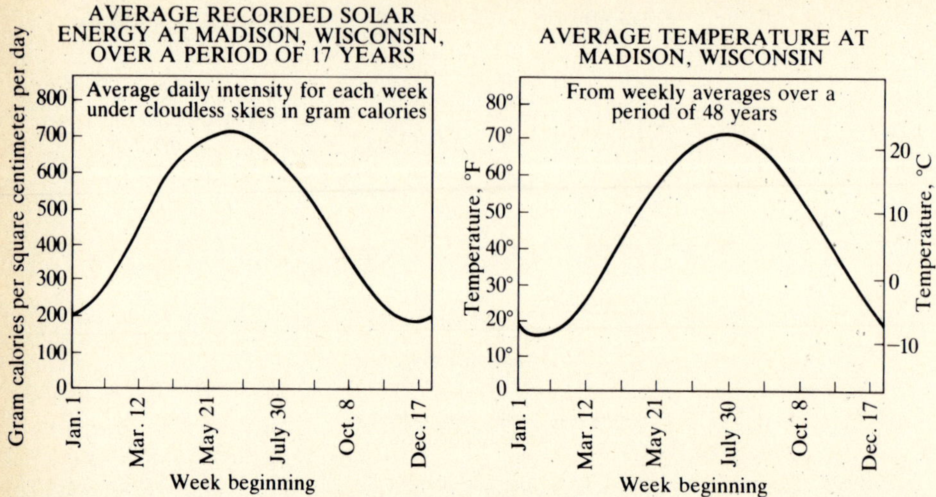

FIGURE 3.4 Maximum and minimum temperatures lag a month or so behind maximum and minimum solar radiation. The solar radiation curve has been smoothed slightly. Source: G. T. Trewartha, *An Introduction to Climate,* 4th ed., McGraw-Hill, 1968. Reproduced with permission.

The seasonal air temperatures also follow closely the annual cycle of incoming solar radiation. The pattern is illustrated in Figure 3.4. The observed lag in peak temperature relative to peak radiation again responds to the delayed peak in the earth's back-radiation. This buffering effect is more significant over oceans or ocean-dominated climates. In such cases, maximum and minimum temperatures occur in August and February (northern hemisphere), respectively. On the other hand, continental climates peak in July and register minimum temperatures in January.

3.4.2 Horizontal Distribution

Figures 3.5 and 3.6 give the time-averaged temperature distribution (maps of equal temperature lines) over the earth for the extreme months of January and July. The temperatures shown have been adjusted to sea level in order to eliminate local topographic effects. Except for the deviations caused by large land masses, the isotherms follow the parallels, which receive equal solar radiation. The effect of continental land masses is particularly important in the northern hemisphere. The winter–summer drift of the isotherms again coincides with the general annual movement of belts of equal radiation induced by planetary motion. The maximum temperature occurs somewhat above the equator, possibly because of the cloud effects in that region. The "greenhouse" effect of the cloudy tropics, together with a more uniform annual radiation budget, makes this region the one with the smallest temperature fluctuations throughout the year.

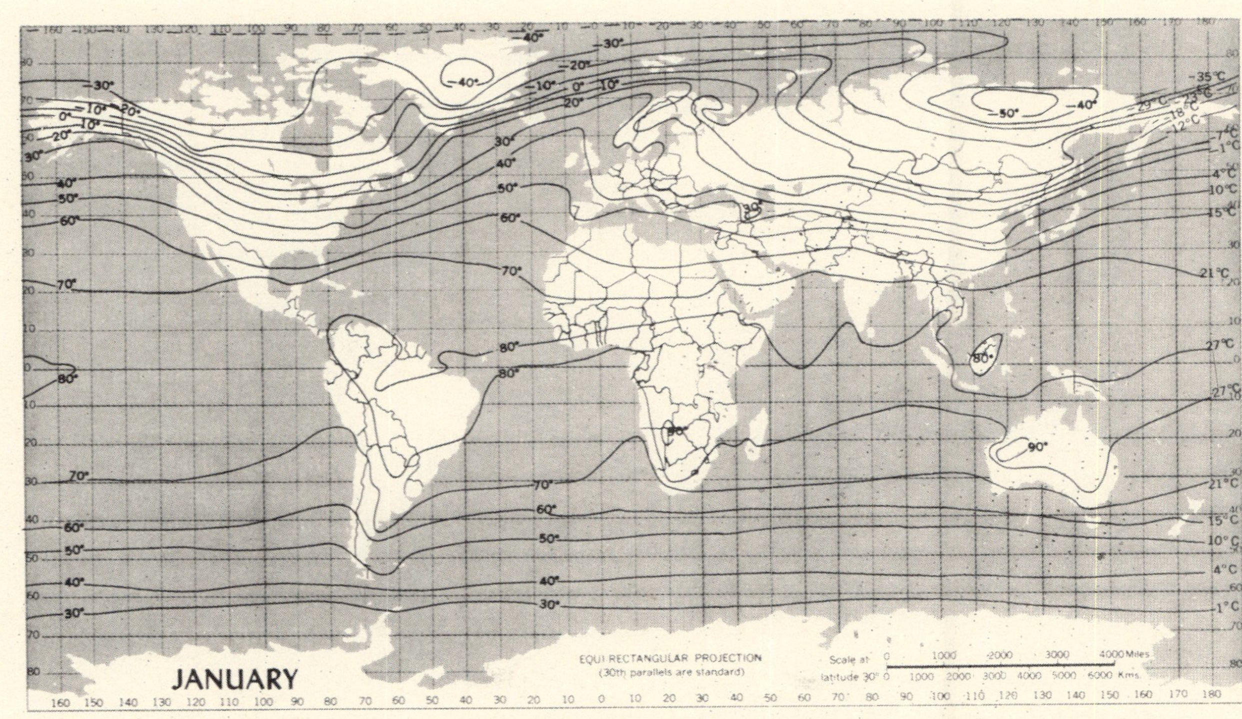

FIGURE 3.5 Average sea-level temperatures, January. Source: G. T. Trewartha, *An Introduction to Climate,* 4th ed., McGraw-Hill, 1968. Reproduced with permission.

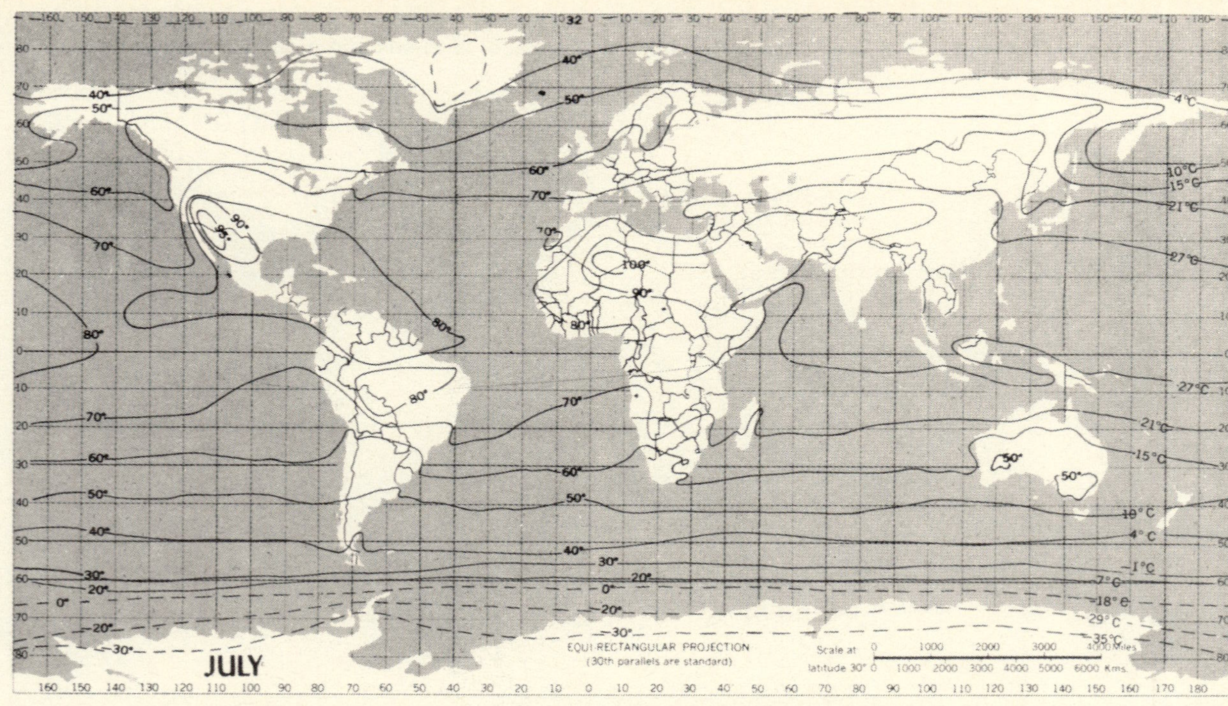

FIGURE 3.6 Average sea-level temperatures, July. Source: G. T. Trewartha, *An Introduction to Climate,* 4th ed., McGraw-Hill, 1968. Reproduced with permission.

If the temperatures expected from radiation received are subtracted from the sea-level isothermal maps shown in Figures 3.5 and 3.6, the results are maps of temperature anomalies or isanomalies of temperature. These point out regions with average temperatures that are unusually high or low. Figures 3.7 and 3.8 are isanomalies for the months of January and July. The features shown reflect the oceans' ability, relative to the land masses, to stay warmer in winter and colder in summer. Most significantly, though, are the effects of ocean currents on air temperature. As will be seen in Section 3.6.2, the unusual winter warmth of western Europe is due to a warm ocean current along its shores. Similarly, the lower temperatures of western Chile are due to cold Antarctic currents. The large land mass of northeastern Asia leads to significant negative and positive temperature anomalies in winter and summer, respectively.

3.4.3 Vertical Temperature Distribution in the Lower Atmosphere

As mentioned previously, the troposphere shows a well-defined gradient of temperature, with the higher temperatures at the earth's surface. As we will soon see, atmospheric temperature is very much dictated by heat from the earth's surface.

The temperature variation in the troposphere is reasonably assumed to be linear (or piecewise linear)

$$T = T_{\circ} - \alpha z, \tag{3.6}$$

where T is ambient temperature at elevation z, and $T_{\circ}$ is surface temperature. The rate of cooling α is called the ambient lapse rate; it usually varies between 5 and 8°C km^{-1}.

The ambient lapse rate plays a major role in establishing the stability or instability of air masses, which in turn lead to no precipitation or precipitation, respectively. Stability here refers to the ability of air masses to rise due to thermal convection in the atmosphere. They will rise only if their temperature is warmer than the surrounding air. Therefore, the rate of cooling of the air mass, relative to the ambient air, is a critical thermal stability factor. This will be discussed in detail in Section 3.8.

Unusually stable weather occurs when the temperature of the air is increasing with elevation rather than decreasing; this condition is called a thermal inversion. In such an atmosphere, any rising (and therefore cooling) (see Section 3.8) air mass will always be heavier than the ambient air and will sink, remaining at low elevations, unable to condense its water vapor significantly.

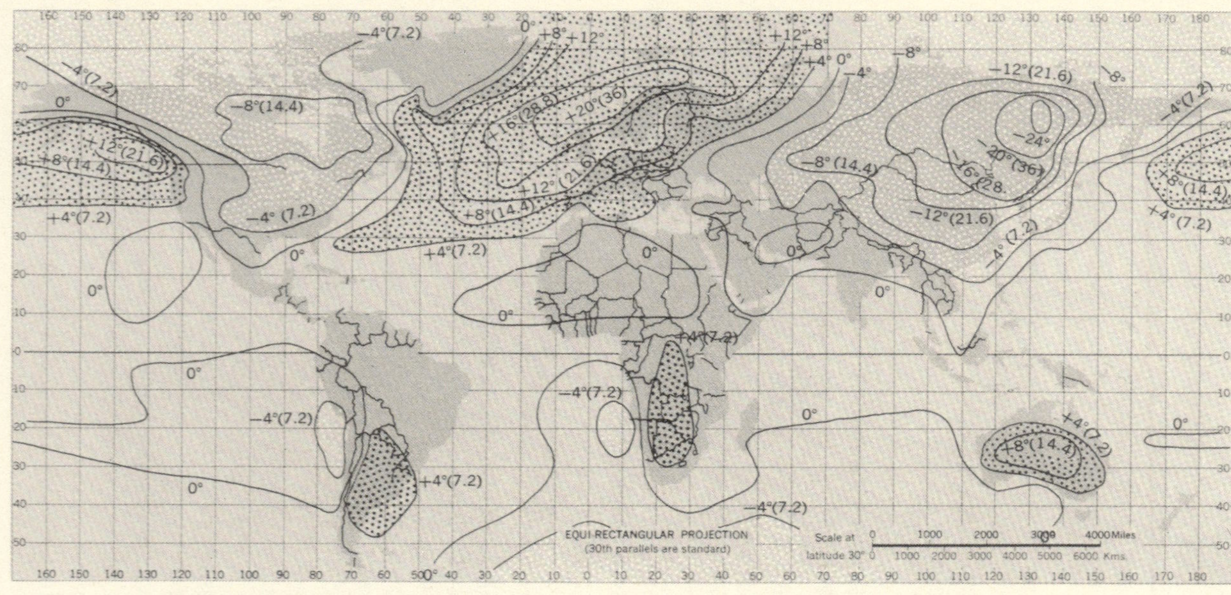

FIGURE 3.7 Isanomalies of temperature, January. Temperatures are in degrees Centigrade with degrees Fahrenheit in parentheses. Source: G. T. Trewartha, *An Introduction to Climate,* 4th ed., McGraw-Hill, 1968. Reproduced with permission.

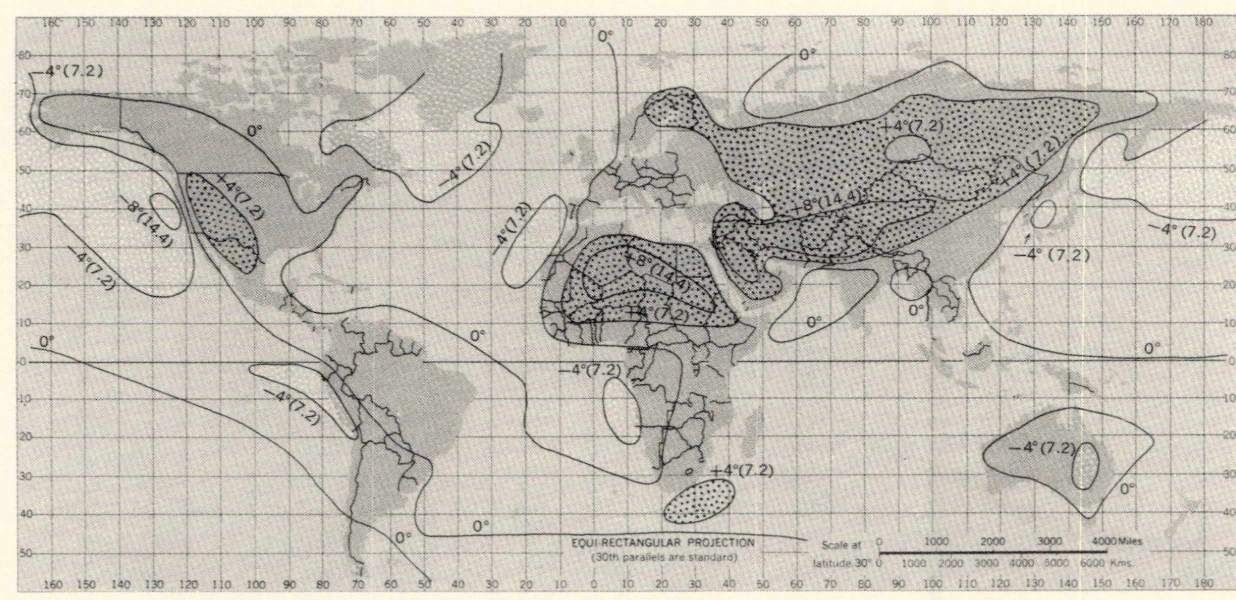

FIGURE 3.8 Isanomalies of temperature, July. Temperatures are in degrees Centigrade with degrees Fahrenheit in parentheses. Source: G. T. Trewartha, *An Introduction to Climate,* 4th ed., McGraw-Hill, 1968. Reproduced with permission.

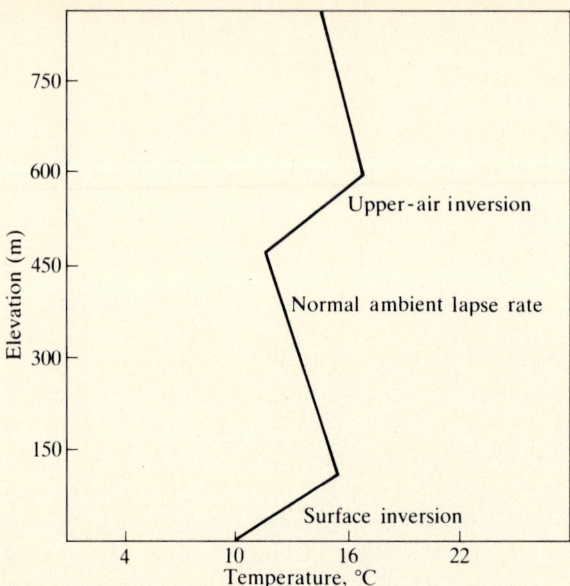

FIGURE 3.9 Surface and upper-air inversion.

Inversions commonly occur near the earth's surface after nights with clear skies, when longwave radiation from the earth can escape through the atmosphere. On a clear night, the land surface cools much faster than upper air layers. The net result is that adjacent air cools and shows increasing temperature with elevation. At some higher elevation the normal ambient lapse rate will take over. A surface inversion is illustrated in Figure 3.9. Inversions are enhanced and more common when the air, in addition to being cloudless, is dry (water vapor is a good radiation absorber); nights are long (as they are in winter); air is calm, so as to avoid turbulent mixing; and the surface is covered with snow, which enhances cooling and prevents soil from heating adjacent air masses. Inversions are more significant over land masses than over oceans. Upper-air inversions may also occur, usually caused by the convergence of air masses of different temperatures or by the subsidence of cold air over centers of diverging air (anti-cyclones, Section 3.6.1). This condition is illustrated in Figure 3.9. Upper inversions can effectively establish a ceiling to cloud development and thermal convection, resulting in little or no precipitation.

3.4.4 Temperature Measurements

Temperature is usually measured by thermometers in degrees Celsius (or Centigrade) (°C). Water freezes at 0°C (32°F) and boils at 100°C (212°F). For correct temperature measurements, the location of the instrument should avoid direct sunlight, otherwise it may register unusually high temperatures.

Thermometers should be reasonably ventilated, but protected from wetness and extreme winds.

Absolute temperature is usually given in degrees Kelvin (°K):

$$°K = °C + 273.15 \tag{3.7}$$

or degrees Rankine:

$$°R = °F + 459.69\,. \tag{3.8}$$

Normally, daily air temperature is measured as a maximum, T_{max}, and a minimum, T_{min}. Reported daily temperature averages are commonly obtained as

$$T_{avg} = \frac{T_{max} + T_{min}}{2}\,. \tag{3.9}$$

The above approximation to the true daily mean is best when the daily temperature cycle is symmetrical. That is rarely the case. Aggregated averages (i.e., weekly, monthly, seasonal, yearly) are obtained by arithmetic averaging of daily quantities over the period of interest.

Average information can be misleading, since it gives no information on the range of temperatures or on the relative frequency of temperatures. This type of information can be obtained by preparing histograms of frequency of occurrence of various temperature ranges. Similarly, the frequency of departure from the mean temperature—or simply, the mean departure—is valuable information.

Computation of monthly (or even daily) means over years of data illustrates the expected variations throughout the year. Obviously, statistical analysis of temperature data is possible. These procedures will be discussed in Chapter 11, within the framework of rainfall and streamflow data analysis.

Temperature information is also sometimes given in terms of degree-days. These are accumulated departures of the daily average temperature from a reference temperature over a preselected number of days N:

$$\text{degree-days} = \sum_{i=1}^{N} (T^{i}_{avg} - T_{ref})\,. \tag{3.10}$$

This concept is commonly used in agriculture to evaluate plant growth. It is assumed that growth occurs as a function of degree-days above a given reference temperature, which is different for each crop. Each crop, in turn, needs a given number of degree-days to achieve maturity.

Similarly, the concept of heating-degree-days is the result of accumulated daily temperatures from 65°F (18.3°C). Usually, the negative of Eq. (3.10) is used so that positive quantities imply temperatures below 65°F (18.3°C). This heating-degree-day indicator is used as a measure of heating-fuel consumption, which is taken to be directly proportional to the indicator.

Another temperature-dependent agricultural measure is the growing season. It is defined as the longest consecutive period of temperatures above freezing — 0°C or 32°F.

Vertical profiles of temperature can be obtained by meteorologic balloons or by aircraft, which carry instruments aloft. Recently, satellite technology has also permitted remote atmospheric temperature measurements. An example is the geostationary meteorologic satellites series, or GOES. The U.S. Department of Commerce (through its National Weather Service), the European community, Japan, and the U.S.S.R. support such systems. The satellites are geostationary because their orbit altitude is such that their speed is synchronous with the earth's rotation. They are always in the same location relative to the earth's surface.

Among the activities of the GOES satellites (and others) are picture-taking in the visible spectrum and in the infrared spectrum. The latter is particularly useful for temperature measurement. It is usually assumed that the radiating bodies are black bodies. Using the Stefan–Boltzmann law seen in Chapter 2, it is then possible to obtain temperature as a function of the intensity of measured longwave, infrared, radiation. This technique can be used to obtain cloud top-temperatures, water-surface temperatures, soil-surface temperatures, etc. The information can potentially be used for determining rainfall rates, moisture contents, and moisture profiles and for detecting other natural or manmade environmental disturbances.

EXAMPLE 3.2

Simple Atmospheric Stability

Assume you have an ambient lapse rate of 6.5°C km^{-1}. An air mass, at an elevation of 4000 m and a temperature of 4°C moves into the area. The ground-level air temperature is 75°F. Will the air mass rise or sink (be unstable or stable) in the new environment?

To answer the above question, we must compare the relative temperatures of the ambient air at 4000 m to that of the air mass. The ground-level air temperature in degrees Centigrade is about 24. With the given ambient lapse rate, the air is 26°C ($6.5 \times 4000/1000$) colder at 4000 m. The ambient temperature at that level is then −2°C ($24 - 26$). Since the incoming air mass is still warmer, it will continue to rise, possibly leading to condensation of the moisture it may be carrying. ◆

3.5 PRESSURE DISTRIBUTION

The daily pressure distribution at sea level is varying and unstable. Nevertheless, upon averaging, there are some clearly discernible patterns of pressure of a semipermanent nature. Figures 3.10 and 3.11 show the average

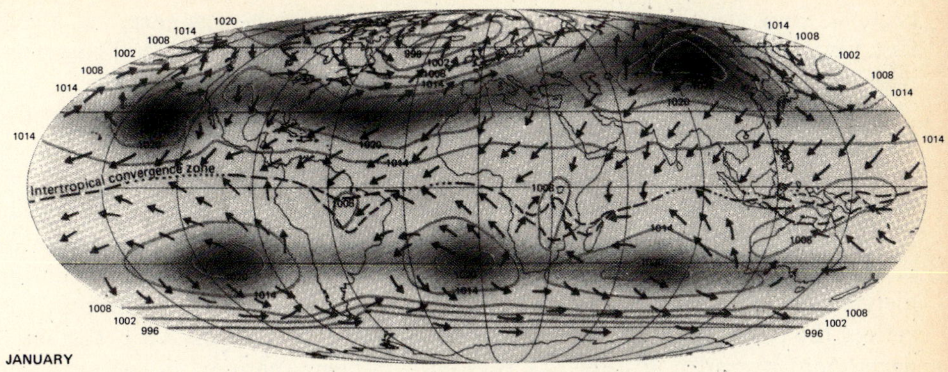

FIGURE 3.10 General distribution of air pressure (in millibars) and atmospheric circulation on a global scale, in January. Source: W. M. Marsh, *Earthscape: A Physical Geography.* Copyright © 1987 by Wiley. Reprinted by permission of John Wiley & Sons, Inc. Map projection by Waldo Tobler.

sea-level pressures (after adjustment for topographic differences) observed in the extreme months of January and July. Pressures are given in millibars—one standard atmosphere is 1013.2 mb. One millibar is equivalent to 100 $\mathrm{N\,m^{-2}}$ (1 $\mathrm{N\,m^{-2}}$ is a Pascal).

Pressure systems seem to be reasonably organized in latitudinal bands, mostly in cells of isobaric conditions spreading over a given latitude. Principal features are the high-pressure cells at the middle latitudes (25 to 35°) in both the northern and southern hemispheres. An equatorial low is found between the two high-pressure centers. Low-pressure belts also occur in the subpolar

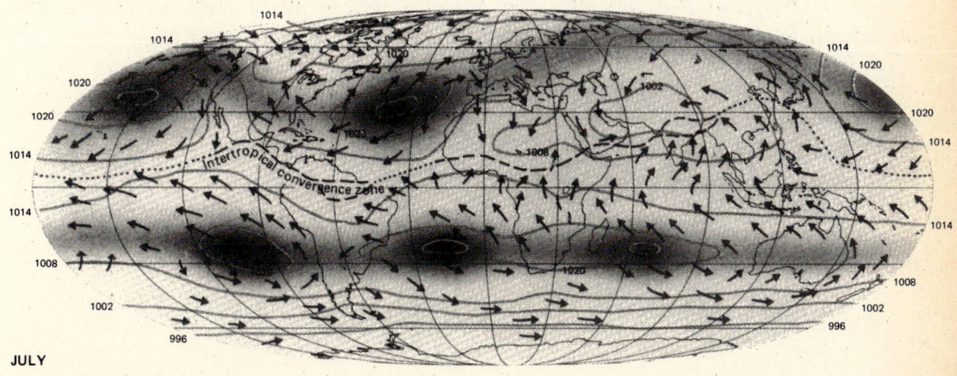

FIGURE 3.11 General distribution of air pressure (in millibars) and atmospheric circulation on a global scale, in July. Source: W. M. Marsh, *Earthscape: A Physical Geography.* Copyright © 1987 by Wiley. Reprinted by permission of John Wiley & Sons, Inc. Map projection by Waldo Tobler.

regions. They are of a cellular nature in the northern hemisphere, but a continuous, beltlike occurrence around the southern polar regions. The actual pressure conditions at the pole are very uncertain and unknown.

The observed mean pressure patterns are not only of a thermal nature. Their origin and the extent to which they are influenced by land masses, oceans, and other factors remains largely unknown. In reading, keep in mind that the pressure features being discussed are average values that are not necessarily discernible on any daily pressure distribution. The patterns of lows and highs migrate northward in July, following the solar radiation distribution. The pressure conditions at higher elevations do not necessarily reflect the sea-level observations. As will be seen in Section 3.6.1, the semipermanent pressure system has related semipermanent wind patterns and associated weather.

There are diurnal pressure fluctuations of small absolute values that can be significant as precursors of weather disturbances. Riehl [1978] discusses how to adjust for these fluctuations (on the order of 1 mb) in order to aid in weather forecasting.

The vertical distribution of pressure is highly variable and weather-dependent. In fact, the height of the 700-, 500-, and 250-mb surfaces are common aids in weather prediction. A representative vertical profile can be obtained by assuming that the atmosphere is hydrostatic, i.e., pressure at any point is proportional to elevation. This is a commonly made first-order approximation. Corrections to the hydrostatic assumption are a function of air velocities. The implied hydrostatic relationship is,

$$\frac{dP}{dz} = -\rho g, \tag{3.11}$$

where P is pressure, z is elevation, g is gravitational acceleration, and ρ is air density. Air in the atmosphere also follows, reasonably well, the ideal gas law, which for a unit mass is

$$P\rho^{-1} = RT', \tag{3.12}$$

where T' is ambient air temperatures in degrees Kelvin and R is the dry-air gas constant in square centimeters per square second per degree Kelvin ($\mathrm{cm^2\,sec^{-2}\,{}^\circ K^{-1}}$).

Alternatively,

$$\rho = \frac{P}{RT'}. \tag{3.13}$$

Combining Eq. (3.13) with Eq. (3.11) yields

$$\frac{dP}{P} = \frac{-g}{RT'}\,dz. \tag{3.14}$$

The above equation can be integrated to obtain

$$P = P_\circ \exp\left(-\frac{1}{R}\int_{z_\circ}^{z} g\frac{dz}{T'}\right), \tag{3.15}$$

where $P_\circ$ is the pressure intensity at elevation $z_\circ$. Using the linear relation between temperature and elevation in the atmosphere (Eq. 3.6) and integrating, results in

$$P = P_\circ\left(\frac{T'}{T_\circ}\right)^{g/R\alpha}, \tag{3.16}$$

which assumes constant g and α values (temperatures in degrees Kelvin).

In the lower stratosphere, T' is many times assumed constant so that by integration of Eq. (3.14), we obtain

$$P = P_\circ \exp\left[-\frac{g(z - z_\circ)}{RT'}\right]. \tag{3.17}$$

Density variations are associated with pressure variation by the ideal gas law. The density profile in the troposphere is then given by

$$\rho = \frac{P_\circ T_\circ^{-g/R\alpha}}{R}(T_\circ - \alpha z)^{(-g/R\alpha)-1}. \tag{3.18}$$

Figure 3.12 plots typical pressure and temperature distributions.

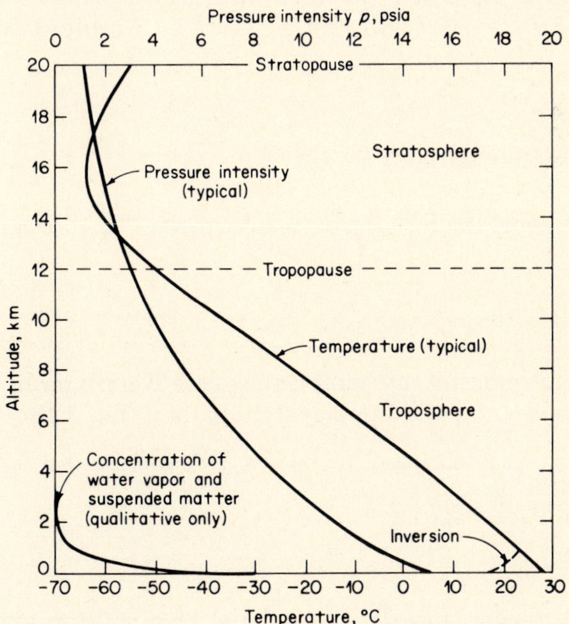

FIGURE 3.12 Typical profiles of water vapor, temperature, and pressure in the atmosphere. Source: Eagleson [1970].

EXAMPLE 3.3

Pressure Distribution in the Atmosphere

A warning commonly given to scuba divers is to avoid flying shortly after diving because of the very quick pressure changes that occur. Assume you are a diver–pilot enjoying yourself in the sunny Caribbean with ground-level air temperatures of 27°C. The ambient lapse rate is 6.5°C km^{-1}. After diving to a 20-m depth you immediately jump into your single-engine plane and fly to 3000 m. What is the total change in pressure you endure? To answer the question you must find the pressure reduction in going from under 20 m of water to atmospheric pressure at sea level and then the reduction in moving 3000 m up into the atmosphere.

The water pressure the diver experiences is nearly hydrostatic; therefore

$$P = \rho_w g z,$$

where z is depth. The density of water ρ_w is about 1 g cm^{-3} and the acceleration of gravity is 9.8 m s^{-2}. Therefore the underwater pressure is

$$\begin{aligned} P &= (1\ \mathrm{g\,cm^{-3}})10^{-3}(\mathrm{kg\,g^{-1}}) \times 10^6(\mathrm{cm^3\,m^{-3}})9.8(\mathrm{m\,s^{-2}}) \times 20(\mathrm{m}) \\ &= 19.6 \times 10^4\ \mathrm{N\,m^{-2}} \\ &= 19.6 \times 10^2\ \mathrm{mb}. \end{aligned}$$

The pressure at 3000 m of elevation in the atmosphere is obtained using Eq. (3.16) with $P_\circ = 1013$ mb and $T_\circ = 27$°C. Ambient temperature at 3000 m is

$$\begin{aligned} T' &= T_\circ - \alpha z \\ &= 27 - 6.5(3000/1000) = 7.5^\circ\mathrm{C}. \end{aligned}$$

The dry-air gas constant R is 2.876×10^6 cm^2 sec^{-2} °K^{-1}. Degrees Kelvin are given by

$$^\circ\mathrm{K} = {}^\circ\mathrm{C} + 273.15,$$

so the lapse rate remains the same in degrees Kelvin as in degrees Centigrade. The pressure at a 3000-m elevation is then (Eq. 3.16)

$$\begin{aligned} P &= 1013\left(\frac{273.15 + 7.5}{273.15 + 27}\right)^{980(\mathrm{cm\,s^{-2}})/2.876\times10^6(\mathrm{cm^2\,s^{-2}\,{}^\circ K^{-1}})6.5({}^\circ\mathrm{K\,km^{-1}})\times10^{-5}(\mathrm{km\,cm^{-1}})} \\ &= 712\ \mathrm{mb} \end{aligned}$$

The change in pressure from sea level to a 3000-m elevation is then 1013 − 712 = 301 mb. Therefore the diver–pilot suffers 2261 mb (1960 + 301) of total pressure change—about 2.23 atm. ◆

3.6 ADVECTION BY WINDS AND OCEAN CURRENTS

Chapter 2 pointed out that the higher latitudes of the globe, both northern and southern hemispheres, had radiation deficits (amounts of outgoing radiation larger than that of incoming radiation), while the lower latitudes had energy surpluses.

The present temperature distributions on the earth are possible only with a major redistribution of energy—the tropics must supply energy to the northern latitudes to achieve the energy balance. Of all the transport processes discussed, advection by winds and ocean currents is the only possible method of energy transfer able to deal with the magnitude of the problem.

3.6.1 Atmospheric Circulation

Atmospheric circulations are generally of thermal origin and related to the earth's rotation. Also, they closely follow the global pressure distribution. A nonrotating earth would result in simply north–south thermal circulation, as shown in Figure 3.13. The warm tropics would induce air masses to rise, leaving a low-pressure "vacuum." The warm masses would circulate along

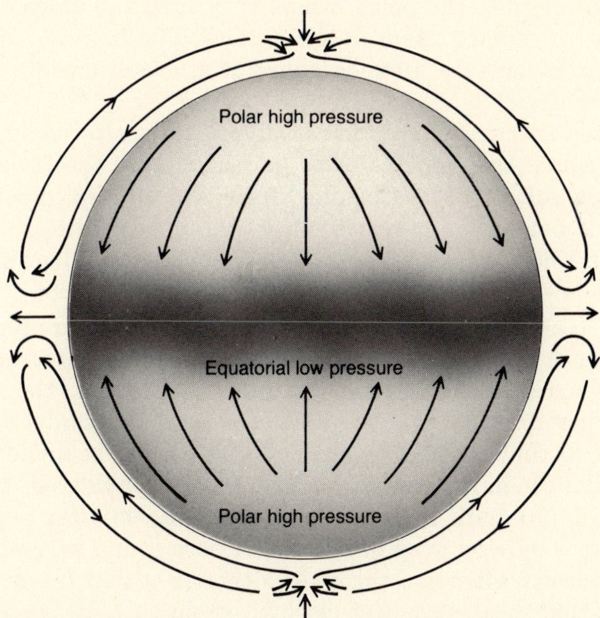

FIGURE 3.13 Atmospheric circulation pattern that would develop on a nonrotating planet. The equatorial belt would heat intensively and would produce low pressure, which would in turn set into motion a gigantic convection system. Each side of the system would span one hemisphere. Source: W. M. Marsh, *Earthscape: A Physical Geography.* Copyright © 1987 by Wiley. Reprinted by permission of John Wiley & Sons, Inc.

FIGURE 3.14 Illustration of the Coriolis effect. Source: W. M. Marsh and J. Dozier, *Landscape: An Introduction to Physical Geography*. Copyright © 1986 by Wiley. Reprinted by permission of John Wiley & Sons, Inc.

the upper atmosphere to the poles, where they take the place of the cold air masses moving to replace them in the tropics. The heavier cold air moves closer to the earth's surface. In such nonrotating-earth models, the poles are high-pressure centers.

The idealized thermal circulation is significantly altered by the effects of the earth's rotation: the Coriolis force and friction of the lower air masses with the earth's surface. The Coriolis effect is really an apparent force. It results from the perception of an observer who moves with a rotating earth and looks at an unattached moving mass. A simple and clear explanation is given by the following example from Marsh and Dozier [1986].*

"To demonstrate the Coriolis effect, imagine that you are playing a game of darts on a large disk rotating in a counter-clockwise direction [Fig. 3.14]. All points on the disk have the same angular velocity (i.e., the same number of revolutions per minute), but those farther from the center have a larger actual velocity, as they travel a greater distance in the same amount of time. In our analysis of the dart game, two facts must be considered: 1) the board is moving in a circular path and will continue to do so after you release the dart; 2) the dart is moving even before you release it and will retain this component of its velocity after you throw it, although it will travel in a straight line. On a disk that rotates counter-clockwise, you will always miss the board to the right, regardless of your position relative to the board." Figure 3.14

*From W. M. Marsh and J. Dozier, *Landscape: An Introduction to Physical Geography*. Copyright © 1986 by Wiley. Reprinted by permission of John Wiley & Sons, Inc.

demonstrates this for three separate positions of the disk. Letter designations in Figure 3.14 are as follows:

R: rotation of the edge of the disk over a time period;

T: movement of the thrower after the dart is released;

B: movement of the dart board after the dart is released;

P: true path of the dart as seen by an observer removed from the rotating disk; and

A: apparent path of the dart as seen by an observer on the rotating disk.

If the disk were rotated clockwise, the Coriolis effect would make the darts veer to the left. The magnitude of the Coriolis effect depends on the angular velocity and the speed of the moving object; it always acts on a plane perpendicular to the axis of rotation.

Marsh and Dozier [1986] also explain why the Coriolis effect increases from effectively zero at the equator to a maximum at the poles. In reality, the angular speed is the same at all locations over the earth, so the net Coriolis force is the same, perpendicular to the axis of rotation (Fig. 3.15). Nevertheless, upon decomposing the force in components parallel and perpendicular to the earth's surface, it is clear that the component parallel to the surface disappears at the equator (Fig. 3.15). It is this component parallel to the earth's surface that affects our perception of wind directions. Therefore, for our purposes, the Coriolis force disappears at the equator. The Coriolis force acts to the right on the northern hemisphere (it rotates counterclockwise relative to the observer) and to the left in the southern hemisphere.

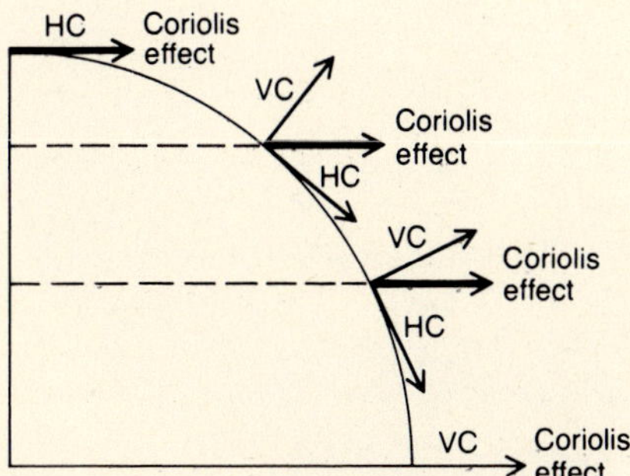

FIGURE 3.15 The Coriolis effect (heavy arrow) is the product of its horizontal and vertical components everywhere except at the poles, where it is equal to the horizontal components, and at the equator, where it is equal to the vertical component. Source: W. M. Marsh and J. Dozier, *Landscape: An Introduction to Physical Geography.* Copyright © 1986 by Wiley. Reprinted by permission of John Wiley & Sons, Inc.

The need to conserve angular momentum also alters the idealized scheme of Figure 3.13. In particular, the circulating cells extending from the equator to the poles are not possible. Also, the deflection given to the winds by the Coriolis force would imply, without preservation of angular momentum, bands of accelerating air masses circling the earth.

The net effect of the Coriolis force and angular momentum is summarized in Figure 3.16, where a circulatory model over a homogeneous smooth sphere is shown. The main features are

1. Convergent winds on the equator of easterly origin. These are the trade winds or "doldrums." These converge in the low-pressure belt called the equatorial convergence zone also called the intertropical (equatorial) convergence zone.
2. At middle latitudes we find the prevailing westerly winds, associated with high-pressure centers.

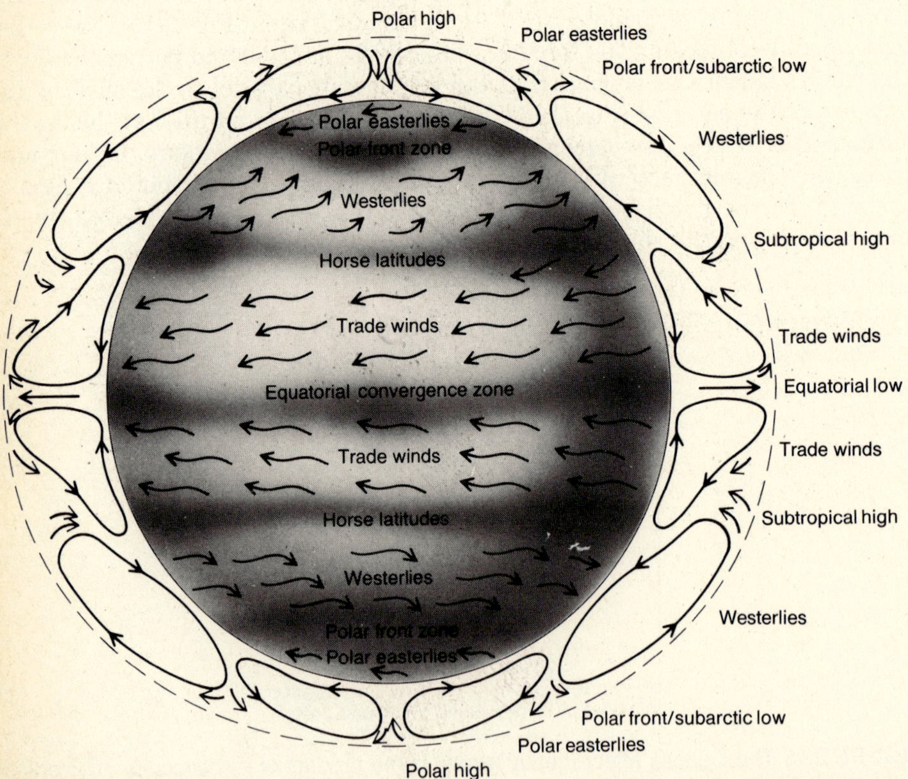

FIGURE 3.16 Idealized circulation of the atmosphere at the earth's surface, showing the principal areas of pressure and belts of winds. Source: W. M. Marsh, *Earthscape: A Physical Geography.* Copyright © 1987 by Wiley. Reprinted by permission of John Wiley & Sons, Inc.

3. Counteracting the westerly winds are highly variable and relatively unknown polar easterly winds.
4. The poleward circulation of tropical air masses is broken into limited gyres, keeping the banded or tubular structure around the earth.

In reality, the topography and variable thermal effects of the land masses and the oceans further alter the pattern of Figure 3.16. Figures 3.10 and 3.11 show the prevailing mean wind patterns in the extreme months of January and July. Besides the general zonal behavior already described, the most obvious feature is the rotating winds around the semipermanent high and low pressures already described in Section 3.5. Rotating winds around low-pressure points are called cyclones. They rotate counterclockwise in the northern hemisphere and clockwise in the southern hemisphere. Winds rotate clockwise around high-pressure areas, anticyclones, in the northern hemisphere. The opposite rotation is observed in the southern hemisphere. As the pressure centers, these cyclones and anticyclones represent averages over long periods and are of a semipermanent nature.

The semipermanent wind structure of Figures 3.10 and 3.11 do much to establish the mean weather patterns over the earth. The tropical easterly winds form a convergence zone of air masses of similar temperatures and densities. This results in ill-defined contact surfaces. Air is displaced upward, to accommodate inflowing masses, resulting in high cloud formations and condensation. This is why the tropics are characterized by a high frequency of precipitation of a showery nature, short duration, and limited extent. This also makes prediction of weather in the tropics fairly difficult. Note that the intertropical convergence zone migrates northward during the northern hemisphere summer. Because of the Coriolis force and low-pressure centers in the middle latitudes in this season, there are some significant components of tropical westerly winds. This behavior explains some of the well-known monsoons, or wind reversals over continental land masses. Monsoons in south Asia are particularly evident in Figures 3.10 and 3.11. In the winter, a large anticyclone sits over central Asia, directing winds in an easterly fashion, toward the Indian Ocean. In the summer, the high-pressure center has migrated northeastward. In its place there is a more southerly (also migrating to the east) low-pressure center. At the same time the intertropical convergence zone moves north. The Coriolis force and the cyclonic formation then conspire to reverse wind direction. Tropical, warm, and moist air flows into southern Asia from the Indian Ocean, resulting in stormy and wet weather.

The high-pressure centers in middle latitudes also affect average weather patterns in other ways. The eastern sides of the anticyclones usually bring cold polar air of diverging nature. This commonly leads to very low and strong temperature inversions that prevent rising air and lead to stable climates with little precipitation. Some of the major desert regions in the world — the Sahara in northwestern Africa, the Chilean-Peruvian Desert in South America, and regions of southwestern United States — are on the eastern side of semipermanent anticyclones.

Winds on the western side of the anticyclones have warmed up and collected moisture from their trip over the tropical oceans. This leads to unstable rising air masses, which yield good amounts of precipitation. This behavior is enhanced by the topographic and thermal lifting induced by land masses on the western side of the anticyclones.

The northern portions of the anticyclones form the prevailing westerly winds of the middle latitudes. These warm winds are confronted by the cold polar easterly winds. The meeting of air masses of different temperatures and densities forces the warm air to rise along extensive well-defined boundaries. These are the stormy frontal systems that will be discussed further in Chapter 4. The warm air masses rising along a front commonly result in constant moderate precipitation of long duration. The poleward migration of the anticyclones during winter accounts for the stormy weather in North America during this season.

The net effect of pressure gradients, Coriolis force, and friction on the winds at the earth's surface is a motion angled relative to the maximum pressure gradient. As we move to higher elevations, the friction effects are reduced and the wind movement is dictated by a balance between Coriolis force and pressure. This type of behavior is called a geostrophic wind. Eagleson [1970] summarizes the mathematical arguments leading to this type of wind. A characteristic of geostrophic winds is that they move almost parallel to the pressure isobars, particularly in regions with important Coriolis effects, such as the high latitudes.

High in the troposphere, at elevations corresponding to 500 mb (on the order of 5.5 km) of pressure and higher, there are high-velocity winds that closely approximate geostrophic behavior; they are called the jet streams. An example is shown in Figure 3.17. Speeds of 160 to 240 km hr^{-1} are common; even higher speeds are achieved at 9000 m and higher elevations.

The jet streams are of a semipermanent nature and play a major role in ground-level wind patterns and weather. The winds exhibit a counterclock-

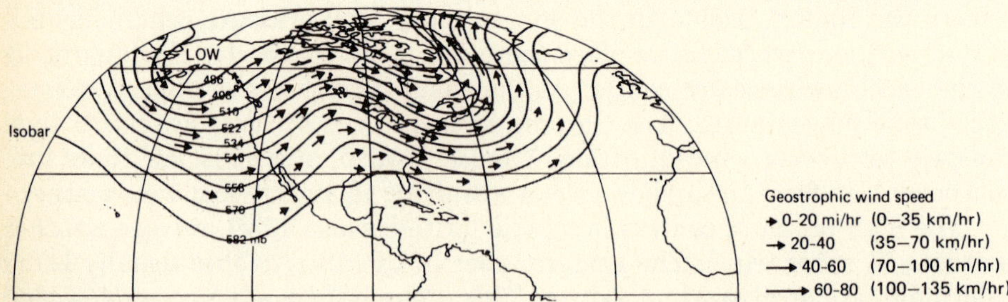

FIGURE 3.17 A geostrophic wind on December 19, 1979, at an altitude of about 5500 m, where pressure is approximately half that at sea level. Note that the wind travels parallel to the isobars. Source: W. M. Marsh and J. Dozier, *Landscape: An Introduction to Physical Geography.* Copyright © 1986 by Wiley. Reprinted by permission of John Wiley & Sons, Inc. Map projection by Waldo Tobler. (After U.S. National Weather Service map.)

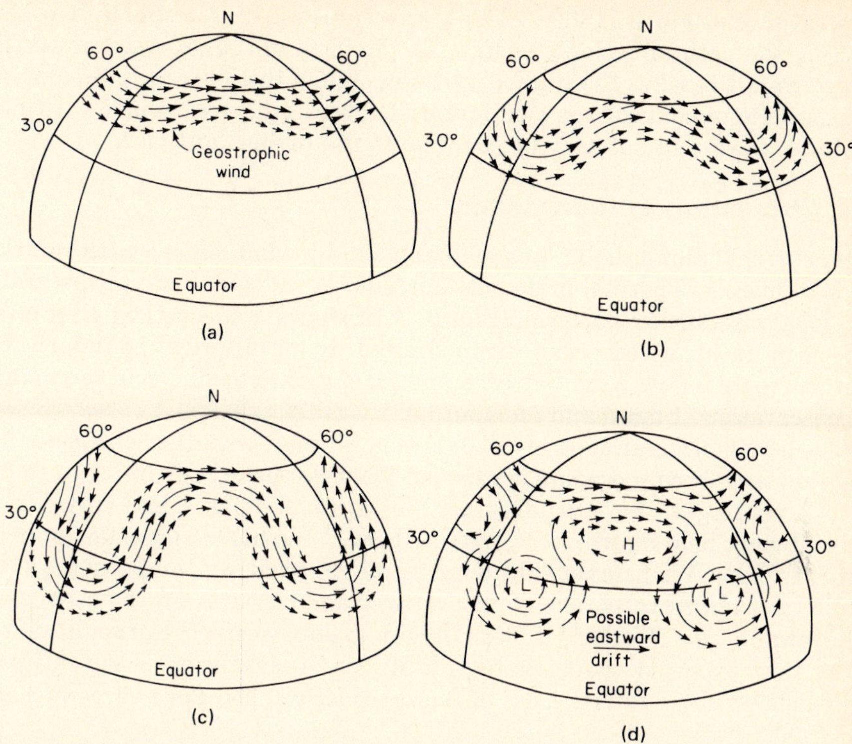

FIGURE 3.18 Upper-air waves on the jet stream, which bring periods of contrasting weather in their wake. The typical sequence, called an index cycle, is shown in the four diagrams. The undulating jet (a) goes into increasingly large oscillations (b and c). North of the jet lies cold polar air, and south of it, warm tropical air. The great oscillations carry polar air into the middle and low latitudes and tropical air into the middle and high latitudes. Finally, the extended waves are cut off, leaving cells of cold air in the south and cells of warm air in the north (d). Source: P. S. Eagleson, *Dynamic Hydrology,* McGraw-Hill, 1970.

wise circulation in the northern hemisphere. The oscillations observed in Figure 3.17 are also very common. These instabilities can be dynamically explained and reproduced in laboratory experiments (Eagleson [1970] and Trewartha and Horn [1980]). At high elevations and high speeds the oscillations may in fact break down into self-contained cells, as in the sequence shown in Figure 3.18.

The jet streams influence climate and weather by moving large masses of cold or warm air with their associated heat contents. Excursions of the jet streams down to the lower latitudes imply the transfer of large, cold polar air masses with obvious effects on temperature. These excursions to low latitudes are generally centered around a high-altitude low-pressure center or trough. Ground-level cyclonic events commonly develop ahead of this trough, in the region where the tropical warm air masses are moving toward the poles

(Trewartha and Horn [1980]). The jet stream plays an important role in directing the path of cyclonic events. At the peak of the jet stream's oscillation, warm air reaches the higher latitudes. When the jet stream breaks into cells, pockets of cold air may be found at lower, otherwise warm latitudes. Similarly, warm regions may be confined to the higher latitudes.

3.6.2 Circulation in the Oceans

Ocean currents and circulation are dominated by wind forces, earth rotation, and land masses. Thermal or density currents are of secondary nature within the global circulation patterns. Figure 3.19 shows a theoretical circulation pattern in an idealized ocean basin limited by an elliptical boundary. The ocean currents follow wind patterns and form gyrations and countercurrents for conservation of mass and momentum.

The actual circulation (Fig. 3.20) is very similar, with the continents limiting flow and forcing oceanic gyrations. The established currents are cold or warm, according to the latitudinal path of the water. The ocean currents are fairly efficient heat distributors, moving heat to the higher latitudes from the tropical zones. With favorable winds and contrasting temperatures, the oceans can significantly alter temperatures and climates on the continents. For example, the North Atlantic current makes western Europe's winters milder than expected, given its high latitudes. A significant ocean influence also exists on the western coast of South America. The Peru current makes

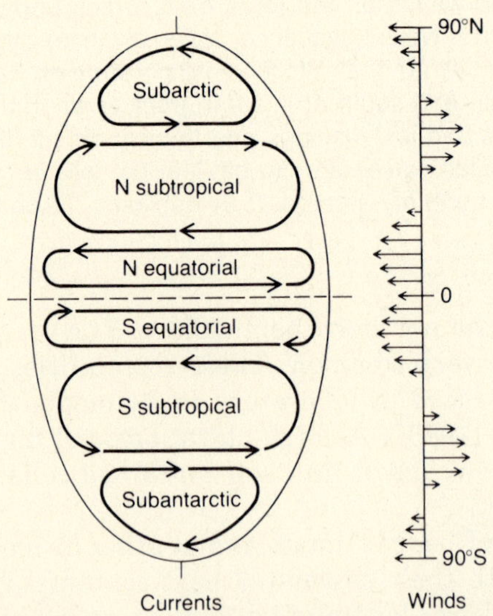

FIGURE 3.19 Surface currents for a hypothetical ocean surrounded by land. Sources: P. K. Weyl, *Oceanography: An Introduction to the Marine Environment*. Copyright © 1970 by Wiley. Reprinted by permission of John Wiley & Sons, Inc. Marsh and Dozier [1986].

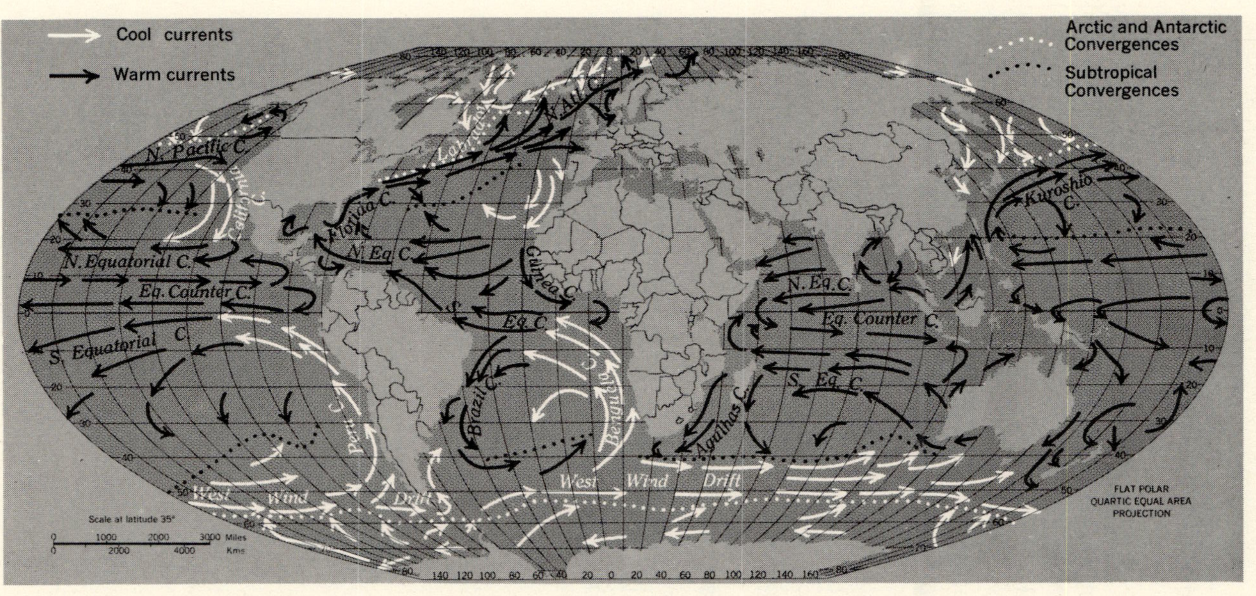

FIGURE 3.20 Generalized scheme of ocean currents. Source: G. T. Trewartha, *An Introduction to Climate,* 4th ed., McGraw-Hill, 1968. Reproduced with permission.

the coast of Chile and Peru much colder than expected. The cool California current effectively reduces summer heat in that state.

3.7 ATMOSPHERIC HUMIDITY

The most variable atmospheric component is humidity. It also plays the starring role in climate and weather determination, with obvious hydrologic effects. Water exists in the atmosphere in all three phases: vapor, liquid, and solid. It is the vapor phase, though, that is most prevalent in the atmosphere. As already seen in Chapter 2, water vapor is a very efficient absorber of radiation and so is very important in the incoming and outgoing radiation balance. Its movement and phase changes are crucial to earthwide heat and mass balance. Upon evaporation and condensation, heat is absorbed and released, respectively. Since both processes rarely occur in the same location, vapor is a carrier of energy and mass from one part of the globe to another. It is the liquid and solid precipitation of vapor that ultimately controls the land-based hydrologic processes that we will study further in this book.

3.7.1 The Phases of Water

The transitions of water between liquid, solid, and gaseous phases are shown in Figure 3.21 as functions of volume, temperature, and pressure. Solid lines represent equilibrium between phases. Dashed lines represent isotherms

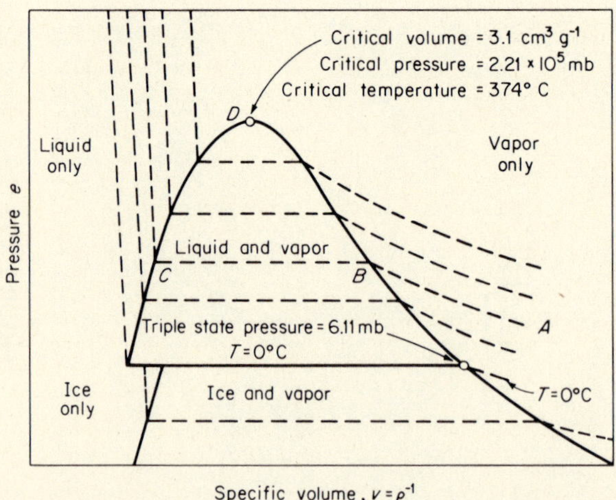

FIGURE 3.21 The phases of water. Source: S. L. Hess, *Introduction to Theoretical Meteorology,* Holt, Rhinehart and Winston. Copyright © 1959 by Seymour L. Hess. Used by permission.

(equal temperature lines). It is important to study the conditions under which condensation and evaporation may occur.

Point A in Figure 3.21 is in the vapor region. Moving along the corresponding line of constant temperature will result in a volume reduction (increasing density) and increasing pressure. At point B, condensation starts, and occurs with minimal changes in pressure as the horizontal line implies. At C, all vapor has condensed. Due to the incompressibility of water, further increases in pressure barely decrease the volume as the nearly vertical dashed lines imply.

A rising air mass suffers changes in ambient temperature as well as pressure changes. Clearly then, the condensation path is not generally so simple.

A particular condensation path—that at constant pressure—results in the definition of dew-point temperature. Moving horizontally (from right to left) in Figure 3.21, along a line of constant pressure while reducing temperature, results in a volume reduction. The temperature at which saturation occurs is called the dew-point temperature (T_d).

A few interesting points in the phase diagram are the critical point and the triple-state points. Only at the critical point are vapor and liquid indistinguishable. At the triple-state point, ice, water, and vapor coexist.

Any phase change (represented in Fig. 3.21 for water) results in a release or absorption of heat. This latent heat does not necessarily change the temperature of the substance, but is necessary to maintain the overall energy balance. Each state has a different level of internal energy and each state change involves some work resulting in density (unit volume) variations. Table 3.2 gives the signs and nomenclature of heat production during the phase changes of water. The sign is positive if heat is absorbed by the changing substance.

The units of latent heat are calories per gram (cal g^{-1}) or joules per kilogram (J kg^{-1}). Latent heat represents the amount of heat exchange required for inducing the state change per gram of substance. Latent heats are a function of temperature.

TABLE 3.2 Latent Heats

STATE 1	STATE 2		SIGN OF L
Liquid	Vapor	L_e = latent heat of evaporation (vaporization)	+
Vapor	Liquid	L_c = latent heat of condensation	−
Ice	Vapor	L_s = latent heat of sublimation	+
Ice	Liquid	L_m = latent heat of melting	+
Liquid	Ice	L_f = latent heat of freezing (fusion)	−

Source: Eagleson [1970].

Common approximations to latent heat are (in calories per gram)

$$\begin{aligned} &\text{Evaporation:} && L_e = 597.3 - 0.57T = -L_c \\ &\text{Sublimation:} && L_s = 677 - 0.07T \\ &\text{Melting:} && L_m = -L_f = 79.7 \end{aligned} \tag{3.19}$$

where T is given in degrees Celsius.

3.7.2 Vapor Pressure and Humidity

Imagine a closed container with equal volumes of water and air at the same temperature. If the air is initially dry, evaporation will take place. Water molecules, vapor, will escape from the water and into the air. Some of the molecules return from the air and condense, but the net effect will be evaporation until the air becomes saturated. Saturation implies that for the given temperature the air is holding the maximum possible amount of water vapor. At this point, the vapor behaves as an ideal gas. From thermodynamics, we know that in a gaseous mixture each component contributes proportionally to the total pressure. So, at equilibrium, the vapor pressure is the maximum possible attainable, or the saturation vapor pressure. At this pressure, water evaporating and condensing through the air–water interface yields zero net transport. In an open system, this balance is hard to attain, particularly where other transport methods—e.g., wind—remove the water vapor from the air. Under such conditions, evaporation will continue.

Applying the ideal gas law to water vapor, we obtain an expression for vapor pressure,

$$e = \rho_v R_v T, \tag{3.20}$$

where e is the vapor pressure (in millibars), ρ_v is vapor density in mass per unit volume ($\mathrm{g\,m^{-3}}$), T is absolute temperature (assumed the same as that of the air it is mixed with) in degrees Kelvin, and R_v is the vapor gas constant.

The vapor gas constant is related to the universal gas constant, $R_\circ$ ($1.9857\ \mathrm{cal\,{}^\circ K^{-1}\,mol^{-1}}$)

$$R_v = \frac{R_\circ}{M_v} = \frac{M}{M_v} R, \tag{3.21}$$

where M_v is the molecular weight of the water vapor, M is the molecular weight of dry air, and R is the dry-air gas constant ($2.876 \times 10^6\ \mathrm{cm^2\,sec^{-2}\,{}^\circ K^{-1}}$). Therefore

$$e = \frac{M}{M_v} \rho_v RT = 1.61 \rho_v RT \tag{3.22}$$

or

$$\rho_v = 0.622 \frac{e}{RT}. \tag{3.23}$$

The vapor density ρ_v is also called the absolute humidity of the atmosphere. The ideal gas assumption is good under most conditions, with the exception of the point of condensation. A curious and useful implication of Eq. (3.23) is that it says that the density of water vapor is 0.622 that of dry air *at the same temperature* T and *pressure* e.

If P represents the total atmospheric pressure (including only dry air and water vapor) then

$$P = P_{\text{dry}} + e. \tag{3.24}$$

The density of the mixture of dry air and vapor is given by

$$\rho_m = \rho_d + \rho_v = \frac{P - e}{RT} + \frac{0.622e}{RT} = \frac{P}{RT}\left(1 - 0.378\frac{e}{P}\right). \tag{3.25}$$

Equation (3.25) shows the interesting result that moist air is less dense than dry air *at the same temperature and pressure*. Pause and think about the meaning of the above statement.

Virtual temperature is defined as that temperature required for dry air to achieve the density of an air mixture (moist air) at the same pressure. Virtual temperature is then

$$T^* = T/(1 - 0.378e/P). \tag{3.26}$$

Relative humidity is defined as the ratio of the vapor density (or pressure) to the saturation vapor density (or saturation vapor pressure) at the same temperature:

$$r = 100\frac{\rho_v}{\rho_s} = 100\frac{e}{e_s}. \tag{3.27}$$

Note that ρ_s and e_s are functions of temperature only.

Specific humidity is the mass of water vapor per unit mass of moist air:

$$q_h = \frac{\rho_v}{\rho_m} = \frac{0.622e}{P - 0.378e} \approx 0.622\frac{e}{P}. \tag{3.28}$$

The mixing ratio is the mass of water per unit mass of dry air:

$$w = \frac{\rho_v}{\rho_d} = \frac{0.622e}{P - e} \approx q_h. \tag{3.29}$$

3.7.3 Measurement and Estimation of Humidity and Vapor Pressure

Saturation vapor pressure can be obtained from tables (see Appendix A) as a function of temperature, dew point, and/or wet-bulb depression (a concept explained later in this section). There are several simple approximations. Over a temperature range of 25 to 55°F (U.S. Army Corps of Engineers [1956])

$$e_s \approx 6.11 + 0.339(T_d - 32) \qquad e_s \text{ in millibars,} \tag{3.30}$$

or

$$e_s \approx 0.180 + 0.01(T_d - 32) \qquad e_s \text{ in inches of mercury,}$$

where T_d is the dew-point temperature given in degrees Fahrenheit. Saturation vapor pressure over water can be approximated within 1% in the range −50 to 55°C (−58 to 131°F) by

$$e_s \approx 33.8639[(0.00738T + 0.8072)^8 - 0.000019|1.8T + 48| + 0.001316] \tag{3.31}$$

(Bosen [1960]), where e_s is in millibars and T is ambient temperature in degrees Celsius.

Saturation vapor pressure over ice and water are not the same, being larger over water than over ice for temperatures below freezing. The ratio of e_s over ice and water is given in tables or approximated by

$$\frac{e_{s,\,\text{ice}}}{e_{s,\,\text{water}}} \approx 1 + 0.00972T + 0.000042T^2, \tag{3.32}$$

where e_s is in millibars and T in degrees Celsius. Accuracy of the above is to within 0.1% of the true ratio for the range 0 to −50°C (32 to −58°F) (Bosen [1961]).

The dew point can be approximated in the temperature range −40 to 50°C (−40 to 122°F) to within 0.3°C by

$$T - T_d \approx (14.55 + 0.114T)x + [(2.5 + 0.007T)x]^3 + (15.9 + 0.117T)x^{14}, \tag{3.33}$$

where T is ambient temperature and x is the complement of relative humidity expressed in decimal form, $x = 1.0 - r/100$.

Relative humidity can be approximated from air temperature and dew point (Bosen [1958]) by

$$r = 100\left(\frac{112 - 0.1T + T_d}{112 + 0.9T}\right)^8. \tag{3.34}$$

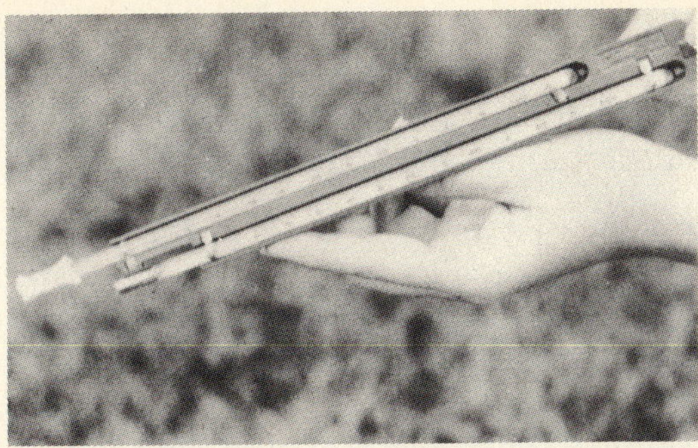

FIGURE 3.22 Sling psychrometer. The wet-bulb thermometer is covered with the gauze sock. Source: W. M. Marsh and J. Dozier, *Landscape: An Introduction to Physical Geography.* Copyright © 1986 by Wiley. Reprinted by permission of John Wiley & Sons, Inc.

Humidity is usually measured with a psychrometer (Fig. 3.22). This is an instrument with two thermometers. One is a wet-bulb thermometer, covered with cloth saturated with water. The other is dry. Ventilated, by rotation or otherwise, the temperature reading of the wet thermometer T_w is lower because of evaporation. The difference between the two readings is called the wet-bulb depression. Appendix A gives a table (Table A.2) relating wet-bulb depression to relative humidity.

A simpler and more convenient humidity measurement can be obtained with a hair hygrometer. These instruments depend on human hair or other organic material that expands when wet and contracts when dry. Although of limited accuracy, these procedures can be useful in remote sensing applications.

EXAMPLE 3.4

Distribution of Moisture and Evaporation

Our diver–pilot is also a runner. He is given the choice of running marathons in two locations of equal temperature ($T = 21$°C). The difference is that in one place the psychrometer shows a wet-bulb depression ($T - T_w$) of 3°C and in the other a depression of 7°C is observed. Both places are equally windy and otherwise similar. Which should be a more comfortable place to run?

To study the situation we should find where evaporation is larger. Given $T = 21$°C and using Eq. (3.31), we find that e_s is about 24.9 mb in both places. The relative humidity in both locations can then be obtained as a function of wet-bulb depression from Table A.2 in Appendix A. The table yields

$$\text{Location 1:} \quad \frac{e}{e_s} = 0.75\,,$$

$$\text{Location 2:} \quad \frac{e}{e_s} = 0.45\,.$$

The corresponding vapor air pressures are then

Location 1: $0.75 \times 24.9 = 18.7\ \text{mb}$,

Location 2: $0.45 \times 24.9 = 11.2\ \text{mb}$.

From Eq. (3.4) we know that evaporation is proportional to the gradient of specific humidity. Given that the locations are similar and that pressure does not change much with elevation, Eq. (3.28) implies that evaporation is effectively proportional to the vapor pressure gradient. We can assume that the marathon runner will effectively be a continuous, unlimited, source of water (sweat) in both locations. Since everything is the same, it is reasonable to state that the vapor pressure at the skin level e_o and the constant of proportionality in Eq. (3.4) are the same in both locations. Due to the amount of water, $e_o \approx e_s$. The ratio of evaporation in both places can then be approximated as

$$\frac{E_2}{E_1} = \frac{K\dfrac{\partial e_2}{\partial z}}{K\dfrac{\partial e_1}{\partial z}} \approx \frac{e_s - e_2}{e_s - e_1} = \frac{24.9 - 11.2}{24.9 - 18.7} = 2.2\,.$$

The runner should be cooler in Location 2 because of the heat removed by the evaporating sweat; nevertheless, he may be better off in Location 1 in terms of energy and body-fluid conservation. ◆

3.7.4 Distribution of Atmospheric Moisture

As stated previously, humidity is a highly variable atmospheric component. Nevertheless, it is possible to make several statements relative to average moisture content. Generally, water vapor, by volume, reduces with elevation as shown in Table 3.3. Temporally, specific humidity increases and decreases seasonally with temperature. The daily variations are masked by local air turbulence that may transfer moisture away from a site, particularly over land areas.

Specific humidity, a measure of absolute water content, is highest in the tropics and low latitudes, sharply decreasing toward the poles (Fig. 3.23). Relative humidity, because of its temperature dependence, shows peaks in the tropics (high moisture and temperature) and in the poles (low moisture with low temperature). Two minima are found in the middle latitudes, coincident with the high-pressure anticyclone regions. This is illustrated in Figure 3.24. Note that the large desert regions exhibit low relative humidity—e.g., the African Sahel region, northwest Mexico, southwestern United States, central Australia, and west central South America—although their specific humidity is relatively high on the average (see Fig. 3.23). Since relative humidity measures not only water content but the ability of the air to

TABLE 3.3 Average Vertical Distribution of Water Vapor in Middle Latitudes

HEIGHT (km)	WATER VAPOR (vol. in %)
0.0	1.3
0.5	1.16
1.0	1.01
1.5	0.81
2.0	0.69
2.5	0.61
3.0	0.49
3.5	0.41
4.0	0.37
5.0	0.27
6.0	0.15
7.0	0.09
8.0	0.05

Source: Landsberg [1958] and Trewartha [1968].

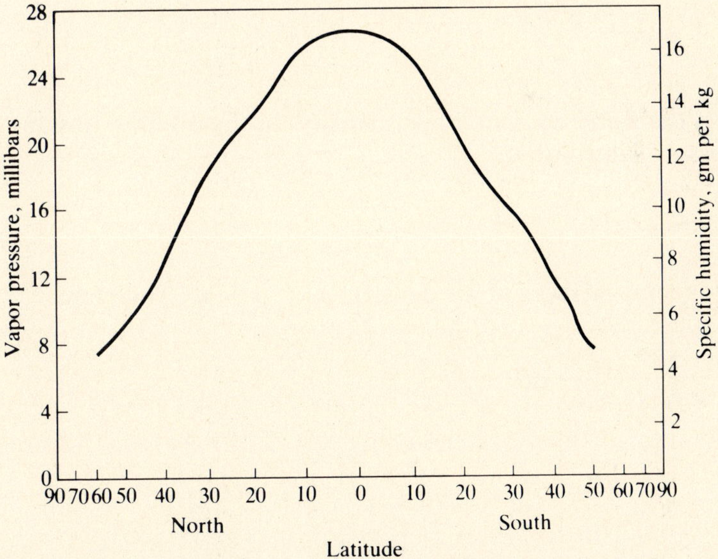

FIGURE 3.23 Distribution by latitude zones of the water vapor content of the air. Specific humidity is highest in equatorial latitudes and decreases toward the poles. Source: G. T. Trewartha, *An Introduction to Climate,* 4th ed., McGraw-Hill, 1968. Reproduced with permission. (After Haurwitz and Austin [1944].)

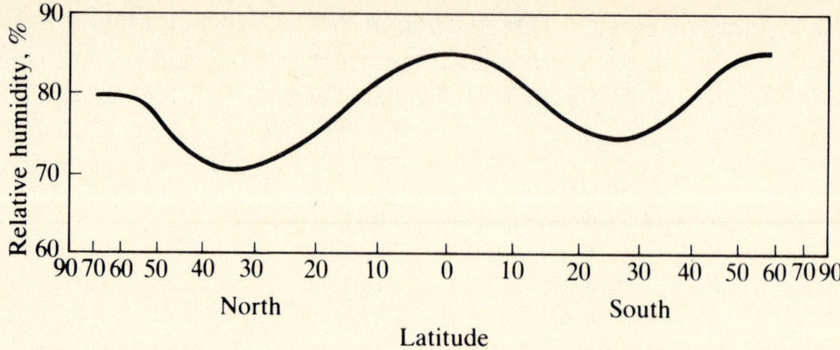

FIGURE 3.24 Distribution of relative humidity by latitude zones. Note that zonal distribution of relative humidity is quite different from distribution of specific humidity. Source: G. T. Trewartha, *An Introduction to Climate,* 4th ed., McGraw-Hill, 1968. Reproduced with permission.

hold moisture, it plays an important role in the condensation process. This will be discussed in Section 3.8.

The knowledge of the vertical and spatial distribution of moisture permits the computation of the potential precipitable water in an area. To compute the total amount of potential precipitable water w_p in a layer between elevations 0 and z, we need to evaluate

$$w_p = \int_0^z \rho_v \, dz \,, \qquad (3.35)$$

where ρ_v is the water content (vapor density) in the column. Assuming hydrostatic pressure distribution,

$$dP = -\rho_m g \, dz \,,$$

where ρ_m is total mixture of air density, $\rho_m = \rho + \rho_d$. Therefore

$$w_p = \frac{1}{g} \int_P^{P_\circ} \frac{\rho_v}{\rho_v + \rho_d} dP \qquad (3.36)$$

$$= \frac{1}{g} \int_P^{P_\circ} q_h \, dP \,, \qquad (3.37)$$

where q_h is the specific humidity. Since $q_h \approx 0.622e/P$,

$$w_p = \frac{0.622}{g} \int_P^{P_\circ} e \frac{dP}{P} \,. \qquad (3.38)$$

The above equation can be integrated analytically if we know the $e - P$ behavior. Numerical integration is always possible by summing over discrete layers that are taken to be of constant pressure. Discretizing Eq. (3.37) and introducing conversion factors leads to

$$\begin{aligned} w_p\,(\text{inches}) &\approx 0.0004 \sum \overline{q}_h\,\Delta P\,, \\ w_p\,(\text{mm}) &\approx 0.01 \sum \overline{q}_h\,\Delta P\,, \end{aligned} \tag{3.39}$$

where $\overline{q}_h$ is the mean specific humidity (in units of grams per kilogram) between a layer with a pressure change ΔP (in millibars). It should be emphasized that for the above potential precipitation to occur, all the moisture must condense and fall to the earth. This would be very rare and difficult. Chapter 4 will discuss the necessary conditions leading to precipitation.

EXAMPLE 3.5

Computation of Precipitable Water

Assume that the vertical pressure distribution and corresponding moisture distribution of a column of atmosphere is

ELEVATION	PRESSURE (mb)	VAPOR PRESSURE (mb)
0.0	1000	22.5
0.5	985	20
1.0	970	18
1.5	955	16
2.0	940	14.5
2.5	920	13
3.0	905	11.5
3.5	890	10.3
4.0	875	9.25
5.0	835	7.4
6.0	800	5.9
7.0	775	4.75
8.0	750	3.8

The precipitable water is obtained by first computing the specific humidity, in grams per kilogram for each layer. Specific humidity would be given by Eq. (3.28) (multiplied by 1000 to convert to grams per kilogram). After the average $\overline{q}_h$ for each layer is obtained, Eq. (3.39) may be used. The following table summarizes the computations.

ELEVATION	q_h	$\overline{q}_h$	ΔP	$\overline{q}_h \, \Delta P$
0	14.0			
		13.3	15	199.5
0.5	12.6			
		12.1	15	181.5
1.0	11.5			
		11.0	15	165.0
1.5	10.4			
		10.0	15	150.0
2.0	9.6			
		9.2	20	184.0
2.5	8.8			
		8.4	15	126.0
3.0	7.9			
		7.6	15	114.0
3.5	7.2			
		6.9	15	103.5
4.0	6.6			
		6.1	40	244.0
5.0	5.5			
		5.1	35	178.5
6.0	4.6			
		4.2	25	105.0
7.0	3.8			
		3.5	25	87.5
8.0	3.2			
		$\sum \overline{q}_h \, \Delta P$	=	1838.5

The potential precipitable water is then

$$w_p \approx 0.01 \times 1838.5 = 18.4 \, \text{mm} \,. \blacklozenge$$

3.8 ATMOSPHERIC STABILITY AND CONDENSATION

Precipitation depends, among other things, on the condensation of atmospheric moisture. For condensation, the moisture-laden air must lower its temperature, increase pressure, or reduce volume according to the phase diagram shown in Figure 3.21. In the atmosphere, condensation generally occurs mostly by temperature reductions resulting from the ascent of air masses.

The initiation of the rising will be further discussed in Chapter 4. Here we will discuss how the rising is enhanced or restrained by convective, thermally induced turbulence.

The stability or instability of an air mass depends on its rate of cooling relative to the atmosphere's lapse rate. The rate of cooling of the air mass depends on whether condensation occurs or not.

3.8.1 Adiabatic Cooling

Assume we have a parcel of dry air that is heated. Apply the first law of thermodynamics under adiabatic (no heat lost) conditions.

$$dQ_\circ = dE + dW , \tag{3.40}$$

where $dQ_\circ$ is net heat received by the system, dE is the increase in internal energy, and dW is the work done by the system on the surroundings.

In expanding gas with no external forces

$$dW = d(PAx) , \tag{3.41}$$

where P is the pressure and Ax is the displaced volume. For small volume changes,

$$dW = P\,dV . \tag{3.42}$$

Equation (3.41) must have consistent units to be used in Eq. (3.40). Dealing with one substance, and dividing by mass, Eq. (3.40) can be expressed in unit mass terms

$$dq_\circ = de + dw . \tag{3.43}$$

In the absence of motion, gravity, electricity, and magnetism, the internal energy is just a function of temperature and volume. Refer now to de as du; then

$$dq_\circ = du + P\,dv , \tag{3.44}$$

where v is the specific volume ρ^{-1}.

For perfect gases, the internal energy is only a function of temperature, and under constant volume conditions, we have

$$(\partial u/\partial T)_v = C_v = \text{specific heat at constant volume.}$$

So,

$$dq_\circ = C_v\,dT + P\,dv , \tag{3.45}$$

where dT is the change in temperature T of the rising air mass. Using the common equation or a perfect gas $Pv = RT$ and differencing,

$$P\,dv = R\,dT - v\,dP , \tag{3.46}$$

where R is the gas constant of dry air in square centimeters per square second per degree Kelvin ($cm^2 s^{-2} {}^\circ K^{-1}$) and T is absolute temperature in degrees Kelvin. Substituting Eq. (3.46) into Eq. (3.45), we then have

$$dq_\circ = (C_v + R)\,dT - v\,dP = (C_v + R)\,dT - RT\frac{dP}{P}, \tag{3.47}$$

but by definition

$$C_v + R = C_p, \tag{3.48}$$

where C_p is the specific heat at constant pressure. Then,

$$dq_\circ = C_p\,dT - RT\frac{dP}{P}. \tag{3.49}$$

Since we are assuming adiabatic rising, $dq_\circ = 0$, which results in

$$\frac{dT}{T} = \frac{R}{C_p}\frac{dP}{P} \tag{3.50}$$

which can be integrated to obtain

$$\frac{T}{T_\circ} = \left(\frac{P}{P_\circ}\right)^{1-n}, \tag{3.51}$$

where $1 - n = R/C_p = (C_p - C_v)/C_p$ or $n = C_v/C_p \approx 1/1.41$ for dry air. $T_\circ$ is absolute temperature at absolute pressure $P_\circ$. $T_\circ$ is also referred to as the potential temperature θ at mean sea level ($P_\circ \approx 1000$ mb); therefore

$$\frac{T}{\theta} = \left(\frac{P}{1000}\right)^{1-n} \tag{3.52}$$

and the potential temperature θ remains constant during an adiabatic process. This implies that on cooling (heating) the air parcel will return to its original, initial, temperature.

If a parcel of dry air is moved vertically, it will expand or contract because of a change in atmospheric pressure. Assume adiabatic conditions; then, using Eq. (3.50) and differentiating with respect to z,

$$\frac{1}{T}\frac{dT}{dz} = \frac{1 - n}{P}\frac{dP}{dz}. \tag{3.53}$$

Recalling Eq. (3.14),

$$\frac{dP}{P} = \frac{-g}{RT'}\,dz,$$

where T' is the absolute temperature of the ambient atmosphere at elevation z, and substituting in Eq. (3.53) results in,

$$\frac{dT}{dz} = \frac{-g(1 - n)}{R} \frac{T}{T'}. \tag{3.54}$$

$dT/dz = \Gamma$ is called the dry adiabatic lapse rate or the rate at which a rising parcel of air cools with elevation under adiabatic conditions. The dry adiabatic lapse rate Γ is approximately 10.0°C km^{-1}.

The relation between Γ and the ambient lapse rate α defines static convective stability in the atmosphere:

If $|\alpha| < |\Gamma|$, rising air cools faster than the atmosphere so once the lifting force is removed, it will sink (since it is denser) to a stable equilibrium.

If $|\alpha| > |\Gamma|$, rising air is always warmer so it keeps rising once given an initial impulse. This is unstable equilibrium.

If $|\alpha| = |\Gamma|$, the rising air parcel will remain in indifferent equilibrium at any elevation.

The above discussion dealt with the rise of dry air. If moist, unsaturated air rises at the adiabatic lapse rate, it will reach a position where the relative humidity becomes 100% and saturation is established. Further cooling by rising will result in condensation. During condensation, latent heat is released (as it is absorbed in evaporation [see Section 3.7.1]) resulting in a warming of the air and a reduction in the lapse rate. The resulting lower rate is called the saturated adiabatic lapse rate Γ'.

This condensation process, which will be discussed in Section 3.8.2, results in precipitation. Precipitation carries heat, breaking the adiabatic assumption. Nevertheless, removed heat is small and the temperature lapse rate is close to Γ' and the process is also called pseudo-adiabatic.

Once condensation occurs, if the parcel of air descends, it will heat up at the fast dry adiabatic rate. This is because moisture has precipitated out of the system. This nonreversibility results in the observed behavior on orographic barriers like mountains where the lee side is usually warmer than the windward side. Figure 3.25 illustrates this.

3.8.2 Condensation by Pseudo-adiabatic Cooling

The following paragraphs use some of the terms defined in past sections to describe analytically the pseudo-adiabatic condensation process, which was defined when discussing the adiabatic lapse rate.

Assume that a unit mass of dry air saturated with ρ_s grams of vapor and with a temperature T and pressure P is forced to rise in the atmosphere. The rising will result in a pressure drop dP and temperature change dT with resulting condensation dw_s, where w_s is the original mixing ratio $w_s = \rho_s/\rho_d =$

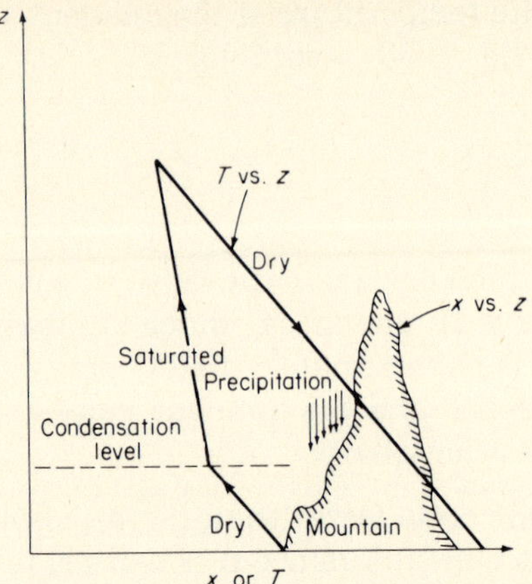

FIGURE 3.25 Warming of moist air through orographic precipitation. Source: Eagleson [1970].

ρ_s, since we are dealing with a unit mass of dry air. Assume, as before, that the latent heat of condensation is entirely absorbed by the 1 g of dry air and not removed by the condensate. An energy balance (Eq. 3.49) of the process will result in

$$L_c dw_s = C_p dT - RT\frac{d(P - e_s)}{P - e_s}. \tag{3.55}$$

In the above equation, the left-hand term is the latent heat of condensation; remember that L_c was previously defined as negative. The first term on the right accounts for internal energy and the second term for work due to volume changes. We have previously seen that

$$w_s = 0.622\frac{e_s}{P - e_s}. \tag{3.56}$$

So, Eqs. (3.55) and (3.56) can be jointly numerically integrated. An approximation of the above is obtained by ignoring e_s relative to P. This results in

$$L_c\frac{dw_s}{T} = C_p\frac{dT}{T} - R\frac{dP}{P}. \tag{3.57}$$

Recall the dry adiabatic temperature–pressure relation,

$$\theta = P_\circ^{1-n}P^{n-1}T, \tag{3.58}$$

where θ is the potential temperature. From the above and using the definition of a total differential,

$$d\theta = \frac{\partial \theta}{\partial P}\, dP + \frac{\partial \theta}{\partial T}\, dT\,, \tag{3.59}$$

it is easy to see that

$$\frac{\partial \theta}{\partial T} = P_\circ^{1-n} P^{n-1}$$

$$\frac{\partial \theta}{\partial P} = P_\circ^{1-n} P^{n-2}(n-1)T \tag{3.60}$$

or

$$\frac{d\theta}{\theta} = \frac{P_\circ^{1-n} P^{n-1}\, dT + P_\circ^{1-n} P^{n-2}(n-1)T\, dP}{P_\circ^{1-n} P^{n-1} T}$$

$$= \frac{dT}{T} + (n-1)\frac{dP}{P}\,. \tag{3.61}$$

Since

$$1 - n = \frac{R}{C_p} \qquad \text{or} \qquad n - 1 = \frac{-R}{C_p}\,,$$

Eq. (3.57) becomes, after using Eq. (3.61),

$$L_c \frac{dw_s}{T} = C_p \frac{d\theta}{\theta} \tag{3.62}$$

by approximating $dw_s/T \approx d(w_s/T)$. Equation (3.62) can be integrated to obtain

$$\frac{\theta}{\theta_e} = \exp\left(\frac{L_c w_s}{C_p T}\right), \tag{3.63}$$

where θ_e is the equivalent potential temperature of a parcel of air after all the moisture has condensed and precipitated out, and all latent heat retained as sensible heat. In using Eq. (3.63), remember that L_c must be taken as a negative number; and hence, $\theta_e > \theta$.

Figure 3.26 shows a diagram of pressure, temperature, lines of potential temperature for dry adiabatic rising, lines of equivalent potential temperature, and lines of water content (mixing ratio). This diagram is obtained using Eqs. (3.56), (3.58), and (3.63), as well as the temperature–pressure relations seen previously.

The pseudo-adiabatic condensation diagram is a very useful tool in the prediction of precipitable moisture. Example 3.6 will serve as illustration.

EXAMPLE 3.6

Illustration of the Use of the Pseudo-adiabatic Diagram

Assume a mass of air at 24°C hits a mountain barrier and is forced to rise. The air has a relative humidity of 75%. The mountain range has an elevation of 1500 m. After clearing the mountain top, the air mass will go back down to sea level, where it started. Using the previous information and the pseudo-adiabatic diagram, it is then possible to obtain precipitable moisture and temperature of the air mass at various locations.

First let us find the mixing ratio. To do that, use Eq. (3.29). The required vapor pressure is obtained using the relative humidity and a table of saturated vapor pressures (or, alternatively, Eq. 3.31). From the tables in Appendix A, the saturated vapor pressure at 24°C can be interpolated to be about 30 mb. With a relative humidity of 75% that implies an actual vapor pressure of 22.5 mb. (0.75×30). From Eq. (3.29) the mixing ratio is then approximately $(0.622 \times 22.5/1000) = 0.014\ \mathrm{g\,g^{-1}}$ or $14\ \mathrm{g\,kg^{-1}}$.

The pseudo-adiabatic diagram is then entered at the intersection of the 1000-mb pressure with the isotherm (solid lines slanting to the right) of 24°C. This is point A in the insert of Figure 3.26. From this point the air rises and cools following the dry adiabat (solid line slanting to the left) that crosses point A. This dry adiabat or isopotential temperature line (Eq. 3.52) also corresponds to 24°C. Rising continues along this curve until the $14\ \mathrm{g\,kg^{-1}}$ mixing ratio line is crossed (mixing ratio lines are dotted, slanting to the right). This occurs fairly quickly, at an elevation of about 600 m. At this point, B in the insert, condensation begins. The condensate releases heat, which reduces the lapse rate of the rising air mass. Further rising occurs along a line of constant equivalent potential temperature (Eq. 3.63), crossing point B. The iso-equivalent, potential temperature lines are shown as dashed curves slanting to the left. Interpolating, we choose a line of $\theta_e \approx 20.5$°C. The isothermal line at the point indicates a temperature of about 18°C. The chosen pseudo-adiabat is then followed to the maximum mountain height of 1500 m. At that point (point C in the insert) the mixing ratio is between 10 and $14\ \mathrm{g\,kg^{-1}}$, approximately $12\ \mathrm{g\,kg^{-1}}$. The air temperature is about 14°C. The implication is that in rising, about 2 g of water per kilogram of rising air has been condensed.

At the lee side of the mountain, the air will descend to sea level. No further condensation occurs, so cooling follows the dry adiabatic line (isopotential temperature line), crossing point C. This corresponds to approximately

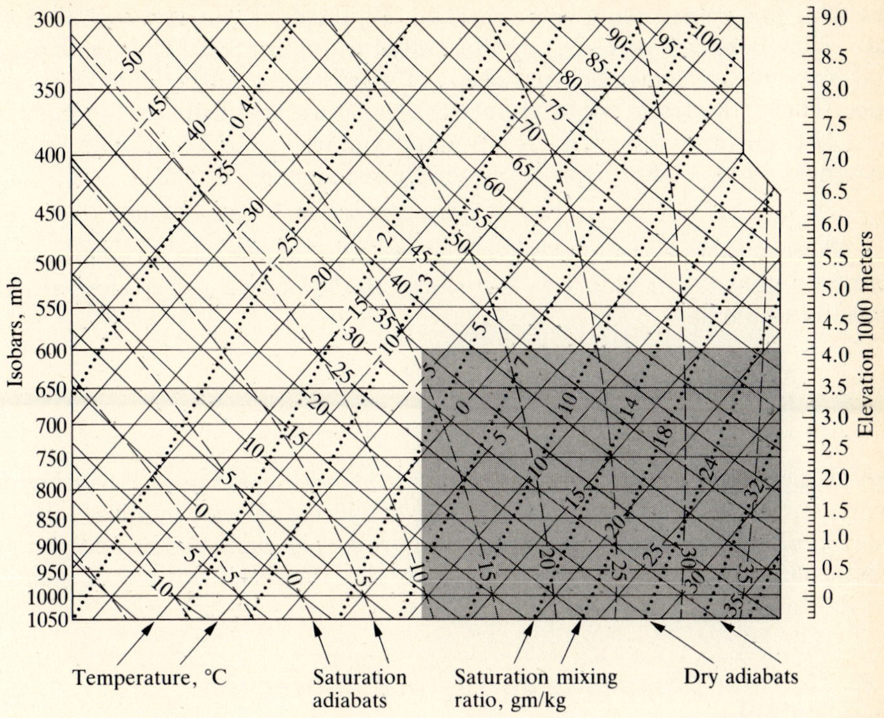

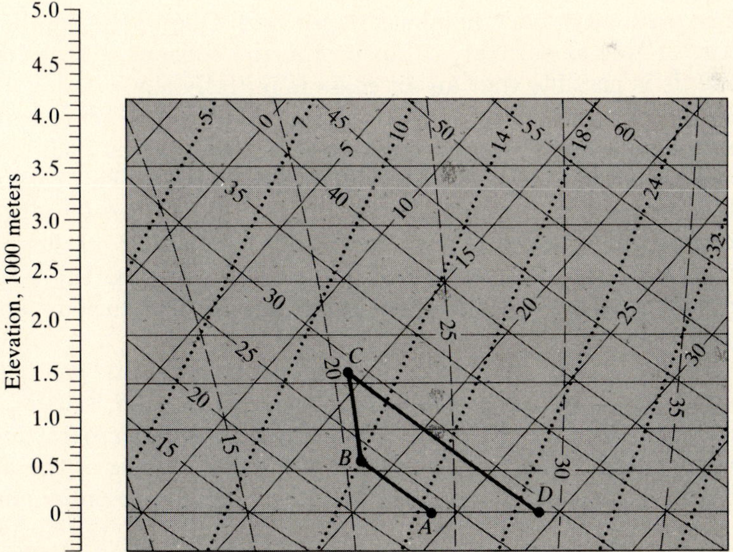

FIGURE 3.26 The pseudo-adiabatic diagram. Source: Adapted from W. M. Marsh, *Earthscape: A Physical Geography.* Copyright © 1987 by Wiley. Reprinted by permission of John Wiley & Sons, Inc.

the 29°C line. When sea level is reached (D), the air is warmer by 5°C (29° − 24°). Such behavior was already illustrated in Figure 3.25. The conditions of this example are somewhat analogous to what occurs on the island of Puerto Rico. There, the trade winds supply the moisture that falls on the northern half of the island, which is closely approximated by a 30- to 35-mile-wide rectangle divided lengthwise by a mountain range. ◆

The calculated precipitable water can be converted to an approximate rainfall intensity per unit area if we knew the velocity v of the rising air. The formula would be

$$p = \frac{\Delta w \gamma_a v}{\gamma_w}, \tag{3.64}$$

where Δw is the condensate in grams per kilogram, γ_a is the specific weight of moist air in kilograms per cubic meter, v is updraft velocity in meters per second, γ_w is the specific weight of water in grams per cubic meter, and p is precipitation rate in meters (of water) per second. Details of this calculation and its limitations will be given in Chapter 4.

3.8.3 Further Comments on Thermal Convection and Stability

In Section 3.8.1 we discussed thermal stability of air masses by comparing the ambient lapse rate to the dry adiabatic lapse rate. Due to the smaller, pseudo-adiabatic (or wet or saturated) lapse rate, the concept of stability must be modified. It is possible that an air mass is initially stable ($|\alpha| < |\Gamma|$) but becomes unstable when forced to rise (for example, by an obstruction) and condensation begins. The heat released upon condensation may be such that the reduced pseudo-adiabatic lapse rate $|\Gamma'|$ is now less than $|\alpha|$. This case is illustrated in Figure 3.27 and is called conditional instability. Condensation can be a very destabilizing process; it acts as a positive feedback on the formation of clouds and precipitation. For convenience in analysis, Table 3.4 gives the pseudo-adiabatic (or saturated adiabatic) lapse rate at various temperatures and pressures.

Up to this point the stability arguments have been based on the behavior of a parcel of air with no vertical dimensions. In reality, though, a mass of air will extend vertically and will have elevation-dependent temperatures and moisture conditions. The net effect is that different portions of the air mass will follow different cooling paths, frequently reducing or enhancing instability.

Trewartha and Horn [1980] give two examples of enhanced instability for rising air masses with extensive vertical development. The first one is illustrated in Figure 3.28. A dry air mass originally extends from point A

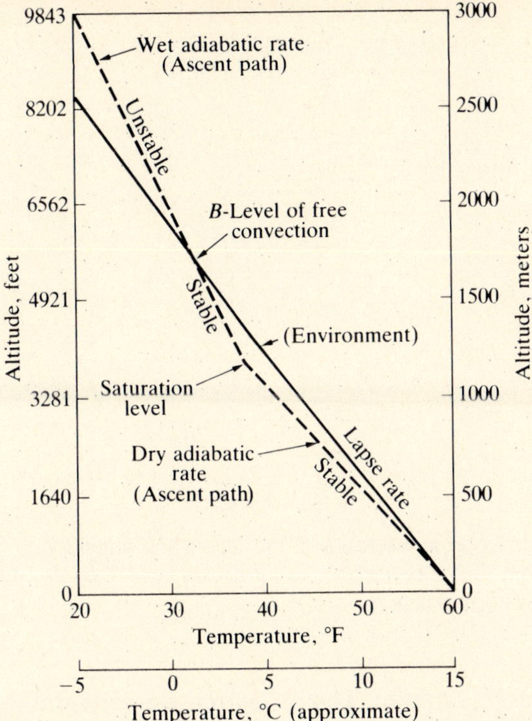

FIGURE 3.27 Conditional instability. Air that was originally stable is made unstable by forced ascent, during which heat of condensation is added. Source: Adapted from G. T. Trewartha, *An Introduction to Climate,* 4th ed., McGraw-Hill, 1968. Reproduced with permission.

TABLE 3.4 Saturated Adiabatic Lapse Rate (°C/100 m)

	PRESSURE (mb)			
TEMPERATURE (°C)	1000	850	700	500
40	−0.30	−0.29	−0.27	
20	−0.43	−0.40	−0.37	−0.32
0	−0.65	−0.61	−0.57	−0.51
−20	−0.86	−0.84	−0.81	−0.76
−40	−0.95	−0.95	−0.94	−0.93

Note: At very cold temperatures, the saturated adiabatic lapse rate is almost equal to the dry adiabatic lapse rate of −0.98°C/100 m. At higher temperatures, the difference is much greater because more water is condensing.

Source: *Understanding Our Atmospheric Environment* by Morris Neilburger et al. Copyright © 1971, 1973, 1983 by W. H. Freeman and Company. Used by permission.

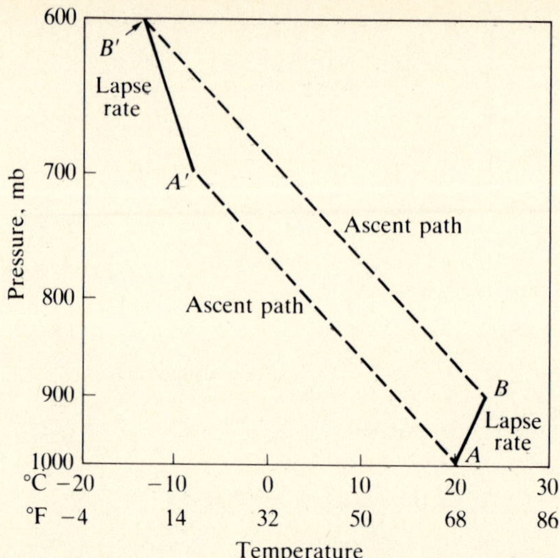

FIGURE 3.28 Effects upon lapse rates of the lifting and subsidence of a thick air mass. Source: Adapted from G. T. Trewartha, *An Introduction to Climate,* 4th ed., McGraw-Hill, 1968. Reproduced with permission.

(1000 mb) to point *B* (900 mb). The air exhibits a very stable temperature inversion, with warmer air aloft. If the mass is forced to rise, points along the line *AB* will follow the dry adiabatic lapse rate to points *A′* and *B′*. There is an obvious expansion of the air column (due to lower pressures), resulting in colder final temperatures for *B′* than for *A′*. The original inversion is now converted to a significant reduction of temperature with height, which is prone to instabilities within the air mass.

A second example is given in Figure 3.29. Here, the original air mass is characterized by larger moisture content in point *A* than in point *B*, a common occurrence. Upon rising (if forced to do so), air in point *A* will begin condensation faster than air in point *B*. At the condensation point, cooling follows the pseudo-adiabatic (also wet or saturated) lapse rate. The net result is that point *B* cools much more than point *A*. The new profile *A′B′* shows a much sharper temperature gradient, which is more apt to support instabilities of rising air parcels within the system.

3.9 SUMMARY

The land, oceans, and atmospheric masses form an indivisible system. They serve as media through which the earth's energy and mass balance are transferred. Reflections of this redistribution activity are long-term climate and local weather conditions, particularly precipitation.

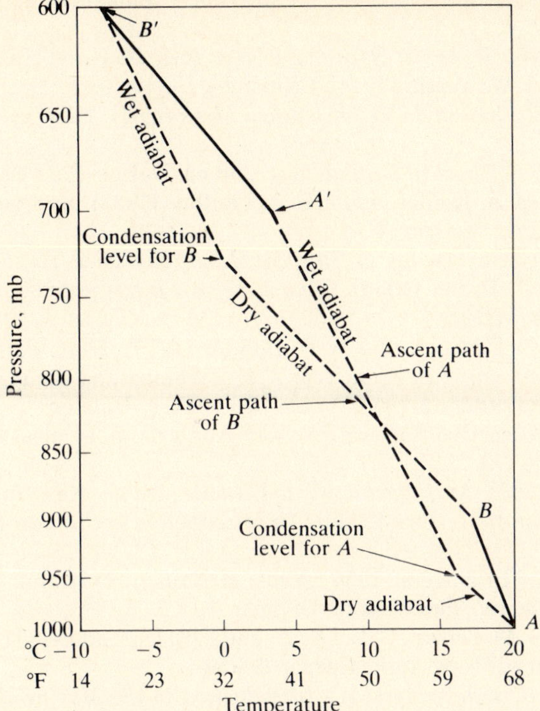

FIGURE 3.29 Moisture effects on convective instability. Source: Adapted from G. T. Trewartha, *An Introduction to Climate,* 4th ed., McGraw-Hill, 1968. Reproduced with permission.

This chapter is intended to provide the necessary meteorologic framework to understand the ultimate origins of the hydrologic processes to be studied next. Obviously, its goal is not to train meteorologists.

Emphasis has been on the earth–atmospheric parameters and patterns that will affect local weather the most. Atmospheric moisture and condensation were given the most attention. Chapter 4 will expand the subject to study how condensation ultimately may lead to precipitation.

REFERENCES

Bosen, J. F. [1958]. "An Approximation Formula to Compute Relative Humidity from Dry Bulk and Dew Point Temperatures." *Monthly Weather Rev.* 86(12):486.

Idem. [1960]. "A Formula for Approximation of the Saturation Vapor Pressure over Water." *Monthly Weather Rev.* 88(8):275.

Idem. [1961]. "Formula for Approximation of the Ratio of the Saturation Vapor Pressure over Ice to That over Water at the Same Temperature." *Monthly Weather Rev.* 92(11):508.

Eagleson, P. S. [1970]. *Dynamic Hydrology.* New York: McGraw-Hill.

Haurwitz, B., and J. W. Austin [1944]. *Climatology.* New York: McGraw-Hill.

Hess, S. L. [1959]. *Introduction to Theoretical Meteorology.* New York: Holt, Rinehart and Winston.

Landsberg, H. [1958]. *Physical Climatology.* 2nd ed. DuBois, Pa.: Gray Printing.

Linsley, R. K., Jr., M. A. Kohler, and J. L. H. Paulhus [1982]. *Hydrology for Engineers.* 3rd ed. New York: McGraw-Hill.

Marsh, W. M. [1987]. *Earthscape: A Physical Geography.* New York: Wiley.

Marsh, W. M., and J. Dozier [1986]. *Landscape: An Introduction to Physical Geography.* New York: Wiley.

Neiburger, M., J. C. Edinger, and W. D. Bonner [1973]. *Understanding Our Atmospheric Environment.* San Francisco: W. H. Freeman and Company.

Oke, T. R. [1978]. *Boundary Layer Climates.* New York: Halsted.

Riehl, H. [1978]. "Objective Surface Pressure Analysis in Tropical Countries." *WMO Bull.* 27(3):175–179.

Trewartha, G. T. [1968]. *An Introduction to Climate.* 4th ed. New York: McGraw-Hill.

Trewartha, G. T., and L. H. Horn [1980]. *An Introduction to Climate.* 5th ed. New York: McGraw-Hill.

U.S. Army Corps of Engineers [1956]. *Snow Hydrology.* Portland, Oreg.: U.S. Army Corps of Engineers, North Pacific Division.

Viessman, W., Jr., J. W. Knapp, G. L. Lewis, and F. E. Harbaugh [1977]. *Introduction to Hydrology.* 2nd ed. New York: Harper & Row.

Weyl, P. K. [1970]. *Oceanography: An Introduction to the Marine Environment.* New York: Wiley.

PROBLEMS

1. Calculate the absolute humidity of an air mass that has a relative humidity of 25% and a temperature of 100°F. (Adapted from P. S. Eagleson, *Dynamic Hydrology,* McGraw-Hill, 1970.)

2. What is the latent heat of vaporization, in calories per gram, for water at (a) 15°C? (b) 77°F?

3. What is the density in kilograms per cubic meter of (a) dry air at 30°C and a pressure of 900 mb? (b) moist air with a relative humidity of 70% at the same temperature and pressure?

4. An air mass at 50°F and 70% relative humidity is forced to go from sea level over a mountain range that is 1000 m high. Estimate the amount of rainfall released for each 30-m thickness of air mass. (Adapted from P. S. Eagleson, *Dynamic Hydrology,* McGraw-Hill, 1970.)

5. For the above problem, what are the temperature and relative humidity of the air mass when it reaches sea level behind the mountain range? (Adapted from P. S. Eagleson, *Dynamic Hydrology,* McGraw-Hill, 1970.)

6. A sounding in the atmosphere indicates temperatures of 16.0, 11.6, and 6.2°C at the 900, 800, and 700 mb levels, respectively. Compute the precipitable water, in millimeters, in the layer between 900 and 700 mb.

7. The ambient atmospheric lapse rate is often taken as $\alpha = 3.3$°F per 1000 ft. Assuming a constant specific humidity throughout the troposphere, calculate the percentage variation in α between dry and saturated atmospheres. (Adapted from P. S. Eagleson, *Dynamic Hydrology*, McGraw-Hill, 1970.)

8. Show that moist air is less dense than dry air. Derive an expression for the density of a mixture of dry air and water vapor. Show that for such a mixture the ideal gas law can be rewritten as

$$p = R\rho T_v = R\rho T(1 + 0.608q),$$

where q is the water vapor mixing ratio of the mixture. T_v is referred to as virtual temperature and represents the temperature dry air would have if it were of the same density as the moist air at the given pressure. (Contributed by Dr. Jorge Ramirez.)

9. In general, atmospheric moisture is confined to the lower portions of the troposphere in the planetary boundary layer (PBL). Construct a hydrostatic atmosphere assuming a quadratic exponential decay of water-vapor mixing ratio with height, as indicated below. Assume a surface pressure of 1000 mb, a surface temperature of 25°C, and a surface relative humidity of 85%. Additionally, assume a constant atmospheric (ambient) lapse rate of 6.5°C km^{-1}. Finally, assume the tropopause is located at 200 mb, beyond which the lapse rate is zero up to 100 mb. Account for the effect of water vapor on density of air.

Let moisture be distributed such that the mixing ratio q_v is given by

$$q_v(z) = q_v(z_o)\exp\left[-\frac{z^2}{H^2}\right],$$

where z is ground elevation and H is the depth of the planetary boundary layer. Assume $H = 1$ km. (Contributed by Dr. Jorge Ramirez.)

10. Using the soundings obtained in Problem 9, answer the following questions:

a) What is the height of the tropopause?

b) How many degrees of cooling at the surface are necessary to bring about saturation (fog formation)? Discuss potential mechanisms producing this effect in nature.

c) Discuss the stability of the atmosphere to vertical displacements of air. What parts of the atmosphere are stable (or unstable) to dry-air displacements? What parts to saturated displacements?

d) If moist air from the PBL is heated so that it rises adiabatically, it will eventually become saturated at the lifted condensation level (LCL). Determine the height of the LCL.

Perform the computations either numerically or with the aid of the pseudo-adiabatic diagram. (Contributed by Dr. Jorge Ramirez.)

11. In a conditionally unstable atmosphere like the one produced in Problem 9, moist air from the PBL is stable to infinitesimal vertical displacements but unstable to finite displacements. This implies that free buoyant convection (cloud formation) will occur if this air is brought to its level of free convection (LFC). An estimate of the energy required to lift air to its LFC can be obtained by computing the area enclosed by the temperature profiles (with elevation) of both the atmosphere and the adiabatically lifted air parcel. Compute the area below the LFC and the area above the LFC. Observe that the total energy released is positive. This energy is released during convective overturning. (Contributed by Dr. Jorge Ramirez.)

12. Consider a parcel of air that is adiabatically displaced a distance Δz, vertically. Discuss the vertical forces acting on it. Derive the vertical momentum equation for this parcel, and discuss its solutions. Determine necessary conditions for stable solutions. (Contributed by Dr. Jorge Ramirez.)

13. Define and describe the following terms. Do it clearly and concisely: (a) dew-point temperature, (b) absolute humidity, (c) latent heat, (d) moist adiabatic or pseudo-adiabatic process, (e) specific humidity, (f) atmospheric convection, (g) convective stability, (h) ambient lapse rate, and (i) dry adiabatic lapse rate.

14. Consider an air mass with a mixing ratio of 12 $\mathrm{g\,kg^{-1}}$ and a temperature of 25°C rising to an elevation of 8 km in the atmosphere. If the cloud system within which this air mass rises was 20% efficient in converting vapor to precipitation, and if the cloud lasts two hours with average updraft velocities of 3 $\mathrm{m\,s^{-1}}$, what would be the total precipitated water?

15. The air in a wind stream over the ocean has the following characteristics: relative humidity, 90%; surface temperature, 25°C; ambient lapse rate, 6°C per 1000 m; and atmospheric pressure, 1020 mb. Three islands in this wind stream have average elevations of 100 m, 400 m, and 1000 m. Would you expect precipitation on any of these islands? Speculate on whether the precipitation would be significant. State your assumptions.

16. This chapter discussed how the saturation vapor pressure e_s behaves as a function of temperature and gave an approximate $e_s(T)$ relation. Assume that temperature decreases at the dry adiabatic lapse rate with elevation and that the pressure distribution is hydrostatic. (a) Develop the $e_s(z)$ relation. (b) Let the surface temperature and pressure be $T_o = 25$°C and $P_o = 1000$ mb. Plot $e_s(z)$ and $p(z)$ for $0 \le z \le 1$ km, and comment on the result. (Contributed by Dr. Angelos Protopapas.)

17. It is possible to fit the saturation vapor pressure versus temperature curve by a function of the form

$$e_s(T) = a(T - b)^{C_p/R},$$

where a and b are constants, R is the gas constant, and C_p is the specific heat at constant pressure, all in proper units.

a) Given the initial temperature T_o, pressure p_o, and mixing ratio w_s of an air parcel, find the temperature T_c and the pressure P_c at which condensation occurs. Assume that the parcel rises adiabatically.

b) If instead of the saturation mixing ratio w_s the initial dew-point temperature T_d of the parcel is known, how does the above result change?

(Contributed by Dr. Angelos Protopapas.)

18. Concepts related to stability criteria for air parcels are based upon the relative magnitude of the ambient lapse rate α as compared with both dry (subsaturated) and saturated adiabatic lapse rates Γ and Γ'. An atmosphere that is stable to subsaturated motions but unstable to saturated ascent is called conditionally unstable (see figure). Generally, cumulus clouds develop in conditionally unstable environments. During the lifetime of a cloud, some of the processes that take place are subsidence and detrainment. Subsidence results from a continuity of mass requirement. Its effect is to take air from the upper troposphere and lower it. During cloud decay, liquid water from the cloud is detrained (mixed turbulently) with subsaturated ambient air. Turbulent mixing will tend to saturate the ambient air.

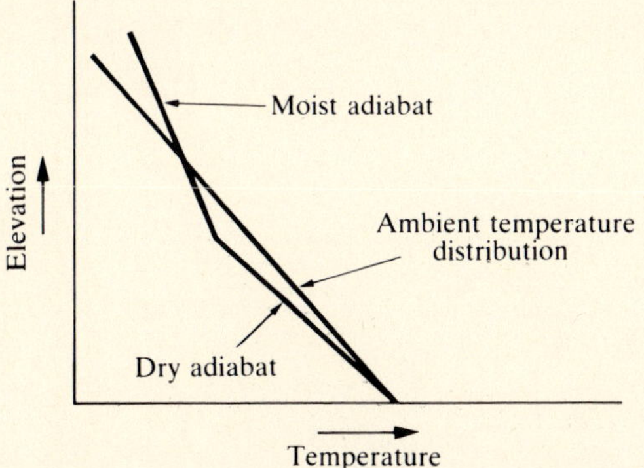

Assume that subsidence takes place adiabatically and that upper air is dry. Also assume that detrained cloud water evaporates instantaneously at constant pressure and that the degree of instability can be measured by the absolute difference in the corresponding lapse rates. Speculate on the stability changes brought about by the above two processes individually. Speculate on the relative magnitude of these effects. What is their likely relative spatial distribution?

19. The atmospheric soundings on April 15, 1988, on the windward side of the Rocky Mountains, are given below.

P (mb)	T (°C)	g kg^{-1}
1004	15.60	7.30
1000	15.60	7.40
850	9.80	7.70
700	0.00	3.80

Find the dew-point temperature of each of the given points on the soundings. The air was forced to go over the mountain range. The recorded soundings at 850 mb on the leeward side were $T = 19.6°\text{C}$, $w = 2.7\ \text{g kg}^{-1}$. From what pressure level on the earlier soundings might this air parcel have come, and how high would it have risen in the interim? Estimate the amount of rainfall released for each 50-m thickness of this air mass. Explain why the temperature on the leeward side is so high. (Contributed by Mr. Shafiqul Islam.)

20. A saturated layer 500 m thick is ascending at $2\ \text{m s}^{-1}$. Its mean pressure is $P = 850$ mb and its mean temperature is 20°C. Estimate the maximum precipitation rate (in millimeters per hour) that can fall from this layer. (Contributed by Mr. Shafiqul Islam.)

Chapter **4**

Precipitation Occurrence and Measurement

4.1 INTRODUCTION

In Chapter 3 we saw that although water vapor is a minor component of the atmosphere, by volume, it is the most important mass input to the hydrologic cycle. Water vapor is the medium by which a large portion of the earthwide mass and energy lack of balance is corrected. We saw that generally the oceans are a major humidity source, particularly tropical zones. The transfer of moisture to higher latitudes with subsequent condensation and latent heat release is an important mechanism of redistribution of energy.

In Chapter 3 we also studied the pseudo-adiabatic condensation process and the atmospheric temperature instabilities that may lead to condensation. But condensation alone does not guarantee precipitation. There are several other requirements for precipitation:

1. Ascending humid air, implying cooling and pressure changes.
2. A change of phase must occur requiring some activation energy to account for the surface energy related to the interface of coexisting phases. This activation energy requirement could be very high unless some nucleating sites are available.
3. Condensate, water drops, or ice crystals must grow sufficiently to counteract cloud updrafts to reach the earth's surface.

This chapter will discuss the above requirements through a brief introduction to the physics of clouds. We will study the form and structure of precipitation as well as its variability in time and space, at local and global scales. The chapter will end with a presentation of some of the issues related to the measurement of precipitation.

4.2 COOLING AND LIFTING PROCESSES

There are four general cooling and lifting mechanisms. Each rarely occurs alone in nature, but some may dominate under certain conditions.

Nonfrontal or horizontal convergence of air into a low-pressure point results in vertical displacement of air, which may lead to condensation and precipitation. Convergence will commonly occur on or near the tropics as northern and southern components of the trade winds, easterlies (see Chapter 3), clash. Horizontal convergence may also occur as western and eastern sides of two adjacent low-pressure cyclones meet.

Extratropical horizontal convergence generates precipitation of moderate intensity—5 to 15 cm over periods of 24 to 72 hours. Those of tropical origin can result in 40 cm in 12 to 24 hours.

Frontal or cyclonic lifting results from convergence of air masses at different temperatures and of different character. They generally occur over temperate regions due to the convergence of polar easterlies and middle-latitudes westerlies. Figure 4.1 shows the concept of a frontal surface. A warm front occurs when warm air impinges on cold air. The warm air rises over the cold air at a relatively gentle slope of 1:100 to 1:400. Because of the slowly rising air masses, warm fronts have associated heralding clouds and precipitation extending 300 to 500 km ahead of the surface front.

Cold fronts occur when cold air moves under a warm air mass forcing the latter upward. A steeper sloping interface (1:25 to 1:100) is observed. Clouds and rainfall are limited to about 80 km ahead of the front.

Frontal storms are steady and extend over a large area. They are generally of moderate intensity with cold fronts causing more variable, tumultuous rainfall of shorter duration.

The development of a cyclonic or frontal weather pattern is depicted in Figure 4.2. Note that the storm develops two fronts, a warm front leading a cold front, which tend to converge. The fronts rotate in a cyclonic pattern around a low-pressure center. The faster-moving cold front finally overtakes the warm front, resulting in an occluded wave. When this occurs the warm, moist air is effectively separated from the cold air.

The cold air is the catalyst that induces (by lifting) condensation and release of energy from the warm air. The moment occlusion occurs, this energy-releasing mechanism stops and the system begins to die.

Cyclonic storms also disappear when they move away from their source of moist, warm air. The moist, warm air masses are of maritime and mostly of tropical origin.

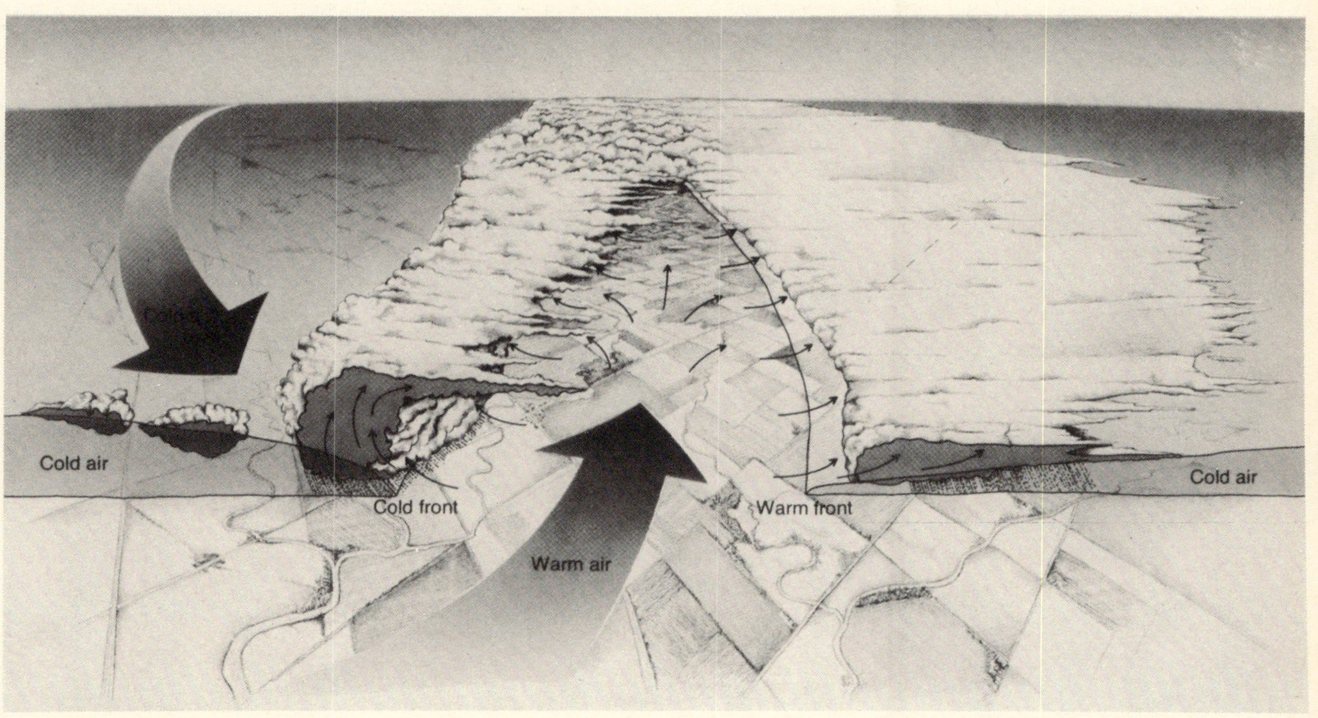

FIGURE 4.1 Cold front, warm front, and associated features of middle latitude cyclone. Turbulence is greatest along the cold front, but the more extensive cloud cover develops along the warm front. (Illustration by William M. Marsh.) Source: W. M. Marsh and J. Dozier, *Landscape: An Introduction to Physical Geography*. Copyright © 1986 by Wiley. Reprinted by permission of John Wiley & Sons, Inc.

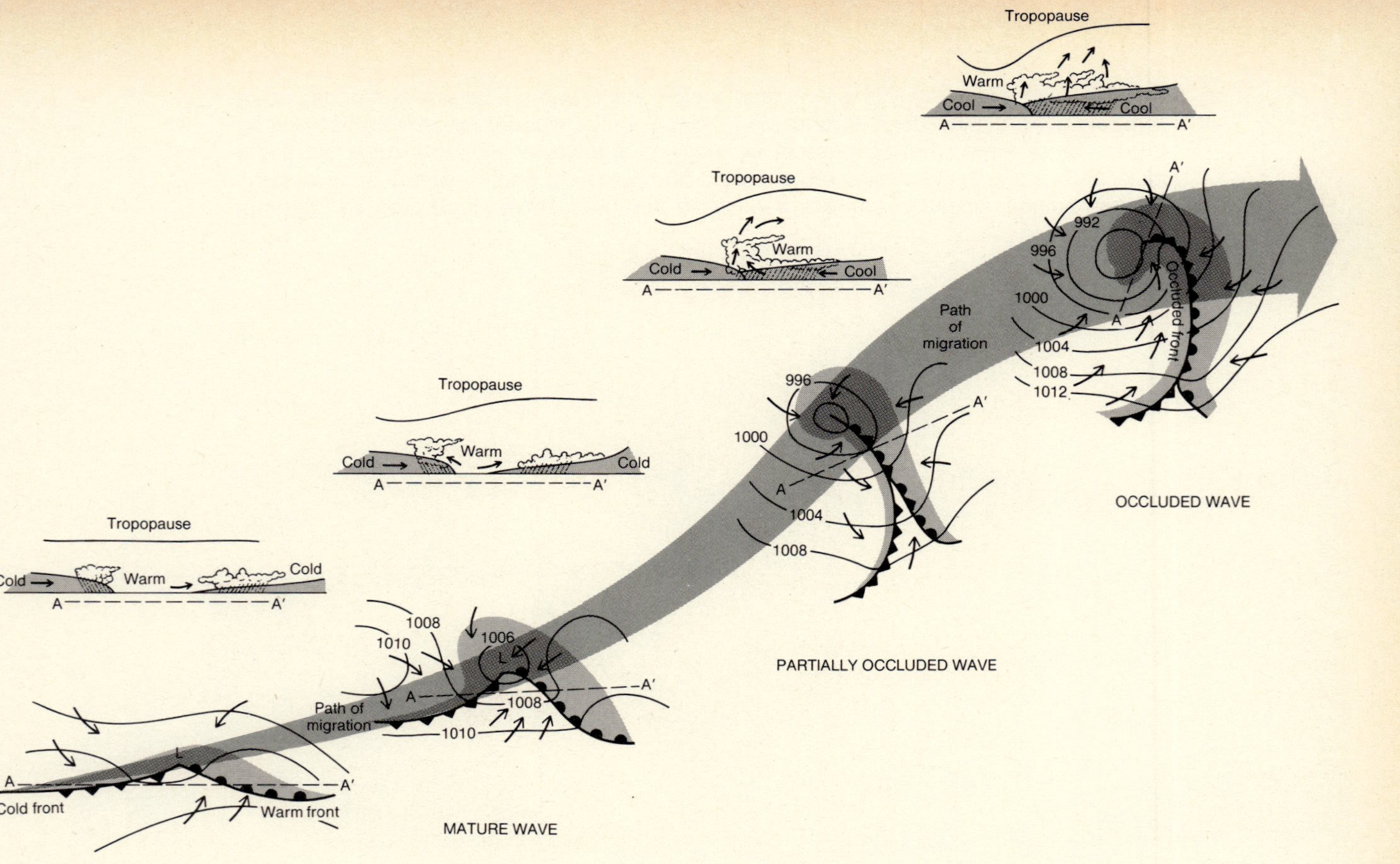

FIGURE 4.2 Development of a cyclone. Source: W. M. Marsh and J. Dozier, *Landscape: An Introduction to Physical Geography*. Copyright © 1986 by Wiley. Reprinted by permission of John Wiley & Sons, Inc. (After Godske et al. [1975].)

The path followed by cyclones was studied by Klein [1957]. Figures 4.3 and 4.4 show the principal tracks for the months of January and July, respectively. The figures agree with the observation that cyclone tracks are very much controlled by the oscillations of the upper atmosphere's jet streams that were discussed in Chapter 3. The fast-moving oscillating streams provide the mechanism to remove the rising air masses of the cyclones, encouraging their development. Like the jet streams, the preferred cyclonic tracks are from west to east, moving south during the northern hemisphere winter and north during the summer.

Orographic cooling occurs when air masses are forced to rise over an obstruction, like a mountain. Figure 4.5 illustrates the process. As discussed in Chapter 3, the end result is condensation and rain in the windward side of the mountain with contrasting dryness in the lee side of the mountain.

Regions of important orographic effects exhibit relatively high precipitation accumulation as well as increased frequency of events. This is an overall average effect; it should not be construed as behavior that can be easily detected by analysis of point precipitation of an event. The spatial variability of point precipitation can often obscure any elevation effects on the intensity or accumulation of an event.

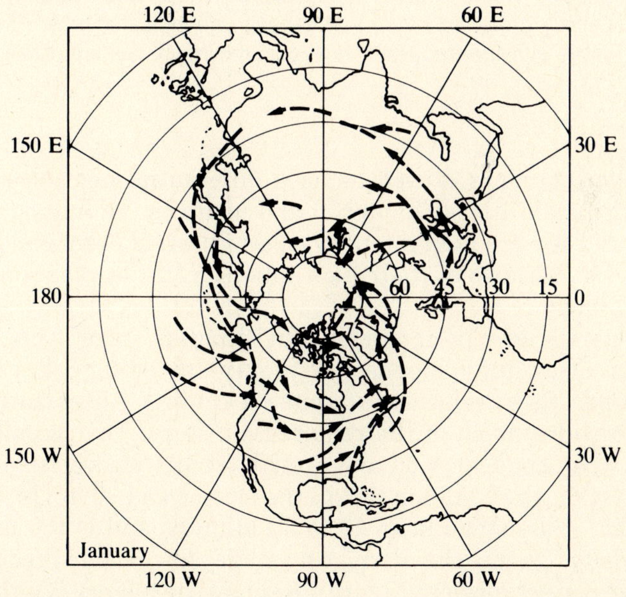

FIGURE 4.3 Principal tracks of northern hemisphere sea-level cyclones in January. Solid lines denote most frequent, well-defined tracks; dashed lines, less frequent and less well-defined tracks. Locally preferred regions of genesis are indicated where tracks begin. Arrowheads end where cyclone frequency is a local minimum. Source: Klein [1957].

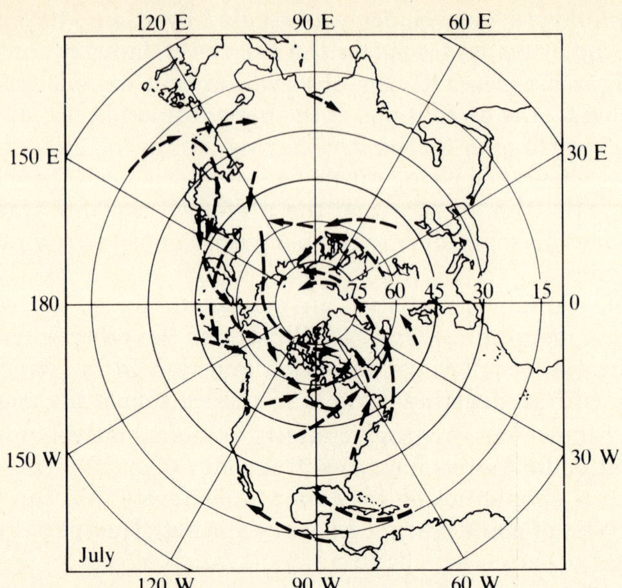

FIGURE 4.4 Principal tracks of northern hemisphere sea-level cyclones in July. Solid lines denote most frequent, well-defined tracks; dashed lines, less frequent and less well-defined tracks. Locally preferred regions of genesis are indicated where tracks begin. Arrowheads end where cyclone frequency is a local minimum. Source: Klein [1957].

Convective cooling is caused by local differential heating of air masses, leading to air instabilities (Chapter 3) and updrafts. Thunderstorms are convective storms. They are high-intensity, short-duration events and are common over the tropics. Because of their high intensity, these events are important in urban hydrology.

Thunderstorms develop in stages. The cumulus stage is characterized by the formation of cumulus clouds (commonly exceeding 8 km in elevation) with strong updrafts at velocities of up to 60 km hr^{-1}. Air entrainment during this period provides much of the required moisture. The second stage is the mature one, when condensation and growth occurs. Updrafts and downdrafts coexist with winds up to 115 km hr^{-1} in the former and 30 km hr^{-1} in the latter (above 1.5 km). This stage lasts 15 to 30 minutes and is the period of peak rainfall intensity. During the dissipating stage, downdrafts predominate until the convective cell disappears. Figure 4.6 shows the history of a thunderstorm.

Convective lifting commonly occurs in conjunction with the other lifting mechanisms. In particular, once condensation occurs and the temperature profiles are unstable, continuing rising and condensation follows the convective pattern.

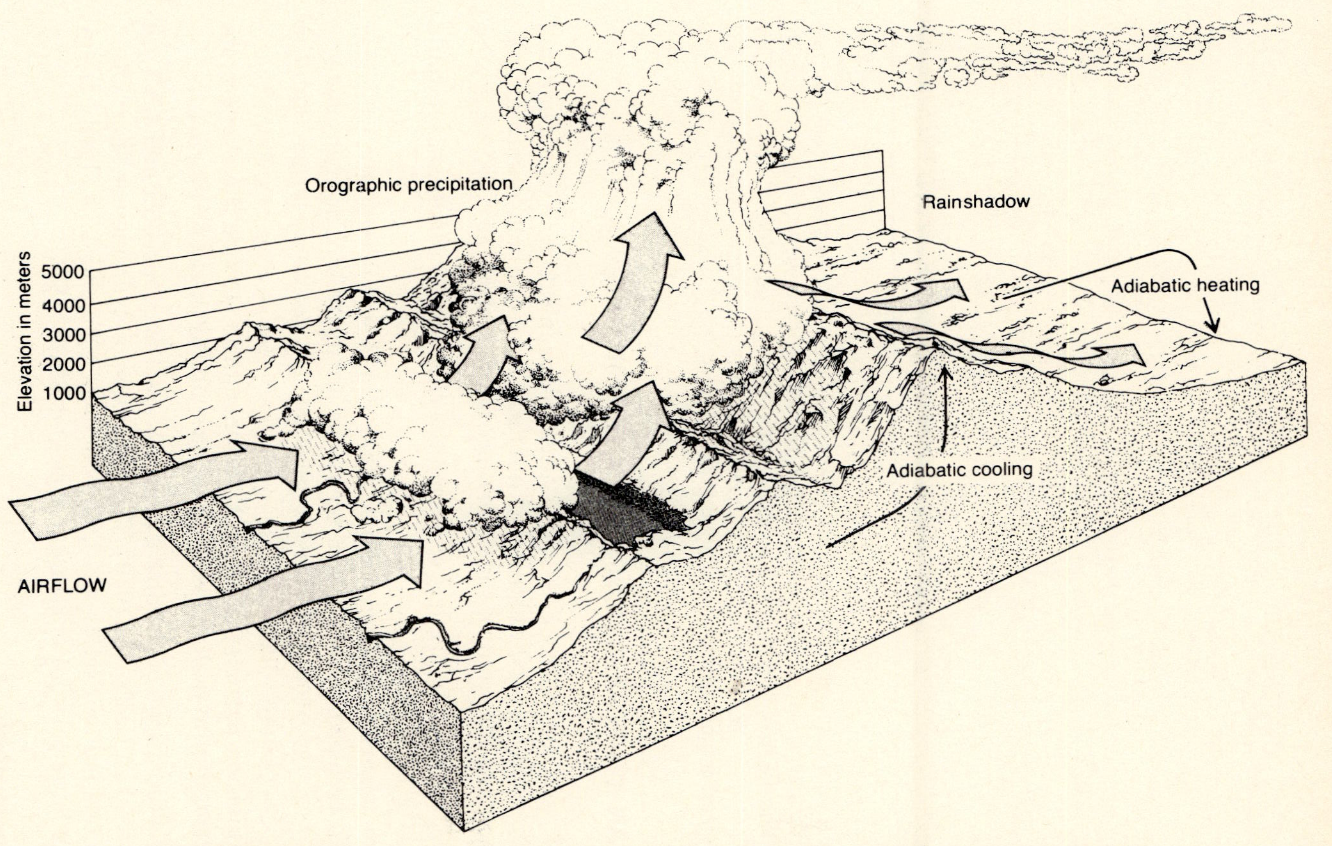

FIGURE 4.5 Orographic precipitation results when moist air is forced over a mountain range. Relative humidity drops when the air descends the leeward slope and heats adiabatically. Source: W. M. Marsh and J. Dozier, *Landscape: An Introduction to Physical Geography*. Copyright © 1986 by Wiley. Reprinted by permission of John Wiley & Sons, Inc.

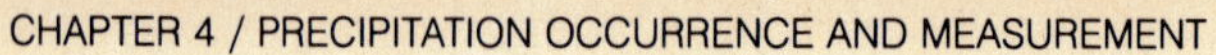

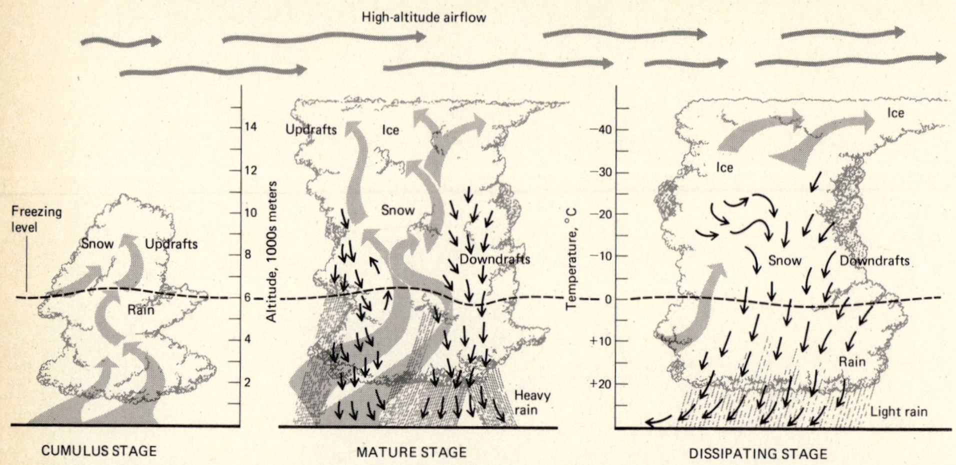

FIGURE 4.6 Formation of a thunderstorm can be described in three stages: cumulus, mature, and dissipating. The first is characterized by updrafts, the second by updrafts, downdrafts, and heavy rain, and the third by downdrafts, and light rain. (Illustration by William M. Marsh.) Source: W. M. Marsh, *Earthscape: A Physical Geography*. Copyright © 1987 by Wiley. Reprinted by permission of John Wiley & Sons, Inc.

4.3 AN INTRODUCTION TO CLOUD PHYSICS

Lifting of moist air masses does not guarantee precipitation. Within the cloud system, there are several complicated interactions of thermodynamic and mechanical nature that control the occurrence of precipitation:

1. Without the catalytic effect of nuclei, the efficiency of condensation would be terribly low.
2. Hydrometeors must grow in size, otherwise they will remain suspended or will be blown off the cloud top by the air updrafts. The hydrometeors must be large enough to achieve terminal fall velocities larger than the velocities of upflowing air.
3. Precipitation particles must be large enough to survive the evaporation occurring in the path between cloud base and ground level.

A brief discussion of the above points follows. Sections 4.3.1 and 4.3.2 are largely based on Eagleson [1970]. The discussion of hydrometeor distribution in the second half of Section 4.3.2, and Sections 4.3.3 and 4.3.4 closely follow and quote large portions of Georgakakos and Bras [1984a].

4.3.1 Nucleation

Condensation under homogeneous conditions requires tremendous saturation conditions and high activation energy. Eagleson [1970] discusses the thermodynamics of nucleation. Any change of phase involves a change in the energy state of the system. The portion of the energy capable of doing work is called free energy. The added free energy involved in condensation of vapor to water or sublimation of vapor to ice is called the activation energy.

Reactions or phase changes move in the direction of minimum free energy. Under conditions of no impurities, and if vapor pressure e is less than saturation, minimum free energy is achieved with zero radius, i.e., no condensation or nucleation of vapor occurs. If $e > e_s$, minimum activation energy can be achieved by no change of phase or by droplet growth beyond a critical radius r^*, which requires a minimum level of activation energy. The rate of nucleation, or drop formation, goes up as the square of the degree of super saturation. Eagleson [1970] states that under conditions of no impurities, the rate of nucleation remains small until e/e_s approaches 4.

The introduction of impurities or nucleating sites reduces the amount of activation energy required for condensation. As stated above, a degree of supersaturation of 4 is demanded by the homogeneous (no impurities) case. In heterogeneous conditions (impurities), the change of phase occurs with e/e_s near 1, a more typical condition in the atmosphere.

In high cloud systems, with below-freezing temperatures, ice particles may be formed by freezing of liquid droplets or by direct sublimation from vapor. Again, the phase changes are aided by nucleating sites or impurities. For example a 1-μm (micron) ice particle is formed from water under homogeneous conditions if the temperature is about −30°C or colder. The same-sized particle may form under heterogeneous freezing at slightly below 0°C.

Homogeneous sublimation of ice from vapor would require temperatures of about −60°C. Under heterogeneous conditions, sublimation of a 0.1-μm particle may occur at temperatures as high as −4°C (see Eagleson [1970] for details leading to previous statements).

Fortunately, the atmosphere is relatively rich in nuclei that assist in phase changes. Nuclei are mostly soil particles (clay), fossil-fuel combustion products, ammonium salts, and seawater salts. Condensation nuclei must be larger than about 0.1 μm.

At elevations where temperatures are below freezing, large-sized nuclei are scarce. As the temperature decreases, though, freezing begins to occur onto smaller particles, and many more potential nucleating sites become active. Nuclei as large as 10 μm, mostly sea salts, can be very effective in inducing condensation under warm clouds (above freezing) conditions.

4.3.2 Growth and Distribution of Precipitation Particles

Before falling, droplets formed by condensation, or ice particles, must grow to a size and weight capable of overcoming updraft velocities in the cloud. Initial growth is achieved by condensation due to supersaturated conditions in the vicinity of the droplets. This mechanism can be reasonably effective in the case of ice particles.

Supersaturation is small in the atmosphere, so it cannot support continuing growth of the type just mentioned. The most effective growth mechanism is called coalescence and results from repeated collisions and aggregations of falling precipitation (mainly water) particles. The coalescence or collection efficiency of particles is measured in terms of effective cross-sectional area swept by a falling particle. Assume there is a large particle of radius r_ℓ and a small particle of radius r_s. The falling large particle will overtake, collide with, and possibly absorb the small particle, if the small particle is within a radius $R = r_\ell + r_s$ of the center of the large particle. The collection efficiency is then defined as

$$E = \frac{R^2}{r_\ell^2}. \tag{4.1}$$

The radius of influence R is $r_\ell + r_s$ only if the dynamics of particle motions do not affect collision probability—i.e., a static analogy. In fact, relative particle motions and sizes may induce or prevent collisions, effectively increasing or reducing R.

Under the static assumption, Eagleson [1970] points out that $E \to 1$ as $r_s/r_\ell \to 0$ and $E \to 4$ as $r_s/r_\ell \to 1$. The maximum efficiencies occur for r_s/r_ℓ ratios larger than about 0.6. Efficiency of collision goes to zero if the large particles are small (20 μm) and the radii ratio is less than 0.3. It also goes to zero if small and large particles are of similar size and have diameters less than about 60 μm. Particles larger than about 80 μm become very efficient at coalescence as the particle distribution becomes uniform ($r_s/r_\ell > 0.9$).

The previous discussion refers to coalescence of water droplets. The growth of snowflakes seems to result mostly from collisions with ice crystals (Eagleson [1970]).

The resulting distribution of hydrometeor sizes resulting from the growth mechanisms is fairly undefined. Nevertheless, it can be inferred (Mason [1971]; Pruppacher and Klett [1978]) that the distribution $N(D)$ of the number of particles per unit volume of diameter within the interval $(D, D + dD)$ is such that it "increases steeply for small D to reach a maximum and possesses a very mild slope for large D" (Georgakakos and Bras [1982]).

Several forms for $N(D)$ satisfying the above characteristics have been suggested (Mason [1971]; Pruppacher and Klett [1978]), but they are all tainted by the difficulties of measuring hydrometeor size in the field. Errors are induced because of inability to measure particles smaller than a given size, inability to measure simultaneously at different locations, or troubles with instrument calibration and reliability under freezing conditions.

Many investigators have then suggested the use of a simple exponential form for the hydrometeor size distribution at a given elevation. The form

$$N(D) = N_o e^{-cD} \tag{4.2}$$

(see Fig. 4.7) has been used by Marshall and Palmer [1948], Gunn and Marshall [1958], Georgakakos and Bras [1984a, b] among others. The parameter c in Eq. (4.2) is the inverse mean diameter at a given level.

The possible objection to Eq. (4.2) is that it implies hydrometeors at diameters approaching zero. The attractive alternative would be to use a distribution starting at zero and peaking somewhere in the small-diameter region. Georgakakos and Bras [1984a] argue that given the acknowledged uncertainties of measuring the number of small hydrometeors, Eq. (4.2) is in fact adequate. Furthermore, they argue that the small-diameter region plays a small role in the macroscopic meteorologic and hydrologic behavior. For example, the water-equivalent mass due to hydrometeors of diameter in the interval $(D, D + dD)$ at a given level per unit volume is

$$X(D) = \rho_\omega N(D)\frac{\pi}{6}D^3 . \tag{4.3}$$

The corresponding rate of mass precipitation would be

$$P(D) = \rho_\omega N(D)\frac{\pi}{6}D^3[V_T(D) - V], \tag{4.4}$$

where ρ_ω is liquid water density, $V_T(D)$ is the terminal velocity of a hydrometeor of diameter D, and V is updraft velocity. Assuming (see Section 4.3.3) that

$$V_T(D) = \alpha D ,$$

where α is a coefficient, and normalizing Eq. (4.2) by $N_\circ$ and Eqs. (4.3) and (4.4) by $\pi\rho_\omega N_\circ/6c^3$ and $\rho_\omega N_\circ\alpha/6c^4$, respectively, results in the curves shown in Figure 4.7.

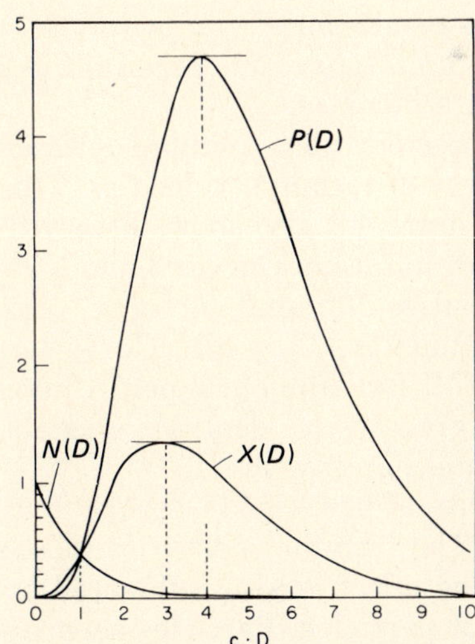

FIGURE 4.7 Normalized number of hydrometeors $N(D)$, water-mass content $X(D)$, and precipitation rate $P(D)$ due to hydrometeors of diameter in the range $(D, D + dD)$ versus the normalized diameter cD. Source: K. P. Georgakakos and R. L. Bras, "A Hydrologically Useful Station Precipitation Model: 1, Formulation," *Water Resources Res*. 20(11):1589, 1984. Copyright by the American Geophysical Union.

The normalized size distribution, liquid moisture content, and precipitation rate are given as a function of the nondimensional hydrometeor size measure cD. The result is that most of the moisture and precipitation is due to diameters well above the mean of $1/c$. The peak liquid water equivalent occurs at $D = 3/c$ and the maximum water-equivalent mass precipitation at $4/c$. Only 1.9% of the total liquid water equivalent is contributed by hydrometeors less than $1/c$ in size. The equivalent figure for the precipitated mass rate is 0.37%.

The mean diameter of the hydrometeors should be larger near the bottom of a cloud and smaller at the top. This is a reflection of the expected nuclei size distribution and the coalescence mechanism. This distribution is very hard to establish because of the sampling difficulties. Georgakakos and Bras [1984a] suggest a simple linear distribution for the value of c.

4.3.3 Terminal Velocities of Hydrometeors

The fall of hydrometeors in the cloud system can be reasonably assumed to occur at their terminal velocities. In quiet air and for an isolated hydrometeor, Pruppacher and Klett [1978] do argue for a constant terminal velocity.

Exact solutions to the equations of motion for a falling hydrometeor do not exist. Nevertheless, experimental and approximate numerical studies of single-particle behavior do exist. Beard [1976] compiled data on free-falling water droplets in air. Droplet size varied from 1 μm to 7 mm. He obtained expressions for velocity V_T as a function of diameter D, particle density ρ_p, and the temperature (T) and pressure (P) of the ambient air. Figure 4.8, curves 1 and 2, show his results for $T = 273.15°\text{K}$, $P = 800$ mb (curve 1); and $T = 293.15°\text{K}$, $P = 1013$ mb (curve 2). Terminal velocity decreases as the temperature and pressure increase.

Contrary to liquid precipitation, solid precipitation particles of the same mass show a wide variety of terminal velocities. This is due to their varied and irregular shapes. Figure 4.8 also shows observed terminal velocities for various types of solid precipitation. Curves 3 and 5 are Beard's [1976] results for solid spheres of densities 500 and 100 $\text{kg}\,\text{m}^{-3}$, respectively. Curves 4, 6, and 7 are simple power functions $V_T = aD^b$. Curve 4 corresponds to results by Locatelli and Hobbs [1974] for "lump graupel." Curves 6 and 7 are for "hexagonal graupel" and aggregates of "dentritic crystals," respectively. The results are also from Locatelli and Hobbs [1974].

In reality, the velocity of particles in groups may be larger than that of isolated hydrometeors. The terminal velocities increase as the particles become closer together. The effect is also influenced by the Reynolds number and the total number of particles. Particles separated by distances larger than 30 or 35 diameters behave very much like single particles.

4.3.4 Evaporation of Precipitating Hydrometeors

Hydrometeors exiting through the cloud base will suffer evaporation throughout their trip to ground level. Evaporation is driven by the difference

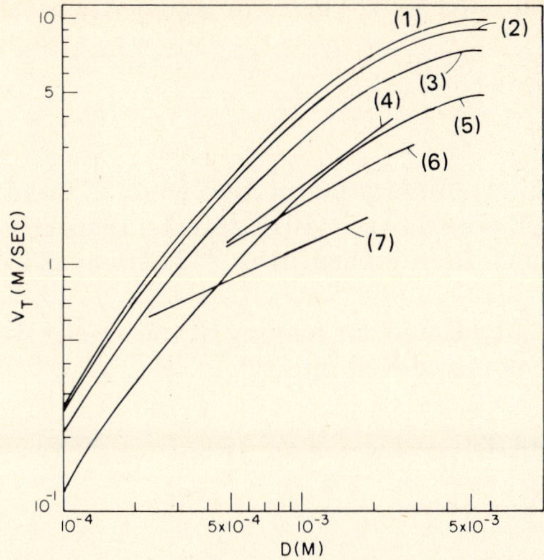

FIGURE 4.8 Terminal velocity as a function of hydrometeor diameter.

Curve 1: Raindrops, $T = 273.15°K$, $P = 800$ mb (Beard [1976]).
Curve 2: Raindrops, $T = 293.15°K$, $P = 1013$ mb (Beard [1976]).
Curve 3: Ice sphere, $\rho = 500$ kg m^{-3}, $T = 273.15°K$, $P = 1013$ mb (Beard [1976]).
Curve 4: Lump graupel (Locatelli and Hobbs [1974]).
Curve 5: Ice sphere, $\rho = 100$ kg m^{-3}, $T = 273.15°K$, $P = 1013$ mb (Beard [1976]).
Curve 6: Hexagonal graupel (Locatelli and Hobbs [1974]).
Curve 7: Aggregates of dendritic crystals (Locatelli and Hobbs [1974]).

Source: K. P. Georgakakos and R. L. Bras, "A Hydrologically Useful Station Precipitation Model: 1, Formulation," *Water Resources Res.*, 20(11):1591, 1984. Copyright by the American Geophysical Union.

between the vapor pressure at particle surface and that in the ambient air. At the particle surface, the vapor pressure can be assumed saturated at the wet-bulb temperature T_w. In the ambient air, the vapor pressure would be equivalent to that at saturation at the dew-point temperature T_d. Since it is expected that $T_w > T_d$, then $e_s(T_w) > e_s(T_d)$, and the vapor-pressure gradient is favorable for evaporation. This evaporation is enhanced by the ventilation effect on the particle moving relative to the surrounding air.

According to Byers [1965] a motionless droplet with surface temperature T_w loses mass at the following rate:

$$\rho_p D \frac{dD}{dt} = \frac{4D^*}{R_v}\left[\frac{e_s(T_d)}{T_\circ} - \frac{e_s(T_w)}{T_w}\right], \tag{4.5}$$

where D is the droplet diameter; t is time; ρ_p is the droplet density; R_v is the gas constant for water vapor (461 J kg^{-1}°K^{-1}); and D^* is the diffusivity of water vapor in air.

Pruppacher and Klett [1978] give the diffusivity as

$$D^* = 2.11 \times 10^{-5} \left(\frac{T_\circ}{T^*}\right)^{1.94} \frac{P^*}{P_\circ}, \tag{4.6}$$

where D^* is in square meters per second when T^* and P^* have values of 273.15°K and 101325 kg m^{-1} s^{-2} (1013.25 mb), respectively. Equation (4.6) is valid for ambient air temperatures $T_\circ$ between 233.15°K (−40°C) and 313.15°K (40°C).

To account for a hydrometeor moving at terminal velocity $V_T(D)$, a ventilation effect factor is introduced into Eq. (4.5)

$$\rho_p D \frac{dD}{dt} = \frac{4D^* f_v(D)}{R_v} \left[\frac{e_s(T_d)}{T_\circ} - \frac{e_s(T_w)}{T_w}\right]. \tag{4.7}$$

Based on experiments, Beard and Pruppacher [1971] suggest the following form for the ventilation factor acting on falling spherical meteors:

$$f_v(D) = 1 + 0.108\, N_{sc}^{2/3} \mathbf{R} \qquad \text{for } N_{sc}^{1/3} \mathbf{R}^{1/2} \le 1.4$$

$$= 0.78 + 0.308\, N_{sc}^{1/3} \mathbf{R}^{1/2} \qquad \text{for } N_{sc}^{1/3} \mathbf{R}^{1/2} > 1.4$$

where Reynold's number $\mathbf{R}$ is defined as

$$\mathbf{R} = \frac{D V_T(D) \rho_a}{\mu} \tag{4.8}$$

and

$$N_{sc} = \frac{\mu}{\rho_a D^*}. \tag{4.9}$$

In the above equations, ρ_a (kg m^{-3}) is the air density at temperature $T_\circ$ and pressure $P_\circ$ and μ is the air dynamic viscosity (kg m^{-1} s^{-1}) at temperature $T_\circ$. The dynamic viscosity can be obtained by

$$\mu = 1.72 \times 10^{-5} \left(\frac{393}{T_\circ + 120}\right) \left(\frac{T_\circ}{273}\right)^{3/2} \tag{4.10}$$

(Rogers [1979]), where the ambient temperature $T_\circ$ is in degrees Kelvin.

The rate of change of the hydrometeor position is given by the terminal velocity $dz/dt = -V_T(D)$. Using the above in Eq. (4.7) yields an expression in terms of elevation and diameter that in turn can be integrated between a final diameter D_f at elevation 0 and an initial diameter at the cloud-base elevation Z_b:

$$\rho_p \int_{D_f}^{D_\circ} \frac{DV_T(D)}{f_v(D)} dD = \int_0^{Z_b} \frac{4D^*}{R_v} \left[\frac{e_s(T_w)}{T_w} - \frac{e_s(T_d)}{T_\circ} \right] dz . \tag{4.11}$$

The left-hand side of Eq. (4.11) is reasonably approximated by a constant multiplying the cube of diameter. The approximation is good for a range of temperatures and pressures, and for liquid/solid precipitation (Georgakakos and Bras [1984a]). Using that approximation on Eq. (4.11) results in

$$C_1(D_\circ^3 - D_f^3) = \int_0^{Z_b} \frac{4D^*}{R_v} \left[\frac{e_s(T_w)}{T_w} - \frac{e_s(T_d)}{T_\circ} \right] dz . \tag{4.12}$$

Georgakakos and Bras [1984a] further approximate Eq. (4.12) by assuming that T_w and $T_\circ$, as well as pressure conditions, do not change over the elevation to cloud base Z_b. Under these isothermal and isobaric conditions, Eq. (4.12) becomes

$$D_f^3 = D_\circ^3 - \frac{1}{C_1} \frac{4D^*}{R_v} \left[\frac{e_s(T_w)}{T_w} - \frac{e_s(T_d)}{T_\circ} \right] Z_b . \tag{4.13}$$

Note that Eq. (4.13) can be expressed as

$$\left(\frac{D_f}{D_\circ} \right)^3 = G(D_\circ)$$

or

$$G(D_\circ) = 1 - \left(\frac{D_c}{D_\circ} \right)^3, \tag{4.14}$$

where D_c is a critical diameter such that particles smaller than D_c completely evaporate on their way to the ground. The critical diameter is

$$D_c = \left[\frac{1}{C_1} \frac{4D^*}{R_v} Z_b \left(\frac{e_s(T_w)}{T_w} - \frac{e_s(T_d)}{T_\circ} \right) \right]^{1/3} . \tag{4.15}$$

If it is assumed that during a storm $T_w = T_\circ$, Eq. (4.13) simplifies further:

$$D_f = \left[D_\circ^3 - \frac{4D^* e_s(T_\circ)}{C_1 R_v T_\circ} (1 - r) Z_b \right]^{1/3}, \tag{4.16}$$

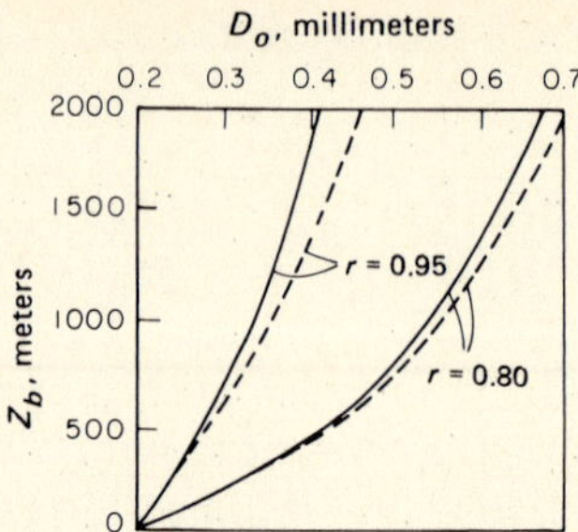

FIGURE 4.9 Initial diameter at cloud base D_o as a function of Z_b for different values of *r*. Solid lines are for numerical integration results in Beard and Pruppacher [1971]. Dashed lines are for Eq. (4.16) with D_f = 0.2 mm. T_o = 273.15°K, and P_o = 765 mb. Source: K. P. Georgakakos and R. L. Bras, "A Hydrologically Useful Station Precipitation Model: 1, Formulation," *Water Resources Res.*, 20(11):1593, 1984. Copyright by the American Geophysical Union.

where r is the fractional relative humidity

$$r = \frac{e}{e_s} = \frac{e_s(T_d)}{e_s(T_o)}.$$

Figure 4.9 shows numerical results (Beard and Pruppacher [1971]) of the initial diameter required to end with a 0.2-mm final hydrometeor diameter after a 2000-m fall. The zero level is the ground. Also shown in the figure are the approximations resulting from Eq. (4.16). Figure 4.9 corresponds to liquid droplets and uses a C_1 value of 7×10^5 kg m^{-3} s^{-1}.

The value of C_1 depends on the nature of the hydrometeor. Georgakakos and Bras [1982] argued that an approximate C_1 value for snow is 1.4×10^5 kg m^{-3} s^{-1}. Figure 4.10 gives the critical diameter D_c required for nonzero

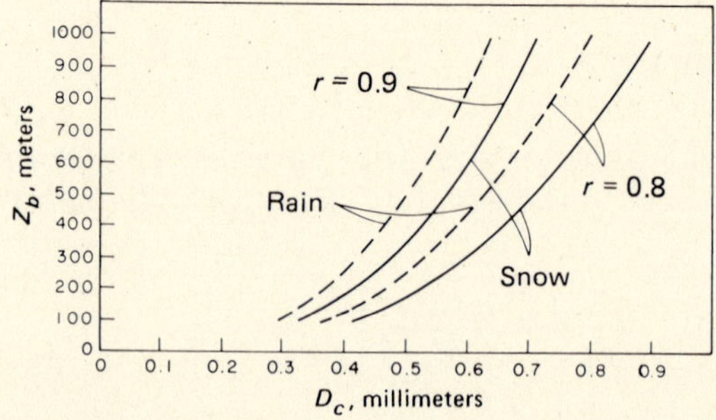

FIGURE 4.10 Initial diameter D_c at cloud base, of the largest completely evaporating hydrometeor in a subcloud layer of depth Z_b, as a function of Z_b for different values of the relative humidity *r*. For raindrops: T_o = 293.15°K, P_o = 1013 mb. For snow particles: T_o = 273.15°K, P_o = 1013 mb. Source: K. P. Georgakakos and R. L. Bras, "A Hydrologically Useful Station Precipitation Model: 1, Formulation," *Water Resources Res.*, 20(11):1594, 1984. Copyright by the American Geophysical Union.

precipitation as a function of cloud-base elevation, relative humidity, and rainfall or snow. The figure results from Eq. (4.15) with $T_w \approx T_\circ$ and the values of C_1 previously discussed.

EXAMPLE 4.1

Evaporation of Hydrometeors

Assume you have the following meteorologic conditions. The average temperature up to cloud base $T_\circ$ is 30°C, the cloud base is at 1000 m, relative humidity is 70%, and average ground pressure is 1000 mb. A drizzle is falling. What is the required water-particle diameter at cloud base to survive evaporation during its fall to ground level? The question really is: What is the critical diameter D_c defined in Eq. (4.15)?

Since we do not have the wet-bulb temperature, we either have to estimate it from $T_\circ$ and relative humidity r or assume that $T_w \approx T_\circ$. This last option is usually not a bad assumption during periods of precipitation. In this case, though, it is not very compatible with the 70% relative humidity. To estimate T_w we may use the tables in Appendix A giving the relative humidity as a function of wet-bulb depression. From Table A.2 for $r = 70\%$ and $T_\circ = 30$°C we get a depression of 4.5°C, which would imply $T_w = 30 - 4.5 = 25.5$°C. The saturated vapor pressure at $T_\circ$ is obtained from Eq. (3.31) as a function of $T_\circ$:

$$\begin{aligned} e_s(T_\circ) &= 33.8639\{[0.00738(30) + 0.8072]^8 \\ &\quad - 0.000019|1.8(30) + 48| + 0.001316\} \\ &= 42.41\,\text{mb} = 4241\,\text{kg}\,\text{m}^{-1}\,\text{s}^{-2}. \end{aligned}$$

Again from Eq. (3.31), we obtain $e_s(T_w)$:

$$\begin{aligned} e_s(T_w) &= 33.8639\{[0.00738(25.5) + 0.8072]^8 \\ &\quad - 0.000019|1.8(25.5) + 48| + 0.001316\} \\ &= 32.62\,\text{mb} = 3262\,\text{kg}\,\text{m}^{-1}\,\text{s}^{-2}. \end{aligned}$$

The saturated vapor pressure at dew-point temperature $e_s(T_d)$ is simply $e(T_\circ) = 0.70(42.41) = 29.69$ mb $= 2969$ kg m^{-1} s^{-2}.

The use of Eq. (4.15) requires a factor C_1, which we take as that given in the text for rain: $C_1 = 7 \times 10^5$ kg m^{-3} s^{-1}. The diffusion factor D^* comes from Eq. (4.6), where $T_\circ$ is given in degrees Kelvin, $T_\circ = 303.15$°K:

$$\begin{aligned} D^* &= 2.11 \times 10^{-5}\left(\frac{303.15}{273.15}\right)^{1.94}\frac{101325}{100000} \\ &= 2.617 \times 10^{-5}\,\text{m}^2\,\text{s}^{-1}. \end{aligned}$$

Now, evaluating Eq. (4.15),

$$D_c = \left[\frac{1}{7 \times 10^5} \frac{4(2.617 \times 10^{-5})}{461} 1000\left(\frac{3262}{298.65} - \frac{2969}{303.15}\right)\right]^{1/3}$$
$$= 7.2 \times 10^{-4}\,\text{m} = 0.72\,\text{mm}.$$

Had we made the assumption that $T_w \approx T_\circ$, the answer would have been 0.68 mm. ◆

4.4 FORMS OF PRECIPITATION

Classification of precipitation forms is generally by size of the hydrometeor and state of the water, usually liquid or solid. Table 4.1 gives a summary of common precipitation types.

The practicing hydrologist-engineer is mostly concerned with rainfall and snow, since they provide the loads to common structures. As will be discussed in later chapters, rainfall is of interest at the place and time of occurrence. On the other hand, snow becomes important during its melting period, which usually occurs long after initial accumulation and is a function of the history of precipitation events.

The occurrence of rainfall or snow and the shape and size of the hydrometeors is very much a function of the meteorologic conditions at the place of formation and of the temperature profile in the atmosphere. Nevertheless,

TABLE 4.1 Forms of Precipitation

NAME	DESCRIPTION	SIZE
Drizzle	Water droplets, low intensity (1 mm hr^{-1})	0.1–0.5 mm
Rain	Water/drops Light: ≤2.5 mm hr^{-1} Moderate: 2.5–7.6 mm hr^{-1} Heavy: >7.6 mm hr^{-1}	>0.5 mm
Glaze	Ice coating, formed by freezing of rain or drizzle	Specific gravity ≈ 0.8
Rime	Opaque, granular ice deposit	Specific gravity ≈ 0.2–0.3
Snow	Ice crystals, hexagonal	Average specific gravity ≈ 0.1
Hail	Balls, irregular ice fragments; convective in nature	5 to over 125 mm; specific gravity ≈ 0.8
Ice pellets	Transparent, translucent ice	<5 mm

Source: Adapted from R. K. Linsley, Jr., M. A. Kohler, and J. L. H. Paulhus, *Hydrology for Engineers*, 3rd ed., McGraw-Hill. Copyright © 1982 by McGraw-Hill. Used by permission.

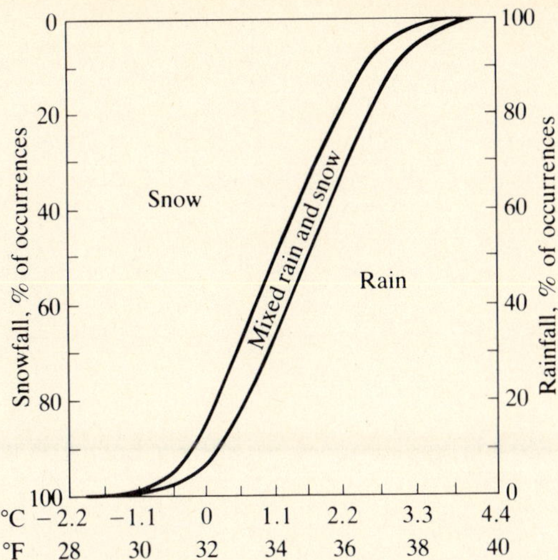

FIGURE 4.11 Frequency of occurrence of rain and snow at various temperatures. Source: U.S. Army Corps of Engineers [1956].

Figure 4.11 shows the frequency of occurrence of snow or rainfall as a function of ground temperature. Clearly, as a rule of thumb, snow occurs when the temperature is below 0 to 1°C.

4.5 STORM STRUCTURE

Any given storm can be classified according to identifiable components and in terms of temporal and spatial characteristics. Houze [1969] has confirmed that storms are composed of well-defined elements of different times and spatial extents.

The microscale is composed of convective cells of short duration, generally about half an hour. The spatial extent of cells is on the order of 5 km in diameter. Convective cells, as thunderstorms, are fairly violent in nature. Resulting rainfall intensities are very high and sporadic. Small cells move randomly, while large cells deviate to the left of upper-layer winds (jet streams) in the northern hemisphere. They do not follow the general storm direction. Cell velocities are on the order of 30 to 50 km hr^{-1}. Cells follow a sequence of growth, maturity, and death typical of convective disturbances.

Mesoscale activity consists of a unit of developing cells, each at different developmental stages and moving in unison is a preferred direction. Houze [1969] defined small and large mesoscale. The large mesoscale covered an area of 2300 to 4700 km^2 while small mesoscale units extended over 150 to 400 km^2. Rainfall intensities are moderate when averaged over the mesoscale area. Figure 4.12 illustrates the movement of a mesoscale with its member cells aging and moving to the left (in the northern hemisphere) of the average

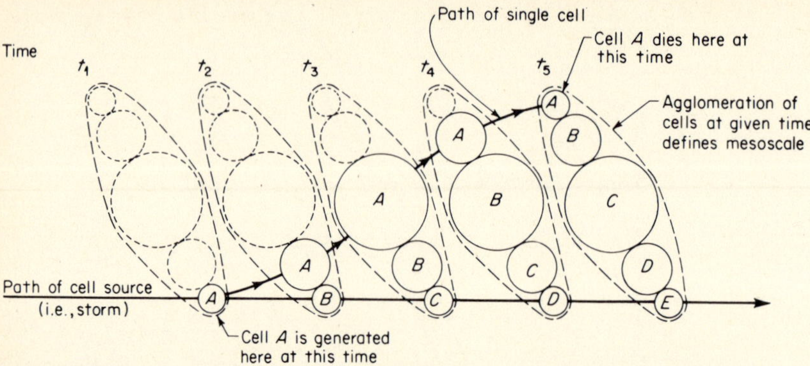

FIGURE 4.12 Cell growth and propagation within a storm. Source: P. S. Eagleson, *Dynamic Hydrology,* McGraw-Hill, 1970.

direction. A thunderstorm is a typical mesoscale unit, composed of an active conglomerate of cells. The familiar rotating arms of a hurricane also correspond to the mesoscale.

The conglomerate of mesoscale units forms the synoptic scale. This is the background climatic disturbance of very large events. The synoptic scale can cover several thousand square kilometers—for example, a typical hurricane with several hundred kilometers in diameter. Frontal climatic events are also within the synoptic scale. Average rainfall intensity over the synoptic area is small. Velocities and mean travel directions are also clearly discernible. Commonly, mesoscale units form bands within the synoptic scale, as illustrated in Figure 4.13.

Houze [1969] studied several events in New England and obtained average characteristics for their components. Some of these results are compiled in Table 4.2.

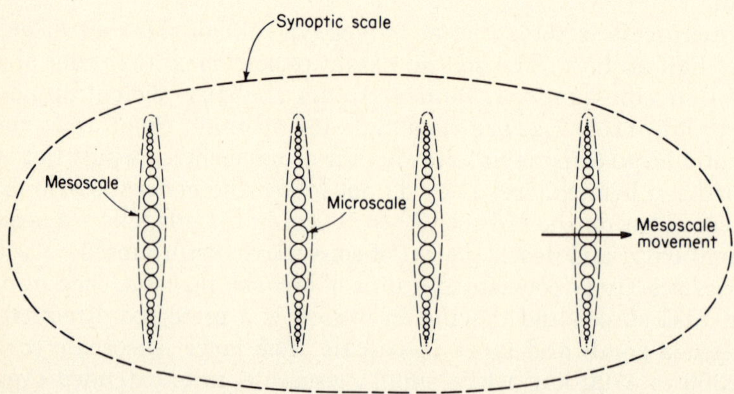

FIGURE 4.13 Typical instantaneous structure of a synoptic-scale event. Source: P. S. Eagleson, *Dynamic Hydrology,* McGraw-Hill, 1970.

TABLE 4.2 Characteristics of General Storms

	SIZE	INTENSITY	DURATION	DENSITY
Synoptic	10^4 to 10^5 mi^2 (26,000 to 260,000 km^2)	0.01 to 0.08 in. hr^{-1} (0.025 to 0.2 cm hr^{-1})	On the order of a few days	
Large mesoscale	900 to 1800 mi^2 (2300 to 4600 km^2)	Approximately twice the intensity of the synoptic level	1 to 12 hr	Covers about 1/3 of the total storm area
Small mesoscale	50 to 150 mi^2 (130 to 390 km^2)	Approximately twice the intensity of the large mesoscale level	1/3 to 3 hr	3 to 6 small mesoscales per large mesoscale or 2 to 5 small mesoscales per 1000 mi^2 (2600 km^2)
Cellular	About 3 mi^2 (8 km^2)	Approximately 2 to 10 times the intensity of the small mesoscale level	Approximately linearly related to cell intensity	1 to 7 cells per small mesoscale

Source: Grayman and Eagleson [1971].

The previous physical description of storms has been used as the basis for rainfall simulation exercises. Among the investigators explicitly using these ideas are Grayman and Eagleson [1971], Amorocho and Slack [1970], Gupta and Waymire [1979], Waymire et al. [1984], and Rodriguez-Iturbe and Eagleson [1987]. The details of those simulations are beyond the scope of this book. Nevertheless, they all rely on the random occurrence of the different storm structures.

Besides the meteorological structure previously discussed, storms at a location can be described by their "exterior" and "interior" statistics. Storm exterior usually refers to total depth of, duration of, and time between storms. These characteristics are generally accepted to be probabilistic in nature. They can be described by location and seasonally dependent probabilistic distributions. The characteristics are generally not statistically independent. Storm depth varies in space and exhibits spatial correlation between points. Depth is related to duration. Large duration is generally associated with larger depths, although the opposite relation exists between average intensity and duration.

Storm interior refers to the time and spatial distribution of intensities throughout a storm. It is observed that for given locations and climatic conditions some type of events exhibit similar histories of rainfall accumulation. The percentage–mass curve is a plot of normalized cumulative depth by normalized duration. Figure 4.14 shows how typical curves are obtained for thunderstorms and cyclonic events. The derivative (slope) of these curves is a graph of intensity versus time. This graph is referred to as the hyetograph

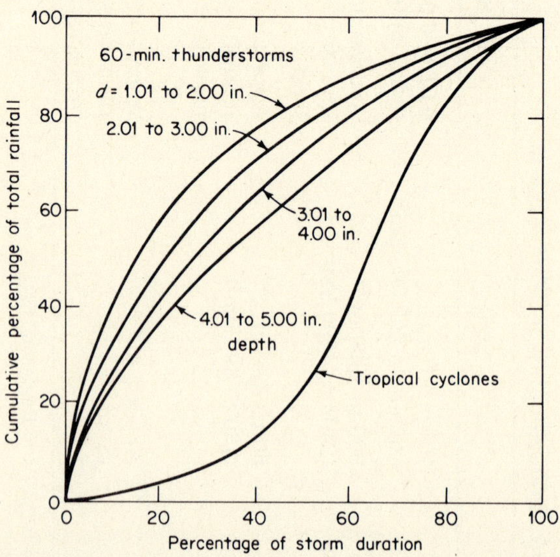

FIGURE 4.14 Typical percentage mass curves of rainfall for thunderstorms and for tropical cyclones. Source: P. S. Eagleson, *Dynamic Hydrology,* McGraw-Hill, 1970.

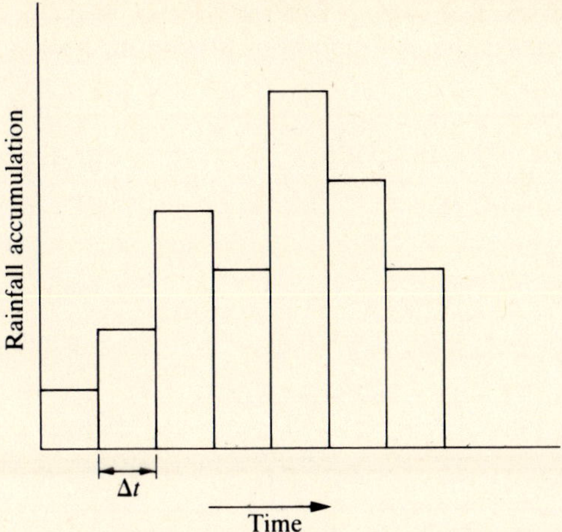

FIGURE 4.15 Typical storm hyetograph in histogram (or bar diagram) form.

and is usually given in histogram form (Fig. 4.15). Note that, according to Figure 4.14, thunderstorms have triangular (or moundlike) hyetographs. Tropical cyclones exhibit intensity diagrams that resemble a distorted bell with long tails to the right.

Rainfall interiors are important in urban hydrology, where histories of rainfall and flow are required. They are also becoming increasingly more important in studying the response of large basins with sophisticated conceptualizations of runoff-generation, mechanisms that will be seen later. Several investigators (Pilgrim et al. [1969]; Marsalek [1978]) have attempted to determine average interiors for design purposes, with some success. Others (Zawadski [1973b]; Bras and Rodriguez-Iturbe [1976]) have hypothesized on the spatial and time structure of storm interiors, suggesting some equivalence between the histories at different points in space, all related by an average velocity of translation for the event.

Spatially, storms generally exhibit one or more centers of maximum depth. The total depth of point rainfall over an area is then a decreasing function of distance from the storm center. Lines of equal precipitation, isohyets, then form close loops around points of maximum precipitation.

Several equations, mostly of local nature, have been suggested to represent this decay from the point of maximum precipitation. Some relations are storm-centered and others are geographically centered.

Mean areal precipitation is usually expressed as a percentage of the storm-center value. The difference between the area-averaged depth and the storm-center value

1. decreases with increasing total rainfall depth, which implies higher uniformity of heavy storms;

TABLE 4.3 Formulas to Compute Average of Precipitation over an Area A or within a Circular Isohyet of Radius x, as a Function of Maximum Precipitation P_m in the Region

AVERAGE PRECIPITATION	AREA
$P_m(1 - 0.14A^{1/4})$	1.6
$P_m - bA^{1/2}$ $(b = 0.03P_m;\ 0.06P_m)$	5–280
$P_m \exp(-kA^n) = P_m \exp(-0.01A^{1/2})$	20–20,000
$2P_m b^{-2}x^{-2}[1 - (1 + bx)\exp(-bx)]$ $(b = 0.0235)$	>100
$(\pi P_m/Aab)[1 - \exp(-Aab/\pi)]$	Any

Note: a and b define scale and ellipticity. When $a = b$, the isohyets are circular.

Source: Gray [1973]. Reproduced by permission of the National Research Council of Canada.

2. decreases with increasing duration, again implying that long storms are more uniform;
3. is greater for convective and orographic precipitation than for cyclonic storms; this behavior is related to points 1 and 2; and
4. increases with increasing area.

Table 4.3 gives some formulas based on storm center, for the areal average of a storm. Rodriguez-Iturbe and Mejia [1974] argue that if the statistical correlation of points in space is known, then it is possible to obtain theoretically the areal average of precipitation from point data.

4.6 MEASUREMENT OF PRECIPITATION

4.6.1 Gages

Precipitation is usually measured at a point by collectors of very simple construction. Essentially, any receptacle with a reasonable opening will serve the purpose of estimating the rainfall volume per unit area accumulated during any given event. In practice, rain gages are of standardized dimensions for quality-control purposes. Each country defines its own standard gage. In the United States, the National Weather Service utilizes the standard 8-in. precipitation gage, which as the name implies, has an 8-in. (20.3-cm)-diameter opening in the collector or receiver. The receiver is like a funnel, feeding another receptacle with one-tenth the cross-sectional area of the receiver. This implies that rainfall depth is amplified 10 times, easing the measurement and increasing its accuracy. Measurements are made with a calibrated stick. Theoretically, the measurement can be made to the nearest 0.25 mm (0.01 in.). The measuring tube is inside another container that will catch any overflow. The whole setup rests on a stand. This technique only records total accumulated depth. The intensity or precipitation rate is lost.

Instruments that measure the precipitation intensity are called recording gages. The most popular of these is the tipping-bucket gage. A 0.01-in. bucket fills with rain, which forces rotation and spillage of the water and exposes another collector. Every rotation is recorded. Knowing the time between rotations and the spillage volume (0.01 in.) permits the rainfall rate calculation.

Weighing-type recording gages measure rate of accumulation in terms of the increasing weight of precipitation. Float-type gages respond continuously to water depth in the float container.

The accuracy of rain-gage information is influenced by many factors. The elevation and exposure of the rain gage clearly play a major part. The recorded precipitation depends on instrument exposure and angle of incidence of rainfall. Gages should not be shadowed by obstructions.

Wind is probably the single most important factor in rain-gage accuracy. Updrafts resulting from air moving up and around the instrument reduce the rainfall catch. A number of studies have attempted to quantify this error and relate it to wind speed. A common approach to reducing wind-induced errors is to provide the gages with wind shields (Larson and Peck [1974]) and positioning them in wind-protected places. Figure 4.16 gives expected catch errors as a function of wind speed.

Snow is commonly measured in the standard 8-in. overflow can. Prevention of ice accumulation in the rim is important to maintain a constant collecting area. Snow accumulation is usually reported in terms of equivalent water depth, the proportionality factor being the density, on the order of 1 cm of rainfall per 10 cm of snow. Snow is also commonly measured in situ by simply recording accumulation with a calibrated staff. The local nature of this type of measurement and the influence of wind and topography on its accuracy are obvious. Snow accumulation can also be measured in terms of the extinction of gamma or other types of radiation (Smith et al. [1965]; Peck and Bissell [1973]; Zotimov [1968]). Satellites can be utilized for estimating snowpack depth and areal coverage. Albedo and heat content are some of the properties that may be correlated to snowpack depth and age (Schneider et al. [1976]; McGinnis et al. [1975]; Peck et al. [1981]; Johnson et al. [1982]).

It is important to emphasize the punctual nature of common precipitation measurements. Spatial variations are important. Ignoring them can introduce considerable error in the computation of total rainfall over an area (Bras and Rodriguez-Iturbe [1976]) and misrepresent the distributed response of a basin.

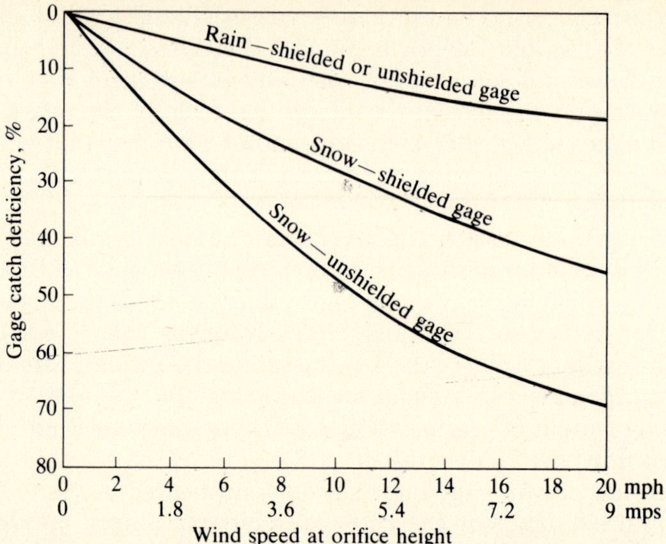

FIGURE 4.16 Effect of wind speed on the catch of precipitation gages. Source: L. W. Larson and E. L. Peck, "Accuracy of Precipitation Measurements for Hydrologic Modeling," *Water Resources Res.*, 10(4):859, 1974. Copyright by the American Geophysical Union.

4.6.2 Radar

The use of radar for precipitation measurement is not a new idea. It has been considered ever since the inception of the instrument. The advantages of radar precipitation measurement may be summarized as

1. Increased time and spatial resolution, in theory a continuous description of the precipitation.
2. Ability to handle all types of precipitation.

Operationally, radar also provides single-site convenience.

In practice, the exclusive use of radar for precipitation measurement is yet to be achieved. Generally, rain gages or other punctual systems (i.e., drop-size measurement apparatus) are required to calibrate radar. Most of the following discussion of the operation of precipitation measuring radar is after Grayman and Eagleson [1971].

Radar emits electromagnetic energy in narrow bands. Upon hitting precipitation it is partially absorbed, scattered, and reflected. Some of the reflected energy returns to the transmitter, the total travel time being $2t$, where t is the time the signal took to reach the target. Electromagnetic waves move at the speed of light. Because of attenuation of the signal during interaction with the storm, the reflected energy must be normalized (adjusted) and amplified to make the comparison of reflection from different ranges possible. The range is the distance from transmitter to target. This procedure is illus-

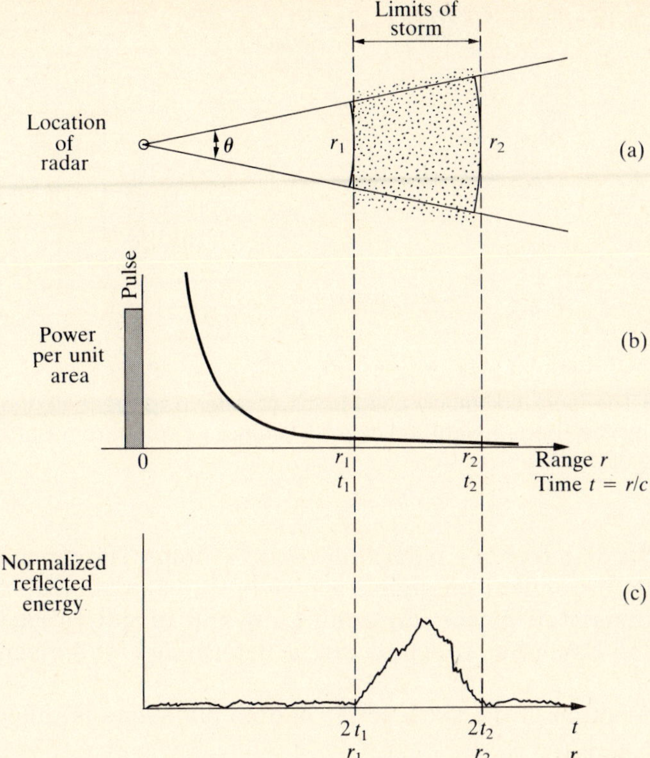

FIGURE 4.17 Essentials of radar detection of rainfall. Source: Grayman and Eagleson [1971].

trated in Figure 4.17. Note that there is always some background noise before and after storm detection.

The radar reflection can be presented in several ways other than the one illustrated in Figure 4.17, which emphasizes the one-dimensional range, or depth of the event. By rotation of the antenna, it is possible to indicate the spatial extent of the precipitation. Vertical tilting is used to quantify the height of the storm. Two common types of displays are the plan position indicator (PPI) and the range height indicator (RHI) shown in Figure 4.18. Present-day weather radars digitize the reflectivity information. Therefore numbers are given for intensities instead of relying of varying-intensity visual displays.

The average returned power of electromagnetic signals is related to precipitation by

$$\overline{P_r} = \frac{C}{r^2} \sum d^6, \tag{4.17}$$

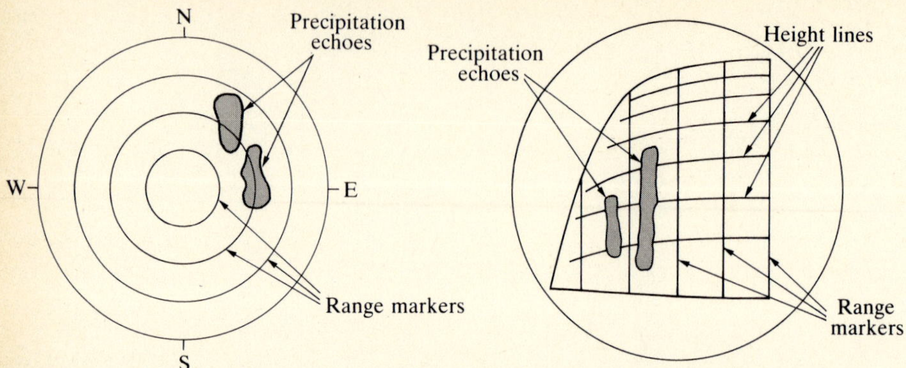

FIGURE 4.18 The more commonly used forms of radar displays. Left: the plan position indicator; right: the range height indicator. Sources: Adapted from Battan [1973] and Barrett and Curtis [1976].

where d is the diameter of a particular particle (drop). The range is r and C is a coefficient "dependent on signal wavelength, beam shape, width, pulse length, transmitted power, antenna gain and target refractive index" (Grayman and Eagleson [1971]). It can be determined for a given equipment configuration.

The summation of the sixth power of drop diameters is called the reflectivity factor Z

$$Z = \int D^6 N(D)\, dD, \tag{4.18}$$

usually expressed in units of millimeters to the sixth power per cubic meter ($mm^6\, m^{-3}$). $N(D)$ is the particle size distribution (see Section 4.3.2). The rainfall rate is related to Z by

$$Z = aR^b, \tag{4.19}$$

which is referred to as the Z–R relationship and is the heart of the procedure. There are two Z–R relationships. The theoretical alternative relates Z, as given in Eq. (4.18), to the rainfall rate, as given by the integration of Eq. (4.4), via the drop-size distribution $N(D)$. A *disdrometer* measures drop-size distribution. The value of parameters a and b vary widely with location, storm type, drop size, etc. Linsley et al. [1982] quote values varying from 15 to 1000 for a, and 1.2 to 3.2 for b, for rainfall intensity R in millimeters per hour. Average values of 200 and 1.6 for a and b, respectively, are usually accepted (Marshall and Palmer [1948]).

For snowflake aggregates, a commonly suggested relationship is (Carlson and Marshall [1972])

$$Z = 2000R^{2.0}. \tag{4.20}$$

The effective reflectivity Z_e measured by the radar itself can be compared with rainfall-intensity measurements from rain gages to obtain an alternative, completely empirical, Z_e–R relationship. Austin [1987] studied storms in New England using the weather radar at the Massachusetts Institute of Technology and suggested the use of $a = 230$ and $b = 1.4$ in a Z_e–R relationship, rather than the Marshall–Palmer values. She divided her data by storm types and concluded that significant improvements could be obtained by using $a = 400$, $b = 1.3$, and $a = 230$, $b = 1.2$ for thunderstorms and convective rain associated with cold fronts, respectively. For drizzle she suggested $a = 100$ and $b = 1.4$. The effect of using different Z–R relationships is illustrated by Wyss [1988] in Figure 4.19.

Calibration with punctual ground data may be in error because of the different nature of the radar measurement. Radar samples a volume at varying elevations, not a point at ground level. Some of the discrepancies may be because

1. precipitation formed at low altitudes may go undetected by radar;
2. detected precipitation at a given height may never reach the ground because of evaporation or winds carrying it to a different location;
3. vertical variability of precipitation type and intensity (i.e., ice versus liquid) may lead to incorrect integration of echoes and incorrect estimates of precipitation; and
4. surface evaporation may lead to false radar echoes.

The problems encountered when the radar signal hits the freezing layer up in the atmosphere, resulting in increased reflectivity, go by the name of bright-

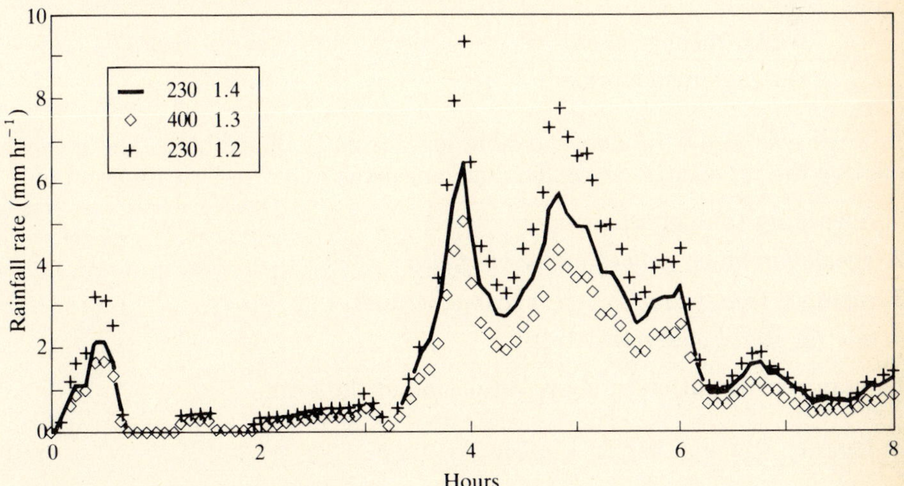

FIGURE 4.19 Areal average rainfall rate measured with radar over the Souhegan River Basin in New England on June 27, 1987. $Z_e = 230R^{1.4}$ (solid line); $Z_e = 400R^{1.3}$ (diamonds); $Z_e = 230R^{1.2}$ (crosses). Source: Wyss [1988].

band effects. Point 4 above is commonly called anomalous propagation. Radar beams have to be pointed upward in order to avoid surface obstructions, or in radar terminology, ground clutter.

There is a conflict between radar-signal attenuation, which favors longer wavelength, and resolution, which favors shorter wavelengths. Radiating at a wavelength greater than 5 cm is recommended. At 5-cm wavelengths, the significant attenuation may occur during intense precipitation. At wavelengths on the order of 10 cm, attenuation is generally not a problem. The common types of weather radar go by the following names: *S* band with 5.77 to 19.3 cm wavelength, *C* band with 7.69 to 9.84 cm of wavelength, and *X* band with 2.75 to 5.77 cm of wavelength.

*Errors in Radar Measurements

Grayman and Eagleson [1971] developed models for sampling errors induced by the use of radar. Radar errors were divided into large- and small-scale errors. Inaccuracies in measuring large area and time-scale events such as storms totals over a large catchment area are called large-scale errors.

This error was represented by the ratio

$$x_t = \frac{\text{radar measurement of storm total}}{\text{true storm total}}.$$

The random variable x_t was taken to be log-normally distributed with mean $\mu_{x_t} = 1.0$ and standard deviation $\sigma_{x_t} = 0.5$. (See Chapter 11 for a review of basic probability theory.)

Small-scale errors are due to local variations in the relation between rainfall intensity and radar reflectivity. The error is measured in terms of the ratio

$$x = \frac{\text{radar measurement}}{\text{rain-gage measurement}}.$$

The ratio x is also a random variable log-normally distributed. The mean is given by the large-scale error. The variance was found to be a function of

1. averaging time of echoes t;
2. area over which echoes are averaged a;
3. distance from radar to storm (range) r; and
4. spatial storm characteristics.

The standard deviation σ_x obeys a relation of the form

$$\log_{10} \sigma_x = c_1 + c_2 \log_{10} t + c_3 \log_{10} a + c_4 r . \tag{4.21}$$

The effects of different storm characteristics were found to be negligible. The range of the parameters of the above equation was $(-1, 0)$, $(-4.38, -0.24)$, $(-0.45, -0.28)$, $(0.01, 0.02)$ for c_1, c_2, c_3, and c_4, respectively.

Rain gages are commonly used for calibrating radar. The use of rain gages has the effect of reducing the variance of the radar measurement error. Given $y = \ln x$, where x is the small-scale error, the mean of the ratio σ_{y_A}/σ_y was found to be

$$\sigma_y^N = 1 - \frac{R_c}{2.3}\left(\frac{\pi}{A}\right)^{1/2}\left[1 - e^{-(2.3/R_c)(A/\pi)^{1/2}}\right], \tag{4.22}$$

where A is rain-gage density, or square kilometers per gage; R_c is distance at which the spatial correlation of rainfall begins oscillating; σ_y is standard deviation of $y = \ln x$ when no calibrating gages are used; σ_{y_A} is the standard deviation of $y = \ln x$ when calibrating gages are available at A square kilometers per gage; and, σ_y^N is the average of the ratio σ_{y_A}/σ_y.

EXAMPLE 4.2

Calibration of Weather Radar

Given below is a spatial correlation function of rainfall accumulation in an area of 10,000 km^2.

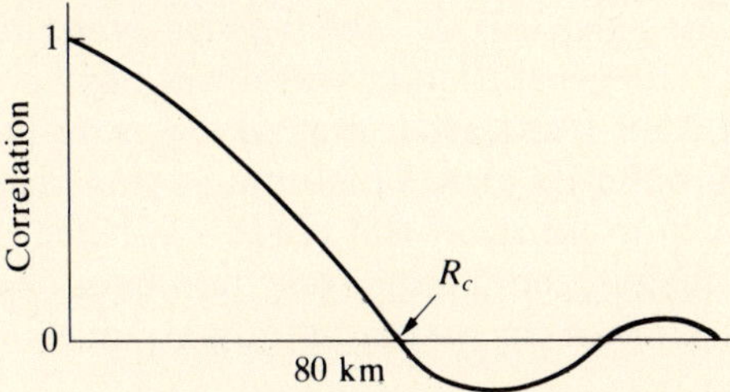

The rainfall in the area is being monitored by radar with a single rainfall gage as a calibrating aid. How many rainfall stations would we need to add in order to decrease the standard deviation of the logarithms of the small-scale error of radar measurements by a factor of 2? To answer the above question, we will make use of Eq. (4.22). With one station, the ratio of calibrated versus noncalibrated standard deviations of the logarithms of radar errors is

$$\sigma_y^N = 1 - \frac{80}{2.3}\left(\frac{\pi}{10{,}000}\right)^{1/2}\left[1 - e^{-(2.3/80)(10000/\pi)^{1/2}}\right] \approx 0.5 .$$

We want to reduce σ_y^N to 0.25. Increasing the number of stations to 10, we get $A = 1000$ km^2 per station, which on substitution into Eq. (4.22) yields

$$\sigma_y^N = 1 - \frac{80}{2.3}\left(\frac{\pi}{1000}\right)^{1/2}\left[1 - e^{-(2.3/80)(1000/\pi)^{1/2}}\right] = 0.22 .$$

Using eight stations, the implied A would be 1250 km^2 per station or

$$\sigma_y^N = 1 - \frac{80}{2.3}\left(\frac{\pi}{1250}\right)^{1/2}\left[1 - e^{-(2.3/80)(1250/\pi)^{1/2}}\right] = 0.24\,.$$

For $A = 1667$, six stations, $\sigma_y^N = 0.27$. Therefore the answer should be seven stations, yielding $A = 1429$ and $\sigma_y^N = 0.25$. ◆

4.6.3 Satellites

The use of meteorologic satellites for weather studies is continuously increasing and should become a dominant sensing technique in the future. Satellite sensors work on the principle that the atmosphere selectively transmits radiation at various wavelengths (see Chapter 2). Most satellites rely on detecting radiation in the visible and thermal infrared wavelengths. The visible wavelengths are on the order of 0.77 to 0.91 μm and the thermal infrared spans 8.0 to 9.2; 10.2 to 12.4; and 17.0 to 22 (Lintz and Simonett [1976]). The satellites produce images from the received radiation. The images are made available in hard copy form and/or digitized for computer processing.

Existing weather satellites are either polar-orbiting or geostationary. The polar-orbiting satellites circle over the poles usually about twice every day at elevations on the order of 1000 km. The geostationary satellites have a high orbit of around 36,000 km such that their translation around the earth is synchronous with the earth's rotation around its axis. This implies that they remain stationary in space relative to a point on the earth's surface. This permits the frequent production of images, commonly every half hour or even down to every five minutes, which is ideal for watching quickly developing weather patterns.

The GOES (Geostationary Operational Environmental Satellite) system of the United States, together with similar satellites of other countries provide an almost continuous view of global weather, except for serious distortions beyond 60° latitude. Figure 4.20 shows the general location of the operational geostationary satellites. The GOES satellites measure albedo (visible light reflection) from 0.5 to 100% and thermal infrared radiation for a temperature range of −93 to 42°C. Their resolution is about 1 km in the visible range and 8 km in the thermal waveband.

With the resolutions given above, GOES satellites are potential tools to define storm events in the mesoscale range. There have been considerable efforts to develop tools to infer precipitation accumulations from remotely sensed satellite information. The implications of such approaches in areas with sparse or no ground-data-collection networks and for improving our appreciation of the spatial extent of precipitation events are obvious.

Ingraham [1980] gives a good literature survey of existing precipitation estimation methods. The technique of Oliver and Scofield [1976] and Scofield and Oliver [1977] is probably the most well-known one. Their procedure is based on subjective and empirical criteria. Using observations of behavior of

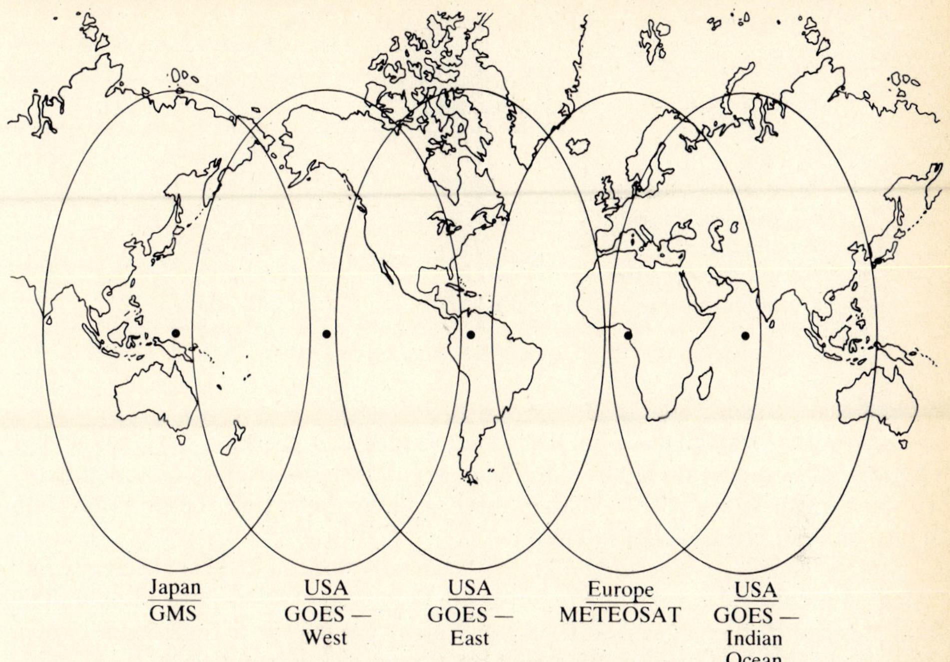

FIGURE 4.20 Five geostationary meteorologic satellites provide global coverage of the weather.

tropical, convective air masses, they suggest a relation between cloud-top temperature (decreases with height) and precipitation. These first estimates are then refined using the rate of growth or contraction of the cloud top and the shape and nature of the clouds.

While studying precipitation over British Columbia, Canada, Ingraham [1980] suggested a simple precipitation model that operates using GOES satellite data. She hypothesized that the rainfall rate per unit area is adequately represented by

$$\frac{dP}{dt} = E\rho_m wV , \tag{4.23}$$

where w is the mixing ratio of air going into the cloud system; ρ_m is the density of the moist air; V is the updraft velocity of the rising moist air masses; and E is an efficiency factor to account for the proportion of vapor that actually condenses in the cloud. Precipitation is given in grams per square meter. If units of depth are required, the equation must be divided by the density of the water. Note that this equation is essentially the same as Eq. (3.64).

In proposing Eq. (4.23), there are obvious shortcuts to the description of cloud physics in Section 4.3. We may mention the assumption that all vapor condenses and precipitates with an efficiency of E, ignoring particle distribu-

TABLE 4.4 Average Efficiencies and Water Content

CLOUD TYPE	EFFICIENCY E (as a fraction)	WATER CONTENT w_c ($g\,m^{-3}$)
Sea fog	—	0.10
Stratus	0.05	0.35
Orographic (Hawaii)	0.05–0.10	0.35
Small cumulus	0.10	0.5–1.4
Cumulus congestus	0.15	1.0
Cumulonimbus	0.20–0.30	2.0–10.0

Source: Mason [1971], Fletcher [1962], and Rogers [1979].

tion, terminal velocities, and the pseudo-adiabatic process. The model also ignores all evaporation below cloud-base level. Precipitation is assumed to occur instantaneously with moisture input and condensation. There is no time delay built into the model behavior.

In Eq. (4.23), ρ_m is assumed known (or easily obtained). The efficiency factor E is a calibration parameter; typical values are given in Table 4.4. The water content may again be fixed at typical values for a region and storm type (Table 4.4) or more adequately measured at ground level (i.e., relative humidity, dew-point temperature). The polar orbiting satellite TIROS-N can also measure and report precipitable water (integrated water content) every six hours.

The GOES satellites can be useful in estimating the updraft velocity V in Eq. (4.23). Ingraham [1980] presents the following arguments. Figure 4.21

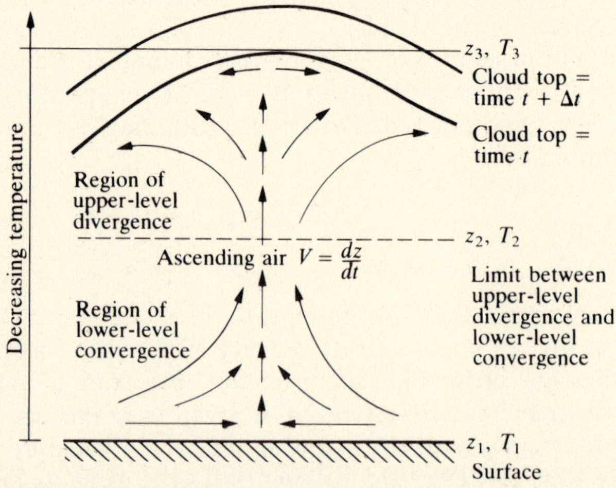

FIGURE 4.21 Schematic diagram of air circulation during a major storm. The horizontal scales of such storms are $0 \sim 10^5$ to 10^7 m, while vertical scales are $0 \sim 10^4$ m. Source: Adapted from Ingraham [1980].

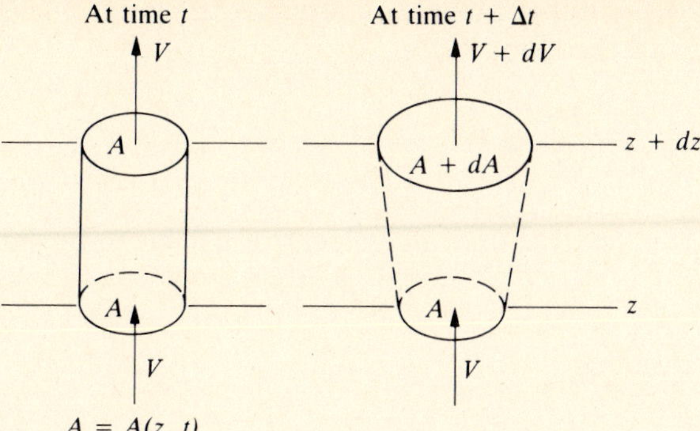

FIGURE 4.22 A column of air between the levels z and $z + dz$ at times t and $t + \Delta t$. Source: Ingraham [1980].

shows a hypothetical cloud system. There is a convergence of air at ground level and a divergence at cloud top. Somewhere in the middle a narrowing point of zero air divergence exists. Figure 4.22 is a simplification of the cloud from the point of zero divergence up at two times—t and $t + \Delta t$.

Conservation of mass M within the unit volume of Figure 4.22 implies,

$$\frac{\partial M}{\partial t} = \frac{\partial}{\partial t}(\rho A\,dz) = \rho VA\,|_z - \rho VA\,|_{z+dz}\,.$$

Expanding the second term in the right-hand side in a Taylor series, keeping only linear terms, leads to

$$\frac{\partial}{\partial t}(\rho A\,dz) = -\frac{\partial}{\partial z}(\rho VA)\,dz\,.$$

Ignoring changes in density, and for a fixed unit volume, the above becomes

$$dz\,\frac{\partial A}{\partial t} = -A\frac{\partial V}{\partial z}\,dz - V\frac{\partial A}{\partial z}\,dz$$

or

$$\frac{\partial A}{\partial t} + V\frac{\partial A}{\partial z} = -A\frac{\partial V}{\partial z}\,.$$

Since $V = dz/dt$, the left-hand term is the total derivative of A with respect to time. If we further assume no acceleration of updrafts (i.e., $(\partial V/\partial t) = 0$), then

$$\frac{1}{A}\frac{dA}{dt} + \frac{dV}{dz} = 0\,, \tag{4.24}$$

where the first term is the horizontal air divergence.

Equation (4.24) can be integrated between the point of zero divergence, z_2 in Figure 4.21, and the cloud top z_3

$$\int_{z_2}^{z_3} dV = -\int_{z_2}^{z_3} \frac{1}{A}\frac{dA}{dt}\, dz\,.$$

Since at z_3 the vertical updraft velocity must be zero, the previous equation reduces to

$$V(z_2) = \int_{z_2}^{z_3} \frac{1}{A}\frac{dA}{dt}\, dz\,,$$

which relates updraft velocity to the growth of the clouds' cross-sectional area (divergence). Since the rising air cools, assume at the adiabatic lapse rate $dT = -\Gamma\, dz$, then the limits of integration can be converted to temperatures:

$$V(z_2) = -\int_{T_2}^{T_3} \frac{1}{\Gamma}\frac{1}{A}\frac{dA}{dt}\, dT\,, \tag{4.25}$$

where T_i is temperature at level i. Substituting in Eq. (4.23) results in an expression relating precipitation rate to the integral of divergence over two temperature layers.

$$\frac{dP}{dt} = -\frac{Ew\rho_m}{\Gamma}\int_{T_2}^{T_3} \frac{1}{A}\frac{dA}{dt}\, dT\,. \tag{4.26}$$

Ingraham [1980] argues that the integral in Eq. (4.26) can be considered linearly proportional to the divergence at the top of the cloud. Lumping the proportionality factor in E, we then obtain

$$\frac{dP}{dt} = -\frac{Ew\rho_m}{\Gamma}\frac{1}{A(T_3,t)}\frac{dA(T_3,t)}{dt}\Delta T\,, \tag{4.27}$$

where the explicit dependence of A on time t and temperature T_3 at the top of the cloud is shown. ΔT is the difference in temperature between levels 2 and 3, $\Delta T = T_3 - T_2$.

If Eq. (4.27) is integrated over a time interval Δt, we obtain the accumulated precipitation during the time interval

$$\Delta P = -\frac{Ew\rho_m}{\Gamma}\ln\left[\frac{A(T_3,t+\Delta t)}{A(T_3,t)}\right]\Delta T\,. \tag{4.28}$$

In theory, Eq. (4.28) is easily evaluated using images from the GOES satellite. All that is required is two consecutive pictures of the satellite from where $A(T_3, t + \Delta t)$ and $A(T_3, t)$ can be estimated.

Ingraham [1980] developed the practical tools to process the GOES visible and infrared images that are needed in evaluating $A(T, t)$. She tested rainfall

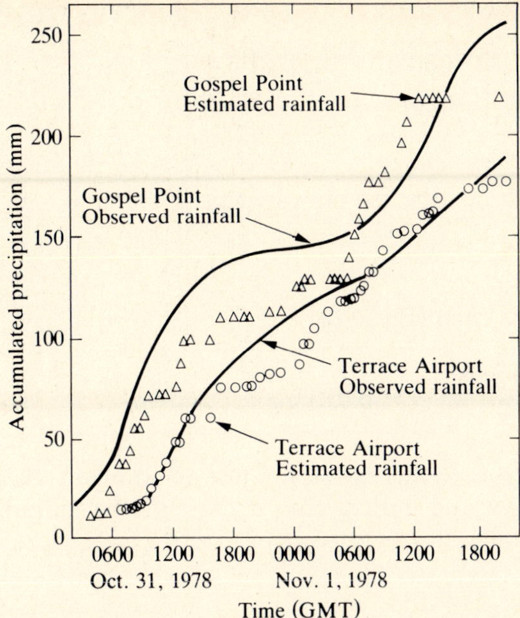

FIGURE 4.23 Estimated and observed precipitation at Gospel Point and Terrace Airport. Source: Ingraham [1980].

estimation in British Columbia, Canada, with good results, as shown in Figure 4.23 for a particular storm at two stations.

EXAMPLE 4.3

Rainfall Estimation from a Satellite

You are following the development of a reasonably stationary thunderstorm using infrared satellite images. The top of the cloud system is measured at −40°C. At the beginning of a half-hour interval of images, the cloud-top area is measured as 100 km^2. At the end of the period the area is 120 km^2. The temperature at the point of zero divergence is estimated at 0°C. The efficiency of storms in the area is usually about 0.2 and the adiabatic lapse rate is taken as 5°C km^{-1}. A meteorologic station below the cloud formation measures a dew-point temperature of 10°C and a ground temperature of 23°C. The pressure is 1000 mb. Can the rainfall accumulation during the period be estimated?

First assume that the air-moisture conditions measured at ground level are representative of moisture supply to the cloud. From Eq. (3.34), we obtain the relative humidity

$$r = \left[\frac{112 - 0.1T + T_d}{112 + 0.9T}\right]^8 = \left[\frac{112 - 0.1(23°\text{C}) + 20°\text{C}}{112 + 0.9(23°\text{C})}\right]^8 \approx 0.83\,.$$

The saturated vapor pressure at $T = 23°C$ is given by Eq. (3.31)

$$\begin{aligned} e_s &\approx 33.8639\{[0.00738(23) + 0.8072]^8 \\ &\quad - 0.000019|1.8(23) + 48| + 0.001316\} \\ &\approx 28.0 \text{ mb}. \end{aligned}$$

Combining the two results, the air vapor pressure is

$$e = (0.83)28.0 = 23.2 \text{ mb}.$$

The mixing ratio is then (Eq. 3.29)

$$w = 0.622 \frac{23.2}{1000 - 27.0} = 0.015 \text{ g g}^{-1}.$$

Approximating ρ_m by the density of dry air at about 73.4°F and 1000 mb, which is 1185 g m^{-3}, and multiplying it by w yields humidity in grams per cubic meter of dry air:

$$0.015 \times 1185 = 17.8 \text{ g m}^{-3}.$$

From Eq. (4.28) the estimated precipitation would then be

$$\begin{aligned} \Delta P &= -\frac{(0.2)(17.8)}{5 \times 10^{-3}}(-40) \ln\left[\frac{120}{100}\right] \\ &= 5.2 \times 10^3 \text{ g m}^{-2} \text{ of water in half hour} \\ &= 5.2 \text{ mm in half hour}. \end{aligned}$$ ◆

4.7 PRECIPITATION DATA ANALYSIS

Precipitation data analysis and handling should be made with knowledge of statistical sampling theory, applicable to any other data-collection experiment. The goal of the hydrologist is to sample and study as much data as required to define sufficiently, both temporally and spatially, the sampled process—i.e., rainfall. The definition of "sufficient" is ultimately a function of the objective and goals for which the data is being used. Many books have been written on hydrologic sampling. Nevertheless, this book must be limited in scope. A few of the common hydrologic data-handling techniques will be mentioned. Wherever possible, brief reference to more-sophisticated techniques will be made, hopefully leading the curious reader to further studies. The techniques to be seen deal with completion of missing data, check on data consistency, estimation of mean areal precipitation, frequency analysis, and network design.

Before going into data analysis techniques, it is useful to emphasize the usual variability of precipitation in time and space. On a global scale, precipitation varies widely. Figure 4.24 gives the mean annual precipitation

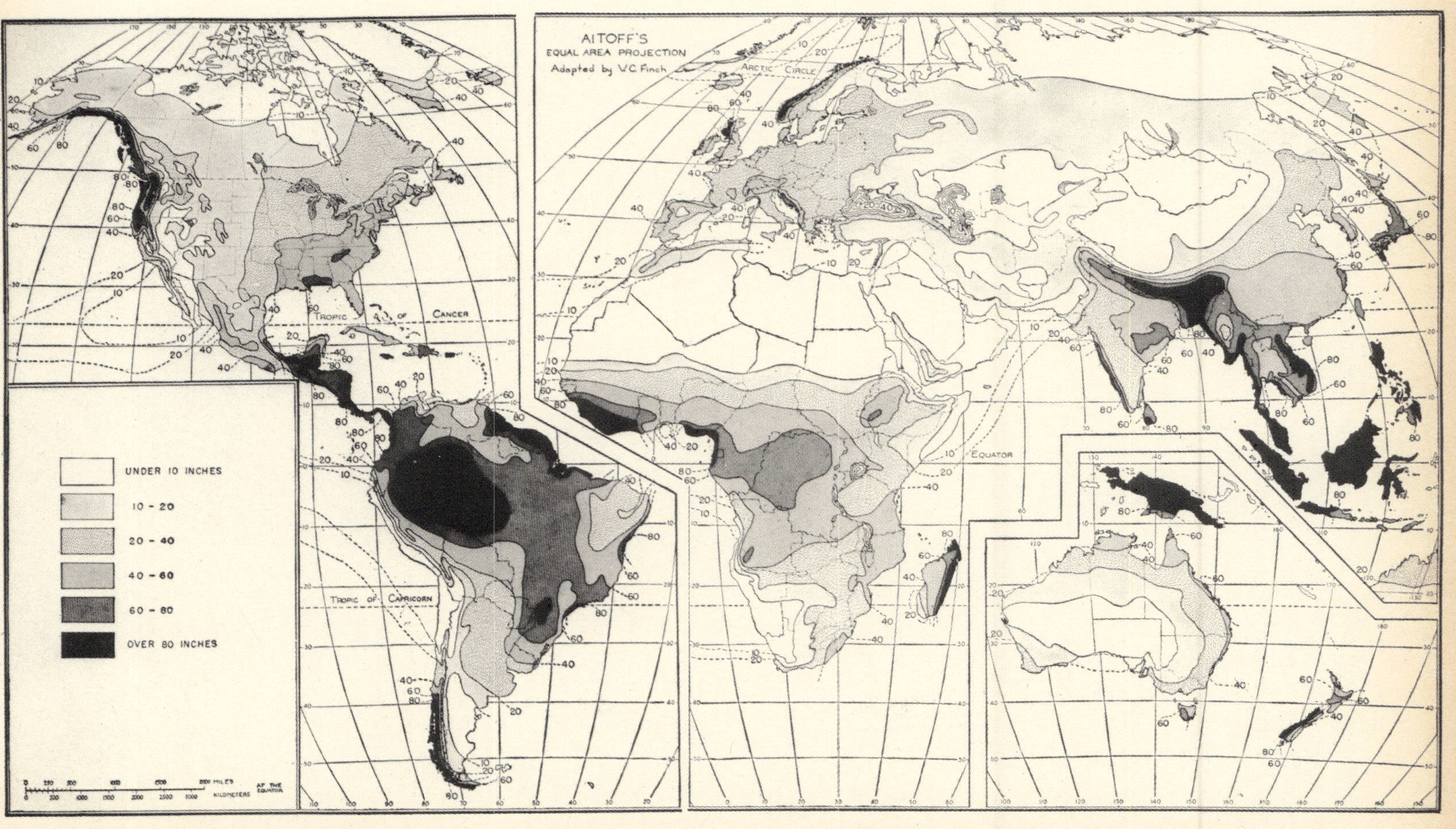

FIGURE 4.24 Terrestrial distribution of mean annual precipitation. Source: G. T. Trewartha, *An Introduction to Climate,* 3rd ed., McGraw-Hill, 1954. Reproduced with permission.

throughout the world. Two of the most important features are the large expanses of desertic regions (occurring almost at all latitudes) and the consistently wet tropics, particularly the Amazon basin in northern South America and the Pacific oceanic regions like Japan, the Philippines, Malaysia, etc.

Spatial variability occurs on a much smaller scale. Figure 4.25 shows isohyets of the annual mean precipitation over the Nile River Basin in the

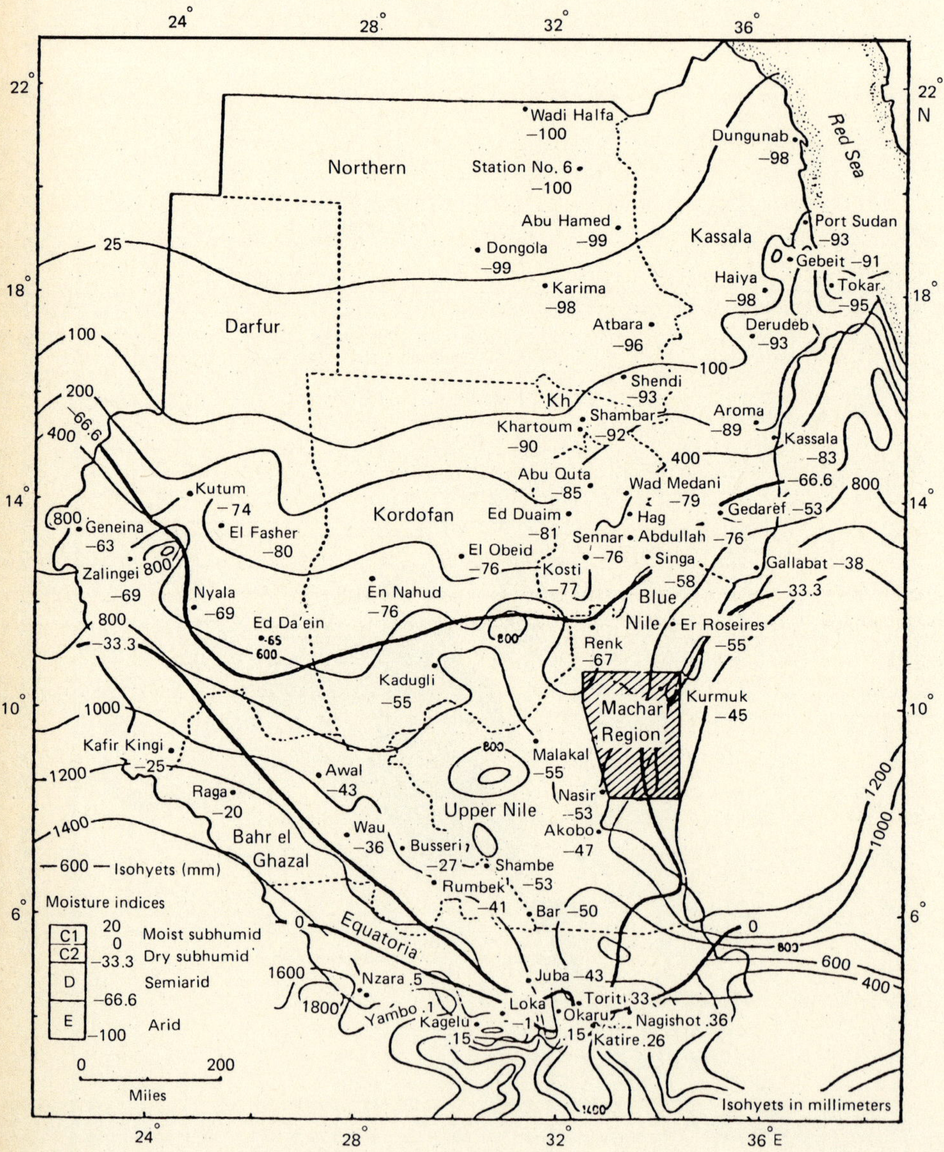

FIGURE 4.25 Rainfall distribution over the Nile River Basin in the Sudan. Source: El-Hemry and Eagleson [1980], after Oliver [1969].

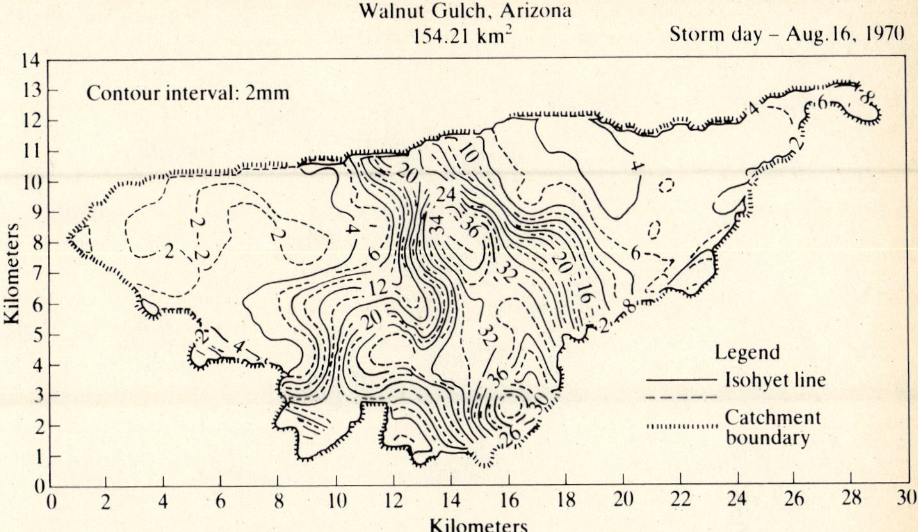

FIGURE 4.26 Total rainfall distribution of a storm over Walnut Gulch, Arizona. Source: Fennessey et al. [1986].

Sudan. Note the sharp drop of precipitation as you move north. At a storm level, the spatial variability can be larger. Figure 4.26 shows a storm over the Walnut Gulch Basin (only 154.2 km^2) in Arizona on August 16, 1970.

In time, precipitation is also highly variable. Some regions show strongly seasonal precipitation and others almost uniform behavior throughout the year. This is evident in Figure 4.27, where the mean monthly accumulations are shown for various parts of the United States. Within a storm, precipitation intensity is quite variable. Figure 4.28 shows the history of a storm over the Souhegan River Basin in New England. The time step is 30 min and the rainfall was measured with weather radar (Wyss [1988]). The basinwide average and the histories at three points in the basin are given. Although the three points are only 13 km apart, the variability in space is evident.

4.7.1 Estimation of Missing Data

The most common method for estimating a missing data point in a rain-gage network is the "normal ratio" method. Three index stations, next to the missing point location, are defined. The missing precipitation value is given by

$$P_4 = \frac{1}{3}\left[\frac{N_4}{N_1}P_1 + \frac{N_4}{N_2}P_2 + \frac{N_4}{N_3}P_3\right], \tag{4.29}$$

where P_4 is the precipitation at the missing location, P_i the precipitation at index station i, and N_i the long-term normal precipitation at station i.

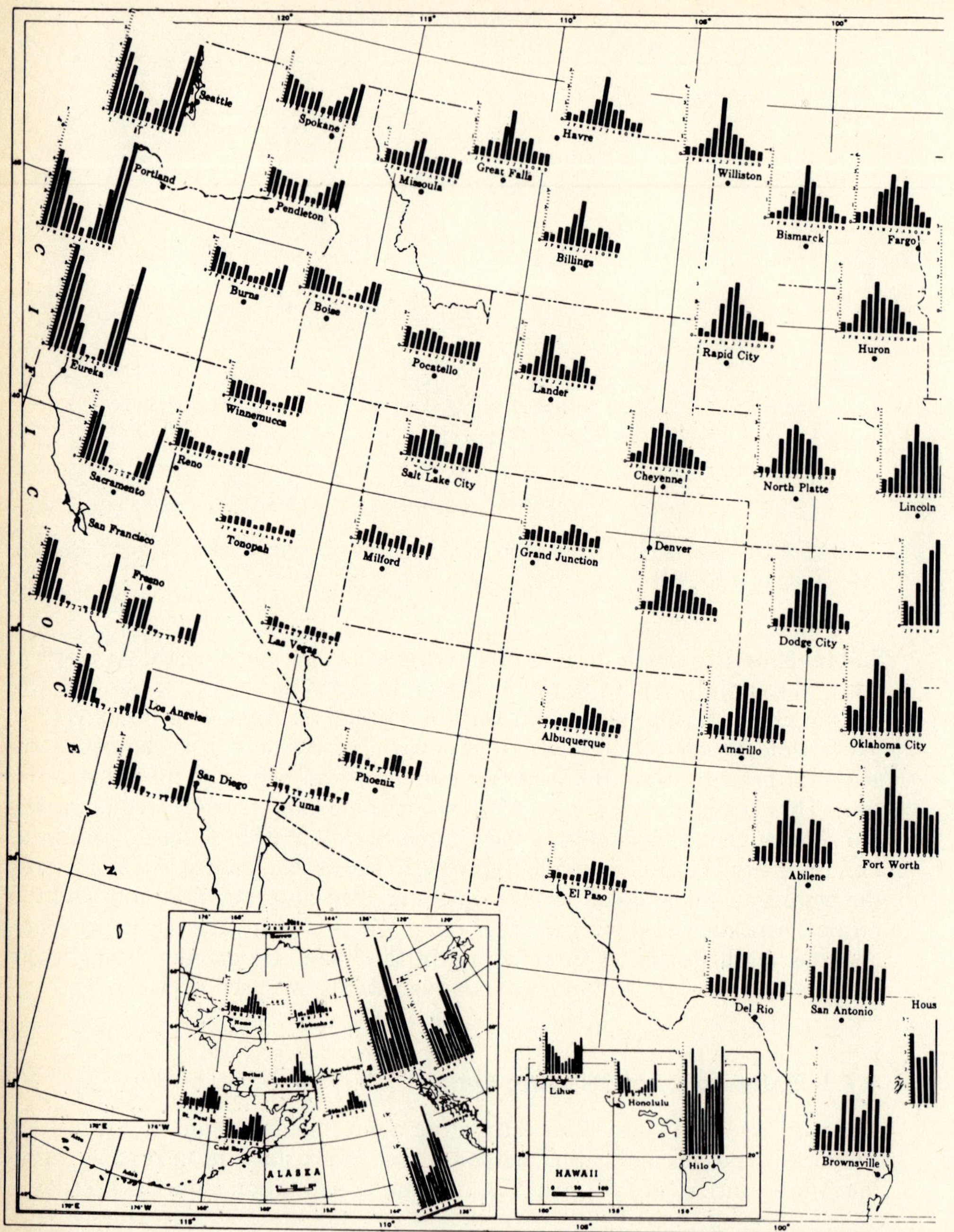

FIGURE 4.27 Monthly distribution of precipitation in the United States, in inches (1 in. = 25.4 mm). Source: Bedient and Huber [1988].

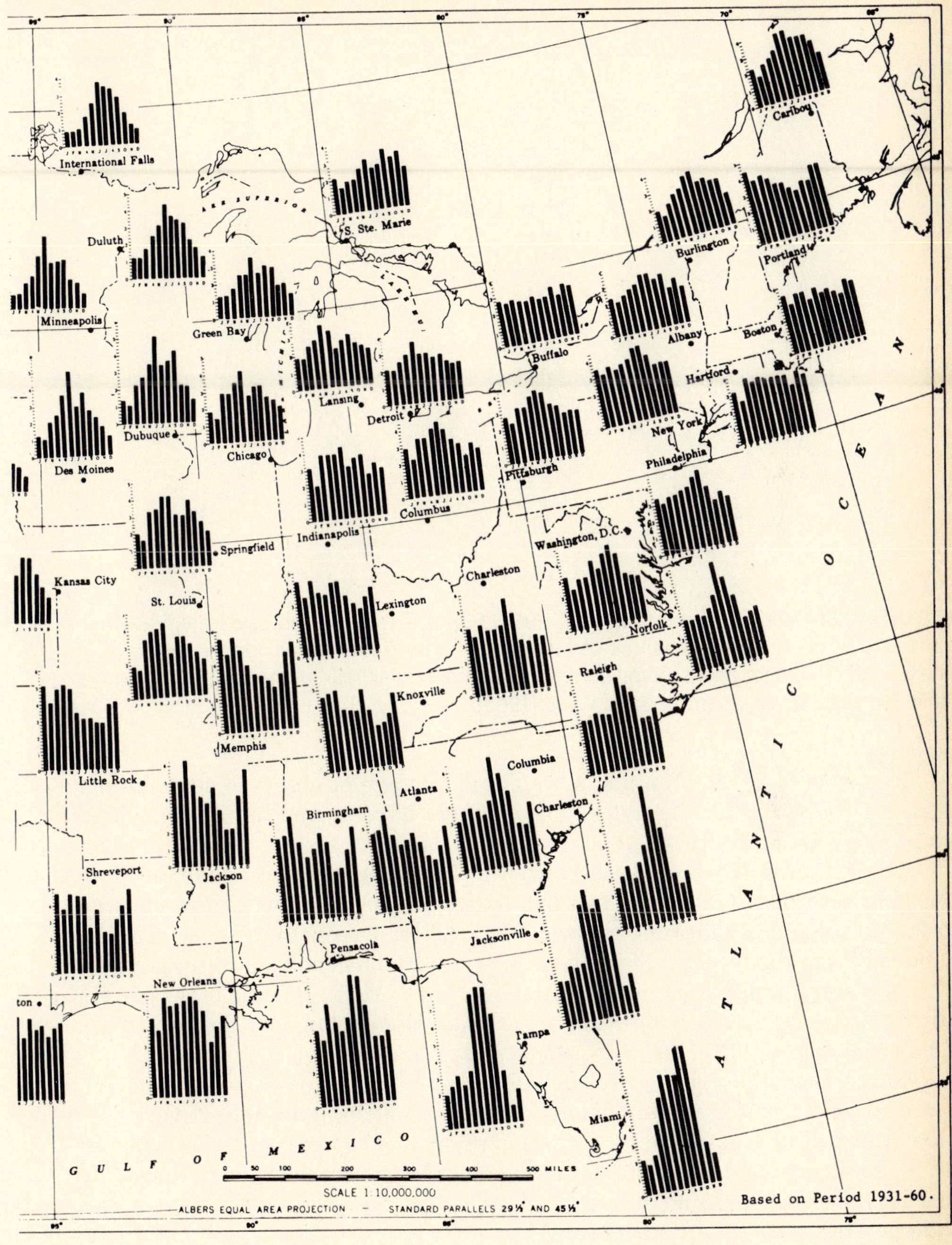
International Falls
Duluth
Minneapolis
Des Moines
Dubuque
Green Bay
S. Ste. Marie
Lansing
Detroit
Chicago
Kansas City
St. Louis
Springfield
Indianapolis
Columbus
Lexington
Charleston
Knoxville
Memphis
Little Rock
Shreveport
Jackson
Birmingham
Atlanta
Columbia
Charleston
Jacksonville
Pensacola
New Orleans
Tampa
Miami
Cariboù
Burlington
Portland
Albany
Boston
Hartford
Buffalo
New York
Philadelphia
Pittsburgh
Washington, D.C.
Norfolk
Raleigh
LAKE SUPERIOR
ATLANTIC OCEAN
GULF OF MEXICO
0 50 100 200 300 400 500 MILES
SCALE 1:10,000,000
ALBERS EQUAL AREA PROJECTION — STANDARD PARALLELS 29½° AND 45½°
Based on Period 1931-60.

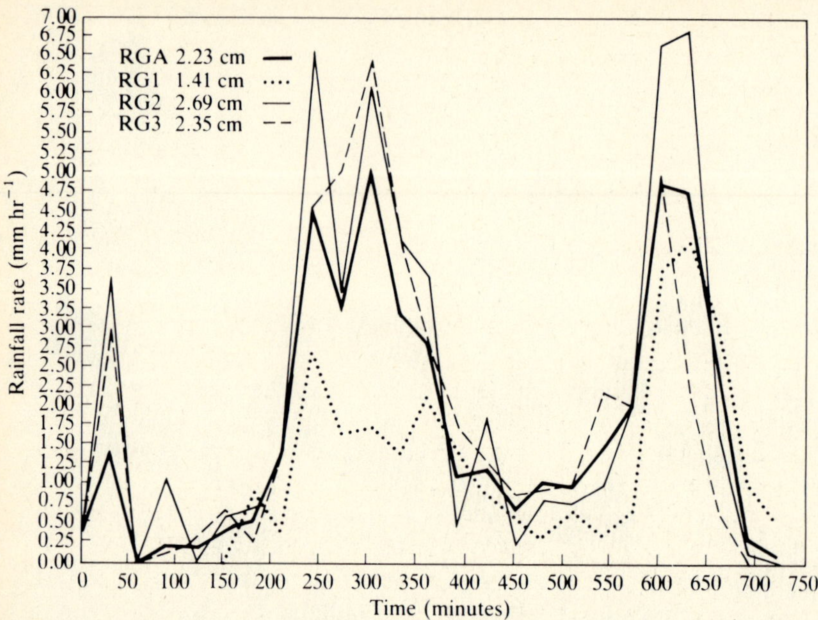

FIGURE 4.28 Rainfall rates on June 27, 1987. Bold lines (RGA) represent areal average rainfall rates over the Souhegan as determined by radar. Total rainfall depth is indicated in the legend. The other three curves correspond to the rainfall rate measured by the radar at selected sites 13 km apart. Source: Wyss [1988].

Equation (4.29) is then just a weighted average of observed surrounding stations. To be useful, the index stations must be near and highly correlated to the location of interest.

The National Weather Service uses another weighted average to estimate missing records. Precipitation at the four stations adjacent to the missing point, in the four quadrants, are weighted by the inverse of the square of the distance from the location of interest. The result is normalized by dividing by the sum of the weights. Other investigators have proposed using other powers of the distance in the weighting algorithm (Shearman and Salter [1975]; Dean and Snyder [1977]; U.S. National Weather Service [1972]).

Regression techniques are clearly applicable to the missing-data problem. Careful consideration of the obvious correlation of the independent variables is required. Investigators have suggested many and varied regression schemes to fill data gaps and complete data series based on complete index series. As methods become more advanced, statements on the goodness of the estimation become more readily available.

Interpolation, both in time and space, plays a major role in completion of missing data. Gandin [1965] has used linear interpolation between adjacent points and derived expressions for the accuracy of the interpolation. Numerous surface-fitting techniques are also available to complete spatially varying data sets. Rhenals-Figueredo et al. [1974] have used double Fourier series and multiquadratic surface to fit rainfall in space. Chua and Bras [1982], Lenton and Rodriguez-Iturbe [1977], Delhomme and Delfiner [1973], and others have used a method called "Kriging" for fitting rainfall and other geophysical processes in space. As the techniques become more sophisticated, they require information like the process spatial correlation.

4.7.2 Consistency Checks

The double-mass analysis is the technique commonly employed to detect changes in data-collection procedures at a given location. These changes may result from changes in instrumentation, changes in observation procedures, changes in gage location and induced (manmade or natural) variations in process characteristics.

The double-mass plot essentially compares, in a diagram, the records of two gages. One of the gages is the standard or base (which could be a combination of gages) while the other is suspect. The accumulated rainfall (or whatever quantity) of the suspected station and the base station(s) are the axes. Under unchanging conditions the two quantities should show a well-behaved (hopefully linear) relation. Any effect that changes the historical behavior of the suspected location without affecting the standard will result in a change of slope in the diagram. This is observed in Figure 4.29. If such a slope change is clear, the reasons for it must be investigated and the record corrected.

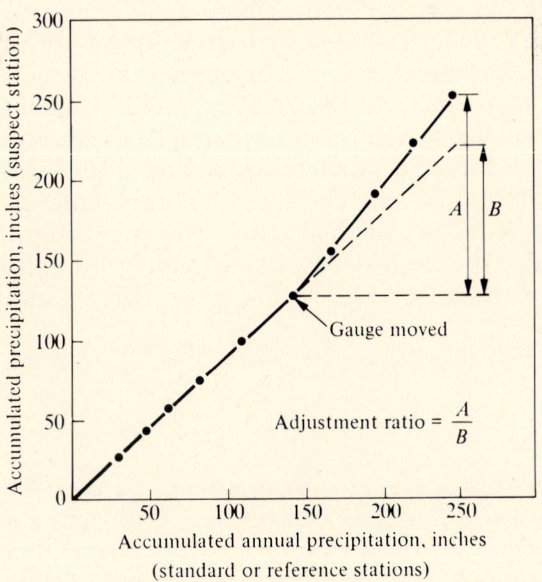

FIGURE 4.29 Illustration of double-mass curve.

4.7.3 Mean Areal Precipitation

There are two common spatial averages used in hydrology, the mean areal precipitation of a storm event and the time-averaged mean areal precipitation over a given period of time. Mathematically, they are defined as

Areal mean of an event:

$$P_1 = \frac{1}{A}\int_A f(x)\,dx \tag{4.30}$$

Long-term areal average:

$$P_2 = \frac{1}{T}\frac{1}{A}\sum_{i=1}^{T}\int_A f(x, t_i)\,dx\,, \quad T \to \infty \tag{4.31}$$

where $f(x)$ is the function describing a storm total accumulation at all points x_i; and $f(x, t_i)$ is a function describing total precipitation at x and period t_i.

Since rainfall observations are generally point values, imperfect for that matter, we do not know the function $f(x)$. Generally, the spatial integration is then approximated by some sort of discrete weighted average. The weights would be one over the number of stations if observations were uniformly distributed and the rainfall process is completely homogeneous in space. This is certainly not the case.

Knowing a spatially varying mean behavior and/or the spatial correlation of the rainfall process, optimal weights could be determined (Lenton and Rodriguez-Iturbe [1977]; Bras and Rodriguez-Iturbe [1976]; Delhomme and Delfiner [1973]). The weights would be optimal in the sense that the mean square error of approximating Eqs. (4.30) and (4.31) is minimized. The mean square error is defined as $E[(P - \hat{P})^2]$, where P is the desired statistic, $\hat{P}$ is its estimate, and E is the expectation (computation of the mean) operator.

Commonly, two different methods to obtain areal averages of storm events are used. The first method is the Thiessen weighting scheme. Figure 4.30 illustrates the method. An area with eight rainfall stations is shown. Rainfall values at each location are also given. The weighting mechanism is of the form

$$\hat{P} = \sum_{i=1}^{N} \rho_i P_i\,, \tag{4.32}$$

where ρ_i is the weight applied to observation P_i. In the Thiessen method, the weight is a measure of rain-gage contributing area. In the procedure, all rain gages are connected, shown in thin lines in the figure. Connecting lines are bisected and extended until they intersect other bisectors. The result is a

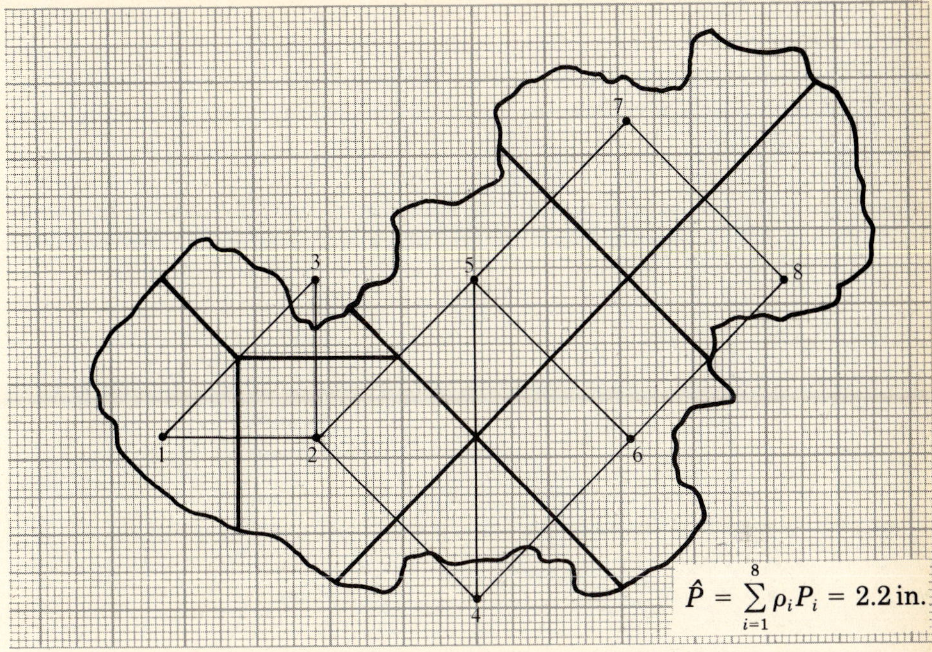

STATION	PRECIPITATION, P_i (INCHES)	THIESSEN WEIGHT, ρ_i
1	1.9	0.105
2	2.3	0.1611
3	2.1	0.0540
4	2.3	0.0705
5	2.2	0.1607
6	2.4	0.1567
7	2.1	0.1560
8	2.2	0.1360

FIGURE 4.30 Illustration of Thiessen coefficients.

polygonal pattern, shown in thick lines in the figure. Each station is surrounded by a closed polygon of given area. The weights ρ_i are given by A_i/A, where A_i is the area of the polygon around station i and A is the total area. The area of each polygon can be estimated by planimeter or any other valid approximation. Here, the area was measured by counting the small squares of the superimposed fine grid. The reader is challenged to repeat the exercise. There is a good chance of proving me wrong! The obtained mean areal precipitation was 2.2 in.

The second common method is the isohyetal method. An isohyetal map is one showing lines of equal precipitation (isohyets). In this method, the weights ρ_i are again A_i/A, where A_i is the area between isohyets. The weighted precipitation values are the average between contiguous curves of equal precipitation. This is illustrated in Figure 4.31. Shown in the figure are isohyets every 0.1 in. for the same storm of Figure 4.30. The areas enclosed by equal precipitation curves are numbered I to XIV. Again, the areas were measured by counting enclosed squares of the superimposed grid. To

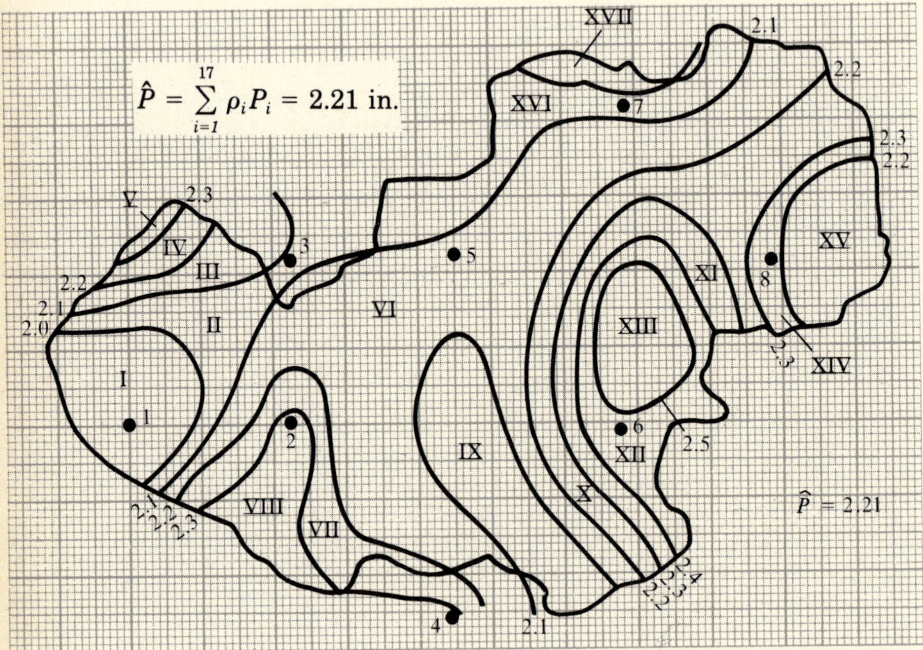

AREA	P_i	ρ_i	AREA	P_i	ρ_i
I	2.05	0.0663	X	2.25	0.0965
II	2.05	0.0426	XI	2.35	0.0468
III	2.15	0.0166	XII	2.45	0.0512
IV	2.25	0.0098	XIII	2.55	0.0417
V	2.35	0.0027	XIV	2.25	0.0237
VI	2.15	0.2952	XV	2.25	0.0563
VII	2.25	0.0444	XVI	2.15	0.0778
VIII	2.35	0.0370	XVII	2.05	0.0138
IX	2.15	0.0776			

FIGURE 4.31 Illustration of the isohyetal method for computing mean areal precipitation.

each area we assigned the mean precipitation of the two boundary isohyets. Where necessary, the basin boundary was given a value 0.1 in. less than the encompassing interior isohyet. The obtained mean areal precipitation was 2.21 in., very close to that resulting from the Thiessen polygon method. Given the uniformity of the storm we are dealing with, this is not surprising.

The isohyetal method is preferable and more accurate. Its main limitation is that it requires enough observations to permit the drawing of contours of equal precipitation. On the other hand, a hydrologist knowledgeable of the typical precipitation patterns in a given area can obtain a better estimate of mean areal precipitation.

Depth–Area–Duration Curves

In Section 4.5 we saw that the area-averaged rainfall depth decreased with increasing area. We also mentioned that as depth and duration increase, the areal average increases and the accumulation generally becomes more uniform in space. It is sometimes useful to quantify these relationships for a given storm or set of storms. The result is the depth–area–duration curve. To obtain this curve for a given storm, we must have the storm history, say at intervals Δt, for a large number of stations within the area. A normal procedure would then be

1. Select intervals of area ΔA such that the total area is given by $A = m\,\Delta A$. Define $A_n = n\,\Delta A$; $n \leq m$.
2. Define the precipitation over the area at all time intervals Δt.
3. For all time intervals Δt, find the maximum mean areal precipitation over a subarea of size A_n, arbitrarily located within the region. In order to maximize the mean areal precipitation over A_n, it is recommended that isohyetal maps of precipitation at time interval Δt be prepared. Repeat this step for all subareas A_n, $n = 1, \ldots, m$.
4. Repeat Step 3 for accumulations over time intervals $2\,\Delta t$, $3\,\Delta t$, and so on, until a period equal to the storm duration is covered.
5. Plot the maximum areal average depth for each period $\ell\,\Delta t$, $\ell = 1, \ldots, L$ (where L is storm duration divided by Δt) against its corresponding A_n.

A typical depth–area–duration curve is shown in Figure 4.32. If the storm has multiple centers, it can be analyzed by centers and the results combined at the plotting stage.

4.7.4 Frequency Analysis

Precipitation, streamflow, evaporation, and all other hydrologic and geophysical processes can be characterized as random occurrences. It is impossible to predict what the future realizations of the processes will be. The analysis of precipitation data should then follow well-established statistical procedures. For example, intensity–frequency–duration (IFD) curves are a

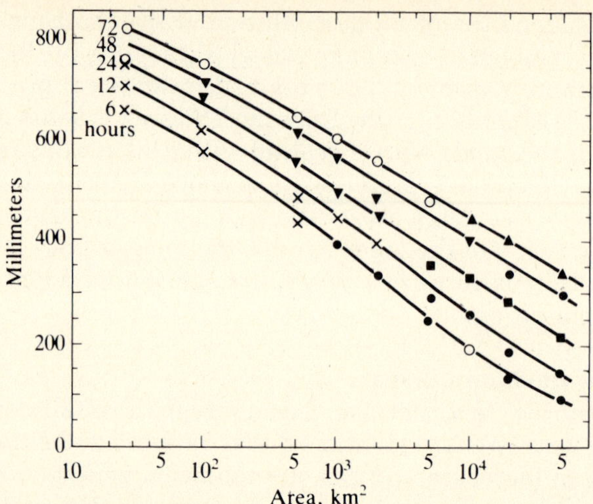

FIGURE 4.32 Diagrammatic presentation of maximum depth–area–duration curves for a catchment. Symbols indicate separate storms. Note the enveloping of data points. Source: A. J. Raudkivi, *Hydrology: An Advanced Introduction to Hydrological Processes and Modelling.* Copyright 1979 by Pergamon Press. Reprinted with permission by the author.

classical rainfall analysis tool that relates the probability of occurrence of storms of given duration and intensity. Probability of occurrence is usually measured in terms of recurrence interval (return period). The recurrence interval of a storm T is the mean time that will go by before an event equaling or exceeding the storm magnitude occurs. Mathematically, it is equal to the inverse of the probability of equaling or exceeding the event in a unit time period. For example, a 50-year-recurrence storm is one that has a probability of being equaled or exceeded in any one year of 1/50. On the average it will take 50 years before that occurs. IFD curves and other probabilistic and statistical measures will be studied in Chapter 11. They are discussed separately in order to make a more complete presentation and to emphasize their applicability to all types of data.

4.7.5 Network Design

As remarked by Lenton and Rodriguez-Iturbe [1974], it is necessary that "all aspects of data management be integrated—the initial collection of data cannot and should not be treated separately from the later stages of data analysis and synthesis." The data management and collection procedure must be defined in terms of the final objectives, goals, and uses of collected information.

Hydrologic-data-collection networks have been divided into various levels (Rodda [1969]). Levels I and II can be related to problems of regional estimation—i.e., there is no clearly defined final goal or use for the collected data. The problem of rainfall monitoring for estimating the total precipitation areal average for a storm event and the problem of finding the long-term (time) mean areal precipitation fall in these two levels. Level III networks are those designed to collect data for a specific, clearly defined, objective, which would imply known net benefits or utility of the data. The problem of rainfall monitoring for use together with a flood forecasting system theoretically fits this framework.

Historically, network design has been strongly influenced by issues of convenience and cost, ignoring the issues of required accuracy. Network design should involve stating the number and location of stations necessary for achieving the accuracy demanded by a given data use and under stated budgetary constraints.

Traditionally, the above objectives were accomplished using heuristic criteria. For example, McKay (Gray [1973]) mentions that for standard precipitation gages a 15-mile separation is adequate for Canadian conditions. Following are a few traditional design criteria and results.

The "index approach" requires the logical condition that sensors have the highest possible correlation with the effects that are being measured. One gage should be located in each "homogeneous" area. Each station should be highly correlated with surrounding effects but uncorrelated among themselves.

Several experiments have been performed on very densely gaged regions. McGuinness [1963] suggests the following formula for Coshocton, Ohio:

$$E = 0.03P^{0.54}G^{0.24}, \tag{4.33}$$

where E is the absolute difference in inches between observed and true average rainfall; P is the rainfall in inches for the "true" dense network; and G is the network density in square miles per gage for a reduced network. The above formula was developed from data of watersheds less than 25 mi^2 but was found to be consistent for larger areas. Being of local origin, extrapolation to other areas is speculative.

Hershfield [1965] suggested that the average spacing between gages should be that required for obtaining a correlation of 0.9 between station values. He related this spacing and correlation to the two-year recurrence, 24-hour duration rainfall and the two-year recurrence, one-hour duration rainfall. Figure 4.33 gives Hershfield's results. Holtan et al. [1962] recommend various rain-gage densities for agricultural areas. These are given in Table 4.5.

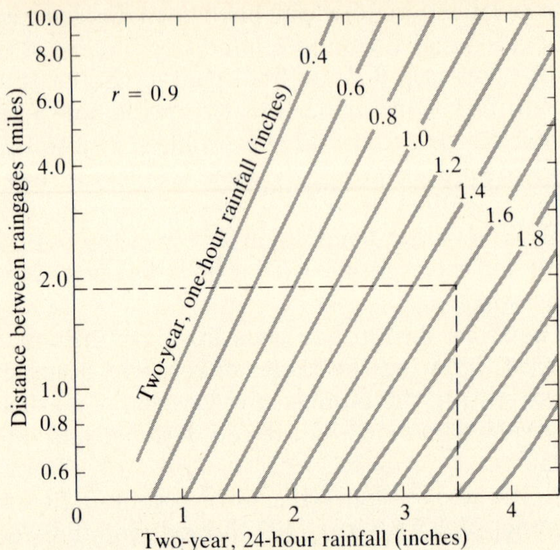

FIGURE 4.33 Diagram for estimating the distance between gages as a function of the two-year, 24-hour and two-year, one-hour rainfall. Source: After Hershfield [1965]. "On the Spacing of Rain Gages", in *Symposium on Design of Hydrological Networks,* vol. 1, pp. 72–81. IAHS Publ. no. 67.

A well-known study in the Muskingum River Basin by the U.S. Weather Bureau [1947] resulted in Figure 4.34, giving the standard error of estimating mean areal precipitation as a function of gage density and total area.

The following minimum densities of precipitation networks have been recommended for general hydrometeorologic purposes (Gray [1970]).

1. Flat regions of temperate, mediterranean, and tropical zones, 600 to 900 km^2 per station.

TABLE 4.5 Number of Rainfall Stations Required

SIZE OF DRAINAGE AREA (ACRES)	GAGING RATIO (mi^2/gage)	MINIMUM NUMBER OF STATIONS
0–30	0.05	1
30–100	0.08	2
100–200	0.10	3
200–500	0.16	1 per 100 acres
500–2500	0.40	1 per 250 acres
2500–5000	1.00	1 per square mile
over 5000	3.00	1 per each 3 mi^2

Source: Holtan et al. [1962].

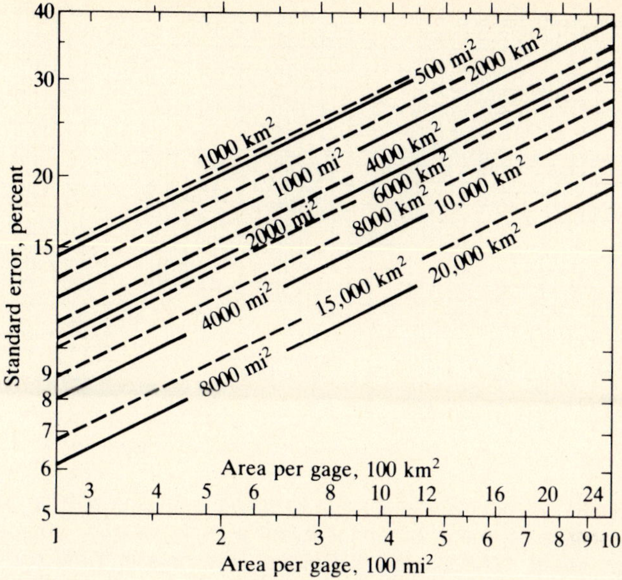

FIGURE 4.34 Standard error of storm precipitation averages as a function of network density and area for the Muskingum Basin. Source: U.S. National Weather Service [1972].

2. For mountainous regions of temperate, mediterranean, and tropical zones, 100 to 250 km² per station.
3. For small mountainous islands with irregular precipitation, 25 km² per station.
4. For arid and polar zones, 1500 to 10,000 km² per station.

*Generalized Network Design

Generalized and theoretical approaches to rainfall network design exist. Although different in techniques and assumptions, all procedures require some knowledge of the rainfall-process spatial correlation. The spatial correlation measures the level linear dependence of precipitation at two points separated a distance v from each other. A correlation of 1 (or -1) will imply perfect linearity. For precipitation, correlation is generally between 0 and 1 and gets smaller as the points are farther apart. A point is perfectly correlated with itself, since v is zero in that case.

Rodriguez-Iturbe and Mejia [1974] developed design curves for the mean square error of estimating the areal average of precipitation (Eq. 4.30) using a random sampling technique. Figure 4.35 gives one such curve, corresponding to random sampling and exponential-type correlation function in space, implying that the correlation falls exponentially with distance between stations. For isotropic, homogeneous random fields, the correlation is just a function of the distance between points v. The results of Figure 4.35 correspond to

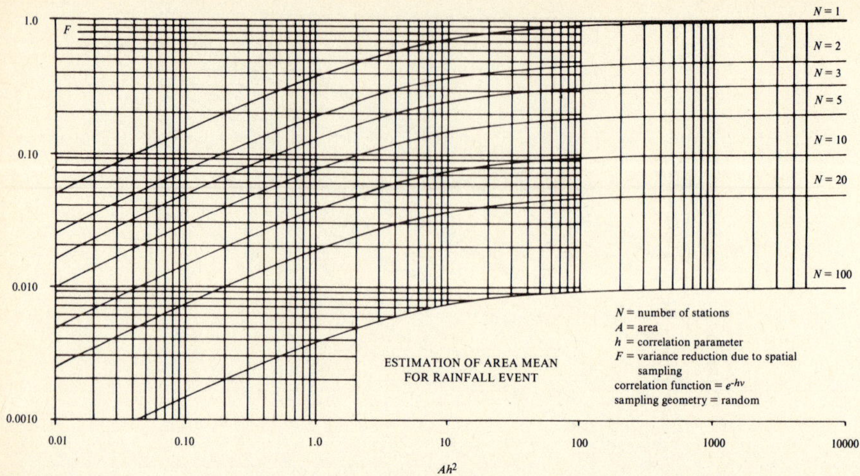

FIGURE 4.35 Variance reduction factor due to spatial sampling with random design used in the estimation of areal mean of rainfall event with $r(v) = e^{-hv}$. Source: I. Rodriguez-Iturbe and J. M. Mejia, "The Design of Rainfall Networks in Time and Space," *Water Resources Res.*, 10(4):725, 1974. Copyright by the American Geophysical Union.

a correlation of the form e^{-hv}, so $v = 0$ implies that rainfall at a point is perfectly correlated with itself. The correlation decreases as the distance between points v increases. The parameter h (km^{-1}) controls the decay in correlation with distance. Random sampling implies that the observations can be anywhere in space. The ratio of mean square error (MSE) to point variance (it is assumed that the process has the same variance everywhere) is given in terms of the number of stations N randomly located in space, and a nondimensional area Ah^2, where h is the parameter of the correlation function:

$$\text{MSE} = F(N;\, Ah^2)\sigma^2. \tag{4.34}$$

Figure 4.36 gives a similar curve for stratified sampling. Stratified sampling refers to random data collection within prespecified strata or regions. Note that the sampling error is smaller under these conditions. Rodriguez-Iturbe and Mejia [1974] assume perfect observations.

In the same work, Rodriguez-Iturbe and Mejia developed curves for evaluating networks designed to obtain the long-term areal average as defined previously (Eq. 4.31). They assume a separable, in time and space, covariance structure of the form $\text{cov}(v, \tau) = \sigma^2 r(v)\rho^\tau$, where σ^2 is the point variance of rainfall, $r(v)$ is the correlation due to distance v between points, ρ is the lag-one serial (time) correlation of data, and τ is the time between data points. The results are that the MSE of estimating the long-term areal average is given by

$$\text{MSE} = F_1(T)F_2(N;\, Ah^2)\sigma^2, \tag{4.35}$$

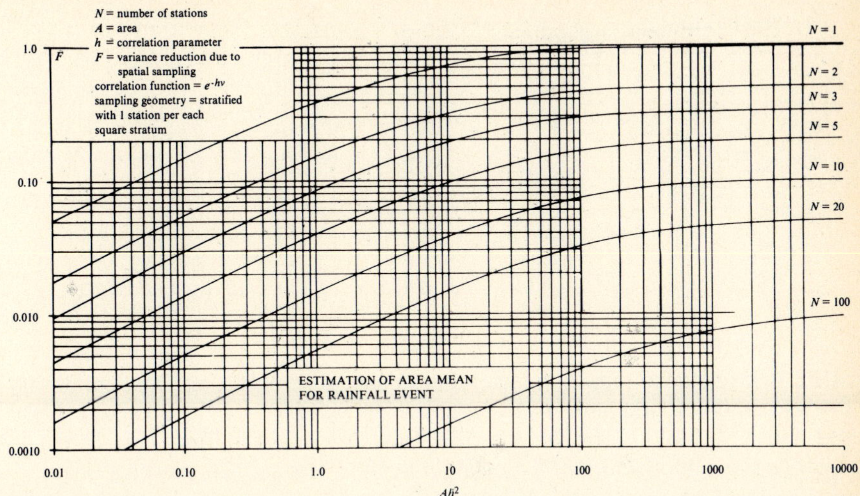

FIGURE 4.36 Variance reduction factor due to spatial sampling with stratified design used in the estimation of areal mean of rainfall event with $r(v) = e^{-hv}$. Source: I. Rodriguez-Iturbe and J. M. Mejia, "The Design of Rainfall Networks in Time and Space," *Water Resources Res.,* 10(4):726, 1974. Copyright by the American Geophysical Union.

where F_1 is a factor function of the number of time periods of observation T and the lag-one autocorrelation of the process. The dependence is shown in Figure 4.37. Figures 4.38 and 4.39 give $F_2(N; Ah^2)$, which is the space-dependent factor, for random and stratified sampling and exponential-type spatial correlation. The variables are the same as for Figures 4.35 and 4.36.

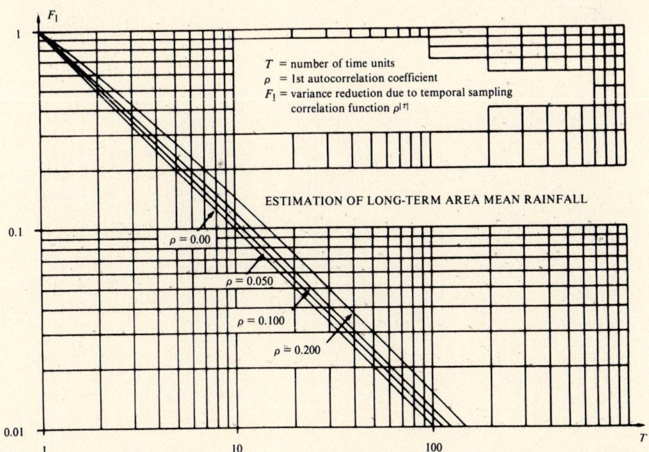

FIGURE 4.37 Variance reduction factor due to temporal sampling used in the estimation of long-term mean areal rainfall. Source: I. Rodriguez-Iturbe and J. M. Mejia, "The Design of Rainfall Networks in Time and Space," *Water Resources Res.,* 10(4):718, 1974. Copyright by the American Geophysical Union.

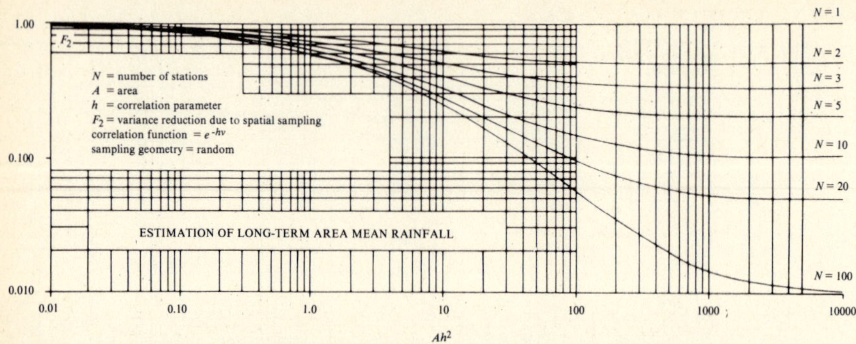

FIGURE 4.38 Variance reduction factor due to spatial sampling with random design used in the estimation of long-term mean areal rainfall with $r(v) = e^{-hv}$. Source: I. Rodriguez-Iturbe and J. M. Mejia, "The Design of Rainfall Networks in Time and Space," *Water Resources Res.*, 10(4):719, 1974. Copyright by the American Geophysical Union.

Bras and Rodriguez-Iturbe [1976] developed a method to handle the systematic sampling condition and instrument error. Systematic sampling implies that stations are given known positions in space. The procedure uses estimation theory and solves for the optimal network to obtain mean areal precipitation of an event by minimizing an objective function of mean square error (accuracy measure) and cost.

Bras and Colon [1978] address the network design for the long-term areal average under the previously mentioned systematic sampling techniques

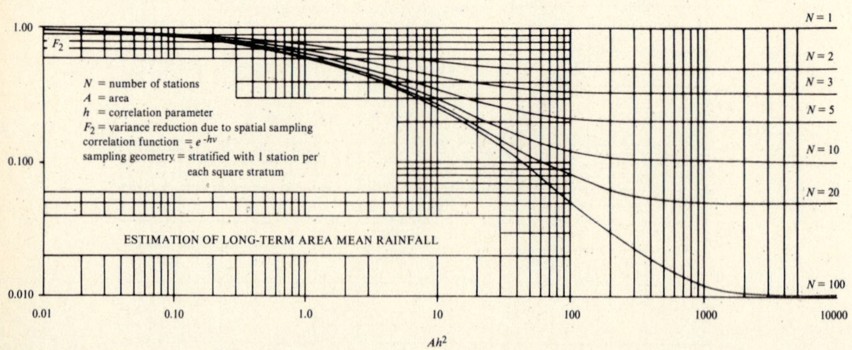

FIGURE 4.39 Variance reduction factor due to spatial sampling with stratified design used in the estimation of long-term mean areal rainfall with $r(v) = e^{-hv}$. Source: I. Rodriguez-Iturbe and J. M. Mejia, "The Design of Rainfall Networks in Time and Space," *Water Resources Res.*, 10(4):721, 1974. Copyright by the American Geophysical Union.

(i.e., not only number but station locations are specified). Estimation theory was used to find the mean square error of estimation expression and include instrument errors in the analysis. The interested reader is referred to Bras and Rodriguez-Iturbe [1985] for a complete view of the above procedures.

EXAMPLE 4.4

Monitoring Network Design

Rodriguez-Iturbe and Mejia [1974] illustrated the network-design exercise with an example from the Central Venezuela region. The region is shown in Figure 4.40, together with the location of its 26 rain gages, over its 30,000-km^2 area. Also shown are the mean annual precipitation isohyets. Table 4.6 gives the stations' annual means and their standard deviation computed as

$$\text{standard deviation} = s = \left[\frac{1}{N-1}\sum_{i=1}^{N}(X_i - \overline{X})^2\right]^{1/2},$$

where N is the number of years of data, X_i is the data point for year i, and $\overline{X}$ is the mean for the particular station,

$$\overline{X} = \frac{1}{N}\sum_{i=1}^{N} X_i .$$

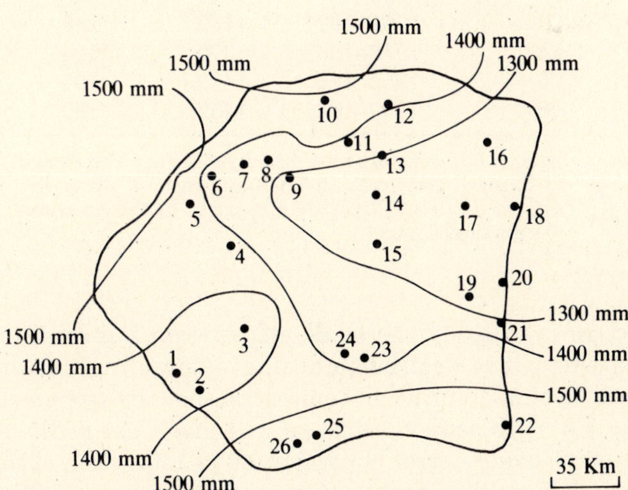

FIGURE 4.40 Central Venezuela region (Portuguese state) used in the example of monitoring network design. Source: I. Rodriguez-Iturbe and J. M. Mejia, "The Design of Rainfall Networks in Time and Space," *Water Resources Res.,* 10(4):715, 1974. Copyright by the American Geophysical Union.

TABLE 4.6 Description of Rainfall Data Used in the Central Venezuela Example

STATION	YEARS OF RECORD	MEAN (mm)	STANDARD DEVIATION (mm)
1	1958–1971	1445	193
2	1955–1971	1412	131
3	1955–1965	1269	234
4	1951–1971	1404	127
5	1956–1971	1500	156
6	1955–1971	1342	189
7	1953–1971	1328	234
8	1948–1971	1294	248
9	1954–1971	1144	195
10	1943–1971	1440	194
11	1956–1971	1341	223
12	1950–1971	1427	261
13	1944–1971	1318	200
14	1958–1971	1296	210
15	1953–1971	1308	225
16	1961–1971	1240	228
17	1961–1971	1255	215
18	1954–1971	1155	206
19	1952–1964	1325	221
20	1958–1971	1252	276
21	1952–1971	1269	246
22	1952–1971	1514	318
23	1961–1971	1462	283
24	1948–1965	1370	194
25	1952–1965	1429	191
26	1946–1965	1452	199

Source: I. Rodriguez-Iturbe and J. M. Mejia, "The Design of Rainfall Networks in Time and Space," *Water Resources Res.* 10(4):721, 1974. Copyright by the American Geophysical Union.

The objective is to study the trade-off between number of stations and years of data when it is desired to design a sampling network that will achieve a given level of accuracy in computing the long-term mean areal precipitation (Eq. 4.31). To achieve this, we will make use of Eq. (4.35), which expresses the mean square error of estimation as a function of the point variance, a factor involving the number of years in the record $F_1(T)$; and a factor involving the number of stations in the region $F_2(N; Ah^2)$.

The procedure requires us to compute the point variance over the region, the lag-one (one-year lag) autocorrelation coefficient, and the spatial correlation. Computing the variance of all records of all stations put together, we get $\sigma^2 = 5.44 \times 10^4 \text{mm}^2$. A study of the correlation between records one year apart indicates that there is no linear statistical relationship between the rainfall accumulation of two adjacent years, i.e., $\rho = 0$. We will assume that the spatial correlation falls exponentially with distance between points v:

$$r(v) = e^{-hv}.$$

The question is what value to give the decay parameter h. In their work, Rodriguez-Iturbe and Mejia [1974] argue that the spatial correlation should be calibrated to a typical distance in the area in question. Their suggestion is to calibrate the correlation between two points chosen randomly in the area. The distance between two random points in an area obeys a precomputable probabilistic distribution. Details are beyond the scope of this work but it suffices to state the following. If λ is the ratio of the sides of a rectangle, then the mean distance between two points of a rectangle of unit area is

Unit Area Rectangle

λ	$\overline{v}$
1	0.5214
2	0.5691
4	0.7137
16	1.3426

Most basins are reasonably approximated by rectangles. The Central Venezuela region is well represented by a rectangle of side ratio $\lambda = 2$. This rectangle would have a diagonal (maximum distance) of 265 km. To obtain the mean distance between two points for the Venezuelan region, we need only scale the unit area rectangle results by the ratio of diagonals. The unit area rectangle with $\lambda = 2$ has diagonal of 1.58; therefore the Central Venezuela region has a mean distance between points of

$$\overline{v} = (265/1.58) \times 0.5691 \approx 100\,\text{km}.$$

Rodriguez-Iturbe and Mejia [1974] computed the sample spatial correlation between points 100 km apart as 0.21. Therefore the correlation function must satisfy

$$r(100) = e^{-h100} = 0.21,$$

which yields $h = 0.0156 \text{ km}^{-1}$ and $Ah^2 = 7.3$.

TABLE 4.7 Variance Reduction Factor Due to Spatial Sampling $F_2(N; Ah^2)$ with $Ah^2 = 7.3$ in the Central Venezuela Example. Exponential Correlation, Random Sampling.

N	$F_2(N; Ah^2)$	N	$F_2(N; Ah^2)$
1	1.00	10	0.37
2	0.65	20	0.33
3	0.54	100	0.31
5	0.43		

Source: I. Rodriguez-Iturbe and J. M. Mejia, "The Design of Rainfall Networks in Time and Space," *Water Resources Res.*, 10(4): 721, 1974. Copyright by the American Geophysical Union.

The variance reduction factors due to spatial sampling, $F_2(N; Ah^2)$ are given in Table 4.7 and come from Figure 4.38 (assuming random sampling). The temporal reduction factor is obtained from Figure 4.37 ($\rho = 0.0$) and is given in Table 4.8.

Combining Tables 4.7 and 4.8, we can estimate the efficiency of different network schemes for the area considered. In the case of one station in operation during 20 years, we can expect a total variance reduction factor of

$$F_1(T) \times F_2(N; 7.3) = 1 \times 0.050 = 0.050\,.$$

In other words, this network will produce an estimate of the long-term areal mean precipitation with a variance on the order of 5% of the variance of the

TABLE 4.8 Variance Reduction Factor Due to Temporal Sampling with $\rho = 0.0$

T	$F_1(T)$	T	$F_1(T)$
1	1.000	15	0.067
2	0.500	20	0.050
3	0.333	30	0.033
5	0.200	50	0.020
7	0.140	75	0.013
10	0.100	100	0.010

Source: I. Rodriguez-Iturbe and J. M. Mejia, "The Design of Rainfall Networks in Time and Space," *Water Resources Res.*, 10(4):722, 1974. Copyright by the American Geophysical Union.

point rainfall process. If we wish to accomplish that type of precision in a lapse of 10 years, we will need

$$F_2(N;\,7.3) = 0.050/F_1(10) = 0.50\,.$$

This corresponds to $N = 4$ stations in the case of random sampling.

It is interesting to observe that the same precision of 0.050 cannot be obtained in a lapse of five years because it will be necessary for

$$F_2(N;\,7.3) = 0.050/0.20 = 0.25\,,$$

which is a value smaller than the asymptotic value of $F_2(N;\,7.3)$ when N goes to infinity. From Figure 4.38 it can be seen that with $Ah^2 = 7.30$ and $F_2(N;\,7.3) = 0.25$, the corresponding value of N is still larger than 100. We thus have the important conclusion that trading time versus space in hydrologic data collection can be done when we do not reduce the time interval too much, but no "miracles" can be expected in short times even from the most dense of all possible networks.

Table 4.9 presents the combined factors $F_1(T) \times F_2(N;\,Ah^2)$ for the example under consideration. This product represents the total reduction in variance relative to variance of point rainfall when the long-term areal mean with N stations during T years is estimated. It can be seen that even for quite a small number of years (like 2, 5, or 10 years), five stations will accomplish

TABLE 4.9 Total Factor of Variance Reduction Due to Temporal and Spatial Sampling $F_1(T) \times F_2(N;Ah^2)$ in the Central Venezuela Region Constructed for the Exponential Correlation Function with a Randomly Designed Network

N	$T = 2$	$T = 5$	$T = 10$
1	0.500	0.200	0.100
2	0.325	0.130	0.065
3	0.270	0.108	0.054
5	0.215	0.086	0.043
10	0.185	0.074	0.037
20	0.165	0.066	0.033
100	0.155	0.062	0.031

Source: I. Rodriguez-Iturbe and J. M. Mejia, "The Design of Rainfall Networks in Time and Space," *Water Resources Res.*, 10(4):722, 1974. Copyright by the American Geophysical Union.

most of the possible reduction in variance, and there is little justification in going over this number. It can also be observed that $F_1(T)$ weights more than $F_2(N; Ah^2)$ in the reduction of the variance of the long-term areal mean; when $T = 5\,\text{yr}$, $F_1(T) = 0.200$, yet an equivalent value of $F_2(N; Ah^2) = 0.200$ cannot be obtained in this example. This shows again that trading time versus space, although it is possible and in some instances necessary, is an expensive proposition. ◆

4.8 SUMMARY

Hydrology has traditionally studied the fluxes of energy and water between the land masses and the atmosphere and oceans. Atmospheric precipitation is, after all, a key element of the hydrologic cycle, the major water input onto the land masses. For too long hydrologists have been content with just measuring precipitation and letting meteorologists be preoccupied with the mechanisms that lead to its formation. This must change. Hydrologists must become knowledgeable about atmospheric processes, particularly those influencing precipitation. This has become apparent as hydrologists realize that most existing precipitation models and meteorologic predictive tools are incompatible with the time and space scales required by representations of the hydrologic land processes of interest. It is also now clearer that the atmospheric and surface processes are interdependent with negative and positive feedbacks that must be understood before any one part becomes explainable. In essence, hydrologists and meteorologists must learn more of each others' efforts and begin to cross the artificial disciplinary barriers that exist.

Chapters 3 and 4, probably a very modest beginning, intended to introduce hydrologists to some of the meteorology that relates to precipitation. A significant amount of space is dedicated to cloud physics in this chapter. Together with the stability issues discussed in Chapter 3, this chapter provides a first-level discussion of all important convective mechanisms that control most precipitation.

This chapter also introduced traditional and modern ways of obtaining and analyzing precipitation data. The issue of monitoring network design was treated at some length, if anything to point out that it pays to think about and plan a monitoring experiment. Unfortunately, in traditional hydrology this most important step is more often than not skipped, commonly resulting in inefficient or even useless data-collection exercises.

Chapter 5 will talk about evaporation, the mechanisms that provide the atmospheric moisture that leads to precipitation. The reader is not to lose sight of the fact that precipitation and evaporation are much more than mass transfers. They are significant energy-transfer mechanisms. During precipitation latent heat is released. Without it there would be little precipitation, since that heat drives the convective mechanisms that lead to further condensation. During evaporation latent heat is absorbed. The fact that evaporated water at a site generally has little influence on precipitation at the same site implies that energy is then transported laterally as well as vertically, a requirement of the energy distributions discussed in Chapter 2.

REFERENCES

Amorocho, J., and J. R. Slack [1970]. "Simulation of Cyclonic Storm Field for Hydrologic Modeling." Presented at the 51st annual meeting of the American Geophysical Union, April 20–24, 1970, Washington, D. C.

Austin, P. M. (1987). "Relation between Radar Reflectivity and Surface Rainfall." *Monthly Weather Rev.* 115:1053–1070.

Barrett, E. C., and L. F. Curtis [1976]. *Introduction to Environmental Remote Sensing.* London: Chapman and Hall.

Battan, L. [1973]. *Radar Observation of the Atmosphere.* Chicago: University of Chicago Press.

Beard, K. V. [1976]. "Terminal Velocity and Shape of Cloud and Precipitation Drops Aloft." *J. Atmos. Sci.* 33:851–864.

Beard, K. V., and H. R. Pruppacher [1971]. "A Wind Tunnel Investigation of the Rate of Evaporation of Small Water Drops Falling at Terminal Velocity in Air." *J. Atmos. Sci.* 28:1455–1464.

Bedient, P. B., and W. C. Huber [1988]. *Hydrology and Floodplain Analysis.* Reading, Mass.: Addison-Wesley.

Braham, R. R. [1965]. "The Aerial Observation of Snow and Rain Clouds." Proceedings of an International Conference on Cloud Physics, May 24–June 1, 1965, Tokyo and Sapparo, Japan.

Bras, R. L., and R. Colon [1978]. "Time Averaged Areal Mean of Precipitation: Estimation and Network Design." *Water Resources Res.* 14(5):8878–8888.

Bras, R. L., and I. Rodriguez-Iturbe [1976]. "Network Design for the Estimation of Areal Mean of Rainfall Events." *Water Resources Res.* 12(6):1185–1195.

Idem. [1985]. *Random Functions and Hydrology.* Reading, Mass.: Addison-Wesley.

Budyko, M. I. [1955]. *Atlas of the Heat Balance.* Leningrad: Gidrometeoizdat.

Budyko, M. I., et al. [1962]. "The Heat Balance of the Surface of the Earth." *Soviet Geograph. Rev. Transl.* 3(5):3–16.

Byers, H. R. [1965]. *Elements of Cloud Physics.* Chicago: University of Chicago Press.

Carlson, P. E., and J. S. Marshall [1972]. "Measurement of Snowfall by Radar." *J. Appl. Meteorol.* 11:494–500.

Chua, S. H., and R. L. Bras [1982]. "Optimal Estimators of Mean Areal Precipitation in Regions of Orographic Influence." *J. Hydrol.* 57(112):23–48.

Corotis, R. B. [1974]. "A Stochastic Rainfall Model and Statistical Analysis of Hydrologic Factors." Evanston, Ill.: Northwestern University. (Ph.D. thesis.)

Coulman, C. E., and J. Warner [1976]. "Aircraft Observations in the Subcloud Layer over Land." Preprints International Cloud Physics Conference, July 26–30, 1976, Boulder, Colorado.

Dean, J. D., and W. M. Snyder [1977]. "Temporally and Areally Distributed Rainfall." *J. Irrig. Drain. Div.* 103(IR2):221–229.

Delhomme, J. P., and P. Delfiner [1973]. "Application du Krigeage a l'Optimisation d'une Campagne Pluviometrique en Zone Aride." *Proc. Symp. Design Water Resource Proj. Inadequate Data.* 2:191–210. Madrid: UNESCO.

Dufour, L., and R. Defay [1963]. *Thermodynamics of Clouds.* Translated by M. Smyth and A. Beer. New York: Academic Press.

Eagleson, P. S. [1967]. "Optimum Density of Rainfall Networks." *Water Resources Res.* 3(4):1021–1033.

Idem. [1970]. Dynamic Hydrology. New York: McGraw-Hill.

El-Hemry, I. I., and P. S. Eagleson [1980]. "Water Balance Estimates of the Machar Marshes." Cambridge, Mass.: MIT Department of Civil Engineering, Ralph M. Parsons Laboratory. (Technical report no. 260.)

Eldridge, R. G. [1957]. "Measurements of Cloud Drop-Size Distributions." *J. Meteorol.* 14:55–59.

Fennessey, N. M., P. S. Eagleson, W. Qinliang, and I. Rodriguez-Iturbe [1986]. "Spatial Characteristics of Observed Precipitation Fields: A Catalog of Summer Storms in Arizona." Cambridge, Mass.: MIT Department of Civil Engineering, Ralph M. Parsons Laboratory. (Technical report no. 307. Vol. 1.)

Fletcher, N. H. [1962]. *The Physics of Rainclouds.* New York: Cambridge University Press.

Fujiwara, M. [1976]. "A Cloud Structure and the Rain Efficiency as Observed by Radars and Raindrop Recorder." Preprints International Cloud Physics Conference, July 26–30, 1976, Boulder, Colorado.

Gandin, L. S. [1965]. *Objective Analysis of Meteorological Fields.* Jerusalem, Israel: Isreal Program for Scientific Translations.

Georgakakos, K. P., and R. L. Bras [1982]. "A Precipitation Model and Its Use in Real-Time River Flow Forecasting." Cambridge, Mass.: MIT Department of Civil Engineering, Ralph M. Parsons Laboratory. (Technical report no. 286.)

Idem. [1984a]. "A Hydrologically Useful Station Precipitation Model: 1, Formulation." *Water Resources Res.* 20(11):1585–1596.

Idem. [1984b]. "A Hydrologically Useful Station Precipitation Model: 2, Case Studies." *Water Resources Res.* 20(11):1597–1611.

Glahn, H. R., and D. A. Lowry [1972]. "The Use of Model Output Statistics (MOS) in Objective Weather Forecasting." *J. Appl. Meteorol.* 11:1203–1211.

Godske, C. L., T. Bergeron, J. Bjerknes, and R. C. Bundgaard [1975]. *Dynamic Meteorology and Weather Forecasting.* Boston: American Meteorological Society.

Gray, D. M., ed. [1973]. *Handbook on the Principles of Hydrology.* Port Washington, N.Y.: Water Information Center.

Grayman, W. M., and P. S. Eagleson [1970]. *A Review of the Accuracy of Radar and Raingages for Precipitation Measurements.* Cambridge, Mass.: MIT Department of Civil Engineering, Ralph M. Parsons Laboratory. (Technical report no. 119.)

Idem. [1971]. *Evaluation of Radar and Raingage Systems for Flood Forecasting.* Cambridge, Mass.: MIT Department of Civil Engineering, Ralph M. Parsons Laboratory. (Technical report no. 138.)

Greene, D. R., and A. F. Flanders [1976]. "Radar Hydrology—The State of the Art." *First Conference on Hydrometeorology.* Fort Worth, Texas, April 20–22, 1976. Boston: American Meteorological Society, pp. 66–71.

Gunn, K. L. S., and J. S. Marshall [1958]. "The Distribution with Size of Aggregate Snowflakes." *J. Meteorol.* 15:452–461.

Gupta, V. K., and E. C. Waymire [1979]. "A Stochastic Kinematic Study of Subsynoptic Space-Time Rainfall." *Water Resources Res.* 15(3):630–636.

Hershfield, D. [1965]. "On the Spacing of Rain Gages." In: *Symp. Design Hydrol. Networks,* vol. 1, pp. 72–81. (IAHS publication no. 67.)

Hobbs, P. V., S. Chang, and J. D. Locatelli [1974]. "The Dimensions and Aggregation of Ice Crystals in Natural Clouds." *J. Geophys. Res.* 79(15):2199–2206.

Hobbs, P. V., and R. A. Houze, Jr. [1976]. "Mesoscale Structure of Precipitation in Extratropical Cyclones." Preprints International Cloud Physics Conference, July 26–30, 1976, Boulder, Colorado.

Holtan, H. N., N. E. Minshall, and L. L. Harrold [1962]. "Field Manual for Research in Agricultural Hydrology." Washington, D.C.: Agricultural Research Service.

Houze, R. A. [1969]. *Characteristics of Mesoscale Precipitation Areas.* Cambridge, Mass.: MIT Department of Meteorology. (S. M. thesis.)

Huff, F. A. [1970]. "Spatial Distribution of Rainfall Rates." *Water Resources Res.* 6(1):254–260.

Ingraham, D. V. [1980]. *Rainfall Estimation from Satellite Images.* Vancouver, British Columbia: Department of Civil Engineering, University of British Columbia. (Ph.D. thesis.)

Johnson, E. R., E. L. Peck, and T. N. Keefer [1982]. "Combining Remotely Sensed and Other Measurements for Hydrologic Areal Averages." Greenbelt, Md.: National Aeronautics and Space Administration. (Report NASA-CR-170457.)

Johnson, E. R., and R. L. Bras [1980]. "Multivariate Short-Term Rainfall Prediction." *Water Resources Res.* 16(1):173–185.

Klein, W. H. [1957]. "Principal Tracks and Mean Frequencies of Cyclones and Anticyclones in the Northern Hemisphere." Washington, D. C.: U.S. Weather Bureau. (Research paper no. 40.)

Larson, L. W., and E. L. Peck [1974]. "Accuracy of Precipitation Measurements for Hydrologic Modeling." *Water Resources Res.* 10(4):857–863.

Lenton, R. L., and I. Rodriguez-Iturbe [1974]. *On the Collection, the Analysis, and the Synthesis of Spatial Rainfall Data.* Cambridge, Mass.: MIT Department of Civil Engineering, Ralph M. Parsons Laboratory. (Technical report no. 194.)

Idem. [1977]. "A Multidimensional Model for the Synthesis of Areal Rainfall Averages." *Water Resources Res.* 13(3):605–612.

Linsley, R. K., Jr., M. A. Kohler, and J. L. H. Paulhus [1982]. *Hydrology for Engineers.* 3rd ed. New York: McGraw-Hill.

Lintz, T., and D. S. Simonett [1976]. *Remote Sensing of Environment.* Reading, Mass.: Addison-Wesley.

Locatelli, J. D., and P. V. Hobbs [1974]. "Fall Speeds and Masses of Some Precipitation Particles." *J. Geophys. Res.* 79(15):2185–2197.

Lowry, D. A., and H. R. Glahn [1976]. "An Operational Model for Forecasting Probability of Precipitation — PEATMOS PoP." *Monthly Weather Rev.* 104:221–232.

Magono, C., and C. W. Lee [1966]. "Meteorological Classification of Natural Snow Crystals." *J. Fac. Sci.* 7(2):321.

Marsalek, J. [1978]. "Research on the Design Storm Concept." New York: ASCE Urban Water Resources Research Program. (Technical memorandum no. 33.)

Marsh, W. M. [1987]. *Earthscape: A Physical Geography.* New York: Wiley.

Marsh, W. M., and J. Dozier [1986]. *Landscape: An Introduction to Physical Geography.* New York: Wiley.

Marshall, J. S., and W. McK. Palmer [1984]. "The Distribution of Raindrops with Size." *J. Meteorol.* 5:165–166.

Mason, B. J. [1971]. *The Physics of Clouds.* 2nd ed. Oxford: Clarendon Press.

McGinnis, D. F., Jr., J. A. Pritchard, and D. R. Wiesnet [1975]. "Determination of Snow

Depth and Snow Extent from NOAA 2 Satellite Very High Resolution Radiometer Data." *Water Resources Res.* 11(6):897–902.

McGuinness, J.L. [1963]. "Accuracy of Estimating Watershed Mean Rainfall." *J. Geophys. Res.* 68(6):4763–4767.

Neil, J.C. [1953]. *Analysis of 1952 Radar and Raingage Data.* Urbana, Ill.: Illinois State Water Survey Meteorologic Laboratories.

Ohtake, T. [1965]. "Preliminary Observations of Size Distribution of Snowflakes and Raindrops at Just Above and Below the Melting Layer." Proceedings of an International Conference on Cloud Physics, May 24–June 1, 1965, Tokyo and Sapporo, Japan.

Oliver, V.J. [1969]. "Problems of Determining Evapotranspiration in the Semi-Arid Tropics Illustrated with Reference to the Sudan." *J. Tropical Geogr.* 29(1):64–74.

Oliver, V.J., and R.A. Scofield [1976]. "Estimation of Rainfall from Satellite Imagery." Proceedings of the Sixth AMS Conference on Weather Forecasting and Analysis, May 10–13, 1976, Albany, N.Y.

Peck, E.L., and V.C. Bissell [1973]. "Aerial Measurement of Snow Water Equivalent by Terrestrial Gamma Radiation Survey." *Bull. Int. Assoc. Hydrol. Sci.* 18(1):47–62.

Peck, E.L., T.N. Keefer, and E.R. Johnson [1981]. "Strategies for Using Remotely Sensed Data in Hydrologic Models." Greenbelt, Md.: National Aeronautics and Space Administration. (Report NASA-CR-66729.)

Pilgrim, D.H., I. Cordery, and R. French [1969]. "Temporal Patterns of Design Rainfall for Sydney, Australia." *Civ. Eng. Trans. Inst. Eng.* April.

Pruppacher, H.R., and J.D. Klett [1978]. *Microphysics of Clouds and Precipitation.* Boston: D. Reidel.

Raudkivi, A.J. [1979]. *Hydrology: An Advanced Introduction to Hydrological Processes and Modelling.* New York: Pergamon.

Rhenals-Figueredo, A.E., I. Rodriguez-Iturbe, and J.C. Schaake, Jr. [1974]. "Bidimensional Spectral Analysis of Rainfall Events." Cambridge, Mass.: MIT Department of Civil Engineering, Ralph M. Parsons Laboratory. (Technical report no. 139.)

Rodda, J.C. [1969]. "Hydrologic Network Design — Needs, Problems and Approaches." Geneva, Switzerland: World Meteorological Organization. (Report no. 12.)

Rodriguez-Iturbe, I., and P.S. Eagleson [1987]. "Mathematical Models of Rainstorm Events in Space and Time." *Water Resources Res.* 23(1):181–190.

Rodriguez-Iturbe, I., and J.M. Mejia [1974]. "The Design of Rainfall Networks in Time and Space." *Water Resources Res.* 10(4):713–728.

Rogers, R.R. [1979]. *A Short Course in Cloud Physics.* 2nd ed. New York: Pergamon.

Schneider, S.R., D.R. Wiesnet, and M.C. McMillan [1976]. "River Basin Snow Mapping at the National Environmental Satellite Service." National Oceanographic and Atmospheric Administration Technical memorandum NESS 83.

Scofield, R.A., and V.J. Oliver [1977]. "A Scheme for Estimating Rainfall from Satellite Imagery." National Oceanographic and Atmospheric Administration Technical memorandum NESS 86.

Shafrir, U., and M. Neiburger [1963]. "Collision Efficiencies of Two Spheres Falling in a Viscous Medium." *J. Geophys. Res.* 68(13):4141–4147.

Shearman, R.J., and P.M. Salter [1975]. "An Objective Rainfall Interpolation and Mapping Technique." *Hydrol. Sci. Bull.* 20(3):353–363.

Smith, J.L., D.W. Willen, and M.S. Own [1965]. "Measurement of Snowpack Profiles with Radioactive Isotopes." *Weatherwise.* 18(6):247–257.

Sulakvelidze, G. K. [1969]. *Rainstorms and Hail.* Translated from Russian by the Israel Program for Scientific Translations, Jerusalem.

Trewartha, G. T. [1954]. *An Introduction to Climate.* 3rd ed. New York: McGraw-Hill.

U.S. Army Corps of Engineers [1956]. *Snow Hydrology.* Portland, Oreg.: U.S. Army Corps of Engineers North Pacific Division.

U.S. Environmental Data Service [1973]. "Monthly Normals of Temperature, Precipitation, and Heating and Cooling Degree Days." *Climatogr. U.S.* No. 81.

U.S. Weather Bureau [1946]. *Manual for Depth–Area–Duration Analysis of Storm Precipitation.* Washington, D.C.: U.S. Weather Bureau.

Idem. [1947]. *Thunderstorm Rainfall.* Washington, D.C.: U.S. Weather Bureau in Cooperation with U.S. Army Corps of Engineers. (Hydrometeorological report no. 5.)

U.S. National Weather Service [1972]. "National Weather Service. River Forecast System, Forecast Procedures." National Oceanographic and Atmospheric Administration Technical memorandum NWS HYDRO. 14:3.1–3.14.

Wallace, J. M., and P. V. Hobbs [1977]. *Atmospheric Science: An Introductory Survey.* New York: Academic Press.

Waymire, E., V. K. Gupta, and I. Rodriguez-Iturbe [1984]. "A Spectral Theory of Rainfall Intensity at the Meso-β Scale." *Water Resources Res.* 20(10):1453–1465.

World Meteorological Organization [1974]. *Guide to Hydrological Practices.* 3rd ed. No. 168. Geneva: World Meteorological Organization.

Wyss, J. [1988]. "Hydrologic Modelling of New England River Basins Using Radar Rainfall Data." Cambridge, Mass.: MIT Department of Earth, Atmospheric and Planetary Sciences. (M. S. thesis.)

Zawadski, I. I. [1972]. *Statistical Studies of Radar Precipitation Patterns.* Montreal: McGill University Department of Meteorology. (Ph.D. thesis.)

Idem. [1973a]. "Errors and Fluctuations of Raingage Estimates of Areal Rainfall." *J. Hydrol.* 18:243–255.

Idem. [1973b]. "Statistical Properties of Precipitation Patterns." *J. Appl. Meteorol.* 12:459–472.

Zotimov, N. V. [1968]. "Investigation of a Method of Measuring Snow Storage by Using the Gamma Radiation of the Earth." *Sov. Hydrol. Sel. Pap.* 3:254–265.

PROBLEMS

1. What is the minimum diameter required for a droplet to survive evaporation under the following conditions: average ground pressure = 1015 mb; average temperature to cloud base = 20°C; relative humidity = 65%; and cloud base at a height of 1000 m?

2. The correlation function of rainfall accumulation at a site starts oscillating at separation distances of 100 km. The area is 5000 km^2. There are 10 stations in the area as well as a radar. What is the expected relative increase in accuracy of using the radar/rain gage system versus using the radar alone?

3. The longwave radiation from the top of a cloud is measured as 0.29 cal cm^{-2} min^{-1} by an environmental satellite. The cloud reaches 10 km in height from ground level and the cloud base is at a 1-km elevation. The mixing ratio of upflowing air is 0.015 $g\,g^{-1}$. In two consecutive pictures half an hour apart the top of the cloud area expands from 200 to 250 km^2. Obtain a reasonable estimate of the precipitation rate. State your assumptions.

4. A hypothetical river basin can be approximated by a 100-by-25-km rectangle. The homogeneous, isotropic correlation function of rainfall accumulation is estimated as

$\bar{v}$ (km)	$r(v)$
0	1
10	0.80
25	0.56
100	0.30
150	0.03

a) Estimate how many rain gages you would need to compute the mean areal precipitation of an event with a mean square error that is 4% of the point variance.

b) Assume that the lag-one annual correlation of rainfall accumulation is 0.2 and that the same spatial correlation is applicable. If you use 10 rain gages, how many years of data would you need to achieve the same accuracy as in part (a) when computing the long-term areal mean of an event?

5. It costs $2,000 per year, annualized, to place and maintain a normal rain gage in a basin 2000 km^2. The rainfall accumulation in the area obeys an exponential correlation function with parameter $h = 0.15\ km^{-1}$. The point variance of a precipitation event is 100 mm^2. As a decision maker, you feel that a unit reduction in mean square error in your estimate of the mean areal precipitation of an event is worth $150.00 annually. What is the number of stations you would locate in the basin if you want to minimize total cost to you?

6. The updraft velocity in a cloud system is related to the pressure and temperature differentials between the bottom and top of the cloud. A reasonable parameterization for updrift velocity (in kilometers per hour) is

$$V_{up} = K\sqrt{\Delta z},$$

where Δz is the difference in elevation between cloud top and base (in kilometers) and $K = 7$. Assume that the raindrop-size distribution is uniform throughout the height of the cloud and is given by Eq. (4.2).

$$N(D) = N_{o}e^{-cD}.$$

With $N_{o} = 1.05\ cm^{-4}$,

$$c = 2\ mm^{-1}.$$

The terminal velocity of drops can be estimated from

$$V_T(D) = \alpha D$$

with $\alpha = 3300\ \text{s}^{-1}$.

A satellite measures the cloud-top temperature as 5°C. The mixing ratio of incoming air at the cloud base is 14 g kg^{-1} and the surface air temperature (sea level) is 25°C.

Estimate a precipitation rate if the "efficiency" of the cloud is 0.2. The solution requires use of the following fact:

$$\int_0^\infty e^{-\alpha x} x^n \, dx = \frac{n!}{\alpha^{n+1}}.$$

7. Total rainfall accumulation from a single storm over an area obeys an exponential correlation function with parameter 0.15 km^{-1}. The point variance (i.e., variance of total depth at any point in the basin) is 100 mm^2. The basin area is 2000 km^2. The "cost" of collecting data to you is a function of the total cost of the rain gages (annualized, or per year) plus $100.00 per every 1 mm^2 of mean square error. If your annualized cost per station is $1,000.00, how many stations (assuming random sampling) would you locate in the basin so as to minimize total cost of estimating the mean areal precipitation of a rainfall event?

8. Consider the rectangular area below. In a particular month precipitation at stations 1, 2, and 3 is $p_1 = 8$ in., $p_2 = 4$ in., $p_3 = 12$ in. A river goes through the area as shown. At stations A and B monthly flows are 4200 acre-ft. and 30,000 acre-ft., respectively. The area is much larger than any existing aquifer and has negligible water-storage capacity. Determine the evapotranspiration in the area during that month. Explain and justify any assumptions. If the spatial correlation function of monthly rainfall is $r(v) = e^{-0.2v}$ (v in miles) and the variance of monthly rainfall at a point is 16 in^2, estimate a reasonable upper limit of the relative variation in your evapotranspiration estimate. (*Hints:* Relative variation = standard deviation/estimated value; there are 2.29×10^{-5} acres per square foot; there are 5280 feet per mile.)

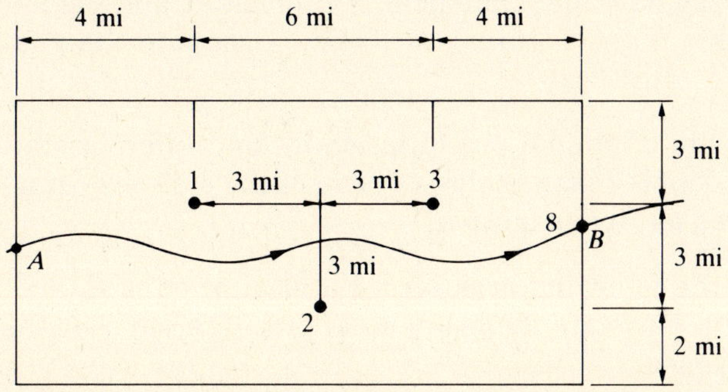

9. Out of a long-term record of precipitation for four adjacent stations you find that records of one station are missing for the summer months, June, July, and August of a given year. For those three months the other three stations recorded the following total depth in millimeters.

	STATION		
	1	2	3
June	55	65	75
July	47	50	45
August	45	40	55

Estimate the missing precipitation if the long-term average precipitation for the three months in all four stations is

	STATION			
	1	2	3	4*
June	60	65	70	67
July	50	55	65	60
August	45	47	60	55

*Station with missing records.

10. An alternative way of estimating a missing record is by weighting observations by the inverse square distance to the point of the missing record. The expression would be

$$P_\circ = \frac{\sum_{i=1}^{N} \frac{1}{r_{\circ i}^2} P_i}{\sum_{i=1}^{N} \frac{1}{r_{\circ i}^2}},$$

where $r_{\circ i}$ is the distance between station i and station $\circ$, which is the one missing a record. Using the data and statement of Problem 9, and the following distances—$r_{41} = 5$ km, $r_{42} = 1$ km, $r_{43} = 10$ km—compute the missing records.

11. Locate at least 15 years of annual precipitation in four different stations in your area. Perform a double-mass analysis of one of the stations using the average of the other three as "index station."

12. The three points in the following area record precipitation of $P_1 = 8$ in., $P_2 = 4$ in., and $P_3 = 12$ in. Estimate the mean areal precipitation over the region using Thiessen polygons.

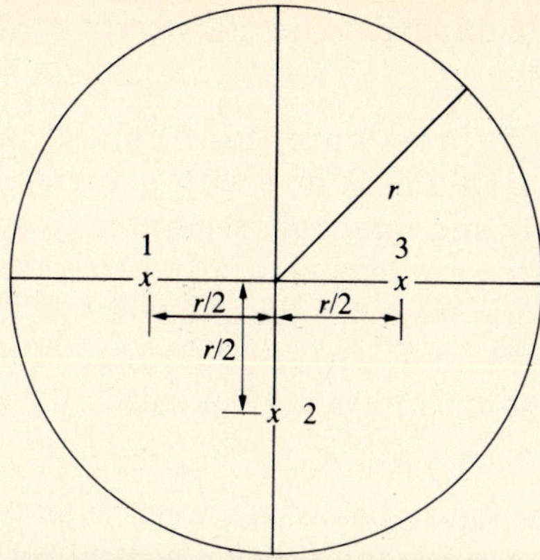

13. Using the contour method compute the mean areal precipitation depth of the storm shown in Figure 4.26.

14. Compute the average rainfall for November in millimeters of depth for the Arno Basin at Florence, Italy, by (a) arithmetic mean, (b) the Thiessen method, and (c) the isohyetal method using the figure below. Comment on the applicability of each method. Given that the Mediterranean is on the west (left in the figure) side of the basin and the elevation increases from left to right, can you explain the variability in the precipitation pattern over the basin?

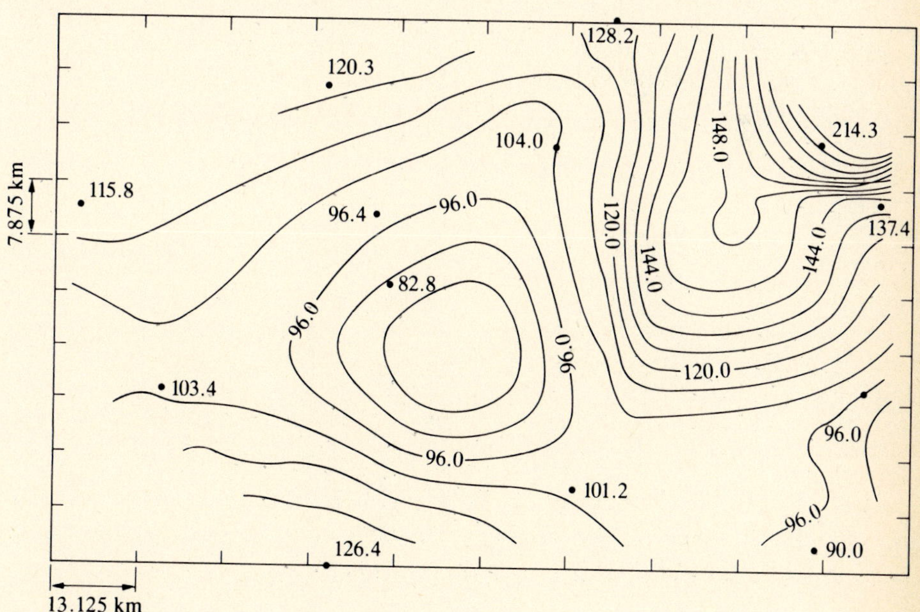

15. Discuss the relationship between depth, duration, and area of rainfall for particular storms.

16. Enumerate the relative advantages and disadvantages of different rainfall-measuring technologies.

17. In an area about 90,000 km^2 in Egypt, there are three annual mean evaporation measurements of 237, 233, and 239 W m^{-2} with a variance of 1600 (W m^{-2})^2. The three points are widely and randomly separated in space. It is estimated that evaporation estimates are exponentially correlated in space with a correlation coefficient ranging from a high of 0.005 km^{-1} to a low of 0.001 km^{-1}. Based on these two extremes, what is the range of mean square error you expect in computing the mean areal evaporation over the whole region?

18. Assume that the gages shown in the figure below are used to estimate mean areal precipitation of events. Also assume they obey a spatial correlation of the exponential type with parameter h = 0.1 km^{-1}. The total area is 500 km^2. The point variance of rainfall is given as 9 in^2. What is the mean square error (MSE) of estimating areal precipitation using the random sampling technique? Assume now that the network is to be used in obtaining long-term annual averages. Serial correlation between annual values is 0.2. How many years of data are needed from the given network (assuming random sampling) to reduce the MSE to 10% of point variance? Why is the random sampling assumption not really applicable in this case?

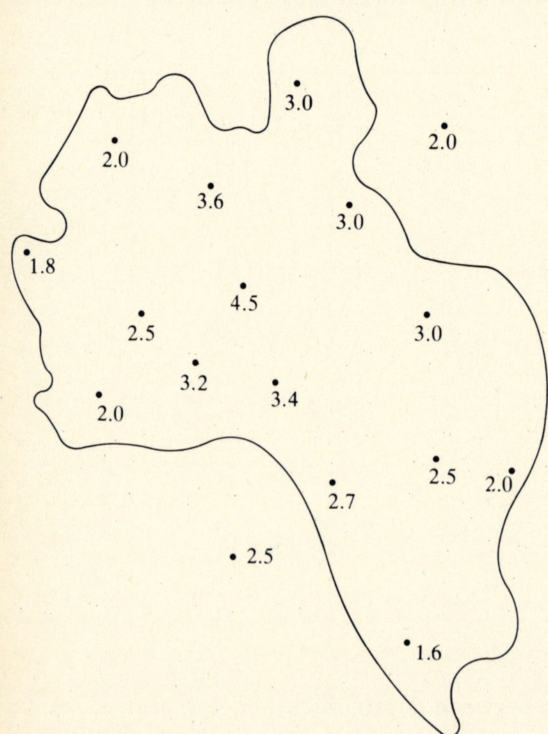

19. Using your knowledge of the types of precipitation and the ways that precipitation occurs, suggest a rain-making machine. Use your imagination! Justify your invention.

20. Using the information from Problem 18, obtain the areal average using two of the formulas relating storm center to mean areal precipitation. Obtain the Thiessen and isohyetal averages.

21. You have a circular area of radius $r = 3$ km. Isohyets are also circular, with centers coinciding with that of the total area. Isohyets correspond to 3 cm at 1-km radius from the center; 2 cm at 2 km; and 1 cm at 3 km. Estimate the mean precipitation over the area.

22. Draw the Thiessen polygons for the station configuration shown below.

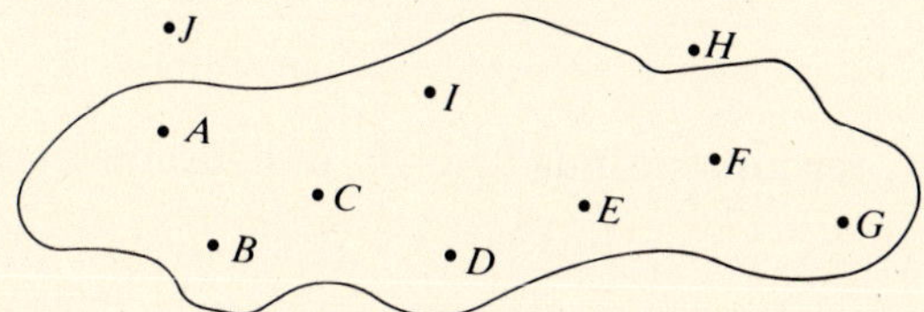

The following information corresponds to the figure.

STATION	THIESSEN AREAS (km^2)	PRECIPITATION (cm)
A	72	3.50
B	34	4.46
C	36	4.28
D	40	5.90
E	76	6.34
F	92	5.62
G	46	5.20
H	40	5.26
I	86	3.86
J	6	3.30

Obtain the average precipitation over the area.

23. An area of 100 km^2 is gaged by three stations as shown in the following figure. The area is a square 10 km by 10 km. After the occurrence of a large cyclonic event, stations 1, 2, and 3 recorded total depths of 106, 152, and 127 mm, respectively. Estimate the mean areal precipitation using the Thiessen polygons method. Assume that cyclonic storms in the region have an exponential correlation function in space with coefficient 0.1 km^{-1}. Estimate a lower bound of the net accuracy (reduction in MSE) *gained* if two additional stations were added. The point variance of total precipitation of an event is 2000 mm^2.

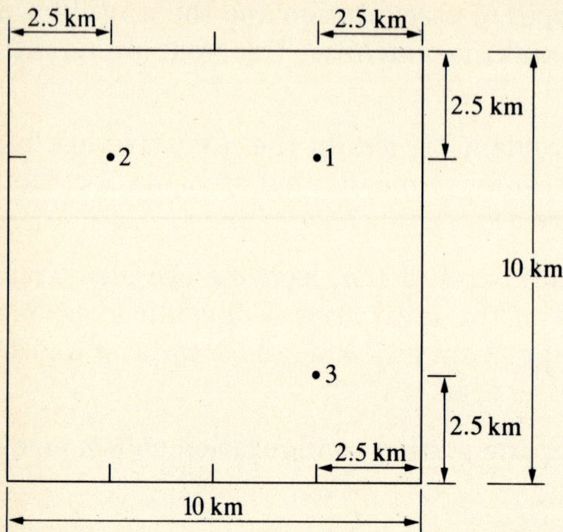

24. Assume that the net velocity, terminal minus updraft, of hydrometeors is of the form

$$V_0(1 - e^{-cD}),$$

where D is diameter; V_0 and c are constants. Assume that the constant c is the same as that appearing in the exponential hydrometeor distribution function. At what value of cD is the rate of mass precipitation per unit volume of air maximized? Using the data on terminal velocities from Figure 4.8 (use curve 1), would the implied values of c make sense? Is this result consistent with the arguments presented in this chapter?

Chapter **5**

Evaporation, Transpiration, Interception, and Depression Storage

5.1 INTRODUCTION

Evaporation refers to the change of water from its liquid to its vapor phase. In the space scales of hydrologic processes, it usually refers to an area-averaged vertical flux of vapor. Evaporation may occur from water bodies, saturated soils, or from unsaturated surfaces. Potential evaporation is the climatically controlled rate of evaporation from a given surface that occurs when the amount and rate of supply of water to the surface is essentially unlimited.

Transpiration is the evaporation occurring from plants' leaves through stomatal openings. Again, given an unlimited rate of water supply at the root zone, the potential transpiration would be a function of climate and, to some extent, plant physiology. The actual transpiration under water-limiting conditions would depend on the ability of the plant to extract moisture from partially saturated soil with limited ability to transmit water.

In science and engineering we frequently use the term evapotranspiration. This is the sum total of evaporation and transpiration. The term responds to the difficulties in separating the two phenomena in the usual situation where vegetative cover is not complete. In such cases transpiration occurs from plants or trees, while evaporation is occurring from soil, water, or other surfaces.

As implied in Eq. (3.4), evaporation is a function of the gradient of vapor-pressure concentration in the vertical direction. That equation came from a turbulent transport analogy to molecular diffusion. In fact, in perfectly still air (and no thermal convection) evaporation would indeed occur as molecular diffusion, at least until the vapor gradient disappears, when evaporation would cease. This closed-container analogy of evaporation, leading to zero evaporation when the air in the container becomes saturated, was described in Section 3.7.2 while defining vapor pressure. In nature, flows are generally turbulent and the transport of vapor is turbulent, as stated by Eq. (3.4). Evaporation in a turbulent medium is a function of the vapor eddy diffusivity as well as of the vapor gradient. This factor is dependent on the degree of turbulence and is in turn largely a function of climatic conditions, mostly wind. We will detail this dependence when we study the Dalton analogy or mass-transfer approach to evaporation in Section 5.2.3.

Evaporation and transpiration can also be obtained by direct or indirect measurement. Direct methods are mostly dominated by point sampling or integrated measurements over small areas, mostly with evaporation pans or lysimeters. These will be described in Section 5.2.6. Indirectly, evaporation or evapotranspiration can be measured by performing a water balance of the region in question, as discussed in Chapter 1. Since latent heat is required for evaporation, an alternative is to do an energy balance of the region in order to calculate evaporation or evapotranspiration indirectly. These last two procedures will be discussed in Sections 5.2.1 and 5.2.2, respectively.

The discussions on evaporation to follow will be grouped in two main sections. Section 5.2 deals with evaporation from free water surfaces. Section 5.3 will concentrate on evapotranspiration, largely from the point of view of agricultural water needs.

Sections 5.4 and 5.5 will describe the processes of interception and surface retention, which in turn provide an opportunity for increased evaporation from a region.

Before entering into detailed descriptions of the evapotranspiration process, we should emphasize its importance in the hydrologic cycle. In Figure 1.2 we saw that 57% of all land precipitation evaporates while oceans evaporate 112% of directly received rainfall. Table 2.2 also indicated that latent heat of condensation and evaporation played a major role in global energy balance. In a semi-arid climate, about 96% of annual precipitation may evaporate (Branson [1976]). Figure 5.1 gives estimates of mean annual evapotranspiration throughout the world. Figure 5.2 shows the mean annual evaporation from shallow lakes and reservoirs over the contiguous United States. Daily evapotranspiration may vary significantly. Figure 5.3 shows the frequency distribution of daily evapotranspiration throughout the year in Davis, California, over a 10-year period (Pruitt et al. [1972]). Values range from about 0 mm to almost 12 mm per day. During a storm, evaporation is reduced to a minimum because of the saturated air conditions; nevertheless, evapotranspiration between storms is usually sufficient to deplete added

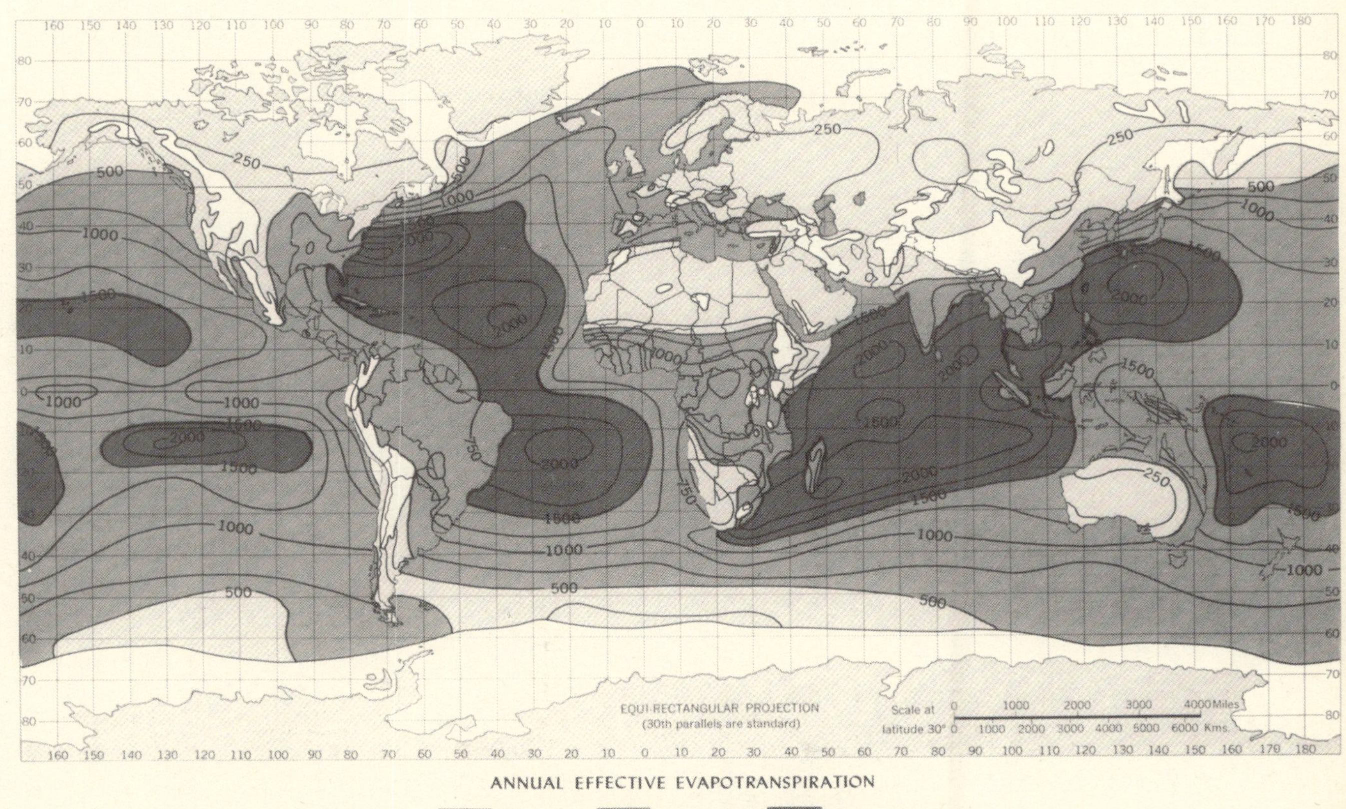

FIGURE 5.1 World distribution of average annual effective evapotranspiration. Source: G. T. Trewartha and L. H. Horn, *Introduction to Climate*, 5th ed., McGraw-Hill, 1980. Reproduced with permission.

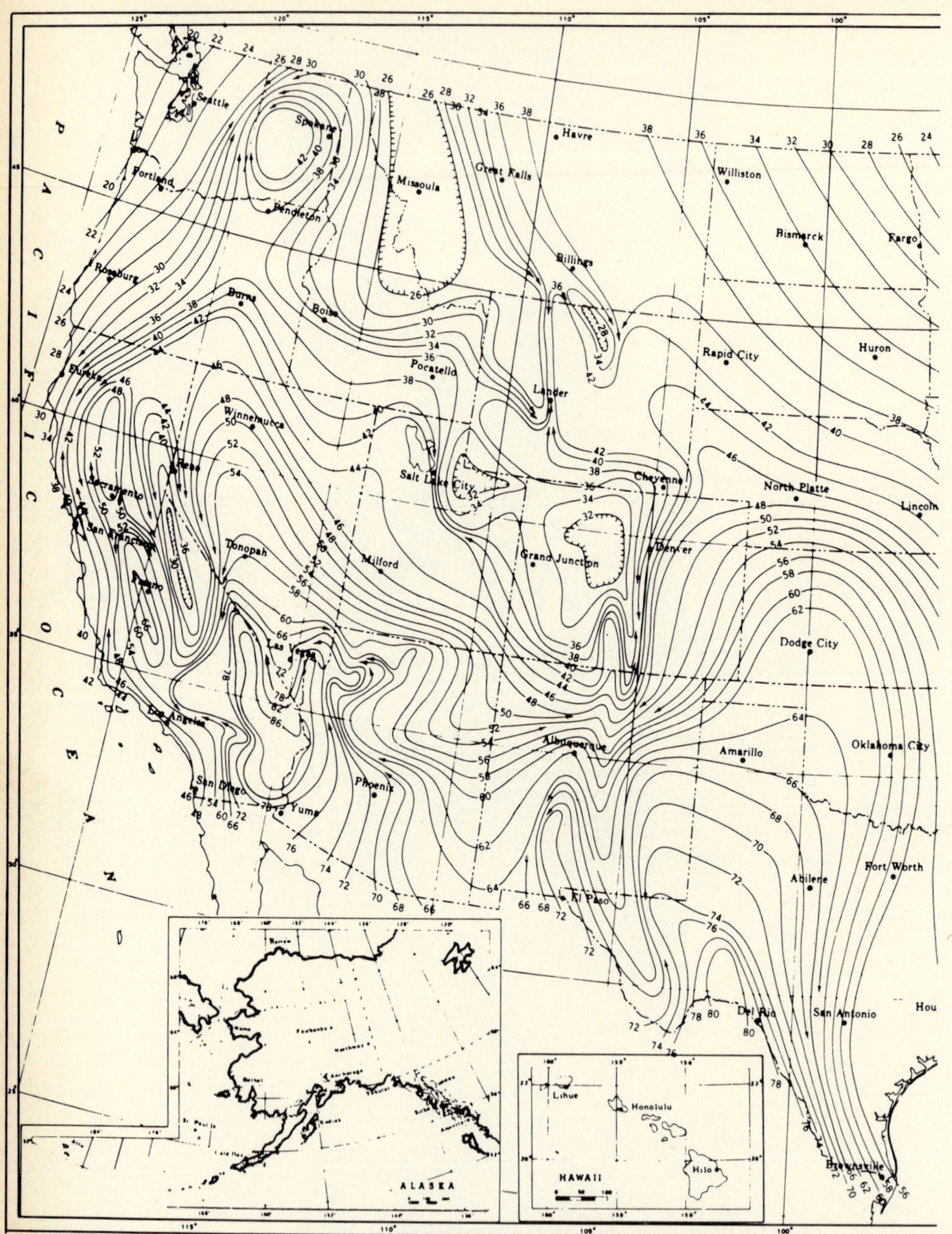

FIGURE 5.2 Average annual evaporation (inches) from shallow lakes. Source: U.S. National Weather Service.

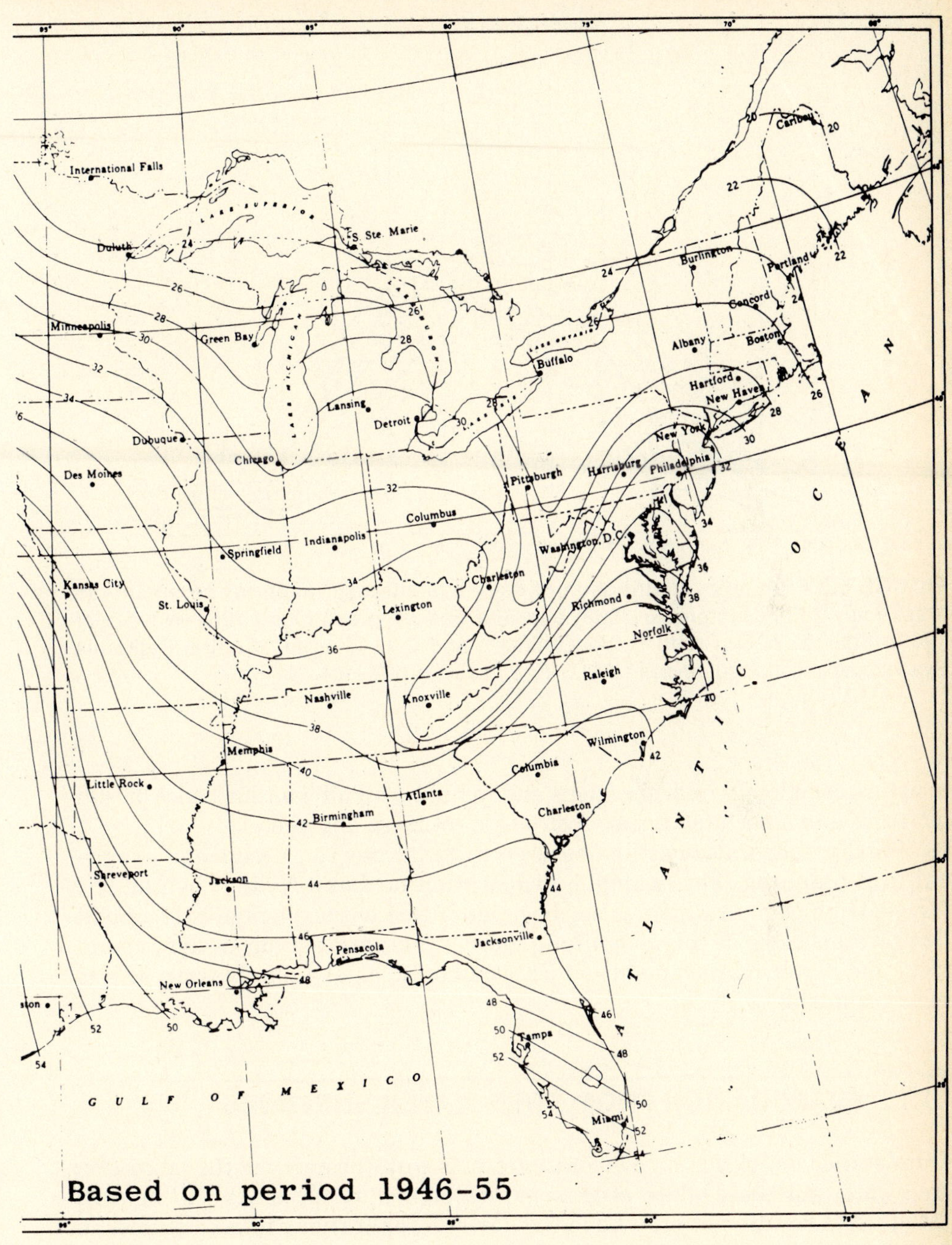
International Falls
Duluth
Minneapolis
Green Bay
S. Ste. Marie
Lake Superior
Lake Michigan
Lake Huron
Lake Ontario
Lake Erie
Lansing
Detroit
Buffalo
Dubuque
Chicago
Des Moines
Springfield
Indianapolis
Columbus
Pittsburgh
Harrisburg
Philadelphia
New York
New Haven
Hartford
Albany
Boston
Concord
Portland
Burlington
Caribou
Kansas City
St. Louis
Charleston
Washington, D.C.
Lexington
Richmond
Norfolk
Raleigh
Nashville
Knoxville
Memphis
Little Rock
Wilmington
Columbia
Atlanta
Birmingham
Charleston
Shreveport
Jackson
Pensacola
Jacksonville
New Orleans
Tampa
Miami
GULF OF MEXICO
ATLANTIC OCEAN
Based on period 1946-55

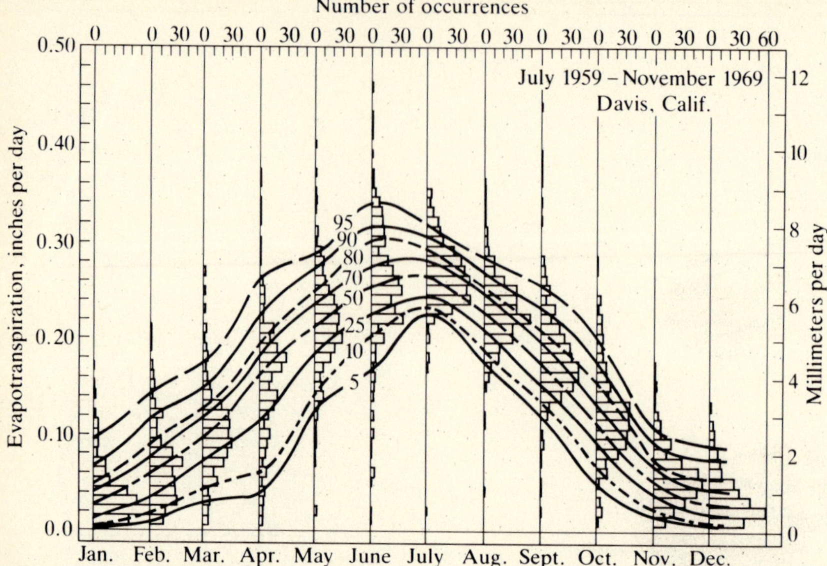

FIGURE 5.3 Frequency distribution of daily evapotranspiration for perennial rye grass (July 1959–September 1963) and Alta Fescue grass (June 1964–November 1969). Source: W. O. Pruitt, S. VonOetinger, and D. L. Morgan, "Central California Evapotranspiration Frequencies," *ASCE Proc*. 98(IR2): 177–184.

moisture completely in arid regions and to have a significant influence on soil moisture and associated future hydrologic response anywhere.

Finally, evaporation estimates are critical in reservoir design and in agricultural planning. For example, evaporation on Lake Nasser, behind the Aswan High Dam in Egypt, is on the order of 15% of the average Nile annual flow. Water losses of this magnitude must influence reservoir design, operation, and water management, which commonly, as at Lake Nasser, affects many countries.

5.2 EVAPORATION FROM FREE WATER SURFACES

Our detailed discussion on evaporation starts with the study of the process on free water surfaces: lakes, oceans, saturated soil surfaces. As will be seen later, these results not only have the obvious applications to lake studies in hydrology, but will have implications in potential evapotranspiration estimates. The important assumption in this section is that water must be available in unlimited amount and rate of supply for evaporation.

5.2.1 Water Balance Method

This method was discussed in Chapter 1 during the discussion of the hydrologic cycle. The hydrologic budget equation written there included an evaporation term. Essentially, if for a given water body, the hydrologist knows the precipitated volume of water, the surface inflow and outflow, the infiltration and exfiltration, and the change in storage over a given period of time, then he or she would be able to calculate the net evaporated and transpired water. This exercise may be possible when applied in well-defined and well-monitored regions and when averaging over relatively long periods of time. A reasonable averaging period for using this technique is one year. Monthly estimates are seriously affected by uncertainty in seasonal values of the elements in the hydrologic budget equation. Daily estimates of evaporation were attempted in the well-documented Lake Hefner, Oklahoma, study (Harbeck et al. [1951]; Harbeck [1954]) with estimated accuracies in the range of 5 to 10%. However, this is an unusual situation. The estimation errors become very large as the magnitudes of the difficult-to-measure terms in the hydrologic balance equation get to be on the order of the evaporation. In summary, the method is difficult and inaccurate under most conditions, particularly for short averaging time periods. Some of the hardest terms to measure are change in storage, seepage, groundwater flow, and advected flows.

EXAMPLE 5.1

Evaporation from Lake Nasser

It was previously stated that evaporation in Lake Nasser is on the order of 15% of the Nile annual flow. This is about 13×10^9 m^3 of evaporated water annually. We can use the density of water and the latent heat of evaporation to convert this quantity to equivalent energy units. The density of water is $10^6\ \mathrm{g\,m^{-3}}$. The latent heat is approximately $600\ \mathrm{cal\,g^{-1}}$ or about $2500\ \mathrm{J\,g^{-1}}$ (there are 4.186 joules per calorie). Therefore the energy expended in evaporation is

$$Q = (13 \times 10^9\ \mathrm{m^3})(10^6\ \mathrm{g\,m^{-3}})(2500\ \mathrm{J\,g^{-1}})$$
$$= 3.25 \times 10^{19}\ \mathrm{J\,yr^{-1}}.$$

There are about 31.536×10^6 seconds in a year; therefore

$$Q = 1.03 \times 10^{12}\ \mathrm{J\,s^{-1}}\ \text{or watts.}$$

A megawatt consists of 10^9 W, so the energy equivalent is

$$Q = 1.03 \times 10^3\ \mathrm{MW}.$$

A relatively large power plant will produce 10^3 MW. In terms of mass, the volume lost to evaporation is about 24% of the actual annual water consumption of about 55×10^9 m^3. ◆

5.2.2 Energy Balance Method

Following the same ideas as in the mass balance, an energy balance can be written for a given area, particularly for a defined water body such as a lake. The equation looks like this:

$$Q_o = Q_s - Q_r + Q_a - Q_{ar} - Q_{bs} + Q_v - Q_e - Q_h, \tag{5.1}$$

where Q_o is the change in system energy; Q_s is the incident solar radiation; Q_r is the reflected solar radiation; Q_a is the incoming longwave atmospheric radiation; Q_{ar} is the reflected longwave radiation; Q_{bs} is the longwave radiation emitted by the water body because of its temperature; Q_v is the net energy advected by moving water; Q_e is the energy used in evaporation; and Q_h is the heat removed from the system into the air as sensible heat.

All terms in the above equation must be given in common units such as calories per unit time, calories per unit area (square centimeters) per unit time, or watts per unit area. The equation ignores chemical and biological heating processes, heat lost by conduction through the bottom of the water body, as well as explicit accounting of possible energy transfers during any condensation.

All the radiation terms can be determined from the equations discussed in Chapter 2 (or others). Preferably, though, radiation is measured at the site of interest. Radiometers exist to measure total incoming or net shortwave and longwave radiation.

In natural water bodies the term Q_h is mostly due to sensible-heat transfer through the water-body surface. This ignores heat fluxes through the ground. The sensible-heat term is difficult to measure; hence, Q_h is commonly related to the energy used in evaporation. Bowen [1926] suggests the use of a ratio of Q_h to Q_e.

$$R = \frac{Q_h}{Q_e} = (0.61 \times 10^{-3}\,{}^\circ\mathrm{C}^{-1})P\frac{(\overline{T}_s - \overline{T}_z)}{(\overline{e}_s - \overline{e}_z)}, \tag{5.2}$$

where P is ambient atmospheric pressure at the water surface; $\overline{T}_s$ is the mean water surface temperature; $\overline{T}_z$ is the mean air temperature at elevation z; $\overline{e}_s$ is saturation vapor pressure at temperature $\overline{T}_s$; and $\overline{e}_z$ is the mean vapor pressure at elevation z. The origin of the Bowen ratio will be explained later.

Using the Bowen ratio, the energy balance equation becomes

$$Q_e(1 + R) = \overline{Q} + (Q_v - Q_o), \tag{5.3}$$

where $\overline{Q}$ is the net radiation (shortwave and longwave). Solving for Q_e in Eq. (5.3) leads to

$$Q_e = \frac{\overline{Q} + (Q_v - Q_o)}{1 + R}, \tag{5.4}$$

where $\overline{Q} = Q_s - Q_r + Q_a - Q_{ar} - Q_{bs}$.

Evaporation is generally expressed in terms of volume per unit area (depth),

$$E = \frac{Q_e}{\rho L_e}, \tag{5.5}$$

where E is evaporation in centimeters of water per day; Q_e is the energy used in evaporation in calories per square centimeter per day; ρ is the density of water in grams per cubic centimeter; and L_e is the latent heat of evaporation in calories per gram.

Depth of evaporation then becomes

$$E = \frac{\overline{Q} + (Q_v - Q_o)}{\rho L_e(1 + R)}. \tag{5.6}$$

Still undefined in Eq. (5.6) are the energy advection term Q_v and the storage term Q_o. To obtain these we need an estimate of the water balance. It should be emphasized that in performing the energy balance, care must be taken to guarantee that mass is conserved.

The advection and storage terms in Eq. (5.6) can be obtained as

$$Q_v - Q_o = \frac{C_p}{\Delta t}(\rho_i V_i T_i + \rho_p V_p T_p - \rho_d V_d T_d - \rho_b V_b T_b - \rho_2 V_2 T_2 + \rho_1 V_1 T_1 - \Delta t \frac{Q_e}{L_e} T_s), \tag{5.7}$$

where C_p is the specific heat at constant pressure of water (≈ 1 cal g^{-1}°C^{-1}); ρ_i is the density of component i; V_i is the volume of component i; and T_i is the absolute temperature of component i. The subscripts are as follows: i, surface inflow; p, precipitation; d, surface outflow; b, seepage flow; 1, water body at the beginning of time interval Δt; and 2, water body at the end of time interval Δt. The last term in Eq. (5.7) accounts for the volume of water removed by evaporation. L_e is the latent heat of evaporation and T_s is a representative water-surface temperature (may be the average). All temperatures are in degrees Centigrade, implying that a reference temperature of 273°K is assumed.

In Eq. (5.7) the temperature of precipitation may be taken as the wet-bulb temperature and that of seepage flow as the temperature of the lowest levels of the water body. If temperature profiles of the water body are available, as well as relationships between water depth and surface area, it may be possible to deal more accurately with the change in storage terms. The depth of the body of water can be discretized in n layers and energy content is calculated for each depth interval:

$$Q_o = \frac{C_p}{\Delta t}\left(\sum_{i=1}^{N} \rho^i \overline{A}{}_1^i T_1^i \Delta h_1^i - \sum_{i=1}^{N} \rho^i \overline{A}{}_2^i T_2^i \Delta h_2^i\right), \tag{5.8}$$

where superscript i refers to the ith layer; Δh_1^i refers to the thickness of layer i at the beginning of the period (subscript 1); and $\overline{A}{}_1^i$ is the average horizontal area covered by layer i at the beginning of the period. For the above technique to be useful, temperature profiles must be available at different locations and the equation preferably applied for subregions. Subregions should be grouped by mean depths, assuming all regions with similar depth will have similar temperature profiles and heat content.

In cases where mass balance is not very accurate, it is possible to attribute a temperature and hence a heat content to a mass closure error.

The energy balance procedure has been used successfully to estimate evaporation on a weekly, monthly, or longer time scale (Kohler [1954]; Harbeck et al. [1958]). It is not recommended for shorter time periods when used in a lumped system as described here. In most cases, its accuracy is higher than that of water balance procedures. As previously stated, radiation measurements are preferred to empirical estimates. Gunaji et al. [1965] have pointed out that results may be more sensitive to longwave radiation estimates than to other terms in the radiation balance. This may simply be a reflection of the importance that the water temperature has throughout the formulation, including its effect on the Bowen ratio. This sensitivity to water temperature may be much less important when the method is used within a context of a dynamic water temperature model, as will be discussed in the following list of advantages and disadvantages of the energy balance method (Helfrich et al. [1982]).

Advantages

1. With good measurements it is usually quite accurate (<10% error).
2. The concept can be implemented and numerically solved to study the spatial nonuniform distribution of temperatures in a lake. This is the basis of hydrothermal models that combine hydraulic and thermal properties of water bodies.
3. With a hydrothermal model the energy balance equations can be solved for short time steps. This dynamic use of the energy balance with a model that computes surface temperatures leads to compensating errors in the

evaporation computation. Too large a surface temperature increases long-wave back-radiation, which in turn works to reduce temperature in the next time period.

Disadvantages

1. Relies on the Bowen ratio concept, and is therefore dependent on the validity of this idea during unstable boundary-layer conditions (see Section 5.2.3).
2. Accuracy may be compromised by use of empirical and/or analytical expressions used to obtain the radiation terms.
3. The time scale of application, when used for a lumped system, should not be less than one week. Results obtained should not be used in computations of time scales smaller than that of original calibration data.

EXAMPLE 5.2

Evaporation Using the Energy Balance Method

Examples 2.1, 2.2, and 2.3 established some hypothetical conditions for a January day in Boston. Let us assume that those conditions are representative of daytime hours for the month of January. What would be the average, daily, shallow (well-mixed) lake evaporation at Boston? Following are a recompilation of the past examples' results and some other necessary data.

MONTHLY AVERAGES	BOSTON
Clear skies, shortwave radiation	0.52 ly min^{-1}
Solar altitude angle, α	24.83 degrees
Pressure over lake	1000 mb
Water-surface temperature T_s	5°C
Air temperature T_z	3°C
Relative humidity at same elevation as T_z	60%

We are to estimate evaporation using the energy budget method (Eq. 5.6). The Bowen ratio was given by

$$R = (0.61 \times 10^{-3}\,°\mathrm{C}^{-1})P\frac{(\overline{T}_s - \overline{T}_z)}{(\overline{e}_s - \overline{e}_z)}.$$

The saturation vapor pressure at temperature T_s can be obtained from Table A.1, in Appendix A. For $T_s = 5°\mathrm{C}$, $e_s \approx 8.72$ mb. At $T_z = 3°\mathrm{C}$, the saturation vapor pressure is 7.59 mb, which, when multiplied by the relative humidity, results in 4.55 mb for e_z. The Bowen ratio then becomes

$$R = (0.61 \times 10^{-3})1000\frac{(5 - 3)}{(8.72 - 4.55)} = 0.29.$$

The specific heat of water at constant pressure is 1.0 cal g^{-1}°C^{-1} at 14.5°C; this value will be used as an approximation at all temperatures for this example. Only at below freezing temperatures is this approximation in serious error.

We will assume that the shallow lake is well mixed (uniform vertical temperature distribution) and that its temperature does not change much during the month. The assumptions lead to no heat-storage changes ($Q_o = 0$). Furthermore, we will assume that water inputs and outputs are in balance and have essentially the same temperature, such that $Q_v = 0$. The energy budget simplifies to

$$Q_e = Q_s - Q_r + Q_a - Q_{ar} - Q_{bs} - Q_h .$$

For Boston, incoming clear-sky shortwave radiation was obtained in Chapter 2 as 0.52 ly min^{-1}. To obtain Q_r we require an estimate of the albedo. For a declination of 24.83 and clear skies, Figure 2.12 gives $A \approx 0.09$, so $Q_r = 0.05$ ly min^{-1}.

The longwave emissivity of the atmosphere is a function of its vapor pressure. Equation (2.35) gave one such relationship. With $e = 4.55$ mb,

$$E_a = 0.740 + 0.0049e = 0.76 .$$

The longwave albedo of water is about 0.03. Therefore

$$Q_a - Q_{ar} = 0.97 E_a \sigma T_z^4$$

with $T_z = 276.15$°K and $\sigma = 0.826 \times 10^{-10}$ cal cm^{-2} min^{-1} °K^{-4}. The above becomes

$$Q_a - Q_{ar} = 0.35 \text{ ly min}^{-1}.$$

The water radiates as a black body, so

$$Q_{bs} = \sigma(278.15)^4 = 0.49 \text{ ly min}^{-1}.$$

At the water temperature $T_s = 5$°C, and using Eq. (3.19)

$$L_e = 597.3 - 0.57T = 594.5 \text{ cal g}^{-1}.$$

With a water density of 1 g cm^{-3}, Eq. (5.6) can now be evaluated:

$$E = \frac{0.52 - 0.05 + 0.35 - 0.49}{1[594.5(1 + 0.29)]} = 4.3 \times 10^{-4} \text{ cm min}^{-1}.$$

If we assume most evaporation does occur during daytime hours and that in January we have about nine hours of sunshine (see Table 5.9), the implied daily evaporation is $E = 0.23 \text{ cm day}^{-1}$. ◆

The Bowen Ratio

Recall the turbulent transport equations of Chapter 3. Evaporation was given as

$$E = -\rho_a K_w \frac{\partial \overline{q}_w}{\partial z}. \tag{5.9}$$

Turbulent heat transport can be similarly expressed as

$$Q_h = -\rho_a C_{p_a} K_h \frac{\partial \overline{T}}{\partial z} \tag{5.10}$$

and momentum transfer as

$$\tau = \rho_a K_m \frac{\partial \overline{U}}{\partial z} = \rho_a U_*^2, \tag{5.11}$$

where ρ_a is moist air density; K_w, K_h, and K_m are turbulent eddy diffusivities for mass, heat, and momentum, respectively; $\overline{q}_w$ is the mean specific humidity (see Chapter 3); $\overline{T}$ is the mean air temperature; z is the vertical coordinate; C_{p_a} is the specific heat of air at constant pressure; and U_* is the apparent "friction velocity." From Eq. (5.11)

$$K_m = U_*^2 \Big/ \left(\frac{\partial \overline{U}}{\partial z}\right). \tag{5.12}$$

Equation (5.10) can be multiplied and divided by K_m to yield

$$Q_h = -\rho_a C_{p_a} \frac{K_h}{K_m} U_*^2 \frac{\partial \overline{T}}{\partial \overline{U}}. \tag{5.13}$$

To proceed further something must be said about the velocity profile in turbulent flow. As air flows over water, a boundary layer forms where the relative velocity at the contact point of air and water is 0 and the mean velocity increases with elevation until the point where boundary effects are lost. This boundary layer can be several hundred meters thick. If a layer develops fully and is stable (surfaces and conditions are not changing), the velocity profile is logarithmic, following the general form

$$\overline{U}(z) = \frac{U_*}{C_1} \ln(z/z_\circ) + C_2. \tag{5.14}$$

Generally, $C_1 = \kappa \approx 0.4$ and $C_2 = 0$, where κ is called the Von Karman's constant. z_o is the roughness height, a characteristic elevation of the surface at which point the velocity is effectively zero.

Equation (5.13) can be approximated in finite differences as

$$Q_h = \rho_a C_{p_a} \frac{K_h}{K_m} U_*^2 \frac{\overline{T}_1 - \overline{T}_2}{\overline{U}_2 - \overline{U}_1}, \tag{5.15}$$

where subscripts 1 and 2 indicate measurements at two heights, and subscript 2 stands for the higher point. Evaluating Eq. (5.14) at two points and subtracting leads to

$$\overline{U}_2 - \overline{U}_1 = \frac{U_*}{\kappa} \ln(z_2/z_1). \tag{5.16}$$

Squaring leads to

$$(\overline{U}_2 - \overline{U}_1)^2 = \frac{U_*^2}{\kappa^2} \ln^2(z_2/z_1). \tag{5.17}$$

Using Eq. (5.17) to express U_*^2 and substituting in Eq. (5.15) results in

$$Q_h = \kappa^2 \rho_a C_{p_a} \frac{K_h}{K_m} (\overline{U}_2 - \overline{U}_1) \frac{(\overline{T}_1 - \overline{T}_2)}{\ln^2(z_2/z_1)}. \tag{5.18}$$

If z_1 is taken as the roughness height z_o, which implies $\overline{U}_1 = 0$, Eq. (5.18) reduces to

$$Q_h = \kappa^2 \rho_a C_{p_a} \frac{K_h}{K_m} \overline{U}_z \frac{(\overline{T}_s - \overline{T}_z)}{\ln^2(z/z_o)}, \tag{5.19}$$

where $\overline{T}_s$ is now the water-surface temperature and $\overline{T}_z$ is the mean temperature at elevation z.

The energy consumed during evaporation is obtained from Eq. (5.9) by multiplying by the latent heat of evaporation. Expressing $\overline{q}_w$ in terms of vapor pressure and assuming reasonably constant pressure (see Chapter 3) results in

$$Q_e = L_e E = -0.622 \frac{L_e \rho_a K_w}{P} \frac{\partial \overline{e}}{\partial z}$$

$$\approx 0.622 \frac{L_e \rho_a K_w}{P} \frac{(\overline{e}_1 - \overline{e}_2)}{(z_2 - z_1)}, \tag{5.20}$$

where P is ambient atmospheric pressure at the water surface and overbars again imply temporal mean values.

We can follow the same procedure used to derive Eq. (5.19) on Eq. (5.20). Start by multiplying and dividing Eq. (5.20) by K_m, as given in Eq. (5.12). Then make use of the velocity profile expression, Eq. (5.14), to eliminate the friction velocity U_*. The result is

$$Q_e = L_e E = 0.622 L_e \kappa^2 \frac{K_w}{K_m} \frac{\rho_a}{P} \overline{U}_z \frac{(\overline{e}_s - \overline{e}_z)}{\ln^2(z/z_o)}. \tag{5.21}$$

Forming the ratio Q_h/Q_e (Eq. 5.19 over Eq. 5.21) yields the Bowen ratio

$$R = \frac{Q_h}{Q_e} = \frac{C_{p_a} P}{0.622 L_e} \frac{K_h}{K_w} \frac{\overline{T}_s - \overline{T}_z}{\overline{e}_s - \overline{e}_z}, \tag{5.22}$$

which Eq. (5.2) gave as

$$R = \frac{Q_h}{Q_e} = (0.61 \times 10^{-3}\,{}^\circ\mathrm{C}^{-1}) P \frac{(\overline{T}_s - \overline{T}_z)}{(\overline{e}_s - \overline{e}_z)}.$$

The above result assumes that turbulent transfer of heat and water vapor are similar. In fact, it implies $K_h \approx K_w$. The equality of eddy diffusivities and the applicability of the Bowen ratio is sometimes questioned, especially when dealing with boundary layers that are not fully developed or stable. Equality of eddy diffusivities simply states that the wind effects on heat and mass transfer are the same. Monin and Yaglom [1971] and Dyer [1974] argue for equal values of K_w and K_h even under unstable conditions. Bowen [1926] suggested $K_h/K_w = 0.92$. In a related conclusion Morgan et al. [1971] gave $K_w/K_m = 1.13$. From Eq. (5.6) it should be clear that evaporation can be very sensitive to the Bowen ratio calculation. This is unfortunate since it in turn depends on quantities that are not always accurately obtained. Errors are to be expected, particularly when $\overline{e}_s$ and $\overline{e}_z$ are of equal magnitude. It is for this reason that parallel computation of surface (water) temperatures with a numerical model can improve the results of an energy balance calculation of evaporation. Although the controversy over the Bowen ratio continues, it is still used to relate Q_e and Q_h given the difficulties in directly measuring Q_h.

5.2.3 Mass-Transfer Methods: The Dalton Law Analogy

From Eq. (5.21) we can see that evaporation (in grams per square meter per second of vapor) can be expressed as

$$E = 0.622 \kappa^2 \frac{K_w}{K_m} \frac{\rho_a}{P} \overline{U}_z \frac{(\overline{e}_s - \overline{e}_z)}{\ln^2(z/z_o)}, \tag{5.23}$$

which states that evaporation is a function of the vapor pressure difference over a given height z above the water surface, and of wind velocity. In stable boundary layers with no major density differences, $K_w \approx K_m$. I must remind the reader that the velocity dependence in Eq. (5.23) was made explicit by mathematical manipulation but always existed implicitly in Eq. (5.9). Equation (5.23) is of the form

$$E = f(\overline{U})(\bar{e}_s - \bar{e}_a), \tag{5.24}$$

where $f(\overline{U})$ is a function of velocity. Equation (5.23) says that there is no evaporation when the wind velocity is zero. This is not the case (Ryan and Harleman [1973]; Brady et al. [1969]; Kohler [1954]) because of free convective heat transfer, given the temperature differences between the water surface and the adjacent air. To account for this problem, Eq. (5.24) is modified so that the evaporation continues even if $f(U) = 0$ when $\overline{U} = 0$:

$$E = [a + f(\overline{U})](\bar{e}_s - \bar{e}_a). \tag{5.25}$$

The form of Eq. (5.25) is the model of many empirical evaporation equations; it is called the Dalton Law model after the eighteenth-century scientist who first suggested the evaporation dependence on vapor pressure gradients.

The development of Eq. (5.23) and therefore the basis of Eq. (5.25) assumed that the boundary layer above the water table was neutrally stable. This permitted the assumption of a constant K_w/K_m value and led to constant coefficients in Eq. (5.25), not dependent on atmospheric stability. Modifications of Eqs. (5.23) and (5.25) are possible in order to make the coefficients of the Dalton's Law equation a function of stability. This may be very important in situations such as cooling-pond analysis, where water temperature is much higher than air temperature and fuels unstable buoyancy effects, which lead to higher evaporation than implied by Eq. (5.23) or (5.25). Other effects can also be included in the evaporation equation. The observed reduction of evaporation downwind over a lake, effectively a reduction with area, may be mentioned. The above points will be discussed in following subsections.

*Effects of Forced Convection

This and the next two subsections are largely based on a presentation by Helfrich et al. [1982] and to a lesser extent on TVA [1972]. Helfrich et al. divided the nonstable-boundary-layer situation into two conditions, One they called forced convection and the other free convection. Forced convection refers to a situation where the production of turbulence by velocity shear dominates over the production or damping of turbulence by buoyancy. This occurs when fast winds flow over water; for example, a cooling pond. Cooling ponds are manmade or controlled reservoirs whose purpose is to induce cooling of heated water discharges from power plants or industries. Cooling occurs as latent heat is used in the evaporation process.

Helfrich et al. [1982] conclude that under forced convection, evaporation takes the general form

$$E = F \cdot (\bar{e}_s - \bar{e}_z), \tag{5.26}$$

where the function F depends on a measure of the relative strength of wind-induced turbulence and buoyancy effects.

A particular form of Eq. (5.26) is

$$E = \frac{\rho_a \kappa^2 \overline{U}_z(\overline{q}_s - \overline{q}_z)}{[\ln(z/z_o) - \psi_m][\ln(z/z_w) - \psi_w]}, \tag{5.27}$$

where z_o is the roughness length where velocity is zero (which increases with turbulence); z_w is a roughness length over which the specific humidity gradient is zero (which decreases with turbulence); and ψ_m and ψ_w are positive constants, increasing with the potential for forced convection. Hence as ψ_m and ψ_w increase, evaporation increases, since the denominator in Eq. (5.27) gets smaller.

*Effects of Free Convection

Free convection occurs when there is no wind over the body of water, but the temperature and vapor content of ambient air and the near-water-surface air are so different that there is a strong buoyancy effect. Making an analogy to convection over a heated plate Shulyakovsky [1969] and Ryan and Harleman [1973] give

$$E = \frac{0.087}{P} \rho_a \alpha \left(\frac{g\beta\Delta T}{\nu\alpha}\right)^{1/3} (\bar{e}_s - \bar{e}_a), \tag{5.28}$$

where $\beta = 1/\rho_a\ (\partial\rho_a/\partial T)$, ν is the viscosity of air, α is the molecular diffusivity of heat, and ΔT is the difference between surface and air temperature.

Instead of using absolute temperature T to compute ΔT, Ryan and Harleman suggest the use of virtual temperature (see Chapter 3)

$$T^* = \frac{\overline{T}}{(1 - 0.378\bar{e}/P)},$$

which is the temperature of dry air if it occurs at the same temperature and pressure as moist air (see Chapter 3). Using virtual temperature explicitly deals with the combined effect of temperature and moisture on buoyancy. So evaporation is now

$$E = A(\Delta T^*)^{1/3}(\bar{e}_s - \bar{e}_a), \tag{5.29}$$

which is a Dalton analogy formula not dependent on wind speed, where A is a calibrating constant.

If forced and free convection effects were to be added, evaporation could be approximated as

$$E = [A(\Delta T^*)^{1/3} + F](\bar{e}_s - \bar{e}_a). \tag{5.30}$$

*Fetch Dependence of Evaporation

Figure 5.4 shows the expected relative behavior of the evaporation rate and the development of a boundary layer over a body of water. Early in the development, the velocity and vapor gradients are sharp, resulting in higher evaporation than when a more stable boundary layer is established. This argument would imply that area-averaged evaporation should be less than point evaporation, particularly for points near the water-body boundaries.

The fetch dependence of evaporation has been argued by Harbeck [1962], Goodling et al. [1976], and Resch and Selva [1979], among others. The conclusion is that E is inversely proportional to the fetch length ℓ

$$E \sim \ell^{-n}, \tag{5.31}$$

where the factor n usually takes values between 0.1 and 0.2. Table 5.1, from Helfrich et al. [1982], summarizes and comments on some of the existing results.

The difficulty with fetch dependence adjustment lies in the varying nature of wind direction and speed and in the irregular shapes of natural and manmade water bodies. Under such conditions the definition of a fetch length may introduce uncertainties that make the calculation questionable.

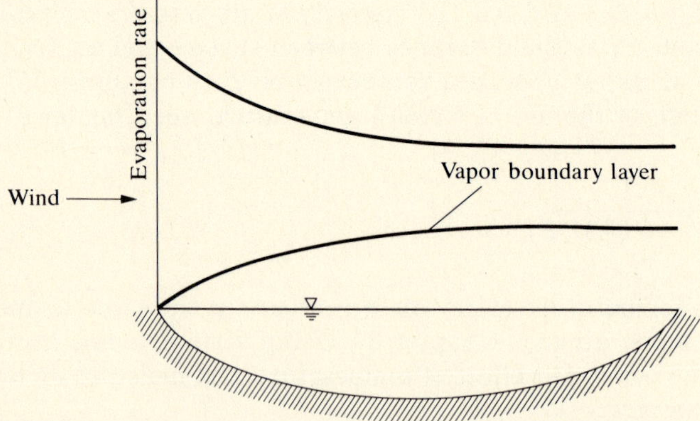

FIGURE 5.4 Evaporation versus fetch. Source: Helfrich et al. [1982].

TABLE 5.1 Evaporation and Fetch Relationships, $E \sim \ell^{-n}$

REFERENCE	n	COMMENTS
Goodling et al. [1976]	0.2	Semitheoretical development
Wegenfeld and Plate [1977]	0.13	Laboratory and field data, $\ell = \sqrt{A}$
Gloyne [1971]	0.12	Circular areas
	0.11	Rectangular strips
Resch and Selva [1979]	0.1	Laboratory data
Harbeck [1962]	0.1	Field data from lakes and reservoirs, $\ell = \sqrt{A}$, A = water-surface area

Source: Helfrich et al. [1982].

5.2.4 Combined Mass-Transfer and Energy Methods: The Penman Equation

Temperature and vapor pressure data necessary to compute evaporation are more often than not collected over land close to the water body of interest. This is in conflict with the requirements of both the energy method (Section 5.2.2) and the mass-transfer methods (Section 5.2.3, Eq. 5.24). Those techniques were developed assuming that the water temperature was available and that a saturated air zone existed over the water, at the water temperature. One way to deal with this issue is the Penman [1948] equation and its modifications (Anderson [1952]; Van Bavel [1966]).

After introducing the concept of a roughness height, Eq. (5.21) stated

$$Q_e = L_e E = L_e B(e_s - e_z), \tag{5.32}$$

where E is evaporation in grams per square meter per second and Q_e is in calories per square meter per second. Overbars are dropped from here on for all variables influenced by turbulent fluctuations, like e, for the sake of notational simplicity. The factor B was a function of velocity and was given by

$$B = 0.622\kappa^2 \frac{K_w}{K_m} \frac{\rho_a}{P} U_z / \ln^2(z/z_o).$$

From the energy balance equation

$$Q_e = Q^* - Q_h, \tag{5.33}$$

where Q^* includes the radiation, advection, and storage terms and Q_h is the sensible-heat term. Using the Bowen ratio concept

$$R = \frac{Q_h}{Q_e} = C_B P \frac{T_s - T_z}{e_s - e_z},$$

where $C_B = C_{p_a} K_h / 0.622 L_e K_w$, in Eq. (5.32) results in

$$Q_h = C_B P L_e B(T_s - T_z) = \gamma L_e B(T_s - T_z), \tag{5.34}$$

where $\gamma = C_B P$. Define the slope of the saturation vapor pressure–temperature relation as

$$\Delta = \frac{e_s - e_{sz}}{T_s - T_z}, \tag{5.35}$$

where e_{sz} and T_z are saturation vapor pressure at elevation z with temperature T_z. Elevation z is a short distance above the surface.

Using Eq. (5.35) in Eq. (5.34) results in

$$Q_h = \frac{\gamma}{\Delta} L_e B(e_s - e_{sz}).$$

Adding and subtracting $(\gamma/\Delta) L_e B e_z$,

$$Q_h = \frac{\gamma}{\Delta} L_e B(e_s - e_z) - \frac{\gamma}{\Delta} L_e B(e_{sz} - e_z), \tag{5.36}$$

where e_z is the vapor pressure at air temperature at an elevation z above the surface.

The first term in Eq. (5.36) is, according to Eq. (5.32), $(\gamma/\Delta)Q_e$. Therefore

$$Q_h = \frac{\gamma}{\Delta} Q_e - \frac{\gamma}{\Delta} L_e B(e_{sz} - e_z). \tag{5.37}$$

Using the energy balance equation (5.33),

$$\begin{aligned} Q_e &= Q^* - Q_h \\ &= Q^* - \frac{\gamma}{\Delta} Q_e + \frac{\gamma}{\Delta} L_e B(e_{sz} - e_z) \\ &= \frac{Q^* + \frac{\gamma}{\Delta} L_e B(e_{sz} - e_z)}{\left(\frac{\gamma}{\Delta} + 1\right)} \\ &= \frac{\frac{\Delta}{\gamma} Q^* + L_e B(e_{sz} - e_z)}{\left(\frac{\Delta}{\gamma} + 1\right)} \end{aligned} \tag{5.38}$$

or

$$E = \frac{1}{L_e}\left[\frac{\dfrac{\Delta}{\gamma}Q^* + L_eB(e_{sz} - e_z)}{\dfrac{\Delta}{\gamma} + 1}\right], \tag{5.39}$$

where E is evaporation in grams of water vapor per unit area per unit time. If divided by the density of water ρ, the result is a volume of water per unit area per unit time.

If Q^* is known, hence not requiring calculation, the modified Penman equation eliminates the need to measure the water-surface temperature. Besides the radiation terms already discussed, the only new component is the term Δ/γ; that is, the ratio of the slope of the $e_s - T$ relation to C_BP. This ratio is given as a function of air temperature T_z in Table 5.2. Van Bavel [1966] shows that only a minor error is incurred in using only the air temperature to obtain this term. Also appearing in Eq. (5.39) is the water vapor deficit at elevation z, which is easily obtained and requires measurements at only one point.

If Q^* is not known or measured, the water-surface temperature would still be needed to compute longwave back-radiation. To avoid this dependence, approximate relations between net radiation exchange and air temperature exist. Section 5.2.6, Eq. (5.48), gives such a relationship.

Most important is that Eq. (5.39) can be manipulated into

$$E = \frac{1}{L_e}\frac{1}{\Delta + \gamma}[Q^*\Delta + L_e\gamma E_a], \tag{5.40}$$

where E_a is now interpreted as any estimate of evaporation (in the same units as E) resulting from a Dalton law analogy. Effectively then, Penman's equation combines the energy and mass-transfer techniques. As will be seen in Section 5.2.6, advection terms in Q^* are many times ignored and accounted for by adjustment factors analogous to the Bowen ratio.

5.2.5 Empirical Equations

Hydrologic practice usually relies on the empirical evaluation of the coefficients of Dalton's law analogy (Eq. 5.25). Helfrich et al. [1982] state that over 100 such evaporation equations have been identified. They point out partial listings in Paily et al. [1974], Ryan and Harleman [1973], and Wunderlich [1972]. All available empirical equations result from the analysis of limited and local data sets. This combined with inconsistent and different data-collection practices always makes the general use of the equations open to suspicion. Some equations were originally calibrated with data collected over the water surface. Others relied on data over land, sometimes not so near to

TABLE 5.2 Δ/γ versus Temperature T, °C

T	Δ/γ	T	Δ/γ	T	Δ/γ	T	Δ/γ	T	Δ/γ	T	Δ/γ
0.0	0.67	10.0	1.23	20.0	2.14	30.0	3.57	40.0	5.70	50.0	8.77
0.5	0.69	10.5	1.27	20.5	2.20	30.5	3.66	40.5	5.83	50.5	8.96
1.0	0.72	11.0	1.30	21.0	2.26	31.0	3.75	41.0	5.96	51.0	9.14
1.5	0.74	11.5	1.34	21.5	2.32	31.5	3.84	41.5	6.09	51.5	9.33
2.0	0.76	12.0	1.38	22.0	2.38	32.0	3.93	42.0	6.23	52.0	9.52
2.5	0.79	12.5	1.42	22.5	2.45	32.5	4.03	42.5	6.37	52.5	9.72
3.0	0.81	13.0	1.46	23.0	2.51	33.0	4.12	43.0	6.51	53.0	9.92
3.5	0.84	13.5	1.50	23.5	2.58	33.5	4.22	43.5	6.65	53.5	10.10
4.0	0.86	14.0	1.55	24.0	2.64	34.0	4.32	44.0	6.80	54.0	10.30
4.5	0.89	14.5	1.59	24.5	2.71	34.5	4.43	44.5	6.95	54.5	10.50
5.0	0.92	15.0	1.64	25.0	2.78	35.0	4.53	45.0	7.10	55.0	10.80
5.5	0.94	15.5	1.68	25.5	2.85	35.5	4.64	45.5	7.26	55.5	11.00
6.0	0.97	16.0	1.73	26.0	2.92	36.0	4.75	46.0	7.41	56.0	11.20
6.5	1.00	16.5	1.78	26.5	3.00	36.5	4.86	46.5	7.57	56.5	11.40
7.0	1.03	17.0	1.82	27.0	3.08	37.0	4.97	47.0	7.73	57.0	11.60
7.5	1.06	17.5	1.88	27.5	3.15	37.5	5.09	47.5	7.90	57.5	11.90
8.0	1.10	18.0	1.93	28.0	3.23	38.0	5.20	48.0	8.07	58.0	12.10
8.5	1.13	18.5	1.98	28.5	3.31	38.5	5.32	48.5	8.24	58.5	12.30
9.0	1.16	19.0	2.03	29.0	3.40	39.0	5.45	49.0	8.42	59.0	12.60
9.5	1.20	19.5	2.09	29.5	3.48	39.5	5.57	49.5	8.60	59.5	12.80
10.0	1.23	20.0	2.14	30.0	3.57	40.0	5.70	50.0	8.77	60.0	13.10

Source: C. H. M. Van Bavel, "Potential Evaporation: The Combination Concept and Its Experimental Verification," *Water Resources Res.* 2(3):467, 1966. Copyright by the American Geophysical Union.

the location of interest. The elevation of temperatures and vapor pressure measurements vary widely. This definitely affects estimates of moisture gradient and of velocity. Some equations modify the Dalton analogy and add nonlinear terms in the wind velocity function or introduce nonlinear dependence on vapor pressure gradients.

It is possible to make adjustments for elevation differences in wind and vapor pressure gradient measurements. The adjustments are based on the logarithmic profiles for wind velocity and humidity. For a neutral, stable atmosphere, the correction from a wind speed measurement U_1 to a value U_2 would be

$$\frac{U_1}{U_2} = \frac{\ln(z_1/z_o)}{\ln(z_2/z_o)}, \tag{5.41}$$

where z_o is the roughness height for wind. Hicks et al. [1977] suggest

$$z_o = \frac{\alpha U_*^2}{g},$$

where the constant α takes a value of 0.009. Helfrich et al. [1982] summarize possible values of z_o. Kohler [1954], using Lake Hefner data, suggested $z_o = 0.0046$ to 0.009 m over water. Harbeck et al. [1958] reported $z_o = 0.00015$ m from Lake Mead data. Ryan and Harleman [1973] used $z_o = 0.001$ m for wind speeds less than 2.25 m s^{-1} and $z_o = 0.005$ m for higher wind speeds. Helfrich et al. [1982] show that an order of magnitude error in z_o (0.001 or 0.0001 m) implies a 4% error in the adjustment of wind speed from 0 m to 2 m in elevation. They also argue that additional corrections on z_o to account for a nonneutral atmosphere are generally not warranted.

Vapor pressure gradient adjustments between two elevations can be made using

$$\frac{e_1 - e_s}{e_2 - e_s} = \frac{\ln(z_1/z_w)}{\ln(z_2/z_w)}, \tag{5.42}$$

where z_w is the roughness height for water vapor, given by Hicks et al. [1977] as

$$z_w = \frac{D_w}{\kappa U_*},$$

where D_w is the molecular diffusivity of water vapor. Helfrich et al. [1982] quote $z_w = 0.000061$ m and 0.0000003 m for the Lake Hefner and Lake Mead studies, respectively. The errors of specification of z_w are comparable to those of wrong z_o specification in the velocity profile equation.

Helfrich et al. [1982] compiled 10 evaporation equations and converted them to 2-m height measurements and to metric units. These are the first

10 equations in Table 5.3, along with a few other alternatives. The reader is referred to Helfrich et al. [1982] for a comparison of the first 10 equations in Table 5.3 in predicting cooling-pond behavior together with a dynamic hydrothermal model. Among their conclusions we can highlight the following general statements:

1. Most equations overestimated evaporation and required adjustments by a fixed factor, effectively a recalibration.
2. Calibrated evaporation equations behave very well when used with data averaged over longer terms than originally intended. For example, the use of daily, weekly, monthly, or annually averaged meteorologic data did not seriously compromise estimates of evaporation for those periods.
3. Uncertainty or errors in parameters or meteorologic data are considerably damped by the feedback effect between temperature and evaporation when the formulas are used within a dynamic hydrothermal model.

EXAMPLE 5.3

Comparison of Empirical Evaporation Formulas

It is useful to study the variable results of the various Dalton-type evaporation formulas of Table 5.3. In order to do that, assume the following daily averaged conditions:

$$U_2 = 1.9 \text{ m s}^{-1}$$

$$e_2 = 22 \text{ mb}$$

$$T_o = 25°\text{C, which implies } e_s \approx 31.6 \text{ mb}$$

Equation (3.19) can be used to obtain the latent heat of evaporation, evaluating at T_o

$$L_e = 597.3 - 0.57T$$
$$= 597.3 - 0.57(25) = 583 \text{ cal g}^{-1}.$$

To be compatible with the units of Table 5.3, the above has to be converted to joules per gram. There are 4.186 J cal^{-1}; therefore, $L_e = 2441 \text{ J g}^{-1}$. Now we can evaluate some of the available formulas.

Lake Hefner:

$$Q_e = 3.75(1.9)(31.6 - 22)$$
$$= 68.4 \text{ W m}^{-2}.$$

In order to convert the above to evaporation in centimeters per day, we first must divide by L_e and by the density of water, 10^6 g m^{-3}.

$$E = \frac{68.4 \text{ J s}^{-1} \text{m}^2}{2441 \text{ J g}^{-1} \, 10^6 \text{ g m}^{-3}} = 0.28 \times 10^{-7} \text{ m s}^{-1}.$$

TABLE 5.3 Empirical Evaporation Equations

The first 10 equations were compiled by Helfrich et al. [1982]. Units are Q_e (in watts per square meter), e (in millibars), T (in degrees Celsius), A (in hectares), and U (in meters per second). Subscripts in U and e indicate the elevation where data were taken. To convert energy flux Q_e to depth of evaporation, divide by latent heat L_e and density of water.

1. Lake Hefner (Kohler [1954]) (on lake data)

$$Q_e = 3.75U_2(e_s - e_2).$$

2. Meyer [1942] (developed for small heated pond water, 5 ft on pond data)

$$Q_e = (7.9 + 2.2U_2)(e_s - e_2).$$

3. Throne [1951] (developed for small heated pond water, 5 ft on pond data)

$$Q_e = (6.6 + 3.75U_2)(e_s - e_2).$$

4. Harbeck [1962] (several lakes, on pond data)

$$Q_e = \frac{5.82U_2}{A^{0.05}}(e_s - e_2).$$

5. Brady et al. [1969] (from moderately heated cooling ponds, originally calibrated with mixed data on and off ponds)

$$Q_e = (6.9 + 0.49U_2^2)(e_s - e_2).$$

6. Rimsha and Donchenko [1957] (thermally loaded river, water-surface data)

$$Q_e = (6.0 + 0.26\Delta T + 3.1U_2)(e_s - e_2)$$
$$\Delta T = T_s - T_2.$$

7. Ryan and Harleman [1973] (cooling pond; originally, pond-edge data at 1 ft)

$$Q_e = [2.26(\Delta T^*)^{1/3} + 3.1U_2](e_s - e_2)$$
$$\Delta T^* = T_2^* - T_1^*.$$

8. Goodling et al., [1976] (theoretical, no data verification)

$$Q_e\left[2.25(e_s - e_2)^{1/3} + \frac{1.52U_2}{R_L^{0.2}}\right](e_s - e_2),$$

where $R_L = (U_2\ell/\nu)$; ℓ = fetch distance; ν = viscosity.

(continued)

TABLE 5.3 (*Continued*)

9. Weisman and Brutsaert [1973] and Weisman [1975] (theoretical, no data verification)

$$Q_e = aU_*\left(\frac{\ell}{z_\circ}\right)^{-n}(q_s - q_a)$$

q_s = specific humidity at water surface temperature

q_a = specific humidity of air

ℓ = fetch length

(a and n are coefficients dependent on stability).

10. Hicks et al. [1975] (cooling-pond data, on pond data)

$$Q_e = \frac{L_e \rho_a \kappa^2 U_2 (q_s - q_z)}{[\ln(z/z_\circ) - \psi_m][\ln(z/z_w) - \psi_w]}$$

(as Eq. 5.27).

The following formulas were adapted from results by TVA [1972]. In making conversions, standard atmospheric pressure was taken as $P_\circ$ = 1013 mb and specific humidity was defined as $q = 0.622\ e/P$. Evaporation E and velocities U are given in meters per second.

11. Penman (from Priestley [1959]) (data on land)

$$E = 1.63 \times 10^{-9}(0.93 + U_2)(e_s - e_2).$$

12. Marciano and Harbeck [1954] (Lake Hefner data) (midlake readings, originally 8 m)

$$E = 1.523 \times 10^{-9} U_2 (e_s - e_2).$$

(Rogers airport data for U_z and e_z, on-lake data for e_s)

$$E = 1.16 \times 10^{-9} U_z (e_s - e_z).$$

(midlake data, 2 m)

$$E = 1.572 \times 10^{-9} U_2 (e_s - e_2).$$

13. Turner [1966] (velocity on lake, vapor pressure on land)

$$E = 2.37 \times 10^{-9} U_2 (e_s - e_2).$$

14. Easterbrook [1969] (Lake Hefner, midlake tower)

$$E = 1.326 \times 10^{-9} U_2 (e_s - e_2).$$

Since there are 86,400 seconds in a day and 10^2 cm m^{-1}, the above is equivalent to an average daily evaporation of

$$E = 0.24 \text{ cm day}^{-1}.$$

Brady et al.:

$$Q_e = [6.9 + 0.49(1.9)^2](31.6 - 22)$$
$$= 83.2 \text{ W m}^{-2}$$

or

$$E = 0.29 \text{ cm day}^{-1}.$$

Penman:

$$E = 0.38 \text{ cm day}^{-1}.$$

Turner:

$$E = 0.37 \text{ cm day}^{-1}.$$

Easterbrook:

$$E = 0.21 \text{ cm day}^{-1}.$$

It is obvious that there can be significant differences between results. To minimize such variability, whenever possible, equations should be used with data of the same nature as that used in their development.

An additional interesting comparison is the evaluation of Eq. (5.23), which has theoretical origins. To do so, assume $K_w \approx K_m$, $\kappa = 0.4$, $P =$ 1000 mb. The density of air at 25°C is about 0.0011 g cm^{-3} or 0.0011×10^6 g m^{-3}. From Table 5.5, z_o is taken as 0.001 cm or 0.001×10^{-2} m. Equation (5.23) then becomes,

$$E = 0.622(0.4)^2 \frac{(0.0011 \times 10^6)}{1000}(1.9)\frac{(31.6 - 22)}{\ln^2(2/0.00001)}$$
$$= 0.013 \text{ g m}^{-2}\text{ s}^{-1} \text{ of vapor.}$$

To convert to centimeters per day of water, we must divide by the density of water, 10^6 g m^{-3}, multiply by 10^2 cm m^{-1} and by 86,400 s day^{-1}. This yields

$$E = 0.11 \text{ cm day}^{-1}.$$

Had we used K_H/K_m equal to 1.13 as stated by Morgan et al. [1971], the result would be

$$E = 0.13 \text{ cm day}^{-1},$$

still considerably less than the empirical equation results. An order of magnitude increase in the roughness height to 0.01 cm will increase the evaporation to $E = 0.17$ cm day^{-1}. ◆

5.2.6 Direct Measurement of Evaporation

The most widely used and proven monitoring technique relies on evaporation pans. These are containers with controlled and monitored water quantities. The pans are exposed to climate and weather in the location of interest. Pan evaporation is then correlated with the expected evaporation from a natural water body.

Evaporation is commonly measured in standardized pans. The most commonly used ones are

1. *U.S. Weather Bureau Class A Land Pan.* This instrument is made of galvanized iron 4 ft (122 cm) in diameter and 10 in. (25.4 cm) deep. It is usually supported on a well-ventilated wooden frame about 15 cm above ground surface.
2. *Bureau of Plant Industry Sunken Pan.* This pan is 6 ft (183 cm) in diameter and 2 ft (61 cm) deep, buried in the ground within 4 in. of the top. The water surface is supposed to be not more than $\frac{1}{2}$ in. above or below ground level.
3. *Colorado Sunken Pan.* This pan is made of unpainted galvanized iron, 3 ft square (91.5 cm) with a depth of 18 in. to 3 ft (45 cm). It is buried in the ground to within 4 in. of its rim.
4. *U.S. Geological Survey Floating Pan.* This instrument is 3 ft (91.5 cm) square and 18 in. (46 cm) deep. It floats in the water body in the center of a raft 14 ft (427 cm) by 16 ft (488 cm).
5. *GGI-3000, the U.S.S.R. Standard Pan.* This galvanized iron instrument is circular (3000 cm^2) and has a depth of about 60 cm. The pan is installed sunk in the ground.
6. *The Young Pan.* This pan is 2 ft (61 cm) in diameter and 3 ft (91 cm) deep. Its main characteristic is that it is covered by a $\frac{1}{4}$-inch-opening screen.

All pans measure more evaporation than natural water bodies because of their smaller size, boundary effects induced by heat transfer through pan material, and wind effects caused by the container itself. This leads to the commonly accepted pan coefficients. The coefficients, which vary with pan type and location, multiply the pan records to obtain actual evaporation estimates. The most commonly used pan in the United States, the Class A pan, shows a relatively stable coefficient. It is near 0.70 under average conditions.

Table 5.4 gives several observed pan coefficients. The Young pan shows coefficient values closer to unity. This is due to a compensating (evaporation reduction) effect of the covering mesh. This effect, though, will vary with mesh size and material. The mesh itself can also lead to maintenance difficulties. The sunken pans—BPI, Colorado, and GGI-3000—also exhibit coefficients near unity. Their drawback is maintenance and possible debris accumulation. Floating pans face the most realistic conditions; nevertheless, their use and access is difficult, and problems with waves are common. The Class A pan shows reasonable variation of the coefficient around a 0.70 value. Coefficients tend to be less variable over time at a given location, but are observed to be reasonably sensitive to location.

Kohler et al. [1955] developed a successful modified Dalton law analogy for land pan evaporation,

$$E_a = (e_s - e_a)^{0.88}(0.42 + 0.0029U_p), \tag{5.43}$$

where E is in millimeters per day, pressures in millibars, and U_p is wind velocity in kilometers per day, at an elevation of 150 mm above the pan rim. Kohler and Richards [1962] used Eq. (5.43) in the Penman formulation (Eq. 5.40) and a pan coefficient of 0.7 to derive Figure 5.5, a graph giving daily shallow-lake evaporation as a function of mean daily air temperature (T_a), mean daily wind velocity, daily solar radiation, and mean daily dew-point temperature (T_d). Obviously, the terms Δ, $\Delta/(\Delta + \gamma)$, net radiation, and vapor pressure gradient are being related to the above-mentioned three variables. Linsley et al. [1982] show how this is done, mostly relying on Eq. (3.31). Their results are

$$\Delta = (0.00815T_a + 0.8912)^7 \qquad T_a \geq -25°\text{C} \tag{5.44}$$

$$\frac{\Delta}{\Delta + \gamma} = \left[1 + \frac{0.66}{(0.00815T_a + 0.8912)^7}\right]^{-1} \tag{5.45}$$

$$\frac{\gamma}{\Delta + \gamma} = 1 - \frac{\Delta}{\Delta + \gamma} \tag{5.46}$$

$$\begin{aligned} e_s - e_a = {} & 33.86[(0.00738T_a + 0.8072)^8 \\ & -(0.00738T_d + 0.8072)^8] \qquad T_d \geq -27°\text{C} \end{aligned} \tag{5.47}$$

$$\begin{aligned} Q^* = {} & 7.14 \times 10^{-3}Q_s + 5.26 \times 10^{-6}Q_s(T_a + 17.8)^{1.87} \\ & + 3.94 \times 10^{-6}Q_s^2 - 2.39 \times 10^{-9}Q_s^2(T_a - 7.2)^2 - 1.02 \end{aligned} \tag{5.48}$$

In the previous equations, all temperatures are in degrees Celsius, pressures in millibars, Q^* in equivalent millimeters of evaporation per day, and

TABLE 5.4 Summary of Pan Coefficients

			SURFACE EXPOSURE		SUNKEN PANS			
LOCATION	YEARS OF RECORD	PERIOD	Class A	X-3[a]	BPI	Young	Colo.	GGI-3000
Davis, Calif.[b]	1966–69	Ann.	0.72	0.71	···	···	···	0.94
Denver, Colo.[c]	1915–16	Ann.	0.67					
	1916	June–Oct.	···	···	0.94			
Felt Lake, Calif.	1955	Ann.	0.77	···	0.91	0.99	0.85	
Ft. Collins, Colo.	1926–28	Apr. –Nov.	0.70	···	···	···	0.79	
Fullerton, Calif.[c]	1936–39	Ann.	0.77	···	0.94	0.98	0.89	
Lake Colorado City, Tex.	1954–55	Ann.	0.72					
Lake Elsinore, Calif.	1939–41	Ann.	0.77	···	···	0.98		
Lake Hefner, Okla.	1950–51	Ann.	0.69	···	0.91	0.91	0.83	
Lake Mead, Ariz.–Nev.[b]	1966–69	Ann.	0.66[d]	0.73[d]	···	···	···	0.71[d]
Lake Okeechobee, Fla.	1940–46	Ann.	0.81	···	···	···	0.98	
Red Bluff Res., Tex.	1939–47	Ann.	0.68					
Salton Sea, Calif.	1967–69	Ann.	0.64	0.64				
Silver Hill, Md.[e]	1955–60	Apr. –Nov.	0.74	···	1.05	···	0.97	
Sterling, Va.[b]	1965–68	Apr. –Nov.	0.69[f]	0.71[f]	···	···	···	1.11[f]
Australia[g]								
Lake Albacutya	···	Ann.	0.79					
Lake Cawndilla	···	Ann.	0.71					
Lake Hindmarsh	···	Ann.	0.74					

Lake Menindee	···	Ann.	0.71					
Lake Pamamaroo	···	Ann.	0.66					
Stephens Cr. Resv.	···	Ann.	0.69					
India								
Poona[b]	1965–68	Ann.	0.69	···	···	···	···	0.78
Israel								
Lod Airport[c]	1954–60	Ann.	0.74					
Sudan								
Khartoum[c]	1960–61	Ann.	0.65					
United Kingdom								
London	1956–62	Ann.	0.70					
U.S.S.R.								
Dubovka[b]	1957–59	May–Oct.	0.64	···	···	···	···	0.91
	1962–67	May–Oct.	0.64	···	···	···	···	0.84
Valdai[b]	1949–53	May–Sept.	0.82	···	···	···	···	0.93
	1958–63	May–Sept.	0.67	···	···	···	···	0.98

[a]Insulated; dimensions approximately the same as for the GGI-3000 pan.

[b]Assuming that evaporation from a tank 5 m (16.5 ft) in diameter is equivalent to lake evaporation.

[c]Assuming that evaporation from a tank 12 ft (3.65 m) in diameter is equivalent to lake evaporation.

[d]Correction for heat flow from soil into 5-m tank would reduce coefficient at least 5%.

[e]Assuming that evaporation (adjusted for heat flow to the soil) from a tank 15 ft (4.57 m) in diameter is equivalent to lake evaporation.

[f]Correction for heat flow from 5-m tank to the soil would increase coefficient a few percent.

[g]The listed coefficients are based on water-budget determinations. Heat-budget determinations for eight additional lakes are cited in Hoy and Stephens [1979]; their average value is appreciably higher.

Source: R. K. Linsley, Jr., M. A. Kohler, and J. L. H. Paulhus, *Hydrology for Engineers*, 3rd ed., McGraw-Hill, 1982. Reproduced with permission.

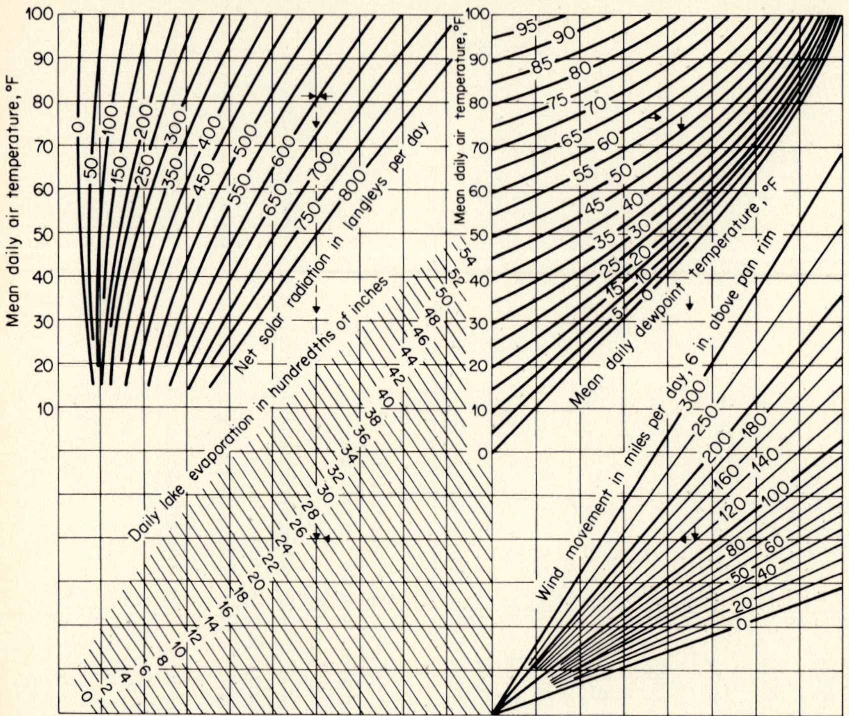

FIGURE 5.5 Shallow-lake evaporation as a function of solar radiation, air temperature, dew point, and wind movement. Sources: U.S. National Weather Service, and Eagleson [1970].

Q_s (shortwave radiation) in calories per square centimeter. Note that while the given equations use metric units, Figure 5.5 uses English units.

Figure 5.5 is not valid for deep lakes, locations and time scales that involve significant advected energy, and/or energy storage changes. Kohler et al. [1955] suggest that the proportion of advected and stored energy used in evaporation is given by a factor α_L. This factor, derived using an approach similar to the Bowen ratio concept, is given in Figure 5.6, or as given by Linsley et al. [1982]:

$$\alpha_L = \left[1 + \frac{0.00066P + (T_s + 273)^3 \times 10^{-8}/(0.177 + 0.00143U_4)}{(0.00815T_s + 0.8912)^7}\right]^{-1}, \quad (5.49)$$

where P is atmospheric pressure in millibars, T_s is water temperature in degrees Celsius, and U_4 is the 4-m wind movement in kilometers per day. Given α_L, the adjusted evaporation is

$$E' = E + \alpha_L(Q_v - Q_o). \quad (5.50)$$

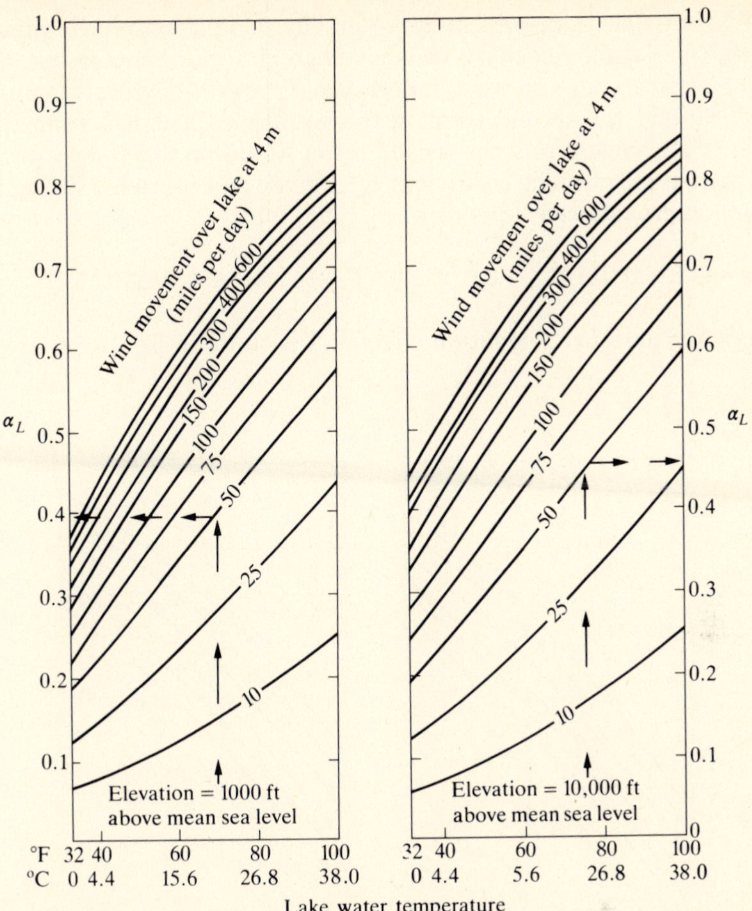

FIGURE 5.6 Proportion of advected energy (into a lake) utilized for evaporation. Source: Gray [1973]. Reproduced by permission of the National Research Council of Canada.

Note that the units of Q_v and Q_o must be in equivalent evaporation units. The evaporation E can be obtained by any method that ignores advection and storage changes, for example, Figure 5.5. Again, Eq. (5.49) is in metric units while Figure 5.6 is its English-units equivalent.

Kohler et al. [1955] also offer a method to correct pan evaporation for sensible-heat transfers through pan walls. This transfer may be out of or into the system, depending on water and air temperatures. Using a basic pan coefficient of 0.7, they then suggest that lake evaporation E is related to pan evaporation E_p by

$$E = 0.7[E_p \pm 0.00064P\alpha_p(0.37 + 0.00255U_p)|T_s - T_a|^{0.88}], \tag{5.51}$$

where evaporation is in millimeters per day, P is atmospheric pressure in millibars; U_p is wind velocity 6 in. above pan rim in kilometers per day, and water temperature T_s and air temperature T_a are in degrees Celsius. When $T_s > T_a$, we add the second term in the equation (heat loss from the pan). When $T_s < T_a$, we subtract the second term. Equation (5.51) does not include lake advection effects. The coefficient α_p is given in Figure 5.7 (using English units) and approximated (Linsley et al. [1982]) by

$$\alpha_p = 0.34 + 0.0117T_s - 3.5 \times 10^{-7}(T_s + 17.8)^3 + 0.0135(U_p)^{0.36}, \tag{5.52}$$

which ignores pressure (elevation effects). The units in Eq. (5.52) are metric.

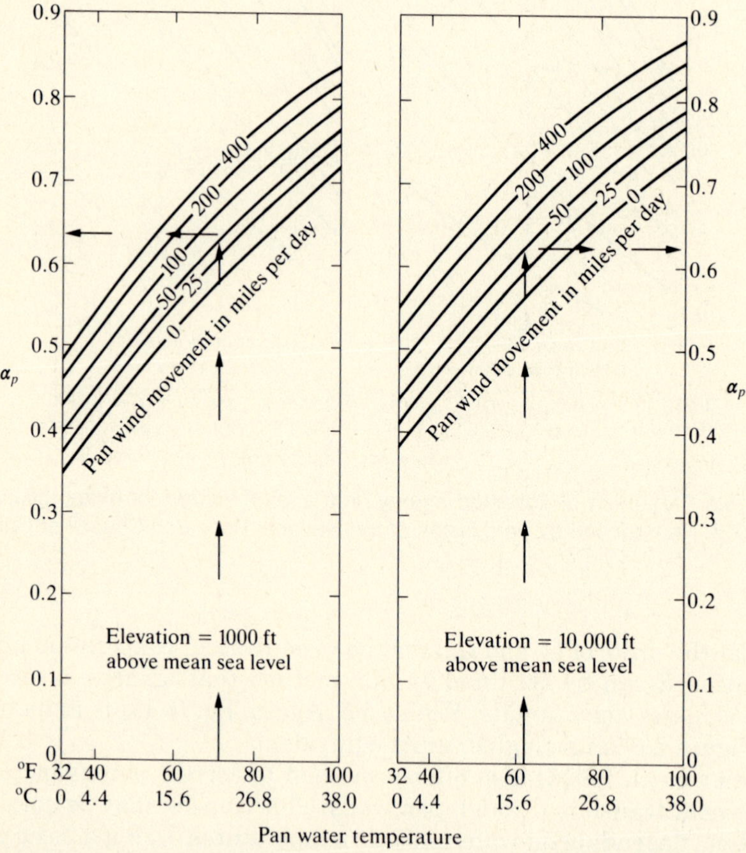

FIGURE 5.7 Proportion of advected energy (into a Class A pan) utilized for evaporation. Source: Gray [1973]. Reproduced by permission of the National Research Council of Canada.

EXAMPLE 5.4

Design of Cooling Pond

As a consultant to a utility, you are asked to estimate the area of the cooling pond required to cool water at 35°C down to 21°C. The power plant discharges at a rate of 40 $m^3 s^{-1}$ at 35°C. Design criteria require that the pond operate while receiving 0.4 ly min^{-1} net shortwave radiation on a clear day when the air temperature is 20°C and the relative humidity is 60%. Wind speed is 200 km per day (2.3 $m s^{-1}$). You can assume that the system is in a steady state, implying no changes in energy storage in the pond. The pond is well mixed, leading to uniform vertical temperature distribution.

The advected energy per unit area of pond is taken to be dominated by the term

$$Q_v = C_p \rho V(T_{in} - T_{out})/A,$$

where V is an assumed constant flow through the system; ρ is the water density; C_p is specific heat; T_{in} and T_{out} are the incoming and outgoing water temperatures, respectively; and A is the area of the pond. Given that C_p = 1 cal g^{-1} °C^{-1} and $\rho = 10^6$ g m^{-3}, we get

$$Q_v = \frac{(1)(10^6)40(35 - 21)}{A} = \frac{560 \times 10^6}{A} \text{ cal s}^{-1}\text{m}^{-2}$$

$$= \frac{3.36 \times 10^{10}}{A} \text{ cal m}^{-2}\text{min}^{-1}.$$

Now it is necessary to compute the longwave-radiation component. Atmospheric radiation depends on vapor pressure as in Example 5.2. Saturation vapor pressure at 20°C is 23.37 mb. With a 60% relative humidity, e = 0.60 × 23.37 = 14. Atmospheric emissivity is then (Eq. 2.35)

$$E_a = 0.74 + 0.0049(14.0) = 0.81.$$

The longwave albedo of water is taken as 0.03, hence

$$Q_a - Q_{ar} = 0.97 E_a \sigma T^4,$$

with T = 293.15°K (20 + 273.15); $\sigma = 0.826 \times 10^{-10}$ cal cm^{-2} min^{-1} °K^{-4}; and E_a as above, we get

$$Q_a - Q_{ar} = 0.48 \text{ ly min}^{-1} = 0.48 \times 10^4 \text{ cal m}^{-2}\text{min}^{-1}.$$

To obtain the outgoing longwave radiation emitted by the water, we need a water temperature that we will take as the average of the input (35°C) and output (21°C) temperatures of the pond, which is 28°C. Emitting as a black body at temperature T_s = 301.15°K, the water radiation is

$$Q_{bs} = \sigma T_s^4 = 0.68 \text{ ly min}^{-1} = 0.68 \times 10^4 \text{ cal m}^{-2} \text{min}^{-1}.$$

Now we have all the components of the energy balance evaporation formula (Eq. 5.6) except for the area in the advection term and the actual evaporation E. We will estimate the latter with the Ryan and Harleman [1973] formula appearing in Table 5.3:

$$Q_e = [2.26(\Delta T^*)^{1/3} + 3.1U_2](e_s - e_2),$$

where T^* is the virtual temperature difference. The virtual temperature adjustment is quite small, so we will use the measured temperatures. Hence, ΔT = 14°C and using $U_2 = 2.3 \text{ m s}^{-1}$ and the saturation vapor pressure at 28°C, which is about 38.1 mb,

$$\begin{aligned} Q_e &= [2.26(14)^{1/3} + 3.1(2.3)](38.1 - 14) \\ &= 303.1 \text{ W m}^{-2} \text{ (J s}^{-1}\text{m}^2) \\ &= 72.4 \text{ cal s}^{-1}\text{m}^2 \\ &= 0.43 \text{ ly min}^{-1}. \end{aligned}$$

The latent heat of vaporization is about 600 cal g^{-1} ≃ 2511 J g^{-1}.

The evaporation in centimeters per minute is then

$$E = \frac{0.43}{(1)600} = 0.0007 \text{ cm min}^{-1} = 7 \times 10^{-6} \text{ m min}^{-1}.$$

Finally, we need the Bowen ratio to evaluate Eq. (5.6):

$$R = (0.61 \times 10^{-3})P\frac{(T_s - T_z)}{(e_s - e_z)} = \frac{0.61(28 - 20)}{(38.1 - 14)} = 0.2.$$

Equation (5.6) becomes

$$7.0 \times 10^{-6} = \frac{(0.4 + 0.48 - 0.68) \times 10^4 + \dfrac{3.36 \times 10^{10}}{A}}{10^6[600(1 + 0.20)]}.$$

Solving for A:

$$A = 1.1 \times 10^7 \text{ m}^2 = 2750 \text{ acres}.$$

Jirka and Harleman [1979] indicate that typical existing or proposed cooling ponds in the United States require about 1 acre per megawatt of electric unit capacity. Naturally, actual values will vary depending on conditions. ◆

5.3 TRANSPIRATION AND EVAPOTRANSPIRATION

Transpiration is the process by which vapor is discharged to the atmosphere through plant stomata. The stomata, intercellular openings in the lower side of leaves, open during daylight (for most plants) to allow in the carbon dioxide necessary for photosynthesis and respiration. Water coming from the roots is also used in this process. A considerable amount of water reaching the stomata is evaporated from the leaf surface. The rate of evaporation is much larger than the rate of consumption in the formation of vegetative matter.

The actual rate of transpiration is a function of the type, stage, and growth of the plants, as well as of soil type and moisture and climatic conditions.

Capillary potentials caused by leaves' intercellular openings and osmotic pressure resulting from the difference in moisture between sap at the roots and surrounding soil favor transpiration. Working against transpiration are gravity and capillary tension in the soil, which increases with soil dryness. If the combined effect of the above processes is that the water supply to the leaves is greater than the evaporative capacity of the atmosphere (in terms of radiant energy, vapor pressure gradient, turbulent advection, etc.), then transpiration is said to be at its potential rate, climate controlled. Otherwise, actual transpiration is controlled by the plant–soil system.

There is considerable argument about the extent to which a plant can control transpiration. When soil moisture is limited, a plant may help the transpiration process in several ways:

1. shifting resources from shoot to root development;
2. reducing its moisture content in search of favorable gradients that would promote water transfers; and
3. controlling stomatal openings.

The mechanisms and the relative importance of the above responses are still a subject of research and controversy. The plant type and stage of growth certainly play a role by defining the degree of leaf surface exposure, orientation, effective surface roughness, radiation absorption, reflection, etc.

Many researchers feel that plant responses are such that the transpiration rate stays at a maximum constant rate (under constant climatic and vegetative cover conditions) until soil moisture is depleted below a certain level, when water intake through the roots is stopped. This point is sometimes assumed to be the wilting point. It is reasonable, though, to think that moisture depletion may become important before the wilting stage. Water intake through the roots may be reduced because of limited osmotic pressures and other concentration-dependent and moisture-gradient-dependent transfer processes.

Climatic and meteorologic conditions play a similar part in transpiration as in evaporation. Vapor pressure gradients, wind, radiation, and other energy and mass-transfer processes still control the rate of vapor removal from the leaves' surfaces.

Evapotranspiration is the name given to the combination of transpiration and evaporation occurring from plants and surrounding land surfaces. This is the term commonly used by hydrologists who are generally interested in the water-mass balance and not in the individual plant consumption. Potential evapotranspiration is the maximum water consumption rate occurring under conditions of unlimited water availability, dense vegetative cover, and normal climatic conditions.

Actual evapotranspiration can be estimated from

1. soil moisture depletion studies on small plots;
2. tanks and lysimeter experiments (lysimeters are usually open-bottom tanks with monitored inputs and outputs);
3. groundwater fluctuations and other mass balance techniques;
4. relations to pan evaporation;
5. soil moisture budgets;
6. energy budgets.

Given full vegetative cover of the soil, potential evapotranspiration is commonly estimated by evaporation from shallow water surfaces, essentially conditions of no significant heat storage. The Penman equation as applied to free water surfaces is commonly taken as a surrogate to potential evapotranspiration, as well as for pan evaporation data. Table 5.5 gives surface roughness assumptions for different vegetative covers to be used with the Penman equation. Table 5.6 gives ratios of observed and calculated evapotranspiration using the Penman equation. The Penman equation can also be modified to account explicitly for evapotranspiration from dry cropped surfaces. The Penman equation can be rewritten (using the definition of the Bowen constant, $C_B = C_{p_a}K_h/0.622L_eK_w$, and $\gamma = C_BP$) as:

$$Q_e = \frac{Q^*\Delta + (\rho_a C_{p_a}/r_a)(e_{sz} - e_z)}{\Delta + \gamma}, \tag{5.53}$$

where C_{p_a} is the specific heat of dry air at constant pressure and r_a is atmospheric diffusion resistance in seconds per centimeter:

$$r_a = \frac{K_m}{K_h}\frac{1}{\kappa^2}\frac{\ln^2(z/z_o)}{\overline{U}_z}.$$

TABLE 5.5 Surface Roughness

SURFACE	WIND SPEED $\overline{U}_z$ at $z = 2$ m ($\mathrm{m\,s^{-1}}$)	ROUGHNESS z_o (cm)
Open water	2.1	0.001
Smooth mud flats	···	0.001
Smooth snow on short grass	···	0.005
Wet soil	1.8	0.02
Desert	···	0.03
Snow on prairie	···	0.10
Mown grass		
1.5 cm	···	0.2
3.0 cm	···	0.7
4.5 cm	2	2.4
4.5 cm	6–8	1.7
Alfalfa		
20–30 cm	1.9	1.4
30–40 cm	1.9	1.3
Long grass		
60–70 cm	1.5	9.0
60–70 cm	3.5	6.1
60–70 cm	6.2	3.7
Maize		
90 cm	···	2.0
170 cm	···	9.5
300 cm	···	22.0
Sugar cane		
100 cm	···	4.0
200 cm	···	5.0
300 cm	···	7.0
400 cm	···	9.0
Brush		
135 cm	···	14.0
Orange orchard		
350 cm	···	50.0
Pine forest		
500 cm	···	65.0
2700 cm	···	300.0
Deciduous forest		
1700 cm	···	270.0

Source: P. S. Eagleson, *Dynamic Hydrology,* McGraw-Hill, 1970.

The resistance factor can be evaluated by taking $K_m \approx K_h$, $\kappa = 0.4$, and using Table 5.5 to define a roughness height z_o. By analogy, Eq. (5.53) is modified to include a soil surface or stomatal diffusion resistance (in seconds per centimeter) that will account for resistance to vapor flux through stomata

TABLE 5.6 24-Hour Tests of the Penman Equation (Eq. 5.53) for Computing Evapotranspiration (E_0) from Various Surfaces (All measurements at z_1 = 2 m)

DATE	SURFACE	UNIFORMITY	HEIGHT, cm	DAYS SINCE IRRIGATION	REMARKS	NET SOLAR RADIATION, ly	AV. TEMPERATURE, °C	AV. VAPOR PRESSURE, mb	AV. WIND SPEED, m sec^{-1}	ROUGHNESS PARAMETER, cm	MEASURED EVAPORATION, E, mm	AV. HOURLY RATIO, E/E_0	DAILY AV. RATIO, E/E_0^*
25/4/61	Open water	Perfect	...	...	....	734	15	4	2.1	0.001	5.85	0.98	
29/4/61	Wet soil	Perfect	...	2	Windy afternoon	748	20	4	1.8	0.02	7.05	0.98	
26/3/63	Alfalfa	Good	20	7	Light breezes	532	17	8	1.6	1.0	5.04	0.89	0.94
21/6/63	Alfalfa	Excellent	30	4	Very windy	761	24	6	1.8	0.7	12.20	1.03	1.04
9/8/63	Alfalfa	Excellent	21	11	Calm, rough surface	625	31	22	1.3	3.0	8.57	0.95	0.86
12/8/63	Alfalfa	Excellent	34	14	Calm	631	32	21	1.2	1.0	8.89	0.97	1.03
12/11/63	Alfalfa	Good	30	15	Calm	310	19	11	1.0	1.0	3.38	0.96	1.09
18/3/64	Alfalfa	Good	25	21	Windy afternoon	479	16	11	2.6	1.0	6.17	0.87	1.06
11/4/64	Alfalfa	Excellent	27	1	Calm	610	10	10	1.0	1.0	7.19	0.99	1.06
18/4/64	Alfalfa	Good	34	10	Partly cloudy, windy	479	24	10	3.8	1.0	10.41	0.93	0.99
21/4/64	Alfalfa	Fair	36	13	Windy afternoon	668	16	9	1.8	1.0	6.62	0.87	0.97
5/6/64	Alfalfa	Excellent	20	7	Light breezes	613	26	13	1.0	1.0	9.06	0.94	0.98
20/7/64	Alfalfa	Good	37	13	Windy afternoon, rough surface	584	32	26	1.7	3.0	11.99	1.05	1.10

Source: C. H. M. Van Bavel, "Potential Evaporation: The Combination Concept and Its Experimental Verification," *Water Resources Res.* 2(3):459, 1966. Copyright by the American Geophysical Union.

and unsaturated soil. The modified equation is

$$Q_e = \frac{Q^*\Delta + (\rho_a C_{p_a}/r_a)(e_{sz} - e_z)}{\Delta + \gamma(1 + r_s/r_a)}. \tag{5.54}$$

Values of r_s will vary with season, soil moisture, and crop type. For pine, r_s varies from 1.0 s cm^{-1} during the moist season to 1.5 s cm^{-1} during the dry season. Alfalfa shows values of 0.3 and 0.8 s cm^{-1} before and after cutting, respectively.

Priestley and Taylor [1972] argued, based on field experiments, that potential evapotranspiration could be approximated as 1.26 times the first term of the Penman equation. This is the term resulting when $e_{sz} = e_z$ and is called the equilibrium rate (Linsley et al. [1982]).

There are several empirical and semi-empirical formulas commonly used to estimate potential evapotranspiration. The Blaney–Criddle (Blaney and Criddle [1950]) equation is probably the most commonly used of the empirical formulas:

$$PE = Kf, \tag{5.55}$$

where PE is the potential evapotranspiration and consumptive use in inches; K is a monthly crop coefficient; $f = (t \times p)/100$ is the monthly consumptive use factor; t is the mean monthly temperature in degrees Fahrenheit; and p is the monthly percentage of daytime hours (as a percentage of yearly total) (Table 5.7).

TABLE 5.7 Monthly Percentage of Annual Daytime Hours (p) for Different Latitudes

LAT. °N	JAN.	FEB.	MAR.	APR.	MAY	JUNE	JULY	AUG.	SEPT.	OCT.	NOV.	DEC.
40	6.76	6.72	8.33	8.95	10.02	10.08	10.22	9.54	8.29	7.75	6.72	7.52
42	6.63	6.65	8.31	9.00	10.14	10.22	10.35	9.62	8.40	7.69	6.62	6.37
44	6.49	6.58	8.30	9.06	10.26	10.38	10.49	9.70	8.41	7.63	6.49	6.21
46	6.34	6.50	8.29	9.12	10.39	10.54	10.64	9.79	8.42	7.57	6.36	6.04
48	6.17	6.41	8.27	9.18	10.53	10.71	10.80	9.89	8.44	7.51	6.23	5.86
50	5.98	6.30	8.24	9.24	10.68	10.91	10.99	10.00	8.46	7.45	6.10	5.65
52	5.77	6.19	8.21	9.29	10.85	11.13	11.20	10.12	8.49	7.39	5.93	5.43
54	5.55	6.08	8.18	9.36	11.03	11.38	11.43	10.26	8.51	7.30	5.74	5.18
56	5.30	5.95	8.15	9.45	11.22	11.67	11.69	10.40	8.53	7.21	5.54	5.89
58	5.01	5.81	8.12	9.55	11.46	12.00	11.98	10.55	8.55	7.10	5.04	4.56
60	4.67	5.65	8.08	9.65	11.74	12.39	12.31	10.70	8.57	6.98	4.31	4.22

Source: Gray [1973]. Reproduced by permission of the National Research Council of Canada.

On a seasonal basis, the equation becomes

$$PE = \sum_{i=1}^{m} K_i f_i = K_s \sum_{i=1}^{m} (t_i \times p_i)/100\,, \qquad (5.56)$$

where K_s is a seasonal crop coefficient and m is the number of months in a season. Typical values of K are given in Table 5.8.

The Thornthwaite [1948] equation gives

$$PE = CT^a, \qquad (5.57)$$

where PE is the monthly potential evapotranspiration in centimeters, C is a coefficient, and T is the mean monthly temperature in degrees Celsius.

The value of a is given by

$$a = 67.5 \times 10^{-8} I^3 - 77.1 \times 10^{-6} I^2 + 0.0179 I + 0.492\,, \qquad (5.58)$$

where

$$I = \text{annual heat index} = \sum_{m=1}^{12} \left[\frac{t_m}{5}\right]^{1.51}.$$

The parameter C depends, among other things, on the percentage of sunshine in a given period. Under the assumption of 12 hours of sunshine each

TABLE 5.8 Consumptive-Use Coefficients (K) for Irrigated Crops in Western United States

ITEM	LENGTH OF GROWING SEASON OR PERIOD	CONSUMPTIVE-USE COEFFICIENTS, SEASONAL (K)	MAXIMUM MONTHLY* (K_i)
Alfalfa	Frost-free	0.85	0.95–1.25
Beans	3 mo	0.65	0.75–0.85
Corn	4 mo	0.75	0.80–1.20
Deciduous orchard	Frost-free	0.65	0.70–0.75
Pasture, grass, hay annuals	Frost-free	0.75	0.85–1.15
Potatoes	3 mo	0.70	0.86–1.00
Small grains	3 mo	0.75	0.85–1.00
Sorghum	5 mo	0.70	0.85–1.10
Sugar beets	$5\frac{1}{2}$ mo		0.85–1.00

*Dependent upon mean monthly temperature and stage of growth of crop.

Source: Gray [1973]. Reproduced by permission of the National Research Council of Canada.

day and 30 days per month, the equation is expressed as

$$PE = 1.62b\left[\frac{10T}{I}\right]^a, \tag{5.59}$$

where b is an adjustment factor to account for the fact that sunshine is not 12 hours a day and the months are not all 30 days in duration. Table 5.9 gives the adjustment coefficient b as a function of month and latitude.

There are many other empirical equations, mainly using temperature and radiation as explanatory variables, to compute potential evapotranspiration. For a good review, the reader is referred to Saxton and Pullman in a monograph edited by Haan et al. [1982].

Actual evapotranspiration, or transpiration, is very much a function of plant development. Saxton and Pullman classify the effects of plants on evapotranspiration into canopy, phenology, root distribution effects, and water stress.

Attempts have been made to quantify the canopy effect through the ratio of leaf area to soil-surface area, called the leaf area index (LAI). Figure 5.8 gives a typical relation between the ratio of actual evaporation to potential evaporation and LAI. Potential evapotranspiration is typically achieved when the LAI is around 3. Since crops, in particular, show quickly varying canopy cover (and LAI), the actual evapotranspiration varies considerably during the growing season.

Plants and crops quickly increase transpiration early in their development and achieve a maximum before or on maturation. With maturation and aging, the transpiration ability is further reduced. This is illustrated in Figure 5.9.

Plant growth also implies root development. As roots grow deeper and spread out, the available moisture for transpiration increases. At the early

TABLE 5.9 Mean Possible Hours of Bright Sunshine Expressed in Units of 30 Days of 12 Hours Each

NORTH LAT.	JAN.	FEB.	MAR.	APR.	MAY	JUNE	JULY	AUG.	SEPT.	OCT.	NOV.	DEC.
0°	1.04	0.94	1.04	1.01	1.04	1.01	1.04	1.04	1.01	1.04	1.01	1.04
10°	1.00	0.91	1.03	1.03	1.08	1.06	1.08	1.07	1.02	1.02	0.98	0.99
20°	0.95	0.90	1.03	1.05	1.13	1.11	1.14	1.11	1.02	1.00	0.93	0.94
30°	0.90	0.87	1.03	1.08	1.18	1.17	1.20	1.14	1.03	0.98	0.89	0.88
35°	0.87	0.85	1.03	1.09	1.21	1.21	1.23	1.16	1.03	0.97	0.86	0.85
40°	0.84	0.83	1.03	1.11	1.24	1.25	1.27	1.18	1.04	0.96	0.83	0.81
45°	0.80	0.81	1.02	1.13	1.28	1.29	1.31	1.21	1.04	0.94	0.79	0.75
50°	0.74	0.78	1.02	1.15	1.33	1.36	1.37	1.25	1.06	0.92	0.76	0.70

Source: Gray [1973]. Reproduced by permission of the National Research Council of Canada.

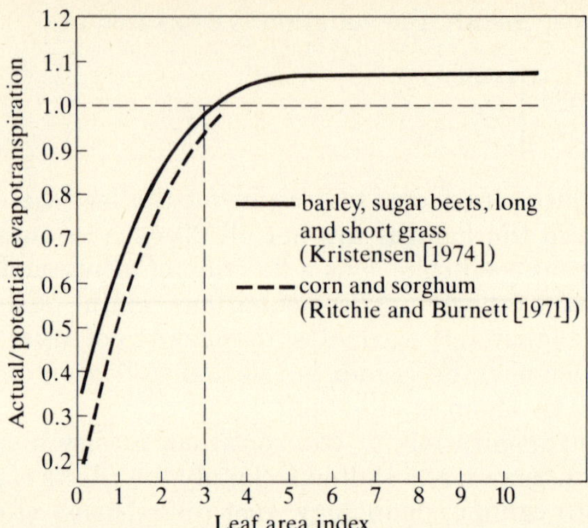

FIGURE 5.8 Relation between leaf area index and the ratio of actual to potential evapotranspiration to define the influence of crop canopy. Source: Haan et al. [1982] by permission from the ASAE.

stages root development limits the water supply to the leaf system. Obviously, maximum crop benefits would be achieved if root development coincided with soil moisture profiles at all times. Mathematical quantification of root development, depth, and density is an active research subject. Table 5.10 gives the distribution in percent of water extraction from various soil depths. The table corresponds to corn planted between May 1 and May 15.

As previously mentioned, a reasonable model of plant transpiration would make actual transpiration a function of soil moisture. As soil moisture

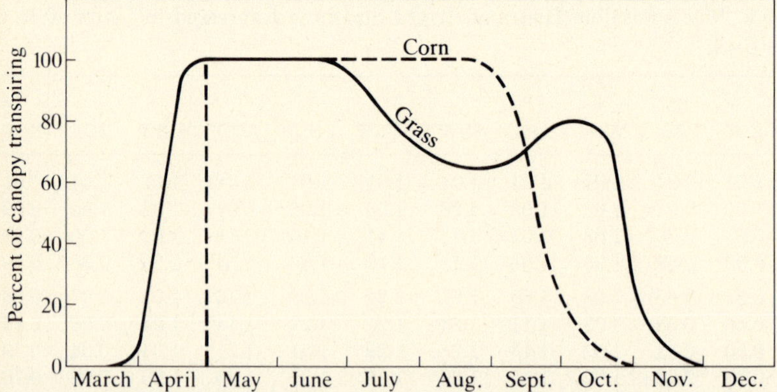

FIGURE 5.9 Phenologic representation of crop transpirability. Sources: Saxton et al. [1974] and Haan et al. [1982] by permission from the ASAE.

TABLE 5.10 Distributions of Water Extraction from Soil Layers by Transpiration, Percentage of Total

DEPTH OF SOIL LAYER, cm	BEGINNING DATE OF EACH DISTRIBUTION*									GRASS, ALL SEASONS
	5/10	6/7	6/14	6/27	7/4	7/11	7/18	7/25	8/1	
0.0– 15.2	100	50	40	35	35	35	35	35	30	35
15.2– 30.5		50	27	25	25	25	25	25	25	30
30.5– 45.7			20	20	18	15	10	8	8	20
45.7– 61.0			13	10	10	8	8	7	7	10
61.0– 76.2				10	7	7	7	5	5	3
76.2– 91.4					5	5	5	5	5	2
91.4–106.7						5	5	5	5	
106.7–121.9							5	5	5	
121.9–137.2								5	5	
137.2–152.4									5	

*Planting date May 1 to May 15.

Source: Saxton et al. [1974] by permission from the ASAE.

decreases, the osmotic potential at the roots gets smaller. At some point the plant will be unable to supply water at the climatic potential rate. This is called a stress condition. Denmead and Shaw [1962] suggested such behavior from experimental evidence. Saxton et al. [1974] suggest the relationship shown in Figure 5.10 for corn in western Iowa. The curves relate the actual to

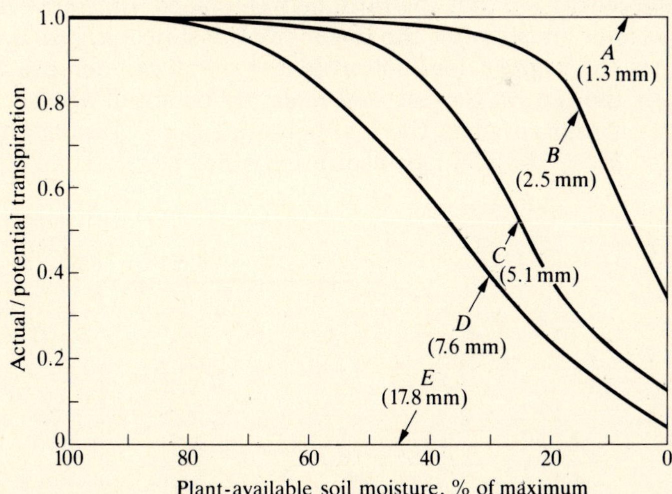

FIGURE 5.10 Moisture–stress relationships used to compute actual transpiration. Curves *A* to *E* represent potential evapotranspiration demand rates (in millimeters per day) with values in parentheses suggested for corn in western Iowa. Sources: Saxton et al. [1974] and Haan et al. [1982] by permission from the ASAE.

potential transpiration rate (stress factor) to soil moisture. Each curve corresponds to a different climatic potential; the smaller the potential, the more resistive is the plant to stress. Neghassi [1974] suggested that the stress behavior can be parameterized by

$$\frac{T_a}{T_p} = \left(\frac{\theta}{\theta_s}\right)^{\lambda} \qquad \theta < \theta^*$$
$$\frac{T_a}{T_p} = 1 \qquad \theta \geq \theta^* \tag{5.60}$$

where θ_s is the saturation soil moisture content (in millimeters), θ is the actual soil moisture (in millimeters), λ is a parameter, and θ^* is a threshold value, dependent on soil crop and climatic conditions, below which the plant is in stress. Soil moistures are commonly measured relative to the permanent wilting point (the moisture value below which the plant dies). Figure 5.11 shows Eq. (5.60). The same type of relationship as Eq. (5.60) has been used to explain actual evapotranspiration (Gardner et al. [1975]; Cordova and Bras [1981]) and evaporation (Saxton et al. [1974]).

In 1948 van den Honert made an analogy between actual transpiration and Ohm's electrical resistance law. He suggested that the volume rate of transpiration per unit width of plant community was directly proportional to the difference between soil moisture potential and leaf moisture potential. It is also inversely proportional to the sum of a resistance to moisture flow in the soil and a resistance to moisture flow in the plant. These resistance and potential concepts were previously discussed. Eagleson [1978] used that idea to support the conclusion that the ratio between plant transpiration T_a to potential evaporation must follow the linear relationship of Figure 5.12. In Figure 5.12 ψ_{1c} is the highest leaf potential the plant can achieve by closing stomata. ψ_{sc} is the critical (lowest) soil moisture potential needed to achieve maximum transpiration when the leaf potential is ψ_{1c}. The maximum ratio T_a/e_p is called k_v, or the plant transpiration efficiency. Linacre et al. [1970]

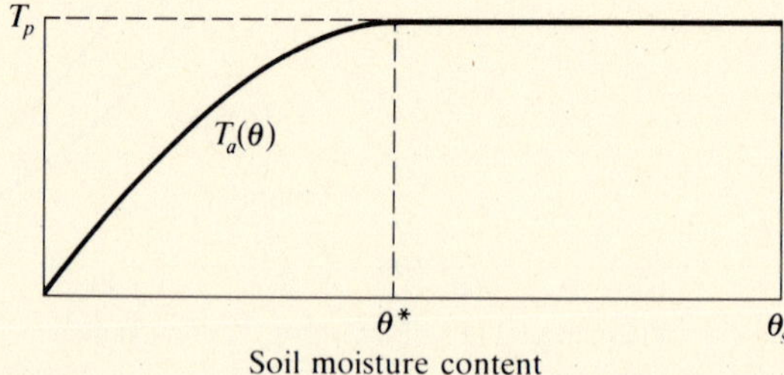

FIGURE 5.11 Relation between the actual transpiration and the soil moisture content at the root zone.

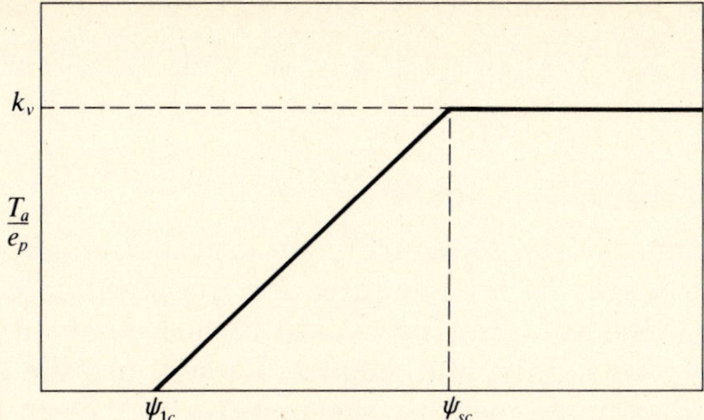

FIGURE 5.12 Steady-state interstorm transpiration. Source: After Eagleson [1978].

have reported k_v values of 0.6 to 2.5. Based on the argument that "evaporation from a plant community never exceeds that of a similar area of wet soil with the same exposure" (Eagleson [1978]), made by Kramer [1969] and Penman [1963], Eagleson hypothesized that k_v is a measure of the ratio of effective leaf transpiring area to vegetated land surface.

All the factors affecting actual evapotranspiration are commonly included in "crop coefficients" like those appearing in the Blaney–Criddle formulation, Eq. (5.55). Figure 5.13 gives the seasonal values of the Blaney–

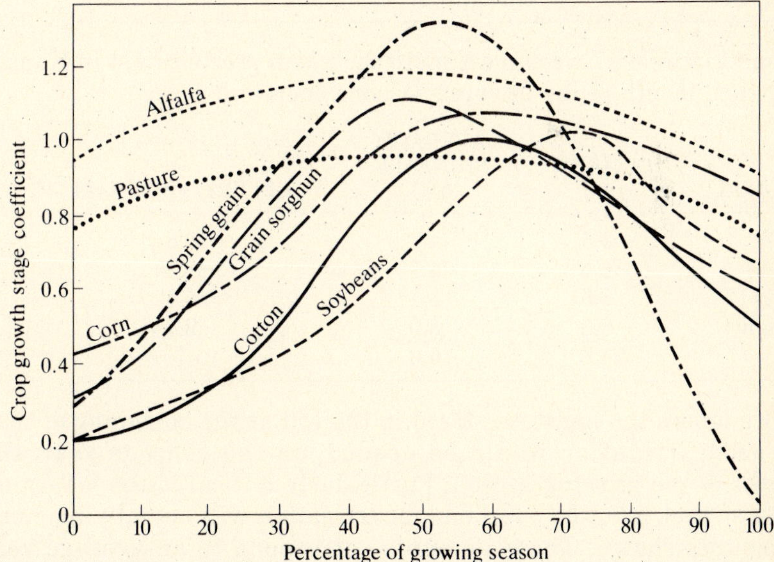

FIGURE 5.13 Seasonal distribution of crop coefficients for application with the Blaney–Criddle equation. Source: Haan et al. [1982] by permission from the ASAE.

Criddle equation for various crops. The local and experimental nature of those coefficients should be considered before extrapolating to other areas.

EXAMPLE 5.5

Estimating Evapotranspiration of Corn

A farmer from Kansas City, Missouri (≈40° latitude), wants to estimate how much irrigation water he will require for his planned corn crop. From Table 5.8 we see that the growing season of corn is about 4 mo; they can be taken as May, June, July, and August. We will use the Blaney–Criddle equation (5.56) to estimate monthly consumptive use. From Figure 5.13 it is possible to obtain reasonable monthly coefficients for the equation. Using the corn curve we can estimate K for May (the first quarter of the growing season) to be about 0.56. Similarly: June, 0.85; July, 1.04; and August, 0.92.

To evaluate Eq. (5.56), we now need the mean monthly temperatures and the monthly percentage of daytime hours. The latter are obtained from Table 5.7. The temperatures are assumed given as shown below.

	MEAN		CONSUMPTIVE USE	
Month	Temperature (°F)	% Daytime hours	K_i	$K_i(t_i \times P_i)/100$ in.
May	63	10.02	0.56	3.5
June	74	10.28	0.85	6.5
July	80	10.22	1.04	8.5
August	72	9.54	0.92	6.3

From Figure 4.27 we can estimate the mean precipitation in Kansas City during each month of the growing season. Therefore

MONTH	DEMAND (in.)	AVERAGE RAINFALL (in.)	−DEFICIT/+SURPLUS
May	3.5	4.3	0.8
June	6.5	4.7	−1.8
July	8.5	3.1	−5.4
August	6.3	3.9	−2.4
	24.8	16.0	

If we ignore the moisture stored in the soil at the beginning of the growing season, irrigation would be needed, on the average year, the last 3 months of the growing season. Particularly critical is the month of July. The water that must be available for irrigation will actually be more than the amounts shown. These amounts correspond to an average year; the farmer will probably want to hedge against the risk of drought, so the systems are usually designed for a more critical rainfall year. Furthermore, water-delivery systems can be very inefficient. Losses of 50% in delivery, for example, through evaporation and seepage in open canals, are not unusual. Finally, additional water may be required to counteract the effect of salts accumulated during agriculture or to leach them. This may demand irrigation even in times of apparent surplus. In actual practice, though, a crop like corn with well-developed and deep root systems, can make substantial use of the moisture stored in the soil prior to the beginning of the growing season. ◆

5.4 EVAPORATION FROM SNOW
(Adapted from Gray [1973])

Computation of evaporation from snow follows the same concepts as discussed in past sections. Possibly, the energy balance is most useful for snow studies; it will be discussed to a larger extent in Chapter 6, when we address snowmelt. The main differences of snow behavior from past surface-water studies are highlighted by Gray [1973] as

1. Snow surfaces have high and variable albedo.
2. Evaporation from snow requires a three-phase change of state from a solid to liquid to gas. The latent heat of sublimation is much higher than the latent heat of melting, so that the latter is a preferred process.
3. Snow exists at temperatures below 0°C. The saturated vapor pressures at these temperatures is about 6.11 mb. In order for evaporation to occur, the air above the snow surface must have a lower vapor pressure. This is particularly difficult to achieve. Since saturated vapor pressure increases with temperature, the maximum relative humidity possible to obtain vapor pressures below 6.11 mb falls very rapidly. For various temperatures, Gray [1973] gives relative humidities as: for 0°C, 99.9%; for 5°C, 70%; for 10°C, 49.7%; for 15°C, 35.7%; and for 20°C, 26%.
4. Given a favorable vapor pressure gradient, evaporation of snow will continue even if the environment does not supply sufficient heat. The necessary heat is taken from the snow itself by cooling it. By doing so the evaporation potential is reduced due to a lower vapor gradient.
5. The conduction and heat absorption properties of snow are different from those of other materials.

Gray [1973] gives three mass-transfer formulas to compute evaporation from snow:

Sverdrup [1946]

$$E = \frac{0.623\rho_a \kappa U_8(e_\circ - e_8)}{P(\ln 800/z_\circ)^2}, \tag{5.61}$$

where κ is 0.4; $z_\circ$ is 0.25 cm; E is evaporation in centimeters per second; ρ_a is the density of air (in grams per cubic centimeter); U_8 is the wind speed at 8 m (centimeters per second); $e_\circ$ is the saturated vapor pressure (in millibars) at snow surface temperature; and e_8 is the air vapor pressure (in millibars) at 8-m elevation. The above equation is valid for favorable vapor pressure gradients and small radiation exchange.

Kuzmin [1957]

$$E = (0.18 + 0.098U_{10})(e_\circ - e_2), \tag{5.62}$$

where E is in millimeters per day; the velocity is in meters per second; and the pressures in millibars.

Central Sierra Snow Laboratory (U.S. Army Corps of Engineers [1956])

$$E = 0.0063(z_a z_b)^{-1/6}(e_\circ - e_a)U_b, \tag{5.63}$$

where E is in inches per day; U_b is the wind speed in miles per hour at elevation z_b feet above the ground; $e_\circ$ is saturation vapor pressure at snow surface temperature in millibars; and e_a is vapor pressure (in millibars) at elevation z_a (in feet).

From previous discussions it should be clear that snow evaporation is rarely a dominant process. It is certainly overshadowed by the significant snowmelt activity.

5.5 INTERCEPTION

Interception refers to the amount of water retained in vegetation. It is usually a function of storm character, vegetative species, density, and season. In heavily forested regions, it may account for up to 25% of annual precipitation.

Most of the interception occurs at the beginning of storms. Many times, it is taken into account by a lump subtraction at the early stages of storm development. The usual expression for total interception is

$$I = S_v + RE\,t_r, \tag{5.64}$$

where I is the total interception for the projected canopy area in units of length, S_v is storage capacity of vegetation for projected area of canopy, usually 0.01–0.05 in. (0.025–0.13 cm), R is the ratio of vegetal surface area to its projected area, E is the evaporation rate during storm from plant surfaces, and t_r is the duration of rainfall.

Some investigators have observed that storage increases with rainfall. To account for that behavior, the following equation is used:

$$I = S_v(1 - e^{-P/S_v}) + RE\,t_r, \tag{5.65}$$

where P is the total precipitation in consistent units with S_v.

In general, interception within a storm is expressed by equations of the form

$$I = a + bP^n. \tag{5.66}$$

Table 5.11 gives typical values of a and b for different types of vegetal cover. Table 5.12 gives typical values for intercepted rainfall by different crops under different rainfall conditions.

It should be obvious that knowledge of interception losses is highly empirical. It can be concluded that within a storm, interception (percentage-wise) is more significant for small accumulations and storms of short duration. The type of vegetation is fairly important as well as the canopy cover. Interception of fresh snow may be large but relatively unimportant in the ultimate snowmelt. Since interception is largely dependent on the retention capacity of vegetation, it becomes more important in regions of infrequent short storms, where the capacity is renewed before the next storm.

Annual interception on the order of 20% of precipitation or more have been reported in the literature for a few particular locations. The reader is referred to Branson et al. [1981] for a more extensive discussion of interception in general.

TABLE 5.11 Evaluation of Constants a, b, and n in an Interception Equation

	INTERCEPTION = $a + bP^n$		
VEGETAL COVER	a	b	n
Orchards	0.04	0.18	1.00
Ash, in woods	0.02	0.18	1.00
Beech, in woods	0.04	0.18	1.00
Oak, in woods	0.05	0.18	1.00
Maple, in woods	0.04	0.18	1.00
Willow, shrubs	0.02	0.40	1.00
Hemlock and pine woods	0.05	0.20	0.50
Beans, potatoes, cabbage, and other small crops grown on hills	0.02h	0.15h	1.00
Clover and meadow grass	0.005h	0.08h	1.00
Forage, alfalfa, vetch, millet, etc.	0.01h	0.10h	1.00
Small grains, rye, wheat, barley	0.005h	0.05h	1.00
Corn	0.005h	0.005h	1.00

Note: Interception is in inches for P in inches. The symbol 'h' refers to the height of plant in feet.

Source: Gray [1973]. Reproduced by permission of the National Research Council of Canada.

TABLE 5.12 Interception under Natural Rainfall for One Square Meter of Area

VEGETATION	PRECIPITATION Inches	PRECIPITATION Character	INTERCEPTION PERCENTAGE
Wheat	0.02	One very light shower	90
	0.06	One very light shower	80
	0.07	One very light shower	72
	0.07	One very light shower	76
	0.24	Two light showers	74
	0.32	One short shower	52
	0.35	One short shower	64
	0.46	One hard shower	46
	0.80	Three showers	51
	1.48	Heavy rain followed by showers and mist	33
Oats	0.11	One light shower	72
	0.15	Several light showers	57
	0.74	Heavy rain followed by light showers	45
Slough grass	0.02	Very light shower	80
	0.06	Light shower	80
	0.07	Light shower	66
	0.07	Light shower	76
	0.38	Hard shower	78
	0.39	Hard shower	67
	0.45	Hard shower	73

Source: Gray [1973]. Reproduced by permission of the National Research Council of Canada.

5.6 DEPRESSION STORAGE

The amount of water that is retained during a storm event, in the surface irregularities of the soil, is called depression storage. Again, it may be taken as a lump quantity, applying it at the beginning of the storm, but generally its total volume is expressed as

$$V = S_d(1 - e^{-(1/S_d)P_e}) = S_d(1 - e^{-KP_e}), \tag{5.67}$$

where V is the volume of water stored, S_d is the maximum storage capacity, and P_e is the rainfall excess (gross rainfall minus infiltration, $I - F$) (see Chapter 8 for a discussion of infiltration).

The rate of depression storage is given by the time derivative of Eq. (5.67):

$$\nu = \frac{dV}{dt} = e^{-KP_e}\frac{dP_e}{dt}, \tag{5.68}$$

where dP_e/dt is the rate of rainfall excess or the rainfall rate minus the infiltration rate, $i - f$.

Therefore

$$\frac{dV}{dt} = (i - f)e^{-KP_e}. \tag{5.69}$$

Equation (5.69) can be used to obtain an expression for the overland flow supply rate,

$$\sigma = i - f - \nu.$$

Using Eq. (5.68):

$$\frac{\sigma}{i - f} = \frac{i - f - \nu}{i - f} = \frac{i - f - (i - f)}{i - f} e^{-KP_e} = [1 - e^{-KP_e}]. \tag{5.70}$$

Figure 5.14 shows a plot of Eq. (5.70). Note that in the limit as rainfall accumulation increases, the ratio of $\sigma/(i - f)$ becomes 1. The vertical line in

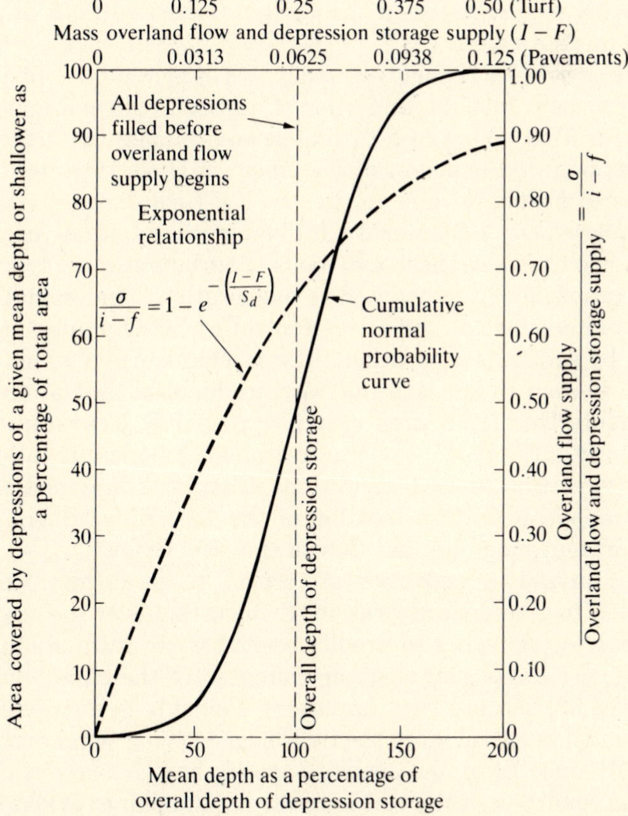

FIGURE 5.14 Depth–distribution curve of depression storage. Source: Reprinted from *J. Sanitary Eng.*, ASCE, March 1959, p. 60, with permission.

the figure represents a step function, where all depressions are filled before overland flow begins. The S-type curve is a suggested middle point between Eq. (5.70) and the step function line.

Note that Figure 5.14 uses values of S_d equal to 0.0625 and 0.25 in. for pavements and turf, respectively.

The overland supply rate given in Eq. (5.70) is the net result after subtraction of infiltration and depression storage from gross rainfall. This overland flow supply rate, as shown in Figure 5.14, becomes runoff and is routed through overland and stream segments resulting in the streamflow hydrograph, a topic of discussion in future chapters.

5.7 SUMMARY

Evaporation is an extremely important component of the hydrologic cycle, yet surprisingly little attention is given to it. In Chapter 4 we mentioned the generally poor coverage of basins with rain gages. Evaporation coverage is far worse, with whole states in the United States without any continuous evaporation measurement.

We have studied evaporation from two major points of view: energy balance and mass transport. Radiation (Chapter 2) plays a major role in the former. Evaporation occurs in a turbulent environment with the transport of vapor proportional to the decreasing gradient of vapor pressure (a measure of water in the atmosphere) and to the level of turbulence, parameterized in terms of wind velocity (Chapter 3). All empirical equations used to compute evaporation are based on those two ideas of turbulent mass transport.

Transpiration is closely related to evaporation. The crucial difference is that the former depends on the active control by living plants (versus the passive control by soils in evaporation). The mechanisms by which plants influence actual transpiration are not well understood, let alone quantified. Unquestionably this is an area of active research. Transpiration through plants is a very large portion of the hydrologic balance. It influences global water balance as well as local conditions. Chapter 1 provided an example of how evapotranspiration in a fraction of the Nile River Basin amounted to 30% of the present mean annual flow of the Nile River.

Evaporation and transpiration are crucial to the farmer, since crop yield is proportional to actual transpiration. Evaporation over the oceans is crucial for the climatologist trying to predict ocean-temperature anomalies (i.e., El Niño) or global climate and weather changes. To the climatologist evaporation and transpiration are very important elements in energy and mass balances. Evaporation is crucial to the hydrologist trying to predict floods from a given rainfall since he or she needs to specify how dry or wet the soil in the basin is. The condition of the soil is mostly a function of evaporation. Evaporation is crucial to the engineer trying to design a reservoir for water supply.

It would be embarrassing if the reservoir never filled or if evaporation from the large surface area of the lake resulted in very serious net yield reductions from the river system. In summary, almost everybody could benefit from improved quantification of evaporation.

Chapter 6 will discuss snow and snowmelt. It should be clear as you proceed that snowmelt analysis is analogous to evaporation analysis. The energy balance is the key to understanding. Energy is required for the change of phase from liquid (or solid) to vapor involved in evaporation. Similarly, energy is needed for the solid phase of water (ice) to melt.

REFERENCES

Anderson, E.R. [1952]. "Energy-Budget Studies, from Waterloss Investigations." Vol. 1. Lake Hefner Studies Technical Report, *U.S. Geol. Surv. Circ.* 229:71–119.

Blaney, H.F., and W.D. Criddle [1950]. "Determining Water Requirements in Irrigated Areas from Climatological and Irrigation Data." U.S. Department of Agriculture Soil Conservation Service. Technical document no. 96.

Blank, H. [1975]. "Optimal Irrigation Decisions with Limited Water." Fort Collins, Colo.: Colorado State University. (Ph.D. dissertation.)

Bowen, I.S. [1926]. "The Ratio of Heat Losses by Conduction and by Evaporation from Any Water Surface." *Physics Rev.* 27:779–787.

Brady, K.D., W.L. Graves, and J.C. Geyer [1969]. "Cooling Water Studies for Edison Electric Institute, Proj. No. RP-49—Surface Heat Exchange at Power Plant Cooling Lakes." Baltimore: Johns Hopkins University.

Branson, F.A. [1976]. "Water Use on Rangelands." Watershed Management on Range and Forest Lands. *Proc. 5th U.S./Australian Rangelands Panel,* Utah Water Research Laboratory, Logan, Utah, pp. 193–209.

Branson, F.A., G.F. Gifford, K.G. Denard, and R.F. Hadley [1981]. *Rangeland Hydrology*. 2nd ed. Dubuque, Iowa: Kendall/Hunt.

Businger, J.A., et al. [1971]. "Flux-Profile Relationships in the Atmospheric Surface Layer." *J. Atmos. Sci.* 28:181–189.

Cordova, J.R., and R.L. Bras [1979]. "Stochastic Control of Irrigation Systems." Cambridge, Mass.: MIT Department of Civil Engineering, Ralph M. Parsons Laboratory. (Technical report no. 234.)

Idem. [1981]. "Physically-Based Probabilistic Models of Infiltration, Soil Moisture and Actual Evapotranspiration." *Water Resources Res.* 17(1):93–106.

Cruff, R.W., and T.H. Thompson [1967]. "A Comparison of Methods of Estimating Potential Evapotranspiration from Climatological Data in Arid and Subhumid Environments." Washington, D.C.: U.S. Geological Survey water supply paper no. 1839–M.

Denmead, O.T., and R.H. Shaw [1962]. "Availability of Soil Water to Plants as Affected by Soil Moisture Contents and Meteorological Conditions." *Agron. J.* 54:385–390.

Dyer, A.J. [1974]. "A Review of Flux-Profile Relationships." *Bound. Layer Meteorol.* 7:363–372.

Eagleson, P. S. [1970]. *Dynamic Hydrology.* New York: McGraw-Hill.

Idem. [1978]. "Climate, Soil, and Vegetation, 4, The Expected Value of Annual Evapotranspiration." *Water Resources Res.* 14(5):731–739.

Easterbrook, C. C. [1969]. "A Study of the Effects of Waves on Evaporation from Free Water Surfaces." Washington, D. C.: U. S. Department of the Interior, Bureau of Reclamation. (Research report no. 18.)

Gardner, W. R., W. A. Jury, and J. Knight [1975]. "Water Uptake by Vegetation. In: D. A. DeVries and N. H. Afgan, eds. *Heat and Mass Transfer in the Biosphere.* Vol. 1. *Transfer Processes in the Environment.* New York: Wiley.

Gloyne, R. W. [1971]. "A Note on the Measurement and Estimation of Evaporation." *Meteorol. Mag.* 100(1189):322.

Goodling, J. S., B. L. Sill, and W. J. McCabe [1976]. "An Evaporation Equation for an Open Body of Water Exposed to the Atmosphere." *Water Resources Bull.* 12(4):843–853.

Gray, D. M., ed. [1973]. *Handbook on the Principles of Hydrology.* Port Washington, N. Y.: Water Information Center.

Gunaji, N. N. [1968]. "Evaporation Investigations at Elephant Butte Reservoir in New Mexico." *Int. Assoc. Sci. Hydrol. Publ.* 78:308–325.

Gunaji, N. N., et al. [1965]. "Evaporation Reduction Investigations — Elephant Butte Reservoir, New Mexico." Engineering Experimental Station, Las Cruces, N. M.: New Mexico State University. (Technical report no. 25.)

Haan, C. T., H. P. Johnson, and D. L. Brakensiek [1982]. "Hydrologic Modeling of Small Watersheds." *Am. Soc. Agr. Eng.* Monograph no. 5.

Hamon, W. R. [1961]. "Estimating Potential Evapo-Transpiration." *Proc. ASCE J. Hydraul. Div.* 87(HY3):107–120. (Paper no. 2817.)

Harbeck, G. E. [1952]. "The Lake Hefner Water Loss Investigations." U. S. Geological Survey circular 229.

Idem. [1954]. "Cummings Radiation Integration in Water-Loss Investigations — Lake Hefner Studies, Technical Report." U.S. Geological Survey professional paper 269.

Idem. [1962]. "A Practical Field Technique for Measuring Reservoir Evaporation Utilizing Mass Transfer Theory." U.S. Geological Survey professional paper no. 272-E.

Harbeck, G. E., et al. [1951]. "Utility of Selected Western Lakes and Reservoirs for Water Loss Studies." U.S. Geological Survey circular 103.

Harbeck, G. E., et al. [1958]. "Water Loss Investigations: Lake Mead Studies." U. S. Geological Survey professional paper 298.

Helfrich, K. R., E. E. Adams, A. L. Godbey, and D. R. F. Harleman [1982]. "Evaluation of Models for Predicting Evaporative Water Loss in Cooling Impoundments." Palo Alto, Calif.: Electric Power Research Institute. (Interim report no. CS-2325.)

Hicks, B. B., and M. L. Wesely [1975]. "An Examination of Some Bulk Formulae Used for Assessing the Performance of Industrial Cooling Ponds." In: *1974 Annual Report of Radiological and Environmental Research Division.* Argonne, Ill.: Argonne National Laboratory. (Report no. ANL75-60, Part IV.)

Hicks, B. B., M. L. Wesely, and C. M. Sheih [1975]. "Eddy-Correlation Measurements over a Cooling Pond with Limited Fetch." In: *1974 Annual Report of Radiological and Environmental Research Division.* Argonne, Ill.: Argonne National Laboratory. (Report no. ANL75-60, Part IV.)

Idem. [1977]. "A Study of Heat Transfer Processes Above a Cooling Pond." *Water Resources Res.* 13(6):901–908.

Hoy, R. D., and S. K. Stephens [1979]. "Field Study of Lake Evaporation— Analysis of

Field Data from Phase 2 Storages and Summary of Phase 1 and Phase 2." Australian Water Resources Council technical paper 41.

Jirka, G. H., and D. R. F. Harleman [1979]. "Cooling Impoundments: Classification and Analysis." *J. Energy Div. ASCE.* 105(EY2):291–309.

Kohler, M. A. [1954]. "Lake and Pan Evaporation in Water Loss Investigations—Lake Hefner Studies, Technical Report." U.S. Geological Survey professional paper 269.

Kohler, M. A., T. J. Nordenson, and W. E. Fox [1955]. "Evaporation from Pans and Lakes." Washington, D. C.: U.S. Department of Commerce, Weather Bureau (Research paper no. 38.)

Kohler, M. A., and M. A. Richards [1962]. "Multicapacity Basin Accounting for Predicting Runoff from Storm Precipitation." *J. Geophys. Res.* 67(13):5187–5197.

Kramer, P. J. [1969]. *Plant and Soil Water Relationships—A Modern Synthesis.* New York: McGraw-Hill.

Kristensen, K. J. [1974]. "Actual Evapotranspiration in Relation to Leaf Area." *Nordic Hydrol.* 5:173–182.

Kuzmin, P. O. [1957]. "Hydrophysical Investigations of Land Waters." *Int. Assoc. Sci. Hydrol.* 3:468–478.

Linacre, E. T., B. B. Hicks, G. R. Sainty, and G. Grauze [1970]. "The Evaporation from a Swamp." *Agr. Meteorol.* 7:375–386.

Linsley, R. K., Jr., M.A. Kohler, and J. L. H. Paulhus [1949]. *Applied Hydrology.* New York: McGraw-Hill.

Idem. [1958]. *Hydrology for Engineers.* 1st ed. New York: McGraw-Hill.

Idem. [1982]. *Hydrology for Engineers.* 3rd ed. New York: McGraw-Hill.

Marciano, T. T., and G. E. Harbeck [1954]. "Mass Transfer Studies in Water Loss Investigations, Lake Hefner Studies." U. S. Geological Survey professional paper 269.

Meyer, A. F. [1942]. "Evaporation from Lakes and Reservoirs." St. Paul: Minnesota Resources Commission.

Monin, A. S., and A. M. Yaglom [1971]. *Statistical Fluid Mechanics: Mechanics of Turbulence.* Vol. 1. Cambridge, Mass.: MIT Press.

Morgan, D. L., W. O. Pruitt, and F. L. Lourence [1971]. "Analysis of Energy, Momentum and Mass Transfers Above Vegetative Surfaces." Davis, Calif.: University of California Department of Water Science and Engineering. (Research and Development technical report ECOM68-G10-F.)

Neghassi, H. M. [1974]. "Crop Water Use in Yield Models with Limited Soil Moisture." Fort Collins, Colo.: Colorado State University. (Ph.D. dissertation.)

Nordenson, T. T., and D. R. Baker [1962]. "Comparative Evaluation of Evaporation Instruments." *J. Geophys. Res.* 67:671–679.

Paily, P. P., E. O. Macagno, and J. F. Kennedy [1974]. "Winter Regime Surface Heat Loss from Heated Streams." Iowa Institute of Hydraulics Research report no. 155.

Panofsky, H. A. [1963]. "Determination of Stress from Wind and Temperature Measurements." *Q. J. R. Meteorol. Soc.* 89:85–94.

Peck, E. L., and R. Farnsworth. "A Dual-Purpose Evaporimeter." U.S. National Weather Service unpublished manuscript.

Penman, H. L. [1948]. "Natural Evaporation from Open Water, Bare Soil and Grass." *Proc. R. Soc. (London) [A].* 193:120–145.

Idem. [1963]. "Vegetation and Hydrology." Harpenden, England: Commonwealth Bureau of Soils. (Technical communication 53.)

Priestley, C. H. B. [1959]. "Turbulent Transfer of the Lower Atmosphere." Chicago: University of Chicago Press.

Priestley, C. H. B., and R. J. Taylor [1972]. "On the Assessment of Surface Heat Flux and Evapotranspiration Using Large Scale Parameters." *Monthly Weather Rev.* 100:81–92.

Pruitt, W. O., S. VonOetinger, and D. L. Morgan [1972]. "Central California Evapotranspiration Frequencies." *ASCE Proc.* 98(IR2):177–184.

Quinn, F. H. [1979]. "An Improved Aerodynamic Evaporation Technique for Large Lakes with Application to the International Field Year for the Great Lakes." *Water Resources Res.* 15(4):935–940.

Rauner, J. L., and O. Teplovom [1965]. "Balanse Listvennogo Lesav Zimnig Period." *Izv. Akad. Navk. U.S.S.R. Ser. Geogr.* No. 4.

Resch, F. J., and J. P. Selva [1979]. "Turbulent Air-Water Mass Transfer Under Varied Stratification Conditions." *J. Geophys. Res.* 84(C7):3205.

Rimsha, V. A., and R. V. Donchenko [1957]. "The Investigation of Heat Loss from Free Water Surfaces in Wintertime" (in Russian). *Trudy Leningrad Gosub-Gidrol.* Inst. 64.

Ritchie, J. T., and E. Burnett [1971]. "Dryland Evaporative Flux in a Sub-Humid Climate: II Plant Influences." *Agron. J.* 63:56–62.

Ryan, P. J., and D. R. F. Harleman [1973]. "An Analytical and Experimental Study of Transient Cooling Pond Behavior." Cambridge, Mass.: MIT Department of Civil Engineering, Ralph M. Parsons Laboratory. (Technical report no. 161.)

Saxton, K. E., H. P. Johnson, and R. H. Shaw [1974]. "Modeling Evapotranspiration and Soil Moisture." *Trans. Am. Soc. Agr. Eng.* 7(4):673–677.

Shulyakovsky, L. G. [1969]. "Formula for Computing Evaporation with Allowance for Temperature of Free Water Surface." *Sov. Hydrol. Selected Papers.* 6:566–573.

Sverdrup, H. V. [1946]. "The Humidity Gradient Over the Sea Surface." *J. Meteorol.* 3:1–8.

Szeicz, G., G. Endrodi, and S. Tajchman [1969]. "Aerodynamic and Surface Factors in Evaporation." *Water Resources Res.* 5(2):380–394.

Tachman, S. [1967]. "Energie und Wasserhaushalt Verschiedener Pflanzenbestande bei Munchen." Munich, Germany: Universitat Munchen. (Ph.D. thesis.)

Tanner, C. B., and W. L. Pelton [1960]. "Potential Evapotranspiration Estimates by the Approximate Energy Balance Method of Penman." *J. Geophys. Res.* 65:3391–3413.

Tennessee Valley Authority [1972]. "Heat and Mass Transfer Between a Water Surface and the Atmosphere." Norris, Tenn.: Tennessee Valley Authority Division of Water Control Planning Engineering Laboratory. (Report no. 14.)

Tholin, A. L., and C. J. Kiefer [1959]. "The Hydrology of Urban Runoff." *Proc. Trans. ASCE. J. San. Eng. Div.* 85(5A2):47–106.

Thornthwaite, C. W. [1948]. "An Approach Toward a Rational Classification of Climate." *Am. Geogr. Rev.* 38:55–94.

Throne, R. F. [1951]. "How to Predict Cooling Lake Action." *Power.* 95:86–89.

Trewartha, G. T., and L. H. Horn [1980]. *An Introduction to Climate.* 5th ed. New York: McGraw-Hill.

Turner, J. F. [1966]. "Evaporation Study in a Humid Region, Lake Michie, N. C." U.S. Geological Survey professional paper 272-G.

Uchijima, Z., and J. L. Wright [1964]. "An Experimental Study of Air in the Corn Plant Air Layer." *Bull. Natl. Inst. Agr. Sci. [A].* 11:19–66.

U.S. Army Corps of Engineers [1956]. "Summary Report of Snow Investigations." Portland, Oreg.: U.S. Army Corps of Engineers. North Pacific Division.

U.S. Department of Agriculture, Soil Conservation Service [1967]. "Irrigation Water Requirements." (Technical report no. 21.)

Van Bavel, C. H. M. [1966]. "Potential Evaporation: The Combination Concept and Its Experimental Verification." *Water Resources Res.* 2(3):455–467.

van den Honert, T. H. [1948]. "Water Transport as a Catenary Process." *Trans. Faraday Soc.* 3:146–153.

Viessman, W., Jr., J. W. Knapp, and T. E. Harbaugh [1972]. *Introduction to Hydrology.* New York: Intext Educational.

Idem. [1977]. *Introduction to Hydrology.* 2nd ed. New York: Harper & Row.

Wegenfeld, P., and E. J. Plate [1977]. "Evaporation from a Water Current Under the Influence of Wind-Induced Waves." IAHR Congress, August 15–19, 1977, Baden, West Germany.

Weisman, R. N. [1975]. "Comparison of Warm Water Evaporation Equations." *J. Hydraul. Div. ASCE.* 101(HY10):1303–1313.

Weisman, R. N., and W. Brutsaert [1973]. "Evaporation and Cooling on a Lake Under Unstable Atmospheric Conditions." *Water Resources Res.* 9(5):1242–1257.

World Meteorological Organization [1966]. "Measurement and Estimation of Evaporation and Evapotranspiration." (WMO technical note no. 83.)

Wunderlich, W. O. [1972]. "Heat and Mass Transfer Between a Water Surface and the Atmosphere." Tennessee Valley Authority Engineering Laboratory. (Lab report no. 14.)

Young, A. A. [1947]. "Some Recent Evaporation Investigations." *Trans. Am. Geophys. Union.* 28:279–284.

PROBLEMS

1. Find the evaporation from a lake using the following information: latitude, 15°N; date, July 1; hour, noon; overcast (cloudy) with 3000-foot cloud-base elevation; smoggy air conditions; water temperature, 18.33°C; emissivity of water, 0.8; cloud base temperature, 20°C; specific heat of water, 1 cal gr^{-1} $°C^{-1}$; shortwave albedo of water, 0.1; temperature of air, 26.67°C; temperature of evaporated water, 32.2°C; and relative humidity of air, 0.85. Ignore cloud effects in net longwave radiation. Assume no precipitation, no seepage, zero net flow, and no net increase in stored energy in the lake. Can you estimate the average daily wind speed at this site?

2. The Machar Marshes lie in Southern Sudan near the confluence of the White and Blue branches of the Nile River. The marshes and surrounding plains cover an area of about 25,000 km^2. Typical yearly characteristics are a mean wind velocity of 0.2 m s^{-1} (2 m above ground surface); vegetation is papyrus with effective roughness of 10 cm; average air temperature at 2 m above ground surface is 28°C; surface temperature is 30°C; relative humidity is 0.51 at 2 m above ground surface; average cloud cover, 0.60; longwave radiation cloud effects coefficient is 1; net shortwave radiation is 0.32 ly min^{-1};

shortwave albedo is 0.2; latent heat of evaporation = 597.3 − 0.57 $(T_s - 0°C)$ cal gr^{-1}; c_1 coefficient in logarithmic wind velocity is $(2.5)^{-1} = 0.4$; $\sigma = 0.826 \times 10^{-10}$ cal cm^{-2} min^{-1} °K^{-4}; dry-air gas constant = $R = 2.876 \times 10^6$ cm^2 s^{-2} °K^{-1}; and saturation vapor pressure $e_s = 33.8639[(0.00738T + 0.8072)^8 - 0.000019|1.8T + 48| + 0.001316]$. Find the yearly potential evapotranspiration from the area. Compare this to the evaporation from a shallow lake in the same area.

3. Consider a small fully mixed pond for which the energy balance is studied. The lake has volume $V = 2500$ acre-ft and surface area $A = 50$ acres. The following table is obtained after analyzing monthly climatic data:

	MARCH	APRIL	MAY	JUNE	JULY
Net solar radiation (ly day^{-1})		364	442	489	477
Atmosphere longwave (ly day^{-1})		603	678	750	796
Back radiation (ly day^{-1})		−695	−771	−848	−907
Evaporative heat flux (ly day^{-1})		−44	−94	−178	−350
Conductive heat flux (ly day^{-1})		156	95	47	−2
Net heat influx (ly day^{-1})					
Pond temperature (°C)	4				

a) Complete the table by computing the net heat flux into the lake and the average lake temperature for April through July.

b) Analyzing many years of data, it is found that the net heat flux into the lake is given by

$$q_n(t) = C \sin(\omega t + \theta),$$

where $\omega = 2\pi/\tau$, τ in days, and C in ly day^{-1}. If at $t = 0$ the lake temperature is T_o, find the variation of this variable during a period τ.

c) For $C = 300$ ly day^{-1}, $\tau = 365$ days, and $\theta = -\pi/4$, sketch the net heat flux in the lake and the temperature during the year.

(Contributed by Dr. Angelos Protopapas.)

4. Consider an agricultural field in which the crop covers 40% of the surface area. Assume that only transpiration occurs from regions covered with vegetation. An irrigation scheme that moistens the whole area is being considered. The particular crop being grown is most efficient when the soil moisture content, as a fraction of total volume, is $\theta = 0.2$. Therefore, irrigation rates are set to keep the soil at this moisture content. The actual evaporation from moist soil is directly proportional to the level of saturation. Using the following parameters compute the rate of evaporation from the soil and the rate of transpiration from the crop. Use this to compute the volume of irrigation water required daily per hectare. You may neglect loss of irrigation water by infiltration to groundwater.

Parameters

Air temperature, 25°C;

Relative humidity, 0.6;

Net radiation exchange for cropped area, 250 ly day^{-1};

Net radiation exchange for bare soil, 300 ly day^{-1};

Latent heat of evaporation, $597.3 - 0.57T$ cal g^{-1} (T in degrees Celsius);

Atmospheric pressure, 1000 mb;

Specific heat of air at constant pressure $C_{p_a} = 0.22$ cal g^{-1} °C^{-1};

Crop vapor flux resistance $r_s = 0.5$ s cm^{-1};

Saturation vapor pressure in millibars for T in degrees Celsius, $e_s = 33.8639[(0.00738T + 0.8072)^8 - 0.000019|1.8T + 48| + 0.001316]$;

Dry air density 1.15×10^{-3} g cm^{-3};

Soil porosity, $n = 0.4$; and

Plant roughness and wind conditions such that the atmospheric diffusion resistance $r_a = 1.6$ s cm^{-1}, the same as bare soil.

5. A location receives 300 ly day^{-1} of net shortwave radiation. The air temperature at the site is 26.6°C and the atmospheric emissivity is 0.74. The Bowen ratio is approximated as 0.1. The observed evaporation is 100 cm yr^{-1}. What would be an estimated average annual temperature of a freshwater lake in the area if on the average the change in energy storage is equal to all the net advected energy in the year?

6. NASA reports that remote sensing information (satellite-borne radar data) leads to a precipitation estimate of 0.4 mm hr^{-1} during the month of August, over a large tropical ocean area. The mean daily air temperature is 70°F; the mean solar radiation is 600 ly day^{-1}; the dew-point temperature is 50°F; and the monthly mean wind speed is 100 mi day^{-1}. Approximate and state assumptions for evaporation over the ocean. Does the estimated precipitation make sense relative to your evaporation estimate? Explain.

7. Consider a reservoir on a river at 30° latitude. The minimum flow occurs during September and is 4000 $ft^3 s^{-1}$ at a mean temperature of 60°F. Construction of a proposed steam plant for power generation requires diversion of 2000 $ft^3 s^{-1}$ of this flow for condenser water. The condenser water undergoes a temperature increase of 25°F. It is proposed to build a reservoir to dispose of the excess heat by evaporation from the reservoir surface. Two geometries have been suggested:

a) The 2000-$ft^3 s^{-1}$ condenser discharge and the 2000-$ft^3 s^{-1}$ unheated water both flow into the heat-exchange reservoir, which has a 4000-$ft^3 s^{-1}$ outflow.

b) Only the 2000-$ft^3 s^{-1}$ condenser discharge passes through the heat-exchange reservoir. The 2000-$ft^3 s^{-1}$ outflow from this reservoir is then recombined with the 2000-$ft^3 s^{-1}$ unheated water.

For each case, estimate the required surface area so that the mean temperature of the recombined streamflow will not exceed 70°F. Assume the reservoir to be in a steady thermal state, the average September air temperature to be 55°F, the relative humidity to average 60%, and the mean wind velocity to be 5 miles per hour. Meteorologic variables are measured 6 in. above the lake surface and the wind velocity vanishes at 0.5 in. above the surface. (From P. S. Eagleson, *Dynamic Hydrology,* McGraw-Hill, 1970.)

8. Derive Eq. (5.21), giving the heat of evaporation in terms of vapor pressure differences, wind velocity, and roughness height.

9. Measurement of evaporation from a vegetated surface yields 0.5 cm day^{-1}. The average wind speed, temperature, and relative humidity 8 ft above the surface are 5 ft s^{-1}, 25°C, and 60%, respectively, while the surface temperature is 28°C. Estimate the magnitude of the effective surface-roughness parameter if the net energy influx is 400 ly day^{-1}. (From P. S. Eagleson, *Dynamic Hydrology,* McGraw-Hill, 1970.)

10. Derive a general form for Eq. (5.49), giving the proportion of advected and stored energy used in evaporation.

11. Cruff and Thompson [1967] performed a comparison of methods yielding potential evapotranspiration. For a growing season from May to October in Los Angeles (about 35°N latitude), they give a mean temperature of 65°F. Using that information, estimate the potential evapotranspiration during the growing season using the Thornthwaite and Blaney–Criddle methods. Assume that the vegetation is alfalfa with a seasonal K value of 0.90.

12. Construct a nomogram for the solution of the Thornthwaite equation (Eq. 5.59). If needed, you can refer to the original reference, Thornthwaite [1948].

13. The growing season of corn in a northeast state extends from May 1 through August 30. The average mean monthly temperatures are May, 50°F; June, 65°F; July, 70°F; and August, 72°F. Estimate the seasonal potential evapotranspiration.

14. Over a 20-year period and using rain gages over a 1000-km^2 terminal lake (no outlets), you compute that the average precipitation over the lake is 45.7 cm. The point variance, σ_p^2, of the precipitation anywhere in any one year is 100 cm^2. Similarly, you have estimated mean evaporation over the lake to be 100 cm, using five evaporation pans with 10 years of data. The point variance of evaporation is 80 cm^2. Average annual lake net inflow is estimated as 50 cm by averaging 20 years of records. The streamflow estimate has a mean square error of 10 cm^2.

Annual precipitation shows a lag-one correlation ρ of 0.1 and an exponential correlation in space that decays with a parameter $h = 0.05$. Evaporation is either more *or* less correlated by a factor of 2 in the parameters than rainfall (you must decide whether it is reasonable to expect that it is more or less correlated both in space and time). What will an estimate of unaccounted losses or inflows into the lake be and what is the mean square error of that estimate?

15. Using the data of Example 5.3 on a lake approximately square in shape, 100 km^2 in area, and air temperature at 2 m of 20°C, find evaporation using equations by Meyer, Throne, Harbeck, Rimsha and Danchenko, Ryan and Harleman, and Goodling et al. (see Table 5.3).

16. Consider a reservoir on a river at 45°N latitude. The minimum flow occurs during September and is 30 $m^3 s^{-1}$ at a mean temperature of 20°C. Construction of a proposed steam plant for power generation requires diversion of 15 $m^3 s^{-1}$ of this flow for condenser water. The condenser water undergoes a temperature increase of 10°C. It is proposed to build a reservoir to dispose of the excess heat by evaporation from the reservoir surface.

Only the 15-$m^3 s^{-1}$ condenser discharge passes through the heat-exchange reservoir. The 15-$m^3 s^{-1}$ outflow from this reservoir is then recombined with 15-$m^3 s^{-1}$ unheated water. Estimate the required reservoir's surface area so that the mean temperature of the recombined streamflow will not exceed 23°C. Assume the reservoir to be in a steady thermal state, the average September air temperature to be 15°C, the relative humidity to average 60%, and the mean wind velocity to be 2 $m s^{-1}$. Also assume 50% cloud cover and smoggy conditions. State any other assumptions you make.

17. Assume that the average summer temperature in July in your home town is 70°F and daily solar radiation is 650 ly day^{-1}. The mean dew-point temperature is 60°F and the mean wind speed is 50 mi day^{-1}. You build a swimming pool 30 ft by 75 ft. What would be the minimum amount of water you would need to add during July to maintain a constant depth?

18. During a month five rainfall events occur over a forested area with a leaf area index of 2.5. Evaporation potential is on the order of 0.5 mm hr^{-1}. The forest has a storage capacity of 0.05 cm. The five storms had the following depths and durations:

STORM	DEPTH (cm)	DURATION (hr)
1	1	7
2	0.5	5
3	1	4
4	2	24
5	0.2	2

Estimate the total amount of throughfall to the ground during the month.

19. Find an expression for the rate of runoff production at a point below a forested canopy. Take into account interception and storage.

20. Over a large region, which of the three depth-distribution curves of Figure 5.14 is more reasonable? Explain.

21. A simple model of soil moisture depletion at the root zone would be

$$\frac{d\theta}{dt} = -P - ET_a ,$$

where P is percolation to deeper soil zones and ET_a is actual evapotranspiration.
Percolation is given by

$$P(\theta) = d\theta^c$$

and similarly, actual evapotranspiration is

$$ET_a = \begin{matrix} ET_P & \theta^* \leq \theta \\ a\theta^b & 0 \leq \theta \leq \theta^*, \end{matrix}$$

where θ^* is a threshold moisture (see Eq. 5.60).

Write integral expressions for the total actual evapotranspiration between an initial time t_i (with initial moisture θ_i) and an arbitrary time t in the future. Carry out the integrals for the case when P and ET_a are linear.

Chapter 6

Snowpack and Snowmelt

6.1 INTRODUCTION

Snowmelt plays a major role in the hydrology of midlatitudes and of rivers originating in high mountains. Even tropical rivers like the Amazon may have significant snowmelt components originating in the Andes. In contrast to rainfall, snowfall has a delayed effect on river flow and hydrology. Accumulations occurring during winter months will become important during the following spring months. Meltwaters are crucial for water supply (i.e., irrigation in the west and midwest of the United States) and can also cause serious floods, particularly when compounded with spring rainfall. In rangelands, snowmelt can be fairly important in the replenishment of soil moisture crucial for crop development and in the recharging of groundwater supplies.

Unfortunately, snow hydrology has lagged somewhat behind other branches of the field. This can probably be attributed to difficulties of field measurements and to the inherent slow response of snowmelt systems. It is difficult to measure and obtain good data on the extensive and highly spatially variable snowfall, particularly given the usually harsh weather conditions that prevail during winter periods. Snow hydrology also requires long-term commitments to study the snowmelt process months after the snowfall, year after year.

Our knowledge of snow hydrology has been significantly advanced by the U.S. Army Corps of Engineers, Cold Regions Research Laboratory, together with the U.S. National Weather Service. Their report (U.S. Army Corps of Engineers [1956]) is still the keystone reference in the field. Anderson (Anderson and Crawford [1964]; Anderson [1968], [1973], [1976], [1978a], and [1978b]) has probably been one of the most influential researchers on the subject, having developed snowmelt components for two of the most popular rainfall-runoff models in existence, the Stanford Watershed Model and the National Weather Service River Forecasting System Model. Other works of significant impact have been those of Winston [1965] and Amorocho and Espildora [1966]. The following sections will follow closely the work of Anderson but will also rely on a report by Laramie and Schaake [1972]. Eagleson [1970], Branson et al. [1981], and Gray [1973] will also be sources of material.

Snowmelt effectively is delayed precipitation. As such it could have been discussed within or immediately after Chapter 4. Many authors put it within the realm of runoff, a concept we really have yet to discuss. Here it is studied after evaporation because, as will be seen, the tools for analysis of snowmelt are analogous to those seen in Chapter 5.

6.2 SNOW ACCUMULATION AND MEASUREMENT

In Chapter 4 we mentioned that the nature of precipitation is very much dependent on the history of hydrometeor development and cloud physics. Nevertheless, Figure 6.1 (also Fig. 4.11) indicates that air temperature is a reasonable index of precipitation type. Chances are that snow will occur for temperatures below 0° to 1°C. The density of new-fallen snow is also a function of multiple factors. Generally, the wetter or warmer the snow is, the denser it will be. Figure 6.2 illustrates this general behavior, relating density to surface air temperature. Note that the density of new-fallen snow ranges

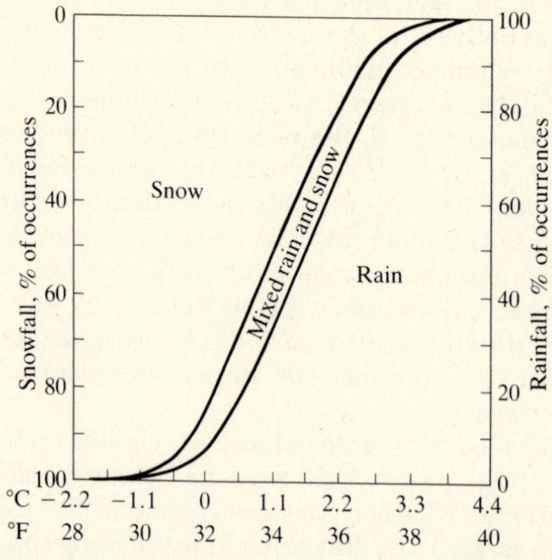

FIGURE 6.1 Frequency of occurrence of rain and snow at various temperatures. Source: U. S. Army Corps of Engineers [1956].

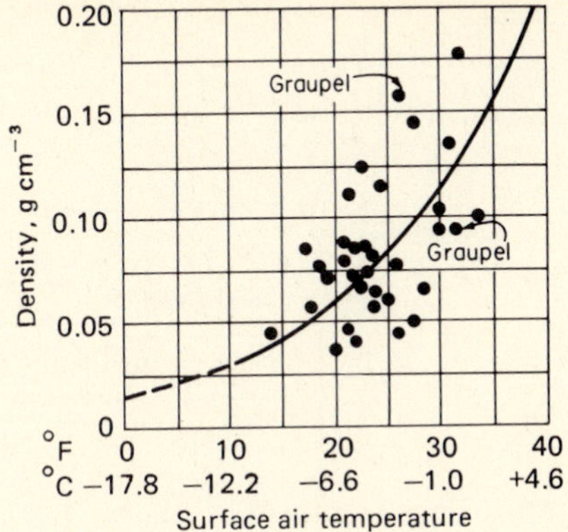

FIGURE 6.2 Density of new-fallen snow. Source: Gray [1973]. Reproduced by permission of the National Research Council of Canada.

from about 0.05 to 0.2 g cm^{-3}, with a concentration near or somewhat below the commonly used value of 0.1 g cm^{-3}.

Snowfall over an area tends to be more uniform than rainfall (Gray [1973]). On the other hand, snow accumulation is largely a function of elevation, slope, exposure, and vegetative cover. Snow accumulation generally increases with elevation because of the combined effect of the prevailing lower temperatures and the increased frequency of precipitation events caused by orographic effects. Figure 6.3 shows a possible distribution of precipitation within a basin. Branson et al. [1981] give results of Meiman [1968], which indicate a 5 to 200% increase in snow accumulation per 1000 ft. The water equivalent of the snowpack (i.e., snow-water equivalent) decreases with slope and exposure and increases with deviation of aspect from the south, in the northern hemisphere (Laramie and Schaake [1972]). Nevertheless, topography is generally too complex to obtain widely applicable relationships. Again from Meiman [1968], a study of seven sites indicated that northern exposure resulted in 3 to 138% more snow accumulation than southern exposure.

Interception of snowfall by vegetation can be fairly effective. In forested areas, this may lead to direct sublimation of snow from the forest canopy. Some of the intercepted snow is redistributed by wind and melt. Very little is known about the amounts involved in this complex process. Most researchers ignore this issue and prefer to treat it as part of the general error in snowfall measurement.

Different surfaces have different snow-retention capacity. Table 6.1 gives snow retention coefficients from Kuzmin [1960]. The snow retention coefficient is the ratio of snow catch in the surface in question to the accumulation in an otherwise virgin soil. Note that forest openings or cuttings are particularly effective snow collectors. Gullies and surface depressions are also similarly effective. This mostly responds to the interaction of wind and topog-

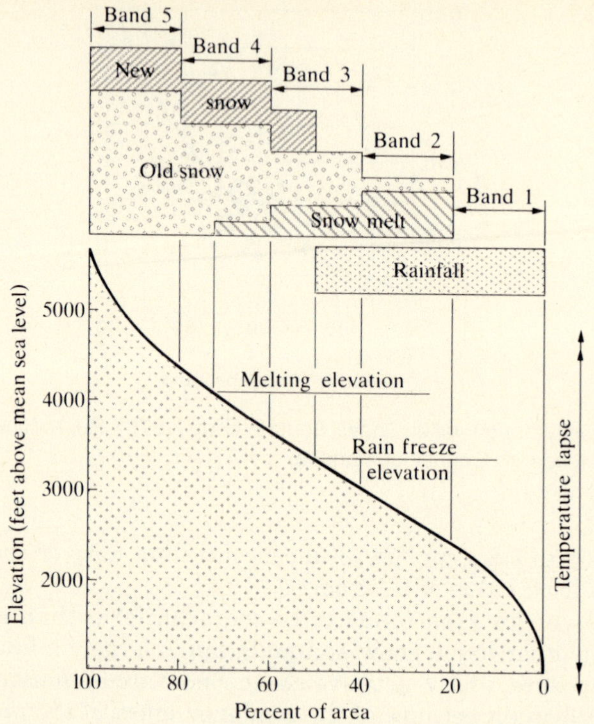

FIGURE 6.3 Various conditions of precipitation (rain or snow) and snowmelt that may occur simultaneously as a function of elevational bands within a basin. Source: U. S. Army Corps of Engineers [1971].

raphy. In forests much of the intercepted snow is blown off and settles in openings where it is shaded from further transport.

Transport by wind is fairly common, particularly after fresh falls of light snow. Given a smooth surface of fresh, light snow, Figure 6.4 may help in

TABLE 6.1 Snow-Retention Coefficients

Open ice surface of lakes	0.4 to 0.5
Arable land	0.9
Virgin soil	1.0
Hilly districts	1.2
Large forest tracts	1.3 to 1.4
River beds	3.0
Rush growth near lakes	3.0
Forest cuttings of a radius of about 100 to 200 m and edges of forests	3.2 to 3.3

Source: Kuzmin [1960].

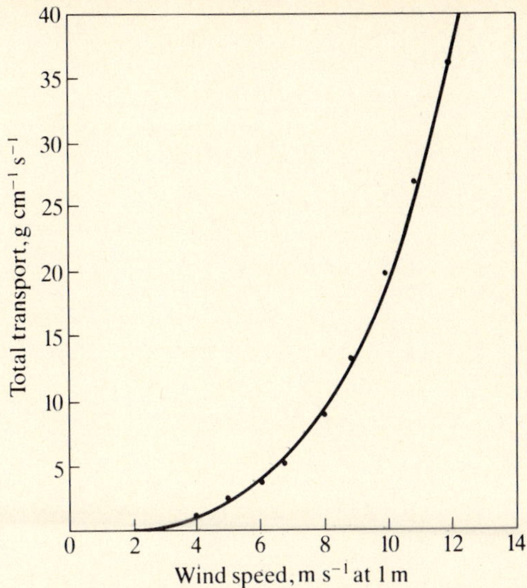

FIGURE 6.4 Total snow transport. Sources: After Komarov [1954] and Branson et al. [1981].

computing total transport. It gives the amount transported in grams per second per centimeter of width perpendicular to the wind, as a function of wind speed. Much of wind-carried snow may sublimate. Schmidt [1970] and Tabler and Schmidt [1973] developed a model of this process and give the distance a particle of a given diameter may travel before it sublimates (Fig. 6.5). The results correspond to a site in Wyoming. They also studied the relative influence of numerous variables in the transport process. Figure 6.6 gives the percent distribution of total snow transport as a function of wind speed, air temperature, incoming radiation, hour of the day, and month. Note that most transport occurs at low temperatures (low-density snow), at night, in February, and at wind speeds above 7.5 m s^{-1}.

Although all the factors discussed up to now conspire to make the quantification of the spatial variability of snow very hard, most locations show persistent patterns of snow coverage and of snow ablation. Figure 6.7 gives typical relations between percent of basin area covered by snow and the ratio of liquid-water equivalent covering the given percent of the basin to the total snowpack liquid-water equivalent when the basin is fully blanketed, A_i. The shape of the curves are related to topographic and snow-accumulation patterns. The meaning of the curves is best explained by Anderson [1978b]:*

> Curve *A* indicates that bare ground appears at a continually increasing rate as the snow cover ablates. Such a curve is typical

*E. A. Anderson, Hydrologic Research Laboratory, National Weather Service/NOAA.

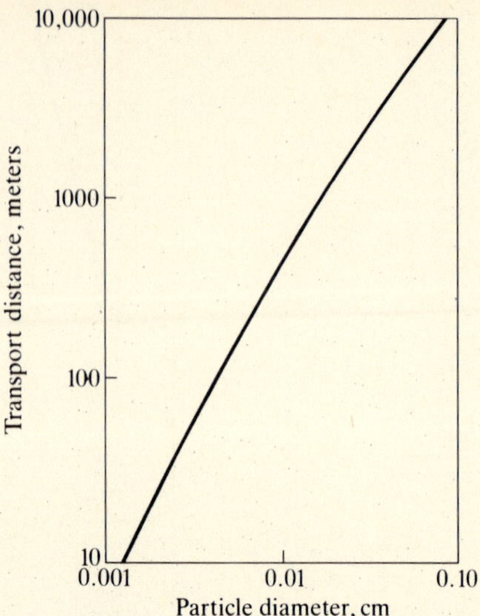

FIGURE 6.5 Snow transport distance as a function of particle diameter, using the mean values for winter conditions in Wyoming during drifting over the 1970–1971 winter. Source: *The Role of Snow and Ice in Hydrology*, vol. 1. Copyright © Unesco/WMO/IAHS. (Article by R. D. Tabler and R. A. Schmidt.)

of areas in which there is variability in accumulation and melt, but the variability is rather evenly scattered over the area. Curve B is similar to curve A in the beginning, but at the lower end the rate at which bare ground appears is reversed. This reversal indicates that a portion of the area accumulates much more snow or has a significantly lower melt rate (or a combination of both factors) than the rest of the area. The reversal may be caused by forested areas with northerly aspects, dense conifer stands within an area with generally mixed cover, or large accumulations of snow in drifts, in ravines, or at the higher elevations. Curve C is similar to curve A in the middle and at the lower end. In the beginning curve C indicates that the areal cover drops off very rapidly when ablation begins. This indicates that a portion of the area accumulates much less snow or has a much higher melt rate (or both) than the remainder. This may be caused by open areas with a southerly aspect, open areas within a forest which consists mainly of conifers, areas which are typically blown free of snow, or little accumulation of snow at lower elevations. Curve D is for an area which can be basically divided into the extremes, i.e., low accumulation and/or high melt rates and high accumulation and/or low melt rates. Normally if such a curve is required, it would be preferable to subdivide the areas and model each portion separately since they are so distinctly different. Both curves C and D generally exist only in areas where the SI value [author's note: value of water

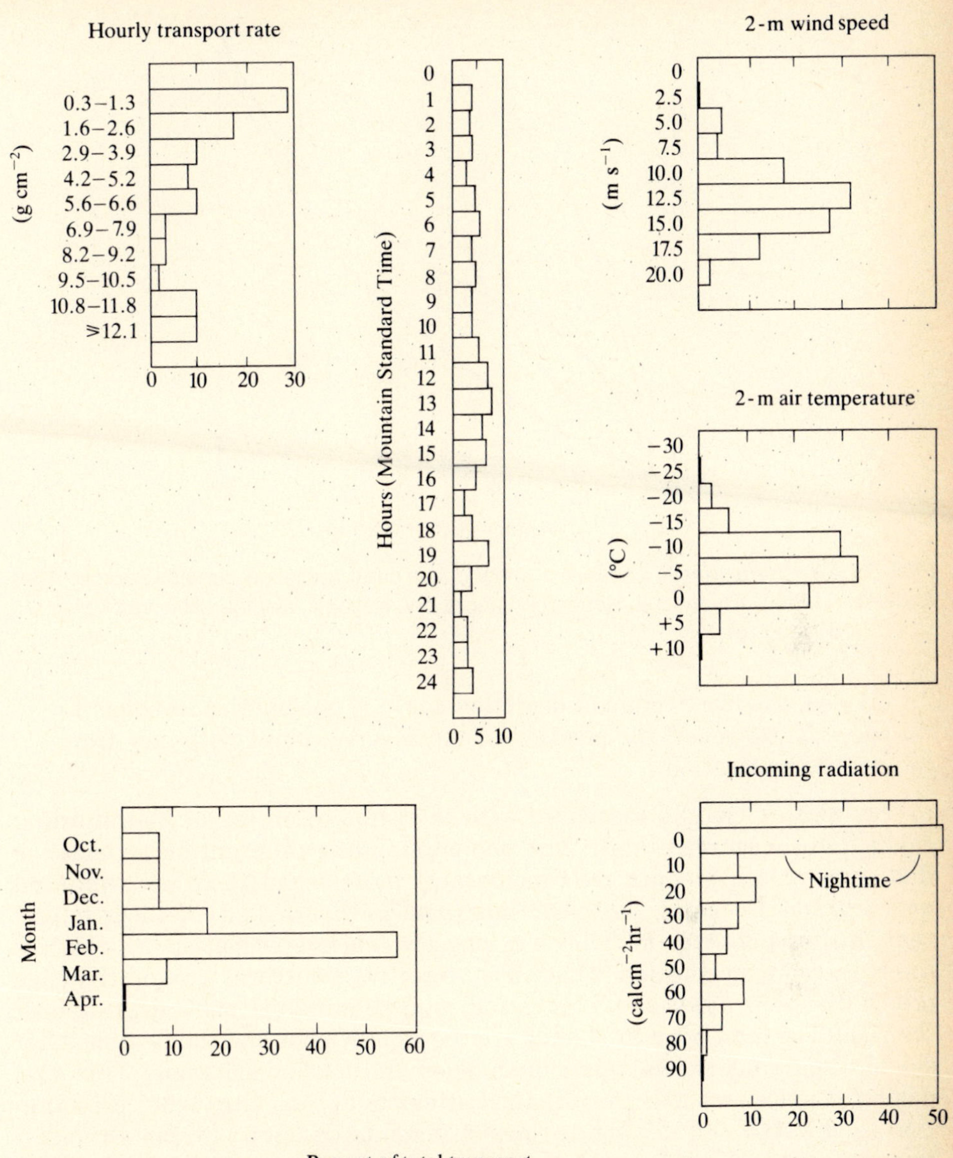

FIGURE 6.6 Percentage distribution of total snow transport per select factor during all drifting events over the 1970–1971 winter. Source: *The Role of Snow and Ice in Hydrology*, vol. 1. Copyright © Unesco/WMO/IAHS. (Article by R. D. Tabler and R. A. Schmidt.)

equivalent when 100% cover is expected] is greater than the largest water-equivalent that occurs during most years. From this discussion and a knowledge of the area, the user should be able to select an initial estimate of the areal depletion curve. If the area is properly sub-

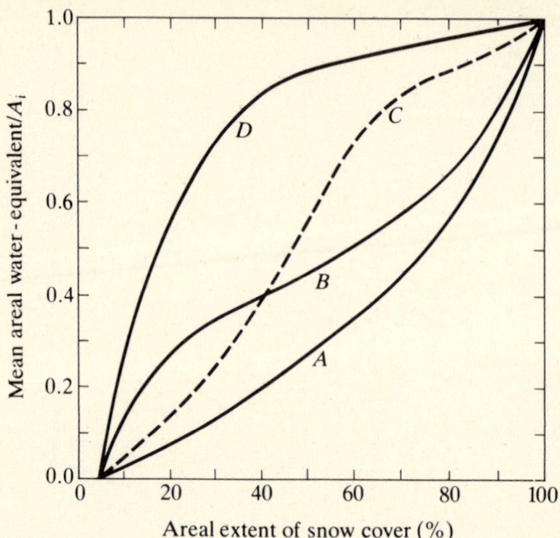

FIGURE 6.7 Characteristic shapes of snow cover areal depletion curves. Source: After Anderson [1978b], Hydrologic Research Laboratory, National Weather Service/NOAA.

divided, the most common depletion curve is one similar in shape to curve B. However, the position of the reversal point will vary from location to location.

Snowfall is usually measured with the same recording or nonrecording gages that measure rainfall. The one outstanding differentiating factor is that snow measurements are considerably more sensitive to wind-induced underestimation errors. This was illustrated in Figure 4.23. Methods to prevent this type of error have been a longstanding preoccupation. Shields are the most common solution. These are protective structures around the gage orifice that slow down and direct wind so as to minimize snow drifting over the gage. The most popular shields are the rigid Nipher and the flexible Alter shield. Figure 6.8 shows a pair of shielded and unshielded gages. This dual gaging practice responds to statistical attempts to correlate "true" precipitation to shielded and unshielded measurements (Hamon [1972]). Other attempts to correct catch errors have involved fences surrounding the gage (Rechard and Larson [1971]; Larson [1971]), and buried installation (Jairell [1975]). Unshielded snow gages may show deficiencies that are three times that of shielded rain gages (Gray [1973]; Larson and Peck [1974]; Larson [1972a, 1972b]).

Once the snow is on the ground, it becomes part of the snowpack, a subject of the following subsection. Snowpack is the amalgamation of old and new snow. Old snow suffers changes induced by added weight (resulting in denser packing), freeze and thaw cycles, and rainfall. The end results are

FIGURE 6.8 Picture of a shielded and unshielded rain gage as used on the Reynolds Creek Experimental Watershed near Boise, Idaho. Source: *Rangeland Hydrology,* 1981, by Branson, Gifford, Renard, and Hadley. Society for Range Management Pub.

changing density, water content, albedo, and other characteristics. It is the melting of the snowpack that is most interesting to hydrologists.

Snowpack is measured in several different ways. Some techniques are geared to estimate the snow-cover water equivalent; others just attempt to get a feeling for the depth of snow. By making assumptions about the density, the latter methods also yield water-equivalent estimates. The following description of the methods is adapted from Gray [1973].

Water Equivalence Sampling Procedures

Snow surveys consist of several core samples along a course. A course is a predetermined geometry of samples, sometimes aligned. The samples are hollow tubes with cutting edges that can be driven into the snowpack. The resulting snow core can be weighed to obtain water content. The method is mostly dependent on the choice of course. The sampling locations should be representative of the region and good indices of whatever the ultimate objective of the water equivalent estimate is. For example, some snow courses may be closely related to runoff or river flow production, others may be good indices of the average water equivalence of snow over the whole region. Errors of this technique are due to both instrument limitations and the inherent bias in selection of a site.

Snow-water equivalent can also be measured with radioisotope techniques. Given a source (gamma radiation from cobalt) and a receiver some distance above, the snow-water equivalent (or density) can be related to net transmitted radiation. Disadvantages of this technique are the cost and the time-consuming procedure of moving source and receiver along a profile. An advantage is the potential of remote aerial readout of radiation from fixed sources. Some instruments have both source and receiver in a single unit, but Gray [1973] argues that these neutron and gamma gages are not accurate.

Direct measurement of snowpack can be achieved with pressure pillows that measure the weight of the accumulated snow above them. These are flexible containers of antifreeze solution. The pressure change in them can be directly related to the weight of the overburden. Their main advantages are that they can be instrumented for telemetering operation and are reliable and fairly easy to operate.

Snow-Depth Sampling Procedures

The ruler or snowstake is the most common instrument for measuring snow depth. It can be fixed or movable. The main consideration of fixed snowstakes is that they are in representative locations and that the stake itself does not affect the pattern of snow accumulation.

Aerial snow markers are stakes with markings that can be read from airplanes. Readings are usually confirmed through photography.

Aerial photography is the most comprehensive method of measuring the spatial extent of snow cover. The spectral distribution of radiation emitted by the snow offers potential opportunities to distinguish between the type of snow cover and its characteristics. Photogrammetry, used with ground truth depth measurements, can be used to obtain fairly accurate estimates of the depth of snow cover over large areas. Photographic remote sensing procedures are limited by weather (cloud cover) and encumbered by forest cover.

6.3 SNOWPACK

The accumulation of snow on the ground — snowpack — suffers a continuous change in properties, or metamorphosis. These changes respond to compaction by weight, percolation of rain or meltwater, freeze–thaw cycles, wind, and other climatic changes. Aging of the snowpack involves a change from crystal to granular structure, an increase in density, a change in albedo, a move toward homogeneous temperature distribution, and an increase in its liquid-water content. When the snowpack becomes nearly homogeneous and contains all the liquid water it can hold against gravity, it is referred to as "ripe" or ready to produce meltwater. In the following subsections, we will detail some of the snowpack properties.

6.3.1 Density

Generally, the density of the snowpack increases with depth and as the accumulation season progresses. Figure 6.9 shows a typical vertical density profile. Figure 6.10 illustrates increased density with time. Typical densities are given in Table 6.2 for snowpacks under various conditions and origins.

Anderson and Crawford [1964] give the following relationship for the reduction in depth of the snowpack due to compaction by new snow

$$\Delta D = \frac{P \times D}{WE}\left(\frac{D}{10}\right)^{0.35}, \tag{6.1}$$

where ΔD is the change in depth; P is the water equivalent of new snow, a function of its density (Fig. 6.2); D is the present snowpack depth; and WE is the water equivalent of the snowpack. The above equation assumes inches as unit of depth. The density of new snow is also parameterized by the same authors as

$$\begin{aligned} \rho_N &= 0.05 + (T_a/100)^2; && \text{for } T_a > 0°\text{F} \\ \rho_N &= 0.05; && \text{for } T_a \leq 0°\text{F}, \end{aligned} \tag{6.2}$$

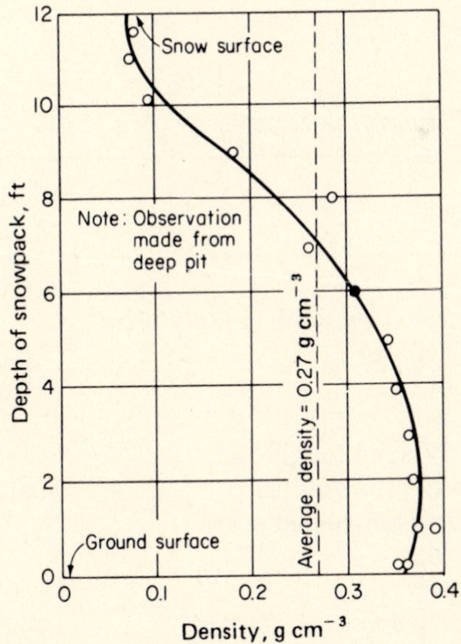

FIGURE 6.9 Density variation in a winter snowpack. Source: U. S. Army Corps of Engineers [1956].

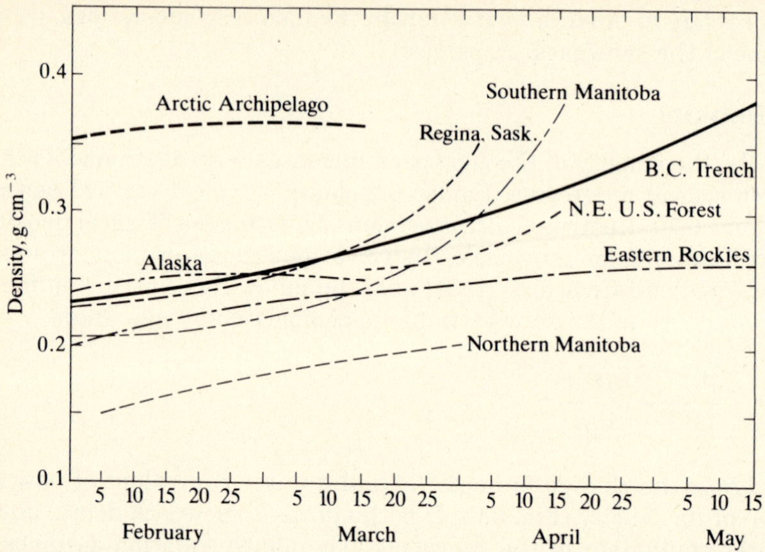

FIGURE 6.10 Seasonal variation in typical snow densities from various geographic areas. Source: McKay and Thompson [1967].

where T_a is air temperature and the density ρ_N is in grams per cubic centimeter. Using Eq. (6.2) would imply that the depth of the new snow is

$$D_N = \frac{P}{\rho_N} \tag{6.3}$$

TABLE 6.2 Snowpack Densities

SNOW TYPE	DENSITY ($g\,cm^{-3}$)
Wild snow	0.01–0.03
Ordinary new snow, immediately after falling in still air	0.05–0.065
Settling snow	0.07–0.19
Settled snow	0.2–0.3
Very slightly wind-toughened, immediately after falling	0.063–0.08
Average wind-toughened snow	0.28
Hard-wind slab	0.35
New firn snow*	0.4–0.55
Advanced firn snow	0.55–0.65
Thawing firn snow	0.6–0.7

*Snow partly consolidated into ice.

Sources: McKay [1968] and Branson et al. [1981].

and the new depth of the snowpack becomes

$$ {}_2D = {}_1D - \Delta D + D_N , \tag{6.4} $$

where ${}_1D$ and ${}_2D$ are the old and new depths, respectively. The water equivalent of the snowpack can also be updated as

$$ {}_2WE = {}_1WE + P \tag{6.5} $$

and the new snowpack density relative to that of water must be

$$ {}_2\rho_p = \frac{{}_2WE}{{}_2D} . \tag{6.6} $$

6.3.2 Cold Content

In order for the snowpack to produce meltwater, it must have a temperature above 0°C. Where this is not the case, a certain amount of heat input is required to raise the temperature before any melt occurs. This threshold energy is called the cold content. To obtain it, we require the snow temperature and density profile,

$$ Q_{cc} = -\int_0^D \rho_p C_s T_p \, dz \text{ cal cm}^{-2} , \tag{6.7} $$

where D is snowpack depth (in centimeters); z (in centimeters) is the vertical coordinate, measured positively upward from ground surface; $\rho_p(z)$ is the snowpack density as a function of depth (grams per cubic centimeter); $T_p(z)$ is the snowpack temperature profile (in degrees Centigrade); and C_s is the snowpack specific heat (calories per gram per degree Centigrade), which takes a value of 0.5 for all practical snow and ice densities. Rarely would $\rho_p(z)$ and $T_p(z)$ be known. At best we can hope to obtain depth-averaged values. The cold content would then become

$$ Q_{cc} \approx -\rho_p C_s D T_p , \tag{6.8} $$

where all quantities are now depth-averaged, and D is taken as a positive snow depth value. Remember, $T_p < 0$°C. Once the snowpack is ripe, it may be reasonable to assume that the cold content is concentrated on any surface ice crust, since any deeper the snowpack is probably still at 0°C. The snow-surface temperature and depth of the ice crust would then be used in Eq. (6.8).

The cold content is commonly given in terms of the depth of water at 0°C, which upon refreezing will raise the snowpack temperature to 0°C. This depth can be expressed as

$$D_{cc} = \frac{Q_{cc}}{\rho_w L_f} \text{ cm}, \tag{6.9}$$

where L_f is the latent heat of freezing (79.7 cal g^{-1}) and ρ_w is the density of water at $T = 0$°C, 1 g cm^{-3}. Using Eq. (6.8),

$$D_{cc} = -\frac{\rho_p}{\rho_w} \frac{C_s DT_p}{L_f}, \tag{6.10}$$

which on substituting values for ρ_w, C_s, and L_f yields

$$D_{cc} = -\frac{\rho_p DT_p}{159.4} \approx -\frac{\rho_p DT_p}{160} \text{ cm}. \tag{6.11}$$

Given new snowfall at a given temperature, its cold content can be obtained as

$${}_N D_{cc} = -\frac{\rho_N D_N T_N}{160}, \tag{6.12}$$

where subscript N implies new snow properties. The snowpack cold content can then be updated by

$${}_2 D_{cc} = {}_1 D_{cc} + {}_N D_{cc}. \tag{6.13}$$

6.3.3 Thermal Quality

The energy required to produce meltwater from the snowpack is equal to the cold content plus the latent heat demanded by the amount of melt produced. This total energy for melting the snowpack is then

$$Q_o = \rho_p D L_{ms} + Q_{cc}, \tag{6.14}$$

where L_{ms} is the latent heat of melting of the snow. The thermal quality is defined as the ratio of Q_o to the energy consumed in producing the same amount of melt from pure ice at 0°C. This latter quantity is

$$Q = \rho_p D L_m, \tag{6.15}$$

where L_m is the latent heat of melting for ice.

The ratio of Eq. (6.14) to Eq. (6.15) is the desired thermal quality θ

$$\theta = \frac{\rho_p D L_{ms} + Q_{cc}}{\rho_p D L_m} = \frac{L_{ms}}{L_m} - \frac{C_s T_p}{L_m}, \tag{6.16}$$

where temperatures are in degrees Centigrade.

For subfreezing snowpacks ($T_p < 0$), $L_{ms} = L_m$; therefore $\theta > 1$. For ripe snowpacks with some water content, $L_{ms} < L_m$ and by definition $T_p = 0$; therefore $\theta < 1$. Eagleson [1970] quotes a typical value of about 0.97 in such cases.

6.3.4 Liquid-Water Content

Much like soil, the snowpack has a given porosity, defined as the percent volume of voids to total volume. Table 6.3 gives some typical snow porosities in comparison with that of sand, gravel, and straw. When the temperature is greater than 0°C, the pores are capable of holding water mostly by capillary and tension forces. The amount of liquid water retained this way is called the liquid-water content, and is defined in terms of the percent by weight of liquid water in the snowpack. There is a maximum value of liquid water content beyond which water will drain by gravity action. This is called the liquid-water holding capacity. Figure 6.11 shows an empirical relationship between liquid-water holding capacity and snow density as suggested by Amorocho and Espildora [1966]. Since the latent heat of melting of snow must be proportional to its water content, we have

$$\frac{L_{ms}}{L_m} = 1 - \frac{W}{100}, \tag{6.17}$$

TABLE 6.3 Air Permeabilities and Porosities of Some Common Porous Media

MATERIAL	POROSITY	AIR PERMEABILITY ($\mathrm{cm\,s^{-1}}$)
Coarse sand (0.75 mm)	0.40	6
Very coarse sand (1.1 mm)	0.39	28
2-mm gravel	0.40	75
Wind-packed snow	0.67–0.80	25–100
Old snow (fine-grained)	0.60–0.77	100–500
Old snow (coarse-grained)	0.50–0.78	200–1000
New snow	0.80–0.87	500–1000
10-mm gravel	0.40	1885
Straw	0.97	5400

Sources: Van Haveren [1975] and Branson et al. [1981].

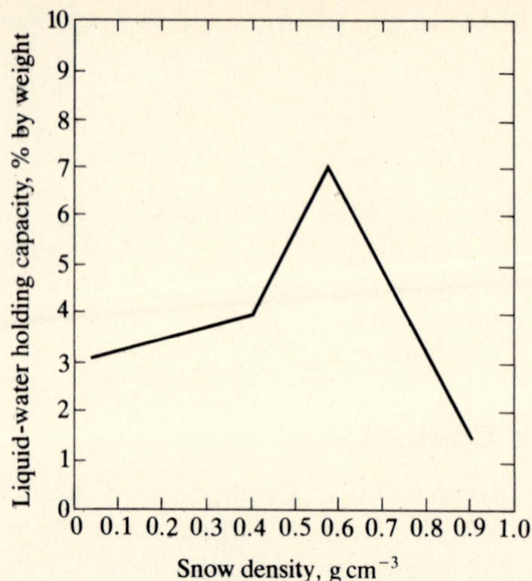

FIGURE 6.11 Variation of the liquid-water holding capacity of snow with snow density. Source: J. Amorocho and B. Espildora, "Mathematical Simulation of the Snow Melting Process," *Water Science and Engineering*, Paper no. 3001, University of California, 1966.

where W is the percent, by weight, of water content. Using Eq. (6.17) in Eq. (6.16), we see that for ripe snowpack ($T_p \approx 0°C$), the thermal quality must then be

$$\theta = 1 - \frac{W}{100} \tag{6.18}$$

or

$$W = 100(1 - \theta). \tag{6.19}$$

6.3.5 Albedo

As the next section will discuss, snowpack melt depends on the available energy. An energy budget as seen in Chapter 5 will then be required. Two of the principal components of that budget are the incoming shortwave radiation and its reflected proportion. Because of its metamorphosis, the snowpack has constantly changing albedo. This was seen in Figure 2.14, repeated here as Figure 6.12. Figure 6.13 relates albedo to the summation of daily maximum temperatures (in degrees Fahrenheit) since the last snowstorm. Clearly there can be drastic reductions in the snowpack's ability to reflect shortwave radiation. As the snowpack ages, more shortwave radiation is absorbed, which in turn increases the available energy for snowmelt.

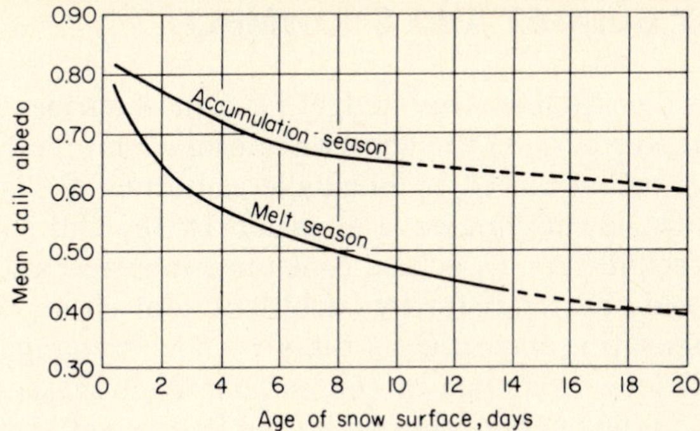

FIGURE 6.12 Time variation in albedo of a snow surface. Source: U. S. Army Corps of Engineers [1956].

Laramie and Schaake [1972] use an expression provided by Anderson [1968] and fit the albedo curves in Figure 6.12 with

$$A = 0.85(0.94)^{(\tau)^{0.58}} \qquad \text{for the accumulation season},$$

$$A = 0.85(0.82)^{(\tau)^{0.46}} \qquad \text{for the melt season},$$

where τ is the number of days since the last storm, or the age of the snow surface.

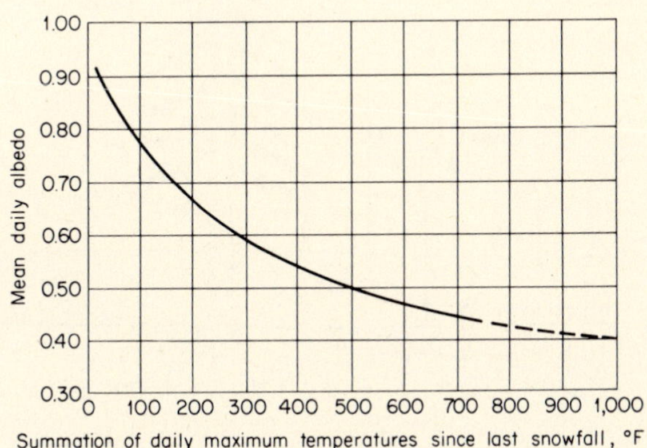

FIGURE 6.13 Variation of albedo of a snow surface with accumulated temperature index. Source: U. S. Army Corps of Engineers [1956].

6.4 ENERGY BUDGET AND SNOWMELT

Figure 6.14 gives a possible energy budget for a snow surface. Most terms are defined as in Chapter 5: Q_s is the shortwave radiation input; Q_r is its reflection of shortwave radiation; Q_a is longwave radiation; Q_{ar} is reflected longwave radiation; Q_{bs} is the longwave emission by the soil surface; Q_v is advected heat in precipitation; Q_e is heat consumed in evaporation and sublimation; Q_h is sensible-heat transfer by turbulent convection; Q_{cd} is heat contributed by condensating vapor; Q_c is the air–snow exchange by conduction; Q_g is the ground–snow exchange by conduction; Q_w is the heat carried away by meltwater; Q_f is the energy released by the freezing of any liquid water in the snowpack; and Q_o will be the change in heat storage in the pack. The conductive heat transfer between the air and snow surface can be neglected

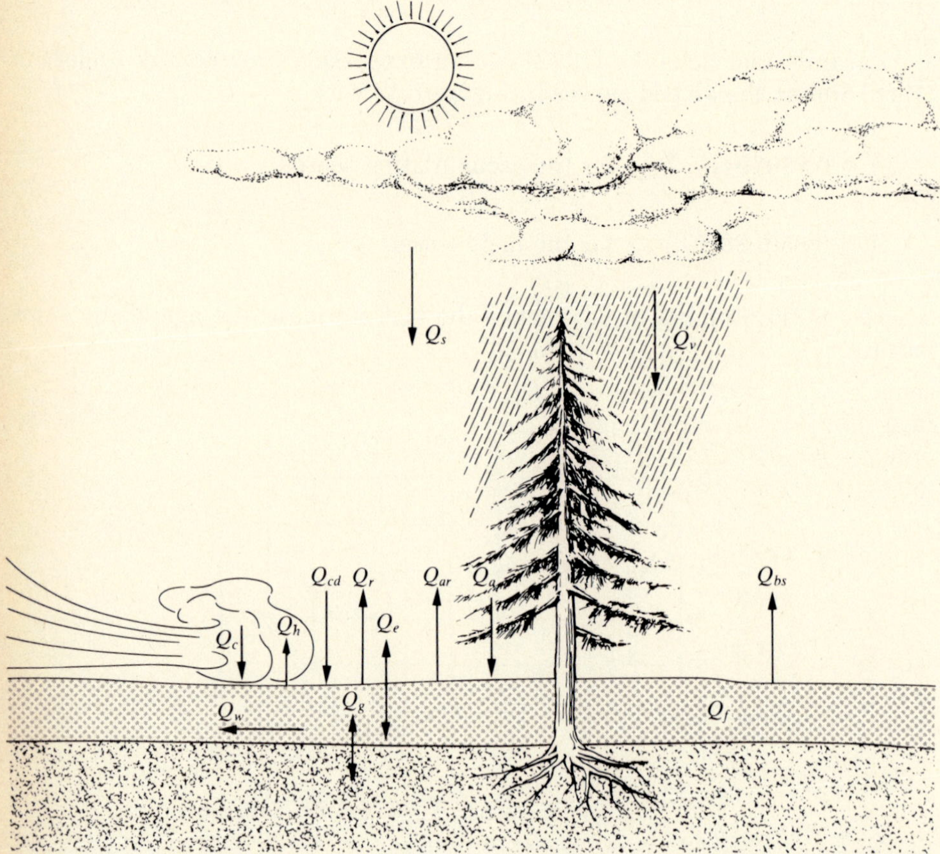

FIGURE 6.14 Schematic representation of heat-transfer components. Source: After Laramie and Schaake [1972].

(Laramie and Schaake [1972]), so $Q_c = 0$. The meltwater leaving the pack will have a temperature near 0°C. Since we are using degrees Centigrade to define heat content, the meltwater is then removing no heat, $Q_w = 0$.

Letting Q^* be the net radiation exchange, the energy budget over a given time interval becomes

$$Q_o = Q^* + Q_v - Q_e - Q_h + Q_{cd} - Q_g + Q_f. \tag{6.20}$$

Keep in mind that all the energy terms must be given in consistent units; for example, calories per square meter per day or calories per square centimeter per hour, and that a mass balance must be satisfied, as in the evaporation computations of Chapter 5.

When the change in the heat storage Q_o is positive and exceeds the existing cold content such that $Q_m = Q_o - Q_{cc}$ is positive, then melt occurs. Given $Q_m > 0$, the total meltwater in units of depth is obtained by dividing the change in heat storage in the snowpack by the latent heat of melting, the density of water, and the thermal quality:

$$H_m = \frac{Q_o}{L_m \rho_w \theta}; \qquad Q_o > Q_{cc} \tag{6.21}$$

with $\rho_w = 1\ \text{g cm}^{-3}$ and $L_m = 80\ \text{cal g}^{-1}$,

$$H_m = \frac{Q_o}{80\theta}\ \text{(centimeters per unit time)}. \tag{6.22}$$

Anderson [1968] points out that except for the Q_g term, Eq. (6.20) represents an energy balance of the upper layer of snowpack. This is true if that layer absorbs the incoming solar radiation, which is the case of layers on the order of 16 cm or so. The terms of Eq. (6.20) will be discussed further in the following subsections.

6.4.1 Net Radiation

The computation of net radiation follows the concepts given in Chapter 2 and does not require further discussion except to state that the longwave albedo of snow is nearly zero and can be neglected.

6.4.2 Advected Heat in Precipitation

The heat carried by incoming precipitation depends on its temperature

$$Q_v = C_P \rho_w P T, \tag{6.23}$$

where P is the water equivalent of precipitation in centimeters per unit time, T is the temperature of precipitation in degrees Centigrade, ρ_w is the density of water ($1\ \mathrm{gr\,cm^{-3}}$), and C_P is the specific heat of precipitation ($0.5\ \mathrm{cal\,g^{-1}\,°C^{-1}}$ if it is snow or $1.0\ \mathrm{cal\,g^{-1}\,°C^{-1}}$ if it is rainfall). The temperature of the precipitation can be taken as the wet-bulb temperature if it is rainfall. Snow occurs if the air temperature is below 0° to 1°C; this temperature can then be assigned to the precipitation. Note that in the case of snow, Q_v is most probably zero or negative and may contribute to the cold content.

6.4.3 Energy Consumed in Evaporation, Condensation, and Sensible-Heat Transfers

These three terms are computed in exactly the same way as presented in our discussion of evaporation (Chapter 5). Evaporation will occur given a decreasing vapor gradient from the snow surface to the overlying air. An inverse gradient will yield condensation. Assuming the turbulent diffusion analogy for transport results in the general Dalton-type equation

$$E = \frac{K_1}{6}\left[\frac{0.622}{P}\right](Z_a Z_b)^{-1/6}(e_s - e_a)U_b, \tag{6.24}$$

where e_a is the vapor pressure at elevation a, e_s is the saturation vapor pressure at the snow surface, U_b is the wind velocity at elevation b, P is pressure, Z_a and Z_b are elevations, and E is depth of condensation or evaporation per unit time, depending on the gradient direction. A sixth-power law for velocity profile has been assumed in Eq. (6.24), allowing e_a and U_b to be measured at different locations. If pressures are measured in millibars, elevations in feet, and velocities in miles per hour, the U.S. Army Corps of Engineers [1956] collapses Eq. (6.24) into

$$E = K_1(Z_a Z_b)^{-1/6}(e_a - e_s)U_b, \tag{6.25}$$

with $K_1 = 0.00635\ \mathrm{in.\,ft^{1/3}\,hr\,day^{-1}\,mb^{-1}\,mi^{-1}}$. E is then in inches per day. For E in centimeters per day, elevations in meters and velocity in kilometers per hour, $K_1 = 0.00651\ \mathrm{cm\,m^{-1/3}\,hr\,day^{-1}\,mb^{-1}\,km^{-1}}$. To convert to energy units, the above should be multiplied by the density of water and by the latent heat involved in the change of state. Anderson [1968] points out that if the snowpack is ripe and melting, then the latent heat of condensation (evaporation) is involved, about $600\ \mathrm{cal\,g^{-1}}$. However, if the snow is not melting, the latent heat of sublimation, involving direct changes between the solid and gaseous phases, should be used. This is about $677\ \mathrm{cal\,g^{-1}}$. In practice, though, only the latent heat of condensation is generally used, since the melting period is the one of interest, and the magnitudes of direct sublimation are such that the error becomes acceptable. Since the latent heat of condensation is 7.5 times that of melting ($600/80 = 7.5$), it is commonly quoted that a unit of condensation yields 8.5 units of melt: itself plus 7.5 units of melt produced by the released $600\ \mathrm{cal\,g^{-1}}$. Evaporation, though, uses $600\ \mathrm{cal\,g^{-1}}$, so it can freeze 7.5 units of liquid-water content for every evaporated one.

Again, based on turbulent diffusion analogy, the sensible-heat transfer becomes proportional to the temperature gradient (as in Chapter 5). The U.S. Army Corps of Engineers [1956] suggests

$$Q_h = L_e K_2 \frac{P_a}{P_o} (Z_a Z_b)^{-1/6} (T_a - T_s) U_b , \tag{6.26}$$

where P_a is the surface atmospheric pressure, P_o is the sea-level atmospheric pressure, T_a is air temperature at elevation Z_a, T_s is temperature at the surface, L_e is the latent heat of evaporation, and U_b is wind velocity at elevation Z_b. For temperatures in degrees Fahrenheit, elevations in feet, and velocities in miles per hour, the constant is $K_2 = 0.00626$ in. ft$^{1/3}$ hr day^{-1} °F^{-1} mi^{-1} (U.S. Army Corps of Engineers [1956]). Q_h is in calories per square centimeter if $L_e = 1524$ ly in.$^{-1}$. With L_e given as 600 cal g^{-1}; temperature in degrees Centigrade; elevations in meters; and velocities in kilometers per hour, the constant K_2 should be 0.00357 cm m$^{1/3}$ hr day^{-1} °C^{-1} km^{-1}.

Anderson [1968] and others simply use, in operational models, empirical equations of the form

$$Q_e = f(U) L_e \rho_w (e_a - e_s) \tag{6.27}$$

$$Q_h = f'(U) L_e \rho_w (T_a - T_s) , \tag{6.28}$$

where $f(U)$ and $f'(U)$ are empirical wind functions. In fact, once Q_e is known, the Bowen ratio concept seen in Chapter 5 would be applicable to obtain Q_h as a function of Q_e. Since the Bowen ratio states

$$\frac{Q_h}{Q_e} = C_B P_a \frac{T_a - T_s}{e_a - e_s} \tag{6.29}$$

substitution of Eq. (6.27) in Eq. (6.29) yields

$$Q_h = C_B P_a f(U) \rho_w L_e (T_a - T_s) , \tag{6.30}$$

with $C_B = 0.61 \times 10^{-3}$ °C^{-1}.

6.4.4 Heat of Conduction from the Soil

The heat flux from the soil to the snowpack is given by

$$Q_g = -K \frac{dT}{dZ} , \tag{6.31}$$

where K is the thermal conductivity of the soil, and dT/dZ is the temperature gradient from soil to snow. This quantity is generally much smaller than the rest of the surface-energy transfers. Laramie and Schaake [1972] use a constant value of 0.17 ly hr^{-1} for this term. Eagleson [1970] estimates even smaller amounts, on the order of 10^{-3} ly day^{-1}. Anderson [1968] assumes that this effect causes 0.01 in. (0.0254 cm) of melt a day, which implies about 2 ly day^{-1}.

6.4.5 Energy Released by Freezing of Liquid-Water Content

If the cold content is positive, the snowpack's temperature is below freezing. The freezing of any existing liquid water will release latent heat. The amount released can be computed as

$$Q_f = \frac{\rho_p D W L_f}{\Delta t}, \tag{6.32}$$

where, as previously defined, W is the liquid-water content (Eq. 6.17), D is the snowpack depth, ρ_p is the snowpack density, L_f is the latent heat of freezing (80 cal g^{-1}), and Δt is the period of computation.

6.5 AIR TEMPERATURE AS AN INDEX OF SNOWMELT

The energy budget procedure requires data commonly not available. Ideally, radiation measurements exist, but generally, we have to rely on the radiation equations of Chapter 2. Even then, the energy budget requires vapor pressure, wind speed, precipitation, and temperature measurements at various elevations. Most locations will have precipitation and air-temperature measurements. In order to deal with the sparcity of data, many empirical procedures have been developed to relate snowmelt, runoff, or energy transfers to easily measured indices. The most common and successful index of snowmelt is air temperature.

Anderson [1978a] presents one of the most successful air-temperature index methods, the one used by the U.S. National Weather Service. The following is adapted from that source.

Snowmelt is divided into rain and no-rain periods. The separation is due to differences in the magnitude of the various energy transfers, knowledge of the dominant transfer procedure during rain on snow periods, and differences in the seasonal variation of melt rates for the two periods.

During rain on snow, it is assumed that the incoming solar radiation is negligible because of overcast conditions, the cloud cover radiates as a black body, with a temperature close to that of air, and there is very high relative humidity. Anderson [1978a] assumes 90% relative humidity.

Under the above assumptions, the energy balance takes the following form.

Net Radiation

$$\begin{aligned} Q^* &= Q_a - Q_{bs} \\ &= \sigma(T_a + 273)^4 - \sigma(T_s + 273)^4 \\ &= \sigma(T_a + 273)^4 - \sigma(273)^4 \\ &= 0.826 \times 10^{-10}(T_a + 273)^4 - 0.45 \text{ cal cm}^{-2}\text{min}^{-1} \\ &= 49.56 \times 10^{-10}(T_a + 273)^4 - 27 \text{ cal cm}^{-2}\text{hr}^{-1}. \end{aligned} \tag{6.33}$$

Advected Heat in Precipitation

$$Q_v = C_P \rho_w P T_a \text{ cal cm}^{-2}\text{hr}^{-1}, \tag{6.34}$$

where C_P is the specific heat and P is precipitation usually given in centimeters per hour.

Condensation and Sensible-Heat Transfers

Since snow is at 0°C, the saturation vapor pressure at the surface is 6.11 mb. The vapor pressure in the air is $0.9e_{\text{sat}}$, using the assumed 90% relative humidity. Using Eq. (6.27) and the Bowen ratio concept,

$$Q_e + Q_h = f(U)\rho_w L_e[(0.9e_{\text{sat}} - 6.11) + C_B P_a T_a], \tag{6.35}$$

where $L_e = 600 \text{ cal g}^{-1}$, pressures are in millibars, $C_B = 0.61 \times 10^{-3}\,°\text{C}^{-1}$, T_a is in degrees Centigrade, $\rho_w = 1 \text{ g cm}^{-3}$, and $f(U)$ is in centimeters per millibar per hour.

The function $f(U)$ is taken as a calibration constant or as $K_1(Z_a Z_b)^{-1/6} U_b$, corresponding to Eq. (6.25), if a wind velocity measurement exists, taking care of unit consistency.

Anderson [1978b] suggests as an initial estimate of the wind function,

$$f(U) = 0.0002U,$$

where U is the wind in kilometers per hour at 1-m elevation and $f(U)$ has units of centimeters per millibar per hour.

Adding Eqs. (6.33) through (6.35), the change of energy available is

$$\begin{aligned} Q_o = {} & 49.56 \times 10^{-10}(T_a + 273)^4 - 27 + PT_a \\ & + f(U)\, 600[(0.9e_{\text{sat}} - 6.11) + 0.61 \times 10^{-3}\, P_a T_a] \end{aligned} \tag{6.36}$$

in calories per square centimeters per hour.

Anderson [1978a] suggests that the saturation vapor pressure in millibars can be estimated from

$$e_{sat} = 2.749 \times 10^8 \exp\left[\frac{-4278.6}{T_a + 242.8}\right], \tag{6.37}$$

with T_a in degrees Centigrade.

The pressure is taken from the following elevation-pressure relationship

$$P_a = 1012.4 - 11.34z + 0.00745z^{2.4}, \tag{6.38}$$

where z is elevation in hundreds of meters.

There are no simplifying assumptions for snowmelt during periods of no rain. During those periods, the complete energy balance must be known or computed. Unfortunately, the variability of the energy exchanges with the conditions of the atmosphere are too large to account explicitly for each term with a simple surrogate such as air temperature. The commonly chosen alternative is to relate empirically snowmelt directly to air temperature. Most equations are of the form

$$M = M_f(T_a - T_b), \tag{6.39}$$

where M is the melt in millimeters of water over a given time period, T_a is air temperature in degrees Centigrade, and M_f is a "melt factor" in millimeters per degrees Centigrade. T_b is a base temperature below which no melt occurs. Generally, T_b is taken as 0°C. Mathematically, this implies that melt occurs only when $T_a > 0$°C, which is not always true. Using the energy budget, it should be easy to see that snowmelt can occur for air temperatures below 0°C. This is particularly true during clear, calm days when solar radiation dominates the energy budget. Conversely, on clear nights, when outgoing longwave radiation is significant, the energy budget may imply that no melt occurs even though air temperature may be above 0°C. These cases, though, are not the dominant ones. Should a given location exhibit a persistent bias in the snowmelt calculation, T_b can be adjusted to something other than 0°C.

Since the melt factor encompasses all energy transfer effects, it must reflect some seasonal variability. Certainly, solar radiation is an important seasonally varying factor in the energy balance. The seasonality of M_f will be directly related to the importance of solar radiation in the energy budget of a location. Anderson [1968, 1978a] indicates that the seasonal variation in the melt factor can be represented by

$$M_f = \frac{M_f^{\max} + M_f^{\min}}{2} + \sin\left[\frac{n2\pi}{366}\right] \times \frac{M_f^{\max} - M_f^{\min}}{2}, \tag{6.40}$$

where $M_f^{\max}$ is a maximum value M_f may take and $M_f^{\min}$ is a corresponding minimum value, and n is the day number beginning with March 21.

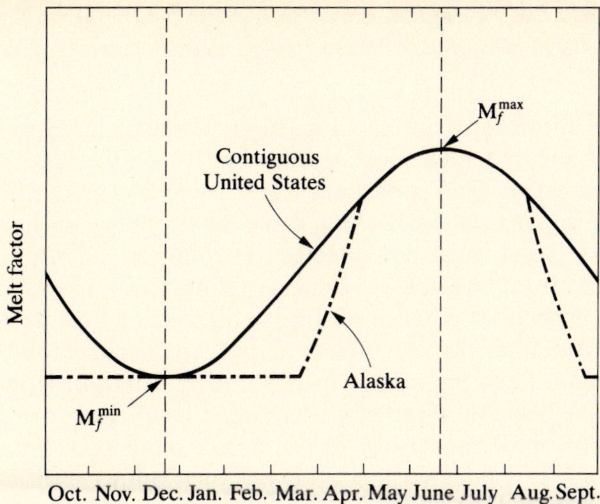

FIGURE 6.15 Seasonal variation in melt factors used during non-rain periods. Source: After Anderson [1978a], Hydrologic Research Laboratory, National Weather Service/NOAA.

Anderson [1978a] points out that the above melt-factor variation proved inadequate for Alaskan conditions. An adjustment was suggested using data from Fairbanks, Alaska:

$$M_f^* = (M_f - M_f^{min})\,F + M_f^{min}, \tag{6.41}$$

where M_f^* is the modified melt factor and F is an adjustment given by

$$\begin{aligned} F &= 0.0 \qquad &&\text{for } X \le 0.48, \\ &= 1.0 \qquad &&\text{for } X \ge 0.7, \\ &= \frac{X - 0.48}{0.22} \qquad &&0.48 < X < 0.7. \end{aligned} \tag{6.42}$$

In the above, X is the decimal fraction of the time between December 21 and June 21. Figure 6.15 shows the nature of Eqs. (6.40) and (6.41).

Parameters M_f^{min} and M_f^{max} are calibration coefficients. They will vary from location to location and will be affected by latitude, forest cover, exposure, wind, and other variables. Following is an excerpt from Anderson [1978b][†] discussing possible values of M_f^{min} and M_f^{max} for use with the U.S. National Weather Service snowmelt model. In reading, keep in mind that the

[†]E. A. Anderson, Hydrologic Research Laboratory, National Weather Service/NOAA.

National Weather Service model operates at 6-hr time intervals. The variables M_f^{min} and M_f^{max} have been substituted for the names used in Anderson's original write-up.

> Surface snowmelt is affected by many climatic and physiographic factors. The factor which most commonly is used to classify melt factors is forest cover. This is because forest cover has a significant effect on many of the variables affecting snow cover energy exchange. Thus, difference in forest cover can be used to explain much of the variation in melt rates from one area to another. Differences in aspect are also important, especially when modeling snowmelt at a point. Over most watersheds or subareas, the effects of differing aspects tend to cancel. Climatic factors are important in explaining differences in melt factors between regions when physiographic conditions are the same. For example, melt factors in arctic areas tend to be smaller than those at lower latitudes with similar physiographic conditions mainly due to lower radiation intensities and relatively little wind during the melt season.
>
> In areas with distinct accumulation and melt seasons, the maximum melt factor is generally more critical than the minimum melt factor since most of the snow melts after March 21. [Table 6.4] gives some typical values for the melt factors as a function of forest cover conditions. Climatic and other physiographic factors will tend to alter these values and should be taken into account when making initial estimates of the melt factors. Melt factors should be increased in areas with predominantly south-facing slopes and reduced in areas with a northerly aspect. Aspect should have more effect in open areas than forested areas. Windy areas typically have higher melt factors than areas where calm conditions prevail. In arctic areas the melt factors should be smaller than those at lower latitudes ([Table 6.4] is based on applications of the model to the contiguous United States).

TABLE 6.4 Typical Values of Melt Factors as Related to Forest Cover for Areas with Distinct Accumulation and Melt Seasons (Units are millimeters per degree Centigrade per 6 hours)

FOREST COVER	M_f^{max}	M_f^{min}
Coniferous forest—quite dense	0.5–0.8	0.2–0.3
Mixed forest—coniferous plus open and/or deciduous	0.8–1.0	0.25–0.4
Predominantly deciduous forest	1.0–1.3	0.35–0.5
Open areas	1.3–2.0	0.5–0.9

Source: Anderson [1978b], Hydrologic Research Laboratory, National Weather Service/NOAA.

In areas with no distinct accumulation and melt seasons, the maximum melt factors are generally similar to those given in [Table 6.4]. However, the minimum melt factors are usually somewhat higher. This is because in such areas the snow cover is normally shallow and mid-winter thaws last long enough to completely ripen the snow surface. In areas with a distinct accumulation season, winter thaw periods are usually brief and the albedo of the snow cover remains relatively high, thus keeping [$M_f^{\max}$] low. [Table 6.4] can also be used to get initial estimates of [$M_f^{\max}$] and [$M_f^{\min}$] at a point. However, even though the point itself might be classified as open, the surroundings should be taken into account. A snow course in a small forest clearing may act similar to a mixed forest, whereas a snow course in a larger opening should be approaching the conditions at a truly open site.

When the air temperature is below 0°C and it is not raining, the model formulated in past paragraphs imply that no snowmelt is occurring. Nevertheless, during those periods of nonzero cold content, the energy exchanges continue, mainly through the air–snow interface. Anderson [1978b] assumes that this energy exchange is proportional to the temperature gradient in the upper part of the snow cover. Since generally no temperature profiles of the snowpack exist, a surrogate measure of the snowpack temperature at some depth is introduced. It is called the antecedent temperature index (ATI) and it is only a function of air temperature.

$$\mathrm{ATI}_2 = \mathrm{ATI}_1 + C_1(T_a - \mathrm{ATI}_1), \tag{6.43}$$

where subscripts 1 and 2 refer to the beginning and end of a time period, respectively, and C_1 is a parameter taking values between 0.1 and 1.0. Temperatures are given in degrees Centigrade. Equation (6.43) mimics the delayed response of the snowpack temperature at some depth to the air temperature. The antecedent temperature index cannot take a value above 0°C since it represents snowpack interior temperature.

Using the difference between T_a and ATI as a measure of the temperature gradient in the snowpack upper sections, Anderson [1978a] uses the following expression for the change in snowpack heat content.

$$\Delta Q = NM_f \cdot (\mathrm{ATI}_1 - T_a), \tag{6.44}$$

where NM_f is a coefficient called the negative melt factor. The negative melt factor is a surrogate to some snow conductivity. Since conductivity is density-dependent and it in turn is seasonally dependent, NM_f should vary as the season progresses. Anderson [1976] suggests using the same seasonal variation as the melt factor:

$$NM_f = \left[\frac{M_f}{M_f^{\max}}\right] \cdot NM_f^{\max}, \tag{6.45}$$

where $NM_f^{\max}$ is the maximum negative melt factor.

The factors C_1 (Eq. 6.43) and NM_f (Eq. 6.44) again are calibration parameters that should be tailored to the conditions of particular locations. Anderson [1978b] also gives guidelines for the values that the parameters may take. Except for the substitution of variable names, the following is quoted from Anderson [1978b]* after adjustment for unit compatibility. The write-up assumes the 6-hr time step of the National Weather Service model.

> Energy exchange during non-melt periods is assumed to be proportional to the temperature gradient defined by the snow surface temperature (approximated by the air temperature) and a temperature at some depth below the surface. The temperature of the snow at some depth below the surface is approximated by an antecedent temperature index, ATI. The parameter $[C_1]$ is used in the computation of ATI. A value of $[C_1]$ above 0.5 essentially gives weight only to air temperatures during the past few 6-hour periods in the computation of ATI. A value of $[C_1]$ below 0.2 gives weight to temperatures over the past 3 to 7 days. Thus, an ATI computed using a high value of $[C_1]$ would correspond to a snow temperature closer to the surface than an ATI based on a low value of $[C_1]$. It seems logical to expect that heat transfer within a deep snow cover would be controlled by a temperature further below the surface than in the case of a shallow cover because of the increased depth and heat storage capacity. Thus, it would be expected that the value of $[C_1]$ used for areas with typically deep snow covers should be smaller than the value of $[C_1]$ used for areas which generally have a shallow snow cover. This has been confirmed by calibration results. It is recommended that a value of $[C_1]$ of 0.5 or greater be used in areas which typically have a relatively shallow snow cover like most of the upper midwest portion of the United States. For areas which generally have a deep snow cover, a value of $[C_1]$ in the range of 0.1 to 0.2 would be appropriate. A value of $[C_1]$ between 0.2 and 0.5 would be reasonable in areas which usually have a moderate amount of snow like much of northern New England.
>
> The steady state equation for heat transfer in a homogeneous snow cover can be expressed as
>
> $$Q = k_e \cdot \frac{\Delta T}{\Delta Z}, \qquad (6.46)$$
>
> where Q equals heat transfer in cal cm$^2 \cdot$ 6 hr^{-1}; k_e equals the effective thermal conductivity of snow (cal °C^{-1} cm^{-1} 6 hr^{-1}); and $\Delta T/\Delta Z$ equals the temperature gradient—difference in temperature, ΔT, over difference in depth, ΔZ (°C $\cdot$ cm^{-1}). A comparison of this equation with [Eq. (6.44)] indicates that the negative melt factor is equal to

*E. A. Anderson, Hydrologic Research Laboratory, National Weather Service/NOAA.

TABLE 6.5 Computed Negative Melt Factors for Various Values of Snow Density and ΔZ (Values are in calories per degree Centigrade per centimeter per 6 hours)

DENSITY, ΔZ	10 cm	20 cm	30 cm
0.3	*1.28*	0.64	0.4
0.4	2.16	*1.12*	0.72
0.5	3.36	1.68	*1.12*

Source: Anderson [1978b], Hydrologic Research Laboratory, National Weather Service/NOAA.

$k_e/\Delta Z$. The value of k_e has been found experimentally to be mainly a function of the density of snow (Anderson [1976]). Thus, the negative melt factor is a function of snow density and ΔZ. [Table 6.5] shows calculated values of the negative melt factor for various values of density and ΔZ. The ΔZ values in [Table 6.5] were selected to reasonably represent depths corresponding to ATI values computed using the recommended initial values of [C_1] for shallow, medium, and deep snow covers, respectively. Since [$NM_f^{\max}$] is the maximum negative melt factor, it should be based on the typical maximum snow density for the area. Maximum snow densities generally occur during the melt season and typically vary from about 0.3 for shallow snow to 0.5 for a very deep snow cover. This suggests that the underlined values in [Table 6.5] would be good initial estimates of [$NM_f^{\max}$]. Since these values are all similar, it suggests that a good default value for [$NM_f^{\max}$ is 1.2 cal °C^{-1} cm^{-1} 6 hr^{-1}]. It should be remembered that in reality heat transfer in a snow cover is not a steady state process, the snow surface temperature is usually not equal to the air temperature, and the depth corresponding to the value of ATI undoubtedly varies with time. Thus, the values given in [Table 6.5] should only be used to suggest reasonable initial values and indicate something about the likely range in values for [$NM_f^{\max}$].

There are other possible changes in the pack heat content besides that given by Eq. (6.44). They are the heat released by freezing liquid water when the snowpack temperature is below 0°C and the heat added by possible snowfall. These two terms were already defined in Eqs. (6.32) (Section 6.4.5) and (6.23) (Section 6.4.2), respectively.

6.6 ROUTING OF MELT THROUGH SNOWPACK

When snowpack is ripe, isothermal at 0°C, and saturated with liquid water (W = liquid-water capacity), any melt or rain will be reflected as runoff from the pack. There are no theoretical studies on how this water transfer (routing)

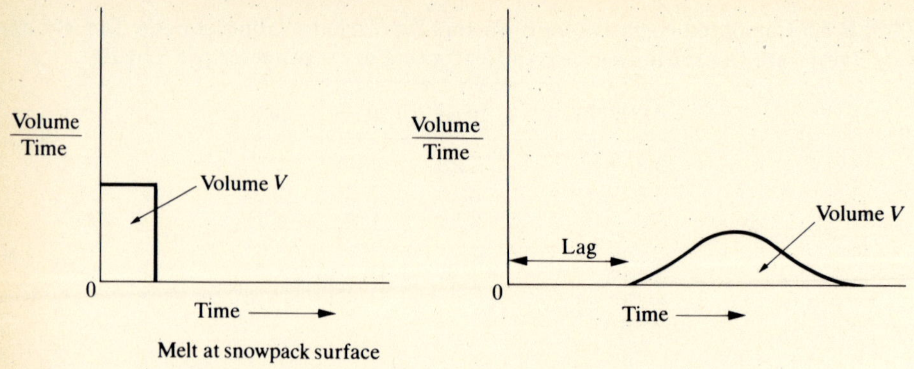

FIGURE 6.16 The effect of lag and attenuation on the routing of snowmelt.

occurs. Anderson [1978a] recommends the use of empirical relationships developed during April and May 1954 using lysimeter data from the Central Sierra Snow Laboratory.

Routing through the snowpack involves a time lag and an attenuation. This type of effect is illustrated in Figure 6.16 for an impulse of melt at the snowpack surface. The time lag of the melt is shown in Figure 6.17 as a function of the ratio of water equivalent of the solid portion of snow cover (in mil-

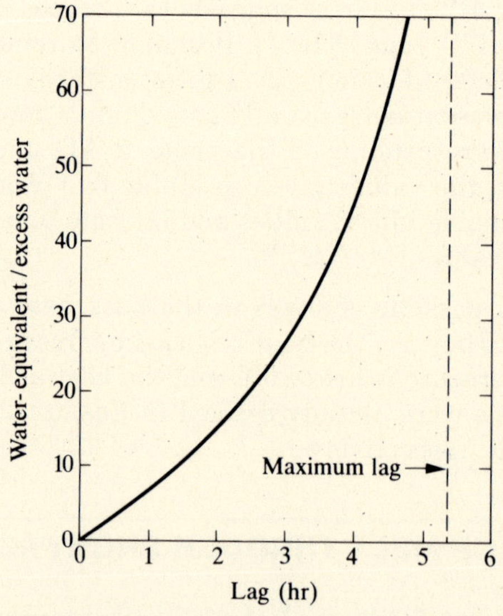

FIGURE 6.17 Lag applied to excess water moving through a snow cover. Source: After Anderson [1978a], Hydrologic Research Laboratory, National Weather Service/NOAA.

limeters) to the excess liquid water (in millimeters per 6 hours). The figure obeys the following expression:

$$L = 5.33 \left[1.0 - \exp\left(\frac{-0.03WE}{E} \right) \right], \tag{6.47}$$

where WE is the water equivalent of the snowpack solid phase and E is the excess water.

Figure 6.18 gives the portion of excess liquid water that drains from storage within the snowpack during a given time interval. It represents the wave attenuation and is given by

$$R = \frac{1.0}{5.0 \exp\left[\dfrac{-500E_{\ell s}}{WE_s^{1.3}} \right] + 1}, \tag{6.48}$$

where R is the normalized runoff or withdrawal rate (per hour); $E_{\ell s}$ is the mean amount of lagged excess liquid water (water in transit) during the current period, given in inches; and WE_s is the mean water equivalent of the solid phase of the snowpack during the time period, in inches. The unfortunate use of English units is due to the original empirical calibration.

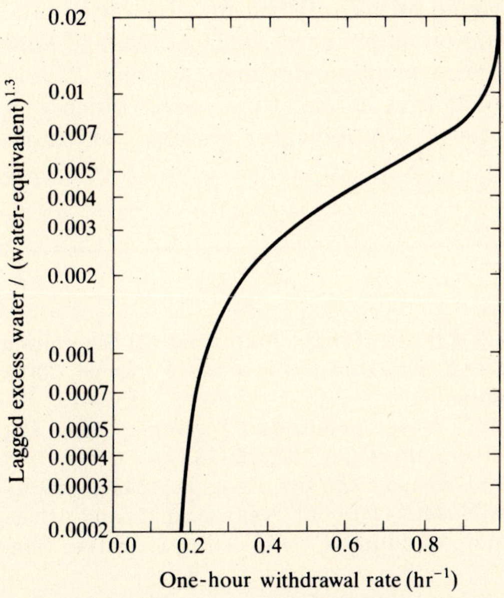

FIGURE 6.18 Attenuation of excess water moving through a snow cover. Source: After Anderson [1978a], Hydrologic Research Laboratory, National Weather Service/NOAA.

In an hour, the total amount of snowpack outflow is then

$$O_s = (S_1 + E_\ell)R\,, \tag{6.49}$$

where O_s is the outflow in millimeters per hour, S_1 is the amount of excess liquid water in storage (in millimeters) in the snowpack at the beginning of the period, and E_ℓ is the amount of lagged liquid water entering the storage during the current time period (in millimeters). Note that the use of the above lag and route procedures requires keeping careful accounting of all liquid water in transit through the snowpack. At the end of a period, the amount of liquid water in storage is

$$S_2 = S_1 + E_\ell - O_s\,. \tag{6.50}$$

6.7 SUMMARY

With snowmelt we have completed the study of the origins and fate of precipitation on the ground surface, before it moves to rivers, lakes, channels, or the ocean. At this point we will move underground, and quantify the relationship between surface waters and water in the soil system, as well as the nature and behavior of the vast amounts of water in the soil (Chapters 7 and 8).

The movement of water into the soil is called the infiltration process; this is one of the main topics of Chapter 8. After the precipitation (rainfall and snowmelt) is acted on by infiltration, the water takes separate paths. That portion moving on or near the surface is called surface runoff. It will ultimately go into river or other drainage systems. The soil water is either held within the soil matrix as moisture, moves quickly in shallow soil profiles, or moves to deeper groundwater systems, which also slowly drain to water bodies.

REFERENCES

Amorocho, J., and B. Espildora [1966]. "Mathematical Simulation of the Snow Melting Process." *Water Science and Engineering*. Paper no. 3001. Davis, Calif.: University of California.

Anderson, E. A. [1968]. "Development and Testing of Snow Pack Energy Balance Equations." *Water Resources Res.* 3(1):19–38.

Idem. [1973]. "National Weather Service River Forecast System—Snow Accumulation and Ablation Model." National Oceanographic and Atmospheric Administration Technical memorandum NWS HYDRO-17. Silver Spring, Md.: National Weather Service.

Idem. [1976]. "A Point Energy and Mass Balance Model of a Snow Cover." National Oceanographic and Atmospheric Administration Technical report NWS 19. Silver Spring, Md.: National Weather Service.

Idem. [1978a]. "Snow Accumulation and Ablation Model." In: *Operational Forecast Programs and Data Components for the National Weather Service River Forecast System Data Management Program, System Documentation.* Silver Spring, Md.: National Weather Service.

Idem. [1978b]. "Initial Parameter Values for the Snow Accumulation and Ablation Model." In: *Operational Forecast Programs and Data Components for the National Weather Service River Forecast System Data Management Program, System Documentation.* Silver Spring, Md.: National Weather Service.

Anderson, E. A., and N. H. Crawford [1964]. "The Synthesis of Continuous Snowmelt Runoff Hydrographs on a Digital Computer." Stanford, Calif.: Stanford University Department of Civil Engineering. (Technical report no. 36.)

Branson, F. A., G. F. Gifford, K. G. Renard, and R. F. Hadley [1981]. *Rangeland Hydrology.* Dubuque, Iowa: Kendall/Hunt.

Eagleson, P. S. [1970]. *Dynamic Hydrology,* New York: McGraw-Hill.

Gerdel, R. W. [1954]. "The Transmissional Water Through Snow." *Trans. Am. Geophys. Union.* 35(3):475–485.

Gray, D. M. [1973]. *Handbook on the Principles of Hydrology.* Port Washington, N.Y.: Water Information Center.

Hamon, R. W. [1972]. "Computing Actual Precipitation." Symposium of the World Meteorological Organization and International Association of Scientific Hydrology, Geilo, Norway, July 31–August 5, 1972.

Jairell, R. L. [1975]. "An Improved Recording Gage for Blowing Snow." *Water Resources Res.* 11(1):34–38.

Komarov, A. A. [1954]. As quoted by Kuzmin [1960], p. 34.

Kuzmin, P. O. [1960]. "Snow Cover and Snow Reserves." *Gidrometeorol. Isdatelsko.* (Translation by the National Science Foundation, Washington, D.C.) pp. 99–105.

Laramie, R. L., and J. C. Schaake, Jr. [1972]. "Simulation of the Continuous Snowmelt Process." Cambridge, Mass.: MIT Department of Civil Engineering, Ralph M. Parsons Laboratory. (Technical report no. 143.)

Larson, L. W. [1971]. "Shielding Precipitation Gages from Adverse Wind Effects with Snow Fences." Laramie, Wyo.: University of Wyoming. (Water Resources no. 25.)

Idem. [1972a]. "Approaches to Measuring 'True' Snowfall." 29th Eastern Snow Conference, Oswego, N.Y., February 3–4, 1972.

Idem. [1972b]. "An Application of the Dual-Gage Approach for Calculation of 'True' Solid Precipitation." 53rd annual meeting of the American Geophysical Union, Washington, D.C., April 17–21, 1972.

Larson, L. W., and E. L. Peck [1974]. "Accuracy of Precipitation Measurements for Hydrologic Modeling." *Water Resources Res.* 10(4):857–864.

McKay, G. A. [1968]. "Problems of Measuring and Evaluating Snow Cover." In: *Snow Hydrology.* Proceedings of a Workshop Seminar, University of New Brunswick, February 28–29, 1968.

McKay, G. A., and H. A. Thompson [1967]. "Snowcover in the Prairie Provinces." Paper presented at the joint meeting of the Canadian Society of Agricultural Engineers and the American Society of Agricultural Engineers. Saskatoon, Sask., June 27–30, 1967.

Meiman, J. R. [1968]. "Snow Accumulation Related to Elevation, Aspect and Forest Canopy." In: *Snow Hydrology.* Proceedings of a Workshop Seminar, University of New Brunswick, February 28–29, 1968, pp. 35–47.

Peck, E. L. [1972]. "Snow Measurement Predicament." *Water Resources Res.* 8(1):244–248.

Rechard, P. A., and L. W. Larson [1971]. "Snow Fence Shielding of Precipitation Gages." *Am. Soc. Civil Eng. Proc.* 97:1427–1439.

Schmidt, R. A., Jr. [1970]. "Sublimation of Wind Transported Snow—A Model." U.S. Forest Service. Research paper no. RM-90.

Tabler, R. D., and R. A. Schmidt [1973]. "Weather Conditions That Determine Snow Transport Distances at a Site in Wyoming." In: *The Role of Snow and Ice in Hydrology,* Symposium on Properties and Processes, Banff, Alberta, September 1972, pp. 118–120.

U.S. Army Corps of Engineers [1956]. "Snow Hydrology." Portland, Oreg.: U.S. Army Corps of Engineers, North Pacific Division.

U.S. Army Corps of Engineers [1971]. "Runoff Evaluation and Streamflow Simulation by Computer." Portland, Oreg.: U.S. Army Corps of Engineers, North Pacific Division.

Van Haveren, B. P. [1975]. "Airflow and Gas Exchange in Snow—Fact or Fiction?" In: *Proceedings of the 43rd Western Snow Conference.* Coronado, Calif., April 23–25, 1975, pp. 21–27.

Winston, W. [1965]. "A Comprehensive Procedure for Evaluating Snow Ablation." In: *Proceedings of the 22nd Eastern Snow Conference.* Vol. 10. Hanover, N.H., February 4–5, 1965.

PROBLEMS

1. The initial conditions of a snowpack were: thickness, 0.5 m; temperature, 15°F; and density, 0.2 g cm^{-3}. A snowfall occurred during which the air temperature was 20°F. The thickness was then measured as 0.8 m.

a) Compute the water equivalent and cold content of the initial snowpack.

b) Estimate the density of the new snow and compute average density, water equivalent, cold content and thermal quality of the resulting snowpack.

c) Estimate the energy per unit area that needs to be applied (by radiation, advection, or conduction) to this snowpack before there is any meltwater runoff. Is the liquid-water holding capacity of the snow significant in this calculation?

2. During the first week of the spring snowmelt, the effective incoming radiation available for snowmelt is 150 ly day^{-1}. Estimate the volume of runoff due to snowmelt during this week, given the following snowpack information: thickness, 1 m; density, 0.4 g cm^{-3}; initial temperature, −5°C; latent heat of freezing, 79.7 cal g^{-1}; and heat capacity of ice, 0.5 cal g^{-1} °C. You may neglect energy transfers between snowpack and the atmosphere or earth.

3. Figure 6.12 shows that the albedo of a fresh snow surface is between 0.70 and 0.80. Snow falling in cities gets dirty quickly. Under otherwise similar conditions, what effect would dirt have on the melting rate? Explain. Can you quantify, for a typical situation, the difference between clean and dirty snow?

4. The mean daily temperature during the month of January in a northern hemisphere city is 40°F with a standard deviation of 7°F. The temperature can be assumed to be normally distributed. Using Figure 6.1, what is the probability of getting snow in any one day?

5. Find a region typically covered with snow during winter time, anywhere in the world. Locate (use your library) topographic and vegetative-cover maps. Suggest a relationship between the percent of area covered by snow and the ratio of liquid-water equivalent covering the given percent of the basin to the total-snowpack liquid-water equivalent when the basin is fully blanketed (see Fig. 6.7). Base your suggestion on issues such as elevation, vegetative cover, topography, aspect, etc.

6. Consider a homogeneous 2-ft snowpack that has a density of 0.5 $\mathrm{g\,cm^{-3}}$ and is in thermal equilibrium at a uniform temperature of −2°C. Rain, having a temperature of 2°C (assume equal to air temperature) begins to fall at a constant rate of 0.1 $\mathrm{in.\,hr^{-1}}$. Wind velocity is 4 $\mathrm{km\,hr^{-1}}$.

a) How long before melt begins?
b) Estimate the rate of runoff (in millimeters per hour) at the foot of the snowpack over some time, assuming that rainfall continues.

(Adapted from P. S. Eagleson, *Dynamic Hydrology*, McGraw-Hill, 1970.)

7. On a clear winter night, the surface of the snowpack is 0°C. The air temperature measured 8 ft above the surface is −10°C and the relative humidity is 70%. Wind velocity is 10 mi per hour, also at 8 ft above the surface. How much condensation or evaporation occurs?

8. On March 31, the air temperature is 5°C. Estimate the maximum snowmelt from a snow-covered open area and from a dense coniferous forest.

9. During the hottest hours of a winter day, the following conditions prevail: latitude, 45°N; time, 11:30 a.m. to 3:30 p.m.; date, December 21; clear sky; snow age, 1 wk; forest canopy density, 0.3; air temperature, 50°F at 4 ft above snow; relative humidity, 20% at 4 ft above snow; wind velocity, 2 miles per hour at 8 ft above snow; snow-surface temperature, 32°F; and surface atmospheric pressure, 1000 mb. At the beginning of the period the snowpack depth and density are 24 in. and 0.5 $\mathrm{g\,cm^{-3}}$, respectively, and the average temperature of the pack is 30°F. Calculate the amount of melt in the 4-hr period. (From P. S. Eagleson, *Dynamic Hydrology*, McGraw-Hill, 1970.)

10. If it were overcast and raining at a rate of 1 $\mathrm{cm\,hr^{-1}}$, what would the approximate melt be for the conditions of Problem 9?

Chapter 7

Groundwater Flow in Saturated Porous Media

7.1 INTRODUCTION

Up to now we have studied elements of the hydrologic cycle occurring on or above the earth's surface. In this chapter we will study the issues related to groundwater. By historical accident and because of the constant move to specialization in all professions, groundwater hydrology has been artificially separated from surface hydrology. The separation has been possible because, as we will see, the time constants involved in groundwater systems are generally longer than those of many common surface-water problems, such as flood forecasting. Nevertheless, problems such as droughts, water supply, irrigation, and water pollution have to be treated with awareness of both the surface and ground waters.

As we saw in Chapter 1 (Table 1.1) groundwater accounts for 62% by volume of available fresh water, four orders of magnitude more than surface waters. Such large water reserves remain mostly untapped, although local exploitation can be great. Many of these water sources cannot be exploited because of the depths involved or because they consist of soil moisture at low levels of saturation, making extraction from the soil impossible.

As surface waters become more and more exhausted, in terms of quantity and quality, groundwater will gain importance. The study of groundwater should not be taken lightly. The field is very well developed, with innumerable specialized books and articles on the subject. In this chapter we intend to introduce the basic concepts of flow in porous media (soil), under saturated conditions. Chapter 8 will deal with unsaturated soil systems.

Flow under saturated conditions will dominate the multidimensional movement of water in deep groundwater deposits or aquifers. We will study the possible forms of these equations and use them to explain the hydraulic behavior of wells and their surrounding soil environment.

7.2 THE SOIL–ROCK PROFILE AND SUBSURFACE WATERS

A common soil–rock system may be divided in four general regions (Fig. 7.1). Sequentially from the surface the classification is as follows (Meinzer [1923]):

1. The soil water zone begins at the ground surface and extends downward, encompassing the root layers. Its depth is variable and dependent on soil type and vegetation. During periods of rainfall (or other water applica-

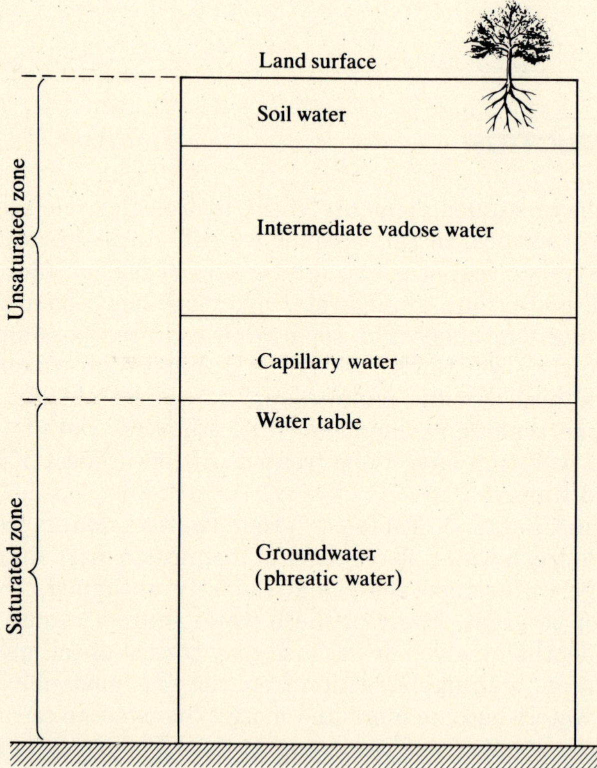

FIGURE 7.1 Classification of soil–rock system. Source: R. H. Brown, A. A. Konoplyantsev, J. Ineson, and V. S. Kovatevsky, *Groundwater Studies.* Copyright © Unesco 1972. Reproduced by permission of Unesco. After Meinzer [1923].

tion such as irrigation), this area may become saturated. It is otherwise in an unsaturated state; part of the soil pores are filled with air.

2. The intermediate zone extends down to the capillary fringe. It is a connecting link to the lower, usually saturated, soil–rock systems. The intermediate zone is unsaturated except during periods of extreme precipitation. If it exists, this zone could be centimeters to tenths (or even hundredths) of meters thick.
3. The capillary zone lies above the lower saturated layers. The name comes from the existence of rising water under capillary forces. Capillarity is a function of the type of soil–rock medium. The rising of the water column may be from a fraction of a meter for sands to tenths of meters for fine clays.
4. The saturated zone has all pores in the soil–rock system filled with water. Formations in this zone that can be exploited for their water content are called aquifers.

Water may exist in all its phases within the soil system. Liquid water appears as hygroscopic, capillary, or gravitational water. Hygroscopic and capillary water are held by molecular forces in thin films around soil particles. The drier the soil and the smaller the interstices between particles, the stronger are the forces holding this water. Hygroscopic water is essentially unavailable, held at negative pressures of 31 to 10,000 bars. Capillary water results when more water is available filling gaps between soil particles but in a discontinuous fashion. This water is held at pressures of 0.33 to 31 bars and can be in direct connection with groundwater or in isolated pockets. Capillary water forms the continuous capillary zone previously mentioned and can be used by plants. If negative pressures have magnitudes greater than about 15 bars, the plant root system cannot extract the water. This level of dryness is called the permanent wilting point and vegetation cannot survive at lower moisture contents. As the moisture increases, a point is reached when gravity is strong enough to counteract the negative pore pressures. This occurs at negative pressures between 0 and 0.33 bars. The maximum amount of water the soil can hold against gravity is called field capacity. Water in excess of field capacity percolates down the soil column, ultimately reaching a zone of saturation bounded by bedrock or some other impermeable material. It is reasonable to expect that soil moisture at some depth in the intermediate layer does not vary much with time. The exact depth will naturally depend on the history of the site, but it is on the order of several meters in most regions. In humid or well-irrigated areas, field capacity is a good moisture assumption for this layer.

The profile in the upper layers after a storm, following a dry period, may look like Figure 7.2. The upper zones are quickly saturated or nearly so, establishing two opposing moisture gradients. Moisture decreases with depth from the newly wetted surface, up to a point when it starts increasing with depth from the previously dry conditions of the upper layers to a constant soil

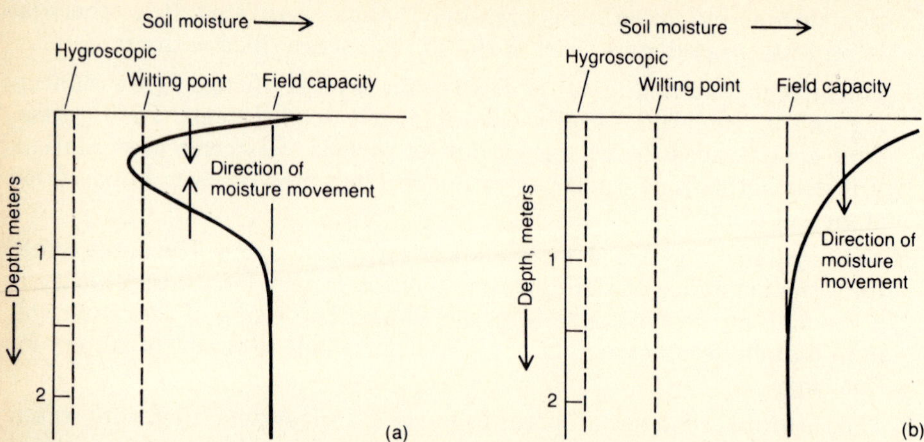

FIGURE 7.2 Soil moisture profiles. (a) Immediately after a drought has been broken by a substantial rainfall—infiltration water has reached 25 cm of soil, and with the upflow of moisture from the lower reservoir, the desiccated layer is being recharged from two directions. (b) After the soil has been recharged in excess field capacity and is incapable of retaining additional water—the surplus is transmitted downward as gravity water. Source: W. M. Marsh and J. Dozier, *Landscape: An Introduction to Physical Geography.* Copyright ©1986 by Wiley. Reprinted by permission of John Wiley & Sons, Inc.

moisture front at some depth. Effectively the dry region is "pinched out" by the two gradients. After sufficient moisture input, a single gradient, with moisture decreasing with depth, is developed and water percolates further down.

After cessation of rainfall, the drying process may look something like Figure 7.3. At time 0 the wet front was moving down. Immediately after the rainfall stops, drying begins in the upper layers. The bottom part of the wet front continues moving down. At some point, no more water moves down and the gradient is completely negative (moisture uniformly increasing with depth). A "dry front" is now established. The lower zones are being depleted in order to satisfy an evaporation potential at the surface. The dry front will continue developing until the capillary forces are unable to move water to the surface (profile 7).

As previously mentioned, exploitable groundwater reservoirs are called aquifers. A saturated formation is exploitable if it can release a considerable amount of water with relative ease at reasonable expense. This is a property dependent on soil types and formation origins. Figure 7.4 shows a typical soil profile illustrating the most common aquifer types; these are confined and unconfined (phreatic or water table) aquifers.

Upon drilling a fully penetrating well through a phreatic aquifer, water will rise to the water-table level, which defines the piezometric surface or head of the system. An unconfined (phreatic) aquifer has a free water surface. This free water surface may be directly connected to a stream or other surface

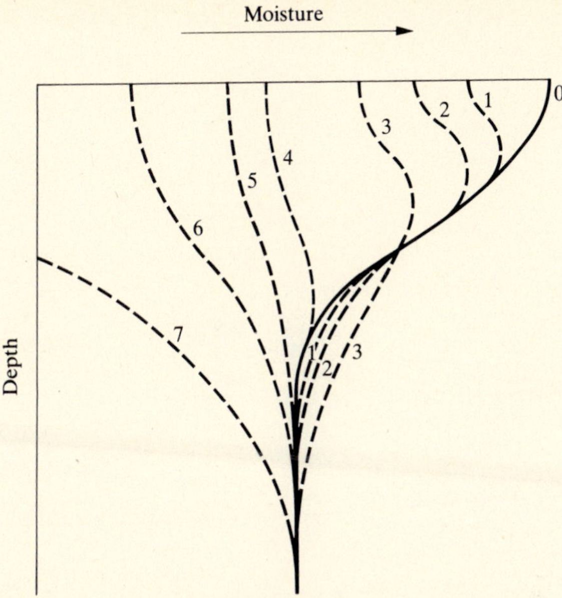

FIGURE 7.3 Hypothetical upper-soil profiles (time progress shown by numbers) during a drying period.

waters. The water in phreatic aquifers comes from direct rainfall recharge over the aquifer, from connections to surface waters, and/or from other aquifers.

Figure 7.4 shows a confining bed separating the phreatic aquifer from a lower confined system. The confining beds may be completely impermeable (aquifuge) or "leaky" (aquiclude). The confined aquifer does not have a free surface. Upon drilling a well through a confined aquifer, water will rise to the piezometric level, which is equal to the elevation above a datum plus the pressure in the aquifer. This piezometric level may be above ground surface, in which case a flowing artesian well results. Confined aquifers recharge through formation outcrops—areas where the soil system is exposed to the surface—or through aquicludes. This type of recharge may be limited. Many confined aquifers contain "fossil waters" deposited in past geologic times.

Whether a rock or soil formation is an aquifer, aquifuge, or aquiclude depends largely on its geologic origins and history. Geologic information tells us much about critical hydraulic properties like permeability and porosity, subjects of detailed study further on in this book. Table 7.1 gives possible rock classifications and their water-bearing potentials.

Aquifers in bedrock are generally not very productive. Sandstone—consolidated sediment deposits—might be expected to have the best aquifer potential. Sandstone and other tightly cemented sedimentary rocks have a propensity to crack and fracture. Fractures or crevices may also develop because of solu-

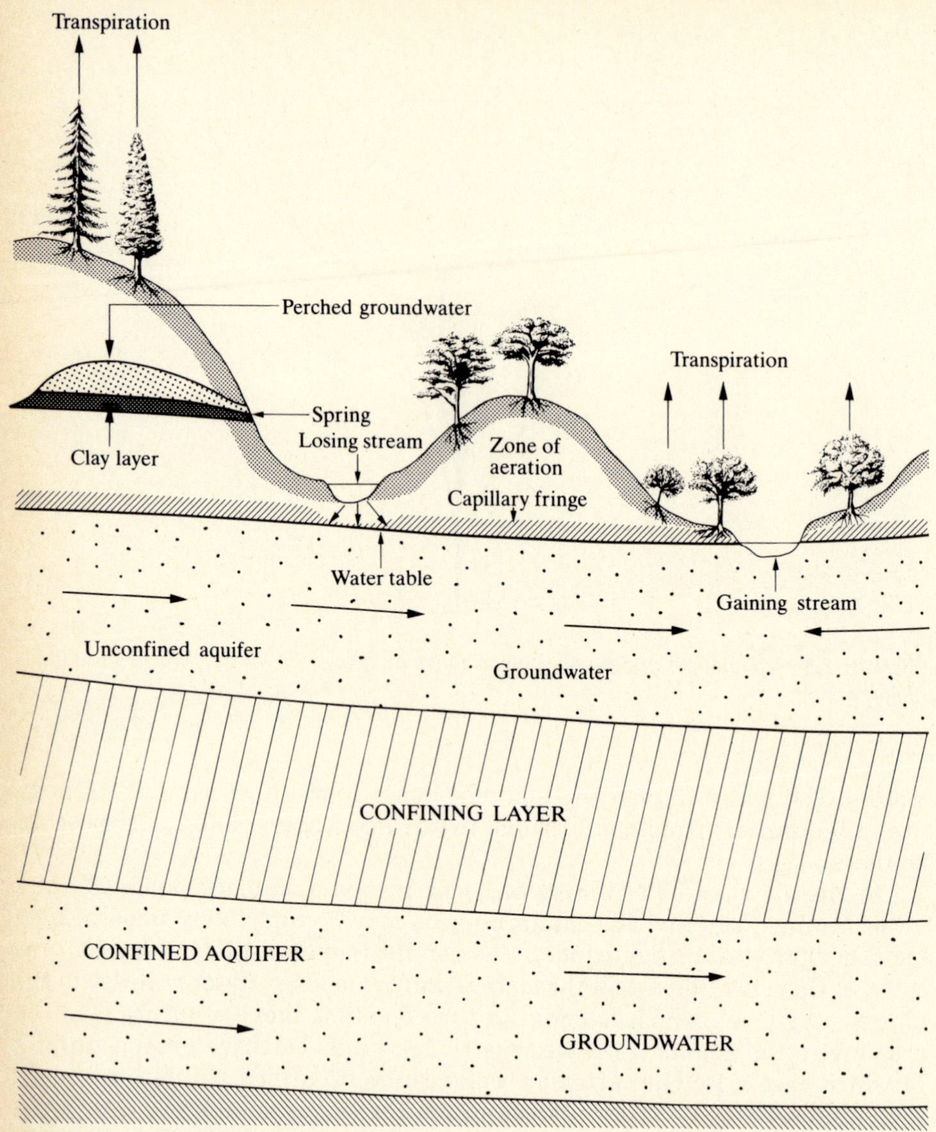

FIGURE 7.4 Schematic cross section showing occurrence of groundwater. Source: R. H. Brown, A. A. Konoplyantsev, J. Ineson, and V. S. Kovatevsky, *Groundwater Studies.* Copyright © Unesco 1972. Reproduced by permission of Unesco.

tion of cementing material. The water-bearing capacity is largely proportional to the degree of fracturing. Sedimentary rocks like dolomite and limestone may have very little inherent permeability through pore openings, but are prone to solution and the development of fractures, crevices, or even cavities. Karst regions throughout the world are tremendously water rich

TABLE 7.1 Rock Classification

ROCK TYPES	POROSITY Primary (grain)	POROSITY Secondary (fracture)[1]	TYPE OF WATER-BEARING UNIT
Sediments, unconsolidated			
Gravel	30–40%		Aquifer
Coarse sand	30–40		Aquifer
Medium to fine sand	30–35		Aquifer
Silt	40–50	Occcasional	Aquiclude
Clay, till	45–55	Rare (mud cracks)	Aquiclude
Sediments, consolidated			
Limestone, dolomite	1–50	Solution joints, planes	Aquifer or aquifuge
Coarse, medium sandstone	<20	Joints, fractures	Aquifer or aquiclude
Fine sandstone, argillite	<10	Joints, fractures	Aquifer or aquifuge
Shale, siltstone	—	Joints, fractures	Aquifuge or aquifer
Volcanic rocks			
Basalt	—	Joints, fractures	Aquifer or aquifuge
Acid volcanic rocks	—		Aquifuge or aquifer
Crystalline rocks			
Plutonic and metamorphic		Weathering and fractures decreasing as depth increases	Aquifuge or aquifer

[1]Rarely exceeds 10%.

Source: R. H. Brown, A. A. Konoplyantsev, J. Ineson, and V. S. Kovatsky, *Groundwater Studies*. Copyright © Unesco 1972. Reprinted by permission of Unesco.

and consist of highly weathered (dissolved) limestone, sometimes forming true underground river systems.

Other bedrock formations of volcanic or crystalline structure have little inherent permeability but could potentially contain and transmit water through cracks and fractures. Another transmission path may be the boundaries between strata of different geologic origin. Folding, faults, weight of overburden, and other sources of geologic stress may conspire to reduce fracturing or to increase it in a given rock formation.

Unconsolidated sediments are generally the best aquifers. These sediment deposits are usually of alluvial or glacial origin. Alluvial sediments are deposited in and around former or present water courses. Since water-carrying capacity depends on the size of particles, alluvial sediments are usually well graded and layered. This enhances their porosity and hydraulic conductivity, making them good aquifers.

The nature of glacial sedimentary deposits depends on their history and chronology relative to glacial formation and movement. For example, glacial fans are much like alluvial fans. They occur downstream of where meltwater emerged from a former glacier. The glacier had to have been stationary for a long time in order to produce extensive useful fans. On the other hand, ground moraines result from the deposit of mixed up material by a quickly receding glacier. This ungraded deposit will usually have poor water-holding and water-transmission potential.

7.3 DARCY'S LAW

The dynamics of flow in a saturated porous medium are described by Darcy's Law. Using an apparatus similar to the one shown in Figure 7.5, where a constant water head is maintained over a medium of thickness ℓ, Darcy [1856] concluded that the volume of flow through the medium is given by

$$\text{Vol} = KA(h_1 + \ell - h_2)t/\ell\,, \tag{7.1}$$

where A equals the cross-sectional area of the medium, t equals time, and K is a proportionality constant. The average velocity of flow over the cross section is then

$$q = K(h_1 + \ell - h_2)/\ell\,. \tag{7.2}$$

More generally, Darcy's Law states that the velocity of flow through a porous medium is directly proportional to the gradient of piezometric head. In one dimension,

$$q = -K\frac{dh}{d\ell}, \tag{7.3}$$

where h equals $z + P/\rho g$, P is pressure, z is elevation, g is gravitational acceleration, ρ is density of water, ℓ is direction of flow, q is velocity (length per time) or discharge per unit cross-sectional area, and K is the hydraulic conductivity (length per time) or "permeability."

As Eq. (7.3) states, flow is in the direction of decreasing piezometric head and is perpendicular to lines of equal head (equipotential lines).

Keep in mind that q is the average velocity over the cross section. Defining $v = q/n_e$, where n_e is effective porosity, we get the average velocity in the actual pores of the cross section. The effective porosity is the ratio of the vol-

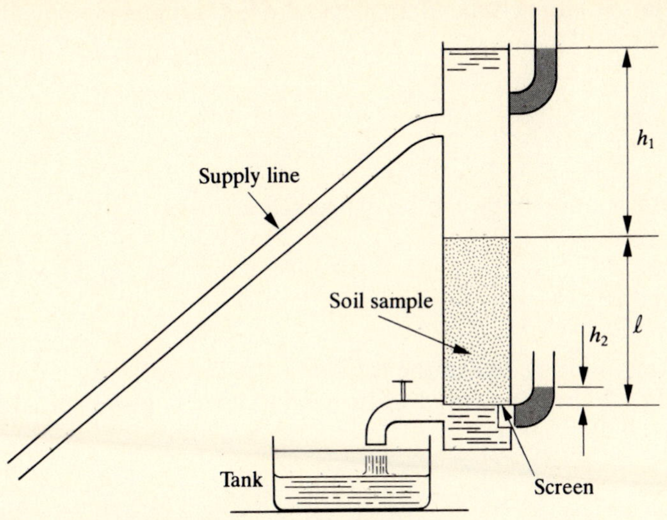

FIGURE 7.5 Darcy's apparatus. Source: Gray [1973].

ume of voids available for flowing water to the volume of the soil. This seepage velocity is greater than q and will correspond to that of a nonreacting, conservative tracer in the porous medium. Darcy's Law can be derived from basic principles of flow in porous media (DeWiest [1965]), and it is applicable under most conditions encountered in practice. Essentially, these are situations where viscous forces predominate over inertial forces leading to laminar flow. The Reynolds number (commonly used in fluid mechanics) is the ratio of inertial to viscous forces and is defined as

$$N_R = \frac{\rho q d}{\mu}, \tag{7.4}$$

where ρ equals water density, q equals velocity of flow, d equals mean particle diameter, and μ equals dynamic viscosity. Darcy's Law is applicable for Reynolds number values less than 1 and has been observed valid for values as high as 10.

The hydraulic conductivity appearing in Eq. (7.3) is a function of both medium and fluid properties. For example, it can be theoretically shown that laminar flow through straight capillary tubes is described by the Hagen–Poiseville equation,

$$q = -\frac{\rho g d^2}{32\mu}\frac{dh}{d\ell}, \tag{7.5}$$

where d equals capillary-tube diameter and $dh/d\ell$ equals piezometric gradient, which implies

$$K = \frac{\rho g d^2}{32\mu}. \tag{7.6}$$

Hydraulic conductivity can be expressed as

$$K = k\frac{\rho g}{\mu}, \tag{7.7}$$

where k is the intrinsic hydraulic conductivity, theoretically a function of the medium only. The intrinsic hydraulic conductivity is also expressed as

$$k = cd^2, \tag{7.8}$$

where c is a proportionality constant function of the medium. Intrinsic hydraulic conductivity has units of length squared (i.e., ft^2, cm^2). To avoid dealing with the usually small values, another unit, the Darcy, is defined as

$$\begin{aligned} 1 \text{ Darcy} &= 0.987 \times 10^{-8} \text{ cm}^2 \\ &= 1.062 \times 10^{-11} \text{ ft}^2. \end{aligned}$$

Hydraulic conductivity has units of velocity. In English units, it is sometimes given in terms of gallons per day per square foot. Table 7.2 shows typical values of intrinsic hydraulic conductivity and hydraulic conductivity (water as fluid) of different soil types. Also in the table are corresponding values of porosities n. Porosity, in a representative elementary volume, is the ratio of the volume of voids to the total volume of the material.

Darcy's equation can be extended to three dimensions. By carefully orienting coordinate axes so as to agree with the preferred direction of flow

TABLE 7.2 Hydraulic Properties of Typical Soils

	k (cm^2)	K (cm s^{-1})	$\psi(1)$ (cm)	n	m	c	d
Clay	4×10^{-10}	3.4×10^{-5}	90	0.45	0.44	7.5	4.3
Silty loam	4×10^{-9}	3.4×10^{-4}	45	0.35	1.2	4.7	2.9
Sandy loam	4×10^{-8}	3.4×10^{-3}	25	0.25	3.3	3.6	2.3
Sand	10^{-7}	8.6×10^{-3}	15	0.20	5.4	3.4	2.2

See Section 8.2 for definitions of $\psi(1)$, m, c, and d.
Source: Entekhabi [1988].

(i.e., commonly the layering of geologic strata), the flux in three orthogonal directions becomes

$$q_x = -K_x \frac{\partial h}{\partial x}; \qquad q_y = -K_y \frac{\partial h}{\partial y}; \qquad q_z = -K_z \frac{\partial h}{\partial z}, \tag{7.9}$$

where K_x, K_y, and K_z are conductivities of all directions. A porous media is called isotropic if $K_x = K_y = K_z$. If the conductivities do not vary from point to point in space, the medium is homogeneous. Mathematically, that implies

$$\frac{\partial K_i}{\partial x} = \frac{\partial K_i}{\partial y} = \frac{\partial K_i}{\partial z} = 0,$$

where K_i is the conductivity in the ith-coordinate direction.

In anisotropic fields, the flow is not perpendicular to the equipotential lines (lines of equal piezometric head). It is nevertheless possible to transform the scale of the x, y, and z axes so that the resulting problem looks mathematically isotropic. The transformations in the three coordinate directions are

$$x' = \sqrt{\frac{1}{K_x}}\,x; \qquad y' = \sqrt{\frac{1}{K_y}}\,y; \qquad z' = \sqrt{\frac{1}{K_z}}\,z. \tag{7.10}$$

The flow across boundaries, perpendicular to the gradient, can be expressed in terms of the new coordinates x', y', and z' as

$$\begin{aligned}
Q_x &= \sqrt{K_y}\sqrt{\frac{1}{K_x}}\sqrt{K_z}\,K_x \frac{\partial h}{\partial x'}\,\Delta y'\Delta z' = \sqrt{K_x K_y K_z}\,\frac{\partial h}{\partial x'}\,\Delta y'\Delta z', \\
Q_y &= \sqrt{K_x}\sqrt{\frac{1}{K_y}}\sqrt{K_z}\,K_y \frac{\partial h}{\partial y'}\,\Delta x'\Delta z' = \sqrt{K_x K_y K_z}\,\frac{\partial h}{\partial y'}\,\Delta x'\Delta z', \\
Q_z &= \sqrt{K_z}\sqrt{K_y}\sqrt{\frac{1}{K_z}}\,K_z \frac{\partial h}{\partial z'}\,\Delta x'\Delta y' = \sqrt{K_x K_y K_z}\,\frac{\partial h}{\partial z'}\,\Delta x'\Delta y'.
\end{aligned} \tag{7.11}$$

Equation (7.11) indicates that the change of scale has resulted in an equivalent isotropic permeability

$$K_\circ = \sqrt{K_x K_y K_z}. \tag{7.12}$$

It is more important to point out that the rotation of axes to obtain Eq. (7.9) is not always possible. In such cases, Darcy's Law takes a more complicated form:

$$q_x = -K_{xx}\frac{\partial h}{\partial x} - K_{xy}\frac{\partial h}{\partial y} - K_{xz}\frac{\partial h}{\partial z},$$

$$q_y = -K_{yx}\frac{\partial h}{\partial x} - K_{yy}\frac{\partial h}{\partial y} - K_{yz}\frac{\partial h}{\partial z}, \tag{7.13}$$

$$q_z = -K_{zx}\frac{\partial h}{\partial x} - K_{zy}\frac{\partial h}{\partial y} - K_{zz}\frac{\partial h}{\partial z},$$

or, using matrix notation,

$$\mathbf{q} = -\mathbf{K}\nabla\mathbf{h}, \tag{7.14}$$

where

$$\mathbf{q}^T = [q_x, q_y, q_z]$$

$$\mathbf{K} = \begin{bmatrix} K_{xx} & K_{xy} & K_{xz} \\ K_{yx} & K_{yy} & K_{yz} \\ K_{zx} & K_{zy} & K_{zz} \end{bmatrix}$$

Matrix **K** is symmetrical and is called the tensor representation of hydraulic conductivity. Readers are referred to Wang and Anderson [1982], Appendix A, for a more extensive, but simple, discussion of anisotropy and tensors.

7.4 MASS BALANCE EQUATIONS—FLOW IN SATURATED POROUS MEDIA

Darcy's equation is the basic expression for momentum balance in flow through porous media. It is combined with continuity or mass balance expressions to obtain full descriptions of the flow field. The derivation of the basic groundwater equations in saturated media follows.

7.4.1 Confined Aquifers

Figure 7.6 defines a unit volume in a saturated medium with no free surface. The fluxes at the center of the unit volume are q_x, q_y, and q_z in the three orthogonal coordinate directions. Writing the net flux in the x direction yields (the following argument is a nonrigorous application of Taylor's series expansion of flux for a small volume)

$$\left\{\left[\rho q_x - \frac{1}{2}\frac{\partial}{\partial x}(\rho q_x)\Delta x\right] - \left[\rho q_x + \frac{1}{2}\frac{\partial}{\partial x}(\rho q_x)\Delta x\right]\right\}\Delta y\,\Delta z = \frac{-\partial}{\partial x}(\rho q_x)\Delta V, \tag{7.15}$$

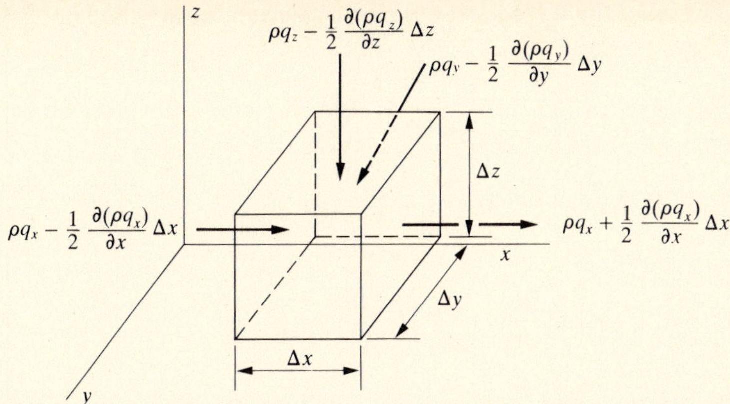

FIGURE 7.6 Mass fluxes in a unit volume of saturated soil. Source: After Gray [1973].

where ρ is the density of water and $\Delta V = \Delta x\,\Delta y\,\Delta z$. As in previous sections, q_x is the flow per unit area perpendicular to the face $\Delta y\,\Delta z$. All other directions yield similar results. When added, they must be equal to the change of fluid mass in the system, defined here as $\partial M/\partial t$. Using the effective porosity to express M and adding the net flux in all directions results in

$$-\left[\frac{\partial}{\partial x}(\rho q_x) + \frac{\partial}{\partial y}(\rho q_y) + \frac{\partial}{\partial z}(\rho q_z)\right]\Delta V = \frac{\partial}{\partial t}(\rho n_e\,\Delta V)\,. \tag{7.16}$$

Under saturated conditions, the change in fluid mass can only be attributed to water density or porosity changes. Changes in water density can be generally neglected when compared to changes in porosity caused by compression of the soil–rock matrix in the vertical direction. Define a coefficient of specific storativity S_o as the volume of water released from storage per unit volume of aquifer per unit change in pressure head. Then, using a hydrostatic pressure assumption, $P = \rho g(h - z_o)$, where h is the piezometric head and z_o is an arbitrary datum, Eq. (7.16) can be simplified to

$$-\left(\frac{\partial q_x}{\partial x} + \frac{\partial q_y}{\partial y} + \frac{\partial q_z}{\partial z}\right) = \frac{S_o}{\rho g}\frac{\partial P}{\partial t} = S_o\frac{\partial h}{\partial t}\,. \tag{7.17}$$

The specific storativity S_o has units of inverse distance. It is a property of the aquifer and is usually taken as a calibration parameter. The form of Eq. (7.17) and the origins of S_o can be obtained using theoretical arguments. Appendix B presents the arguments used by Jacob [1949] to arrive at Eq. (7.17).

Under steady-state conditions, Eq. (7.17) becomes

$$\frac{\partial q_x}{\partial x} + \frac{\partial q_y}{\partial y} + \frac{\partial q_z}{\partial z} = \nabla q = 0\,, \tag{7.18}$$

where q is a vector of velocities in all directions and ∇ is the divergence operator

$$\nabla = \frac{\partial}{\partial x} + \frac{\partial}{\partial y} + \frac{\partial}{\partial z}.$$

Substitution of Darcy's expression for the fluxes, Eq. (7.9), in Eq. (7.18) results in

$$\frac{\partial}{\partial x}\left(K_x \frac{\partial h}{\partial x}\right) + \frac{\partial}{\partial y}\left(K_y \frac{\partial h}{\partial y}\right) + \frac{\partial}{\partial z}\left(K_z \frac{\partial h}{\partial z}\right) = 0. \tag{7.19}$$

For an isotropic and homogeneous medium the hydraulic conductivities can be taken out of the derivatives and divided out, resulting in the well-known Laplace equation,

$$\frac{\partial^2 h}{\partial x^2} + \frac{\partial^2 h}{\partial y^2} + \frac{\partial^2 h}{\partial z^2} = \nabla^2 h = 0. \tag{7.20}$$

The Laplace equation is common to all problems of potential flow in fluid mechanics as well as in problems of electricity, magnetism, and thermal fields. Note that the terms of the Laplace equation are the curvature of the piezometric surface. A nonzero total curvature implies failure of the homogeneity and isotropy assumptions.

7.4.2 Unconfined Aquifer

The unconfined-aquifer problem is complicated by the presence of a free surface. Although Eq. (7.17) would still be valid in regions of complete saturation, it must be solved with variable boundary conditions associated with the phreatic surface. The boundary condition is one of atmospheric pressure over a variable (dynamic) interface between the water and air. It must also include the potential input due to direct aquifer recharge. The analysis is not simple, the interested reader is referred to Bear [1972].

An alternative simplified approach is to integrate the equations in the vertical and deal with a two-dimensional system that assumes horizontal flow and that there are no fluid particles moving along the free surface gradient. This is called the Dupuit approximation.

7.4.3 Horizontal-Plane Flow and the Dupuit Approximation

Figure 7.7 shows a vertical, full-depth column of a phreatic aquifer. A useful assumption is that the slope of the phreatic surface is small in comparison to the total depth, so that vertical components of flow can be ignored. The Dupuit approximation then states: $q_z = 0$ and $\partial q_y / \partial z = \partial q_x / \partial z = 0$. Assuming horizontal bottom and constant hydraulic conductivity over depth, the

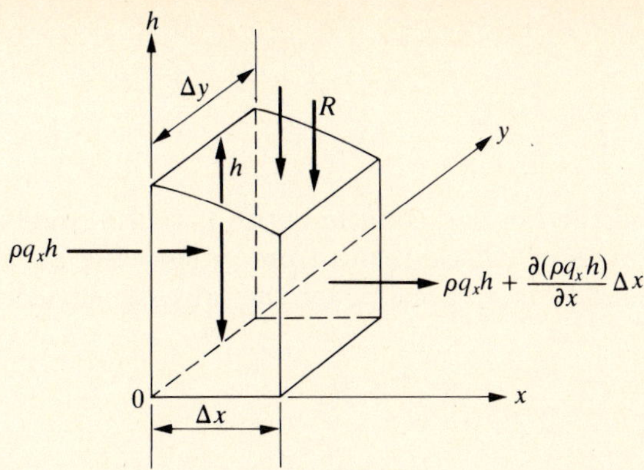

FIGURE 7.7 Control volume for the derivation of the Dupuit approximation.

mass flux in the x direction is (see Fig. 7.7, again using a Taylor expansion concept)

$$\rho q_x h \, \Delta y - \left[\rho q_x h \, \Delta y + \frac{\partial(\rho q_x h)}{\partial x} \Delta x \, \Delta y \right] = -\frac{\partial}{\partial x}(\rho q_x h) \, \Delta x \, \Delta y . \tag{7.21}$$

Adding a similar result in the y direction and equating to the change in mass leads to

$$-\left[\frac{\partial}{\partial x}(\rho q_x h) + \frac{\partial}{\partial y}(\rho q_y h) \right] \Delta x \, \Delta y + \rho R \, \Delta x \, \Delta y = \frac{\partial}{\partial t}(\rho n_e \, \Delta V) + n_e \frac{\partial h}{\partial t} \rho \, \Delta x \, \Delta y . \tag{7.22}$$

The term $\frac{\partial}{\partial t}(\rho n_e \, \Delta V)$ represents changes in mass due to changes in porosity and density of the matrix. This term is several orders of magnitude smaller than the term $n_e \frac{\partial h}{\partial t} \rho \, \Delta x \, \Delta y$, which accounts for the change in the surface elevation. The first term in the right-hand side of Eq. (7.22) is therefore ignored. The last term on the left-hand side accounts for the recharge rate per unit area of the aquifer R. Introducing Darcy's expressions for velocities q_x, q_y, and q_z and assuming that water is incompressible results in

$$\frac{\partial}{\partial x}\left(K_x h \frac{\partial h}{\partial x} \right) + \frac{\partial}{\partial y}\left(K_y h \frac{\partial h}{\partial y} \right) = n_e \frac{\partial h}{\partial t} - R . \tag{7.23}$$

The steady-state result for a homogeneous isotropic medium is

$$\frac{\partial}{\partial x}\left(h \frac{\partial h}{\partial x} \right) + \frac{\partial}{\partial y}\left(h \frac{\partial h}{\partial y} \right) = -\frac{R}{K}$$

or

$$\frac{1}{2}\left[\frac{\partial^2 h^2}{\partial x^2} + \frac{\partial^2 h^2}{\partial y^2}\right] = \frac{1}{2}\nabla^2 h^2 = -\frac{R}{K}. \tag{7.24}$$

The above is called the Poisson equation with potential field h^2.

A more general result would result from integrating Eq. (7.17) with valid upper boundary conditions to account for the phreatic surface. The result is

$$\frac{\partial}{\partial x}\left(T_x \frac{\partial h}{\partial x}\right) + \frac{\partial}{\partial y}\left(T_y \frac{\partial h}{\partial y}\right) = S\frac{\partial h}{\partial t} - R, \tag{7.25}$$

where T is the transmissivity defined for an unconfined system as

$$T_i = \int_{z_1}^{h} K_i(z)\, dz \tag{7.26}$$

and for a confined system as

$$T_i = \int_{z_1}^{z_2} K_i(z)\, dz, \tag{7.27}$$

where z_1 is the bottom of the aquifer, z_2 is the top of a confined aquifer, and $K(z)$ represents the permeability variation with depth. Note that in a confined aquifer of constant permeability and depth b, $T = bK$.

In an unconfined system, S in Eq. (7.25) is the effective porosity n_e and in a confined aquifer

$$S = \int_{z_1}^{z_2} S_o(z)\, dz, \tag{7.28}$$

where S is the nondimensional storativity coefficient.

Usually the integrals in Eqs. (7.26), (7.27), and (7.28) are replaced by sums over finite layers of the aquifer thickness.

$$T = \sum_{j=1}^{N} K^j \Delta z^j, \tag{7.29}$$

$$S = \sum_{j=1}^{N} S^j \Delta z^j, \tag{7.30}$$

where the superscript represents the jth layer of N and Δz^j the thickness of the jth layer.

Given the definition of T_i, Eq. (7.25) is clearly nonlinear in h for the unconfined aquifer. Nevertheless, both unconfined and confined situations in two dimensions collapse to the same equation, for the homogeneous isotropic case with a horizontal bottom by defining the following potentials:

$$\text{unconfined:} \quad \phi = \frac{h^2}{2}$$

$$\text{confined:} \quad \phi = h$$

for which the generalized equation is

$$\frac{\partial^2 \phi}{\partial x^2} + \frac{\partial^2 \phi}{\partial y^2} = C_1 \frac{\partial \phi}{\partial t} - \frac{R}{C_2}, \tag{7.31}$$

where

	UNCONFINED	CONFINED
C_1	$\frac{n_e}{Kh_\circ}$	$\frac{S}{T}$
C_2	K	T

and $h_\circ$ is some representative hydraulic head around which Eq. (7.23) is linearized.

To study hydraulics of wells, the cylindrical-coordinates form of Eq. (7.31) is used:

$$\frac{1}{r} \frac{\partial}{\partial r} \left(r \frac{\partial \phi}{\partial r} \right) + \frac{1}{r^2} \frac{\partial^2 \phi}{\partial \theta^2} = C_1 \frac{\partial \phi}{\partial t} - \frac{R}{C_2}. \tag{7.32}$$

The source (sink) term R in Eq. (7.32) can also be used to represent pumping, recharge, or leakage across aquifer boundaries. The latter condition generally refers to aquifers with leaky aquitards as one or more confining layers. The vertical leakage across a semipervious stratum (aquiclude or aquitard) can be approximated as

$$R = \frac{K_v}{B}(H_\circ - h), \tag{7.33}$$

where K_v is the vertical permeability of the aquitard, B is the thickness of the layer, and $H_\circ$ is the time-invariant head externally imposed on the leaky layer. In using Eq. (7.32) in (7.33), it is assumed that the aquitard cannot store and release water in response to changes in internal pressure.

7.4.4 Initial and Boundary Conditions

To solve the saturated groundwater flow equations seen in the past sections, we need boundary, initial conditions, and the geometry of the problem. There are three general types of boundary conditions.

1. Prescribed head $h = h(x, y, z, t)$. As implied, this potential is not only a function of space but may also be a function of time. These are called Dirichlet boundary conditions.
2. Neumann boundary conditions prescribe the flux or gradient of head along the boundary $q = q(x, y, z, t)$. It could also be a function of time. An impervious boundary is a no-flux condition, $\nabla h(x, y, z) = 0$.
3. A semipervious boundary is called a Cauchy condition. It establishes flux as a function of some external head (usually fixed) and the local potential $q = q[H_o, h(x, y, z, t)]$. The relationship is usually linear:

$$q = f_1(x, y, z, t) + f_2(x, y, z, t)h \,.$$

Initial conditions usually specify the head throughout the aquifer at the beginning of the period of interest. They are a snapshot of conditions at relative-time 0.

7.4.5 Linearity and the Superposition Principle

Equation (7.31) is a linear differential equation on the potential ϕ. Keep in mind that the definition of this potential depends on the nature of the problem under analysis. A property of linear systems is the additivity of solutions. If ϕ_1 and ϕ_2 are solutions to Eq. (7.31), then $\phi = \phi_1 + \phi_2$ is a solution. This principle of superposition of solutions permits the decomposition of large complicated problems into a series of smaller, simpler problems. For example, solutions to complicated input (i.e., recharge, pumping) patterns can be obtained by decomposing the pattern into simple parts, solving the corresponding problems, and adding the results. Similarly, difficult boundary and initial conditions can be achieved by adding carefully selected subproblems. This is the basis of the method of images to be discussed in Section 7.5.5. Remember, though, that boundary conditions must be linear (i.e., not dependent on the state) in order to be able to add solutions.

EXAMPLE 7.1

One-Dimensional Island

A classic example is the aquifer model of an island of infinite length. Jacob [1943] used this model to simulate groundwater in Long Island, New York. An idealization of the situation is shown in Figure 7.8. Because of the infinite length, flow is assumed one dimensional with $\partial h/\partial y = 0$. A condition of no flow across the impervious bottom is also assumed. The aquifer is homogeneous.

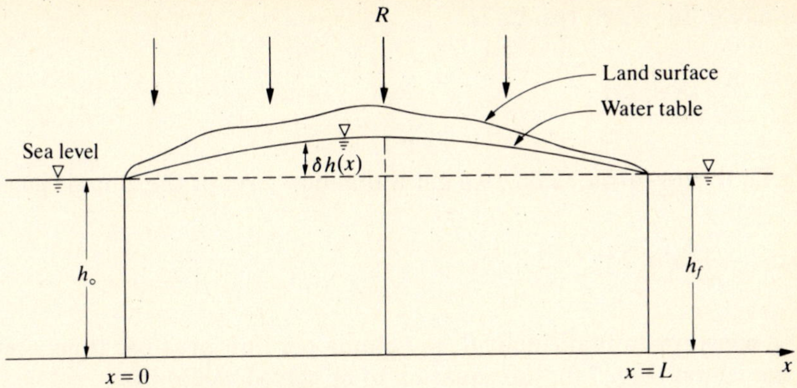

FIGURE 7.8 An island of infinite extent.

Equation (7.24) is applicable for the no-recharge case,

$$\frac{\partial^2 h^2}{\partial x^2} = 0. \tag{7.34}$$

Integrating twice,

$$h^2 = ax + b. \tag{7.35}$$

Imposing the boundary condition at $x = 0$, $h = h_\circ$, leads to $b = h_\circ^2$ or

$$h^2 = ax + h_\circ^2. \tag{7.36}$$

Similarly, at $x = L$, $h = h_f$,

$$h_f^2 = aL + h_\circ^2$$

or

$$a = \frac{h_f^2 - h_\circ^2}{L}. \tag{7.37}$$

The head then obeys,

$$h^2 = \frac{h_f^2 - h_\circ^2}{L}x + h_\circ^2. \tag{7.38}$$

From Darcy's Law the discharge per unit width must be

$$Q = -Kh\frac{\partial h}{\partial x}, \tag{7.39}$$

which using Eq. (7.38) results in

$$Q = \frac{K}{2L}(h_o^2 - h_f^2). \tag{7.40}$$

Using Eq. (7.40) in Eq. (7.38), we get a common form of the Dupuit parabola,

$$h^2 = \frac{-2Q}{K}x + h_o^2. \tag{7.41}$$

If a constant rainfall input R, in volume per unit area per time, occurred over the island (Fig. 7.8), the equation to be solved would be

$$\frac{\partial^2 h^2}{\partial x^2} = -\frac{2}{K}R. \tag{7.42}$$

A first integration yields

$$\frac{\partial h^2}{\partial x} = -\frac{2}{K}Rx + c_1.$$

A second results in

$$h^2 = -\frac{Rx^2}{K} + c_1 x + c_2. \tag{7.43}$$

Imposing boundary conditions, at $x = 0$, $h = h_o$; therefore,

$$c_2 = h_o^2.$$

At $x = L$, $h = h_f$, or

$$h_f^2 - h_o^2 = -\frac{RL^2}{K} + c_1 L;$$

therefore,

$$c_1 = \frac{\left(h_f^2 - h_o^2 + \frac{RL^2}{K}\right)}{L}. \tag{7.44}$$

Finally, then,

$$h^2 = -\frac{Rx^2}{K} + \frac{\left(h_f^2 - h_\circ^2 + \frac{RL^2}{K}\right)}{L} x + h_\circ^2. \tag{7.45}$$

Flow per unit width is given by

$$Q = -Kh\frac{\partial h}{\partial x}$$

or

$$Q = -K\left(-\frac{Rx}{K} + \frac{h_f^2 - h_\circ^2 + \frac{RL^2}{K}}{2L}\right) \tag{7.46}$$

or at $x = L$,

$$Q = \frac{RL}{2} + \frac{h_\circ^2 - h_f^2}{2L}K. \tag{7.47}$$

If $h_f = h_\circ$, Eq. (7.45) yields

$$h^2 - h_\circ^2 = \frac{-Rx^2}{K} + \frac{RL^2}{LK}x = \frac{R}{K}(Lx - x^2).$$

In such a case, the problem is symmetrical. At $x = L/2$, the above equation gives the maximum head

$$h^2 - h_\circ^2 = \frac{R}{K}\left(\frac{L^2}{2} - \frac{L^2}{4}\right) = \frac{RL^2}{4K}. \tag{7.48}$$

An alternative is to deal with the problem with equations analogous to those of confined aquifers. Figure 7.8 can help in the formulation. Define δh as the excess head above elevation $h_\circ$, the boundary value. Also assume that $\delta h \ll h_\circ$. The unconfined aquifer equation

$$\frac{\partial^2 h^2}{\partial x^2} + \frac{2}{K}R = 0$$

can also be expressed as

$$\frac{\partial}{\partial x}\left(h\frac{\partial h}{\partial x}\right) + \frac{R}{K} = 0.$$

Substituting $h = h_\circ + \delta h$,

$$\frac{\partial}{\partial x}\left[(h_\circ + \delta h)\frac{\partial \delta h}{\partial x}\right] + \frac{R}{K} = 0,$$

and ignoring the high-order term in δh, $\frac{\partial}{\partial x}\left[\delta h \frac{\partial \delta h}{\partial x}\right]$, we get

$$\frac{\partial^2 \delta h}{\partial x^2} + \frac{R}{h_\circ K} = 0, \tag{7.49}$$

where $h_\circ K$ is the "transmissivity" of the aquifer under this first-order linearization procedure.

Defining $x = 0$ at the center of the symmetrical domain and integrating twice yields

$$\delta h = \frac{-R}{h_\circ K}\frac{x^2}{2} + c_1 x + c_2. \tag{7.50}$$

The "groundwater divide" implies that at $x = 0$ there is no flow in the x direction. This implies a boundary condition of the form

$$\frac{\partial \delta h}{\partial x} = 0 = \frac{-R}{h_\circ K}x + c_1 \bigg|_{x=0}$$

or $c_1 = 0$. At $x = L/2$, $\delta h = 0$; therefore

$$0 = \frac{-RL^2}{8h_\circ K} + c_2$$

or

$$c_2 = \frac{RL^2}{8h_\circ K}.$$

Substituting in Eq. (7.50) above,

$$\delta h = \frac{R}{2h_\circ K}\left(\frac{L^2}{4} - x^2\right). \tag{7.51}$$

For illustration purposes, assume the following values of the parameters: $R = 0.1 \text{ cm hr}^{-1} = 0.1 \times 10^{-2} \text{ m hr}^{-1}$, $L = 500$ m, $h_\circ = 100$ m, and

$K = 10^{-3}$ cm s^{-1} = 3.6×10^{-2} m hr^{-1}. According to Eq. (7.48), the maximum (center) head would be

$$h^2 - (100)^2 = \frac{(0.1) \times 10^{-2}}{4(3.6 \times 10^{-2})}(500)^2,$$

$$h = 108.3 \text{ m}.$$

Using the first-order linearization given by Eq. (7.51), the maximum h would be

$$h_\circ + \delta h \,|_{x=0} = 100 + \frac{0.1 \times 10^{-2}}{2(100)3.6 \times 10^{-2}}\left[\frac{(500)^2}{4}\right]$$

$$= 108.68 \text{ m}. \blacklozenge$$

EXAMPLE 7.2

Superposition of Solutions

The superposition principle can be used to decompose large problems into groups of simpler problems whose solutions are added. The results of Example 7.1 can be used to illustrate the concept. The problem of rainfall over an island can be broken in two parts.

First, we make use of the homogeneous (no recharge) solution with the boundary conditions implied in Figure 7.8. The solution was seen to be Eq. (7.40)

$$Q_1 = \frac{K}{2L}(h_\circ^2 - h_f^2).$$

To account for the added rainfall, we solve the complementary problem (with different boundary conditions)

$$\frac{\partial^2 h^2}{\partial x^2} + \frac{2R}{K} = 0,$$

$$h(0) = 0,$$

$$h(L) = 0.$$

Integrating twice (Eq. 7.43) resulted in

$$h^2 = -\frac{Rx^2}{K} + c_1 x + c_2$$

and imposing boundary conditions, we get for $x = 0$, $c_2 = 0$, and for $x = L$, $c_1 = RL/K$. Therefore,

$$h^2 = \frac{-Rx^2}{K} + \frac{R}{K} Lx .$$

The discharge at $x = L$ would be

$$Q_2 = -Kh \frac{\partial h}{\partial x}\bigg|_{x=L} = \frac{-K}{2}\left[\frac{-2Rx}{K} + \frac{RL}{K}\right]\bigg|_L$$
$$= \frac{RL}{2}.$$

Adding the two solutions,

$$Q = Q_1 + Q_2 = \frac{K}{2L}(h_\circ^2 - h_f^2) + \frac{RL}{2},$$

which is the same as Eq. (7.47). ◆

EXAMPLE 7.3

Leaky Aquifer

Imagine a leaky aquifer as shown in Figure 7.9. Overlying the zone of saturation is a semipervious layer of thickness B and hydraulic conductivity K_v. The flow through this wall is (Eq. 7.33)

$$q(x) = K_v \frac{H_\circ - h(x)}{B}; \ (\text{cm s}^{-1}),$$

where $H_\circ$ is a fixed external head, also equal to the head at $x = 0$. At $x = L$ the head is equal to the aquifer depth b. In steady state what would be the

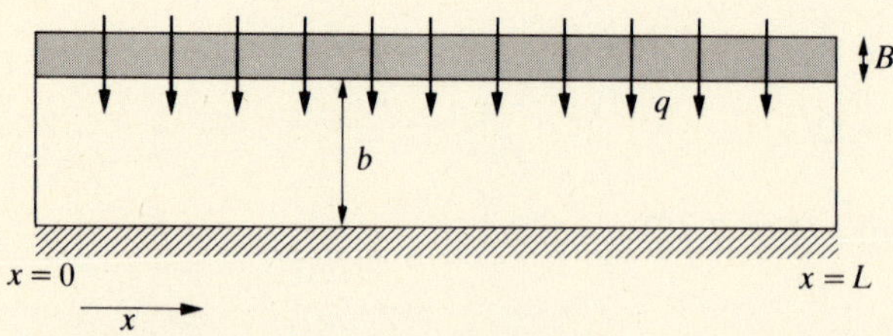

FIGURE 7.9 A confined aquifer with a leaky upper layer.

piezometric head distribution? Assuming one-dimensional flow, we could write (Eq. 7.31)

$$\frac{\partial^2 h}{\partial x^2} + \frac{q}{Kb} = 0\,, \tag{7.52}$$

where K is the hydraulic conductivity of the aquifer. Substituting the expression for q leads to

$$\frac{\partial^2 h}{\partial x^2} + \frac{K_v(H_\circ - h)}{KbB} = 0\,. \tag{7.53}$$

Let $Z = h - H_\circ$ and substitute in Eq. (7.53)

$$\frac{\partial^2 Z}{\partial x^2} - \frac{Z}{M^2} = 0\,, \tag{7.54}$$

where $M = (KbB/K_v)^{1/2}$.

The solution to Eq. (7.54) is of the form

$$Z = C_1 \exp(x/M) + C_2 \exp(-x/M)\,. \tag{7.55}$$

Using the boundary conditions, at $x = L$, $h(L) = b$ or $Z(L) = b - H_\circ$. At $x = 0$, $h(0) = H_\circ$, which implies $Z(0) = 0$. Substituting this last condition in Eq. (7.55),

$$0 = C_1 + C_2\,,$$

and using the former condition

$$b - H_\circ = C_1 \exp(L/M) + C_2 \exp(-L/M)$$

or

$$C_1 = -C_2 = (b - H_\circ)/[\exp(L/M) - \exp(-L/M)]\,. \tag{7.56}$$

Equation (7.55) now becomes

$$h - H_\circ = C_1[\exp(x/M) - \exp(-x/M)]\,, \tag{7.57}$$

with M as in Eq. (7.54) and C_1 given by Eq. (7.56). As an example, let $K_v = 10^{-6}$ cm s^{-1}, $K = 10^{-4}$ cm s^{-1}, $b = 100$ m, $B = 1$ m, $H_\circ = 125$ m, and $L = 1000$ m. The coefficients become $M = 100$ and $C_1 = -1.13 \times 10^{-3}$.

A plot of Eq. (7.57) with the above parameters is given in Figure 7.10. The effect of relatively high leakage is reflected on the flatness of the piezometric head distribution up to 600 m along the aquifer. ◆

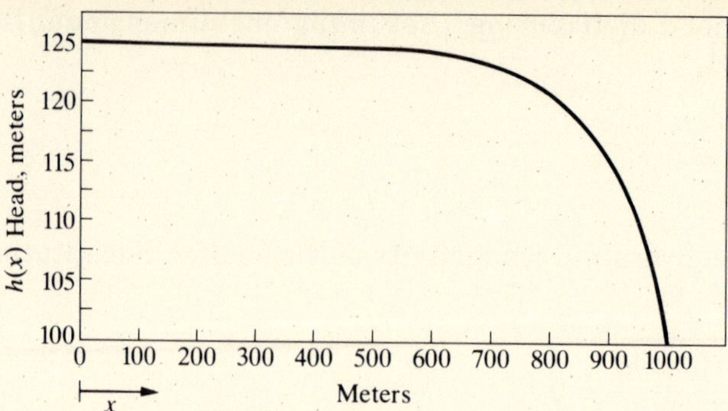

FIGURE 7.10 Solution to leaky aquifer.

You can find in the literature analytical solutions to problems much more complicated than those of the past three examples. Two such results follow. The details of the actual solution have been eliminated as inappropriate for the goals of this book.

EXAMPLE 7.4

Two-Dimensional Model of a Small River Valley

Toth [1962] studied a series of small tributary valleys in the Red Deer River system of central Alberta, Canada. Based on information about hydrology, geology, physiography, topography, and groundwater levels, he argued that the groundwater systems in the small tributary valleys could be represented as in Figure 7.11. The main characteristics are two-dimensional flow in a cross section, impermeable no-flow barriers at the top and bottom of the valley (groundwater divides), impermeable horizontal bedrock, and a piezometric surface that followed the topography. He also assumed that the river was not a significant sink of groundwater and could be ignored.

Toth [1962] then solved the steady-state equations of flow in a square region (height $z_\circ$, length s) approximating the more trapezoidal reality. The problem statement was

$$\frac{\partial^2 h}{\partial x^2} + \frac{\partial^2 h}{\partial z^2} = 0 \tag{7.58}$$

with the following boundary conditions:

$$h = g(z_\circ + cx) \quad \text{at } z = z_\circ \quad \text{for } 0 \leq x \leq s$$

$$\partial h/\partial x = 0 \quad \text{at } x = 0 \text{ and } s \quad \text{for } 0 \leq z \leq z_\circ$$

$$\partial h/\partial z = 0 \quad \text{at } z = 0 \quad \text{for } 0 \leq x \leq s$$

where c is a constant and g is gravitational acceleration.

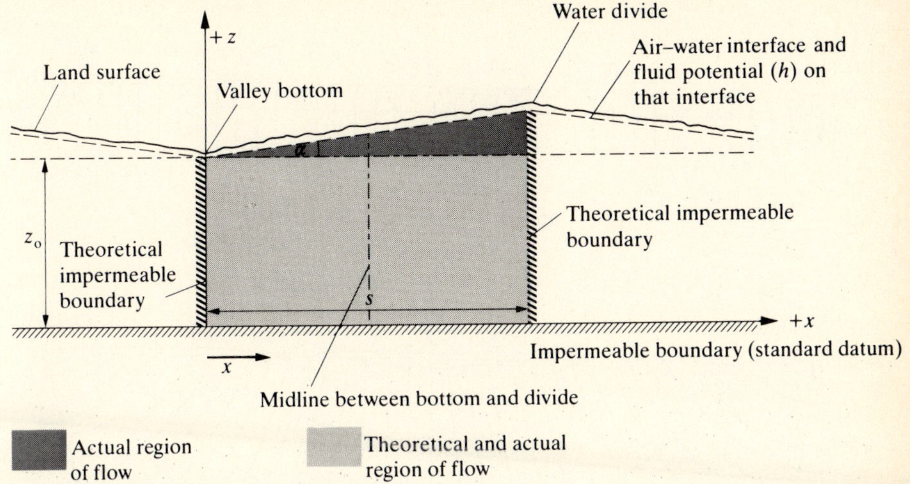

FIGURE 7.11 Cross section of a valley, showing real and theoretical boundaries and flow regions. Source: J. Toth, *J. Geophys. Res.* 67(11):4379, 1962. Copyright by the American Geophysical Union.

The solution to the linear partial differential equation is found, by separation of variables, to be

$$h(x, z) = g\left(z_\circ + \frac{cs}{2}\right) - \frac{4gcs}{\pi^2}\sum_{m=0}^{\infty} \frac{\cos[(2m + 1)\pi x/s]\cosh[(2m + 1)\pi z/s]}{(2m + 1)^2 \cosh[(2m + 1)\pi z_\circ/s]}. \tag{7.59}$$

Figure 7.12 diagrams Eq. (7.59). Equipotential (equal head) lines are drawn together with the perpendicular streamlines. Such a diagram is called a flow net and is a common graphical solution procedure and representation of potential flow fields obeying Laplace's equation. Two solutions are given, one for a 500-ft-deep valley (152.4 m) and one for a 10,000-ft-deep (3048.8 m) valley. The width of the flow tubes (between streamlines) is inversely proportional to the flow velocity since, by construction, flow tubes carry the same increment of discharge. Therefore, in Figure 7.12, flow is more intense near the surface particularly close to the flow divides.

Toth [1962] used the above result to argue that the valley can be divided, at its midpoint, in recharge (water-receiving) and discharge areas. Furthermore, the discharge is significant throughout the whole valley bottom, not only at the stream. This type of behavior could be used to argue that when rainfall comes, certain lower reaches are quickly saturated and contribute most of the surface flowing waters to the river. This is called the "partial area" runoff contributing concept. The related concepts of infiltration and runoff will be discussed in Chapter 8, including runoff by soil profile saturation.

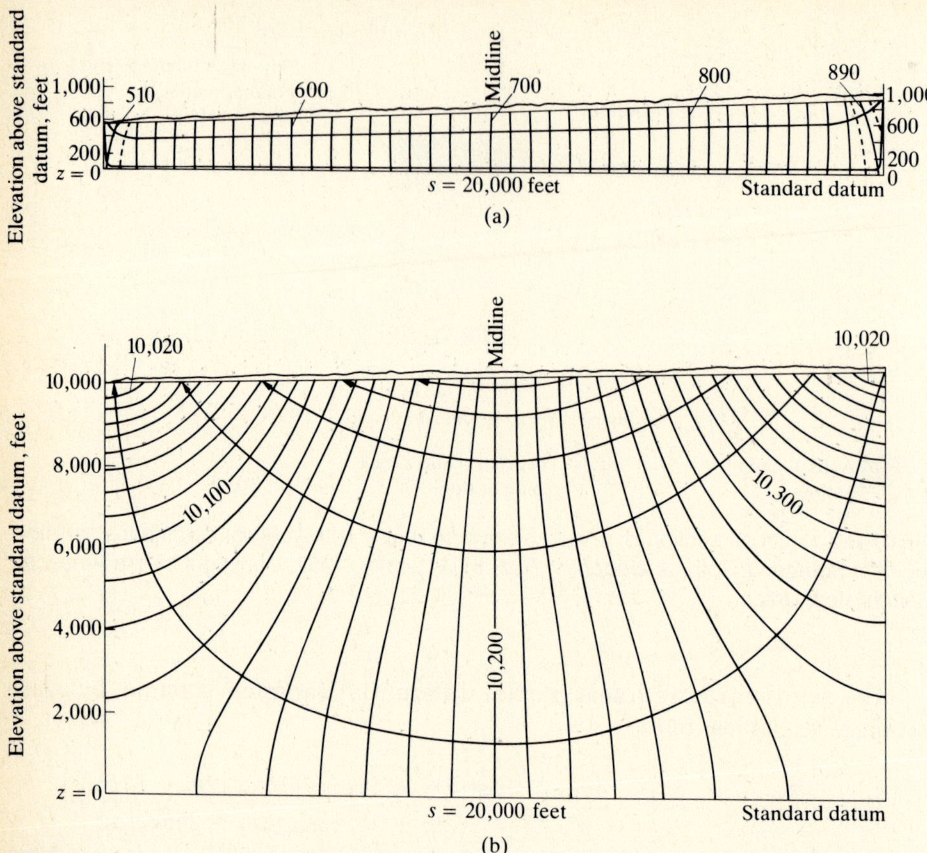

FIGURE 7.12 Two-dimensional distributions and flow patterns for different depths to the horizontal impermeable boundary. Source: J. Toth, *J. Geophys. Res.* 67(11):4380, 1962. Copyright by the American Geophysical Union.

Toth [1963] expands the presented approach to study groundwater flow under more complicated phreatic surface descriptions. He was attempting to account for the highly variable topographic relief in a valley floor. The reader is urged to study this illuminating example further. ◆

EXAMPLE 7.5

Surface–Groundwater Interaction

Cooper and Rorabaugh [1963] studied the interaction between surface streams and aquifers. They idealized the situation as in Figure 7.13 and analytically studied the response of the aquifer to a passing flood in the stream. The aquifer could be of finite or semi-infinite (no impermeable valley-wall effect) extent.

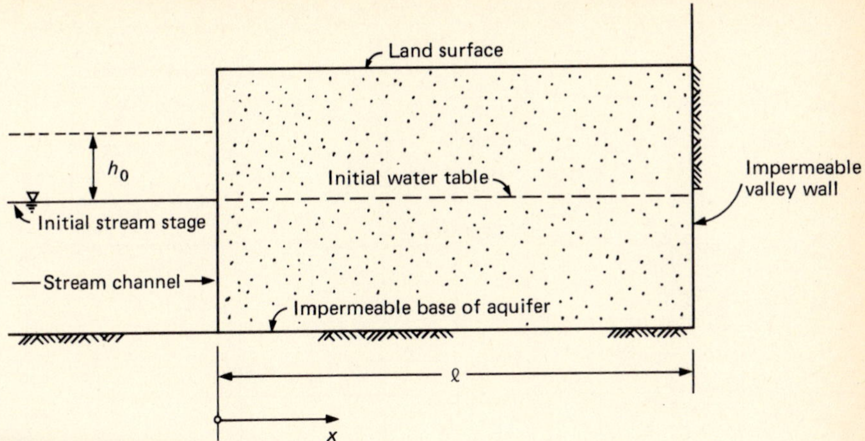

FIGURE 7.13 Idealized interaction between a stream and an aquifer. Sources: Cooper and Rorabaugh [1963], and Eagleson [1970].

The passing flood was represented by a general time-varying head at $x = 0$ of the form,

$$h(0, t) = Nh_\circ e^{-\delta t}(1 - \cos \omega t), \qquad \text{for } 0 \le t \le \tau, \tag{7.60}$$

where $h_\circ$ is the maximum height of the wave above the initial water table, δ is a factor to control the symmetry of the wave, N depends on δ and insures that all flood waves peak at $h_\circ$, and τ is the duration of the wave. Clearly, the problem is one of non-steady-state, time-varying, interest. For $0 \le t \le \tau$, Cooper and Rorabaugh defined h as the deviation of the head from the original steady-state problem and solved the following one-dimensional problem:

$$\frac{\partial^2 h}{\partial x^2} - \frac{S}{T}\frac{\partial h}{\partial t} = 0 \tag{7.61}$$

$$h(x, 0) = 0 \qquad \text{for } 0 \le x \le \ell$$

$$\frac{\partial h(\ell, t)}{\partial x} = 0 \qquad \text{for } t \ge 0$$

and $h(0, t)$ according to Eq. (7.60). The above is a first-order linearization of the unconfined groundwater flow equations. The factors S and T depend on the linearization.

For $t \ge \tau$ some of the boundary conditions change:

$$h(x, \tau) = h_{t \le \tau}(x, \tau) \qquad \text{for } 0 \le x \le \ell$$

$$h(0, t) = 0 \qquad \text{for } t \ge \tau$$

$$\frac{\partial h(\ell, t)}{\partial x} = 0 \qquad \text{for } t \ge \tau,$$

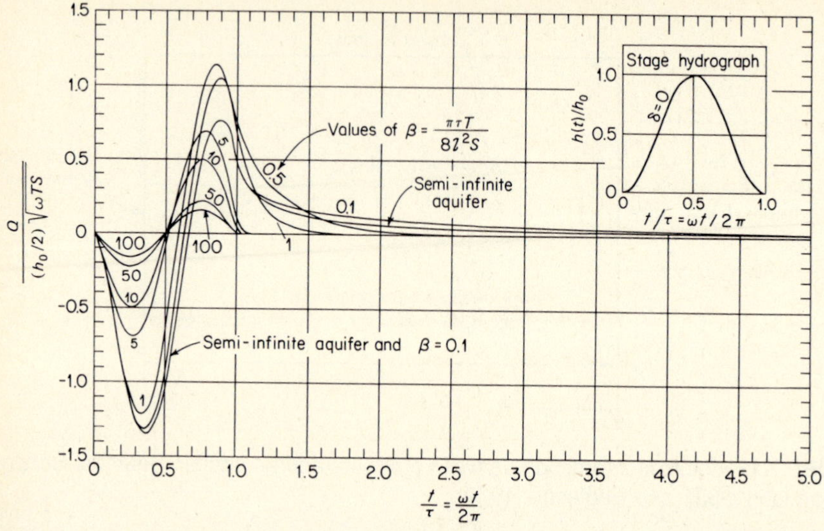

FIGURE 7.14 Groundwater flow into stream resulting from stage oscillation defined by Eq. (7.60) when $\delta = 0$. Sources: Cooper and Rorabaugh [1963], and Eagleson [1970].

where $h_{t\leq\tau}(x, \tau)$ is the solution to Eq. (7.61) at time τ. Cooper and Rorabaugh only solved Eq. (7.61) and used the principle of superposition seen in Section 7.4.5 to obtain the full solution. Laplace transform techniques were used to solve the partial differential equation.

The solutions are best visualized in Figures 7.14 and 7.15. Both figures correspond to a symmetrical flood wave (stage hydrograph with $\delta = 0$). Curves are shown for various values of $\beta = \pi\tau T/8\ell^2 S$.

A value of β at or near zero corresponds to an aquifer of semi-infinite extent. T and S in the ordinate are transmissivity and the nondimensional storativity coefficient. Figure 7.14 shows the variation in time of flow into the stream. Negative values imply that water is going into the aquifer. Note that for values of $\beta = 10$ or higher, the flow is in phase with the flood wave. Water goes in during the rising limb and out during the recession in an almost symmetrical manner. This is not true for the semi-infinite aquifer. In that case flow and the wave are out of phase. The banks receive water during 70% of the event. It releases water at a much slower rate.

The solution for the volume of water absorbed by the aquifer (bank storage) is given in Figure 7.15. Limited-dimension aquifers release as much water as they absorb, almost at the same rate. An infinite aquifer retains a considerable amount of water for effectively an infinite time. ◆

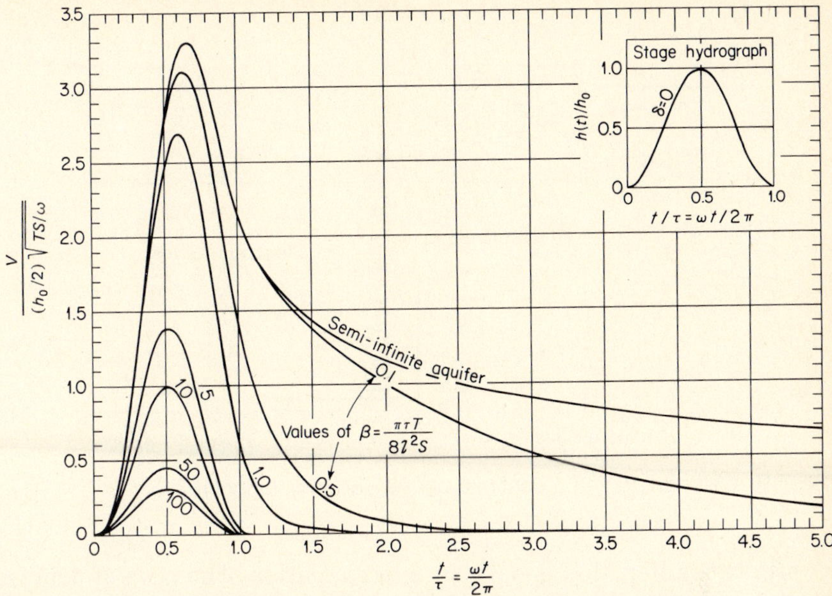

FIGURE 7.15 Bank storage resulting from stage oscillation defined by Eq. (7.60) when $\delta = 0$. Sources: Cooper and Rorabaugh [1963], and Eagleson [1970].

7.5 HYDRAULICS OF WELLS

The following sections will study the behavior of wells, generally from a somewhat idealized perspective. Assumptions will include isotropy, homogeneity, and, generally, full penetration of the aquifer by the well. Although restrictive, these simplifications will not detract from the tremendous usefulness of the results. As we will see, the study of well behavior will provide ways of estimating aggregate parameters of aquifers. Therefore, wells are not only a way to take out or add water to aquifers, but they are also instruments of the study of groundwater flow parameters.

7.5.1 Steady-State Solution of a Fully Penetrating Well in a Confined Aquifer

Imagine a single well penetrating a confined aquifer of depth b, as shown in Figure 7.16. The well is being pumped at a constant rate Q. After some pumping time (theoretically, infinite time; practically, a finite time), the originally horizontal piezometric head forms a time-invariant cone of depression. (*Note:*

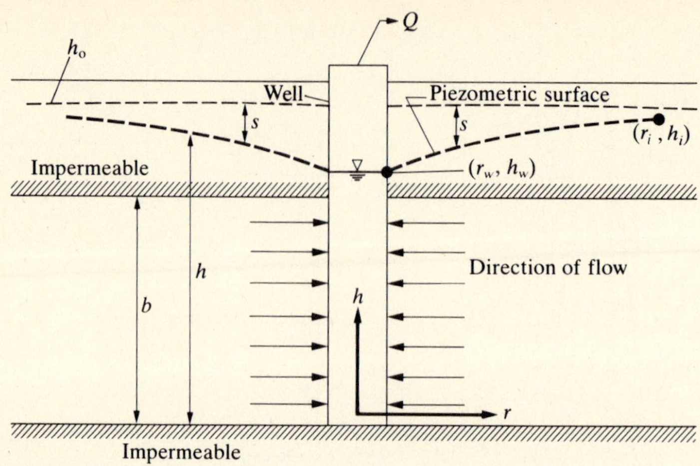

FIGURE 7.16 Well penetrating a confined aquifer. Source: Gray [1973].

In an infinite aquifer, this is only an approximation. The cone of depression continues to expand, only very slowly, as time goes to infinity.) In a homogeneous, isotropic aquifer of infinite extent, this cone is radially symmetric. At the well radius r_w, the head is taken as h_w, the level to which the water rises in the well. In fact, this is not true in practice; the water level will be lower than h_w and the effective well radius will be larger than r_w. This behavior accounts for energy losses occurring through the graded gravel packing and screens usually found around correctly built wells (Gray [1973], Brown et al. [1972]). The head at a radius r_i away from the well is h_i. Flow is horizontal, given the parallel confining layers and the complete penetration of the well.

The steady-state saturated groundwater problem just described must obey Laplace's equation, which in its cylindrical-coordinates version is

$$\frac{1}{r}\frac{\partial}{\partial r}\left[r\frac{\partial h}{\partial r}\right]+\frac{1}{r^2}\frac{\partial^2 h}{\partial \theta^2}=0\,. \tag{7.62}$$

Since flow is radial, irrotational,

$$\frac{\partial h}{\partial \theta}=0\,,$$

leading to the one-dimensional statement

$$\frac{1}{r}\frac{\partial}{\partial r}\left[r\frac{\partial h}{\partial r}\right]=0\,. \tag{7.63}$$

Integrating once,

$$r \frac{\partial h}{\partial r} = c_1 .$$

A second integration yields

$$h = c_1 \ln r + c_2 . \tag{7.64}$$

Boundary conditions must now be imposed. At $r = r_w$, $h = h_w$; therefore $h_w = c_1 \ln r_w + c_2$. At $r = r_i$, $h = h_i$; therefore $h_i = c_1 \ln r_i + c_2$.

Simultaneous solution of the above results in

$$c_1 = \frac{h_i - h_w}{\ln\left(\dfrac{r_i}{r_w}\right)}$$

$$c_2 = h_w - \frac{(h_i - h_w) \ln[r_w]}{\ln\left(\dfrac{r_i}{r_w}\right)} .$$

Finally,

$$h = \left[\frac{h_i - h_w}{\ln\left(\dfrac{r_i}{r_w}\right)}\right] \ln r + h_w - \left[\frac{h_i - h_w}{\ln\left(\dfrac{r_i}{r_w}\right)}\right] \ln r_w . \tag{7.65}$$

Using Darcy's equation, flow into the well through any of the concentric cylinders, at radius r, around the well is given by

$$Q = -2\pi r b K \frac{\partial h}{\partial r} , \tag{7.66}$$

where discharge into the well is defined as negative.

From Eq. (7.65),

$$\frac{\partial h}{\partial r} = \frac{h_i - h_w}{\ln\left[\dfrac{r_i}{r_w}\right]} \frac{1}{r} . \tag{7.67}$$

Substituting into Eq. (7.66) leads to the following discharge magnitude into the well (negative sign is ignored):

$$Q = 2\pi bK \frac{h_i - h_w}{\ln\left(\dfrac{r_i}{r_w}\right)} = 2\pi T \frac{h_i - h_w}{\ln\left(\dfrac{r_i}{r_w}\right)}. \tag{7.68}$$

Note that if the pumping rate is known and water-level observations are available at the site and at a point a distance r_i away (observation well), then Eq. (7.68) can be solved for the transmissivity T of the aquifer. This parameter-estimation problem is generally called the inverse problem in hydrology. Also note that the choice of a finite head value at a given radial distance for this infinite domain problem leads to a solution that is not valid when r_i goes to infinity. As it stands, $h_i - h_w$ would illogically also go to infinity.

7.5.2 Steady-State Solution of a Fully Penetrating Well in an Unconfined Aquifer

Steady-state flow into a well in an unconfined aquifer would require the solution of the homogeneous form of Eq. (7.32), but now with $\phi = h^2/2$. Again, note that this steady-state solution can at best approximate reality. As expected, the integration and results will parallel those of Section 7.5.1. An alternative approach uses geometric considerations. Figure 7.17 shows the typical unconfined flow into a well. Assuming the Dupuit assumptions hold, the flow across the concentric cylinders surrounding the well must be equal to each other and equal to the well discharge. At the cylinder a distance r from the well center (defining flow into the well as negative),

$$Q = -2\pi rhK \frac{\partial h}{\partial r}. \tag{7.69}$$

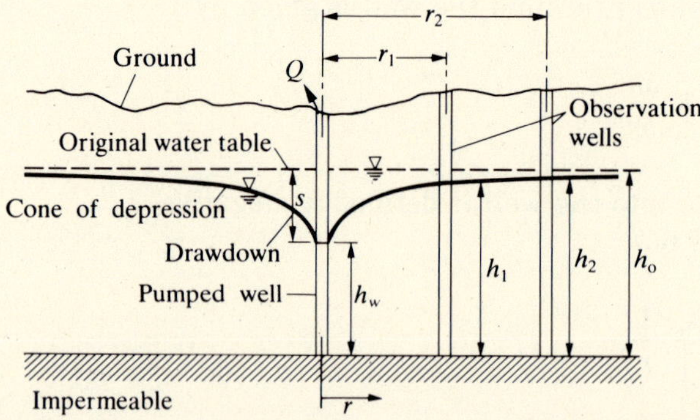

FIGURE 7.17 Schematic diagram of a fully penetrating well in an unconfined aquifer.

Integrating Eq. (7.69),

$$\int_{r_1}^{r_2} Q \frac{\partial r}{r} = -2\pi K \int_{h_1}^{h_2} h \partial h ,$$

leads to the following discharge magnitude (ignoring the negative sign):

$$Q = \frac{\pi K(h_2^2 - h_1^2)}{\ln\left(\dfrac{r_2}{r_1}\right)}. \tag{7.70}$$

Equation (7.70) can again be solved for hydraulic conductivity, given Q and two observations of well-water elevation at two different points of radius r_1 and r_2.

7.5.3 Unsteady Flow in Wells

As previously seen, the solution of unsteady flow in wells must satisfy Eq. (7.32), which for radial flow and no recharge becomes

$$\frac{1}{r}\frac{\partial}{\partial r}\left(r\frac{\partial \phi}{\partial r}\right) = C_1 \frac{\partial \phi}{\partial t}. \tag{7.71}$$

For a single well in a homogeneous, isotropic, confined aquifer of infinite extent, Theis [1935] gave the solution of Eq. (7.71) as

$$s = \frac{Q}{4\pi T} W(u), \tag{7.72}$$

where s is drawdown, the distance from the original piezometric surface to the new surface (drawdown curve) at a point r from the center of the pumping well (see Fig. 7.17 for definition); T is transmissivity; and Q is well constant discharge. The term $W(u)$ is the so-called well function

$$W(u) = \int_u^{\infty} \frac{e^{-u}}{u} du , \tag{7.73}$$

where

$$u = \frac{r^2 s}{4Tt}, \tag{7.74}$$

where r is the radial distance in feet, S is the storativity coefficient, t is the duration of pumping, and T is the transmissivity.

TABLE 7.3 Values of $W(u)$

N \ u	Nx10^{-15}	Nx10^{-14}	Nx10^{-13}	Nx10^{-12}	Nx10^{-11}	Nx10^{-10}	Nx10^{-9}	Nx10^{-8}	Nx10^{-7}	Nx10^{-6}	Nx10^{-5}	Nx10^{-4}	Nx10^{-3}	Nx10^{2}	Nx10^{-1}	N
1.0.	33.9616	31.6590	29.3564	27.0538	24.7512	22.4486	20.1460	17.8435	15.5409	13.2383	10.9357	8.6332	6.3315	4.0379	1.8229	0.2194
1.1	33.8662	31.5637	29.2611	26.9585	24.6559	22.3533	20.0507	17.7482	15.4456	13.1430	10.8404	8.5379	6.2363	3.9436	1.7371	.1860
1.2	33.7792	31.4767	29.1741	26.8715	24.5689	22.2663	19.9637	17.6611	15.3586	13.0560	10.7534	8.4509	6.1494	3.8576	1.6595	.1584
1.3	33.6992	31.3966	29.0940	26.7914	24.4889	22.1863	19.8837	17.5811	15.2785	12.9759	10.6734	8.3709	6.0695	3.7785	1.5889	.1355
1.4	33.6251	31.3225	29.0199	26.7173	24.4147	22.1122	19.8096	17.5070	15.2044	12.9018	10.5993	8.2968	5.9955	3.7054	1.5241	.1162
1.5	33.5561	31.2535	28.9509	26.6483	24.3458	22.0432	19.7406	17.4380	15.1354	12.8328	10.5303	8.2278	5.9266	3.6374	1.4645	.1000
1.6	33.4916	31.1890	28.8864	26.5838	24.2812	21.9786	19.6760	17.3735	15.0709	12.7683	10.4657	8.1634	5.8621	3.5739	1.4092	.08631
1.7	33.4309	31.1283	28.8258	26.5232	24.2206	21.9180	19.6154	17.3128	15.0103	12.7077	10.4051	8.1027	5.8016	3.5143	1.3578	.07465
1.8	33.3738	31.0712	28.7686	26.4660	24.1634	21.8608	19.5583	17.2557	14.9531	12.6505	10.3479	8.0455	5.7446	3.4581	1.3098	.06471
1.9	33.3197	31.0171	28.7145	26.4119	24.1094	21.8068	19.5042	17.2016	14.8990	12.5964	10.2939	7.9915	5.6906	3.4050	1.2649	.05620
2.0	33.2684	30.9658	28.6632	26.3607	24.0581	21.7555	19.4529	17.1503	14.8477	12.5451	10.2426	7.9402	5.6394	3.3547	1.2227	.04890
2.1	33.2196	30.9170	28.6145	26.3119	24.0093	21.7067	19.4041	17.1015	14.7989	12.4964	10.1938	7.8914	5.5907	3.3069	1.1829	.04261
2.2	33.1731	30.8705	28.5679	26.2653	23.9628	21.6602	19.3576	17.0550	14.7524	12.4498	10.1473	7.8449	5.5443	3.2614	1.1454	.03719
2.3	33.1286	30.8261	28.5235	26.2209	23.9183	21.6157	19.3131	17.0106	14.7080	12.4054	10.1028	7.8004	5.4999	3.2179	1.1099	.03250
2.4	33.0861	30.7835	28.4809	26.1783	23.8758	21.5732	19.2706	16.9680	14.6654	12.3628	10.0603	7.7579	5.4575	3.1763	1.0762	.02844
2.5	33.0453	30.7427	28.4401	26.1375	23.8349	21.5323	19.2298	16.9272	14.6246	12.3220	10.0194	7.7172	5.4167	3.1365	1.0443	.02491
2.6	33.0060	30.7035	28.4009	26.0983	23.7957	21.4931	19.1905	16.8880	14.5854	12.2828	9.9802	7.6779	5.3776	3.0983	1.0139	.02185
2.7	32.9683	30.6657	28.3631	26.0606	23.7580	21.4554	19.1528	16.8502	14.5476	12.2450	9.9425	7.6401	5.3400	3.0615	.9849	.01918
2.8	32.9319	30.6294	28.3268	26.0242	23.7216	21.4190	19.1164	16.8138	14.5113	12.2087	9.9061	7.6038	5.3037	3.0261	.9573	.01686
2.9	32.8968	30.5943	28.2917	25.9891	23.6865	21.3839	19.0813	16.7788	14.4762	12.1736	9.8710	7.5687	5.2687	2.9920	.9309	.01482
3.0	32.8629	30.5604	28.2578	25.9552	23.6526	21.3500	19.0474	16.7449	14.4423	12.1397	9.8371	7.5348	5.2349	2.9591	.9057	.01305
3.1	32.8302	30.5276	28.2250	25.9224	23.6198	21.3172	19.0146	16.7121	14.4095	12.1069	9.8043	7.5020	5.2022	2.9273	.8815	.01149
3.2	32.7984	30.4958	28.1932	25.8907	23.5881	21.2855	18.9829	16.6803	14.3777	12.0751	9.7726	7.4703	5.1706	2.8965	.8583	.01013
3.3	32.7676	30.4651	28.1625	25.8599	23.5573	21.2547	18.9521	16.6495	14.3470	12.0444	9.7418	7.4395	5.1399	2.8668	.8361	.008939
3.4	32.7378	30.4352	28.1326	25.8300	23.5274	21.2249	18.9223	16.6197	14.3171	12.0145	9.7120	7.4097	5.1102	2.8379	.8147	.007891
3.5	32.7088	30.4062	28.1036	25.8010	23.4985	21.1959	18.8933	16.5907	14.2881	11.9855	9.6830	7.3807	5.0813	2.8099	.7942	.006970
3.6	32.6806	30.3780	28.0755	25.7729	23.4703	21.1677	18.8651	16.5625	14.2599	11.9574	9.6548	7.3526	5.0532	2.7827	.7745	.006160
3.7	32.6532	30.3506	28.0481	25.7455	23.4429	21.1403	18.8377	16.5351	14.2325	11.9300	9.6274	7.3252	5.0259	2.7563	.7554	.005448
3.8	32.6266	30.3240	28.0214	25.7188	23.4162	21.1136	18.8110	16.5085	14.2059	11.9033	9.6007	7.2985	4.9993	2.7306	.7371	.004820
3.9	32.6006	30.2980	27.9954	25.6928	23.3902	21.0877	18.7851	16.4825	14.1799	11.8773	9.5748	7.2725	4.9735	2.7056	.7194	.004267
4.0	32.5753	30.2727	27.9701	25.6875	23.3649	21.0623	18.7598	16.4572	14.1546	11.8520	9.5495	7.2472	4.9482	2.6813	.7024	.003779
4.1	32.5506	30.2480	27.9454	25.6428	23.3402	21.0376	18.7351	16.4325	14.1299	11.8273	9.5248	7.2225	4.9236	2.6576	.6859	.003349
4.2	32.5265	30.2239	27.9213	25.6187	23.3161	21.0136	18.7110	16.4084	14.1058	11.8032	9.5007	7.1985	4.8997	2.6344	.6700	.002969
4.3	32.5029	30.2004	27.8978	25.5952	23.2926	20.9900	18.6874	16.3848	14.0823	11.7797	9.4771	7.1749	4.8762	2.6119	.6546	.002633
4.4	32.4800	30.1774	27.8748	25.5722	23.2696	20.9670	18.6644	16.3619	14.0593	11.7567	9.4541	7.1520	4.8533	2.5899	.6397	.002336
4.5	32.4575	30.1549	27.8523	25.5497	23.2471	20.9446	18.6420	16.3394	14.0368	11.7342	9.4317	7.1295	4.8310	2.5684	.6253	.002073
4.6	32.4355	30.1329	27.8303	25.5277	23.2252	20.9226	18.6200	16.3174	14.0148	11.7122	9.4097	7.1075	4.8091	2.5474	.6114	.001841
4.7	32.4140	30.1114	27.8088	25.5062	23.2037	20.9011	18.5085	16.2959	13.9933	11.6907	9.3882	7.0860	4.7877	2.5268	.5979	.001635
4.8	32.3929	30.0904	27.7878	25.4852	23.1826	20.8800	18.5774	16.2748	13.9723	11.6697	9.3671	7.0650	4.7667	2.5068	.5848	.001453
4.9	32.3723	30.0697	27.7672	25.4646	23.1620	20.8594	18.5568	16.2542	13.9516	11.6491	9.3465	7.0444	4.7462	2.4871	.5721	.001291
5.0	32.3521	30.0495	27.7470	25.4444	23.1418	20.8392	18.5366	16.2340	13.9314	11.6289	9.3263	7.0242	4.7261	2.4679	.5598	.001148
5.1	32.3323	30.0297	27.7271	25.4246	23.1220	20.8194	18.5168	16.2142	13.9116	11.6091	9.3065	7.0044	4.7064	2.4491	.5478	.001021
5.2	32.3129	30.0103	27.7077	25.4051	23.1026	20.8000	18.4974	16.1948	13.8922	11.5896	9.2871	6.9850	4.6871	2.4306	.5362	.0009086
5.3	32.2939	29.9913	27.6887	25.3861	23.0835	20.7809	18.4783	16.1758	13.8732	11.5706	9.2681	6.9659	4.6681	2.4126	.5250	.0008086
5.4	32.2752	29.9726	27.6700	25.3674	23.0648	20.7622	18.4596	16.1571	13.8545	11.5519	9.2494	6.9473	4.6495	2.3948	.5140	.0007198
5.5	32.2568	29.9542	27.6516	25.3491	23.0465	20.7439	18.4413	16.1387	13.8361	11.5336	9.2310	6.9289	4.6313	2.3775	.5034	.0006409
5.6	32.2388	29.9362	27.6336	25.3310	23.0285	20.7259	18.4233	16.1207	13.8181	11.5155	9.2130	6.9109	4.6134	2.3604	.4930	.0005708
5.7	32.2211	29.9185	27.6159	25.3133	23.0103	20.7082	18.4056	16.1030	13.8004	11.4978	9.1953	6.8932	4.5958	2.3437	.4830	.0005085
5.8	32.2037	29.9011	27.5985	25.2959	22.9934	20.6908	18.3882	16.0856	13.7830	11.4804	9.1779	6.8758	4.5785	2.3273	.4732	.0004532
5.9	32.1866	29.8840	27.5814	25.2789	22.9763	20.6737	18.3711	16.0685	13.7659	11.4633	9.1608	6.8588	4.5615	2.3111	.4637	.0004039
6.0	32.1698	29.8672	27.5646	25.2620	22.9595	20.6569	18.3543	16.0517	13.7491	11.4465	9.1440	6.8420	4.5448	2.2953	.4544	.0003601
6.1	32.1533	29.8507	27.5481	25.2455	22.9429	20.6403	18.3378	16.0352	13.7326	11.4300	9.1275	6.8254	4.5283	2.2797	.4454	.0003211
6.2	32.1370	29.8344	27.5318	25.2293	22.9267	20.6241	18.3215	16.0189	13.7163	11.4138	9.1112	6.8092	4.5122	2.2645	.4366	.0002864
6.3	32.1210	29.8184	27.5158	25.2133	22.9107	20.6081	18.3055	16.0029	13.7003	11.3978	9.0952	6.7932	4.4963	2.2494	.4280	.0002555
6.4	32.1053	29.8027	27.5001	25.1975	22.8949	20.5923	18.2898	15.9872	13.6846	11.3820	9.0795	6.7775	4.4806	2.2346	.4197	.0002279
6.5	32.0898	29.7872	27.4846	25.1820	22.8794	20.5768	18.2742	15.9717	13.6691	11.3665	9.0640	6.7620	4.4652	2.2201	.4115	.0002034
6.6	32.0745	29.7719	27.4693	25.1667	22.8641	20.5616	18.2590	15.9564	13.6538	11.3512	9.0487	6.7467	4.4501	2.2058	.4036	.0001816
6.7	32.0595	29.7569	27.4543	25.1517	22.8491	20.5465	18.2439	15.9414	13.6388	11.3362	9.0337	6.7317	4.4351	2.1917	.3959	.0001621
6.8	32.0446	29.7421	27.4395	25.1369	22.8343	20.5317	18.2291	15.9265	13.6240	11.3214	9.0189	6.7169	4.4204	2.1779	.3883	.0001448
6.9	32.0300	29.7275	27.4249	25.1223	22.8197	20.5171	18.2145	15.9119	13.6094	11.3068	9.0043	6.7023	4.4059	2.1643	.3810	.0001293
7.0	32.0156	29.7131	27.4105	25.1079	22.8053	20.5027	18.2001	15.8976	13.5950	11.2924	8.9899	6.6879	4.3916	2.1508	.3738	.0001155
7.1	32.0015	29.6989	27.3963	25.0937	22.7911	20.4885	18.1860	15:8834	13.5808	11.2782	8.9757	6.6737	4.3775	2.1376	.3668	.0001032
7.2	31.9875	29.6849	27.3823	25.0797	22.7771	20.4746	18.1720	15.8694	13.5668	11.2642	8.9617	6.6598	4.3636	2.1246	.3599	.00009219
7.3	31.9737	29.6711	27.3685	25.0659	22.7633	20.4608	18.1582	15.8556	13.5530	11.2504	8.9479	6.6460	4.3500	2.1118	.3532	.00008239
7.4	31.9601	29.6575	27.3549	25.0523	22.7497	20.4472	18.1446	15.8420	13.5394	11.2368	8.9343	6.6324	4.3364	2.0991	.3467	.00007364
7.5	31.9467	29.6441	27.3415	25.0389	22.7363	20.4337	18.1311	15.8286	13.5260	11.2234	8.9209	6.6190	4.3231	2.0867	.3403	.00006583
7.6	31.9334	29.6308	27.3282	25.0257	22.7231	23.4205	18.1179	15.8153	13.5127	11.2102	8.9076	6.6057	4.3100	2.0744	.3341	.00005886
7.7	31.9203	29.6178	27.3152	25.0126	22.7100	20.4074	18.1048	15.8022	13.4997	11.1971	8.8946	6.5927	4.2970	2.0623	.3280	.00005263
7.8	31.9074	29.6048	27.3023	24.9997	22.6971	20.3945	18.0919	15.7893	13.4868	11.1842	8.8817	6.5798	4.2842	2.0503	.3221	.00004707
7.9	31.8947	29.5921	27.2895	24.9869	22.6844	20.3818	18.0792	15.7766	13.4740	11.1714	8.8689	6.5671	4.2716	2.0386	.3163	.00004210
8.0	31.8821	29.5795	27.2769	24.9744	22.6718	20.3692	18.0666	15.7640	13.4614	11.1589	8.8563	6.5545	4.2591	2.0269	.3106	.00003767
8.1	31.8697	29.5671	27.2645	24.9619	22.6594	20.3568	18.0542	15.7516	13.4490	11.1464	8.8439	6.5421	4.2468	2.0155	.3050	.00003370
8.2	31.8574	29.5548	27.2523	24.9497	22.6471	20.3445	18.0419	15.7393	13.4367	11.1342	8.8317	6.5298	4.2346	2.0042	.2996	.00003015
8.3	31.8453	29.5427	27.2401	24.9375	22.6350	20.3324	18.0298	15.7272	13.4246	11.1220	8.8195	6.5177	4.2226	1.9930	.2943	.00002699
8.4	31.8333	29.5307	27.2282	24.9256	22.6230	20.3204	18.0178	15.7152	13.4126	11.1101	8.8076	6.5057	4.2107	1.9820	.2891	.00002415
8.5	31.8215	29.5189	27.2163	24.9137	22.6112	20.3086	18.0060	15.7034	13.4008	11.0982	8.7957	6.4939	4.1990	1.9711	.2840	.00002162
8.6	31.8098	29.5072	27.2046	24.9020	22.5995	20.2969	17.9943	15.6917	13.3891	11.0865	8.7840	6.4822	4.1874	1.9604	.2790	.00001936
8.7	31.7982	29.4957	27.1931	24.8905	22.5879	20.2853	17.9827	15.6801	13.3776	11.0750	8.7725	6.4707	4.1759	1.9498	.2742	.00001733
8.8	31.7868	29.4842	27.1816	24.8790	22.5765	20.2739	17.9713	15.6687	13.3661	11.0635	8.7610	6.4592	4.1646	1.9393	.2694	.00001552
8.9	31.7755	29.4729	27.1703	24.8678	22.5652	20.2626	17.9600	15.6574	13.3548	11.0523	8.7497	6.4480	4.1534	1.9290	.2647	.00001390
9.0	31.7643	29.4618	27.1592	24.8566	22.5540	20.2514	17.9488	15.6462	13.3437	11.0411	8.7386	6.4368	4.1423	1.9187	.2602	.00001245
9.1	31.7533	29.4507	27.1481	24.8455	22.5429	20.2404	17.9378	15.6352	13.3326	11.0300	8.7275	6.4258	4.1313	1.9087	.2557	.00001115
9.2	31.7424	29.4398	27.1372	24.8346	22.5320	20.2294	17.9268	15.6213	13.3217	11.0191	8.7166	6.4148	4.1205	1.8987	.2513	.000009988
9.3	31.7315	29.4290	27.1264	24.8238	22.5212	20.2186	17.9160	15.6135	13.3109	11.0083	8.7058	6.4040	4.1098	1.8888	.2470	.000008948
9.4	31.7208	29.4183	27.1157	24.8131	22.5105	20.2079	17.9053	15.6028	13.3002	10.9976	8.6951	6.3934	4.0992	1.8791	.2429	.000008018
9.5	31.7103	29.4077	27.1051	24.8025	22.4999	20.1973	17.8948	15.5922	13.2896	10.9870	8.6845	6.3828	4.0887	1.8695	.2387	.000007185
9.6	31.6998	29.3972	27.0946	24.7920	22.4895	20.1869	17.8843	15.5817	13.2791	10.9765	8.6740	6.3723	4.0784	1.8599	.2347	.000006439
9.7	31.6894	29.3868	27.0843	24.7817	22.4791	20.1765	17.8739	15.5713	13.2688	10.9662	8.6637	6.3620	4.0681	1.8505	.2308	.000005771
9.8	31.6792	29.3766	27.0740	24.7714	22.4688	20.1663	17.8637	15.5611	13.2585	10.9559	8.6534	6.3517	4.0579	1.8412	.2269	.000005173
9.9	31.6690	29.3664	27.0639	24.7613	22.4587	20.1561	17.8535	15.5509	13.2483	10.9458	8.6433	6.3416	4.0479	1.8320	.2231	.000004637

Sources: Wenzel [1942] and Gray [1973].

The well function is tabulated and given in Table 7.3. A diagram of $W(u)$ versus u is shown in Figure 7.18. The well function is also expressed as

$$W(u) = -0.577216 - \ln u + u - \frac{u^2}{2 \times 2!} + \frac{u^3}{3 \times 3!} + \cdots. \tag{7.75}$$

Given T and S and using Table 7.3, Eq. (7.72) is easily solved to describe the drawdown curve at any point in space and time after the beginning of pumping.

The solution will approach the steady-state solution as time progresses and the drawdown cone develops. Theoretically, the drawdown is asymptotically zero at infinite distance from the well.

In most situations, the hydrologist is ignorant of the aquifer parameters and must carry out field tests to determine transmissivity and storativity. These tests require pumping from a well and observing drawdown s in time and space and later solving for T and S in the Theis equation (Eq. 7.72). Such a solution is not explicit and is usually obtained graphically.

For constant Q, S, and T, we see from Eq. (7.72) that $\ln s$ = constant + $\ln W(u)$. Similarly, from Eq. (7.74) we get $\ln(r^2/t)$ = constant + $\ln u$. Therefore, if drawdown s is plotted versus r^2/t, the resulting curve will be similar in shape, but not in scale and origin, to the plot of $W(u)$ versus u. The graphical solution of the Theis equation consists of superimposing a plot of s versus r^2/t (obtained from well-pumping tests) on a plot of $W(u)$ versus u and finding a match point that defines corresponding pairs of $W(u)$, s, and r^2/t, u. This is illustrated in Figure 7.19.

Using the matched points, the transmissivity is obtained using Eq. (7.72):

$$T = \frac{Q}{4\pi s} W(u). \tag{7.76}$$

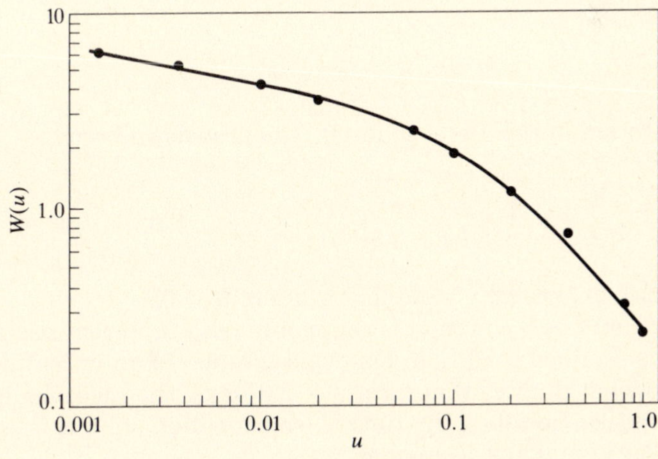

FIGURE 7.18 Type curve—$W(u)$ versus u. Source: Gray [1973].

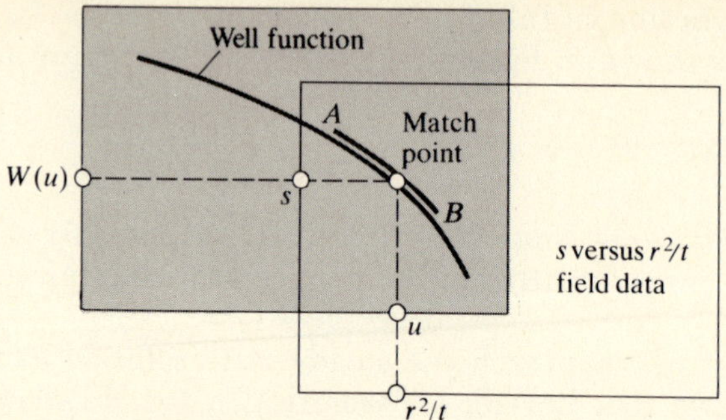

FIGURE 7.19 Illustration of graphic solution of Theis equation. Source: D. K. Todd, *Groundwater Hydrology,* 1st ed. Copyright © 1959 by Wiley. Reprinted by permission of John Wiley & Sons, Inc.

Storativity comes from Eq. (7.74):

$$S = \frac{4Ttu}{r^2}. \tag{7.77}$$

With a redefinition in Eq. (7.71) of ϕ as $h^2/2$ and C_1 as n_e/KH_o, it should be clear that the Theis equation is also a solution for $(H_o^2 - h^2)/2$ in an unconfined aquifer, where H_o is the original phreatic surface level. For small changes in head from the original levels, the Theis equation for direct drawdown s is still valid in an unconfined aquifer with transmissivity defined as $T = KH_o$.

Modified Theis Equation

For small values of u, the well function becomes nearly logarithmic and approximated by

$$W(u_2) - W(u_1) = -2.3(\log_{10} u_2 - \log_{10} u_1). \tag{7.78}$$

Upon substitution in the Theis equation, the drawdown becomes

$$s_2 - s_1 = \frac{2.3Q}{4\pi T} \log_{10}\left(\frac{r_1^2/t_1}{r_2^2/t_2}\right). \tag{7.79}$$

Equation (7.79) is generally valid for values of $u < 0.1$.

The modified Theis equation is commonly used in unconfined aquifers as well as in the confined condition. For typical values of an unconfined aquifer, $u \approx 0.23/t$, which implies that after $t = 2.3$ days, the modified equation is valid. For confined aquifers, the time is even smaller, since S is generally at least one order of magnitude smaller.

If observations are made at a single point at different times, the modified Theis equation becomes

$$s_2 - s_1 = \frac{2.3Q}{4\pi T}\log_{10}\left(\frac{t_2}{t_1}\right). \tag{7.80}$$

Equation (7.80) implies that drawdown versus time will plot on semilog paper as a straight line in the range of validity of the equation. Such a time–drawdown curve is given in Figure 7.20. If t_2/t_1 is 10 (one log cycle), then T is easily obtained as

$$T = \frac{2.3Q}{4\pi\Delta s}. \tag{7.81}$$

Extrapolating from the first two terms of Eq. (7.75), $W(u)$ becomes 0 when $u = 0.562$. This is the condition of zero drawdown, under the small u approximation. Using

$$S = \frac{4uTt}{r^2}$$

and substituting $u = 0.562$,

$$S = \frac{2.25Tt_\circ}{r^2}, \tag{7.82}$$

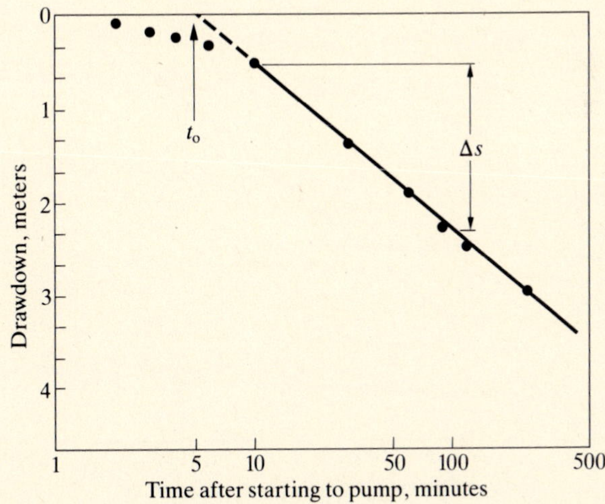

FIGURE 7.20 Illustration of a time–drawdown curve for use with the modified Theis equation.

where t_o is the time of zero drawdown resulting from the linear extension of the S versus t semilog plot, as shown in Figure 7.20. Once t_o is identified, S is easily obtained.

Observations at different points at the same time lead to

$$s_2 - s_1 = \frac{2.3Q}{2\pi T}\log\left(\frac{r_1}{r_2}\right), \tag{7.83}$$

which again plots linearly on semilog paper (s vs. r). Such plots are called distance–drawdown plots. As shown in Figure 7.21, again over one log cycle in distance, transmissivity is given by

$$T = \frac{2.3Q}{2\pi\Delta s}.$$

The storage coefficient is obtained from

$$S = \frac{2.25Tt}{r_o^2},$$

where r_o is the extrapolated distance where drawdown is 0.

Note the absolute value of the slope of the distance–drawdown curve is twice that of the time–drawdown curve, which allows you to obtain one from the other knowing only one point on the line.

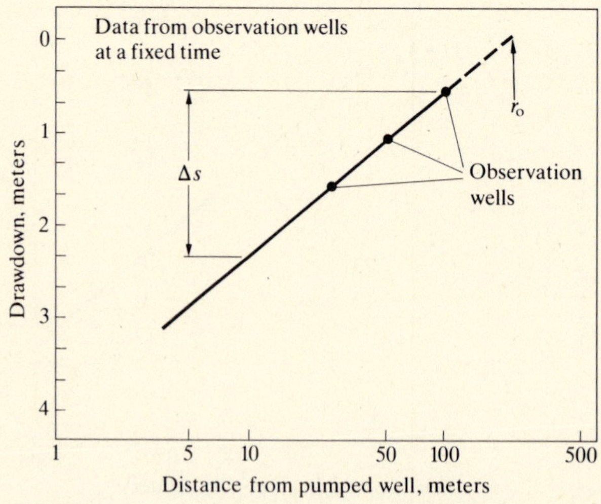

FIGURE 7.21 Illustration of a distance–drawdown curve for use with the modified Theis equation.

7.5.4 Wells in Leaky Aquifers

Hantush and Jacob [1955] developed a procedure analogous to the Theis equation to deal with nonsteady well flow in a leaky aquifer. The solution is just like Eq. (7.72), but a new well function, dependent on aquiclude thickness and conductivity, is defined. The inverse problem is complicated by requiring the additional inference of the vertical hydraulic conductivity of the confining layer.

7.5.5 Superposition of Wells and the Method of Images

Section 7.4.5 already discussed the linearity of the basic groundwater equations. Linearity implies that the potential distribution in an aquifer with multiple wells (a well field) must be equal to the sum of the potential distributions resulting from each single well in the field. For example, if N wells pumping at rates Q_i, $i = 1, \ldots, N$ exist in a field (Fig. 7.22), then the drawdown at any point in time and space is given by

$$s(t) = \sum_{i=1}^{N} s_i(t) = \sum_{i=1}^{N} \frac{Q_i}{4\pi T} W(u_i), \tag{7.84}$$

where

$$u_i = \frac{r_i^2 S}{4T(t - t_{oi})}.$$

In the above, r_i is the distance between the location of interest and well i, Q_i is the pumping rate of well i; t_{oi} is the time since the beginning of pumping in well i, and t is the time since the first pumping activity anywhere in the field.

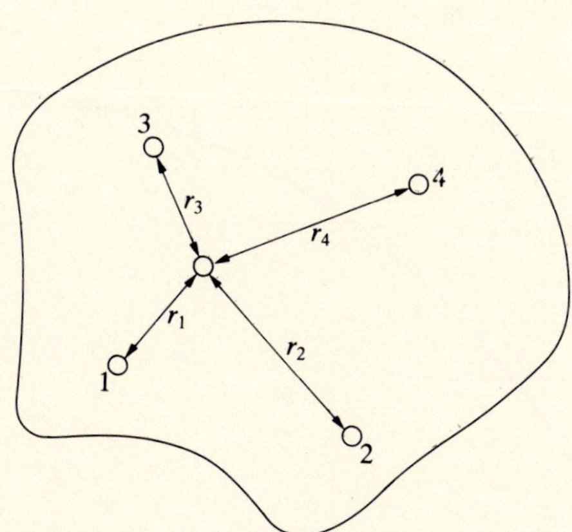

FIGURE 7.22 A well field.

Linearity also permits the use of the Theis equation to find the groundwater response to a well with a variable pumping rate, as in Figure 7.23. The variable pumping rate is substituted by piecewise constant segments. By linearity, the aquifer response up to time t_1 corresponds to a well pumping at rate Q_1. At time t_2 the response is the sum of a well pumping at rate Q_1 plus a well pumping at rate $Q_2 - Q_1$. In general, the potential drawdown at any time t is

$$s(t) = \sum_{i=1}^{n} \frac{Q_i - Q_{i-1}}{4\pi T} W(u_i) \qquad \text{for } t_{n-1} < t \le t_n$$
$$u_i = \frac{r^2 S}{4T(t - t_{i-1})} \tag{7.85}$$

and $Q_o = t_o = 0$.

Linearity and the concept of superposition is also used in the Method of Images. This is a tool used to build difficult boundary conditions by adding solutions to imaginary wells. The classic situations are a river intersecting an aquifer and an impervious boundary limiting an aquifer.

Figure 7.24 shows the first case. The intersecting river imposes a constant head boundary condition at the point $x = 0$. Without the river, the cone of depression of the real well may look like the lower dashed curve in the figure. The constant head boundary raises the piezometric surface. This effect can be achieved by locating a mirror-image well a distance ℓ on the other side of the river. This well will be assumed to be recharging the aquifer at the same rate that the real well is pumping. The upper dashed curve in the figure is the cone of influence (this time accretion rather than depression) of the image well. Adding both solutions, a constant head is achieved at the river boundary and the result corresponds to the desired (solid curve) cone of depression.

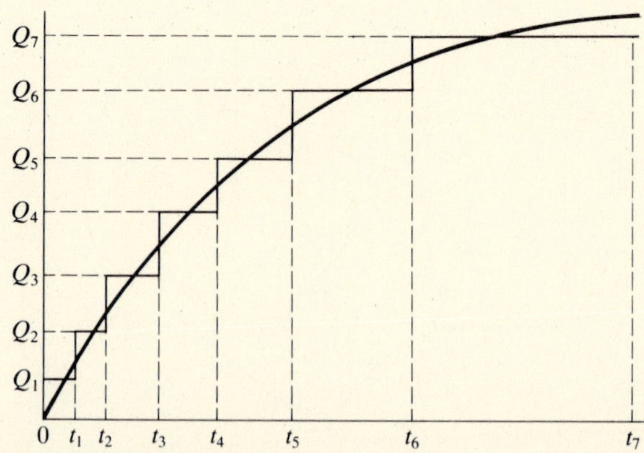

FIGURE 7.23 Variable pumping rate represented by piecewise constant segments.

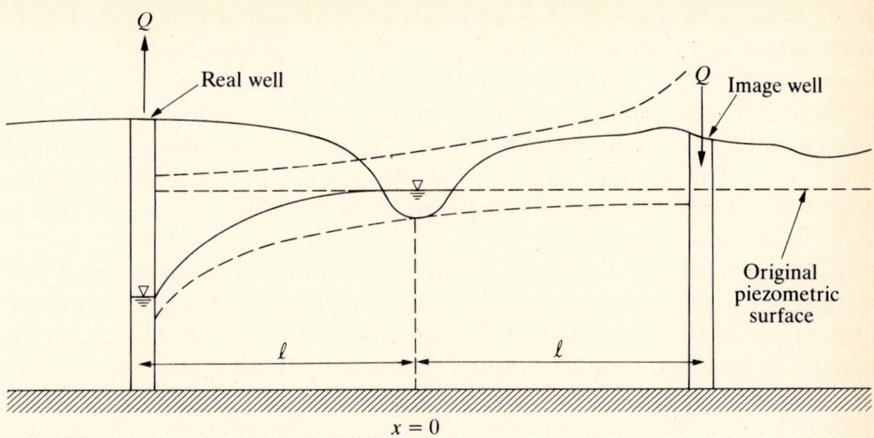

FIGURE 7.24 The method of images: constant head boundary case.

The second case is shown in Figure 7.25. At point $x = 0$, there is an impermeable boundary, the real (solid curve) cone of depression is horizontal at $x = 0$, $\partial h/\partial x = 0$. The curve as a whole is also depressed, reflecting the limited extent of the water source. To achieve this behavior a pumping image well is placed a distance ℓ on the other side of the boundary. Adding the two cones of depression with opposing slopes results in the desired no-flow condition at $x = 0$ as well as the lower piezometric head throughout.

7.5.6 Aquifer Tests

Past subsections have addressed the inverse problem, determining parameters of the aquifer. Pumping tests are the most common procedure to achieve this. Similarly, it is possible to use the time history of the recovery of an

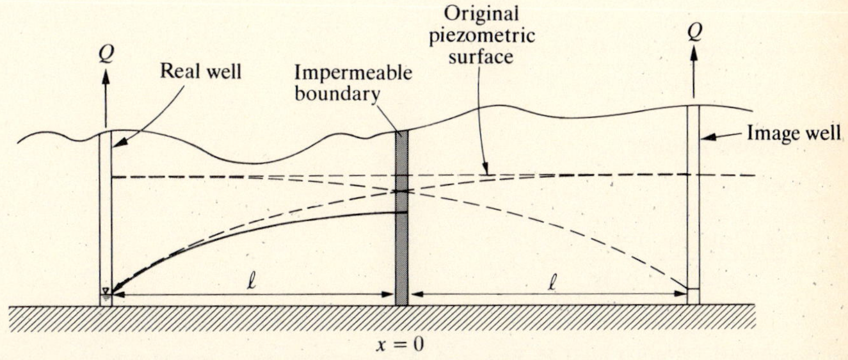

FIGURE 7.25 The method of images: no flow boundary case.

aquifer after pumping ceases to determine soil properties. Note that the recovery period solution would correspond to the superposition of a recharge well on the pumping well from the time that pumping stops. Solutions like the modified Theis equation (Eq. 7.80) would still be valid by redefining s as a recovery from a previous drawdown level and t as the time since pumping stopped. Gray [1973] gives examples of this approach. The U.S. Department of the Navy [1962] also gives many empirical formulas to use the recovery history in the inverse problem.

Another common deviation from the formulations presented is the fact that wells do not always fully penetrate the aquifer. An approach in Brown et al. [1972] attributed to Verigin [1961] is to modify the steady-state solutions to confined and unconfined aquifers by adding seepage resistance factors caused by partial penetration and by wells not fully screened throughout their length. We quote the equations and explanation from Brown et al. [1972].* Note that they originate from the steady-state solutions given in Sections 7.5.1 and 7.5.2 except that use is made of drawdown s instead of head h. Also, r_∞ is defined as a large distance where the drawdown is effectively zero. Subscripts w, 1, and 2 refer to the pumped well and two observation wells, respectively. H is the average saturated thickness of the unconfined aquifer and b is that of the confined aquifer.

Confined Aquifer

Pumped well:

$$K = \frac{Q_w[\ln(r_\infty/r_w) + 0.5\xi_0]}{2\pi b s_w}. \tag{7.86}$$

Pumped well and one observation well:

$$K = \frac{Q_w[\ln(r_1/r_w) + 0.5(\xi_0 - \xi_1)]}{2\pi b(s_w - s_1)}. \tag{7.87}$$

Two-observation wells:

$$K = \frac{Q_w[\ln(r_2/r_1) + 0.5(\xi_1 - \xi_2)]}{2\pi b(s_1 - s_2)}. \tag{7.88}$$

Unconfined Aquifer

Pumped well:

$$K = \frac{Q_w[\ln(r_\infty/r_w) + 0.5\xi_0]}{\pi(2H - s_w)s_w}. \tag{7.89}$$

*From *Groundwater Studies* by R. H. Brown, A. A. Konoplyantsev, J. Ineson, and V. S. Kovatevsky. Copyright © Unesco 1972. Reproduced by permission of Unesco.

Pumped well and one observation well:

$$K = \frac{Q_w[\ln(r_1/r_w) + 0.5(\xi_0 - \xi_1)]}{\pi(2H - s_w - s_1)(s_w - s_1)}. \tag{7.90}$$

Two observation wells:

$$K = \frac{Q_w[\ln(r_2/r_1) + 0.5(\xi_1 - \xi_2)]}{\pi(2H - s_1 - s_2)(s_1 - s_2)}. \tag{7.91}$$

[In the above equations,] ξ_0, ξ_1, and ξ_2 are seepage resistances for partially penetrating wells at the pumped locations, nearer, and more-distant observation wells, respectively. The resistance parameter ξ_0 for the pumped well takes into account the effects of both partial penetration and well construction.

Values of ξ_0, ξ_1, and ξ_2 can be determined from Table 7.4, the two relevant parameters being (b/r), where b is the thickness of the confined aquifer and r the distance from the pumped well to the observation well (r_1 or r_2) or the radius of the pumped well (r_w), respectively; and (ℓ/b), where ℓ is the unlined open part of the well or screen.

For unconfined conditions of groundwater flow, the value of b is taken as $[H - (s_w/2)]$ and similarly, $[\ell - (s_w/2)]$ is used instead of ℓ.

The values of ξ given in Table 7.4 are adopted when the unlined or screened part of the well is near the top or bottom of the aquifer. If this part of the well is near the middle part of the aquifer, ξ should be multiplied by 1.5 when $\ell/b = 0.3$ and by 0.7 when $\ell/b = 0.5$.

TABLE 7.4 Values of the Seepage Resistance Function ξ

	$\left(\frac{\ell}{b}\right)$				
$\left(\frac{b}{r}\right)$	0.1	0.3	0.5	0.7	0.9
0.5	0.00391	0.00297	0.00165	0.000546	0.000048
1	0.122	0.0908	0.0494	0.0167	0.0015
3	2.04	1.29	0.656	0.237	0.0251
10	10.4	4.79	2.26	0.879	0.128
30	24.3	9.23	4.21	1.69	0.334
100	42.8	14.5	6.55	2.67	0.528
200	53.8	17.7	7.86	3.24	0.664
500	69.5	21.8	9.64	4.01	0.846
1000	79.6	24.9	11.0	4.58	0.983
2000	90.6	28.2	12.4	5.19	1.12

Source: R. H. Brown, A. A. Konoplyantsev, J. Ineson, and V. S. Kovatsky, *Groundwater Studies*. Copyright © Unesco 1972. Reproduced by permission of Unesco.

Partial penetration of observation wells is considered to be insignificant if the observation wells are located at distances greater than the equivalent aquifer thickness from the pumped well.

With some risk, the value of r_∞ may be estimated and used in the appropriate equations. Some examples of possible values of r_∞ are given in Table 7.5.

EXAMPLE 7.6

Unsteady Flow in Wells: Theis Equation

A well is being pumped at a constant rate of 1500 $m^3\,day^{-1}$. At an observation well 100 m away, we measured the following history (Table 7.6) of drawdown versus time. Assume the aquifer is confined and the well is fully penetrating. What are the transmissivity and storativity of the aquifer?

To solve the inverse problem, we will use the Theis equation and the matching-point graphical approach. Figure 7.26 shows a plot of drawdown s versus r^2/t. Figure 7.27 is a plot of a portion of the well function $W(u)$ versus u with Figure 7.26 superimposed in gray. Superimposing one figure on another, maintaining parallel axes, we note that with a small shift the curves overlap very well. Along the overlapping segments, we choose the following matching points:

$$W(u) = 3.35 \qquad u = 0.02$$

$$s = 0.4 \text{ m} \qquad r^2/t = 2.7 \times 10^5 \text{ m}^2\,\text{day}^{-1}$$

Transmissivity is obtained from Eq. (7.76) as

$$T = \frac{1500(3.35)}{4\pi(0.40)} = 999.7 \text{ m}^2\,\text{day}^{-1}$$

TABLE 7.5 Approximate Values of r_∞

TYPE OF SEDIMENT	GROUNDWATER CONDITIONS	EXTENT OF CONE OF DEPRESSION r_∞ (METERS)
Fine- and medium-grained sands	Confined	250–500
	Unconfined	100–200
Coarse-grained sands and gravel-pebble beds	Confined	750–1500
	Unconfined	300–500
Fissured rocks	Confined	1000–1500
	Unconfined	500–1000

Source: R. H. Brown, A. A. Konoplyantsev, J. Ineson, and V. S. Kovatsky, *Groundwater Studies*. Copyright © Unesco 1972. Reproduced by permission of Unesco.

TABLE 7.6 Observation Well Drawdown History and Computations

s (m)	t (min)	r^2/t (m^2 day^{-1})
0.02	1	1.50×10^7
0.14	5	2.90×10^6
0.21	10	1.50×10^6
0.25	15	1.00×10^6
0.29	20	7.10×10^5
0.33	30	5.00×10^5
0.41	60	2.40×10^5
0.46	90	1.60×10^5
0.49	120	1.21×10^5
0.54	180	8.00×10^4
0.58	240	6.00×10^4
0.60	300	4.81×10^4
0.63	360	4.00×10^4
0.64	420	3.43×10^4
0.65	480	3.00×10^4
0.69	600	2.40×10^4
0.73	840	1.70×10^4
0.79	1440	1.00×10^4

and storativity from Eq. (7.77):

$$S = \frac{4(999.7)(0.02)}{2.7 \times 10^5} = 2.96 \times 10^{-4}. \ \blacklozenge$$

EXAMPLE 7.7

Unsteady Flow in Wells: Modified Theis Equation

The previous example can be solved using the modified Theis procedure. Figure 7.28 shows the corresponding time–drawdown plot, s versus t. After about 20 minutes of pumping, the points follow a straight line on the semilogarithmic paper. Extending the line to the zero drawdown point yields a corresponding time $t_\circ = 1.9$ minutes or 1.32×10^{-3} days. Over a logarithmic cycle, from 100 to 1000 minutes, the drawdown is shown to be $\Delta s = 0.270$ m. This is all the information required to use Eqs. (7.81) and (7.82). The computations are given in the figure, yielding $T = 1017$ m^{2}day^{-1} and $S = 3 \times 10^{-4}$. ◆

EXAMPLE 7.8

Unsteady Flow in Wells: An Alternate Approach

Chow [1952] suggested a procedure that avoids some of the graphical pitfalls and difficulties of the past two examples. From the time–drawdown curve of

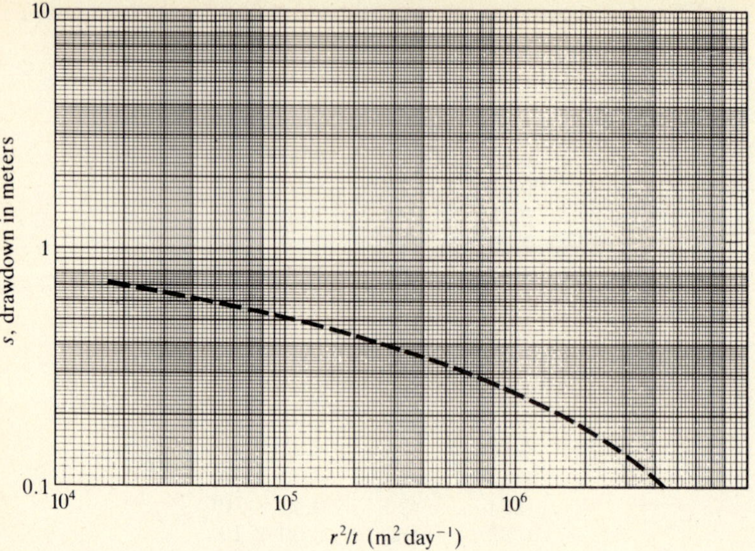

FIGURE 7.26 Plot of drawdown versus r^2/t.

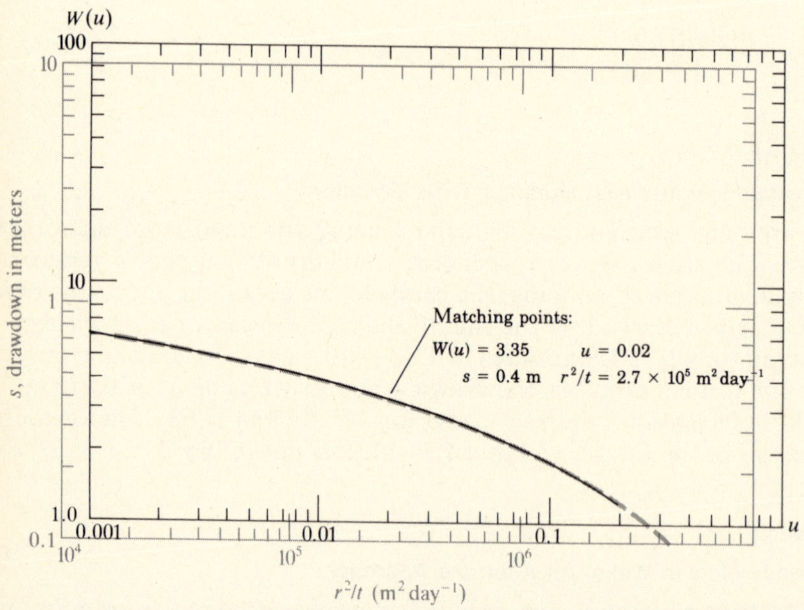

FIGURE 7.27 Plot of $W(u)$ versus u with Figure 7.26 superimposed in gray.

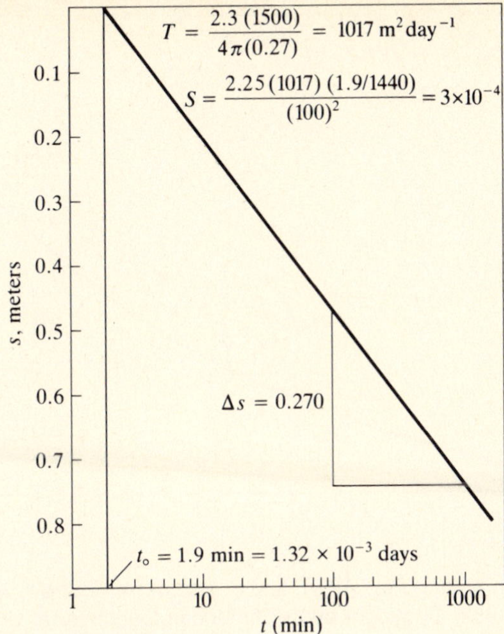

FIGURE 7.28 Plot of drawdown versus time.

Figure 7.28 select a point s. Around that point find the drawdown change in a log cycle and form the function

$$F(u) = \frac{s}{\Delta s}. \tag{7.92}$$

Use Figure 7.29 and the value of $F(u)$ to find a corresponding value of u and $W(u)$, which together with the chosen s, t, and r can be used in Eqs. (7.76) and (7.77) to find S and T. As an example select the point $s = 0.6$ and $t =$ 300 minutes in Figure 7.28. We already saw that a log cycle around that point shows a 0.270-m drawdown. Therefore

$$F(u) = \frac{s}{\Delta s} = \frac{0.60}{0.27} = 2.22\,.$$

With the above result and using Figure 7.29, we get $W(u) = 5$ and $u = 0.004$, which results in $T = 995$ and $S = 3.3 \times 10^{-4}$ when used in Eqs. (7.76) and (7.77). ◆

It should be pointed out that the data of the past three examples were generated with values of $T = 1000\ \mathrm{m^2\,day^{-1}}$ and $S = 0.0003$.

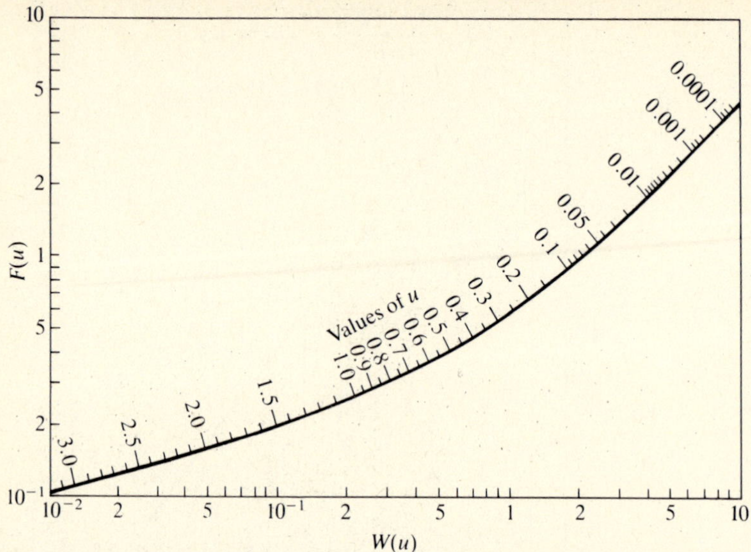

FIGURE 7.29 Relation among *F*(*u*), *W*(*u*), and *u*. Source: V. T. Chow, *Trans. Am. Geophys. Union.* 33:397–404, 1952. Copyright by the American Geophysical Union.

EXAMPLE 7.9

Recovery Analysis

As previously stated, data from the well recovery period can be used to estimate aquifer parameters. Figure 7.30 illustrates what happens after pumping stops. Define s' as the drawdown after cessation of pumping, s as the drawdown that would have occurred had pumping continued, t^* as the time pumping stopped, t as absolute real time, and $t' = t - t^*$ as the time since pumping stopped. Using the modified Theis equation concept (combine Eqs. 7.72 and 7.75), drawdown at an observation well had pumping continued would have been

$$s = \frac{Q}{4\pi T}\left(-0.577 - \ln\frac{r^2S}{4Tt}\right).$$

By superposition, the drawdown after pumping has to be equivalent to that resulting from adding a recharging well $-Q$ at time t^*. Therefore

$$s' = \frac{Q}{4\pi T}\left(-0.577 - \ln\frac{r^2S}{4Tt}\right) - \frac{Q}{4\pi T}\left(-0.577 - \ln\frac{r^2S}{4Tt'}\right).$$

The recovery $s - s'$ is then

$$s - s' = \frac{Q}{4\pi T}\left(-0.577 - \ln\frac{r^2S}{4Tt'}\right), \tag{7.93}$$

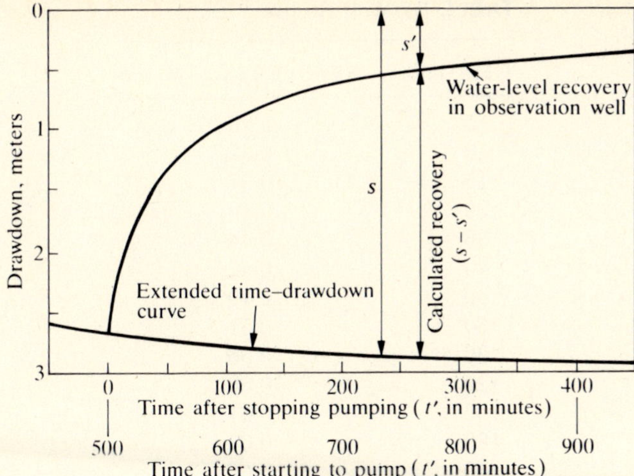

FIGURE 7.30 Illustration of a well recovery after pumping.

which has the same form of the modified Theis equation during pumping. At two times, and assuming that s is nearly constant (pumping stopped only after a long time of activity)

$$s_2' - s_1' = \frac{Q}{4\pi T} \ln\left(\frac{t_2'}{t_1'}\right). \tag{7.94}$$

Because of Eq. (7.94) a plot of recovery $s - s'$ versus time must be nearly linear for large times (small u) on semilogarithmic paper.

Assume the discharge is $Q = 2000\ \text{m}^3\,\text{day}^{-1}$ at a well in a confined aquifer. After four hours pumping is stopped, resulting in the history of recovery $(s - s')$ given in Table 7.7, corresponding to an observation well 75 m away. The drawdown at four hours was 1 m.

Figure 7.31 gives the plot of recovery versus time. The zero recovery time is 0.5 minute, obtained by extrapolation. Over one logarithmic cycle (10 to 100 minutes) the recovery was 0.25. The figure shows the calculations for T and S using Eqs. (7.81) and (7.82). The results $T = 1464\ \text{m}^2\,\text{day}^{-1}$ and $S = 2.03 \times 10^{-4}$ agree very well with those used to generate that data ($t = 1500$, $S = 2 \times 10^{-4}$). ◆

EXAMPLE 7.10

Method of Images

Figure 7.32 shows the plan view of a proposed excavation site. It is planned to build four equal, fully penetrating dewatering wells of 0.5 m diameter each to lower the water table 5 m everywhere within the excavation. As an estimate, the original phreatic surface is taken horizontally at level $H_\circ = 100$.

TABLE 7.7 Well Recovery Data

TIME (min)	RECOVERY ($s - s'$) (m)
5	0.25
10	0.32
15	0.37
20	0.40
30	0.44
40	0.47
50	0.49
60	0.51
90	0.56
120	0.59
150	0.61
180	0.63
240	0.66
300	0.68
360	0.70
720	0.78
1440	0.85

The river is also maintained at that head value. The phreatic aquifer has a permeability of 2×10^{-5} m s^{-1} (1.73 m day^{-1}). What would be an estimate of the discharge Q required in each well to achieve the dewatering objective? As a first approximation, we will assume steady-state conditions. By symmetry and geometry, it could be argued that the maximum head will occur somewhere between points c and d, the latter being the center of the well field.

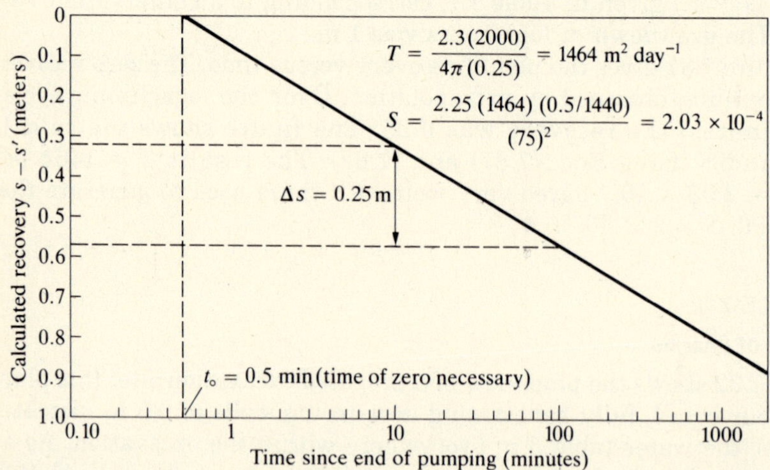

FIGURE 7.31 Well recovery versus time.

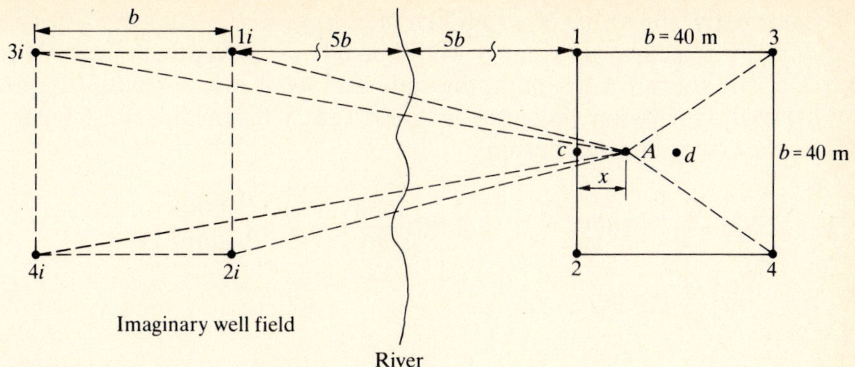

FIGURE 7.32 Plan view of dewatering system.

To account for the effect of the river, we must use the method of images. Figure 7.32 also shows the required image wells, all recharging.

Equation (7.70) gave the steady-state solution for a single well in a phreatic aquifer. Now we will further assume that the cone of depression does not extend beyond 1000 m. It must be emphasized that assumptions of that type are difficult and dangerous. The single well expression is then

$$H_\circ^2 - h^2 = \frac{\ln\left(\dfrac{1000}{r_i}\right)Q}{\pi K}.$$

The total effect of all real and imaginary wells on a point A must be

$$H_\circ^2 - h_A^2 = \frac{Q}{\pi K}\left[\sum_{j=1}^{4} \ln\left(\frac{1000}{r_j^r}\right) - \sum_{j=1}^{4} \ln\left(\frac{1000}{r_j^i}\right)\right],$$

where r_j^i are distances to imaginary wells from point A, and r_j^r are distances to real wells from point A. Since $10b \gg b/2$, the distance to imaginary wells are approximated:

$$r_1^i = r_2^i \approx 10b + x,$$

$$r_3^i = r_4^i \approx 11b + x,$$

where x is the horizontal distance from the left edge of the excavation to point A.

For real wells,

$$r_1^r = r_2^r = \sqrt{\left(\frac{b}{2}\right)^2 + x^2},$$

$$r_3^r = r_4^r = \sqrt{(b - x)^2 + \left(\frac{b}{2}\right)^2}.$$

Theoretically, the value of x could be found by maximizing the expression for $H_\circ^2 - h_A^2$. For a faraway river, x would be $b/2$. For simplicity, we will first locate A at the center of the pumping field and let $x = b/2$. Using the previously derived expression (keeping in mind that it is inconsistent with the very faraway river approximation),

$$H_\circ^2 - h_A^2 = \frac{Q}{\pi K}\left[4 \ln\left(\frac{1000}{28.28}\right) - 2 \ln\left(\frac{1000}{420}\right) - 2 \ln\left(\frac{1000}{460}\right)\right]$$
$$= \frac{Q}{\pi K}(10.97).$$

For a 5-m drawdown, $h_A = 95$; therefore

$$Q = [(100)^2 - (95)^2]\frac{\pi(1.73)}{10.97} = 483 \text{ m}^3\,\text{day}^{-1}.$$

Had we assumed $x = 0$,

$$(H_\circ^2 - h_A^2) = \frac{Q}{\pi K}\left[2 \ln\left(\frac{1000}{20}\right) + 2 \ln\left(\frac{1000}{44.72}\right) - 2 \ln\left(\frac{1000}{400}\right) - 2 \ln\left(\frac{1000}{440}\right)\right]$$
$$= \frac{Q}{\pi K}(10.56)$$

or

$$Q = 502 \text{ m}^3\,\text{day}^{-1}.$$

The above are fairly high pumping rates. ◆

7.6 SUMMARY

The study of groundwater hydrology has had particular relevance during the past 10 years due to our "sudden" realization that groundwater supplies have been contaminated through decades of neglect and uncontrolled disposal of waste on and in the ground. There has been a profusion of literature, including many textbooks, dedicated exclusively to the subject of groundwater hydrology. This chapter and Chapter 8 are not intended to compete with those far more complete treatments, but they do provide a fairly self-contained initiation to the subject, permitting some practice, providing basic understanding, and serving as foundation for the studies. The continuing student is referred to textbooks by Bear [1972], Freeze and Cherry [1979], Todd [1959], Bouwer [1978], and DeMarsily [1986] as samples of available didactic literature.

There are several themes important to the study of groundwater that cannot be treated in this type of introductory textbook. One is the whole field of numerical solutions to the groundwater flow equations and resulting aquifer models for both quantity and quality of water in porous media. A good reference is Wang and Anderson [1982]. Pinder and Gray [1977] provide a more specific and advanced view of the subject. A related issue is the identification of parameters (inverse problem) of aquifer models. This is the spatially distributed analog of the simple well tests discussed in this chapter. A review of the subject is provided by Yeh [1986].

Groundwater moves within a very heterogeneous medium. Hence, the common assumptions of homogeneity and isotropy are generally crude approximations. Since it is impossible to deterministically characterize the spatial variability of aquifer and soil properties, a lot of research interprets soil as a medium with random properties. Hence, groundwater flow and contaminant transport are also random processes that must be characterized with probabilistic distributions or their moments (i.e., mean, variance, etc.). The mathematically advanced reader can refer to Gelhar [1984, 1986, 1987] to start this adventure.

Chemical and biologic interactions in groundwater are another topic of current research, with little known and far more to learn. In the same category is the characterization of flow in fractured porous or nonporous media.

This chapter does provide the basic concepts of groundwater hydrology. The equations of flow in saturated porous media given here are the cornerstone of all other topics. Their solution for the simple cases of wells in confined and unconfined aquifers are useful tools in practice. Many of the other simple solutions developed as examples are first, and sometimes good, approximations to real problems of groundwater flow. It is useful to end this summary by making reference to the problems of groundwater flow in a hillslope and aquifer–stream interaction discussed in Examples 7.4 and 7.5, respectively. These should serve as reminders that the division of hydrology into surface and subsurface processes is very often artificial and wrong.

REFERENCES

Bear, J. [1972]. *Dynamics of Fluids in Porous Media*. New York: American Elsevier.

Bouwer, H. [1978]. *Groundwater Hydrology*. New York: McGraw-Hill.

Brown, R. H., A. A. Konoplyantsev, J. Ineson, and V. S. Kovatevsky [1972]. *Groundwater Studies*. Paris: UNESCO.

Busch, F. K., and L. Luckner [1972]. *Geohydraulik*. Leipzig, East Germany: VEB Deutscher Verlag Fur Grunstoffindustrie.

Chow, V. T. [1952]. "On the Determination of Transmissibility and Storage Coefficients from Pumping Test Data." *Trans. Am. Geophys. Union.* 33:397–404.

Cooper, H. H., Jr., and M. I. Rorabaugh [1963]. "Ground-Water Movements and Bank Storage Due to Flood Stages in Surface Streams." U.S. Geological Survey water supply paper no. 1536-J.

Darcy, H. [1856]. *Les Fontaines Publiques de LaVille de Dijon.* Paris: V. Oalmont.

DeMarsily, G. [1986]. *Quantitative Hydrogeology: Groundwater Hydrology for Engineers.* New York: Academic Press.

DeWiest, R. J. M. [1965]. *Geohydrology.* New York: Wiley.

Dupuit, J. [1863]. *Etudes Theoriques et Practiques sur le Mouvement des Eaux dans les Canaux D'ecouverts et à Travers les Terrains Perméables.* 2nd ed. Paris: Dunod.

Dyck, S. [1978]. *Angewandte Hydrologie.* Teil 2. Berlin, East Germany: VEB Verland Fur Bauwesse.

Eagleson, P. S. [1970]. *Dynamic Hydrology.* New York: McGraw-Hill.

Idem. [1978a]. "Climate, Soil and Vegetation 3: A Simplified Model of Soil Moisture Movement in the Liquid Phase." *Water Resources Res.* 14(5):722–729.

Idem. [1978b]. "Climate, Soil, and Vegetation 5: A Derived Distribution of Storm Surface Runoff." *Water Resources Res.* 14(5):741–748.

Entekhabi, D. [1988]. Personal communication.

Ferris, J. G., et al. [1962]. "Theory of Aquifer Tests." U.S. Geological Survey water supply paper no. 1536-E.

Freeze, R. A., and J. A. Cherry [1979]. *Groundwater.* Englewood Cliffs, N.J.: Prentice-Hall.

Gelhar, L. W. [1984]. "Stochastic Analysis of Flow in Heterogeneous Porous Media." In: J. Bear and M. Y. Corapcioglu, eds. *Fundamentals of Transport Phenomena in Porous Media.* Dordrecht, The Netherlands: Martinus Nijhoff.

Idem. [1986]. "Stochastic Subsurface Hydrology from Theory to Applications." *Water Resources Res.* 22(9):Suppl:355–1455.

Idem. [1987]. "Stochastic Analysis of Solute Transport in Saturated and Unsaturated Porous Media." In: J. Bear and M. Y. Corapcioglu, eds. *Advances in Transport Phenomena in Porous Media.* Dordrecht, The Netherlands: Martinus Nijhoff.

Gray, D. M. [1973]. *Handbook on the Principles of Hydrology*, Port Washington, N. Y.: Water Information Center.

Hantush, M. S., and J. E. Jacob [1955]. "Non Steady Radial Flow in an Infinite Leaky Aquifer." *Trans. Am. Geophys. Union.* 36:95 – 100.

Jacob, C. E. [1943]. "Correlation of Ground-Water Levels and Precipitation on Long Island, New York." *Trans. Am. Geophys. Union.* 24:564–573.

Idem. [1949]. "Flow of Groundwater." In: H. Rouse, ed. *Engineering Hydraulics.* New York: Wiley.

Johnson, E. E. [1959]. "Factors Affecting Permeability." *The Drillers Journal.* January–February. (St. Paul: E. E. Johnson.)

Idem. [1961]. "Analyzing Water Level Recovery Data." *The Drillers Journal.* May–June, p. 9. (St. Paul: E. E. Johnson.)

Marsh, W. M., and J. Dozier [1986]. *Landscape: An Introduction to Physical Geography.* New York: Wiley.

Meinzer, O. E. [1923]. "Outline of Ground-water Hydrology with Definitions." U.S. Geological Survey water supply paper no. 493.

Pinder, G. F., and W. G. Gray [1977]. *Finite Element Simulation in Surface and Subsurface Hydrology*. New York: Academic Press.

Soil Conservation Service [1968]. *Hydrology*. (Suppl. A to Sec. 4, *Engineering Handbook*.) Washington, D. C.: U.S. Department of Agriculture.

Theis, C. V. [1935]. "The Relation Between the Lowering of the Piezometric Surface and the Rate and Duration of Discharge of a Well Using Ground-Water Storage." *Trans. Am. Geophys. Union*. 16:519–524.

Thiem, G. [1906]. *Hydrologische Methoden*. Leipzig, East Germany: Gebhardt.

Todd, D. K. [1959]. *Groundwater Hydrology*. 1st ed. New York: Wiley.

Toth, J. [1962]. "A Theory of Groundwater Motion in Small Drainage Basins in Central Alberta, Canada." *J. Geophys. Res*. 67(11):4375–4387.

Idem. [1963]. "A Theoretical Analysis of Groundwater Flow in Small Drainage Basins." *J. Geophys. Res*. 68(16):4795–4813.

U.S. Department of the Navy [1962]. *Design Manual, Soil Mechanics, Foundations, and Earth Structures*. Washington, D.C.: Government Printing Office. (Publication no. NAVDOCKS DM-7.)

Verigin, N. N. [1961]. "Methody Opredelenniya fil'tratsionnykh Svoistv Gornykh Porod (Methods of Determining Filtration Characteristics of Porous Strata)." Moscow: Gosstroiizdat.

Viessman, W., Jr., J. W. Knapp, G. L. Lewis, and T. E. Harbaugh [1977]. *Introduction to Hydrology*. 2nd ed. New York: Harper & Row.

Wang, H. F., and M. P. Anderson [1982]. *Introduction to Groundwater Modelling*. San Francisco: Freeman.

Wenzel, L. K. [1942]. "Methods of Determining Permeability of Water-Bearing Materials, with Special Reference to Discharging-Well Methods." U.S. Geological Survey water supply paper no. 887.

World Meteorological Organization [1974]. *Guide to Hydrological Practices*. 3rd ed. Geneva: World Meteorological Organization. (WMO report no. 168.)

Yeh, W.-G. [1986]. "Review of Parameter Identification Procedures in Groundwater Hydrology: The Inverse Problem." *Water Resources Res*. 22(2):95–108.

PROBLEMS

1. Three wells located 1000 m apart extract water from the same horizontal aquifer. Well A is south of Well B, and Well C is east of the line AB. The surface elevations of A, B, and C are 95, 110, and 135 m, respectively. The depth to water in A is 5 m; in B, 30 m; and in C, 35 m. Determine the direction of groundwater flow through the triangle ABC and calculate the head gradient.

2. The instrument shown in the following illustration is a variable-head permeameter. The tube, in the right of area a, is initially full of water. A discharge occurs out of the soil in the left bucket, and the head $h(t)$ decreases with time. Find an expression for the varying head $h(t)$ and for hydraulic conductivity in terms of the head at different times, by computing $Q(t)$ from conservation of mass and from Darcy's Law.

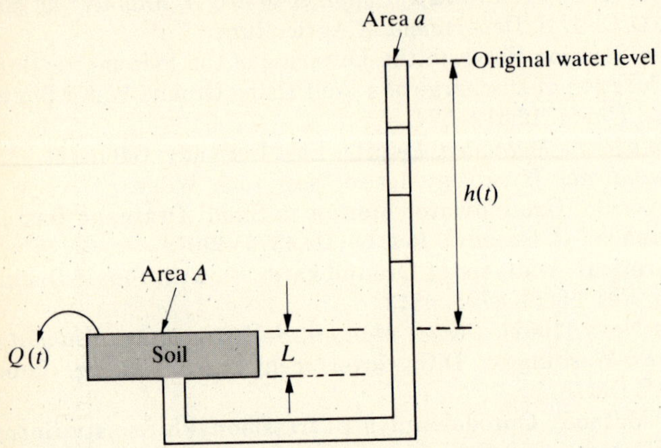

3. A single well is pumping at 300,000 gal day^{-1}. The following data are observed in a well 50 ft from the pumped well. The aquifer is artesian.

Time (min)	3	5.5	8	10	29	40	65	100
Drawdown (ft)	0.2	0.6	1.2	1.5	4	4.8	6	7

a) Using the modified Theis equation, obtain the value of S (storage coefficient) and T (transmissivity) in U.S. gallons per day per foot. Use three-cycle semilog paper.

b) What is the slope of the distance–drawdown relationship?

c) What is T if you assume steady-state drawdown of 7 ft?

4. The time–drawdown data for an observation well located 296 ft from a pumped artesian well (500 gal min^{-1}) are given in the following table. Find the coefficient of storage and the transmissivity in gallons per hour per foot of the aquifer by the Theis method. Use 3-by-5-cycle log paper.

TIME (hr)	DRAWDOWN (ft)	TIME (hr)	DRAWDOWN (ft)
1.9	0.28	9.8	1.09
2.1	0.30	12.2	1.25
2.4	0.37	14.7	1.40
2.9	0.42	16.3	1.50
3.7	0.50	18.4	1.60
4.9	0.61	21.0	1.70
7.3	0.82	24.4	1.80

5. A plan view of a well field is as shown below. The aquifer is confined, of depth 100 ft and permeability $K = 1300\ \text{gal day}^{-1}\,\text{ft}^{-2}$. The well has been pumped for a long time at a rate $Q = 1000\ \text{gal min}^{-1}$. The two observation wells at $r_1 = 50$ ft and $r_2 = 150$ ft indicate a difference in drawdown of 1.349 ft. What is the distance ℓ to the impervious wall?

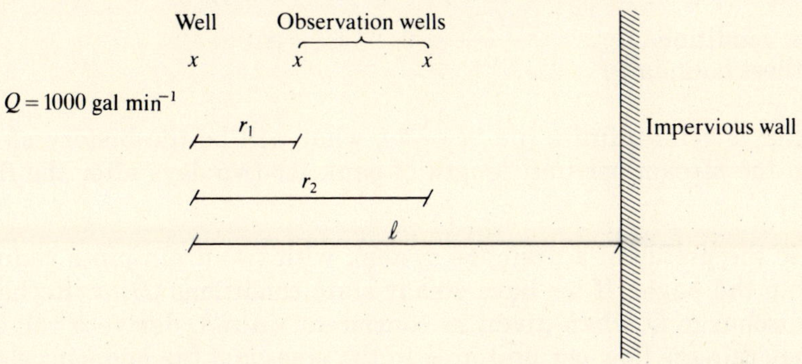

6. Time–drawdown data collected at a distance of 100 ft from a well pumping at a rate of $192 \times 10^3\ \text{ft}^3\,\text{day}^{-1}$ are as follows:

t (min)	s (ft)	t (min)	s (ft)
1	3.8	10	10.0
2	5.2	20	12.2
3	6.2	40	14.0
4	7.0	80	15.8
5	7.6	100	16.4
6	8.3	300	19.0
7	8.8	500	20.2
		1000	21.6

Calculate the transmissivity and storativity. Solve by any procedure desired.

7. A well starts pumping at a rate of $1000\ \text{m}^3\,\text{day}^{-1}$. After one day the pumping rate is increased to $2000\ \text{m}^3\,\text{day}^{-1}$. The aquifer is confined and has the following properties: $T = 1400\ \text{m}^2\,\text{day}^{-1}$ and $S = 10^{-4}$. What is the drawdown at an observation well 1000 m from the pumping well, three days after the initiation of pumping?

8. A stream–aquifer interaction is schematized in Figure 7.13. A flood wave goes by (in the stream), imposing a variation in head, at the bank, of the form:

$$h(0, t) = \frac{h_\circ}{2}\left(1 - \cos\frac{2\pi t}{\tau}\right), \qquad 0 \leq t \leq \tau.$$

The duration of the flood τ is 24 hr and peaks at 10 m elevation (h_o). The storativity of the aquifer is 0.001 and the transmissivity 1000 $m^2 day^{-1}$. Assume the initial conditions,

(no initial storage)	$h(x, 0) = 0$	for $0 \leq x \leq \ell$
(no-flow condition at farthest boundary)	$\dfrac{\partial h(\ell, t)}{\partial x} = 0$	for $t \geq 0$

If the aquifer is semi-infinite (i.e., $\ell \rightarrow \infty$), what will be the history of discharge into the stream per unit length of bank for two days after the flood begins?

9. Below is the profile of a drainage system, with two drain pipes running parallel (into the page). If we have steady-state conditions (*Hint:* Recharge R equals discharge Q when given in consistent units.), derive a relation between the drainage flow per unit area in the pipes and the midpoint elevation m'.

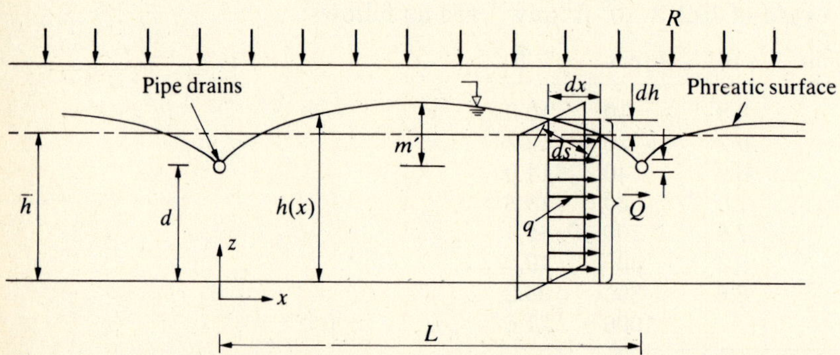

10. Given the following data: $Q = 60{,}000$ $ft^3 day^{-1}$, $T = 650$ $ft^2 day^{-1}$, $t = 30$ days, $r = 1$ ft, and $S = 6.4 \times 10^{-4}$, find the drawdown s. Assume this to be a nonequilibrium problem.

11. Determine the seepage loss through a clay layer under an earth-filled dam when $K = 10^{-4}$ $m s^{-1}$, $H = 20$ m, $L = 50$ m, and $b = 10$ m.

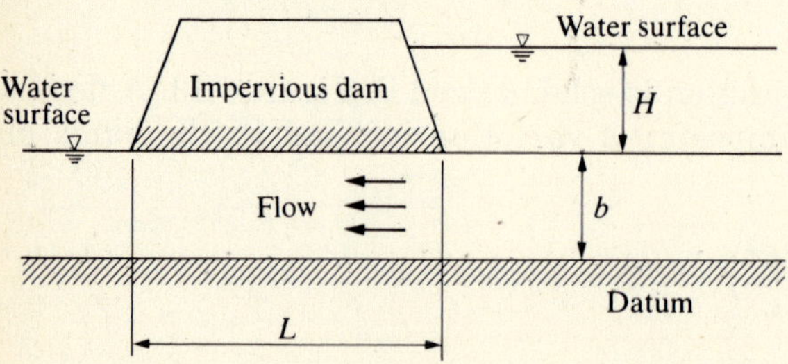

12. The following data is observed in an observation well 50 ft from a pumped well, $Q = 200 \text{ gal day}^{-1}$, in an artesian aquifer.

Time (min)	3	5.5	8	10	29	40	65	100
S (ft)	0.2	0.6	1.2	1.5	4	4.8	6	7

Using the modified Theis equation, obtain the value of S (storage coefficient) and T (transmissivity) in U.S. gallons per day per foot.

13. A 12-in. diameter well is pumped at a uniform rate of 500 gal min^{-1} while the observations of drawdown are made in a well 100 ft distant. Values of t and s (drawdown) observed together with values of r^2/t are given below. Find T and S for the aquifer, and estimate the drawdown in the observation well at the end of one year of pumping.

t (hr)	1	2	3	4	5	6	8	10	12	18	24
s (ft)	0.6	1.4	2.4	2.9	3.3	4.0	5.2	6.2	7.5	9.1	10.5
$\frac{r^2}{t}\left(\frac{\text{ft}^2}{\text{day}} \times 10^{-5}\right)$	2.4	1.2	0.8	0.6	0.5	0.4	0.3	0.24	0.2	0.13	0.1

14. Using the modified Theis method, find the transmissivity and storativity coefficient for the data of Problem 13.

15. Below is a diagram illustrating the long-standing positions and conditions of a pond and a river relative to a confined aquifer.

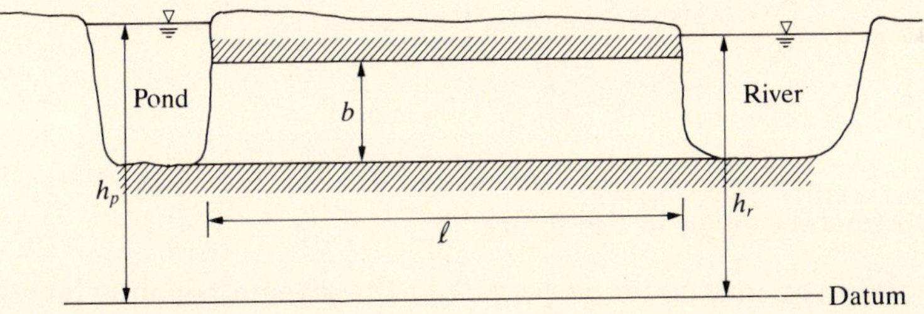

A chemical company suddenly dumps a conservative chemical substance into the pond. Assuming that the chemical instantly mixes with the pond water, approximate how long it would take for the contaminant to get to the river. You can assume that the contaminant moves with the water, does not get absorbed in the soil, and dispersion is minimal. Also, the pond and river are infinitely long. This corresponds to a worst-case (fastest transport) scenario of the bulk of the contaminant. $\ell = 3$ km, n (porosity) $= 0.3$, $b = 10$ m, $T = 0.1 \text{ m}^2 \text{s}^{-1}$, h_p (pond) $= 301$ m, and h_r (river) $= 300$ m. What do you expect would have been different in the approach and solution if the aquifer had been unconfined?

16. A confined aquifer extends between two rivers as shown below.

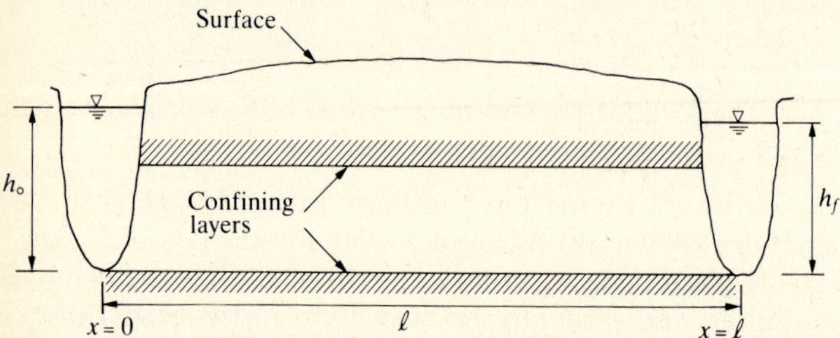

A fully penetrating well is drilled in the middle of the two rivers and pumps steadily Q gal min^{-1}. Assuming the cone of influence from the well is much less than the distance between the rivers, sketch the total head distribution from $x = 0$ to $x = \ell$.

If the field transmissivity is 10,000 gal day^{-1} ft^{-1}, $Q = 200$ gal min^{-1}, $h_f = 88$ ft, $h_o = 83$ ft, $\ell = 10{,}000$ ft, and $r_\infty = 250$ ft, compute the steady-state head 50 ft to the right of the well.

17. A hillside drains to a stream as shown below.

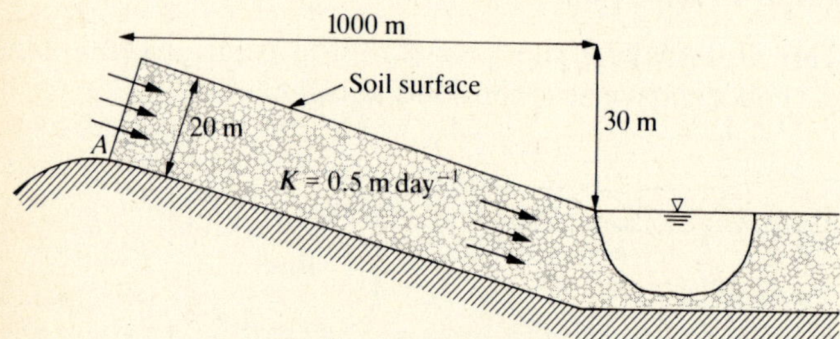

Using the parameters given in the figure:

a) Calculate the maximum seepage into the stream per unit length of the stream (when the entire soil is saturated).

b) What is the maximum runoff rate at the base of the hillside if it rains steadily at a rate of 0.3 cm day^{-1}? Express this maximum at a percentage of the rainfall rate.

c) If the soil is a silty loam, how long would it take the center of mass of a contaminant injected instantaneously at point A to reach the stream?

(Contributed by Laurens van der Tak, based on class notes from the University of California at Davis.)

18. Consider the unconfined flow through a stratified aquifer, as shown in the figure below. Determine the rate of flow Q per unit width of the aquifer, using the Dupuit assumptions. Neglect the presence of a seepage face. (Contributed by Laurens van der Tak, based on class notes from the University of California at Davis.)

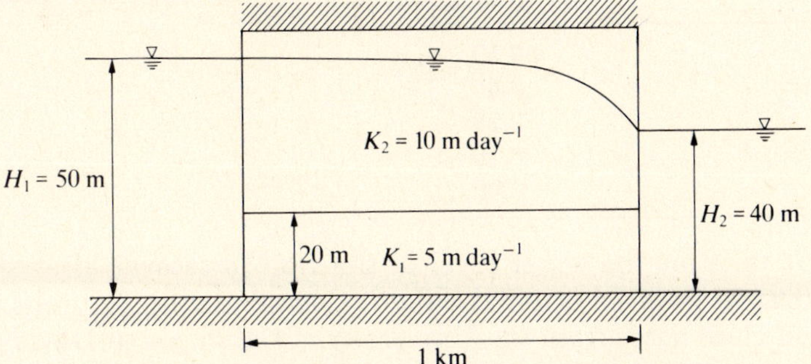

19. An aquifer with transmissivity T of 2000 ft^2 day^{-1} is confined by an aquitard 10 ft thick. The phreatic aquifer above the aquitard has a constant piezometric head h_s as shown in the figure below. At an upstream point, the piezometric head in the confined aquifer is 10 ft above that in the phreatic aquifer, and at a point 10,000 ft down gradient, the head excess in the confined aquifer is 4 ft. Determine the hydraulic conductivity of the aquitard and the rate of upward leakage through that aquitard at the upstream point. Determine the distance downstream L where the leakage ε is negligible [$\varepsilon(x = L) = 0.01 \times \varepsilon(x = 0)$]. (Contributed by Laurens van der Tak, based on homework notes from Prof. Lynn W. Gelhar at MIT.)

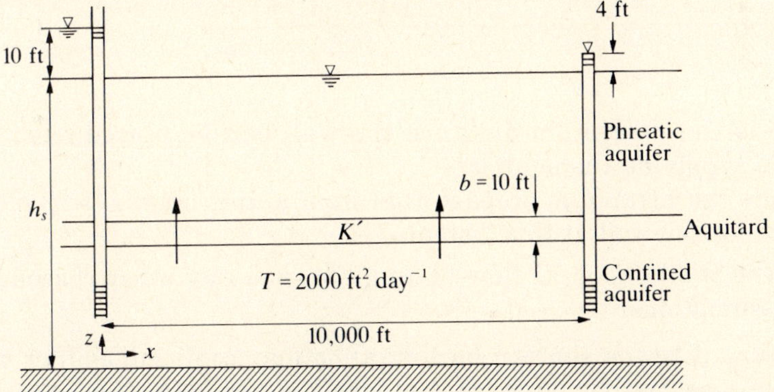

20. If a well fully penetrating a confined aquifer produces 60 gal min^{-1} with a 10-ft drawdown, compute the rate of discharge with a 15-ft drawdown. Assume steady-state conditions and neglect well losses. (Contributed by Laurens van der Tak, based on class notes from the University of California at Davis.)

21. Determine the values of T and K indicated by a test of a confined aquifer of thickness 50 m, in which a constant pumped discharge of 1500 $m^3\,day^{-1}$ for 0.25 day is shut off. The drawdown during recovery is indicated in the table below.

TIME SINCE PUMPING STOPPED (days)	DRAWDOWN DURING RECOVERY (m)
0.02	0.59
0.05	0.40
0.10	0.28
0.25	0.16
0.50	0.09

How long does it take for the water level in the well to recover after shutdown to about 0.01 m from its original value? (Contributed by Laurens van der Tak, based on class notes from the University of California at Davis.)

22. A well is to be placed in an unconfined aquifer as shown below and pumped at a rate of 30,000 $gal\,day^{-1}$. Environmental conservation requirements specify that the maximum steady-state drawdown outside a radius of 50 ft from the well can be 2 ft.

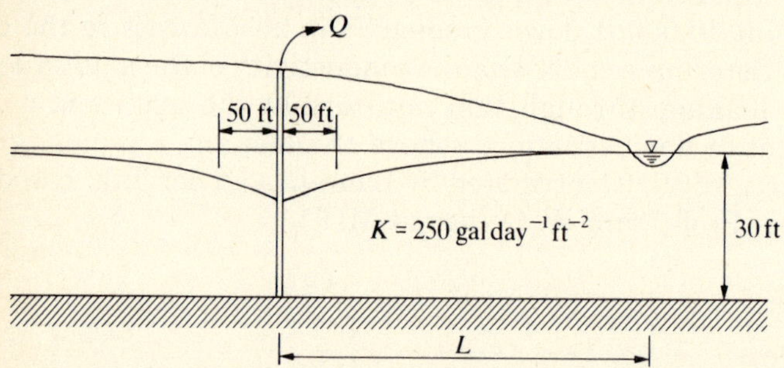

a) What is the maximum distance the well can be placed from the stream supplying the aquifer?

b) What is the maximum seepage rate from a unit length (1 ft) of the stream for the well at this distance?

You must make the horizontal flow assumption and any other reasonable simplifying assumptions.

23. The following diagram shows a horizontal bottom, confined, aquifer with boundary conditions fixed by large lakes at levels h_o and h_f ($h_o > h_f$). A cylindrical slug of contaminant is moving left to right without mixing with the groundwater. In order to prevent it from reaching Lake 2, you decide to dig a well, 50 cm in diameter, in the middle of the contaminant slug and pump. Assuming steady state, estimate how much you would need to pump to stagnate (stop) the point x at the edge of the plume from moving. Will this be enough to prevent contaminant from reaching Lake 2? (The aquifer parameters are $L = 3000$ m, $h_\circ = 103$ m, $\ell = 300$ m, $h_f = 100$ m, $K = 10^{-3}$ cm s^{-1}, and $b = 100$ m.) (*Hint:* Flow is in the direction of decreasing head and solutions are linear.)

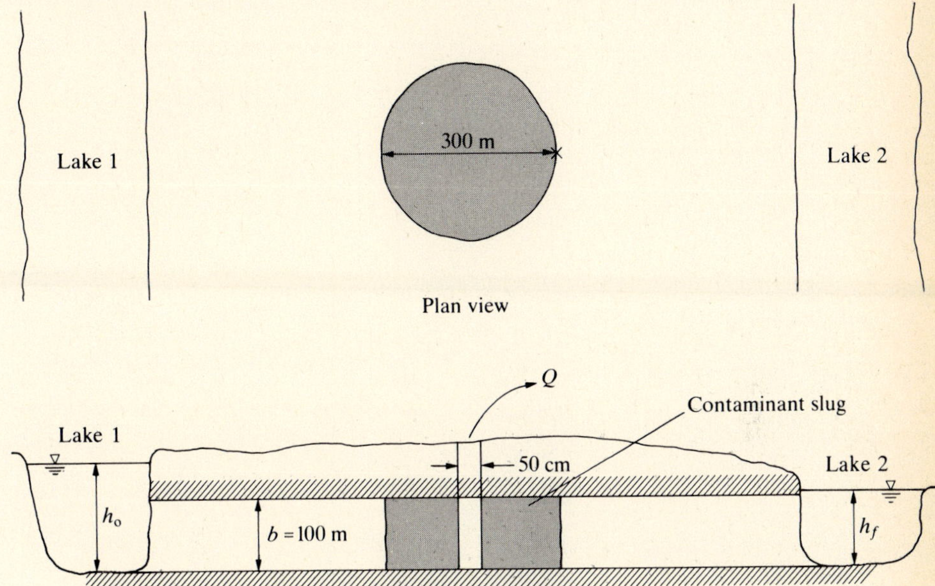

Chapter 8

Flow in Unsaturated Porous Media and Infiltration

8.1 INTRODUCTION

Chapter 7 discussed the occurrence and transport of water in saturated media. Equally important is the water found in the unsaturated or vadose zone of the soil. This water is the direct source of moisture for vegetation and hence is invaluable to food production and to the planet ecology. Furthermore, this zone is the link between surface and underground hydrologic processes. Water evaporated or transpired from this unsaturated region supplies a large portion of the atmospheric moisture so important to climate and meteorology. The water in the unsaturated zone also controls the amount of precipitation that will enter the soil or remain on the surface. The path taken by precipitation will determine the nature of the dominating hydrologic processes in a region.

The transfer of water from the atmosphere to the soil is called infiltration. Infiltration is the most crucial element of the description of the transformation of rainfall into streamflow. In the following sections we will describe the nature of flow in unsaturated porous media and use this knowledge to discuss the infiltration process. The related concept of runoff will also be studied.

8.2 FLOW IN UNSATURATED POROUS MEDIA

As a soil–rock matrix dries, the interconnectivity between pores becomes irregular and discontinuous, since air substitutes water in many locations. Furthermore, in contrast with saturated flows where gravitational forces play a dominant role, molecular forces become extremely important. Strong negative capillary pressures develop in the air–water interfaces. These pressures change with the effective pore sizes of the soil, and therefore depend on the intrinsic structure of the material and on the degree of saturation. The drier the material, the smaller and more discontinuous are the pores containing water, and the stronger the capillary forces.

The change in moisture content not only affects the forces, but obviously hinders the path of water through pores. The hydraulic conductivity is a function of soil moisture. The problem is further complicated by the fact that under unsaturated conditions, the flux of water is in both liquid and vapor forms. This flux depends not only on gravitational and capillary potentials, but on vapor density and temperature gradients throughout the soil. All these complicated issues are beyond the scope of this book. The reader is referred to Milly [1982] for a concise explanation and complete references. Nevertheless, if we assume isothermal conditions and keep incompressible assumptions for the vapor phase, we can suggest that the moisture flux obeys the following analogy to Darcy's Law:

$$q_x = -K_x(\theta)\frac{\partial h}{\partial x}; \qquad q_y = -K_y(\theta)\frac{\partial h}{\partial y}; \qquad q_z = -K_z(\theta)\frac{\partial h}{\partial z}, \tag{8.1}$$

where the hydraulic conductivities are now explicitly dependent on volumetric soil moisture (volume of water over total volume of soil) θ, and h is now the piezometric head in terms of capillary potential:

$$h = \frac{P_c}{\gamma} + z = \psi(\theta) + z, \tag{8.2}$$

where z is defined as being positive upward from an arbitrary datum within the soil column.

The matrix or capillary potential $\psi(\theta)$ is a function of moisture θ and is measured in centimeters. It is negative relative to atmospheric pressure. P_c is also called the suction or tension pressure.

Substituting Eq. (8.2) into (8.1) leads to

$$q_x = -K_x(\theta)\frac{\partial\psi(\theta)}{\partial x} \tag{8.3a}$$

$$q_y = -K_y(\theta)\frac{\partial\psi(\theta)}{\partial y} \tag{8.3b}$$

$$q_z = -K_z(\theta)\frac{\partial\psi(\theta)}{\partial z} - K_z(\theta), \tag{8.3c}$$

where the last term (Eq. 8.3c) accounts for the gravitational effect on flow. Note that a negative q_z value implies flow vertically downward in the soil system.

The use of Eq. (8.3) is hampered because there is no one-to-one relationship between K, ψ, and θ. This is called the hysteresis effect and is illustrated in Figure 8.1. The relationship between matrix potential and soil moisture is not the same during wetting and drying cycles. In fact, the actual function is very much dependent on the history of wetting and drying events. This behavior is partly explained by the fact that during wetting the filling of small-diameter pores is aided by capillary forces, while during drying the same forces act to delay their emptying. Furthermore, given the tortuosity of channels formed by interconnecting pores, it is to be expected that the location of air pockets and discontinuities will vary widely from one type of event to another.

The quantification of hysteresis is difficult. Only a few attempts to conceptualize this behavior exist (Milly [1982]). Generally, the hysteresis is ignored, an assumption that becomes more reasonable in the isothermal case we discuss here (Rogers and Klute [1971], Mualem [1977]). Eagleson [1978a] uses results of Brooks and Corey [1966], Burdine [1958], and his own further development to recommend the following relationships:

$$\psi(s) = \psi(1)s^{-1/m} \tag{8.4}$$

$$k(s) = k(1)s^{c} \tag{8.5}$$

$$K(s) = K(1)s^{c} \tag{8.6}$$

$$c = (2 + 3m)/m, \tag{8.7}$$

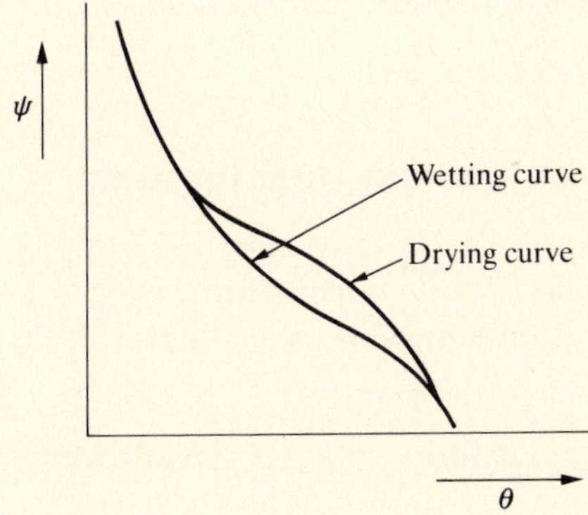

FIGURE 8.1 Illustration of hysteresis loop in the $\psi(\theta)$ versus θ relationship.

TABLE 8.1 Hydraulic Properties of Typical Soils

	k (cm^2)	K (cm s^{-1})	$\psi(1)$ (cm)	n	m	c	d
Clay	4×10^{-10}	3.4×10^{-5}	90	0.45	0.44	7.5	4.3
Silty loam	4×10^{-9}	3.4×10^{-4}	45	0.35	1.2	4.7	2.9
Sandy loam	4×10^{-8}	3.4×10^{-3}	25	0.25	3.3	3.6	2.3
Sand	10^{-7}	8.6×10^{-3}	15	0.20	5.4	3.4	2.2

See Section 8.2 for definitions of $\psi(1)$, m, c, and d.
Source: Entekhabi [1988].

where $s = \theta/n$ is the degree of saturation (n is porosity), m is the pore-size distribution index of the soil, c is the pore-disconnectedness index (a measure of tortuosity, or the ratio of the actual path to the straight-line path between two points), and $k(1)$ is the saturated intrinsic hydraulic conductivity. Table 7.2, repeated here as Table 8.1, gives values of $\psi(1)$, m, and c that are generally representative of typical soils. It should be emphasized that these values result from fits of Eqs. (8.4)–(8.7) to reported properties of soils in laboratory situations. Any particular soil, in the laboratory or the field, can exhibit parameters varying considerably from the given values. For example, soils in nature can exhibit porosities that are 0.1 larger, in all categories, than these laboratory preparations. Similarly, the range of hydraulic conductivities span several orders of magnitude around the values shown. The pore distribution index m could be considerably smaller (one order of magnitude). In summary, Table 8.1 should be used only as an example of soil hydraulic parameter values within reasonable range and consistent with each other.

Note also that $\psi(1)$ is nonzero only in the sense of the empirical fit to Eq. (8.6). Models to represent $K(s)$ and $\psi(s)$ exist that are compatible with a theoretical value $\psi(1) = 0$. van Genuchten [1980] and Wösten and van Genuchten [1988] discuss one such model.

Figures 8.2 and 8.3 show the fit of Eqs. (8.4) and (8.6) to cohesive and noncohesive soils. Figure 8.4 shows the behavior of Eq. (8.7).

8.2.1 Conservation of Mass in Unsaturated Porous Media

Assuming incompressible flow (ρ = constant), we can substitute Eq. (8.3) in the basic mass-conservation equation (see Chapter 7, Eq. 7.17), and obtain

$$\frac{\partial}{\partial x}\left[K_x(\theta)\,\frac{\partial\psi(\theta)}{\partial x}\right] + \frac{\partial}{\partial y}\left[K_y(\theta)\,\frac{\partial\psi(\theta)}{\partial y}\right] + \frac{\partial}{\partial z}\left[K_z(\theta)\,\frac{\partial\psi(\theta)}{\partial z}\right] + \frac{\partial}{\partial z}K_z(\theta) = \frac{\partial\theta}{\partial t}. \tag{8.8}$$

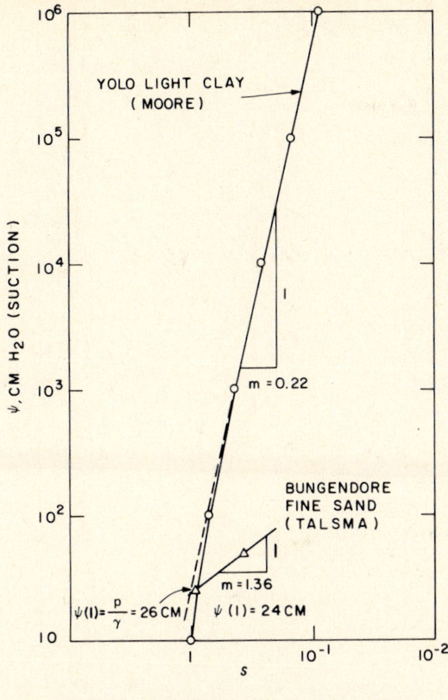

FIGURE 8.2 Matrix potential (water at 15°C). Source: P. S. Eagleson, "Climatic, Soil and Vegetation 3: A Simplified Model of Soil Moisture Movement in the Liquid Phase," *Water Resources Res.* 14(5):724, 1978. Copyright by the American Geophysical Union.

Using the assumed single-value function of $\psi(\theta)$, the above can also be expressed as

$$\frac{\partial}{\partial x}\left[D_x(\theta)\,\frac{\partial\theta}{\partial x}\right] + \frac{\partial}{\partial y}\left[D_y(\theta)\,\frac{\partial\theta}{\partial y}\right] + \frac{\partial}{\partial z}\left[D_z(\theta)\,\frac{\partial\theta}{\partial z}\right] + \frac{\partial}{\partial z}K_z(\theta) = \frac{\partial\theta}{\partial t}, \tag{8.9}$$

where

$$D_i(\theta) = K_i(\theta)\,\frac{\partial\psi(\theta)}{\partial\theta} \tag{8.10}$$

and is called the diffusivity in the ith direction.

For an isotropic medium, diffusivities in all three directions are the same. Eagleson [1978a] uses Eqs. (8.4) and (8.6) to obtain

$$D(\theta) = \frac{|\psi(1)|K(1)}{nm}\,s^d, \tag{8.11}$$

where $d = c - (1/m) - 1 = (c + 1)/2$ (see Fig. 8.4 for form of d).

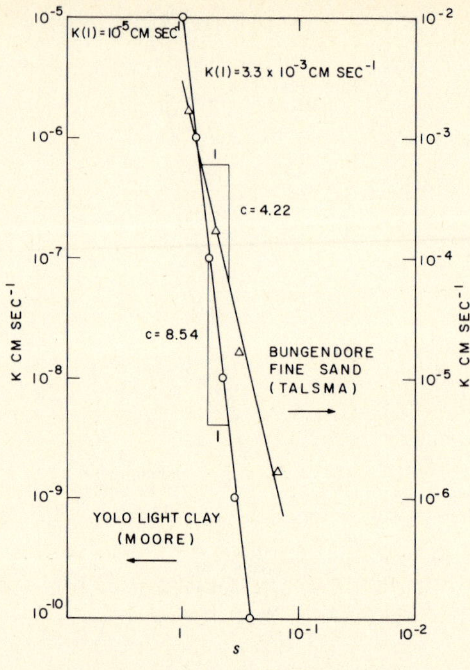

FIGURE 8.3 Hydraulic conductivity (water at 15°C). Source: P. S. Eagleson, "Climate, Soil and Vegetation 3: A Simplified Model of Soil Moisture Movement in the Liquid Phase," *Water Resources Res.* 14(5):724, 1978. Copyright by the American Geophysical Union.

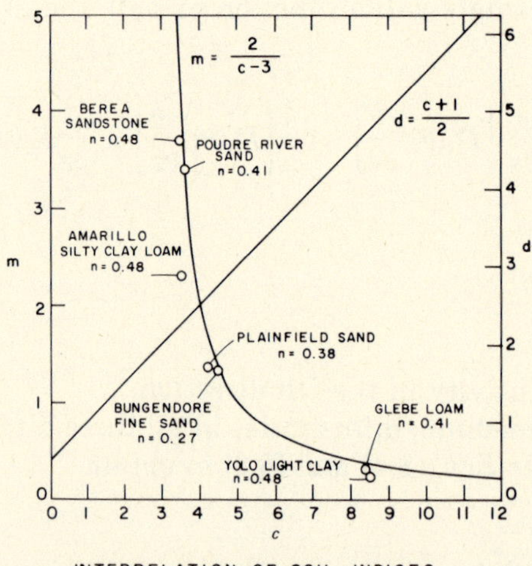

FIGURE 8.4 Interrelation of soil indices. Source: P. S. Eagleson, "Climate, Soil and Vegetation 3: A Simplified Model of Soil Moisture Movement in the Liquid Phase," *Water Resources Res.* 14(5):724, 1978. Copyright by the American Geophysical Union.

8.3 INFILTRATION AND EXFILTRATION

Infiltration is the movement of surface waters into the soil matrix. Since the upper zones of the soil are generally unsaturated, the infiltration process is described by Eq. (8.8). The reality of flow into the ground is three-dimensional. Nevertheless, it is commonly assumed that one-dimensional vertical flow dominates, which leads to the following equation:

$$\frac{\partial \theta}{\partial t} = \frac{\partial}{\partial z}\left[D(\theta)\frac{\partial \theta}{\partial z} + K_z(\theta)\right]. \tag{8.12}$$

Even the analytical solution of Eq. (8.12) for general boundary and initial conditions escapes us. Milly [1982] gives a complete numerical solution, including the vapor phase and temperature effects. Some analytical results are possible for particular assumptions, boundary conditions, and initial conditions. A common initial assumption is to ignore the gravitational term $K_z(\theta)$ in Eq. (8.12). Eagleson [1970] points out that this will be reasonable if $K_z(\theta)$ is uniform with depth or if the capillary potential is so large as to dominate Eq. (8.12). He states that Gardner and Mayhugh [1958] and Gardner [1959] argue that those conditions will be met when the level of saturation is low, like in the beginning of infiltration or the end of exfiltration. The equation can be further simplified if the diffusion term is assumed constant:

$$\frac{\partial \theta}{\partial t} = D\frac{\partial^2 \theta}{\partial z^2}. \tag{8.13}$$

Equation (8.13) describes infiltration as well as the exfiltration process. The latter is when water moves vertically up and out of the soil, fueled by a dry soil surface. The drying of the upper soil zones is due to surface evaporation. The equation also corresponds to a general diffusion formulation with wide applications in heat transfer through solids and in contaminant transport in water bodies. As such, it has been widely studied and a variety of solutions exist, corresponding to many boundary and initial conditions. The classic reference on the heat equation is Carslaw and Jaeger [1959].

Following the presentation by Eagleson [1970], the solution of Eq. (8.13) is

$$\frac{\theta - \theta_0}{\theta_i - \theta_0} = \mathrm{erf}\left[\frac{|z|}{2(Dt)^{1/2}}\right] \tag{8.14}$$

when the initial and boundary conditions are

$$\theta = \begin{cases} \theta_i & z \le 0;\ t = 0 \\ \theta_0 & z = 0;\ t > 0 \end{cases} \tag{8.15}$$

and the soil system is semi-infinite, extending to great depths. The conditions of Eq. (8.15) state that the soil column is initially at a uniform constant moisture content θ_i and a different constant moisture θ_0 is maintained at the surface, for all times greater than zero. In Eq. (8.14) "erf" represents the error function:

$$\operatorname{erf}(x) = \left(\frac{4}{\pi}\right)^{1/2} \int_0^x \exp(-y^2)\,dy, \tag{8.16}$$

which is tabulated in most handbooks of mathematics and in Carslaw and Jaeger [1959]. The form of Eq. (8.14) is shown in Figure 8.5. Note that Figure 8.5 implies that $\theta = \theta_i$ for values of $|z|/2(Dt)^{1/2}$ greater than 2. So the approximate depth of the moisture front, defined as the position where the moisture is still essentially that of the initial conditions, is

$$|z| = 4(Dt)^{1/2}. \tag{8.17}$$

The depth of the front estimate would be in error by the effect of the gravitational term, which will approximately add $K_z(\theta)t$ to the depth. Diaz-Granados et al. [1983] show by simulation that this total depth rarely exceeds 2 m for a variety of soils and climates. From Eqs. (8.3c) and (8.10), the vertical flux of

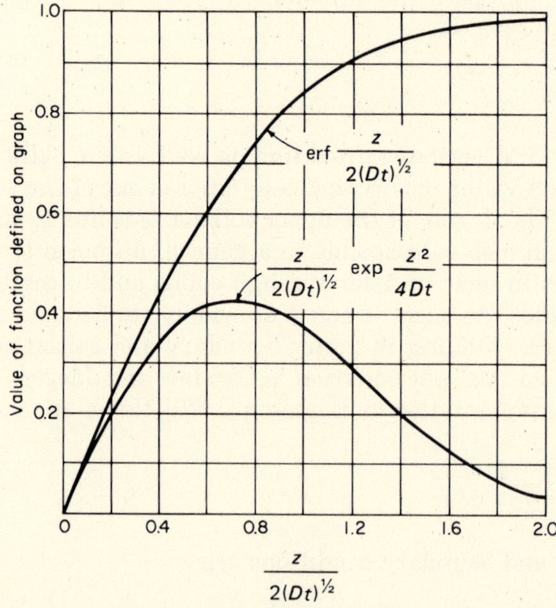

FIGURE 8.5 Functions for one-dimensional diffusion into a semi-infinite region. Source: H. S. Carslaw and C. Jaeger, *Conduction of Heat in Solids*, 2nd ed., 1959, Oxford University Press.

water at the surface is

$$q_z = -D \frac{\partial \theta}{\partial z}\bigg|_{z=0} = f, \tag{8.18}$$

which from Eq. (8.14) yields

$$f = (\theta_i - \theta_0)\left(\frac{D}{\pi t}\right)^{1/2}; \tag{8.19}$$

f is called the infiltration rate or capacity and has units of velocity.

If we assume that the gravitational effects linearly influence the infiltration rate, then

$$f = (\theta_i - \theta_0)\left(\frac{D}{\pi t}\right)^{1/2} - K(\theta_0)\,. \tag{8.20}$$

Note that f above has a direction-dependent definition. Positive values of f correspond to exfiltration; negative values are infiltration. Infiltration increases with decreasing θ_i, increasing θ_0, and increasing $K(\theta_0)$. Similarly, the infiltration rate increases with the square root of the diffusion coefficient, but it is inversely proportional to the square root of time, implying that infiltration decreases as time increases. All these are expected and observed results.

Eagleson [1970] also shows that the flux (Eq. 8.18) must also obey the original differential equation (Eq. 8.13). A general solution can be expressed:

$$f = (A \cos mz + B \sin mz)e^{-D\ell^2 t} + \text{constant}\,. \tag{8.21}$$

The above solution allows the imposition of boundary values for f at $z = 0$,

$$\begin{aligned} f &= f_c \qquad z = 0\,, t = \infty \\ f &= f_0 \qquad z = 0\,, t = 0\,, \end{aligned} \tag{8.22}$$

which leads to (for $z = 0$)

$$f = f_c + (f_0 - f_c)e^{-D\ell^2 t}, \tag{8.23}$$

where ℓ^{-1} is a characteristic length dependent on the value of z, f_c is an asymptotic value for infiltration, at infinite time, and $f_0 > f_c$ is an initial value. Equation (8.23) has the form of an empirical formulation called the Horton equation. It will be discussed in detail in Section 8.3.1. The advantage of the Horton form is that it gives a finite initial infiltration value. Equation (8.19) gives $f = \infty$ at $t = 0$. Figure 8.6 shows the behavior of both derived infiltration formulas.

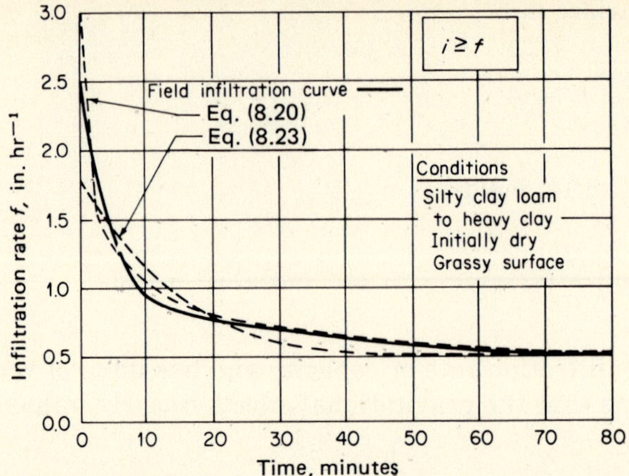

FIGURE 8.6 Comparison of observed and best-fit theoretical curves of infiltration capacity. Source: K. K. Watson, "A Note on the Field Use of a Theoretically Derived Infiltration Equation," *J. Geophys. Res.* 64(10):1614, 1959. Copyright by the American Geophysical Union.

Ignoring gravitational effects, the variable diffusivity equation for infiltration would be

$$\frac{\partial \theta}{\partial t} = \frac{\partial}{\partial z}\left[D(\theta)\frac{\partial \theta}{\partial z}\right]. \tag{8.24}$$

A possible solution to the above involves the so-called Boltzmann transformation

$$\eta = \frac{z}{2}(D_0 t)^{-1/2},$$

where D_0 is a constant diffusivity assumed at the surface $z = 0$. The boundary and initial conditions expressed in terms of η are

$$\theta = \theta_i; \qquad \eta \to -\infty$$

$$\theta = \theta_0; \qquad \eta = 0$$

The use of the Boltzmann transformation converts Eq. (8.24) into an ordinary differential equation (Eagleson [1970]). The solution is of the form (adding gravity)

$$f = (\theta_i - \theta_0)\left(\frac{\overline{D}}{\pi t}\right)^{1/2} - K(\theta_0), \tag{8.25}$$

which looks like Eq. (8.20), but now $\overline{D}$ is an effective diffusivity over the range of possible moisture values $[\theta_0, \theta_i]$. Crank [1956] suggested that $\overline{D}$ be given the following forms:

Infiltration:

$$\overline{D} = \frac{5}{3}(\theta_0 - \theta_i)^{-5/3} \int_{\theta_i}^{\theta_0} (\theta - \theta_i)^{2/3} D(\theta)\, d\theta. \tag{8.26}$$

Exfiltration:

$$\overline{D} = 1.85(\theta_i - \theta_0)^{-1.85} \int_{\theta_0}^{\theta_i} (\theta - \theta_i)^{0.85} D(\theta)\, d\theta. \tag{8.27}$$

If the variability of diffusivity with moisture is taken to be of the form (Gardner [1959])

$$\frac{D(\theta)}{D(\theta_0)} = \exp[A(\theta - \theta_0)], \tag{8.28}$$

then it is possible to relate $\overline{D}/D(\theta_0)$ to $D(\theta_0)/D(\theta_i)$ for the infiltration case [or $D(\theta_i)/D(\theta_0)$ for the exfiltration case] via Eqs. (8.26) and (8.27). $D(\theta_i)$ is the diffusivity at moisture level θ_i, the initial value. Gardner [1959] obtains this relationship, which is shown in Figure 8.7. The hysteresis effect becomes

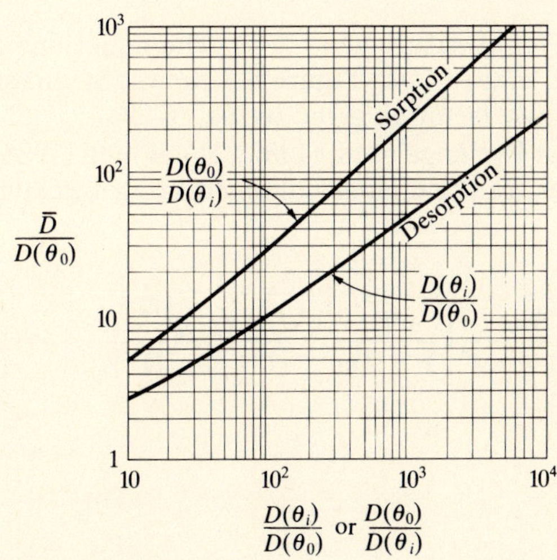

FIGURE 8.7 Weighted mean diffusivity for sorption and desorption. Source: Adapted from W. R. Gardner, "Solutions of the Flow Equation for the Drying of Soils and Other Porous Media," *Soil Sci. Soc. Am. J.* 23(3):183–195, May–June 1959. By permission of the Soil Society of America, Inc.

obvious; the effective diffusivity is higher during sorption (infiltration) than during desorption (exfiltration) for the same relative boundary and initial moisture conditions.

Assuming either saturated ($s_0 = 1$) or completely dry ($s_0 = 0$) boundary conditions, and using Eq. (8.11), Eagleson [1978a] obtained the following expressions for dimensionless effective diffusivities:

Infiltration:

$$\frac{3mn\overline{D}}{5K(1)\psi(1)} = \phi_i(d, s_i)$$
$$= (1 - s_i)^d \left\{ \frac{1}{d + 5/3} + \sum_{n=1}^{d} \frac{1}{d + [(5/3) - n]} \binom{d}{n} \left(\frac{s_i}{1 - s_i} \right)^n \right\}. \tag{8.29a}$$

Exfiltration:

$$\frac{mn\overline{D}}{K(1)\psi(1)} = \phi_e(d) = \left[1 + 1.85 \sum_{n=1}^{d} (-1)^n \binom{d}{n} \frac{1}{1.85 + n} \right]. \tag{8.30a}$$

In the above,

$$\binom{d}{n} = \frac{d!}{n!(d - n)!}.$$

s_i is the initial level of saturation (θ_i/n). The equations are only valid for integer values of $d = (c + 1)/2$. Figure 8.8 shows the dimensionless effective infiltration diffusivity factor as given by Eq. (8.29a).

Using results of Parlange et al. [1985], Entekhabi [1988] derived new and simpler expressions for the dimensionless effective diffusivities:

Infiltration:

$$\phi_i(m, s_i) = \frac{3\pi}{10(1 - s_i)^2} \left[\frac{m}{1 + 4m} + \frac{m^2 s_i^{(1/m + 4)}}{(1 + 4m)(1 + 3m)} - \frac{ms_i}{1 + 3m} \right]. \tag{8.29b}$$

Exfiltration:

$$\phi_e(m) = \frac{2\pi m}{3(1 + 3m)(1 + 4m)}. \tag{8.30b}$$

Philip [1960] also suggested an exact solution to the variable diffusivity problem, with gravity effects, when initial and boundary conditions corre-

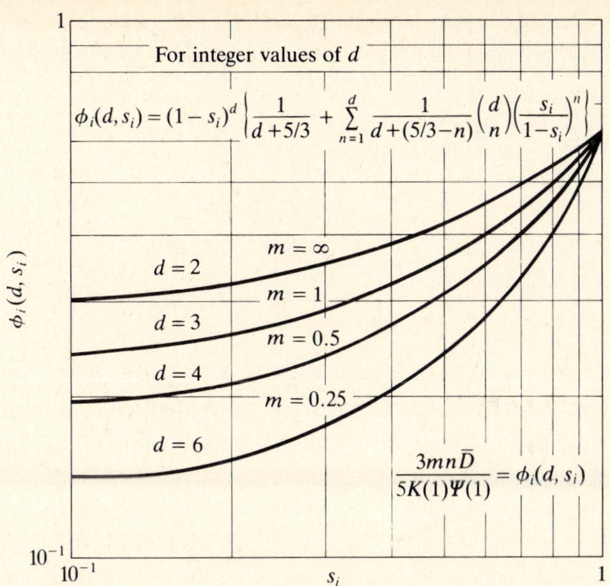

FIGURE 8.8 Dimensionless infiltration diffusivity. Source: P. S. Eagleson, "Climate, Soil and Vegetation 3: A Simplified Model of Soil Moisture Movement in the Liquid Phase," *Water Resources Res.* 14(5):727, 1978. Copyright by the American Geophysical Union.

spond to Eq. (8.15) but with $\theta_0 = n$, the surface is saturated. The solution is in the form of a series expansion. The first three terms of the series reduce to

$$f_i(t) \approx \frac{1}{2} S_i t^{-1/2} + A_i \,, \tag{8.31}$$

where f_i is the (positive) infiltration rate. Parameters A_i and S_i are

$$S_i = 2(\theta_0 - \theta_i)[\bar{D}/\pi]^{1/2} \tag{8.32}$$

$$A_i = \frac{1}{2}[K(\theta_0) - K(\theta_i)]. \tag{8.33}$$

Using Eq. (8.29) for the effective diffusivity, Eagleson [1978a] obtains useful expressions for A_i and S_i in terms of soil parameters and initial soil moisture:

$$A_i = \frac{1}{2} K(1)(1 + s_i^c) - W \tag{8.34}$$

$$S_i = 2(1 - s_i)\left[\frac{5nK(1)\psi(1)\phi_i(d, s_i)}{3m\pi}\right]^{1/2}, \tag{8.35}$$

where W is an adjustment for possible capillary rise flux, to be discussed later.

Following Philip's approach, Eagleson [1978c] also obtained an exfiltration equation for a dry ($\theta_0 = 0$) surface and initial moisture θ_i:

$$f_e(t) \approx \frac{1}{2} S_e t^{-1/2} - MT_a + W\,, \tag{8.36}$$

where

$$S_e = 2s_i^{1+d/2} \left[\frac{nK(1)\psi(1)\phi_e(d)}{\pi m} \right]^{1/2}; \qquad s_0 = 0\,. \tag{8.37}$$

The term MT_a is an approximate way of dealing with moisture extraction by vegetation. M is the vegetated fraction of land surface or canopy density and T_a is the transpiration rate in centimeters per second. The transpiration rate is sometimes expressed as

$$T_a = k_v e_p\,, \tag{8.38}$$

where $k_v = T_a / e_p$ is the transpiration efficiency of the vegetation and e_p is the potential evaporation (climatically controlled) of the site.

8.3.1 Empirical Infiltration Equations

In practice, infiltration is usually calculated from one of the innumerable empirical infiltration equations available. Most of them mimic the following behavior. Under given soil type and antecedent moisture conditions, there will be an initial infiltration rate f_0. This rate will decrease as more water is infiltrated, finally achieving a constant rate, or ultimate infiltration capacity f_c. This infiltration capacity rate occurs when the soil is saturated. Under steady state (no storage change), it will be equal or less than the rate at which water percolates and flows into deep groundwater systems (aquifers). The parameters f_0, f_c, and the decay of infiltration capacity are functions of the soil, moisture conditions, vegetation, rainfall intensity, and soil surface conditions. For example, the behavior of a given soil may be different under different storms because of surface sealing or crusting caused by the impact of raindrops.

Horton Infiltration Equation [1939, 1940]
The Horton formulation was already given in Eq. (8.23). It takes the form

$$f = f_c + (f_0 - f_c)e^{-\alpha t}, \tag{8.39}$$

where in practice f_0, f_c, and α are parameters to be estimated from data. The initial infiltration capacity and the decay rate depend on soil and antecedent conditions.

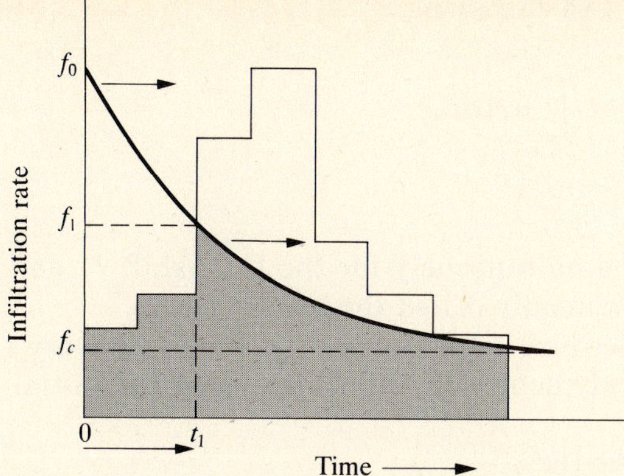

FIGURE 8.9 Infiltration-rate curve when initial rainfall rate is higher than initial infiltration rate.

Figure 8.9 shows the Horton infiltration equation as applied to a given rainfall event. It may be argued that at point t_1, where rainfall first exceeds infiltration, the actual infiltration rate will be larger than that given by f_1 in the figure. This is so because f_1 assumes that the infiltration rate has decayed from f_0 as a function of increased soil moisture, which is given by the area under the f curve between time 0 and t_1. This inconsistency results because the Horton, like the Philip's, equation assumes that the surface is saturated all the time, hence there is an unlimited supply of moisture. To account for this discrepancy, the procedure shown in Figure 8.10 is commonly used with the Horton and any other time-dependent infiltration equation.

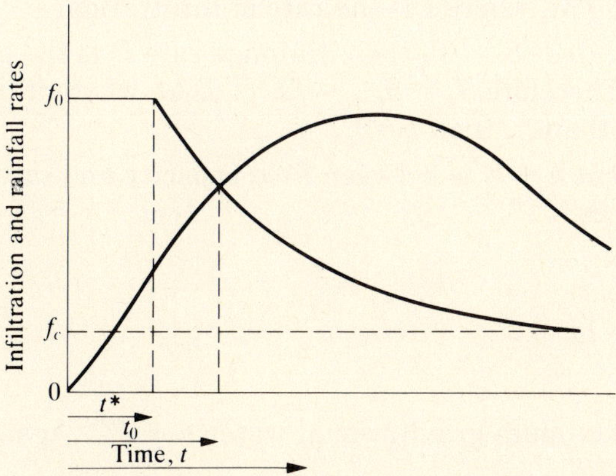

FIGURE 8.10 Illustration of computation of the effective infiltration rate.

The following two equations,

$$\int_{t^*}^{t_o} f(t - t^*)\, dt = \int_0^{t_o} i(t)\, dt\,,$$

$$f(t_o - t^*) = i(t_o)\,, \tag{8.40}$$

should be solved simultaneously for the time shift t^* and t_o (see Fig. 8.10). The time, t^*, is commonly called the ponding time.

The above procedure would need to be repeated every time $i < f$. Nevertheless, this is rarely done in practice except for the initial time shift.

Huggins–Monke [1966]

Other suggested equations try to avoid the problem of computing surface saturation conditions by making infiltration a function of soil moisture. The Huggins–Monke equation is an example:

$$f = f_c + A\left(\frac{S}{n}\right)^P, \tag{8.41}$$

where A and P are coefficients, n is the total porosity of the upper soil layer, and S is storage potential of the soil upper zone or layer, initially n minus antecedent moisture content.

Equation (8.41) must be solved iteratively and in sequence by accounting for all infiltrated and drained water. Let F be the total volume of infiltrated water. At the beginning of the storm $F = 0$ and S is given by porosity n minus antecedent moisture. As time progresses, the storage potential must be found in terms of infiltrated and drained water.

1. If $n - S$ is presently below field capacity f_c, then the drainage rate is 0, so $S_t = S_{t-1} - f\Delta t$, where f is the rate of infiltration.
2. If soil is saturated ($S = 0$), then drainage rate D is the same as the infiltration rate. Therefore $S_t = S_{t-1} - f\Delta t + D\Delta t$, where if the rainfall intensity is larger than f_c, then $f = f_c$.
3. If water content $n - S$ is between field capacity and saturation, drainage rate is taken as

$$D = f_c\left(1 - \frac{S}{G}\right)^3,$$

where G is maximum gravitational water $n - FC$. Again,

$$S_t = S_{t-1} - f\Delta t + D\Delta t\,.$$

Antecedent Precipitation Methods

Antecedent precipitation is a concept used as a moisture surrogate. It is usually computed by accounting for precipitation and moisture decay over a past period of time. The accounting takes the form shown in Figure 8.11. Between time periods, moisture decreases as

$$AP_t = AP_0 K^t, \tag{8.42}$$

where K is a decay rate.

When precipitation occurs, the total precipitation amount is added to the existing moisture conditions to define a new antecedent precipitation level. This concept has led to infiltration equations such as

$$f = f_c + (f_0 + f_c)e^{-\beta(AP_{30}+I)}, \tag{8.43}$$

where β is a decay parameter, AP_{30} is antecedent precipitation computed over a 30-day period, and $I = \int_0^t i(t)$.

Several other empirical relations exist that give infiltration or runoff as a function of the antecedent precipitation index. Most of these procedures relate, through regressions, infiltration or runoff not only to the antecedent precipitation index but to time of the year, storm duration, storm intensity, and surface retention (Linsley et al. [1949], Sittner et al. [1969]).

Green–Ampt Model [1911]

This equation is based on a Darcy-type water flux. Infiltration has to be proportional to the total gradient, including a suction effect. Therefore

$$f = \frac{K_s(H + \psi + L_F)}{L_F}, \tag{8.44}$$

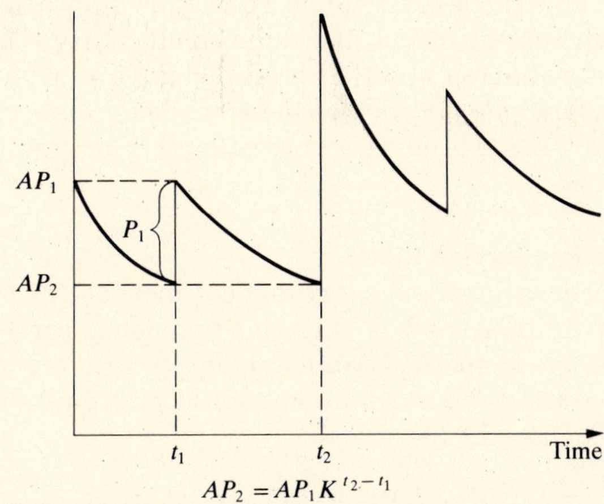

FIGURE 8.11 Schematic representation of the behavior of antecedent precipitation index.

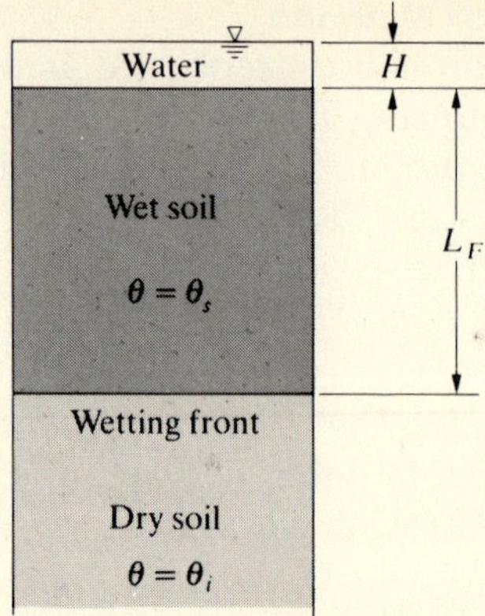

FIGURE 8.12 The Green–Ampt model assumes piston flow with a sharp wetting front between the infiltration zone and soil at the initial water content. The wet zone increases in length as infiltration progresses. Source: Haan et al. [1982].

where H is some level of ponding on the surface, ψ is a suction effect due to dryness at lower levels, and L_F is the increasing depth of the wet front. This situation is shown in Figure 8.12; it is assumed that the wet front moves as a piston.

If H is assumed small, Eq. (8.44) can be expressed as

$$f = K_s + \frac{K_s S \psi}{F}, \tag{8.45}$$

where F is the total infiltrated water given by $(\theta_s - \theta_i)L_F = SL_F$, and S is the initial moisture (as a fractional volume) deficit of the soil column. Note that as F increases, f approaches K_s, which is a hydraulic conductivity usually taken as less than the saturated hydraulic conductivity. The parameters of the Green–Ampt equation should be empirically evaluated using data. Nevertheless, considerable efforts have been made to relate them to soil properties. Haan et al. [1982] provide a good summary of results related to the Green–Ampt model.

Soil Conservation Service [1968]

The Soil Conservation Service empirical method for obtaining runoff (infiltration) over finite areas has enjoyed tremendous popularity because of this agency's attempts to specify parameters for various regions in the United States. This procedure gives the volume of total precipitation minus infiltration (runoff, see Section 8.3.2)

$$P - F - I_a = R_s = \frac{(P - I_a)^2}{P - I_a + S}; \qquad \begin{matrix} P \geq I_a \\ S \geq I_a + F \end{matrix}, \tag{8.46}$$

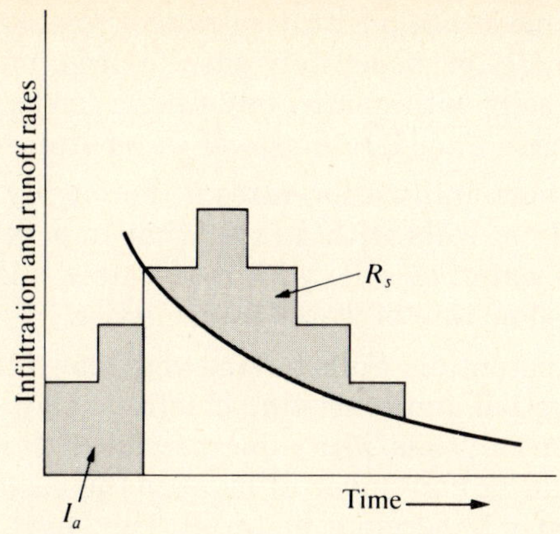

FIGURE 8.13 Implied behavior of infiltration and runoff by the Soil Conservation Service method.

where I_a is an initial retention volume and S is the potential maximum surface retention. The method was developed and calibrated in English units, so Q, P, I_a, and S are in inches. Figure 8.13 shows the implied behavior. The initial abstraction I_a is commonly taken as $I_a = 0.2S$, which leads to

$$R_s = \frac{(P - 0.2S)^2}{P + 0.8S}. \tag{8.47}$$

The retention volume is given by

$$S \text{ (inches)} = \frac{1000}{CN} - 10, \tag{8.48}$$

where CN is called the curve number, a parameter dependent on soil type, use, and antecedent moisture conditions. Curve numbers are given for various soil types, conditions, and locations throughout the United States. Soils are classified in the following way (Viessman et al. [1977]):*

A. (Low runoff potential.) Soils having high infiltration rates even if thoroughly wetted and consisting chiefly of deep well to excessively drained sands or gravels. They have a high rate of water transmission.

*Excerpt from p. 618 of *Introduction to Hydrology* by Warren Viessman, Jr., John W. Knapp, Gary L. Lewis, and Terrence E. Harbaugh. Copyright © 1977, 1972 by Harper & Row, Publishers, Inc. Reprinted by permission of Harper & Row, Publishers, Inc.

B. Soils having moderate infiltration rates if thoroughly wetted and consisting chiefly of moderately deep to deep, moderately well to well-drained soils with moderately fine to moderately coarse textures. They have a moderate rate of water transmission.

C. Soils having slow infiltration rates if thoroughly wetted and consisting chiefly of soils with a layer that impedes the downward movement of water, or soils with moderately fine to fine texture. They have a slow rate of water transmission.

D. (High runoff potential.) Soils having very slow infiltration rates if thoroughly wetted and consisting chiefly of clay soils with a high swelling potential, soils with a permanent high water table, soils with a claypan or clay layer at or near the surface, and shallow soils over nearly impervious material. They have a very slow rate of water transmission.

Table 8.2 gives curve numbers CN for various soil-cover complexes and soil groups. These numbers correspond to antecedent moisture conditions normal for the annual maximum flood, so-called condition II. Condition I implies a dry watershed, but not yet at wilting point. Condition III is when the soil is nearly saturated. The values of Table 8.2 can be adjusted for conditions I and III using Table 8.3. Ultimately, though, the curve number is nothing but a fitting parameter and the user should not become overconfident nor blindly trust the above attempts to regionalize results. McCuen [1982] discusses the use of the Soil Conservation Service runoff model in detail.

Constant Infiltration Indexes

In very large areas, for slow responses, and for storms of long duration, the time distribution of infiltration may not be very important. Under these conditions, we may assume that infiltration occurs at a constant rate. The ϕ index is the uniform distribution of total infiltration throughout storm duration. Total infiltration is computed as the difference between precipitation and runoff. The W index is the same idea but total infiltration is computed taking surface storage into account, not lumped into infiltration as in the ϕ index.

8.3.2 Storm Runoff

Effective rainfall or runoff is traditionally defined as the net liquid water supplied to channels at time scales comparable to the duration of the storm after evaporation, interception, surface retention, infiltration, and percolation to underlying aquifers. Except for small storms in heavily forested areas, where evaporation can be significant, infiltration is usually seen as controlling the amount of water available for storm runoff, particularly storm surface runoff, which is water delivered to the channels as overland flow.

TABLE 8.2 Runoff Curve Numbers for Hydrologic Soil-Cover Complexes

LAND USE OR COVER	TREATMENT OR PRACTICE	HYDROLOGIC CONDITION	HYDROLOGIC SOIL GROUP			
			A	B	C	D
Fallow	Straight row	—	77	86	91	94
Row crops	Straight row	Poor	72	81	88	91
	Straight row	Good	67	78	85	89
	Contoured	Poor	70	79	84	88
	Contoured	Good	65	75	82	86
	Contoured and terraced	Poor	66	74	80	82
	Contoured and terraced	Good	62	71	78	81
Small grain	Straight row	Poor	65	76	84	88
		Good	63	75	83	87
	Contoured	Poor	63	74	82	85
		Good	61	73	81	84
	Contoured and terraced	Poor	61	72	79	82
		Good	59	70	78	81
Closed seed	Straight row	Poor	66	77	85	89
Legumes[a] or	Straight row	Good	58	72	81	85
Rotation meadow	Contoured	Poor	64	75	83	85
	Contoured	Good	55	69	78	83
	Contoured and terraced	Poor	63	73	80	83
	Contoured and terraced	Good	51	67	76	80
Pasture or range		Poor	68	79	86	89
		Fair	49	69	79	84
		Good	39	61	74	80
	Contoured	Poor	47	67	81	88
	Contoured	Fair	25	59	75	83
	Contoured	Good	6	35	70	79
Meadow		Good	30	58	71	78
Woods		Poor	45	66	77	83
		Fair	36	60	73	79
		Good	25	55	70	77
Farmsteads		—	59	74	82	86
Roads						
Dirt[b]		—	72	82	87	89
Hard surface[b]		—	74	84	90	92

[a]Close-drilled or broadcast.

[b]Including right of way.

Source: Soil Conservation Service, "Hydrology," Section 4, *National Engineering Handbook*, March 1985.

TABLE 8.3 Curve Numbers for Different Soil Conditions

	CORRESPONDING *CN*s	
CN FOR AMC II	AMC I	AMC III
100	100	100
95	87	98
90	78	96
85	70	94
80	63	91
75	57	88
70	51	85
65	45	82
60	40	78
55	35	74
50	31	70
45	26	65
40	22	60
35	18	55
30	15	50
25	12	43
20	9	37
15	6	30
10	4	22
5	2	13

AMC denotes antecedent moisture condition.

AMC I. Lowest runoff potential. Soils in the watershed are dry enough for satisfactory plowing or cultivation.

AMC II. The average condition.

AMC III. Highest runoff potential. Soils in the watershed are practically saturated from antecedent rains.

Source: After Soil Conservation Service, "Hydrology," Section 4, *National Engineering Handbook,* March 1985.

The infiltration-rate formulas given above give the soil-controlled potential absorption rate. It should be clear that when the rainfall rate i is less than the infiltration rate, the actual infiltration is the rainfall itself. The soil cannot absorb more than is available.

Most of the infiltration equations seen, like the Philip equation, assume that the soil surface is saturated. This is rarely the case; it will take some time (ponding time) before saturation at the surface occurs. In the interim, the absorbed rainfall (infiltration) reduces the soil infiltration capacity.

Figure 8.14 illustrates the occurrence of runoff using the Philip equation (Eq. 8.31) and assuming that the storm is of uniform intensity i and duration

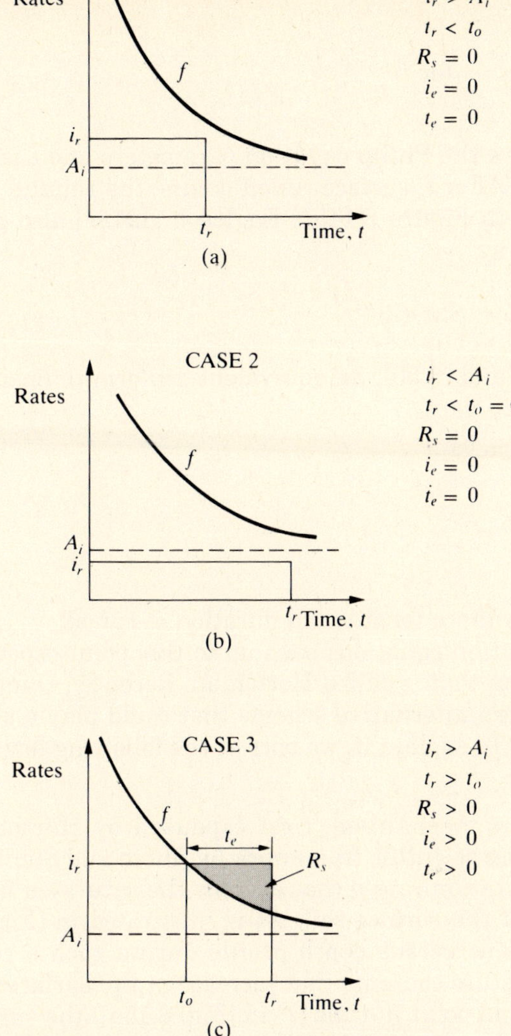

FIGURE 8.14 Storm characteristics versus infiltration capacity.

t_r. There is no surface runoff if (a) the duration of the storm t_r is less than that required to saturate the soil surface (Fig. 8.14a), or (b) the intensity of the storm is less than the minimum infiltration capacity rate A_i (Fig. 8.14b). Figure 8.14(c) shows the occurrence of surface runoff. Initially, the infiltration capacity is greater than the rainfall intensity. During this period there is no surface runoff, all rainfall is infiltrated, and the soil moisture is continuously increasing. Surface runoff will occur only after rainfall i is equal to the infiltration rate f. This occurs at a time t_o. Eagleson [1978c] approximates this "ponding" time as

$$t_o \approx \frac{S_i^2}{2(\bar{i} - A_i)^2} \qquad \text{for } \bar{i} > A_i\,, \tag{8.49}$$

where S_i and A_i are the Philip equation parameters and $\bar{i}$ is a time-averaged rainfall intensity. After t_o surface runoff occurs, the rainfall excess over infiltration is shaded in Figure 8.14(c). Eagleson [1978c] also gives the surface runoff volume as

$$R_s \approx (i - A_i)t_r - S_i(t_r/2)^{1/2}. \tag{8.50}$$

Using Eqs. (8.49) and (8.50), an equivalent uniform-intensity effective rainfall event is defined as

$$t_e = t_r - t_o\,, \tag{8.51}$$

$$i_e = \frac{R_s}{t_e}\,, \tag{8.52}$$

where i_e and t_e are the intensity and duration of runoff.

All the infiltration equations seen up to this point espouse a mechanism of runoff production that is called Hortonian. Recently, Dunne [1978] (among others) suggested an alternative scheme that could play a significant role in runoff production. To explain it, we borrow the following few paragraphs from Freeze [1980]:*

> The classic mechanism, first espoused by Horton [1933] and placed in a more scientific framework by Rubin and Steinhardt [1963], is for a precipitation rate p that exceeds the saturated hydraulic conductivity K^s of the surface soil. As is illustrated in [Figure 8.15a], a moisture content versus depth profile during such a rainfall event will show moisture contents that increase at the surface as a function of time. At some point in time (t^3 in [Fig. 8.15a]) the surface becomes saturated, and an inverted zone of saturation begins to propagate downward into the soil. It is at this time [Fig. 8.15] that the infiltration rate drops below the rainfall rate and overland flow is generated. The time t^3 is called the ponding time. The necessary conditions for the generation of overland flow by the Horton mechanism are (1) a rainfall rate greater than the saturated hydraulic conductivity of the soil, and (2) a rainfall duration longer than the required ponding time for a given initial moisture profile.

*From R. A. Freeze, "A Stochastic Conceptual Analysis of Rainfall–Runoff Process on a Hillslope," *Water Resources Res.* 16(2):394–395, 1980. Copyright by the American Geophysical Union.

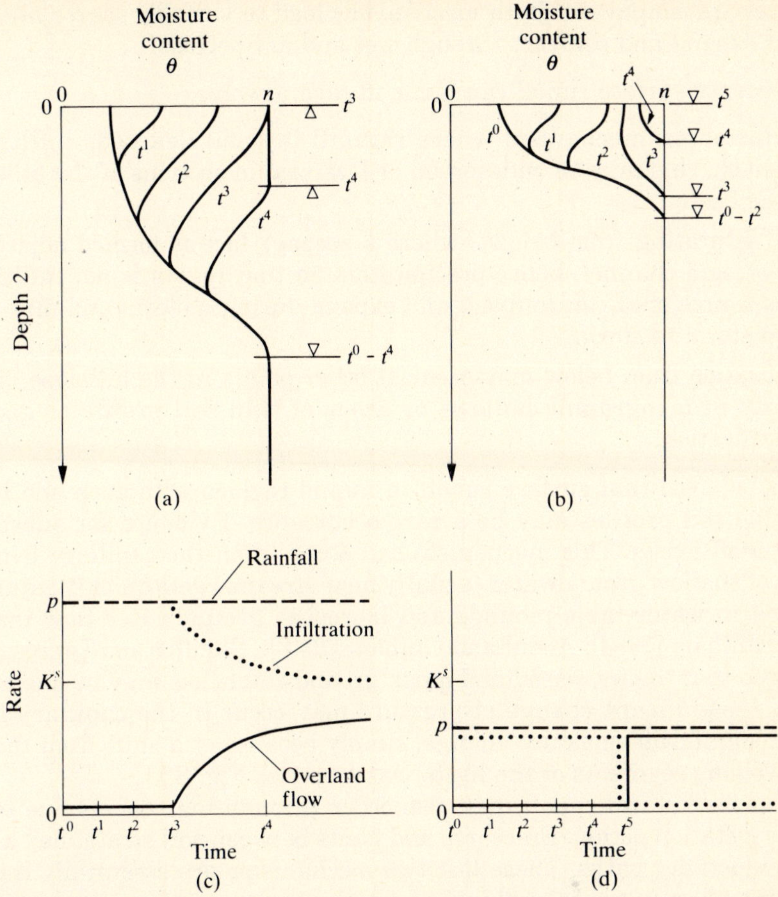

FIGURE 8.15 Moisture content versus depth profiles for (a) the Horton mechanism and (b) the Dunne mechanism. Overland flow generation for (c) the Horton mechanism and (d) the Dunne mechanism. Source: R. A. Freeze, "A Stochastic Conceptual Analysis of Rainfall–Runoff Process on a Hillslope," *Water Resources Res.* 16(2):395, 1980. Copyright by the American Geophysical Union.

The second mechanism as described by Dunne [1978] is illustrated in [Figure 8.15b and d]. In this case $p < K^s$, and the initial water table is shallow. Surface saturation occurs because of a rising water table; ponding and overland flow occur at time t^5 when no further soil moisture storage is available.

The Horton mechanism is more common on upslope areas. The Dunne mechanism is more common on near-channel wetlands. Horton overland flow is generated from partial areas of the hillslope where surface hydraulic conductivities are lowest. Dunne overland flow is generated from partial areas of the hillslope where water tables are shallowest. Both mechanisms lead to variable source areas that expand and contract through wet and dry periods.

In essence surface runoff can occur in several ways:

1. A Hortonian mechanism where rainfall intensity exceeds infiltration capacity. This may be widespread or localized in sections of the hillslopes of the basin.
2. By "saturating from below," where a seepage face is formed adjacent to the stream channel, hence precipitation on that region is not infiltrated. This source area can contract and expand during a storm and may differ from storm to storm.
3. Saturation from below may occur at other points in the hillslope, particularly in topographic hollows or areas of thin soil profile or geologic stratification.

It is believed that surface runoff in humid regions with deep and highly permeable soil profiles may be a rare occurrence. Evidence for subsurface storm runoff exists. This mechanism can occur when the capillary fringe in regions of shallow groundwater (usually near streams) gets quickly saturated, resulting in water-table mounds and increased groundwater flow into the stream (Gillham [1984]; Abdul and Gillham [1948]). Stauffer and Dracos [1986] also argue that the increased and quick groundwater flow may be due to hysteresis. A significant change of pressure may occur in the capillary fringe without significant moisture change, simply because of a shift from the drying to wetting segments of the hysteresis loop (see Fig. 8.1).

Subsurface storm runoff may also occur through macropores due to animal or vegetation action. Fractures and joints between soil strata may also be high conductivity paths. These last two mechanisms are essentially the concept of interflow in traditional hydrologic thinking.

The quantification of when, how, and how much each runoff mechanism will contribute to streamflow remains a subject of active research in hydrology. O'Brien [1983] presents a very complete review of the literature on the subject. Sections of his work are quoted here.*

PARTIAL AREA RUNOFF

Betson [1964] is generally credited with initiating the partial area concept in which a fairly small, yet consistent area of a watershed is assumed to contribute overland flow to the main drainage network. Betson used a non-linear mathematical model which incorporated Horton's [1939] infiltration capacity function to analyze the runoff from a number of basins in Tennessee. The basins were located in areas of steeply sloping terrain, ranged in size from 3.7 acres to

*Reprinted from *Journal of Boston Society of Civil Engineers* Section A.S.C.E., 69(2):303–319,

32.7 square miles, and included open pasture as well as more complexly vegetated areas resulting from diverse agricultural practices. The percent of area contributing runoff for the 14 basins studied was found to range from 5% to 36% with an average, less extremes, of 22%. Further verification of these results for one watershed was provided by small gaged subplots located approximately midway between the stream and the divide. Runoff from these sub-plots was usually less than 0.01 inches and was seldom recorded as occurring from all three plots during a given storm. Significantly, the basin with a contributing area of 86% represented an extreme form of land use. It was completely denuded over two-thirds of the area and was intricately dissected by a deeply incised, thoroughly integrated system of gullies. Betson concluded that in the geographic area of the study, and under normal land use practices, storm runoff frequently occurs from only a small part of the watershed area. Given this conclusion, it is clear why infiltration capacity as measured in the field versus that determined from rainfall-runoff data yielded very different values.

Ragan [1967] provided further insight into the partial area concept through a detailed analysis of a 619 foot length of second order stream segment flowing through a 114 acre forested watershed in Vermont. The watershed was underlain by 80 feet of glacially deposited sands and was monitored by 54 piezometers, 42 observation wells, an interception structure to measure subsurface flow and gages to measure the inflow from 8 seeps located along the stream. Maximum precipitation recorded was 1.32 inches with a maximum observed intensity of 6 inches/hour. For these conditions Ragan concluded that the 'contributing area' for overland flow did not exceed 3% of the total watershed, and that the bulk of the water entered the channel through the seeps as ground-water inflow.

The variable source concept, a variant of the partial area concept, was first presented by the U.S. Forest Service (1961), the Tennessee Valley Authority [1965], Hewlett and Hibbert [1967], and further advanced by Dunne, Moore and Taylor [1975]. This concept as developed by Dunne, et al., [1975] holds that runoff is generated from direct precipitation onto areas that *are saturated by a rising water table.* Runoff produced by this process has two components: (1) precipitation which, unable to penetrate the saturated soils, becomes direct runoff, and (2) subsurface water which, upon rising to the surface, is discharged to run overland to a stream. This latter component, termed return flow (Dunne and Black [1970]), provides a mechanism for the rapid discharge of subsurface water to stream channels and is observed to be sensitive to rainfall intensity.

In a detailed study of a 10 acre portion of an experimental watershed in Danville, Vermont, Dunne and Black [1970] observed the results of numerous natural and artificial (sprinkler produced) rainstorms. The 0.6 acre instrumented portion of the basin was grassed pastureland with slopes that ranged from 30% to 100%. In one storm, 1.83 inches of rain, falling within 34 minutes, followed by one half hour 2.41 inches of artificial rain to produce an event estimated to have a return period of between fifty to several hundred years (Dunne and Black [1970]). Yet, for this event no overland flow as observed on the hillslopes and measurements from the gaged subplots showed that the flood peaks were the result of variable source runoff from saturated areas along the valley bottoms.

Generally, such saturated areas are found in valley bottoms, along streams, and in swales, but various subsurface conditions can also cause saturated zones to occur in topographically high regions of a basin. The area of saturation depends on the season and expands with increases in storm size, hence the origin of the term 'variable source.' Dunne and Black [1970] have noted that basins generating variable source runoff respond rapidly to precipitation events and display the same type of relationship to rainfall and watershed conditions as are recognized for Hortonian overland flow. Consequently a superficial analysis might yield the false impression that Hortonian runoff was occurring in such basins.

SUBSURFACE RUNOFF

The unsaturated zone, lying above the water-table and commonly called the zone of aeration, may also supply considerable amounts of water to the storm hydrograph. Hewlett and Hibbert [1963] were among the first to call attention to the possibility that water draining from the unsaturated zone could, in certain watersheds be the primary source of baseflow. Working at Coweeta (North Carolina) with a 45 foot long concrete trough to produce, in effect, an inclined soil column on a 40 percent slope, the authors found that water was discharged within 1.5 days from the larger soil pores at a high rate, but continued to drain at a lower rate for the next 80 days from the entire soil mass. Moreover, the rate of discharge over time could be described by exponential decay functions which were distinct for the two phases. Subsequent studies (Hewlett and Nutter [1970]) with a 200 foot long soil model, representing a segment of a 38 acre watershed, led to the conclusion that 'subsurface' flow produced the flood peak in the watershed. This conclusion also held for a 20.3 inch rainfall occurring over a 5 day period (a 100-year storm event) which produced no overland flow from the soil model or the basin. Throughout his research, Hewlett has stressed the importance of a belt of saturation, lying along stream channels, and varying in width in response to rainfall, as the critical zone from which subsurface water and groundwater emerge to form a flood peak (cf. Hewlett and Hibbert [1967]).

In forested areas of the Allegheny-Cumberland Plateau region, Whipkey [1969] determined that subsurface discharge (often called interflow) accounted for up to 60% of the stormflow for 130 separate events simulated by a sprinkler system. Interestingly, the subsurface component was the greatest in fine textured soils and appeared to be the result of flow through biological and structural openings in the soil profile.

Corbett, et al., [1975] simulated rainfall on selected portions of a 19.5 acre, highly instrumented watershed, to determine the sources of storm runoff. The researchers noted virtually no surface runoff and concluded that the hydrograph peak was primarily the result of 'subsurface' flow from both the upper and lower slopes with the lower slopes contributing slightly more water. Beasley [1976] used interception trenches to determine that subsurface flow from the upper slopes of a forested watershed can contribute significantly to storm hydrographs where permeable soils overlie impermeable deposits. Beasley noted that flow from the subsurface zone peaked at about the same time as channel flow and theorized that the implied rapid drainage could only occur if water traveled through macrochannels formed by decayed roots. Similar findings have been reported by Mosley [1979].

8.3.3 Actual Evaporation

Between storms, the net evaporation becomes the important process and the opposite analogy to runoff. Figure 8.16 illustrates the situation. A constant climatically controlled mean potential evaporation is assumed during the interstorm period of duration t_b. This evaporation rate is $\bar{e}_p$. The soil capacity to exfiltrate is represented, for example, by Eq. (8.36). As long as the soil is ca-

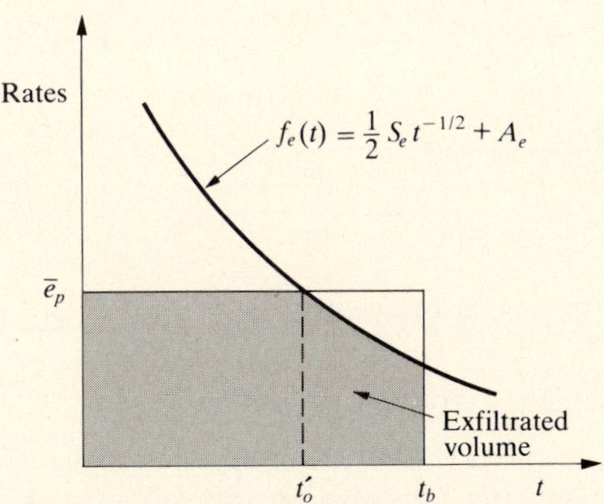

FIGURE 8.16 Representation of the exfiltration process. Source: P. S. Eagleson, "Climate, Soil and Vegetation 4: The Expected Value of Annual Evapotranspiration," *Water Resources Res.* 14(5):733, 1978. Copyright by the American Geophysical Union.

pable of delivering water in excess of $\bar{e}_p$, then the climatic potential evaporation is the actual evaporation. At some time t_o', though, the soil moisture has been depleted to the extent that the soil cannot deliver the climatic demand. The actual evaporation is then soil controlled by Eq. (8.36) and shown shaded in the figure. The time t_o' is also given by Eagleson [1978a, b, and c] as

$$t_o' = \frac{S_e^2}{2\bar{e}_p^2(1 + Mk_v - W/\bar{e}_p)}\left[1 - M + \frac{M^2 k_v + (1 - M)W/\bar{e}_p}{2(1 + Mk_v - W/\bar{e}_p)}\right]. \tag{8.53}$$

The interstorm processes will continue until the next storm arrives.

EXAMPLE 8.1

Variable Diffusivity Infiltration

Assume we have a sandy loam with $K(1) = 3.43 \times 10^{-3}$ cm s^{-1}, $\psi(1) = 39.6$ cm, $n = 0.25$, $m = 1$, and $d = 3$. What would the history of infiltration be if we assume (a) constant diffusivity evaluated at the linear average of the boundary condition and initial condition, or (b) that diffusivity is a function of soil moisture? The conditions to be used are

$$\theta_0 = \theta_s; \qquad s_0 = 1,$$

$$\theta_i = 0.50\theta_s; \qquad s_i = 0.5.$$

The constant diffusion solution is Eq. (8.20). The variable diffusion case is Eq. (8.25), where the effective diffusivity $\overline{D}$ is used.

With $\theta_0 = \theta_s = n = 0.25$, then $\theta_i = 0.125$. A simple average of moisture is then $\bar{\theta} = 0.19$ or $\bar{s} = 0.75$. Using Eq. (8.11), the diffusivity evaluated at the average is

$$\begin{aligned} D(\bar{\theta}) &= \frac{|\psi(1)|K(1)}{nm}\bar{s}^d \\ &= \frac{39.6(3.43 \times 10^{-3})}{0.25 \times 1}(0.75)^{3.0} \\ &= 0.23 \text{ cm}^2\text{s}^{-1}. \end{aligned}$$

Assuming $K_0 = K(1) = 3.43 \times 10^{-3}$ cm s^{-1}, Eq. (8.20) leads to

$$f_i = -0.03t^{-1/2} - 3.43 \times 10^{-3} \text{ cm s}^{-1}.$$

Table 8.4 gives the resulting f_i values for different times.

For the variable diffusion case, we must evaluate $\overline{D}$ in Eq. (8.25). For $s_i = 0.5$, $d = 3$, and $m = 1$, Figure 8.8 results in $\phi_i(d, s_i) = 0.35$, which, using Eq. (8.29a), leads to $\overline{D} = 0.32$.

TABLE 8.4 Infiltration Capacity Resulting from Variable- and Constant-Diffusivity Assumptions and the Philip's Equation

	f_i (cm s^{-1})		
t (sec)	Constant D	Variable D	Philip's
900	4.4×10^{-3}	4.8×10^{-3}	4.4×10^{-3}
1800	4.1×10^{-3}	4.4×10^{-3}	3.7×10^{-3}
2700	4.0×10^{-3}	4.2×10^{-3}	3.3×10^{-3}
3600	3.9×10^{-3}	4.1×10^{-3}	3.1×10^{-3}
4500	3.9×10^{-3}	4.0×10^{-3}	3.0×10^{-3}
5400	3.8×10^{-3}	3.9×10^{-3}	2.8×10^{-3}
6300	3.8×10^{-3}	3.9×10^{-3}	2.8×10^{-3}
7200	3.8×10^{-3}	3.9×10^{-3}	2.7×10^{-3}

The infiltration is then

$$f_i = -0.04t^{-1/2} - 3.43 \times 10^{-3}\ \text{cm s}^{-1}$$

and the results are also in Table 8.4. ◆

EXAMPLE 8.2

Philip's Equation

For the same conditions of the previous example, we can obtain the rate of infiltration implied by Eagleson's [1978a] version of the Philip's equation (Eq. 8.31).

Using Eq. (8.34) and ignoring capillary rise,

$$\begin{aligned} A_i &= \frac{1}{2} K(1)\,(1 - s_i^c) \\ &= \frac{1}{2}\,(3.43 \times 10^{-3})\,(1 + 0.5^5) = 1.77 \times 10^{-3}, \end{aligned}$$

where Eq. (8.7) was used to compute $c = 5$. We will evaluate S_i using Eq. (8.35) and Figure 8.8 [$\phi(d, s_i) \approx 0.35$].

$$\begin{aligned} S_i &= 2(1 - s_i)\left[\frac{5nK(1)\psi(1)\phi_i(d, s_i)}{3m\pi}\right]^{1/2} \\ &\approx 0.08\,. \end{aligned}$$

Therefore

$$f \approx 0.08t^{-1/2} + 1.77 \times 10^{-3}.$$

The tabulated values are also given in Table 8.4. In Table 8.4, note that the variable-diffusion case yields slightly larger infiltration, consistent with the larger diffusion coefficient. Philip's equation results in smaller infiltration rates because of its smaller gravity term. ◆

EXAMPLE 8.3

Hortonian Runoff

It is useful to study the Hortonian runoff potential of typical soils, say those given in Table 8.1. Initial soil moisture at the site is $s_i = 0.3$. Assume it rains for $t_r = 5$ hr at 1 cm hr^{-1} (2.78×10^{-4} cm s^{-1}) and that the surface saturates quickly, $s_0 = 1$. To compute runoff, we first have to compute the ponding time. This is given in Table 8.5 for the various soils. (Equation 8.29b was used to obtain diffusivities.) As the table indicates, $t_0 > t_r$ for silty loam, sandy loam, and sand. No surface runoff will occur on those soils for the given storm. Runoff will occur on clay. Using Eq. (8.50) the runoff volume in centimeters is computed and tabulated in Table 8.5. Results imply that 52% of rainfall infiltrates in clay and everything infiltrates in all other soils. High surface runoff is not easy to achieve! ◆

EXAMPLE 8.4

Soil Conservation Service Formula

Let us look at the Soil Conservation Service (see Section 8.3.1) runoff expression. The volume of infiltration must be

$$F = P - I_a - R_s.$$

If $I_a = 0.2S$, the above becomes

$$F = P - 0.2S - \frac{(P - 0.2S)^2}{P + 0.8S}.$$

TABLE 8.5 Ponding Time and Runoff Resulting from Different Soils and a Storm of 1 cm hr^{-1} for Five Hours

	CLAY	SILTY LOAM	SANDY LOAM	SAND
$\phi_i(d, s_i)$[a]	0.197	0.248	0.273	0.279
S_i (Eq. 8.35)	0.025	0.034	0.043	0.037
A_i[b] (Eq. 8.34)	1.70×10^{-5}	1.70×10^{-4}	1.72×10^{-3}	4.37×10^{-3}
t_o (hours) (Eq. 8.49)	1.3	13.8	∞	∞
Runoff (cm) (Eq. 8.50)	2.3	0	0	0

[a]Approximated using Eq. (8.29b).

[b]Capillary rise is assumed zero.

The maximum infiltration possible is found by taking the limit of F as $P \to \infty$. To do that we use L'Hospital's rule and find

$$\lim_{P\to\infty} F = \lim_{P\to\infty} \frac{(P + 0.8S)P - 0.2S(P + 0.8S) - (P - 0.2S)^2}{P + 0.8S}$$

$$= \lim_{P\to\infty} \frac{\partial(\text{numerator})/\partial P}{\partial(\text{denominator})/\partial P}$$

$$= \frac{2P + 0.8S - 0.2S - 2P + 0.4S}{1} = S.$$

If we now look at the Horton equation, the infiltrated volume is

$$F = \int_0^t f(t)\,dt = f_c t + \frac{f_0 - f_c}{\alpha}[1 - e^{-\alpha t}].$$

For very large t (somewhat analogous to large P) the above becomes

$$F \approx f_c t + \frac{f_0 - f_c}{\alpha},$$

which can be equated to the Soil Conservation Service result only if $f_c = 0$. Doing that,

$$F = S = \frac{f_0}{\alpha} = \frac{1000}{CN} - 10.$$

Given any other storm of normal duration t where the infiltrated volume $F' \approx P - R_s$ is known, we can write

$$F' = \frac{f_0}{\alpha}[1 - e^{-\alpha t}] = S[1 - e^{-\alpha t}]$$

or

$$\alpha = -\ln\left(1 - \frac{F'}{S}\right)\Big/ t.$$

The only reason to make this analogy of the Horton formulation to the Soil Conservation Service formula is to obtain a reasonable distribution of infiltration in time, if that is necessary. The ponding time estimate is I_a/i, where

i is the storm intensity. To illustrate, take a common curve number $CN = 75$. Assume that in a three-hour storm of 2-in. depth, 1.2 in. infiltrated. Then,

$$S = \frac{1000}{75} - 10 = 3.33$$

$$\alpha = -\ln\left(1 - \frac{1.2}{3.33}\right)\Big/3 = 0.15 \text{ hr}^{-1}$$

$$f_0 = \alpha S = 0.15[3.33] = 0.5 \text{ in. hr}^{-1}. \blacklozenge$$

8.4 PERCOLATION AND CAPILLARY RISE

Percolation is defined as the transfer of water through the intermediate zone of Figure 7.1. Percolation ultimately leads to aquifer recharge. Percolation would obey the unsaturated flow equations as infiltration but multidimensional effects and gravitational forces are more inportant. The boundary conditions of the solution could also be difficult. At the upper end there should be fairly fast variability of moisture due to infiltration. At the lower end the position of the groundwater affects capillary fringe and moisture conditions. If the groundwater is fairly deep, then moisture in the intermediate zone should be reasonably constant in time and space. As previously stated, it is probably close to field capacity. Accepting this steady-state and uniform moisture distribution implies that the one-dimensional analogy to the percolation flux is

$$P = K(\theta), \tag{8.54}$$

where $K(\theta)$ is the vertical hydraulic conductivity. Given the form of $K(\theta)$, Eq. (8.6), a common empirical solution is to suggest

$$P = a\theta^b, \tag{8.55}$$

where a and b are calibration parameters.

Capillary rise was studied by Eagleson [1978b] using some results by Gardner [1958]. One-dimensional steady flow is assumed, leading to

$$\frac{d}{dz}\left[K(\theta)\,\frac{d\psi(\theta)}{dz} + K(\theta)\right] = 0. \tag{8.56}$$

The term in brackets is an apparent constant fluid velocity. So Eq. (8.56) implies, after one integration,

$$K(\theta)\left[\frac{d\psi(\theta)}{dz} + 1\right] = -W. \tag{8.57}$$

Using Eqs. (8.4) and (8.6), $\psi(\theta)$ and $K(\theta)$ are related by

$$K(\theta) = a\psi(\theta)^{-b}, \tag{8.58}$$

where

$$a = K(1)\,[\psi(1)] \tag{8.59}$$

$$b = mc\,. \tag{8.60}$$

With the above relationship, Eq. (8.57) can be integrated to obtain W as a function of the matrix potential at the soil surface. The lower integrating limit is $\psi(1)$ at the saturated groundwater depth and the upper limit is some $\psi(s)$ at the surface $z = 0$. If the soil is assumed dry at the surface, and $\psi(0)$ is very large relative to $\psi(1)$, the results asymptotically approach

$$\frac{W}{K(1)} = B\left[\frac{\psi(1)}{z}\right]^{mc}, \tag{8.61}$$

where B is fitted by Eagleson [1978b] to experimental values given by Gardner [1958] as

$$B = 1 + \left[\frac{3}{2}\Big/(mc - 1)\right]. \tag{8.62}$$

The above equation loses accuracy for shallow groundwater. Table 8.6 shows the exfiltration velocity W in centimeters per second for the soil groups of Table 8.1 and various groundwater depths.

TABLE 8.6 Exfiltration Velocities, in Centimeters per Second, for Various Soil Groups and Groundwater Depths

z (cm)	CLAY	SILTY LOAM	SANDY LOAM	SAND
50	N/A[a]	N/A[a]	1.0×10^{-6}	~0
100	N/A[a]	5.0×10^{-6}	2.7×10^{-10}	~0
150	1.0×10^{-5}	5.1×10^{-7}	~0	~0
200	4.0×10^{-6}	1.0×10^{-7}	~0	~0
250	1.9×10^{-6}	2.8×10^{-8}	~0	~0
300	1.1×10^{-6}	1.0×10^{-8}	~0	~0
350	6.4×10^{-7}	4.3×10^{-9}	~0	~0

[a]N/A denotes not applicable because $\psi(1) \geq z$.

8.5 SUMMARY

Flow in unsaturated porous media is a highly nonlinear problem due to the dependence of hydraulic conductivity and matrix potential on the soil moisture distribution. The quantification of that relationship, particularly its hysteretic behavior, has been a focus of attention of many researchers. The joint existence of gaseous (vapor) and liquid phases of water (or other compounds) further complicates the problem. Here we have not dealt with those issues. For the sake of simplicity, this chapter has emphasized one-dimensional solutions to the unsaturated flow equations, ignoring multiphase conditions and temperature effects. The one-dimensional solution is best suited to deal with the infiltration problem, i.e., determining the flux of water at the surface of the soil. We have seen several possible theoretical approaches as well as empirical formulations that mimic expected behavior.

The heterogeneity of natural soils, particularly the common stratification, may lead to significant three-dimensional movement of moisture as it percolates down through the soil. In problems like the migration of contaminants leaching through a heterogeneous unsaturated medium, this multidimensional effect may be very important. Due to our inability to quantify dependence of the soil properties on moisture in three-dimensional detail, the problem is best handled using stochastic conceptualizations of the porous medium. The advanced reader is referred to Yeh et al. [1985a, b, and c] and Mantoglou and Gelhar [1987a, b, and c] for details on the approaches.

The analytical, one-dimensional, infiltration equations presented in this chapter are truly valid at a point. Extrapolation to the behavior over a large area (i.e., river basin) is a conceptualization. Infiltration integrated over an area with varying soil properties is a problem still being actively researched (Dagan and Bresler [1983] and Bresler and Dagan [1983a and b]). For this reason the use of empirical infiltration equations, calibrated to local conditions, remains the safest operational tool. Unfortunately, data do not always permit adequate calibration.

Infiltration is the link between surface and subsurface processes. Hence, it is the switch that controls storm runoff. This chapter presented the various runoff production mechanisms that may be active during a storm. The traditional Hortonian mechanism leads to surface runoff when rainfall intensity exceeds the infiltration rate. It has been found, though, that surface runoff sometimes occurs only when the soil column is saturated, i.e., saturated from below, and hence is usually observed near streams, hollows, or areas with high water tables. This behavior is not contradictory to Horton's conceptualization. In Hortonian runoff, there is also a zone of saturation, beginning at ponding. In essence a perched saturated zone is formed. Saturation from below is an extreme, but compatible, scenario.

Subsurface storm runoff, it is now argued, accounts for most river flow in humid regions of the world with well-developed, very pervious, upper soil zones. In such regions, surface runoff may indeed be limited only to extremely high-intensity storms. Subsurface storm runoff commonly occurs through macropores (Beven and Germann [1982]). These are small fractures or channels resulting from boring animals or past root activity. The boundary between two soil types, particularly between a highly pervious and an impervious soil, is also a preferred flow path. Quantification and prediction of subsurface storm runoff still escapes theoretical treatment. The classical porous media view presented in this chapter would not be applicable for flow through macropores and fractures.

REFERENCES

Abdul, A. S., and R. W. Gillham [1984]. "Laboratory Studies on the Effects of the Capillary Fringe on Streamflow Generation." *Water Resources Res.* 20(6):691–698.

Beasley, R. S. [1976]. "Contribution of Subsurface Flow from the Upper Slopes of Forested Watersheds to Channel Flow." *Soil Sci. Soc. Am. J.* 40:955–957.

Betson, R. S. [1964]. "What is Watershed Runoff?" *J. Geophys. Res.* 69(8):1541–1552.

Beven, K., and P. Germann [1982]. "Macropores and Water Flow in Soils." *Water Resources Res.* 18(5):1311–1325.

Bodman, G. B., and E. A. Colman [1943]. "Moisture and Energy Conditions During Downward Entry of Water into Soils." *Soil Sci. Soc. Am. J.* 7:116–122.

Bresler, E., and G. Dagan [1983a]. "Unsaturated Flow in Spatially Variable Fields 2. Application of Water Flow Models to Various Fields." *Water Resources Res.* 19(2):421–428.

Idem. [1983b]. "Unsaturated Flow in Spatially Variable Fields 3. Solute Transport Models and Their Application to Two Fields." *Water Resources Res.* 19(2):429–435.

Brooks, R. H., and A. T. Corey [1966]. "Properties of Porous Media Affecting Fluid Flow." *J. Irrig. Drainage Div. A.S.C.E.* IR2:61–88.

Burdine, N. T. [1958]. "Relative Permeability Calculations from Pore Size Distribution Data." *Trans. A.I.M.E.* 198:71–78.

Carslaw, H. S., and C. Jaeger [1959]. *Conduction of Heat in Solids.* 2nd ed. Fair Lawn, N.J.: Oxford University Press.

Corbett, E. S., W. E. Sopper, and J. A. Lynch [1975]. "Watershed Response to Partial Area Applications of Simulated Rainfall." Tokyo, Japan: l'Association Internationale des Sciences Hydrologiques, pp. 63–73. (Publication no. 117.)

Crank, J. [1956]. *The Mathematics of Diffusion.* New York: Oxford University Press.

Dagan, G., and E. Bresler [1983]. "Unsaturated Flow in Spatially Variable Fields 1. Derivation of Models of Infiltration and Redistribution," *Water Resources Res.* 19(2):413–420.

Diaz-Granados, M., J. B. Valdes, and R. L. Bras [1983]. "A Derived Flood Frequency Distribution Based on the Geomorphoclimatic IUH and the Density Functions of Rainfall Excess." Cambridge, Mass.: MIT Department of Civil Engineering, Ralph M. Parsons Laboratory. (Technical report no. 292.)

Dunne, T. [1978]. "Field Studies of Hillslope Flow Processes." In: M. J. Kirby, ed. *Hillslope Hydrology.* New York: Wiley-Interscience, pp. 227–293.

Dunne, T., and R.D. Black [1970]. "An Experimental Investigation of Runoff Production in Permeable Soils." *Water Resources Res.* 6(2):478–490.

Dunne, T., T.R. Moore, and C.H. Taylor [1975]. "Recognition and Prediction of Runoff Producing Zones in Humid Regions." *Hydrol. Sci. Bull.* 20(3):305–327.

Eagleson, P.S. [1970]. *Dynamic Hydrology.* New York: McGraw-Hill.

Idem. [1978a]. "Climate, Soil and Vegetation 3: A Simplified Model of Soil Moisture Movement in the Liquid Phase." *Water Resources Res.* 14(5):722–729.

Idem. [1978b]. "Climate, Soil and Vegetation 4: The Expected Value of Annual Evapotranspiration." *Water Resources Res.* 14(5):731–739.

Idem. [1978c]. "Climate, Soil, and Vegetation 5: A Derived Distribution of Storm Surface Runoff." *Water Resources Res.* 14(5):741–748.

Entekhabi, D. [1988]. Personal communication.

Freeze, R.A. [1980]. "A Stochastic Conceptual Analysis of Rainfall–Runoff Process on a Hillslope." *Water Resources Res.* 16(2):391–408.

Freeze, R.A., and J.A. Cherry [1979]. *Groundwater.* Englewood Cliffs, N.J.: Prentice-Hall.

Gardner, W.R. [1958]. "Some Steady-State Solutions of the Unsaturated Moisture Flow Equation with Application to Evaporation from a Water Table." *Soil Sci.* 85(4):228–232.

Idem. [1959]. "Solutions of the Flow Equation for the Drying of Soils and Other Porous Media." *Soil Sci. Soc. Am. J.* 23(3):183–195.

Gardner, W.R., and M.S. Mayhugh [1958]. "Solutions and Tests of the Diffusion Equation for the Movement of Water in Soil." *Soil Sci. Soc. Am. J.* 22:197–201.

Gillham, R.W. [1984]. "The Capillary Fringe and Its Effect on Water Table Response." *J. Hydrol.* 67:307–324.

Green, W.H., and G. Ampt [1911]. "Studies of Soil Physics. Part I — The Flow of Air and Water Through Soils." *J. Agric. Sci.* 4:1–24.

Haan, C.T., H.P. Johnson, and D.L. Brakensick [1982]. *Hydrologic Modeling of Small Watersheds.* St. Joseph, Mich.: American Society of Agricultural Engineers.

Hewlett, J.D., and A.R. Hibbert [1963]. "Moisture and Energy Conditions within a Sloping Soil Mass During Drainage." *J. Geophys. Res.* 68(4):1081–1087.

Idem. [1967]. "Factors Affecting the Response of Small Watersheds to Precipitation in Humid Areas." In: W.E. Sopper and H.W. Lull, eds. *Forest Hydrology.* Elmsford, N.Y.: Pergamon, pp. 275–290.

Hewlett, J.D., and W.L. Nutter [1970]. "The Varying Source Area of Streamflow from Upland Basins." Paper presented at the Symposium on Interdisciplinary Aspects of Watershed Management. Montana State University, Bozeman, pp. 65–83.

Horton, R.E. [1933]. "The Role of Infiltration in the Hydrologic Cycle." *Trans. Am. Geophys. Union.* 14:446–460.

Idem. [1939]. "Analysis of Runoff Plot Experiments with Varying Infiltration Capacity." *Trans. Am. Geophys. Union.* Part IV:693–711.

Idem. [1940]. "An Approach Toward a Physical Interpretation of Infiltration Capacity." *Soil Sci. Soc. Am. J.* 5:399–417.

Huggins, L.F., and E.J. Monke [1966]. "The Mathematical Simulation of Hydrology in Small Watersheds." Lafayette, Ind.: Purdue University Water Resources Center. (Technical report no. 1.)

Linsley, R.K., M.A. Kohler, and J.L.H. Paulhus [1949]. *Applied Hydrology.* New York: McGraw-Hill.

Mantoglou, A., and L. W. Gelhar [1987a]. "Capillary Tension Head Variance, Mean Soil Moisture Content, and Effective Specific Soil Moisture Capacity of Transient Unsaturated Flow in Stratified Soils." *Water Resources Res.* 23(1)P:47–56.

Idem. [1987b]. "Effective Hydraulic Conductivities of Transient Unsaturated Flow in Stratified Soils." *Water Resources Res.* 23(1):57–67.

Idem. [1987c]. "Stochastic Modeling of Large-Scale Transient Unsaturated Flow Systems." *Water Resources Res.* 23(1):37–46.

Marsh, W. M., and J. Dozier [1986]. *Landscape: An Introduction to Physical Geography.* New York: Wiley.

McCuen, R. H. [1982]. *A Guide to Hydrologic Analysis Using SCS Methods.* Englewood Cliffs, N.J.: Prentice-Hall.

Milly, P. C. D. [1982]. "Moisture and Heat Transport in Hysteric, Inhomogeneous Porous Media: A Matrix-Head-Based Formulation and a Numerical Model." *Water Resources Res.* 18(3):489–498.

Mosley, M. P. [1979]. "Streamflow Generation in a Forested Watershed, New Zealand." *Water Resources Res.* 15(4):795–806.

Mualem, Y. [1977]. "Extension of the Similarity Hypothesis Used for Modelling the Soil Water Characteristics." *Water Resources Res.* 13(4):773–780.

O'Brien, A. L. [1983]. "Alternative Approaches to Understanding Runoff in Small Watersheds." *J. B.S.C.E.* 69(2):303–319.

Parlange, J. Y., M. Vauclin, R. Haverkamp, and I. Lisle [1985]. "Note: The Relation between Desorptivity and Soil-Water Diffusivity." *Soil Sci.* 139(5):458–461.

Philip, J. R. [1960]. "General Method of Exact Solution of the Concentration Dependent Diffusion Equation." *Aust. J. Phys.* 13(1):1–12.

Idem. [1969]. "The Theory of Infiltration." In: V. T. Chow, ed. *Advances in Hydroscience.* Vol. 5. New York: Academic Press, pp. 215–296.

Ragan, R. M. [1967]. "An Experimental Investigation of Partial Area Contributions." In: *Hydrological Aspects of the Utilization of Water,* Volume II of the *Proceedings of the General Assembly of Bern,* pp. 241–251. (IAHS publication no. 76.)

Rogers, J. S., and A. Klute [1971]. "The Hydraulic Conductivity-Water Content Relationship During Non-Steady Flow Through a Sand Column." *Soil Sci. Soc. Am. J.* 35(5):695–700.

Rubin, J., and R. Steinhardt [1963]. "Soil Water Relations During Rain Infiltration I: Theory." *Soil Sci. Soc. Am. J.* 27:246–251.

Sittner, W. T., C. E. Schauss, and J. C. Monroe [1969]. "Continuous Hydrograph Synthesis with an API-Type Hydrologic Model." *Water Resources Res.* 5(5):1007–1022.

Soil Conservation Service [1968]. "Hydrology." Supplement A to Section 4, *National Engineering Handbook.* Washington, D.C.: U.S. Department of Agriculture. (Also March 1985.)

Stauffer, F., and T. Dracos [1986]. "Local Infiltration into Layered Soil and Response of the Water Table Experiment and Simulation." *Frontiers in Hydrology.* Littleton, Colo.: Water Resources Publications.

Tennessee Valley Authority [1965]. "Area-Stream Factor Correlation, a Pilot Study in the Elk River Basin." *Bull. Int. Assoc. Sci. Hydrol.* 10(2):22–37.

Todd, D. K. [1980]. *Groundwater Hydrology.* 2nd ed. New York: Wiley.

U.S. Forest Service [1961]. "Some Ideas about Storm Runoff and Baseflow." Annual Report of the Southeastern Forest Experiment Station. Washington, D.C.: U.S. Forest Service, pp. 61–66.

van Genuchten, M. T. [1980]. "A Closed-form Equation for Predicting the Hydraulic Conductivity of Unsaturated Soils." *Soil Sci. Soc. Am. J.* 44:892–898.

Viessman, W. W., J. W. Knapp, G. L. Lewis, and T. E. Harbaugh [1977]. *Introduction to Hydrology.* 2nd ed. New York: Harper & Row.

Watson, K. K. [1959]. "A Note on the Field Use of a Theoretically Derived Infiltration Equation." *J. Geophys. Res.* 64(10):1614.

Whipkey, R. Z. [1969]. "Storm Runoff from Forested Catchments by Subsurface Routes." *Int. Assoc. Sci. Hydrol.* 85:773–779.

World Meteorological Organization [1974]. *Guide to Hydrological Practices.* 3rd ed. Geneva: World Meteorological Organization. (WMO report no. 168.)

Wösten, J. H. M., and M. T. van Genuchten [1988]. "Using Texture and Other Soil Properties to Predict the Unsaturated Soil Hydraulic Functions." *Soil Sci. Soc. Am. J.* 52:1762–1770.

Yeh, T.-C. J., L. W. Gelhar, and A. L. Gutjahr [1985a]. "Stochastic Analysis of Unsaturated Flow in Heterogeneous Soils 1: Statistically Isotropic Media." *Water Resources Res.* 21(4):447–456.

Idem. [1985b]. "Stochastic Analysis of Unsaturated Flow in Heterogeneous Soils 2: Statistically Anisotropic Media with Variable α." *Water Resources Res.* 21(4): 457–464.

Idem. [1985c]. "Stochastic Analysis of Unsaturated Flow in Heterogeneous Soils 3: Observations and Applications." *Water Resources Res.* 21(4):447–456.

PROBLEMS

1. A soil has the following properties: $m = 1.36$, $c = 4.47$, $K(1) = 3.3 \times 10^{-3}$ cm s^{-1}, $\psi(1) = 24$ cm, and $n = 0.2$. The soil is initially completely saturated (after a long storm, maybe) and is subjected to evaporation, which, it is assumed, maintains the surface at a constant degree of saturation $s = \theta_0/n = 0.2$. Estimate the time taken for soil at a depth of 2 m to dry to a moisture content $s = \theta/n = 0.6$. Ignore gravity and state other assumptions.

2. The infiltration rate for excess rain on a small area was observed to be 4.5 in. hr^{-1} at the beginning of a rainfall, and it decreased exponentially to an equilibrium of 0.5 in. hr^{-1} after 10 hr. A total of 30 in. of water infiltrated during the 10-hr interval. Determine the value of α in Horton's equation $f = f_c + (f_0 - f_c)e^{-\alpha t}$.

3. The following figure illustrates infiltration during a storm of constant intensity i and duration t_r. If the infiltration $f_i(t)$ is given by the Philip's equation,

$$f_i(t) = \frac{1}{2} S t^{-1/2} + A,$$

find expressions for the times t' and t_o.

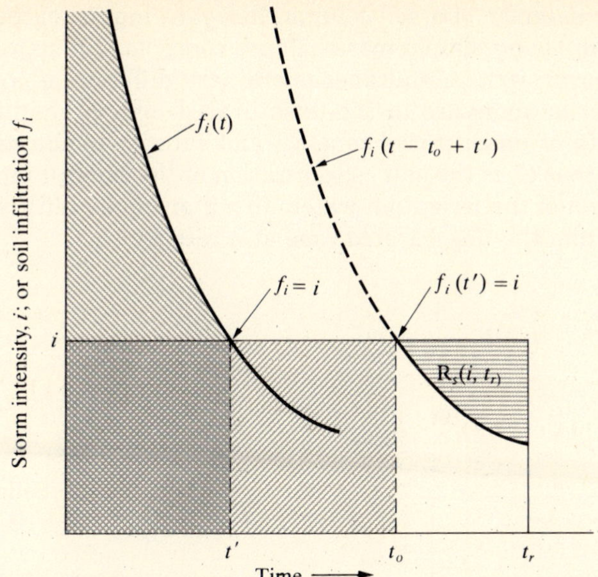

The coefficients S (sortivity) and A (gravitational infiltration rate) can be approximated by

$$S = 2\left(1 - \frac{\theta_i}{n}\right)\left|\frac{5nK(1)\psi(1)\phi(d, \theta_i)}{3m\pi}\right|^{1/2}$$

$$A = \frac{1}{2}K(1)\left[1 + \left(\frac{\theta_i}{n}\right)^c\right] - W,$$

where n = porosity, $K(1)$ = saturated hydraulic conductivity, $\psi(1)$ = saturated matrix potential, $\phi(d, \theta_i)$ = infiltration diffusivity function, W = capillary rise from water table, m = pore-size distribution index of soil, d = diffusivity index of soil, c = pore-connectivity index of soil, and θ_i = initial soil moisture content (at beginning of event).

A given soil has parameter values $n = 0.35$, $K(1) = 1.25\ \text{mm hr}^{-1}$, $\psi(1) = 190$ mm, $d = 5.5$, $m = 0.286$, $c = 10$, $W = 0$ (for deep water table), and $\theta_i = 0.15$. What is the rainfall rate below which no runoff will ever occur from the given soil, irrelevant of storm duration?

4. During a four-hour storm with 2.5 in. of rainfall, 1.5 in. infiltrated. What are the implied α and f_0 coefficients or the Horton infiltration equation if the curve number is 75? Plot the implied infiltration rate.

5. Below is a schematic diagram of a soil column. Relevant inputs, outputs, and variables are defined. Using simple mass balance concepts and approximations to the fluxes shown, write simultaneous ordinary differential equations (nonlinear) to describe moisture in the unsaturated zone θ, height of saturated zone h, salinity of unsaturated zone S_1, and salinity of the saturated region S_2. Assume that C_1 is the salt concentration of the rainfall input, C_2 is the salt concentration of the irrigation water, all water inputs infiltrate, and roots absorb water from the unsaturated zone at a rate $R(\theta, S_1)$.

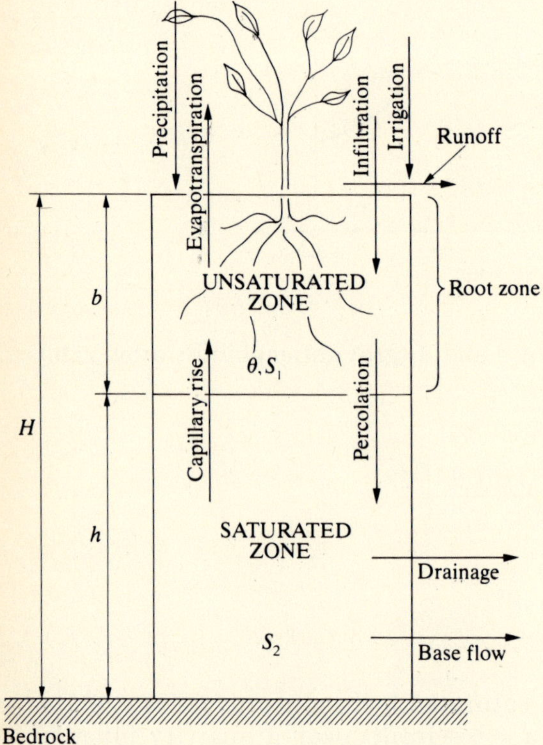

To the extent possible expand all terms in the equation—i.e., express them as functions of the variables θ, S_1, S_2, and h.

6. a) Assume a soil has a constant diffusivity $D = 1 \text{ m}^2 \text{day}^{-1}$ and porosity $n = 0.4$. If the initial moisture content is $\theta_i = 0.1$ and a 12-hr storm maintains a saturated condition at the surface, sketch the soil moisture profile at the end of the storm, indicating at what depths degrees of saturation of 80%, 50%, and 30% will be found.

b) Sketch qualitatively how you might expect the above profile to change for D decreasing with depth.

c) Sketch qualitatively how you might expect the profile to change for D dependent on θ, given, for example, by Eq. (8.11), where D increases with θ.

7. Two different soils are characterized by the following parameters:

	m	c	$K(1)$, cm s^{-1}	$\psi(1)$, cm	θ_r	θ_s
Fine sand	1.36	4.47	3.3×10^{-3}	24	0.15	0.48
Light clay	0.22	12.10	1.0×10^{-5}	26	0.05	0.38

The initial soil moisture is $\theta_i = 0.20$. For the two soils

a) Compute the infiltration as a function of time for constant $\theta = \theta_s$ at the surface.

b) Compute the exfiltration as a function of time for constant $\theta = \theta_r$ (residual soil moisture) at the surface.

c) Plot the results of parts (a) and (b).

(Contributed by Dr. Angelos Protopapas.)

8. The pressure potential ψ profile over depth z is described by the expression

$$\alpha z = -\ln[e^{-\alpha\psi} + b] - \alpha\psi + c\,,$$

where α, b, and c are constant parameters in proper units, and the vertical coordinate z is positive upward. The soil hydraulic conductivity is given by an exponential form

$$K(\psi) = K_s e^{\alpha\psi} \qquad (\alpha > 0)\,.$$

a) What is the constant c if the water table is located at $z = 0$?

b) Find the profile of the moisture flux q as a function of depth.

(Contributed by Dr. Angelos Protopapas.)

9. For a Yolo clay soil the following parameters are given: $K(1) = 1.73$ cm day^{-1}, $\psi(1) = 19$ cm, $m = 0.286$, $c = 10.0$, and $n = 0.35$.

a) Find the capillary rise flux if the water table is located at a 200-cm depth.

b) Find the effective diffusivity for exfiltration from this soil if the initial water content is $s_i = 0.70$.

(Contributed by Dr. Angelos Protopapas.)

10. Derive Eq. (8.49), ponding time, and Eq. (8.50), Hortonian runoff, when using the Philip equation.

11. An industrial park is to be constructed in 50 acres of well-developed woods in a soil of low runoff potential. After development, 30 acres will be covered with buildings, parking places, and roads, all impervious. Ten acres will be terraced landscaping and the remaining will be woods. The 100-year recurrence design storm is 6 in. in 12 hr. Estimate the runoff volume before and after construction.

12. Assume that storms are of constant intensity i_r and duration t_r. Infiltration is given by Philip's equation, evaluated at the initial moisture before the storm s_i. Percolation is given by $K(1)s_i^c$. If no evaporation occurs, show that the moisture at the end of the storm can be approximated by

$$\begin{aligned} s_f &= s_i + [i_r t_r - K(1)s_i^c t_r]/nz, && 0 \le t_r \le t_0 \\ &= s_i + [i_r t_0 + S(t_r^{1/2} - t_0^{1/2}) + A(t_r - t_0) - K(1)s_i^c t_r]/nz, && t_r > t_0, \end{aligned}$$

where z is a characteristic depth over which the mass balance is performed, A and S are the Philip's equation parameters, and t_0 is ponding time. (*Hint:* Write an ordinary differential equation for moisture over depth z.)

13. If exfiltration is given by Eq. (8.36), derive equations, similar to those of Problem 12, for the mean saturation level s_i at the end of a dry period of duration t_b, given that the initial saturation is s_p.

14. In Problems 12 and 13 what is a reasonable value for the characteristic depth z?

15. In Chapter 5 we say that actual transpiration may be related to soil moisture. Assume that relationship to be linear:

$$\frac{T_a}{T_p} = \frac{\theta}{\theta^*} \qquad \text{for } \theta < \theta^*$$

$$\frac{T_a}{T_p} = 1 \qquad \text{for } \theta > \theta^*.$$

The presence of salinity in the root zone will reduce transpiration in the plant. It has been suggested that the response of crops may be expressed as a function of a "total" moisture potential defined as the sum of osmotic and matrix potentials

$$\psi^* = \Pi(c) + \psi(\theta),$$

where ψ^* is the "total" soil moisture potential and $\Pi(c)$ is the osmotic potential, a function of the average salt concentration c. $\psi(\theta)$ is the matrix potential. The osmotic potential is commonly approximated as

$$\Pi(c) = Kc,$$

where K is a constant. Find an approximate expression for actual transpiration as a function of θ and c. In other words if

$$\frac{T_a}{T_p} = f(\theta, c),$$

find the function $f(\theta, c)$.

16. Hillslopes in a river basin can be classified as concave (upward) or convex, and diverging or converging (see figure). Which type of hillslope is more prone to runoff production of the variable-source (saturation from below) type and why?

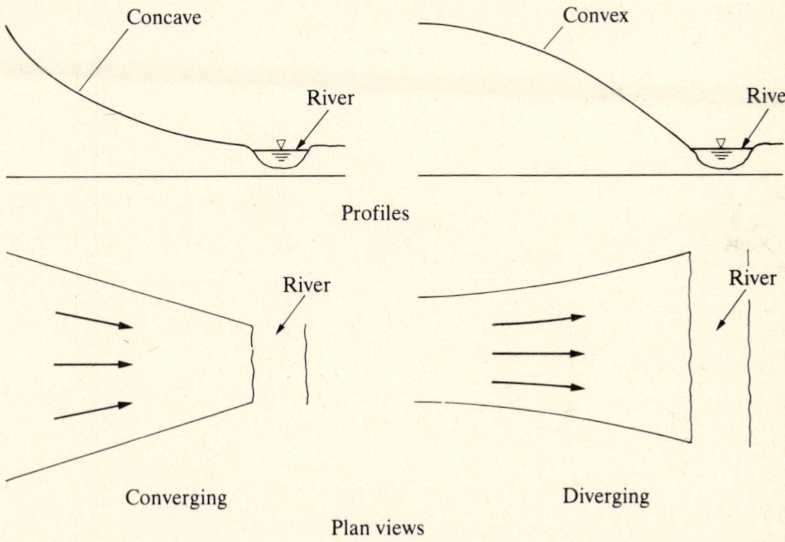

17. Approximately equating the infiltration rates given by the Green–Ampt and Horton models, find an expression for the depth of a moisture front as a function of time in terms of f_0, f_c, α, and ψ.

18. Rainfall intensity is given by an exponential function of the form $u = 6e^{-0.3t}$ cm hr^{-1}, where t is in hours. Infiltration is given by Horton's equation with $f_0 = 4.5$ cm hr^{-1} and $f_c = 2$ cm hr^{-1}, and α is 0.1 hr^{-1}. What is the total infiltration over a 10-hr storm?

Chapter 9

The Hydrograph and Simple Rainfall–Discharge Relationships

9.1 THE HYDROGRAPH

In Chapter 8 we introduced the concept of runoff. Each point on a surface has the potential to contribute to the total runoff input into the channels composing the river network. Other components of total runoff may follow subsurface paths. The nature of the storages, delays, and time of travel in each of the surface and subsurface paths of runoff is different. Discharge in the channels results from the integration of flow from all runoff sources. The distribution of discharge as a function of time in the channel is called the hydrograph.

The shape of a hydrograph for any given stream is a function of total available overland flow supply, subsurface flow, groundwater flow, slope of the overland and stream segments, roughness characteristics of flow elements, and geometry of channels. Figure 9.1 shows the various ways that storm intensity may influence hydrograph shape. Figure 9.1(a) corresponds to low-intensity storms that essentially result in no overland flow, only an increased groundwater flow (base flow). Figure 9.1(b) is a case of low- to medium-intensity conditions, where the system responds to flow through the upper layers of the soil profile (sometimes called interflow) or subsurface storm runoff. Interflow is superimposed on base flow and is faster to respond

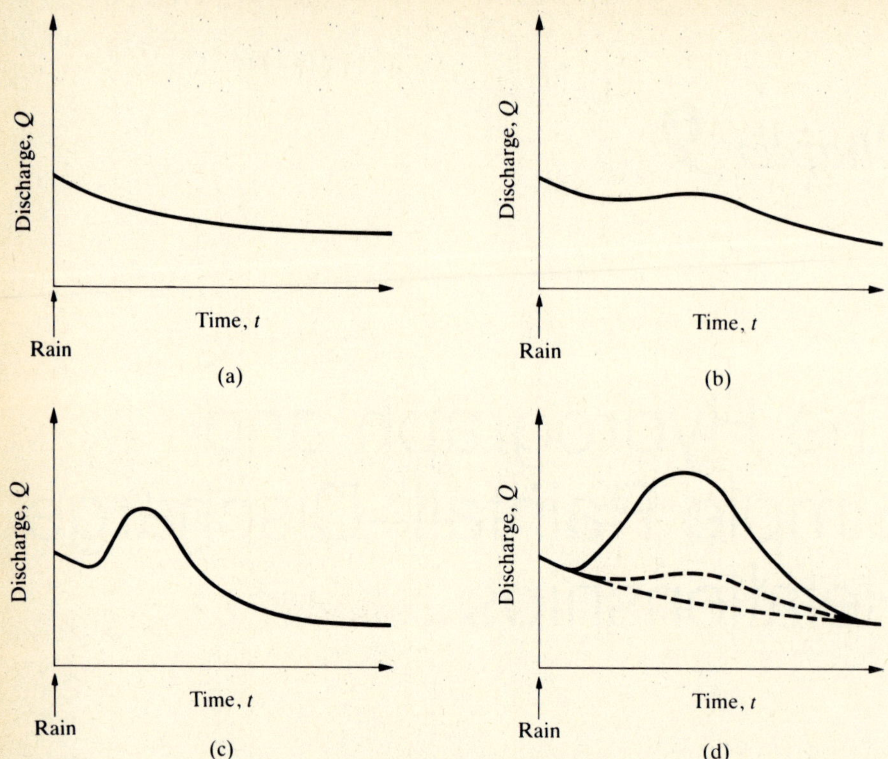

FIGURE 9.1 Hydrographs resulting from storms of different intensities. Source: Adapted from Viessman et al. [1977].

than the latter. Figure 9.1(c) shows a case of pure overland flow or runoff from a variable-source area. Figure 9.1(d) is a combination of all hydrograph components.

Figure 9.2 shows the effects of localized rainfall and different basin shapes on the streamflow hydrograph. Characteristics such as peak discharge and time to peak are dependent on the combined geometry of the basin and the storm. The hydrograph can be subdivided into the rising limb, the peak or crest, and the receding limb. The rising limb and the peak are functions of history and rainfall intensity as well as of basin characteristics. The details of this behavior will be discussed later.

The recession or receding limb is strongly related to storage and change in storage in the basin after the storm stops. Generally this recession curve takes the form

$$Q_t = Q_0 K^t, \tag{9.1}$$

where K is a recession constant and t is the time between Q_0 and Q_t. In fact, K is rarely a constant throughout the whole recession and may vary with

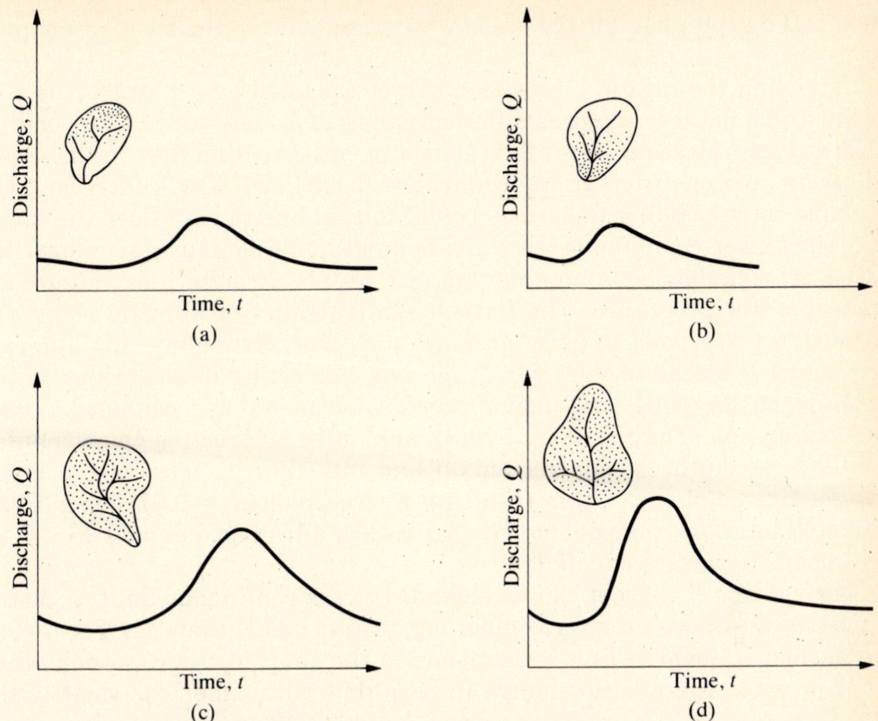

FIGURE 9.2 Interaction between basin shape and storm coverage in producing the hydrograph. Source: Adapted from Viessman et al. [1977].

season and soil-moisture conditions. Taking K as a constant, the change in storage in the basin is given by

$$-dS = Q_t \, dt = Q_0 K^t \, dt. \tag{9.2}$$

Integrating Eq. (9.2) results in

$$\begin{aligned} -\int_{S_0}^{S_t} dS &= \int_{t_0}^{t} Q_t \, dt = \int_{t_0}^{t} Q_0 K^t \, dt, \\ S_0 - S_t &= \frac{Q_t - Q_0}{\ln K}. \end{aligned} \tag{9.3}$$

Equation (9.3) gives the change in storage between any two times as a function of the discharge at those two times. Using those equations between t_1 and t_∞ when $S_{t_\infty} = 0$ and $Q_{t_\infty} = 0$ results in

$$S_{t_1} = \frac{-Q_{t_1}}{\ln K}, \tag{9.4}$$

which is the expression for the amount of storage left in the basin at any one time.

Extending the idea of a recession curve explained by Eq. (9.1), it is possible to argue for at least three different values of K: one value corresponding to the storage release from surface elements, another from flow in the upper soil layers, and one from groundwater flow (base flow). K will increase from the value corresponding to surface runoff to that due to base flow, the latter being the slower-changing process. If a hydrograph is plotted on semilogarithmic paper, segments behaving like Eq. (9.1) will be straight lines. Figure 9.3 illustrates this procedure. The flattest straight-line segment ($K = 0.97$) is assumed to correspond to base (groundwater) flow. Extending this line to a point under the peak of the hydrograph and connecting it with a line to the beginning of the rising limb (end of previous recession) approximates a base-flow hydrograph. The resulting hydrograph, after subtracting the computed base flow, accounts for surface runoff and interflow. Extending the lower straight-line segment of this resulting hydrograph ($K = 0.67$) leads to an estimated interflow component. The net result, a hydrograph with $K = 0.13$, corresponds to surface runoff.

The varying K concept can be extended to the continuous limit. A plot of $Q_{t+\tau}$ versus Q_t, where τ is a convenient lag, results in diagrams like Figure 9.4. The dashed, 45-degree line corresponds to the limit of the recession when $K = 1$ or when there is no change in base flow with time. The slope of the

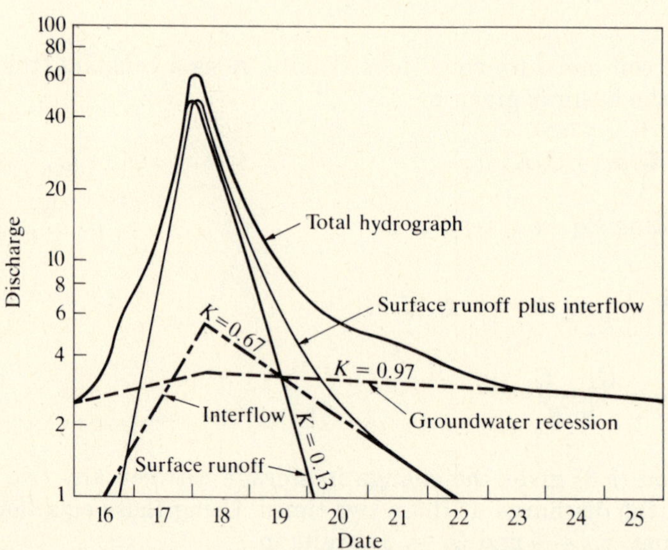

FIGURE 9.3 Semilogarithmic plotting of a hydrograph, showing method of recession analysis. Source: R. K. Linsley, Jr., M. A. Kohler, and J. L. H. Paulhus, *Hydrology for Engineers,* 3rd ed., 1982, McGraw-Hill. Reproduced with permission.

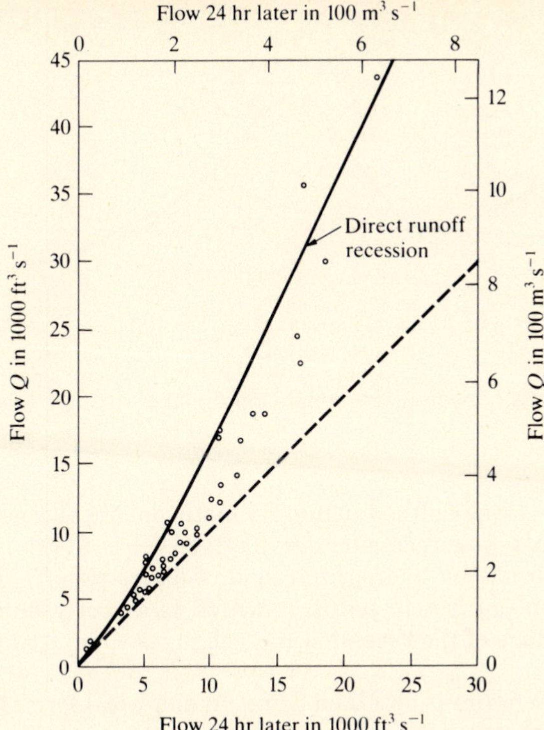

FIGURE 9.4 Recession curve in the form Q_τ versus $Q_{t+\tau}$ for the American River at Fair Oaks, California. Source: R. K. Linsley, Jr., M. A. Kohler, and J. L. H. Paulhus, *Hydrology for Engineers,* 3rd ed., 1982, McGraw-Hill. Reproduced with permission.

envelope of the plotted points is a limit of the K values corresponding to base flow and interflow. Note that Figure 9.4 defines a recession curve that could be used in hydrograph generation or analysis.

9.2 HYDROGRAPH SEPARATION

The previous discussion implies several methods of hydrograph analysis, including hydrograph separation into surface runoff, interflow, and base flow, or simply into direct runoff and base flow. Direct runoff is defined as the portion of the hydrograph that responds relatively quickly, and is clearly related to a given storm event. We will now discuss several techniques of hydrograph separation, which are acknowledged as empirical, if not arbitrary.

Figure 9.5 helps explain some of the methods. Dotted line 1 to 2 separates base flow from direct runoff by simply extending a horizontal line between points 1 and 2, where point 1 is the beginning of the rising limb. Base

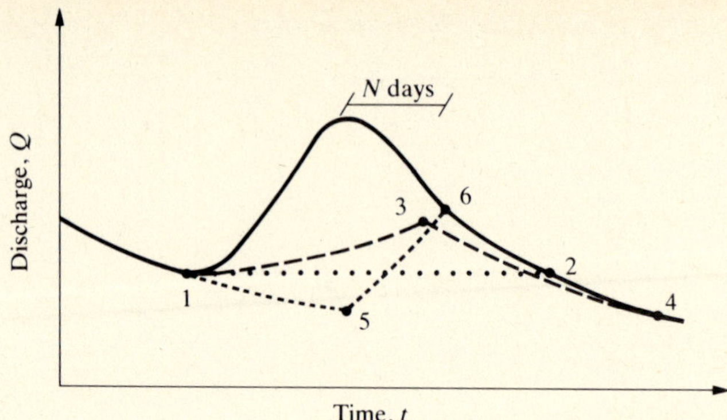

FIGURE 9.5 Methods of hydrograph separation. Source: Adapted from Viessman et al. [1977].

hydrograph 1–3–4 was defined using the methodology discussed in the previous section. A base flow recession curve (given K) is fitted. Point 4 is where the fit deviates from the hydrograph (end of direct runoff). The fitted curve is extended from point 4 to point 3, which is directly beneath the point where the curvature of the recession limb changes. Curve 1 to 3 is arbitrarily defined.

Point 6 in the figure is obtained using an empirical formula (Linsley et al. [1949]),

$$N = A^{0.2}, \tag{9.5}$$

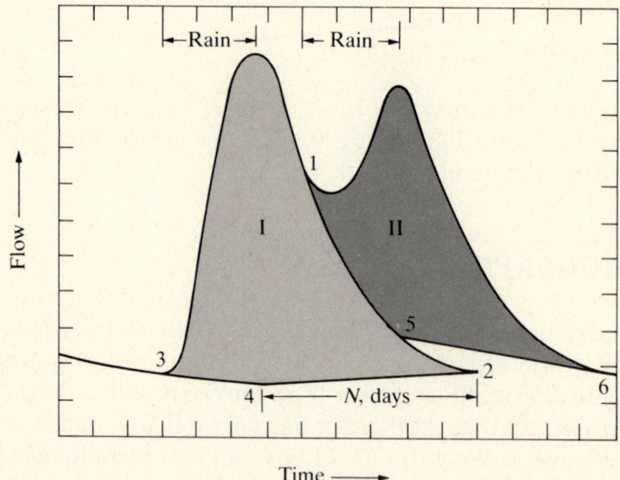

FIGURE 9.6 Separation of complex hydrograph, using recession curve. Source: R. K. Linsley, Jr., M. A. Kohler, and J. L. H. Paulhus, *Hydrology for Engineers*, 3rd ed., 1982, McGraw-Hill. Reproduced with permission.

where N is the time from peak to the beginning of base flow in days and A is the area of the basin in square miles; therefore, the area in square kilometers should be multiplied by 0.39. Point 6 can be used in several ways to define base flow. First, the recession preceding the hydrograph rise may be extended to a point 5 below the hydrograph peak. Points 5 and 6 are then connected by a straight line. The direct-runoff hydrograph is whatever is left after subtracting the defined base flow.

Base-flow separation from more complicated hydrographs is illustrated in Figure 9.6. Segment 1 to 2 is obtained by extending the recession of the first peak to a point 2, defined by Eq. (9.5). Segment 1–2 defines two different events. Base-flow separation within them is achieved with any of the previously described methods.

9.3 STREAMFLOW MEASUREMENTS

Streamflow is generally obtained by measuring stage (water-surface elevation) at calibrated locations where stage–discharge relations or rating curves permit obtaining the latter. A simple stage–discharge relation is shown in Figure 9.7. The relations commonly obey equations of the form

$$Q = a(y - b)^c, \tag{9.6}$$

where a and c are parameters, y is the elevation or stage, Q is discharge, and b is the elevation of zero flow. Variability in cross section and the nature of flow makes it convenient to define curves for low-, normal-, or high-flow conditions (Fig. 9.7).

The stage–discharge relation must be defined for relatively stable river "control sections." Backwater effects near the controls may require stage–discharge relations that are also a function of water-surface slope.

Under uniform flow conditions and gradually varying flow, the well-known Chezy and Manning equations are stage–discharge relations. The Chezy equation is

$$Q = AC\sqrt{RS}, \tag{9.7}$$

where C is the roughness coefficient, A is the cross-sectional area of flow, R is the hydraulic radius A/P, P is the wetted perimeter, and S is the energy slope (bottom slope in uniform flow or approximately the water-surface slope for nonuniform, slowly varying conditions). The equation shows the relation among discharge, slope, and stage (since for a given cross-sectional area, the hydraulic radius is a function of stage).

Another stage–discharge relation is the Manning equation,

$$Q = \frac{1}{n} AR^{2/3}S^{1/2}, \tag{9.8}$$

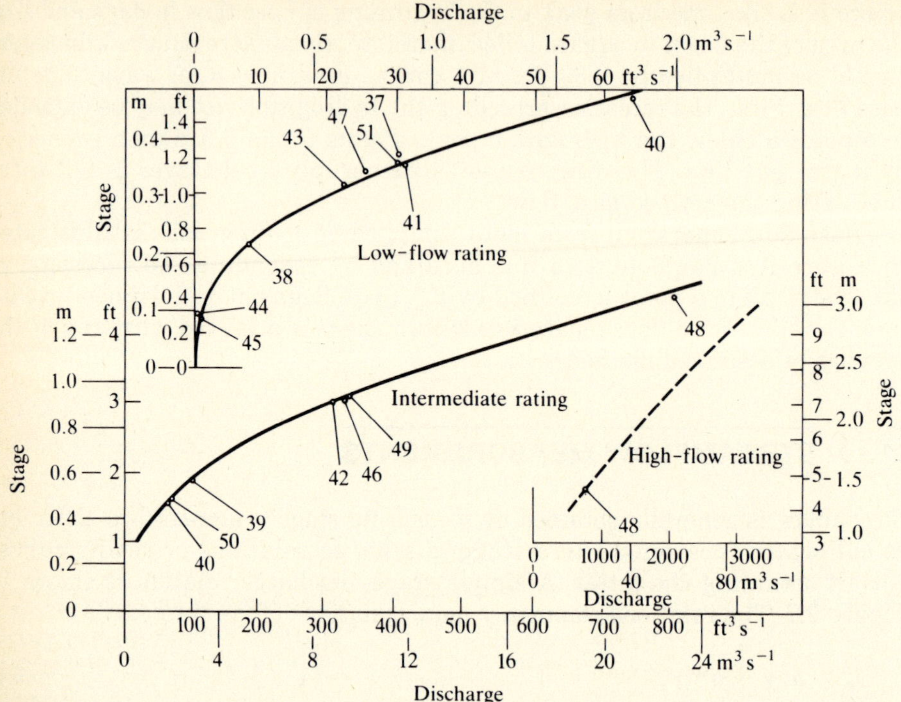

FIGURE 9.7 A simple stage–discharge relation. Source: R. K. Linsley, Jr., M. A. Kohler, and J. L. H. Paulhus, *Hydrology for Engineers*, 3rd ed., 1982, McGraw-Hill. Reproduced with permission.

where Q is discharge in cubic meters per second, A is cross-sectional area of flow in square meters, n is the Manning roughness coefficient, and R is hydraulic radius in meters.

The previous equations may be used to illustrate the nature of possible adjustments to account for backwater effects, or other periods where the energy slope S does not coincide with the bottom slope. Figure 9.8 shows two stations close to one another along a river with stages y_1 and y_2. From Eqs. (9.7) and (9.8) we know that discharge is proportional to S^m, hence we can write

$$\frac{Q}{Q^\circ} \approx \left(\frac{S}{S^\circ}\right)^m \approx \left(\frac{\Delta y}{\Delta y^\circ}\right)^m, \tag{9.9}$$

where m is an exponent, Q is the discharge at station 2 when the surface slope is S, and Q° is the uniform flow discharge that would occur at station 2, with stage y_2, when the energy slope is equal to the bottom slope S°. Δy is the difference in stage between adjacent stations: $\Delta y = y_1 - y_2$. Coefficient m should be close to 1/2. Δy° corresponds to the uniform, normal, flow condi-

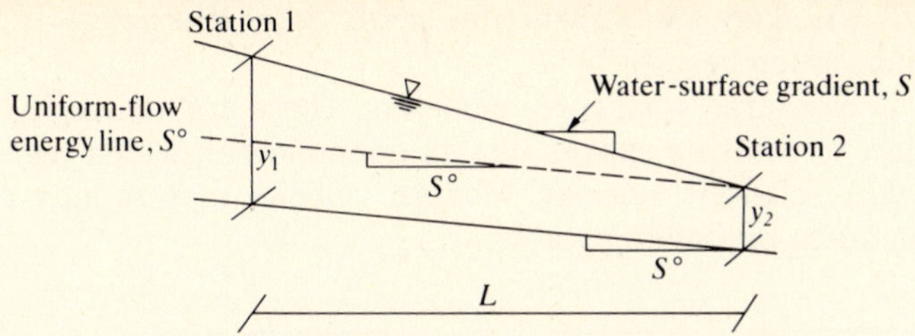

FIGURE 9.8 Water-surface slope between two stations.

tions when the energy slope is the bottom slope. Once the above relationship is developed from data, it can be used to make adjustments to the normal rating-curve results. That is, Q° is first obtained from a stage measurement at the primary station (in this case, y_2), assuming uniform flow. Given a reading of stage at the auxiliary station, Δy can be computed and used with Q° and Δy° in Eq. (9.9) to obtain the adjusted discharge Q.

Stage measurement techniques are discussed in other references (World Meteorological Organization [1974]; Linsley et al. [1982]). Instruments that respond continuously or discretely to surface elevation are used. The stage–discharge calibration results from experimentation at a location. Discharge is obtained by measuring velocities across a river section. Because of expected velocity distributions with depth and across the channel, velocities are usually measured at 0.2 and 0.8 of the total depth (or 0.6 when only one measurement is made) in different river subsections. No subsection should account for more than 10% of the total flow. Figure 9.9 shows this procedure. Average velocity in each subsection is the average between the 0.2 and 0.8 depth values. The discharge per subsection is the corresponding average

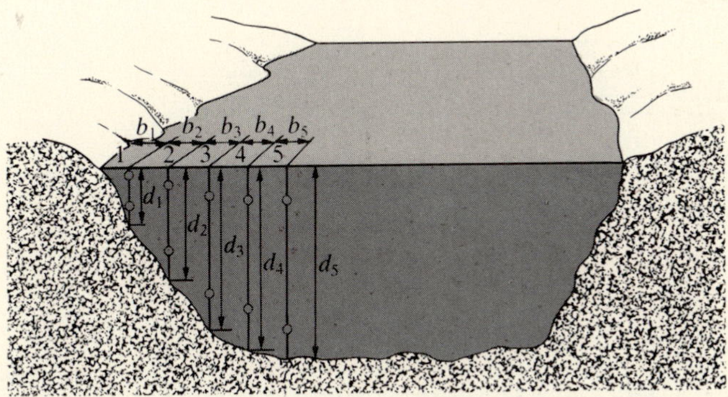

FIGURE 9.9 View of a stream cross section showing location of points of observation. Source: World Meteorological Association, *Guide to Hydrological Practices,* 3rd ed., World Meteorological Association, 1974.

velocity multiplied by the subsection area. Total discharge is the sum of individual subsection results.

Velocity is measured with current meters. These are generally instruments with propellers or rotating spokes like an anemometer, which respond proportionally to water-current velocity. Modern velocity meters may also rely on electric inductance by flowing water.

9.4 RAINFALL–DISCHARGE RELATIONSHIPS

Streamflow responds to precipitation inputs, rainfall, or snowmelt. During a rainfall event, the hydrograph is dominated by surface runoff and/or flow in the upper soil layers. Hydrologists have always been concerned with obtaining discharge from rainfall. The interest is not only on total volumes but on a description of the transformation of the rainfall history (hyetograph) to the streamflow history (hydrograph). These functions or rainfall–discharge relationships are required for the design of hydraulic structures, for describing historical basin behavior, for predicting response when the basin topography or land use changes (i.e., urbanization), or for predicting future discharges (i.e., flood forecasting).

Figure 9.10 gives the theoretical shape of a hydrograph resulting from a storm of constant effective intensity and long duration t_r completely covering a basin. The effective storm intensity is the actual precipitation rate minus the rate at which some of the water becomes unavailable for runoff because of evaporation, detention, or infiltration into deep, slow-responding, soil layers. As the storm starts, the channel responds slowly (but almost immediately) to precipitation falling on the channel itself or on nearby areas. As time passes, rain falling on more remote overland regions affects the channels and moves along these to the outlet. During this period, the hydrograph rises sharply.

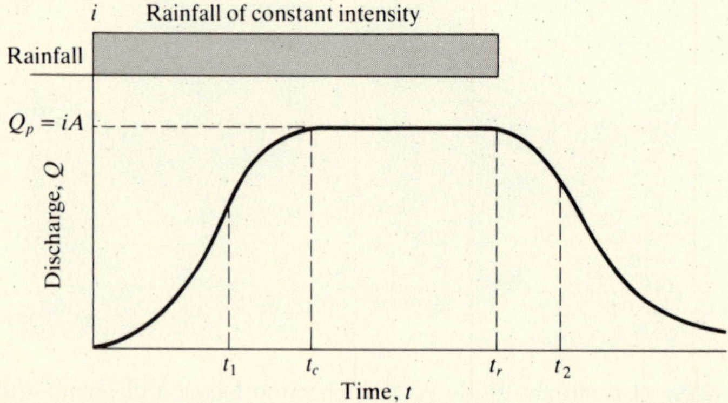

FIGURE 9.10 Hydrograph corresponding to a long storm.

The majority of the basin is contributing to outlet discharge by time t_1. At time t_c (time of concentration) the whole basin is responding. All storage capacity in the basin is used at this point and the input is equal to the output. The peak of the hydrograph must then be iA, where i is the storm's effective intensity and A is the basin area. As long as rain continues, the discharge remains constant. When the storm stops at time t_r, the discharge depends on water being released from storage in channels, upper soils, and overland surfaces. Since the amount released may be considerable, the outflow may remain fairly high for a period after t_r, until the bulk of the storages (usually channels and storage in river banks) is nearly depleted. This can be interpreted as the time that it takes for the outlet to "learn" that it stopped raining everywhere. At this point the inflection at time t_2 occurs. The bulk of the recession is then due to slower-responding storages, like further bank storage and groundwater (base flow). We have already discussed the nature of this recession.

9.4.1 Peak Discharge Formulas—The Rational Formula

Many have attempted to relate peak discharge to rainfall and topographic characteristics such as area, stream length, and slope of the region (Nash [1959]; Sherman [1932b]; Taylor and Schwartz [1952]). The reader can find a considerable number of existing equations, mostly local in nature, in any hydrology handbook (Chow [1964]; Gray [1973]). An expression that has been and is still commonly used in design of urban drainage systems is the rational formula, stated as

$$Q_p = CIA\,, \tag{9.10}$$

where Q_p is the peak hydrograph discharge, A is the area, I is the rainfall intensity, and C is a coefficient. The coefficient C is nondimensional (usually between 0.5 and 0.8) if I and A are given in compatible units (i.e., meters per second and square meters, respectively, leading to Q in cubic meters per second) or if I is in inches per hour and A is in acres, which yields discharge in cubic feet per second (approximately). To be valid, the rainfall intensity in Eq. (9.10) must be of duration equal to or larger than the time of concentration of the basin, which ensures that Q_p is the largest peak obtained for a storm of the given intensity. Usually a given probability of occurrence is assigned to the chosen storm. By construction, the implication is that the resulting Q_p will have the same probability of occurring in any one year (Schaake et al. [1967]). This concept of probability and the associated idea of recurrence will be seen again in Chapter 11. The coefficient C accounts for infiltration and permanent storage, taken as constant throughout the storm duration.

The rational formula is a design tool (limited for that matter) adequate to handle, at best, extreme events. As long as I is defined for a duration equal to the concentration time, the rational formula gives reasonable results.

In practice, the rational formula is mostly used in small (not larger than a few hundred acres [1 acre is about 4000 m^2]) urban areas for designing storm sewer systems. Sewers are designed for a given annual probability (usually 0.1 to 0.04) of failure, defined as times when flow exceeds capacity. Drainage pipes, gutters, and inlets should accept the flows resulting from events with the above annual probabilities of occurrence. As stated previously, it is assumed (not generally correctly, because of the effects of antecedent and moisture conditions) that the peak discharge has the same probability of occurring as the corresponding storm.

EXAMPLE 9.1

The Design of a Storm Sewer System

Traditionally, urban areas are drained by systems of pipes (sewers) intended to prevent surface flooding and remove water from the region as quickly as possible. This philosophy may lead to downstream flooding and water-quality problems, issues of so-called urban hydrology (Overton and Meadows [1976]; Bras and Perkins [1975]). A complete description of classical design procedures can be found in the Engineering Practice Manual No. 37 of the American Society of Civil Engineers [1970]. Here we will illustrate how the rational formula and uniform flow assumptions are used in design practice.

Figure 9.11(a) shows a plan view of an urbanized area with a common gridded street pattern. The area topography and the system of streets and gutters induces the flow pattern shown by arrows. The region is divided into four subareas $C1$, $C2$, $C3$, and $C4$, according to drainage patterns. Runoff from these areas is collected by inlets $I1$, $I2$, $I3$, and $I4$, respectively. A pipe $P1$ connects inlets $I1$ and $I3$. Similarly, $P2$ connects $I2$ and $I4$. The combined waters from catchments $C1$ and $C3$ are then routed through pipe $P3$ to join outflow from $P2$ and $C4$ going into pipe $P4$. A schematic diagram of the interconnections is shown in Figure 9.11(b). Assume each of the catchment areas has an area of 40,000 m^2. The design will be done for a storm that has a 0.04 probability of occurring in any one year. Chapter 4 presented the concept of intensity–frequency–duration curves. This relationship between mean rainfall intensity and duration for a given recurrence can be parameterized as

$$i = \frac{a}{b + t_r}, \tag{9.11}$$

where a and b are constants. Assume that a and b have values of 3 cm and 0.17 hr, respectively, and that i is in centimeters per hour and t_r is in hours. To continue the analysis, we need values of C for each of the subareas and estimates of the time of concentration in each of them. We will assume $C = 0.7$ for all subareas. The time of concentration is taken as 10 minutes for all subareas (note that in practice subareas should be kept to similar times of concentration and size).

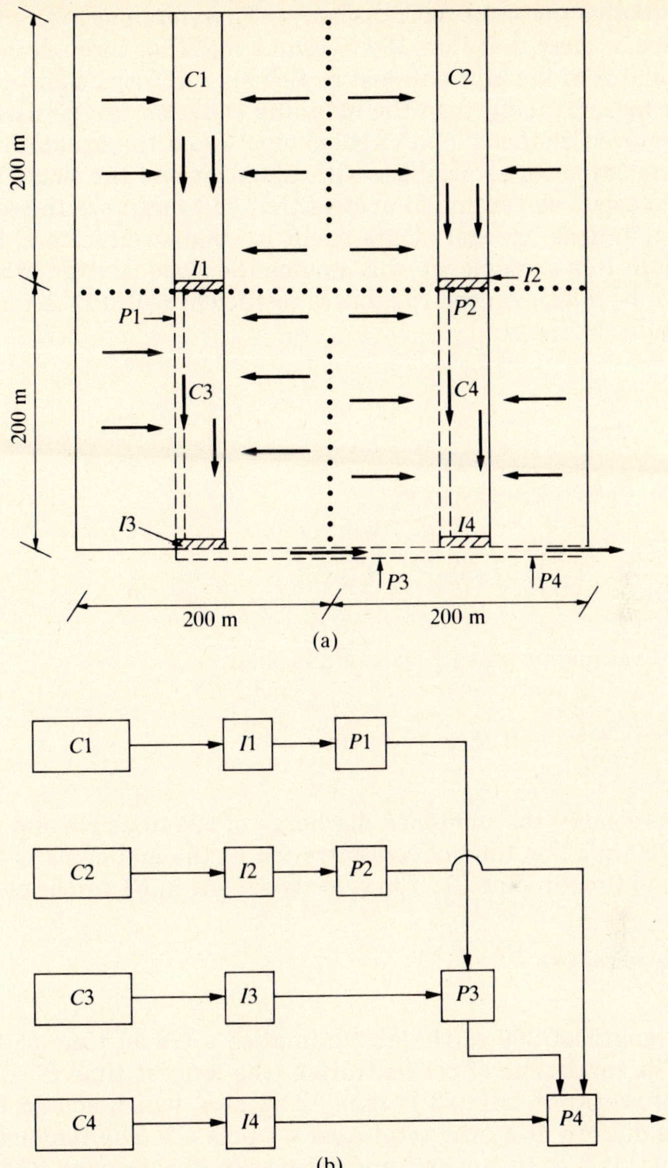

FIGURE 9.11 Plan view of urban storm sewer system.

To compute flow into inlet 1, we first find the rainfall intensity corresponding to a 10-minute (t_c for $C1$) duration. Using the previously given equation, we get $i = 8.9\ \text{cm hr}^{-1}$ or $i = 8.9 \times 10^{-2}\ \text{m hr}^{-1}$. Using Eq. (9.10),

$$Q_p = (0.7)(8.9 \times 10^{-2})(40{,}000) = 2492\ \text{m}^3\,\text{hr}^{-1} = 0.7\ \text{m}^3\,\text{s}^{-1}.$$

Inlet 2 will receive the same peak discharge. Therefore pipes $P1$ and $P2$ have to be designed to carry that flow. If we assume that flow through the pipe will be uniform and that the pipe will just be flowing full (not under pressure but as a free surface channel), then the Manning equation can be used to determine the required diameter of a circular pipe. From the equation it is clear that there are two design variables—the diameter and the slope. In practice the goal is to select slopes and diameters that will minimize the total cost of excavation and pipes. Steeper slopes result in smaller diameters, but deeper excavations. In this example we will assume the slope is given. For pipes $P1$ and $P2$, let it be 0.005. A good roughness coefficient is 0.015. A just-full pipe has a hydraulic radius of

$$R = \frac{A}{P} = \frac{\pi r^2}{2\pi r} = \frac{r}{2}.$$

Therefore Eq. (9.8) takes the form

$$Q_p = \frac{1}{0.015}\pi r^2 \left(\frac{r}{2}\right)^{2/3} (0.005)^{1/2} = 9.3 r^{8/3}.$$

The required radius for pipes $P1$ and $P2$ is then

$$r_1 = r_2 = \left(\frac{Q_p}{9.3}\right)^{3/8} = 0.38 \text{ m} = 38 \text{ cm}.$$

Pipe $P3$ has to carry the combined discharge of subareas $C1$ and $C3$ with a total of 80,000 m^2. The time of concentration at the end of $P1$ is that of $C1$ plus the travel time in pipe $P1$. The velocity of flow in $P1$ can be computed as

$$V = \frac{Q_p}{\pi r^2} = 1.5 \text{ m s}^{-1}.$$

For a pipe length of 200 m the above implies a travel time of 133 sec or 2.22 min. The total time of concentration (the longest time of travel of all contributing areas) at inlet $I3$ is then 12.22 min, which should be used to compute the outflow from the total area $C1$ plus $C3$. The implied intensity (from Eq. 9.11) is 8×10^{-2} m hr^{-1} and the design flow for pipe $P3$ is then

$$\begin{aligned} Q_p &= CIA = 0.7(8 \times 10^{-2})80{,}000 = 4480 \text{ m}^3 \text{hr}^{-1} \\ &= 1.24 \text{ m}^3 \text{s}^{-1}. \end{aligned}$$

If pipe $P3$ has a slope of 0.003 and $n = 0.015$, the required radius is

$$r_3 = \left(\frac{Q_p}{7.2}\right)^{3/8} = 0.52 \text{ m} = 52 \text{ cm}.$$

Pipe $P4$ drains all the areas for a total of 160,000 m^2. Converging into this pipe are the flows of pipes $P2$ and $P3$ and of subarea $C4$. To obtain the design intensity, we use the longest of the two travel paths $C1$–$P1$–$P3$ and $C2$–$P2$, which is clearly the former with a computed time of concentration of $12.22 + 2.28 = 14.5$ min. This leads to an intensity of $7.3 \times 10^{-2}\ \mathrm{m\,hr^{-1}}$ and a design discharge of

$$Q_p = CIA = 0.7(7.3 \times 10^{-2})160{,}000 = 8176\ \mathrm{m^3\,hr^{-1}}$$
$$= 2.3\ \mathrm{m^3\,s^{-1}}.$$

Using Manning's equation with a given slope of 0.002 and $n = 0.015$ yields

$$r_4 = \left(\frac{Q_p}{5.9}\right)^{3/8} = 0.7\ \mathrm{m} = 70\ \mathrm{cm}.$$

The use of the rational formula for design is a generally conservative assumption. In the above we ignored storage effects in pipes as well as time delays. Assuming that all subareas peak simultaneously amounts to using an improbable worst condition. For larger areas with significant pipe or other storage capacity, design should be based on procedures giving the whole history of discharge. Nonuniform flow conditions can also be important in large complicated pipe networks. ◆

9.4.2 The Unit Hydrograph

The peak discharge formulas fail to give any information about the time development of discharge. They do not give the full hydrograph or simply imply a symmetrical hydrograph with the obtained peak. The simplest and most successful of the rainfall–discharge formulations that describe the full hydrograph is the unit hydrograph concept, introduced by Sherman [1932a].

The basic hypothesis of this approach is that the river basin responds linearly to effective rainfall. Linearity implies that if

$$I = aI_1 + bI_2,$$

then

$$Q = aQ_1 + bQ_2,$$

where I_i and Q_i are corresponding inputs and outputs. The unit hydrograph is defined as the discharge produced by a unit volume of effective rainfall of a given duration. Linearity then implies that the basin discharge corresponding to a storm of different effective depth (volume per unit area) but of the same duration would simply be the scaled unit hydrograph. The scaling factor is the new storm depth.

For every duration of effective rainfall, there is a unit hydrograph. Besides linearity, the concept further assumes that rainfall is uniformly distributed in time and space. This limits the application of the technique to uniform storms over relatively small areas.

Unit hydrographs are developed from pairs of corresponding rainfall–discharge events. The criteria used are the following:

1. Select individual events with simple structures, as uniform in time and space as possible.
2. Use hydrograph separation to eliminate base flow.
3. Define duration of effective rainfall resulting in the obtained direct runoff. This duration should preferably be between 10 and 30% the time of rise of the hydrograph.
4. Select events, if possible, with direct runoff volume near 1 cm.
5. Select as many pairs of rainfall–discharge events as possible.
6. All selected events should have equal effective storm duration, $\pm 25\%$.

A unit hydrograph is obtained for each data pair by dividing the ordinates of the direct runoff hydrograph by the effective rainfall volume. The result is a hydrograph of unit volume whose ordinate is in inverse time units.

EXAMPLE 9.2

The Unit Hydrograph

Figure 9.12 gives the precipitation and discharge history in the Bird Creek Basin in Oklahoma for May 1958. The basin has an area of 2344 km^2. Both precipitation and discharge data are given in units of volume per unit area (millimeters) per 6 hr. That is, the ordinate shown at any time is really the accumulation over the last 6-hr period. This is emphasized in the discrete hyetographs superimposed on two of the storms. We will derive the unit hydrographs for pairs of data corresponding to the first and fourth storms in the month. These two are the most significant events on record. Note that both rainfall events have about 18 hr (three time steps) of significant rainfall, so we will be deriving the 18-hr unit hydrograph. In doing so, we are assuming that the whole basin was affected by the events and that the events are of uniform intensity. The last point is clearly not true.

The first step is to find the direct runoff. Figure 9.12(b) gives an estimate of the base-flow hydrograph obtained using one of the empirical procedures of Section 9.2. It is interesting to note that Eq. (9.5) is not valid for this case, since it would lead to an impossibly large value of N. Table 9.1 gives the values of precipitation, discharge, base flow, and direct runoff for storm 1. Table 9.2 does the same for storm 4.

Since the discharge data are given so that each value is the accumulation of the past 6 hr, the total direct runoff is simply the sum of ordinates. Storm 1 produced 7.48 mm and storm 4 produced 10.06 mm of direct runoff. As indi-

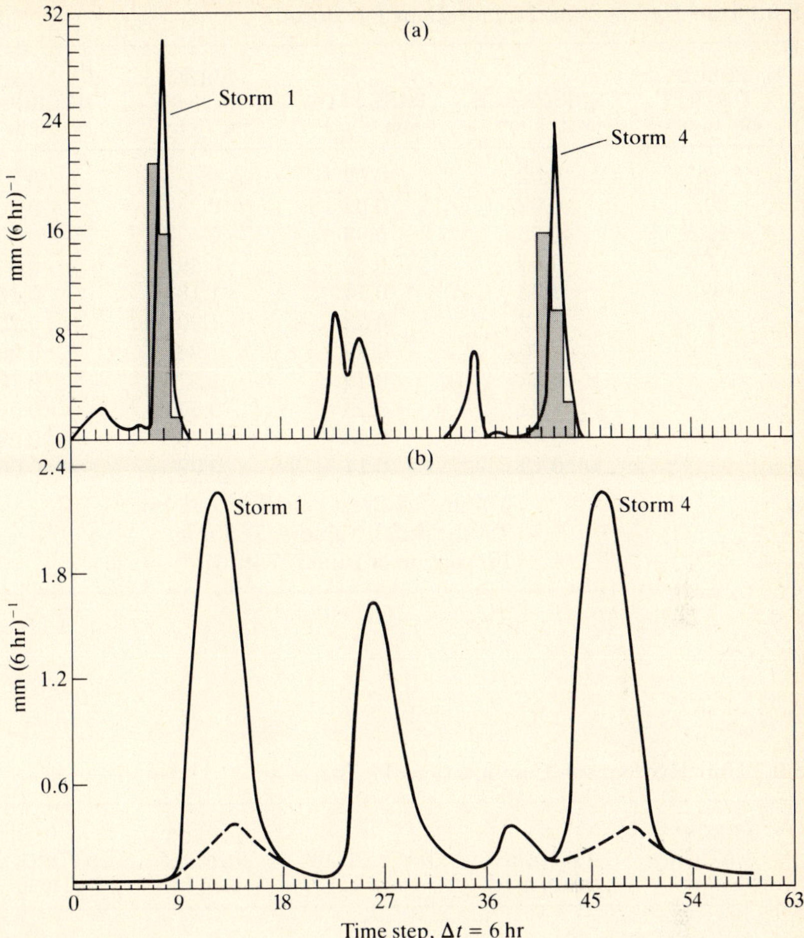

FIGURE 9.12 Precipitation (a) and flow (b) in the Bird Creek Basin, Oklahoma, for the month of May 1958.

cated in Tables 9.1 and 9.2, storm 4 produced proportionally more runoff. The unit hydrograph is obtained by dividing each ordinate of the direct-runoff hydrograph by its total volume. The two results, with ordinates of (6 hr)$^{-1}$ are shown in Tables 9.1 and 9.2 and Figure 9.13.

The fact that the two unit hydrographs are not equal is a reflection of the accuracy of the procedure, the assumptions on the nature of the storms, and the failure of the linearity assumption. Such results are the rule rather than the exception in practice. The hydrologist should compute several unit hydrographs and average them. But be careful! Averaging the ordinates of a given time will result in a distorted, flattened unit hydrograph. We should average characteristics. For example, averages of the time to peak; the peak; the

TABLE 9.1 Unit Hydrograph Computations for Storm 1

TIME (hr)	PRECIP-ITATION mm (6 hr)$^{-1}$	DISCHARGE mm (6 hr)$^{-1}$	BASE FLOW mm (6 hr)$^{-1}$	DIRECT RUNOFF mm (6 hr)$^{-1}$	UNIT HYDROGRAPH (6 hr^{-1})
42	0	0.02	0.02	0	0
48	20.5	0.02	0.02	0	0
54	15.5	0.05	0.05	0	0
60	1.5	0.58	0.09	0.49	0.07
66		1.32	0.14	1.18	0.16
72		2.12	0.22	1.90	0.25
78		2.23	0.29	1.94	0.26
84		2.23	0.29	1.27	0.17
90		0.81	0.29	0.52	0.07
96		0.39	0.23	0.16	0.02
102		0.19	0.17	0.02	0.003

Volume of direct runoff 7.48 mm
Total rainfall volume, 37.5 mm
Proportion of runoff, 7.48/37.5 = 0.20

TABLE 9.2 Unit Hydrograph Computations for Storm 4

TIME (hr)	PRECIP-ITATION mm (6 hr)$^{-1}$	DISCHARGE mm (6 hr)$^{-1}$	BASE FLOW mm (6 hr)$^{-1}$	DIRECT RUNOFF mm (6 hr)$^{-1}$	UNIT HYDROGRAPH (6 hr^{-1})
240	0				0
246	0.6				0
252	15.4	0.15	0.15	0	0
258	9.5	0.29	0.16	0.13	0.01
264	2.6	1.27	0.17	1.10	0.11
270		1.85	0.20	1.65	0.16
276		2.22	0.23	1.99	0.20
282		2.22	0.27	1.95	0.19
288		1.85	0.30	1.55	0.15
294		1.39	0.35	1.04	0.10
300		0.81	0.30	0.51	0.05
306		0.39	0.25	0.14	0.01
312		0.19	0.19	0.00	0.00

Volume of direct runoff 10.06 mm
Total rainfall volume, 28.10
Proportion of runoff, 0.36

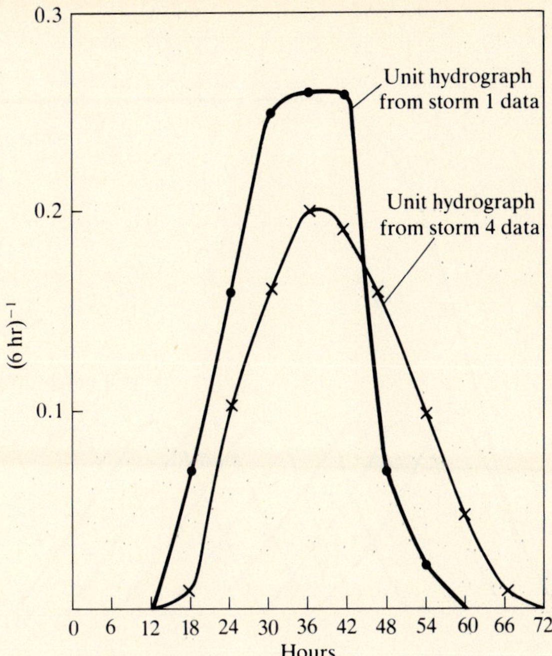

FIGURE 9.13 Unit hydrographs corresponding to storms in Figure 9.12.

width of the hydrograph at the 25, 50, and 75% of the peak; and the duration of the hydrograph could be obtained, and a curve with unit volume (in this example, 1 mm) drawn through the resulting points. ◆

Derivation of Different-Duration Unit Hydrographs — The *S* Curve

Each unit hydrograph corresponds to a given effective rainfall duration. The user of the unit hydrograph theory either computes from data unit hydrographs of various durations to obtain flexibility in handling different types of storms, or makes full use of the linearity assumption and computes the so-called S hydrograph or curve.

The S hydrograph is the expected response of an infinite sequential series of fixed duration storms of unit volume. According to unit hydrograph theory, this response must be the infinite summation of unit hydrographs of the given duration. This is shown in Figure 9.14. The D-hour S hydrograph is obtained by adding corresponding D-hour unit hydrographs, each lagged by D. The resulting S-shaped curve peaks and stabilizes after the summation of T/D unit hydrographs, where T is the time length of the base of the original D-hour unit hydrograph. Since the effective intensity of the assumed uniform

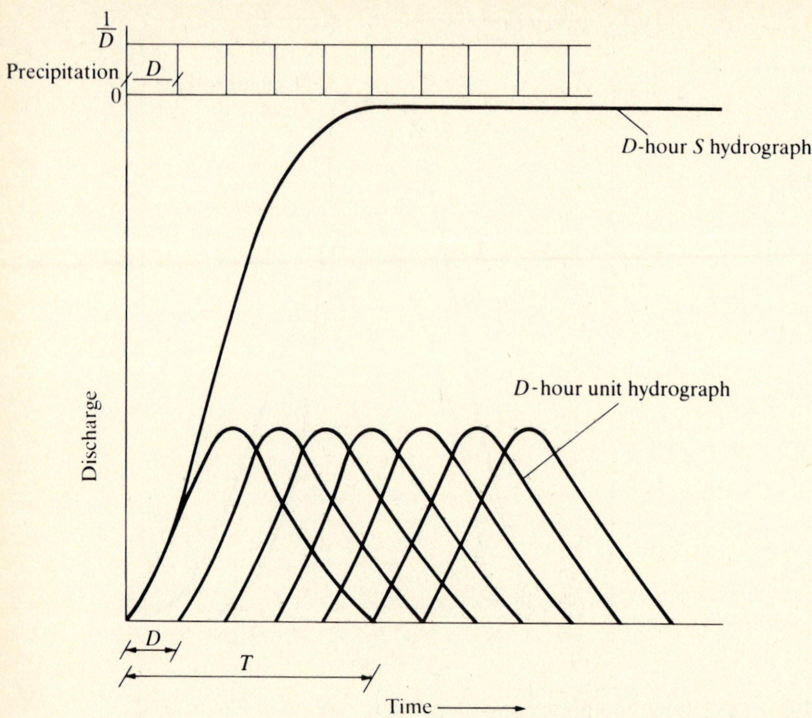

FIGURE 9.14 Computation of *S* hydrograph. Source: Adapted from Viessman et al. [1977].

volume storm has to be $1/D$, the S hydrograph peaks and stabilizes at $Q = 1/D$ in volume per unit area per time or A/D in units of volume per time. The area of the basin is A.

The D-hour-duration S hydrograph can be used to obtain the unit hydrograph corresponding to any other duration t. Using the linearity property, the procedure is

1. Lag S curve by t hours.
2. Subtract the ordinates of the displaced curves; the result is a hydrograph of volume t/D.
3. Normalize resulting hydrograph to volume 1 by multiplying by D/t.

The procedure as illustrated in Figure 9.15 results in the t-hour unit hydrograph.

In practice, the S hydrograph can be easily obtained without requiring the direct summation of T/D unit hydrographs. Column 2 of Table 9.3 is the 18-hr (three time steps) unit hydrograph computed for storm 1 in Example 9.2. Column 3 is obtained by copying the first three ordinates of the unit hydrograph (since the first 18 hr of the S curve are the same as the 18-hr

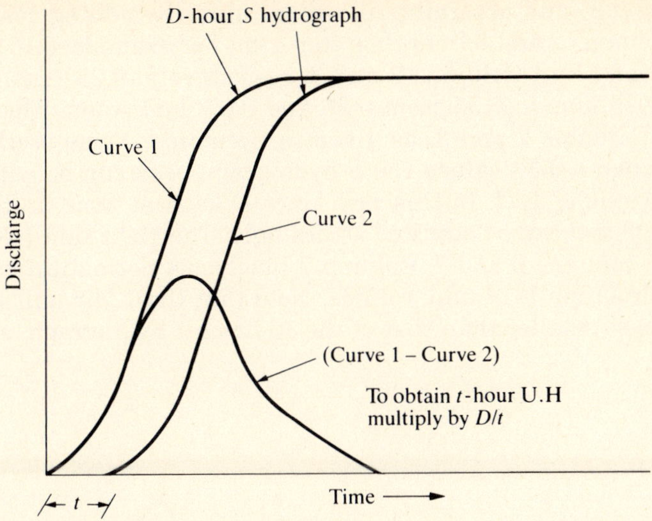

FIGURE 9.15 Obtaining a *t*-hour unit hydrograph from the *D*-hour *S* hydrograph. Source: Adapted from Viessman et al. [1977].

TABLE 9.3 Computation of the *S* Curve and a 12-Hr Unit Hydrograph from an 18-Hr Unit Hydrograph

1	2	3	4	5	6	7
Time Step $\Delta t = 6$ hr	18-Hr Unit Hydrograph ($3\Delta t$)	Intermediate Result	18-Hr *S* Curve	Lagged *S* Curve	Column 4 Minus Column 5	12-Hr Unit Hydrograph (18/12) Times Column 6
1	0		0		0	0
2	0		0		0	0
3	0.07		0.07	0	0.07	0.11
4	0.16	0	0.16	0	0.16	0.24
5	0.25	0	0.25	0.07	0.18	0.27
6	0.26	0.07	0.33	0.16	0.17	0.26
7	0.17	0.16	0.33	0.25	0.08	0.12
8	0.07	0.25	0.32	0.33	0.00	0.00
9	0.02	0.33	0.33*	0.33	0.00	0.00
10	0.003	0.33	0.33	0.33		
11	0	0.32	0.33*	0.33		
12	0	0.35	0.33*	0.33		
13	0	0.33	0.33	0.33		

*Adjusted value.

unit hydrograph) and obtaining all other values by adding corresponding values in Columns 2 and 3 from time step 4 on. For example, 0.16 comes from adding 0.16 and 0 in Columns 2 and 3, respectively; 0.25 results from the next pair, 0.25 plus 0; 0.33 from 0.26 and 0.07; and so on. The S curve is the sum of Columns 2 and 3, as given in Column 4. Some oscillations are observed around a 0.33 value. The S hydrograph peak can be adjusted to its theoretical value of $1/D$. In this case since D is three time units, the peak should be 0.33 and can be adjusted accordingly. The 12-hr unit hydrograph is computed in columns 6 and 7. Column 7 involves a normalization to ensure that the hydrograph is of unit volume. Note that the 12-hr unit hydrograph has a time base smaller than that of the 18-hr unit hydrograph, as expected.

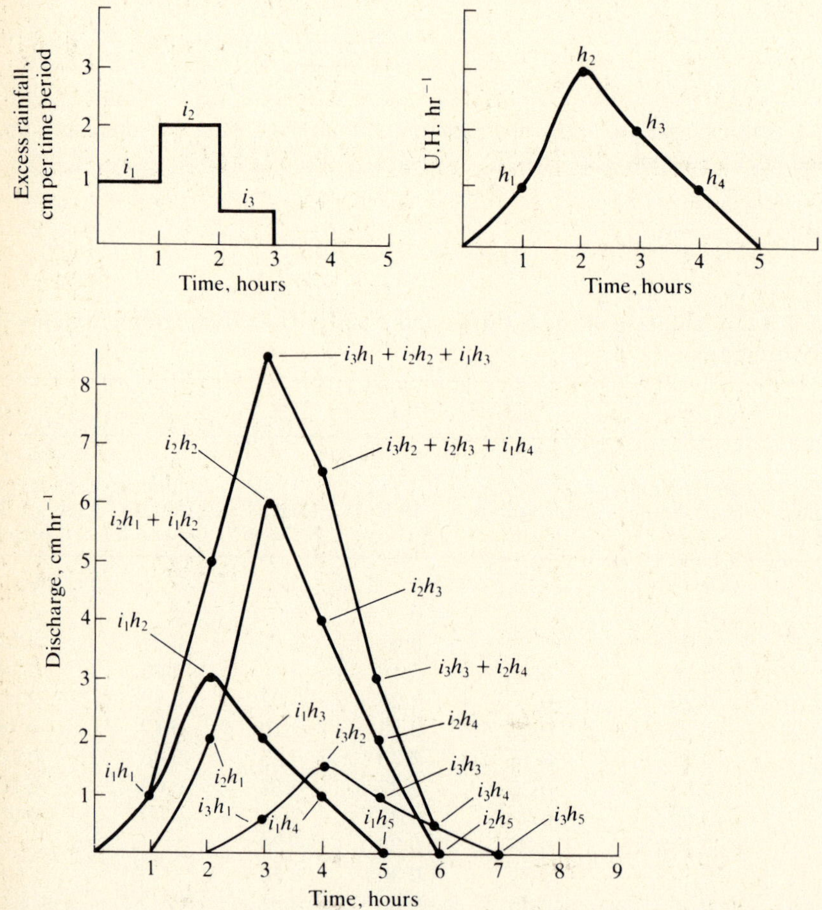

FIGURE 9.16 Hydrograph from a compound storm as obtained with the unit hydrograph. (a) Excess rainfall; (b) unit hydrograph; and (c) discharge hydrograph. Source: Adapted from Viessman et al. [1977].

*Mathematical Formulation of the Unit Hydrograph

The response of a linear basin to any rainfall input is proportional to the unit hydrograph, as discussed previously. This proportionality can be generalized to compound storms, not only to a single uniform event.

Figure 9.16 shows a compound storm extending over three time periods. The unit hydrograph corresponding to one time period of effective rainfall can be used to obtain the response of the compound storm. Based on the ordinates shown in Figure 9.16, it should by now be clear that

$$
\begin{aligned}
i_1h_1 &= Q_1\\
i_2h_1 + i_1h_2 &= Q_2\\
i_3h_1 + i_2h_2 + i_1h_3 &= Q_3\\
i_3h_2 + i_2h_3 + i_1h_4 &= Q_4\\
i_3h_3 + i_2h_4 + 0 &= Q_5\\
i_3h_4 + 0 + 0 &= Q_6
\end{aligned}
\tag{9.12}
$$

The system of equations shown as Eq. (9.12) is over-determined with respect to the unit hydrograph ordinates h_i ($i = 1, \ldots, 4$). It can be expressed in matrix form as

$$\mathbf{IH} = \mathbf{Q}, \tag{9.13}$$

where

$$\mathbf{I} = \begin{bmatrix} i_1 & 0 & 0 & 0\\ i_2 & i_1 & 0 & 0\\ i_3 & i_2 & i_1 & 0\\ 0 & i_3 & i_2 & i_1\\ 0 & 0 & i_3 & i_2\\ 0 & 0 & 0 & i_3 \end{bmatrix}$$

$$\mathbf{H} = \begin{bmatrix} h_1\\ h_2\\ h_3\\ h_4 \end{bmatrix}$$

$$\mathbf{Q} = \begin{bmatrix} Q_1\\ Q_2\\ Q_3\\ Q_4\\ Q_5\\ Q_6 \end{bmatrix}$$

Note that the order n of the output vector is given by

$$n = j + i - 1,$$

where n is the number of output hydrograph ordinates, j is the number of unit hydrograph ordinates, and i is the number of excess-rainfall periods.

A common problem in hydrology is that of model identification, essentially the determination of the ordinates of the unit hydrograph, h_is. Due to the generally over-determined condition of Eq. (9.13), the solution is not $\mathbf{H} = \mathbf{I}^{-1}\mathbf{Q}$; since the inverse of $\mathbf{I}$ does not exist, $\mathbf{I}$ is a non-square matrix of rank less than n.

This condition is bypassed by using a pseudo-inverse concept. Multiplying by $\mathbf{I}^T$ (transpose of $\mathbf{I}$)

$$\mathbf{I}^T\mathbf{I}\mathbf{H} = \mathbf{I}^T\mathbf{Q}. \tag{9.14}$$

The matrix $\mathbf{I}^T\mathbf{I}$ is of rank n and has an inverse. Therefore

$$\mathbf{H} = (\mathbf{I}^T\mathbf{I})^{-1}\mathbf{I}^T\mathbf{Q}. \tag{9.15}$$

Equation (9.15) also amounts to an unconstrained least squares estimation of the vector $\mathbf{H}$.

The above approach will generally result in values of $\mathbf{H}$ that are unstable and not well behaved. Negative unit hydrograph coordinates, volumes less than 1, and tail oscillations in the obtained unit hydrograph are common. These instabilities are due to non-perfect observations of $\mathbf{I}$ and $\mathbf{Q}$ and to the non-perfect linear assumption. The model identification problem should really be a constrained optimization of the form

$$\begin{aligned} &\min_{\mathbf{H}} \; F(\mathbf{Q}_{\text{obs}} - \mathbf{I}\mathbf{H}) \\ &\text{subject to} \quad \mathbf{H} \geq 0 \\ &\qquad\qquad\quad \mathbf{C}\mathbf{H} = 1, \end{aligned} \tag{9.16}$$

where $F(\cdot)$ is the objective function of the error between observed discharge $\mathbf{Q}_{\text{obs}}$ and predicted model discharge $\mathbf{IH}$.

The optimization is constrained to positive values of $\mathbf{H}$ and to unit volume expressed in terms of a linear operation on the vector $\mathbf{H}$, $\mathbf{CH}$. For an objective function of the general form

$$\text{Min}(\mathbf{Q}_{\text{obs}} - \mathbf{I}\mathbf{H})\mathbf{V}^{-1}(\mathbf{Q}_{\text{obs}} - \mathbf{I}\mathbf{H})^T, \tag{9.17}$$

the problem becomes one of quadratic programming and is solved in the literature (Natale and Todini [1976a]; Eagleson et al. [1966]).

The system of equations given in Eq. (9.12) can also be expressed as

$$Q_i = \sum_{j=1}^{i} i_j h_{i-j+1}, \tag{9.18}$$

which is the discrete form of the so-called convolution integral. This convolution or Duhamel integral will become very important in our study of continuous linear systems in hydrology.

EXAMPLE 9.3

Derivation of Unit Hydrographs Using Least Squares Procedures

Singh [1976] derived the unit hydrographs corresponding to the four storms given in Table 9.4. The table gives the effective precipitation and the corresponding discharge. He used two procedures, one based on solving Eq. (9.16) when the objective was to minimize the absolute value of the error $\mathbf{Q}_{\text{obs}} - \mathbf{IH}$; the other was equivalent to Eq. (9.15). Method 1 imposed nonnegativity and volume constraints on the unit hydrograph. A constraint requiring a monotonically decreasing recession was also added. Linear programming was the optimizing tool used. Method 2 imposed no constraints. Figure 9.17 gives the resulting unit hydrographs. Differences in the methods were observed only in storms 1 and 4. Note the tail oscillations and negative unit hydrograph ordinates occurring with method 2. As Figure 9.18 shows, the derived unit hydrographs reproduce the observed discharges quite well. ◆

9.4.3 Synthetic Unit Hydrographs

Determination of the unit hydrograph depends on the availability of input and output data. For areas where these are not available, hydrologists have developed techniques to relate parts of the hydrograph to physical basin characteristics. The hydrograph characteristics of interest are the peak, time to peak, duration of corresponding effective rainfall, and time base of the unit hydrograph. Knowing these points, curves with unit volume could be traced, which would be a fair approximation of the desired system response function.

The first and the most commonly used synthetic unit hydrograph is that developed by Snyder [1938], modified by Taylor and Schwartz [1952], and used by the U.S. Army Corps of Engineers [1959]. As most of these procedures are based on empirical relations obtained through analysis of data from several basins, Snyder [1938] relates the time from the centroid of the rainfall to the peak of the unit hydrograph to geometrical characteristics of the basin by

$$t_\ell = C_t(LL_c)^{0.3}, \tag{9.19}$$

TABLE 9.4 Study Storms over North Branch Potomac River near Cumberland, Maryland

STORM 1 Δt = 4 hr		STORM 2 Δt = 4 hr		STORM 3 Δt = 4 hr		STORM 4 Δt = 4 hr	
R (in.)	Q (in. hr^{-1})	R (in.)	Q (in. hr^{-1})	R (in.)	Q (in. hr^{-1})	R (in.)	Q (in. hr^{-1})
0.12	0.003	0.36	0.001	0.12	0.0005	0.20	0.0015
0.88	0.011	0.84	0.005	0.48	0.0017	0.04	0.0050
0.80	0.036	0.92	0.023	0.60	0.0049	0.32	0.0080
1.00	0.090	0.04	0.052		0.0151	0.08	0.0095
0.24	0.140	0.12	0.083		0.0256	0.72	0.0110
3.04	0.140	0.24	0.088		0.0238	0.44	0.0270
	0.110	2.52	0.081		0.0174	0.12	0.0500
	0.082		0.070		0.0136	0.84	0.0720
	0.057		0.061		0.0111	0.12	0.0850
	0.037		0.047		0.0091	0.52	0.0890
	0.025		0.037		0.0073	3.40	0.0845
	0.014		0.028		0.0060		0.0830
	0.008		0.020		0.0046		0.0835
	0.004		0.014		0.0035		0.0690
	0.002		0.009		0.0024		0.0520
	0.001		0.005		0.0016		0.0360
	0.760		0.003		0.0011		0.0260
	× 4		0.002		0.0005		0.0190
	= 3.04 in.		0.001		0.0002		0.0140
			0.630		0.1500		0.0100
			× 4		× 4		0.0070
			= 2.52 in.		= 0.60 in.		0.0040
							0.0025
							0.0010
							0.0005
							0.8500
							× 4
							= 3.40 in.

Note: 1 in. = 0.0254 m; 1 in. hr^{-1} = 0.0254 m hr^{-1}.

Source: Singh [1976]. Reproduced by permission of the American Water Resources Association.

where t_ℓ is time to peak in hours, L is length of the main stream channel in miles, and L_c is length of the main stream channel to a point opposite the basin centroid in miles. English units are used as a result of the original calibration.

The factor LL_c attempts to parameterize the shape of the basin in terms of a length (L) and width (L_c) measure. The time to peak is a function of basin shape, but following the linear assumption is not assumed to be a function of rainfall intensity. The factor C_t represents variations in watershed

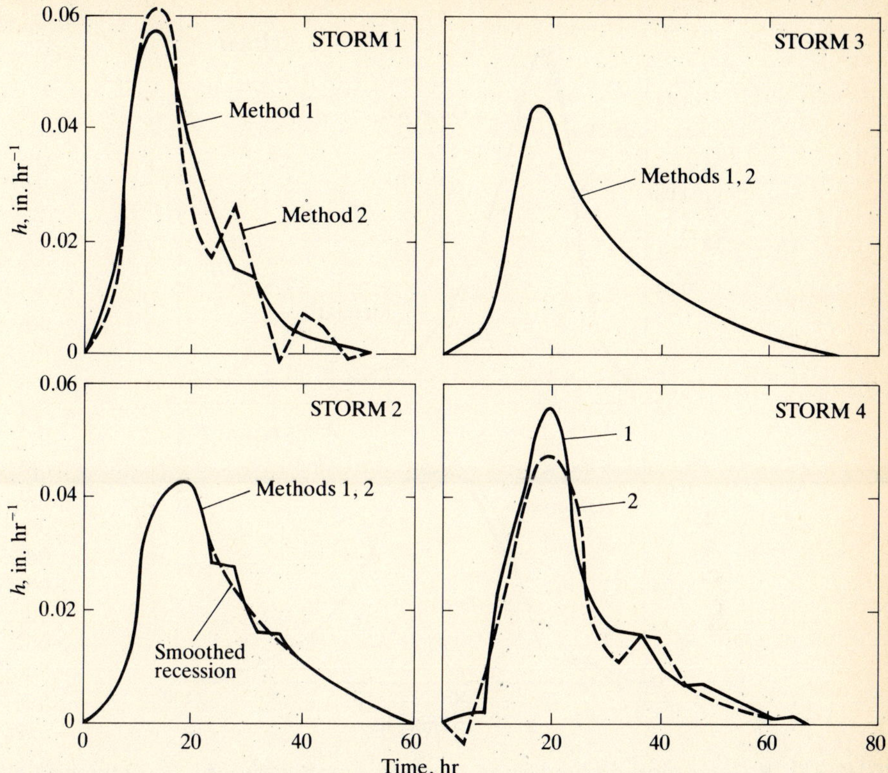

FIGURE 9.17 Unit hydrographs obtained from constrained (method 2) and unconstrained (method 1) estimation procedures. Source: Singh [1976]. Reproduced by permission of the American Water Resources Association.

slopes and storage and usually varies between 1.8 and 2.2 (for English units). The time to the peak is generally measured from the centroid of the rainfall input.

Given the time to the peak defined by Eq. (9.19), the duration of corresponding effective rainfall is obtained by

$$t_r = \frac{t_\ell}{5.5}, \tag{9.20}$$

where t_r is the effective rainfall duration in hours.

The peak discharge is

$$Q_p = 640\frac{C_p A}{t_\ell}, \tag{9.21}$$

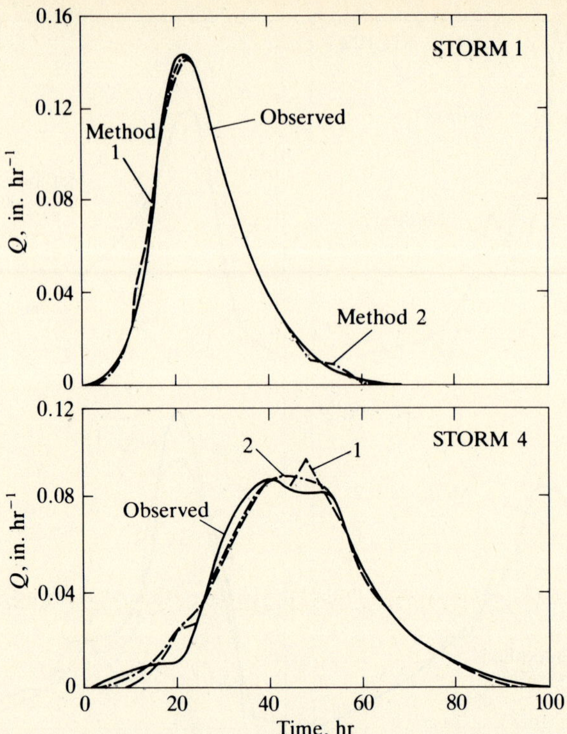

FIGURE 9.18 Given and derived direct surface runoff hydrographs obtained using unit hydrographs from the two methods and rainfall data for storms 1 and 4 (1 in. hr^{-1} = 0.0254 m hr^{-1}). Source: Singh [1976]. Reproduced by permission of the American Water Resources Association.

where Q_p is the peak discharge of unit hydrograph in cubic feet per second, A is the area in square miles, and C_p is the coefficient accounting for retention and storage, usually varying between 0.4 and 0.8.

The unit hydrograph time base is given as

$$T = 3 + \frac{t_\ell}{8}, \tag{9.22}$$

where T is in days. Equation (9.22) is adequate for large areas but highly inadequate for fast-responding small basins. In such cases, three to five times the time to peak t_ℓ is a good approximation for T.

Equations (9.19) and (9.20) limit the user to a hydrograph corresponding to a fixed duration of effective rainfall. To obtain a different-duration hydrograph, Snyder suggests using

$$t_{\ell R} = t_\ell + 0.25(t_R - t_r), \tag{9.23}$$

where t_R is the new desired effective rainfall duration in hours, $t_{\ell R}$ is the new time to peak corresponding to new effective rainfall duration, and t_ℓ and t_r are given by Eqs. (9.19) and (9.20).

Equations (9.19) through (9.23) define points in the unit hydrograph that can be used to trace a volume 1 curve. This sketching exercise is aided by Figure 9.19, where widths at 50 and 75% of peak flow are given as a function of peak flow per unit area. As a rule of thumb, the obtained widths should be divided in a 1:2 ratio between the rising and receding portions of the sketched unit hydrograph.

The U.S. Soil Conservation Service (Mockus [1957]) bases their synthetic unit hydrograph procedures on a nondimensional unit hydrograph developed from studies of many basins in the United States. The dimensionless hydrograph is shown in Figure 9.20 and in Table 9.5. To use this method, t_p, the time from the beginning of the storm to peak discharge, in hours, is obtained by

$$t_p = \frac{D}{2} + t_\ell, \tag{9.24}$$

where D is the duration of effective rainfall and t_ℓ is the lag time from a rainfall centroid to peak discharge.

Lag time t_ℓ can be obtained using localized relations between t_ℓ and A as

$$\begin{aligned} t_\ell &= 1.44A^{0.6} \quad \text{Texas}, \\ t_\ell &= 0.54A^{0.6} \quad \text{Ohio}, \end{aligned} \tag{9.25}$$

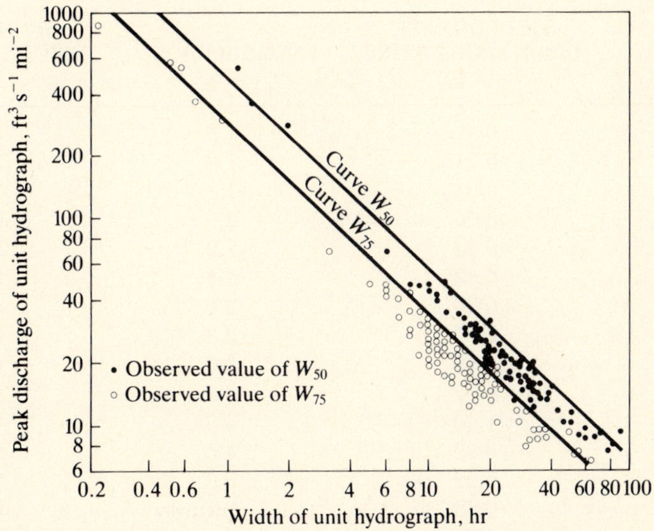

FIGURE 9.19 Unit hydrograph width at 50 and 75% of peak flow. Source: U.S. Army Corps of Engineers [1959].

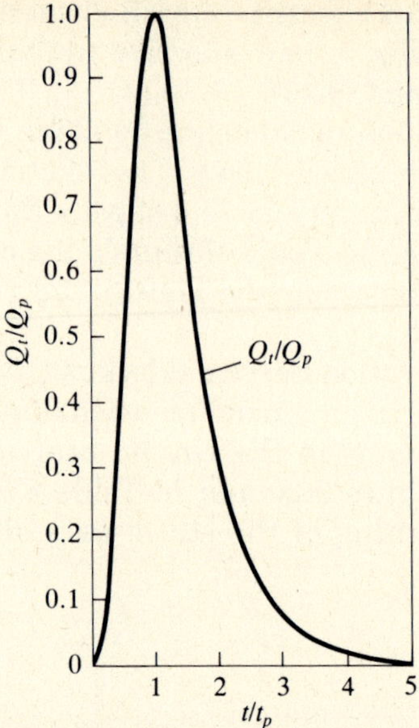

FIGURE 9.20 Dimensionless unit hydrograph of the Soil Conservation Service. Source: After Mockus [1957].

TABLE 9.5 Ratios for the Basic Dimensionless Hydrograph of the SCS

TIME RATIOS, t/t_p	HYDROGRAPH DISCHARGE RATIOS, Q/Q_p	TIME RATIOS, t/t_p	HYDROGRAPH DISCHARGE RATIOS, Q/Q_p
0	0	1.5	0.66
0.1	0.015	1.6	0.56
0.2	0.075	1.8	0.42
0.3	0.16	2.0	0.32
0.4	0.28	2.2	0.24
0.5	0.43	2.4	0.18
0.6	0.60	2.6	0.13
0.7	0.77	2.8	0.098
0.8	0.89	3.0	0.075
0.9	0.97	3.5	0.036
1.0	1.00	4.0	0.018
1.1	0.98	4.5	0.009
1.2	0.92	5.0	0.004
1.3	0.84	Infinity	0
1.4	0.75		

Source: Gray [1973]. Reproduced by permission of the National Research Council of Canada.

where A is in square miles or by relations such as Eq. (9.19), generalized to include the effects of slope

$$t_\ell = a\left(\frac{LL_c}{\sqrt{S}}\right)^b, \tag{9.26}$$

where S is the average main channel slope. Relations of the form given by Eq. (9.26) are shown in Figure 9.21.

The peak discharge results from

$$Q_p = \frac{484A}{t_p}, \tag{9.27}$$

where Q_p is discharge in cubic feet per second, A is area in square miles, and t_p is time to peak from beginning of rainfall in hours. The constant 484 results from the conversion of square miles to acres and an assumption that the storm duration $t_r = 1.67t_p$. The reader is referred to McCuen [1982] for more detail on the Soil Conservation Service (SCS) unit hydrograph.

The Gray synthetic unit hydrograph has the form of a gamma function. Following Gray [1973], it is assumed that the amount of contributing area to the discharge hydrograph increases proportionally to a power of time

$$A(t) \propto t^x. \tag{9.28}$$

Discharge is also proportional to contributing area. Therefore

$$Q(t) \propto t^x. \tag{9.29}$$

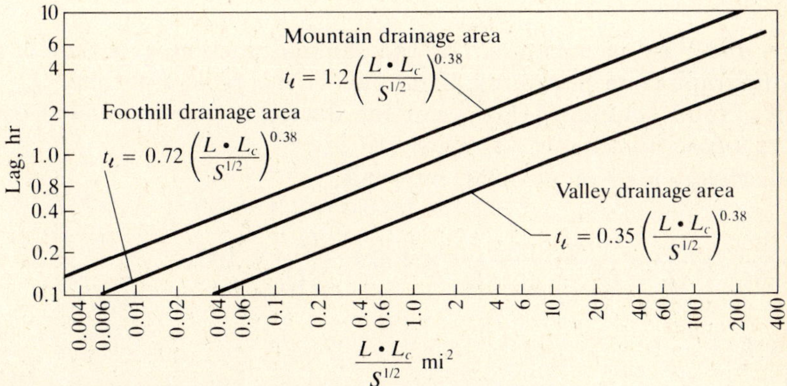

FIGURE 9.21 Relation between lag time and basin properties.

We have seen previously that the recession of the hydrograph can be expressed as

$$Q(t) = Q_0 K^t$$

$$\ln Q(t) = \ln Q_0 + t \ln K = \ln Q_0 - kt \tag{9.30}$$

or

$$Q(t) \propto e^{-kt}. \tag{9.31}$$

Combining Eqs. (9.29) and (9.31),

$$Q(t) \propto t^x e^{-kt} \qquad \text{or} \qquad Q(t) = Bt^x e^{-kt}. \tag{9.32}$$

Integrating Eq. (9.32),

$$\text{Vol} = \int_0^t Q(t)\, dt = \int_0^t Bt^x e^{-kt}\, dt. \tag{9.33}$$

Let $x = m - 1$ and $x = kt$, carrying on the integration to evaluate constant B and rearranging will result in

$$Q(t) = \frac{\text{Vol}\; k^m e^{-kt} t^{m-1}}{\Gamma(m)}. \tag{9.34}$$

The last equation is in the form of a gamma probability density function (see Chapter 11).

For small watersheds, Gray [1961a, 1962] suggested a dimensionless unit hydrograph similar to Eq. (9.34).

$$Q_{t/t_p} = \frac{25\gamma^q e^{-\gamma t/t_p} (t/t_p)^{q-1}}{\Gamma(q)}, \tag{9.35}$$

where γ and q are parameters, Γ is the gamma function, t_p is time of rise of the hydrograph from beginning of rainfall to the peak, t is time, Q_{t/t_p} is the percent of total volume of flow occurring during a time increment equal to $0.25t_p$ at a specific value of t/t_p (Gray [1973]).

If discharge is given at $0.25t_p$ intervals,

$$\frac{\%\ \text{flow}}{0.25t_p} = \frac{Q_i}{Q_1 + Q_2 + \cdots + Q_n} \times 100 = \frac{Q_i}{\Sigma Q_i} \times 100. \tag{9.36}$$

Therefore

$$Q_i = \frac{\%\ \text{flow}/0.25t_p}{100} \frac{\text{A}}{0.25t_p} P_n, \tag{9.37}$$

where P_n is the net storm rainfall depth and

$$\frac{AP_n}{0.25t_p} = \Sigma Q_i\,.$$

The parameters q and γ are given by

$$q = \gamma + 1$$
$$\frac{t_p}{\gamma} = a\left(\frac{L}{\sqrt{S_c}}\right)^b, \tag{9.38}$$

where L is the length of the main stream and S_c is the mean percentage slope of the main channel. If t_p/γ is in minutes (γ is dimensionless), then

$$\frac{t_p}{\gamma} = \frac{1}{2.676/t_p + 0.0139}. \tag{9.39}$$

Equation (9.38) plots as a straight line on log–log paper with parameters a and b varying with location. For L in miles and t_p in minutes, some typical values for a and b, respectively, are 11.4 and 0.531, Ohio; 7.4 and 0.498, Nebraska and Iowa; 9.27 and 0.562, Illinois, Montana, and Wisconsin. Figures 9.22 and 9.23 plot Eqs. (9.38) and (9.39), respectively. The general

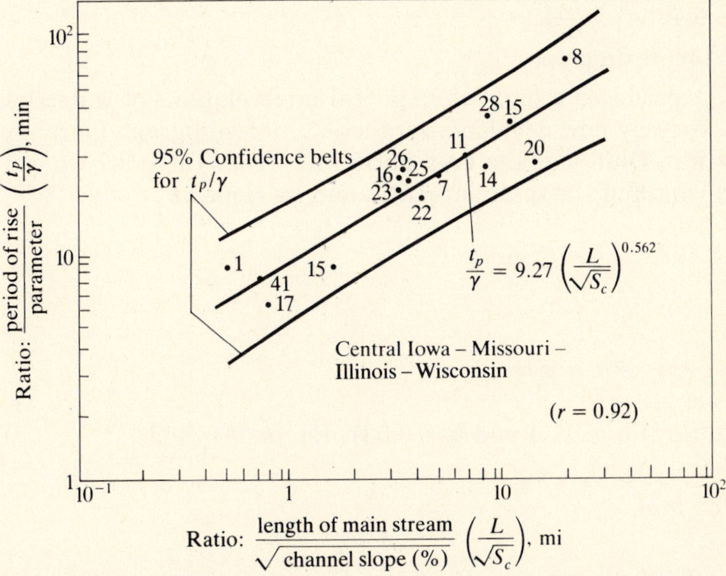

FIGURE 9.22 Relation of storage factor t_p/γ and watershed parameter $L/\sqrt{S_c}$, for watersheds in central Iowa–Missouri–Illinois–Wisconsin. Source: D. M. Gray, "Synthetic Unit Hydrographs for Small Watersheds," *J. Hydraul. Div. A.S.C.E.* 87(HY4):33–54, 1961.

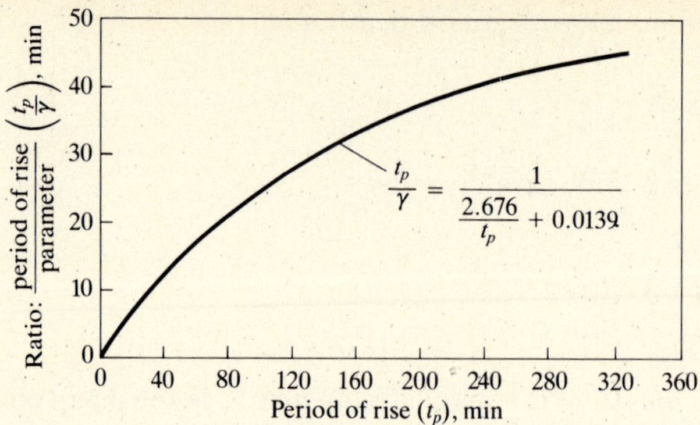

FIGURE 9.23 Relation of storage factor t_p/γ and period of rise t_p. Source: D. M. Gray, "Synthetic Unit Hydrographs for Small Watersheds," *J. Hydraul. Div. A.S.C.E.* 87(HY4): 33–54, 1961.

procedure to use Gray's method is then to

1. Obtain t_p/γ from Eq. (9.38) (Fig. 9.22).
2. With this value enter Figure 9.23 (Eq. 9.39) and obtain t_p.
3. Use Eq. (9.35) or Table 9.6 and Eq. (9.37) to obtain discharges at any t/t_p.

EXAMPLE 9.4

Synthetic Unit Hydrograph

Table 9.7 reproduces a table of empirical interrelations of watershed characteristics. We may now develop a synthetic unit hydrograph for a hypothetical river basin in Ohio. Let the area be 10 mi^2. From Table 9.7 we can approximate the length of the main stream L and its slope as

$$L = 1.40A^{0.568} = 5.2 \text{ mi}$$

and

$$S_c = 1.57L^{-0.662} = 0.53\%.$$

Using values of $a = 11.4$ and $b = 0.531$, Eq. (9.38) yields

$$\frac{t_p}{\gamma} = 32 \text{ min}.$$

From Eq. (9.39) we get that the time of rise t_p is 154 min. This implies that the unit hydrograph to be developed corresponds to a storm of 38.5 min

TABLE 9.6 Dimensionless Graph Coordinates for Different Values of the Parameter γ

	% FLOW/$0.25t_p$[a]								
t/t_p	$\gamma = 2.0$	$\gamma = 2.5$	$\gamma = 3.0$	$\gamma = 3.5$	$\gamma = 4.0$	$\gamma = 4.5$	$\gamma = 5.0$	$\gamma = 5.5$	$\gamma = 6.0$
0.000	0.0	0.0	0.0	0.0	0.0	0.0	0.0	0.0	0.0
0.125	1.2	0.8	0.5	0.3	0.2	0.1	0.1	0.1	0.1
0.375	6.6	6.3	5.8	5.2	4.7	4.2	3.7	3.2	2.8
0.625	11.2	12.0	12.6	13.0	13.3	13.6	13.6	13.6	13.5
0.875	13.3	14.9	16.4	17.7	18.9	20.0	21.0	22.0	22.9
1.000	13.5	15.3	16.8	18.2	19.5	20.8	21.9	23.0	24.1
1.125	13.4	15.0	16.5	17.8	19.0	20.1	21.1	22.1	23.1
1.375	12.1	13.3	14.2	14.9	15.6	16.1	16.6	16.9	17.2
1.625	10.3	10.7	11.1	11.2	11.2	11.1	10.9	10.7	10.5
1.875	8.3	8.2	8.0	7.7	7.3	6.8	6.4	6.0	5.5
2.125	6.5	6.0	5.5	5.0	4.4	3.9	3.4	3.0	2.6
2.375	4.9	4.3	3.6	3.1	2.5	2.1	1.7	1.4	1.1
2.625	3.6	2.9	2.3	1.8	1.4	1.0	0.8	0.6	0.5
2.875	2.6	2.0	1.4	1.0	0.8	0.6	0.4	0.3	0.2
3.125	1.9	1.3	0.9	0.6	0.4	0.3	0.2	0.1	
3.375	1.3	0.9	0.5	0.3	0.2	0.1	0.1		
3.625	1.0	0.6	0.3	0.2	0.1				
3.875	0.6	0.3	0.2	0.1					
4.125	0.4	0.2	0.1	0.1					
4.375	0.3	0.1	0.1						
4.625	0.2	0.1							
4.875	0.1	0.1							
5.125	0.1								
5.375	0.1								
Sum[b]	100.0	100.0	100.0	100.0	100.0	100.0	100.0	100.0	100.0

[a]Rounded to the nearest 0.10%.

[b]Sums do not include peak percentages.

Source: D. M. Gray, "Synthetic Unit Hydrographs for Small Watersheds," *J. Hydraul. Div. A.S.C.E.* 87(HY4):50, 1961.

duration, $t_p/4$. Parameter γ must be 4.8. The unit hydrograph peaks at $t/t_p = 1$. From Table 9.6, for $\gamma = 4.8$, we get % flow/$0.25t_p = 21.5$. Since the unit hydrograph corresponds to a net storm depth of 1 unit, say 1 cm, we can use Eq. (9.37) to obtain the unit hydrograph peak:

$$Q_p = \frac{21.5}{100}\,\frac{10\text{ mi}^2(2.59\text{ km}^2\text{ mi}^{-2})(10^6\text{ m}^2\text{ km}^{-2})(1\times 10^{-2}\text{ m})}{0.25(154)\times 60}$$

$$= 24\text{ m}^3\text{ s}^{-1}.\ \blacklozenge$$

TABLE 9.7 Interrelations of Watershed Characteristics

REFERENCE	RELATIONSHIP[a]	CORRELATION	GEOGRAPHIC LOCATION
Gray [1961], Taylor and Schwartz [1952]	$L = 1.40A^{0.568}$	0.97	Ill., Iowa, Mo., Nebr., No. Calif., Ohio, Wisc., No. and Middle U.S.
Gray [1961a]	$L_{ca} = 0.54L^{0.95}$ or $L_{ca} = 0.90A^{0.56}$	0.99	Ill., Iowa, Mo., Nebr., No. Calif., Ohio, Wisc.
Langbein and others [1947]	$L_{ca} = 0.74A^{0.55}$		340 watersheds from Northeastern U.S.
Gray [1961a]	$S_c = 15.95L^{-0.610}$	0.94	North Carolina
	$S_c = 2.61L^{-0.770}$	0.93	Nebr., Western Iowa
	$S_c = 1.57L^{-0.662}$	0.97	Ohio
	$S_L = 0.86s_1^{0.67}$	0.96	Iowa
	$A/A_c = 0.61$	$C_v = 23\%$	Ill., Iowa, Mo., Nebr., No. Calif., Ohio, Wisc.
	$D/L' = 0.68$	$C_v = 20\%$	Ill., Iowa, Mo., Nebr., No. Calif., Ohio, Wisc.

[a] L = length of main stream from gaging station to outermost point (mi); A = plane area of the watershed enclosed within the topographic divide (mi^2); L_{ca} = distance along the main stream from the gaging station to a point on the stream nearest the center of mass of the area (mi); S_c = slope of the main stream, i.e., slope of a line drawn along the longitudinal section of the main channel in such a manner that the area between the line and a horizontal line drawn through the channel outlet elevation is equal to the area between the channel grade line and the same horizontal line (%); S_L = mean land slope of a watershed as determined by the grid-intersection method (ft/ft); s_1 = average slope of a number of first order streams (ft/ft) which are all streams that have no upstream tributaries and are sources of the network (see Chapter 12); D = diameter of circle with equal area as the basin (mi); A_c = area of the circle with equal perimeter of the basin (mi^2); L' = maximum length of basin parallel to the principal drainage lines (mi); and C_v is a coefficient of variation.

Source: Gray [1973]. Reproduced by permission of the National Research Council of Canada.

9.5 THE INSTANTANEOUS UNIT HYDROGRAPH

A unit hydrograph of arbitrary duration was seen to be derived from a subtraction and normalization of two displaced S curves. Having the 1-hr S curve, a dt-hour unit hydrograph can be obtained by

$$\frac{S(t) - S(t - dt)}{dt}.$$

If we allow dt to become infinitely small, the equation becomes

$$\frac{S(t) - S(t - dt)}{dt} \xrightarrow[dt \to 0]{} \frac{dS}{dt}. \tag{9.40}$$

Therefore dS/dt defines the response of a unit volume storm instantaneously occurring over the area (a Dirac-delta function). This response function is called the instantaneous unit hydrograph (IUH).

The instantaneous unit hydrograph $h(t)$ defines a linear system through the convolution integral

$$\begin{aligned} Q(t) &= \int_{-\infty}^{\infty} h(\tau) I(t - \tau)\, d\tau \\ &= \int_{-\infty}^{\infty} I(\tau) h(t - \tau)\, d\tau, \end{aligned} \tag{9.41}$$

where $Q(t)$ is output of the system at time t, $I(t)$ is input to the system, and $h(t)$ is a response function. The convolution is a restatement of the superposition principle for continuous systems. Note that the response function is defined through the identity

$$h(t) = \int_{-\infty}^{\infty} \delta(t - \tau) h(\tau)\, d\tau, \tag{9.42}$$

where $\delta(t - \tau)$ is the Dirac-delta function:

$$\delta(t - \tau) = \begin{cases} \infty & \text{for } t = \tau \\ 0 & \text{for } t \neq \tau \end{cases}$$

and

$$\int_{-\infty}^{\infty} \delta(\tau)\, d\tau = 1.$$

In hydrology and most real and causal systems, there is no negative time nor system response to future inputs. Therefore the convolution equation for a river basin becomes

$$Q(t) = \int_{0}^{t} I(\tau) h(t - \tau)\, dt. \tag{9.43}$$

Equation (9.43) corresponds graphically to Figure 9.24. Any continuous input is interpreted as a series of infinitesimal impulses of finite intensity that respond according to the IUH. In analogy to the unit hydrograph, each

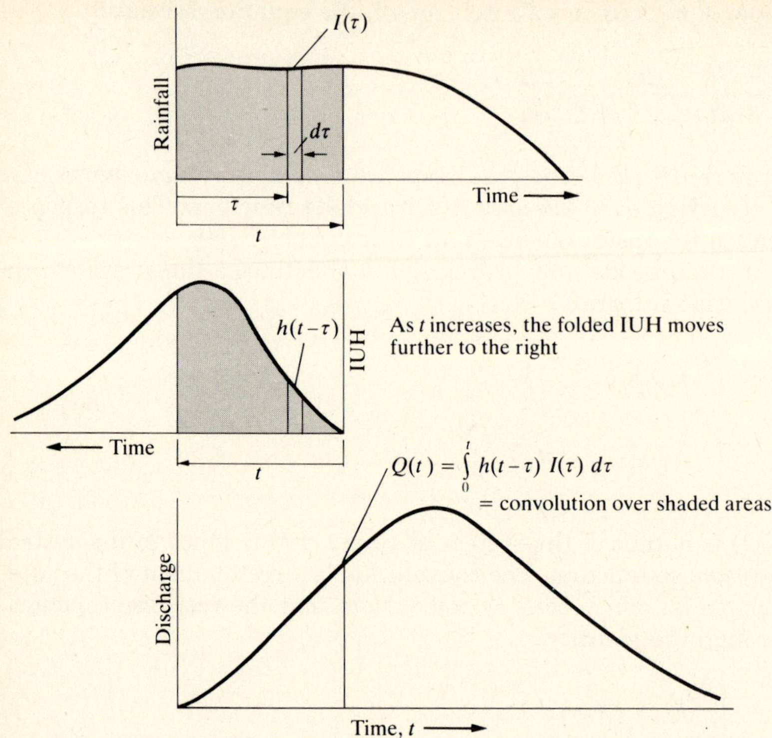

FIGURE 9.24 The concept of convolution.

lagged response is added to the others to obtain the global response of the compound input. The convolution integral represents that summation. Graphically, it amounts to translating the mirror image of the folded function (argument $t - \tau$) under that of the unfolded one (argument τ) and at every translation (corresponding to a time t) multiplying ordinates and adding. Note that by definition the time base of the IUH is the time of concentration of the basin.

Identification of $h(t)$ is a perennial topic that lends itself to innumerable mathematical techniques developed to deal with continuous linear systems. A detailed discussion of all different model identification procedures is beyond the scope of this work. The next few sections will simply touch on some of the methods, hopefully leading the curious to deeper studies.

*9.5.1 Fourier Series

Mathematical expansion of functions in terms of polynomials of orthogonal functions is useful because the coefficients of the expansion are easily identified. The most common expansions of periodic functions are Fourier series.

Most periodic functions $f(t)$ can be expressed as

$$f(t) = \frac{1}{2}A_0 + \sum_{k=1}^{\infty}\left(A_k \cos\frac{k2\pi t}{T} + B_k \sin\frac{k2\pi t}{T}\right), \tag{9.44}$$

where T is period of the function and A_k and B_k are coefficients of sine and cosine terms of increasing frequencies.

$$\begin{aligned} A_k &= \frac{2}{T}\int_{-T/2}^{T/2} f(t)\cos\left(\frac{k2\pi t}{T}\right) dt \\ B_k &= \frac{2}{T}\int_{-T/2}^{T/2} f(t)\sin\left(\frac{k2\pi t}{T}\right) dt\,. \end{aligned} \tag{9.45}$$

None of the hydrologic functions in Eq. (9.43) are periodic. To apply Fourier series, they must be assumed periodic with a period equal to or greater than

$$T = T_i + T_u\,, \tag{9.46}$$

where T is the time base of the output hydrograph, T_i is the time base of the input hydrograph, and T_u is the time base of the unit hydrograph. This periodic assumption is represented in Figure 9.25. It can then be shown (see Bras and Rodriguez-Iturbe [1985] for details of this procedure) that Eq. (9.43) is equivalent to

$$Q(t) = \int_{t-T}^{t} I(\tau)h(t-\tau)\,d\tau\,, \tag{9.47}$$

where each of the intervening functions can be expressed as

$$\begin{aligned} Q(t) &= \sum_{r=0}^{\infty} A_r \cos r\frac{2\pi t}{T} + \sum_{r=1}^{\infty} B_r \sin r\frac{2\pi t}{T} \\ I(\tau) &= \sum_{n=0}^{\infty} a_n \cos n\frac{2\pi\tau}{T} + \sum_{n=1}^{\infty} b_n \sin n\frac{2\pi\tau}{T} \\ h(t-\tau) &= \sum_{m=0}^{\infty} \alpha_m \cos m\frac{2\pi(t-\tau)}{T} + \sum_{m=1}^{\infty} \beta_m \sin m\frac{2\pi(t-\tau)}{T}\,. \end{aligned} \tag{9.48}$$

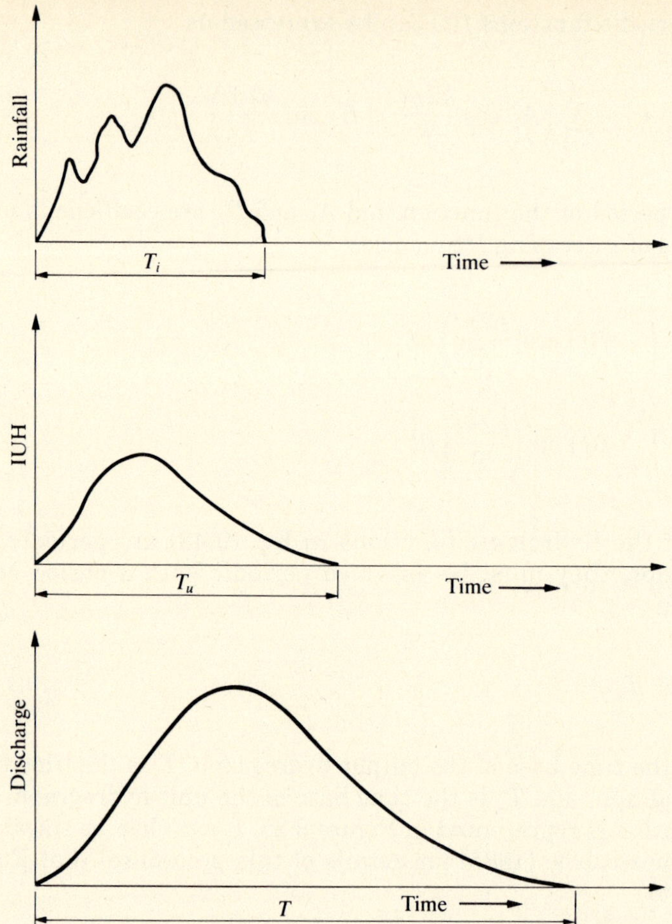

FIGURE 9.25 A periodic interpretation of the instantaneous unit hydrograph.

Substitution of Eqs. (9.48) into Eq. (9.47) and interchanging integrals and summations results in an expression for $Q(t)$ in terms of an infinite summation of integrals. For each pair of m and n, the following products appear:

$$
\begin{aligned}
&\text{a)}\quad a_n\alpha_m \int_{t-T}^{t} \cos n\frac{2\pi\tau}{T} \cos m\frac{2\pi(t-\tau)}{T}\,d\tau \\
&\text{b)}\quad a_n\beta_m \int_{t-T}^{t} \cos n\frac{2\pi\tau}{T} \sin m\frac{2\pi(t-\tau)}{T}\,d\tau \\
&\text{c)}\quad b_n\alpha_m \int_{t-T}^{t} \sin n\frac{2\pi\tau}{T} \cos m\frac{2\pi(t-\tau)}{T}\,d\tau \\
&\text{d)}\quad b_n\beta_m \int_{t-T}^{t} \sin n\frac{2\pi\tau}{T} \sin m\frac{2\pi(t-\tau)}{T}\,d\tau
\end{aligned}
\tag{9.49}
$$

Because of the orthogonality of the sine and cosine functions over the period T, the above are zero except when $m = n$. In that case, the integrals are equal to

$$\begin{aligned}
&\text{a)}\quad \frac{T}{2}a_n\alpha_n \cos n\frac{2\pi t}{T}\\
&\text{b)}\quad \frac{T}{2}a_n\beta_n \sin n\frac{2\pi t}{T}\\
&\text{c)}\quad \frac{T}{2}b_n\alpha_n \sin n\frac{2\pi t}{T}\\
&\text{d)}\quad -\frac{T}{2}b_n\beta_n \cos n\frac{2\pi t}{T}
\end{aligned} \tag{9.50}$$

$Q(t)$ is then expressed as

$$\begin{aligned}
Q(t) = \frac{T}{2}\alpha_0 a_0 + \sum_{n=1}^{\infty}\frac{T}{2}(a_n\alpha_n - b_n\beta_n)\cos n\frac{2\pi t}{T}\\
+ \sum_{n=1}^{\infty}\frac{T}{2}(a_n\beta_n + b_n\alpha_n)\sin n\frac{2\pi t}{T}.
\end{aligned} \tag{9.51}$$

By analogy to Eq. (9.48), then

$$\begin{aligned}
\frac{A_0}{2} &= \frac{Ta_0\alpha_0}{2}\\
A_n &= \frac{T}{2}(a_n\alpha_n - b_n\beta_n) \qquad n \geq 1\\
B_n &= \frac{T}{2}(a_n\beta_n + b_n\alpha_n)
\end{aligned} \tag{9.52}$$

or

$$\begin{aligned}
\alpha_0 &= \frac{1}{T}\frac{A_0}{a_0}\\
\alpha_n &= \frac{2}{T}\frac{a_nA_n + b_nB_n}{a_n^2 + b_n^2} \qquad n \geq 1\\
\beta_n &= \frac{2}{T}\frac{a_nB_n - b_nA_n}{a_n^2 + b_n^2}.
\end{aligned} \tag{9.53}$$

Given the Fourier series expansion for a known pair of input and output functions, we can obtain the coefficients of the Fourier series expansion of the response function in terms of the known input and output coefficients. This procedure was first suggested by O'Donnell [1960].

The technique or idea used in Fourier series can be extended to any series expansion of orthogonal functions. There are several of these types of functions. Expansions by Laguerre polynomials is discussed by Dooge [1973].

*9.5.2 Fourier and Laplace Transforms

A Fourier transform pair is defined as

$$f(\omega) = \frac{1}{2\pi}\int_{-\infty}^{\infty} f(t)e^{-i\omega t}\,dt \tag{9.54a}$$

$$f(t) = \int_{-\infty}^{\infty} f(\omega)e^{i\omega t}\,d\omega, \tag{9.54b}$$

where $f(\omega)$ is the representation of $f(t)$ in the frequency (ω) domain. The Laplace transform

$$f(s) = \int_{0}^{\infty} e^{-st}f(t)\,dt \tag{9.55}$$

has also an inverse transform that is considerably more complicated.

Taking Fourier transforms of

$$Q(t) = \int_{0}^{t} I(\tau)h(t-\tau)\,d\tau \tag{9.56}$$

will result in

$$Q(\omega) = 2\pi I(\omega)H(\omega), \tag{9.57}$$

where $Q(\omega)$, $I(\omega)$, and $H(\omega)$ are Fourier transforms of the corresponding real time outputs, inputs, and response function.

From Eq. (9.57),

$$H(\omega) = \frac{Q(\omega)}{2\pi I(\omega)}, \tag{9.58}$$

where $H(\omega)$ is complex with a real and imaginary part. These can be decomposed as

$$\begin{aligned} H_R(\omega) &= \frac{I_R(\omega)Q_R(\omega) + I_I(\omega)Q_I(\omega)}{I_R^2(\omega) + I_I^2(\omega)} \\ H_I(\omega) &= \frac{I_R(\omega)Q_I(\omega) + I_I(\omega)Q_I(\omega)}{I_R^2(\omega) + I_I^2(\omega)}, \end{aligned} \tag{9.59}$$

where subscripts R and I represent real and imaginary parts [$I(\omega)$ has been redefined as $2\pi I(\omega)$].

Since $h(t)$ is real, $H_R(\omega)$ is even, which allows a back-transform of the form

$$h(t) = \int_0^\infty H_R(\omega)\cos(\omega t)\,d\omega\,. \tag{9.60}$$

Linear systems are ideally suited for treatment in the frequency domain, i.e., with Fourier transforms. There are many other results that can be useful in the analysis or synthesis of hydrologic signals. The interested reader is referred to Bras and Rodriguez-Iturbe [1985] for a detailed treatment of the problem. Among the results are alternative procedures for the identification of the IUH Fourier transform, $H(\omega)$. Define, respectively, the autocorrelation and cross correlation of functions by

$$\psi_{ff}(\tau) = \int_{-\infty}^{\infty} f(t)f(t+\tau)\,d\tau\,,$$

$$\psi_{fg}(\tau) = \int_{-\infty}^{\infty} f(t)g(t+\tau)\,d\tau\,.$$

Their Fourier transforms, via Eq. (9.54), are $\Phi_{ff}(\omega)$ and $\Phi_{fg}(\omega)$, respectively. It can be shown, by operating on Eq. (9.56),

$$\Phi_{QQ}(\omega) = 4\pi^2|H(\omega)|^2\Phi_{II}(\omega) \tag{9.61}$$

and

$$\Phi_{IQ}(\omega) = 2\pi H(\omega)\Phi_{II}(\omega)\,, \tag{9.62}$$

where $|H(\omega)|^2 = H_R^2(\omega) + H_I^2(\omega)$.

The previous two equations can be used to obtain estimates of $H(\omega)$.

In summary, the power of transform methods lies in that they reduce a complicated integral equation to a linear algebraic form. Their problem is the possibly complicated inverse transformation required to obtain the time domain expression of the response function.

9.5.3 Moments and Cumulants (After Dooge [1973])

The last few subsections have shown how certain transformations convert the convolution equation into easily handled algebraic relations. The operation of taking moments of a linear system have the same simplifying characteristics.

The Rth moment of a function, which has been normalized to unit area, around the origin is

$$\mu_R'(f) = \int_{-\infty}^{\infty} f(t)t^R\,dt\,. \tag{9.63}$$

Similarly, the Rth moment around the center of mass (the first moment around the origin) is

$$\mu_R(f) = \int_{-\infty}^{\infty} f(t)\,(t - \mu_1')^R\, dt. \tag{9.64}$$

Any moment is related to the derivatives of a generating function that is the Fourier or bilateral Laplace transform of the original function. Using the bilateral Laplace transform,

$$F(s) = \int_{-\infty}^{\infty} f(t) \exp(-st)\, dt. \tag{9.65}$$

The first derivative of the Laplace transform is given by

$$\frac{d}{ds}[F(s)] = -\int_{-\infty}^{\infty} f(t)t \exp(-st)\, dt. \tag{9.66}$$

The Rth derivative is

$$\frac{d^R[F(s)]}{ds^R} = (-1)^R \int_{-\infty}^{\infty} f(t)t^R \exp(-st)\, dt. \tag{9.67}$$

The latter evaluated at the origin $s = 0$ results in

$$\left.\frac{d^R[F(s)]}{ds^R}\right|_{s=0} = (-1)^R \int_{-\infty}^{\infty} f(t)t^R\, dt = (-1)^R \mu_R'(f). \tag{9.68}$$

Equation (9.68) then relates the Rth moment around the origin of a function to the Rth derivative of the Laplace transform of the function at $s = 0$.

The Laplace transform of the convolution equation,

$$y(t) = \int_0^t h(\tau)x(t - \tau)\, d\tau, \tag{9.69}$$

results in

$$Y(s) = X(s)H(s) \tag{9.70}$$

(by analogy to the procedure shown in Section 9.5.2 for the Fourier transform).

The Rth moment about the origin of the output is

$$\begin{aligned}\mu_R'(y) &= (-1)^R \frac{d^R}{ds^R}[Y(s)]\Big|_{s=0} \\ &= (-1)^R \frac{d^R}{ds^R}[X(s)H(s)]\Big|_{s=0}\end{aligned} \tag{9.71}$$

Equation (9.71) can be expanded by using product differentiation rules, resulting in

$$\mu_R'(y) = (-1)^R \sum_{K=0}^{K=R} \binom{R}{K} \frac{d^K}{ds^K}[X(s)]\bigg|_{s=0} \frac{d^{R-K}}{ds^{R-K}}[H(s)]\bigg|_{s=0}$$

$$= \sum_{K=0}^{K=R} \binom{R}{K} \mu_K'(x)\mu_{R-K}'(h), \tag{9.72}$$

where $\binom{R}{K} = \frac{R!}{(R-K)!K!}$. A similar relationship exists for moments around the center of mass. Equation (9.72) leads to a very convenient additive relation between the centroids of inputs, outputs, and response function in a linear system. Relations between higher moments can also be obtained from Eq. (9.72) (an equivalent equation yields moments around the center of mass). For example,

$$\begin{aligned}
\mu_1'(y) &= \mu_1'(x) + \mu_1'(h) \\
\mu_2'(y) &= \mu_2'(x) + 2\mu_1'(x)\mu_1'(h) + \mu_2'(h) \\
\mu_2(y) &= \mu_2(x) + \mu_2(h) \\
\mu_3(y) &= \mu_3(x) + \mu_3(h)
\end{aligned} \tag{9.73}$$

Unfortunately, the additive convenience for the moments around the center of mass end with the third order.

To obtain the additive results throughout, a new parameter called a cumulant is defined. The Rth cumulant is the Rth derivative at the origin of the logarithm of the Laplace transform of a function

$$K_R(f) = (-1)^R \frac{d^R}{ds^R}[\log F(s)]\bigg|_{s=0}. \tag{9.74}$$

In a linear system,

$$Y(s) = X(s)H(s)$$

or

$$\log Y(s) = \log X(s) + \log H(s). \tag{9.75}$$

Upon differentiation and evaluation at $s = 0$, this results in the additive relation for cumulants

$$K_R(y) = K_R(x) + K_R(h). \tag{9.76}$$

It can be shown that

$$\begin{aligned} K_1(f) &= \mu_1'(f) \\ K_2(f) &= \mu_2(f) \\ K_3(f) &= \mu_3(f) \\ K_4(f) &= \mu_4(f) - 3\mu_2^2(f) \end{aligned} \tag{9.77}$$

The usefulness of moment transformations is that any function is completely defined by all its moments. The Laplace transform of a function is given by

$$F(s) = \sum_{K=0}^{K=\infty} (-1)^K \mu_K'(f) \frac{s^K}{K!} \tag{9.78}$$

as a function of all the moments of the function. Theoretically, then, given the moments, the inverse Laplace transform could be found for any function.

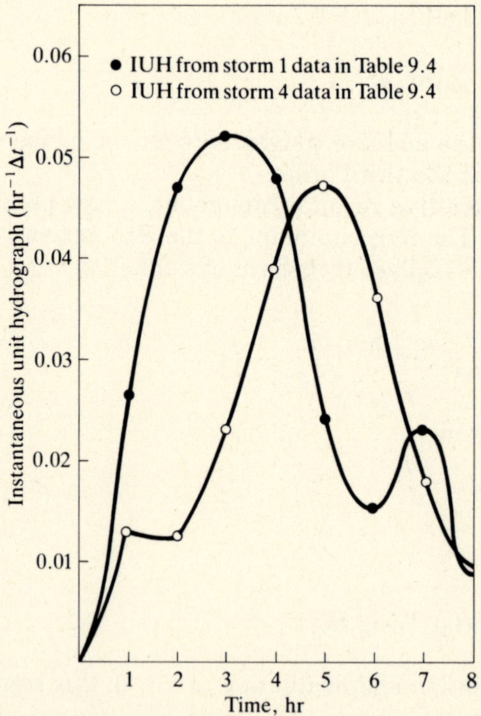

FIGURE 9.26 Instantaneous unit hydrographs obtained by O'Donnell's procedure for the North Branch Potomac River near Cumberland, Maryland.

The moments of the response function are easily obtained in terms of input and output moments using Eqs. (9.72), (9.73), or (9.76).

Fortunately, many functions are completely defined with a limited number of moments (i.e., normal, log normal, exponential, gamma). Furthermore, the higher moments can usually be ignored without seriously altering the shape of the function.

In the following section, the problem is further simplified by using conceptual instantaneous unit hydrographs. These are functional forms whose parameters are easily identified using the method of moments.

EXAMPLE 9.5

IUH Identification Using Fourier Series

Table 9.4 gave a series of inflow–outflow pairs in the North Branch Potomac River near Cumberland, Maryland (Singh [1976]). The area is 2266 km^2 in size. Using the Fourier series expansion procedure suggested by O'Donnell [1960], we get the IUHs shown in Figure 9.26 for storms 1 and 4. To agree with the data, the units of the IUH must be $hr^{-1}\,\Delta t^{-1}$, where Δt is 4 hr. Note the oscillations and negative values at the tail of the computed instantaneous unit hydrograph. ◆

9.6 CONCEPTUAL INSTANTANEOUS UNIT HYDROGRAPHS

Possibly the simplest conceptual IUH results from the assumption that a river basin behaves like a linear reservoir. Figure 9.27 schematizes the linear reservoir responding with discharge $Q(t)$ and receiving input $I(t)$. The definition of the linear reservoir is that

$$S(t) = KQ(t) \tag{9.79}$$

or that storage in the reservoir is proportional to discharge (and vice versa) at all times. Equation (9.79) is the dynamics equation of the simple model.

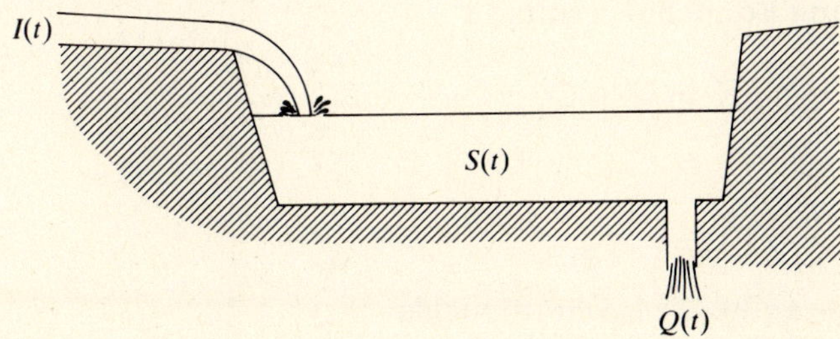

FIGURE 9.27 A conceptual storage reservoir. Source: Evans et al. [1972].

Mass balance states

$$I(t) = Q(t) + \frac{dS}{dt}. \tag{9.80}$$

Combining Eqs. (9.77) and (9.80) results in

$$I(t) = Q(t) + K\frac{dQ(t)}{dt} \tag{9.81}$$

or

$$\frac{dQ(t)}{dt} + \frac{Q(t)}{K} = \frac{I(t)}{K}. \tag{9.82}$$

For constant I, the solution to the above linear, ordinary differential equation is of the form

$$Q(t) = Ce^{-t/K} + I. \tag{9.83}$$

Setting initial conditions

$$Q(0) = 0$$

gives $C = -I$.

Therefore the discharge from a linear reservoir under a constant input I is

$$Q(t) = I(1 - e^{-t/K}). \tag{9.84}$$

Assume now that I terminates at t_0. The mass balance and dynamic Eqs. (9.79) and (9.80) now become

$$Q(t) = -K\frac{dQ(t)}{dt} \tag{9.85}$$

with $Q = Q_0$ at $t = t_0$.

Integrating Eq. (9.85) results in

$$t - t_0 = \tau = -K \ln Q + C' \tag{9.86}$$

or

$$Q = Ce^{-\tau/K}.$$

Imposing $Q = Q_0$ at $t = t_0$ ($\tau = 0$) yields

$$Q(t) = Q_0 e^{-\tau/K}. \tag{9.87}$$

Equation (9.87) gives the discharge resulting after the input I ends. To obtain the instantaneous unit hydrograph of the linear reservoir, impose the definition that the IUH is the response to a unit impulse of infinitesimal duration. By construction, the initial storage S_0 is given by

$$S_0 = KQ_0$$

or

$$Q_0 = \frac{S_0}{K}. \tag{9.88}$$

Since the impulse is of unit volume $S_0 = 1$, and since it ends immediately after application, then Eq. (9.87) applies or

$$h(t) = Q(t) = Q_0 e^{-t/K} = \frac{S_0}{K} e^{-t/K} = \frac{1}{K} e^{-t/K}. \tag{9.89}$$

Equation (9.89) is the form of the instantaneous unit hydrograph of a linear reservoir.

Convoluting any input with the system response given in Eq. (9.89) will result in the corresponding discharge. This procedure is schematized in Figure 9.28. Its use is better illustrated rederiving Eqs. (9.84) and (9.87) as a response of a linear reservoir to a constant-intensity storm of duration t_0.

For a time $t < t_0$, the convolution results in

$$\begin{aligned} Q(t) &= I \int_0^t \frac{1}{K} e^{-(t-\tau)/K} \, d\tau \\ &= \frac{I}{K} e^{-t/K} \int_0^t e^{\tau/K} \, d\tau, \end{aligned} \tag{9.90}$$

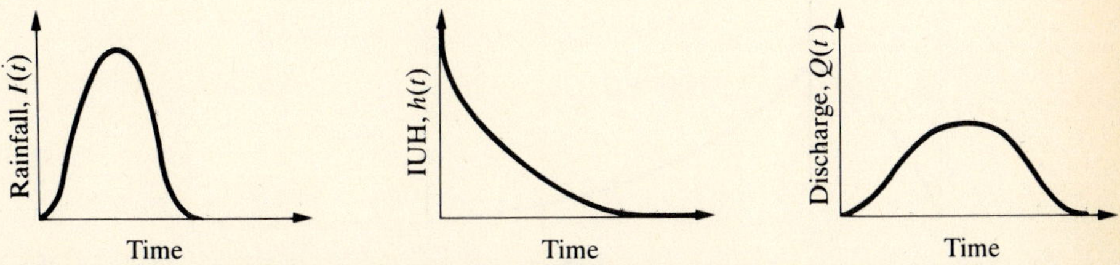

FIGURE 9.28 Discharge $Q(t)$ resulting from the convolution of an input $I(t)$ and a system response function $h(t)$. Source: Evans et al. [1972].

which leads to

$$\begin{aligned} Q(t) &= \frac{I}{K} e^{-t/K} K(e^{t/K} - 1) \\ &= I(1 - e^{-t/K}). \end{aligned} \tag{9.91}$$

Equation (9.91) is then the rising limb of the hydrograph.

For times $t > t_0$ (after storm stops),

$$\begin{aligned} Q(t) &= \int_0^{t_0} I \frac{1}{K} e^{-(t-\tau)/K} \, d\tau \\ &= I \int_0^{t_0} \frac{1}{K} e^{-(t-\tau)/K} \, d\tau . \end{aligned}$$

Letting $u = t - \tau$, $du = -d\tau$,

$$\begin{aligned} Q(t) &= -\frac{I}{K} \int_t^{t-t_0} e^{-u/K} \, du \\ &= \frac{I}{K} \int_{t-t_0}^{t} e^{-u/K} \, du \\ &= I[e^{-(t-t_0)/K} - e^{-t/K}]. \end{aligned} \tag{9.92}$$

Equation (9.92) is the recession part of the hydrograph and corresponds to Eq. (9.87). The resulting hydrograph is shown in Figure 9.29.

The linear reservoir model is a one-parameter (K) model. This parameter is particularly easy to obtain using the method of moments, since for the linear

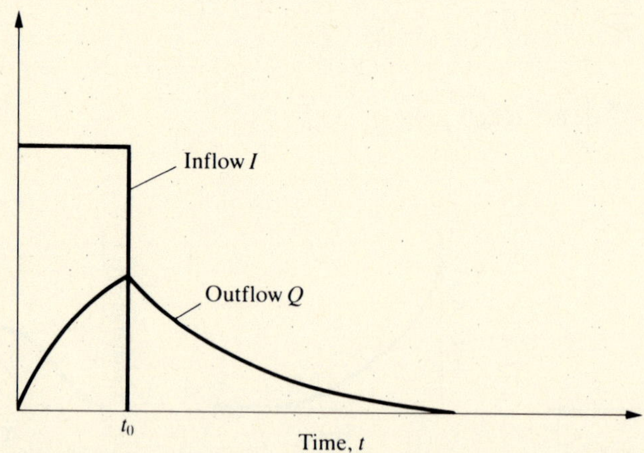

FIGURE 9.29 Outflow hydrograph resulting from a constant input to a linear reservoir. Source: Evans et al. [1972].

reservoir it is easy to show

$$\begin{aligned} \mu_1'(h) &= K \\ \mu_2(h) &= K^2 \\ \mu_3(h) &= 2K^3 \\ \mu_4(h) &= 6K^4 \end{aligned} \tag{9.93}$$

In general, for the linear reservoir

$$\mu'(h) = R!K^R \tag{9.94}$$

and

$$K_R(h) = (R - 1)!K^R. \tag{9.95}$$

Since only one parameter is needed, it is usually computed using the additive first-moment relationships. Lower moments are statistically more stable when computed from discrete, error-prone data. For this reason, lower moments are usually preferred for parameter estimation.

Other conceptual instantaneous unit hydrographs have been suggested. The simplest development beyond the linear reservoir is the combination of the latter with a linear channel. A linear channel is one that simply lags the input

$$Q(t) = I(t - T), \tag{9.96}$$

where T is a prespecified lag.

The combination of a linear channel and a linear reservoir results in the so-called lag and route models schematized in Figure 9.30 and whose IUH is expressed as

$$h(t) = \frac{1}{K}e^{-(t-T)/K}. \tag{9.97}$$

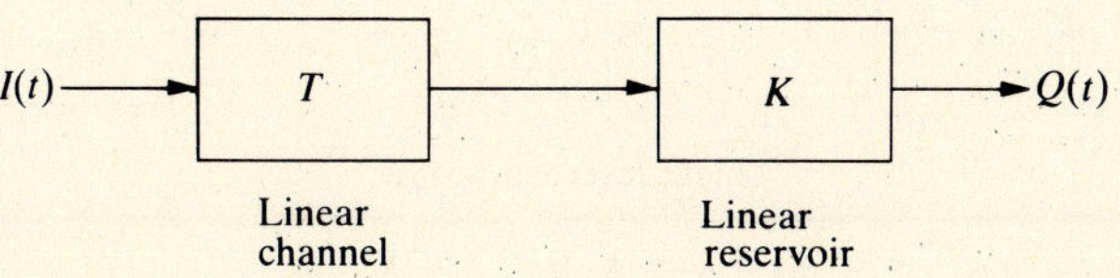

FIGURE 9.30 Block diagram of a lag and route model. Source: Evans et al. [1972].

The moments of the lag and route model are

$$\begin{aligned}
\mu_1'(h) &= T + K \\
\mu_2(h) &= K^2 \\
\mu_3(h) &= 2K^3 \\
\mu_4(h) &= 6K^4
\end{aligned} \tag{9.98}$$

Being a two-parameter model, usually the first two moments are used in parameter estimation.

Possibly the most well-known conceptual IUH is the two-parameter Nash [1957, 1959] model. Shown in Figure 9.31, it consists of a series of n linear reservoirs; the output of one being the input of the one immediately downstream. Clearly, the form of the Nash model can be obtained by convoluting n times the upstream inflow with the downstream linear reservoir. This is analogous in probability theory to the density function resulting from the addition of exponentially distributed terms (see Chapter 11). The result is a gamma function. The Nash model is

$$h(t) = \frac{1}{K}\left(\frac{t}{K}\right)^{n-1} \frac{1}{\Gamma(n)} e^{-t/K}, \tag{9.99}$$

where $\Gamma(\cdot)$ is the gamma function [$\Gamma(n) = (n - 1)!$ for integer n values], K is time constant of the added linear reservoirs, and n is number of linear reservoirs added in series.

The moments of the two-parameter Nash model are

$$\begin{aligned}
\mu_1'(h) &= nK \\
\mu_2'(h) &= n(n + 1)K^2 \\
\mu_2(h) &= nK^2 \\
\mu_3(h) &= 2nK^3 \\
\mu_4(h) &= 6nK^4
\end{aligned} \tag{9.100}$$

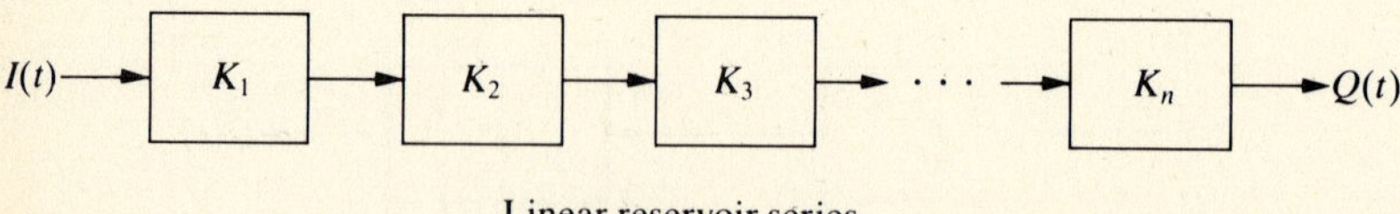

FIGURE 9.31 Block diagram of a Nash model. Source: Evans et al. [1972].

In general, the cumulants are

$$K_R = n(R - 1)!K^R. \tag{9.101}$$

Being the consecutive convolution of many linear reservoirs, the moments of the Nash model are related to those of the linear reservoir by the simple relations discussed in the previous section. Had the Ks been different in each serial linear reservoir, the first moment of the resulting model would have been the addition of the individual Ks.

Other conceptual IUHs resulting from combinations of the linear reservoir, lag and route, and Nash models exist. For example, the Diskin [1964] model (Fig. 9.32) results from the parallel operation of Nash models with the input distributed among the parallel systems. The IUH expression is

$$h(t) = \frac{\alpha}{K_1(n_1-1)!}\left(\frac{t}{K_1}\right)^{n_1-1} e^{-t/K_1} + \frac{1-\alpha}{K_2(n_2-1)!}\left(\frac{t}{K_2}\right)^{n_2-1} e^{-t/K_2}. \tag{9.102}$$

The first moment is given by

$$\mu_1'(h) = \alpha n_1 K_1 + (1-\alpha) n_2 K_2. \tag{9.103}$$

It is pertinent to finish this subsection with mention of the first conceptual IUH ever used, actually before the development of the linear system theory discussed here.

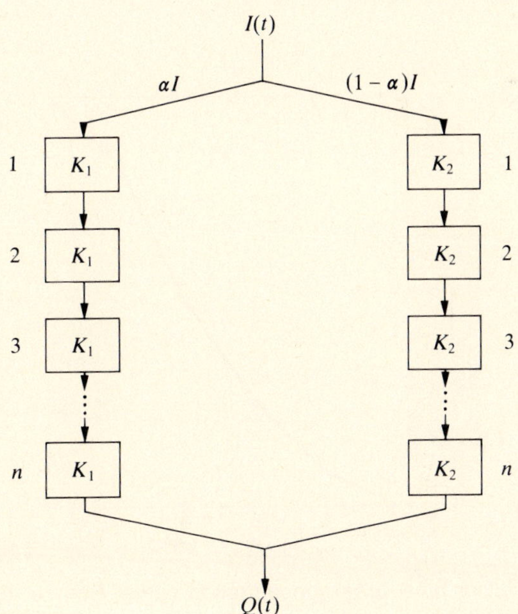

FIGURE 9.32 Diskin's parallel Nash model configuration. Source: Evans et al. [1972].

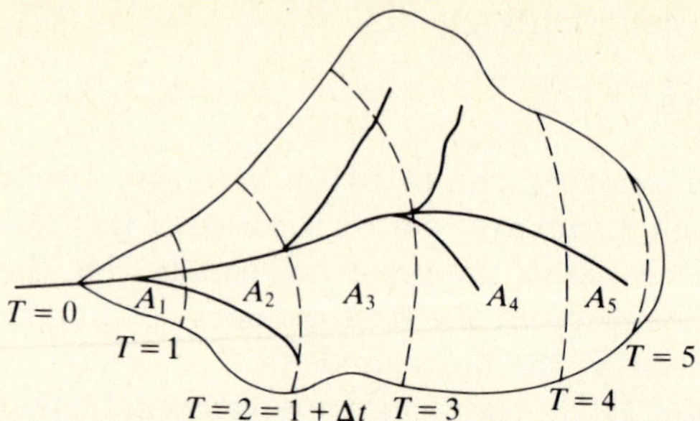

FIGURE 9.33 Basin map with isochrones. Source: World Meteorological Organization, *Guide to Hydrological Practices,* 3rd ed., World Meteorological Organization, 1974.

As an extension to the rational formula, early hydrologists introduced the concept of a time–area curve. In Figure 9.33, a basin is mapped by isochrones or lines of equal travel time. Any point in a given isochrone takes the same time to reach the basin outlet. Using this map, a diagram of cumulative area contributing to runoff versus time could be drawn. The curve is similar to an S curve when normalized by total area. The normalized curve reaches 100% when the total area is contributing (Fig. 9.34). The derivatives or differences of this curve constitute an instantaneous unit hydrograph. Its time base is the time of concentration of the basin and its convolution with any input results in a corresponding output.

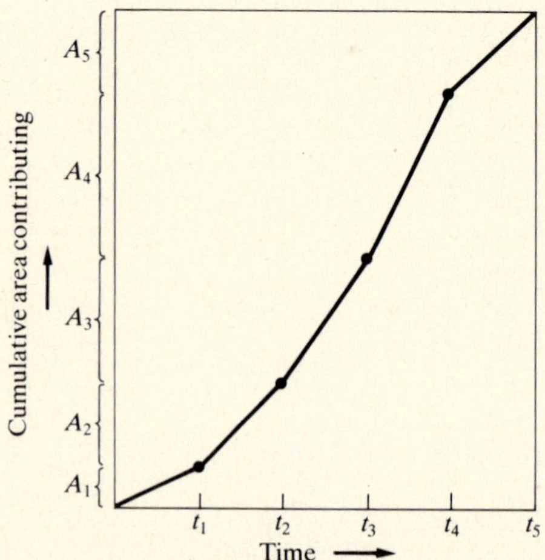

FIGURE 9.34 Cumulative time–area concentration curve. Source: Gray [1973]. Reproduced by permission of the National Research Council of Canada.

Combinations of time–area curves with other conceptual IUHs are common. Time–area curves are commonly routed through linear reservoirs. Resulting conceptual IUHs depend on the shape of the time–area curve but its moments are related (through convolution) as discussed in Section 9.5.3.

9.7 SUMMARY

The hydrograph is the output from a river basin. One of the main tasks of the hydrologist has been to predict the hydrograph from knowledge of the input (i.e., rainfall, snowmelt) given information about initial conditions (i.e., soil moisture, groundwater levels) and about physical properties (i.e., soils, vegetation, topography) of the basin. Let there be no doubt that the hardest part of this exercise is computing infiltration, evaporation, interception, detention, etc. Finding the runoff or "effective rainfall" as referred to in this chapter is the key step to hydrograph prediction. This step occupied much of the previous chapters and will continue to haunt us in the foreseeable future.

This chapter dealt with producing the hydrograph given the runoff. The approach taken is simple. Rather than reductionist (looking at the detail of flow in every corner of the basin), it is holistic or systematic. The basin is looked at as a system whose many parts interact to produce an aggregated-response hydrograph. Our interest has been not in the behavior of the many parts but in the conceptualization of the total response in terms of what we know about inputs (rainfall, snowmelt) to the system (river basin) and outputs (hydrographs).

A lot of time was given to the linear response functions, the unit hydrograph, and the instantaneous unit hydrograph. Basins are not linear. All students of fluid mechanics and hydrology will agree that the response in the basin is truly nonlinear. Nonlinearities arise in the dependence of channel flow on depth of flow. Nonlinearities arise in the dependence of runoff production mechanisms on rainfall input, depth, and intensity. Nevertheless, the linear analogy is not only didactically useful but has many times been proven to be a good approximation. This statement is relative to some of the major uncertainties we have about the nature of the input and runoff production.

The conceptualizations seen in this chapter are also "lumped." The whole river basin is assumed to be responding as a point, with a characteristic response function. This lumped view again responds to a desire for simplicity. But it is also a reflection of the dearth of data that hydrologists commonly face. Generally, a basin will be sampled by a single stream gage at the outlet and commonly one or two rain gages. Soil, vegetation, and evaporation data are generally minimal. Given the data sources, it is wise to limit models to a few parameters, hence the popularity of the lumped approaches.

There are more complicated conceptual and reductionists models, many nonlinear, all with a lot of parameters. As observation technology and information sources improve, hydrologists will and should rethink their conceptualizations for calculating basin response. Clearly radar rainfall data, remote sensing of soil moisture and land fluxes, increased stream gaging, and digitized maps of basins — all becoming more readily available — can support more detailed spatially distributed models of the river basin.

REFERENCES

American Society of Civil Engineers [1970]. *Design and Construction of Sanitary Storm Sewers.* New York: American Society of Civil Engineers. (A.S.C.E. Manuals and Reports on Engineering Practice, no. 37.)

Bernard, M. M., R. L. Gregory, and C. E. Arnold [1932]. "Runoff–Rational Runoff Formulas." *Trans. A.S.C.E.* 96:10–38.

Bras, R. L., and F. E. Perkins [1975]. "Effects of Urbanization on Catchment Response." *J. Hydraul. Div. A.S.C.E.* 101:HY3.

Bras, R. L., and I. Rodriguez-Iturbe [1985]. *Random Functions and Hydrology.* Reading, Mass.: Addison-Wesley.

Chow, V. T., ed. [1964]. *Handbook of Applied Hydrology.* New York: McGraw-Hill.

Clark, C. O. [1945]. "Storage and the Unit Hydrograph." *Trans. A.S.C.E.* 69:1419–1447.

Cordova, J. R., and I. Rodriguez-Iturbe [1983]. "Geomorphoclimatic Estimation of Extreme Flow Probabilities." *J. Hydrol.* 65(1/3):159–173.

DeWiest, R. J. M. [1965]. *Geohydrology.* New York: Wiley.

Diaz-Granados, M., R. L. Bras, and J. B. Valdes [1985]. "Infiltration Aspects in the Geomorphologic IUH." Cambridge, Mass.: MIT Department of Civil Engineering, Ralph M. Parsons Laboratory. (Technical report no. 293.)

Diaz-Granados, M., J. B. Valdes, and R. L. Bras [1984]. "A Physically Based Flood Frequency Distribution for Ungaged Catchments." *Water Resources Res.* 20(7):995–1002.

Diskin, M. H. [1964]. "A Basic Study of the Linearity of the Rainfall–Runoff Process in Watersheds." Urbana, Ill.: University of Illinois. (Ph.D. thesis.)

Dooge, J. C. I. [1956]. "Synthetic Unit Hydrographs Based on Triangular Inflow." Iowa City, Iowa: State University of Iowa. (M.S. thesis.)

Idem. [1959]. "A General Theory of the Unit Hydrograph." *J. Geophys. Res.* 64:241–256.

Idem. [1973]. "Linear Theory of Hydrologic Systems." Washington, D.C.: U.S. Department of Agriculture. (Technical bulletin no. 1468.)

Eagleson, P. S., R. Mejia-R, and F. March [1966]. "Computation of Optimum Realizable Unit Hydrographs." *Water Resources Res.* 2(4):755–764.

Evans, D., B. Harley, and R. L. Bras [1972]. "Application of Linear Routing Systems to Regional Groundwater Problems." Cambridge, Mass.: MIT Department of Civil Engineering, Ralph M. Parsons Laboratory. (Technical report no. 155.)

Gray, D. M. [1961a]. "Interrelationships of Watershed Characteristics." *J. Geophys. Res.* 66(4):1215–1223.

Idem. [1961b]. "Synthetic Unit Hydrographs for Small Watersheds." *J. Hydraul. Div. A.S.C.E.* 87(HY4):33–54.

Idem. [1962]. "Derivation of Hydrographs for Small Watersheds from Measurable Physical Characteristics." Ames, Iowa: Iowa State University Agriculture and Home Economics Experimental Station Research. (Bulletin no. 506.)

Gray, D. M., ed. [1973]. *Handbook on the Principles of Hydrology.* Port Washington, N.Y.: Water Information Center, Inc.

Gupta, V. K., E. Waymire, and C. T. Wang [1980]. "A Representation of an Instantaneous Unit Hydrograph from Geomorphology." *Water Resources Res.* 16(5):855–862.

Hansen, T., and H. P. Johnson [1964]. "Unit Hydrograph Methods Compared." *Trans. Am. Soc. Agric. Eng.* 4:448–451.

Henderson, F. M. [1963]. "Some Properties of the Unit Hydrograph." *J. Geophys. Res.* 68(16):4785–4793.

Hickok, R. B., R. V. Keppel, and B. R. Rafferty [1959]. "Hydrograph Syntheses for Small Watersheds for Small Arid-Land Watersheds." *Agric. Eng.* 40:608–611, 615.

Horton, R. E. [1932]. "Drainage Basin Characteristics." *Trans. Am. Geophys. Union.* 13:350–361.

Idem. [1945]. "Erosional Development of Streams and Their Drainage Basins: Hydrophysical Approach to Quantitative Morphology." *Bull. Geol. Soc. Am.* 56:275–370.

Kirshen, D. M., and R. L. Bras [1983]. "The Linear Channel and Its Effect on the Geomorphologic IUH." *J. Hydrol.* 65:175–208.

Langbein, W. B. [1940]. "Channel Storage and Unit Hydrograph Studies." *Trans. Am. Geophys. Union.* 21:620–627.

Langbein, W. B., et al. [1947]. "Topographic Characteristics of Drainage Basins." U.S. Geological Survey water supply paper no. 968-C.

Lee, M. T., and J. W. Delleur [1976]. "A Variable Source Area Model of the Rainfall–Runoff Process Based on the Watershed Stream Network." *Water Resources Res.* 12(5):1029–1036.

Leopold, L. B., and T. Maddock [1953]. "The Hydraulic Geometry of Stream Channels and Some Physiographic Implications." U.S. Department of the Interior Geological Survey professional paper no. 252.

Linsley, R. K., Jr., M. A. Kohler, and J. L. H. Paulhus [1949]. *Applied Hydrology.* New York: McGraw-Hill.

Idem. [1982]. *Hydrology for Engineers.* 3rd ed. New York: McGraw-Hill.

McCuen, R. H. [1982]. *A Guide to Hydrologic Analysis Using SCS Methods.* Englewood Cliffs, N.J.: Prentice-Hall.

Mitchell, W. D. [1948]. *Unit Hydrographs in Illinois.* Springfield, Ill.: Illinois Department of Public Works and Bridges, Division of Waterways.

Mockus, V. [1957]. "Use of Storm and Watershed Characteristics in Synthetic Hydrograph Analysis and Application." Washington, D.C.: U.S. Department of Agriculture Soil Conservation Service.

Nash, J. E. [1957]. "The Form of the Instantaneous Unit Hydrograph." International Association for Scientific Hydrology, Assemblée Générale de Toronto TOME III, pp. 114–121.

Idem. [1959]. "Systematic Determination of Unit Hydrograph Parameters." *J. Geophys. Res.* 64:111–115.

Natale, L., and E. Todini [1976a]. "A Stable Estimator for Linear Models 1." *Water Resources Res.* 12(4):667–671.

Idem. [1976b]. "A Stable Estimator for Linear Models 2." *Water Resources Res.* 12(4):672–676.

O'Donnell, T. [1960]. "Instantaneous Unit Hydrograph Derivation by Harmonic Analysis." *Int. Assoc. Sci. Hydrol.* 51:546–557.

Overton, D. E., and M. Meadows [1976]. *Stormwater Modeling.* New York: Academic Press.

Pilgrim, D.H. [1977]. "Isochrones of Travel Time and Distribution of Flood Storage from a Tracer Study on a Small Watershed." *Water Resources Res.* 13(3):587–595.

Rodriguez-Iturbe, I., G. Devoto, and J.B. Valdes [1979]. "Discharge Response Analysis and Hydrologic Similarity: The Interrelation between the Geomorphologic IUH and the Stream Characteristics." *Water Resources Res.* 15(6):1435–1444.

Rodriguez-Iturbe, I., M. Gonzalez-Sanabria, and G. Caamaño [1982a]. "On the Climatic Dependence of the IUH: A Rainfall–Runoff Analysis of the Nash Model and Geomorphoclimatic Theory." *Water Resources Res.* 18(4):887–903.

Rodriguez-Iturbe, I., M. Gonzalez-Sanabria, and R.L. Bras [1982b]. "A Geomorphoclimatic Theory of the Instantaneous Unit Hydrograph." *Water Resources Res.* 18(4):877–886.

Rodriguez-Iturbe, I., and J.B. Valdes [1979]. "The Geomorphic Structure of Hydrologic Response." *Water Resources Res.* 15(6):1409–1420.

Schaake, J.C., Jr. [1965]. "Synthesis of the Inlet Hydrograph." Baltimore: Johns Hopkins University Department of Sanitary Engineering and Water Resources. (Technical report no. 3.)

Schaake, J.C., Jr., J.C. Geyer, and J.W. Knapp [1967]. "Experimental Examination of the Rational Method." *J. Hydraul. Div. A.S.C.E.* 93:HY6.

Sherman, L.K. [1932a]. "Stream-Flow from Rainfall by the Unit Graph Method." *Eng. News Rec.* 108:501–505.

Idem. [1932b]. "The Relation of Runoff to Size and Character of Drainage Basins." *Trans. Am. Geophys. Union.* 13:332–339.

Idem. [1940]. "The Hydraulics of Surface Runoff." *Civil Eng.* 10:165–166.

Idem. [1949]. "The Unit Hydrograph Method." In: O. Meinzer, ed. *Hydrology.* New York: Dover, pp. 514–526.

Singh, K.P. [1976]. "Unit Hydrographs—A Comparative Study." *Water Resources Bull.* 12(2):381–391.

Smart, J.S. [1972]. "Channel Networks." *Adv. Hydrosci.* 8:305–346.

Snyder, F.F. [1938]. "Synthetic Unit Graphs." *Trans. Am. Geophys. Union.* 19:447–454.

Soil Conservation Service [1972]. "Hydrology." *National Engineering Handbook,* Section 4. Washington, D.C.: U.S. Department of Agriculture.

Stall, J.B., and Y.S. Fok [1968]. "Hydraulic Geometry of Illinois Streams." Urbana, Ill.: Water Resources Research Center. (Illinois water survey report no. 15.)

Strahler, A.N. [1957]. "Quantitative Analysis of Watershed Geomorphology." *Trans. Am. Geophys. Union.* 38(6):913–920.

Idem. [1964]. *Geology—Part II, Handbook of Applied Hydrology.* New York: McGraw-Hill.

Surkan, A.J. [1969]. "Synthetic Hydrographs: Effects of Network Geometry." *Water Resources Res.* 5(1):112–128.

Taylor, A.B., and H.E. Schwartz [1952]. "Unit Hydrograph Lag and Peak Flow Related to Basin Characteristics." *Trans. Am. Geophys. Union.* 33:235–246.

U.S. Army Corps of Engineers [1959]. "Flood-Hydrograph Analysis and Computations, Engineering, and Design Manual." Washington, D.C.: U.S. Government Printing Office. (Publication no. EM 1110-2-1405.)

Valdes, J.B., Y. Fiallo, and I. Rodriguez-Iturbe [1979]. "A Rainfall–Runoff Analysis of the Geomorphologic IUH." *Water Resources Res.* 15(6):1421–1434.

Viessman, W., Jr., J.W. Knapp, G.L. Lewis, and T.E. Harbaugh [1977]. *Introduction to Hydrology.* 2nd ed. New York: Harper & Row.

Wang, C. T., V. K. Gupta, and E. C. Waymire [1981]. "A Geomorphologic Synthesis of Nonlinearity in the Surface Runoff." *Water Resources Res.* 17(3):545–554.
World Meteorological Organization [1974]. *Guide to Hydrological Practices.* 3rd ed. Geneva, Switzerland: World Meteorological Organization. (Publication no. 168.)

PROBLEMS

1. An architectural design firm is planning a large development on a 100-acre lot, presently in its natural state. State law requires that any new development will not increase peak runoff discharges from the area. As a hydrologic consultant, you are called in for a meeting and asked for a quick estimate of the required volume of a detention pond from which water can be pumped downstream at a rate not higher than the peak natural discharge. You are given the intensity–frequency–duration curves below: the natural state runoff coefficient is 0.5; the developed state runoff coefficient is 0.8; the time of concentration is 30 min; and the design recurrence is 50 yr. What is your detention-volume estimate?

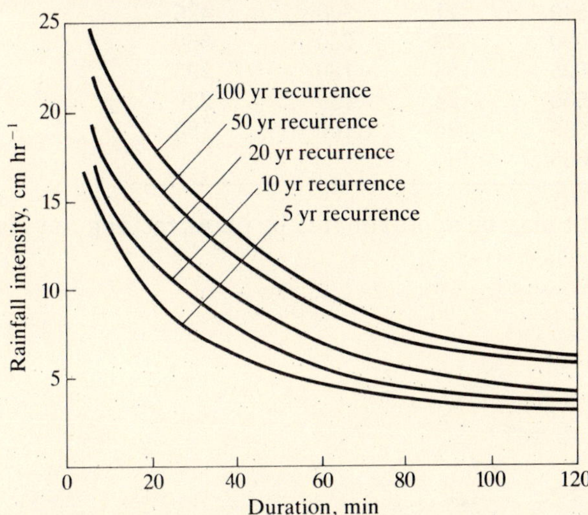

2. For the following data, use Gray's method to construct a unit hydrograph: drainage area = 27 mi^2, length = 8 mi, and S_c (slope) = 2.8%.

3. The instantaneous unit hydrograph for a given area is

$$\text{IUH} = \beta e^{-\beta t}.$$

Gross rainfall on the area is given by

$$I(t) = I_0 e^{\alpha t}.$$

The only loss of precipitation is that due to infiltration, which obeys Horton's equation with $f_c = 0$ and parameters f_0 and α. Infiltration starts coincidentally with rainfall.

What is the equation of the resulting hydrograph? For runoff to occur, what must be the relation between α and β?

4. The input in a system is given by 2, 6, 1. The output is 0, 4, 14, 8, 1, 0. Find the unit response function of the system using the matrix inversion solution and by direct substitution in the resulting set of simultaneous equations.

5. Using the data in the table below, fit a linear reservoir and a Nash model as an instantaneous unit hydrograph using the method of moments.

DAY	INFLOW	OUTFLOW	DAY	INFLOW	OUTFLOW
1	97	86	14	630	660
2	135	100	15	600	645
3	200	140	16	575	630
4	325	200	17	530	610
5	440	300	18	480	585
6	550	378	19	420	535
7	630	465	20	360	490
8	680	540	21	300	425
9	690	590	22	230	360
10	691	625	23	180	295
11	680	650	24	130	230
12	665	658	25	100	180
13	650	665			

6. A first-order catchment may be approximated by plane rectangular slopes draining into a channel.

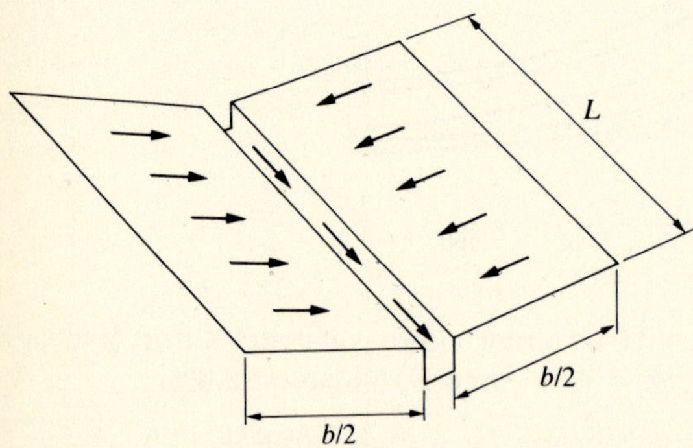

The velocity of overland flow may be approximated as a constant v_o and the velocity in the channel as a constant v_c.

a) Obtain the diagram of cumulative percent of area contributing to runoff as a function of time, assuming an instantaneous input (i.e., time–area curve).

b) Obtain a diagram of cumulative length of stream contributing to the outlet as a function of time (assuming an instantaneous lateral inflow).

c) What are the instantaneous unit hydrographs for the areas draining into the stream, the stream, and the basin as a whole?

7. A simple rainfall–runoff model assumes overland flow velocities of 0.1 m s^{-1} and subsurface flow velocities of 5 m day^{-1}. You have a river basin that can be approximated by the rectangle below. All water, overland and subsurface, enters the stream, which flows at a velocity of 1 m s^{-1}.

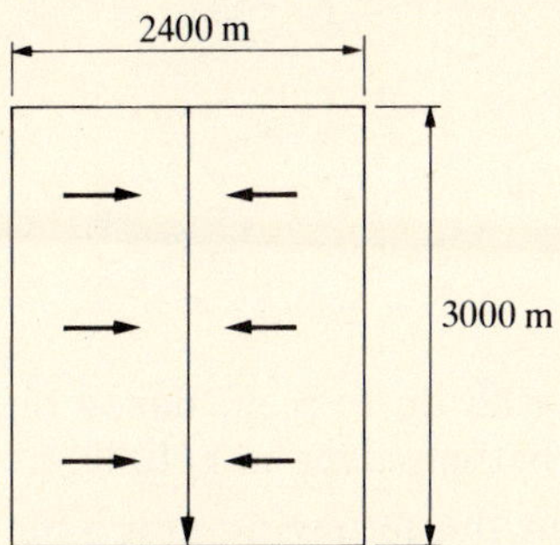

a) Compute the instantaneous unit hydrograph for overland flow.

b) Compute the instantaneous unit hydrograph for subsurface flow.

c) Consider a storm where 20 mm of rain falls in 2 hr and model infiltration using Horton's equation with $f_0 = 20$ mm hr^{-1}, $f_c = 2$ mm hr^{-1}, and $\alpha = 1.5$ hr^{-1}. Compute the hydrograph due to overland and the hydrograph due to subsurface flow. Use approximations where appropriate.

d) This basin lies in a region where there is rarely (take it as never) more than a month (30 days) between storms, which are typically like the storm described in part (c). By using a series of 2-hr storms of 20 mm a month apart, estimate the minimum flow from this basin.

e) A developer wishes to build a shopping center (roofs, parking lots, etc.) over 300 ha of this area. Regard the developed area as essentially impervious and estimate the effect this has on storm runoff from the storm in part (c).

f) A community downstream relies on the minimum flow for water supply. Estimate the effect of the development on the minimum flow.

g) To satisfy environmental-impact requirements, the developer has to ensure that the flow is never less than the minimum prior to development. He also has to ensure that the flood peak is not increased. He intends to do this by building a reservoir to retain floods and store water for supplementing low flows. Determine the storage volume and flood reserve that this reservoir should hold.

h) In the light of your results think about and write a few qualitative remarks about the following issues: irregularly shaped basin, flow velocities dependent on flow, evaporation/transpiration, randomness in storm arrivals, benefits (in terms of smaller storage requirement) of rainfall and flow prediction, seasonality (i.e., separate wet and dry seasons), and intensity–duration relations to rainfall.

(Contributed by David Tarboton.)

8. Consider the channel cross section shown.

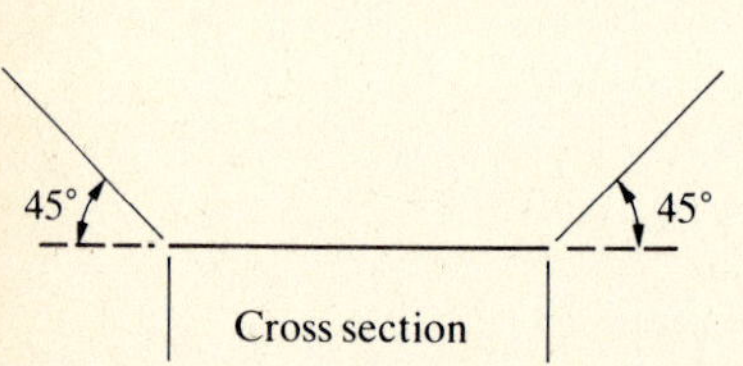

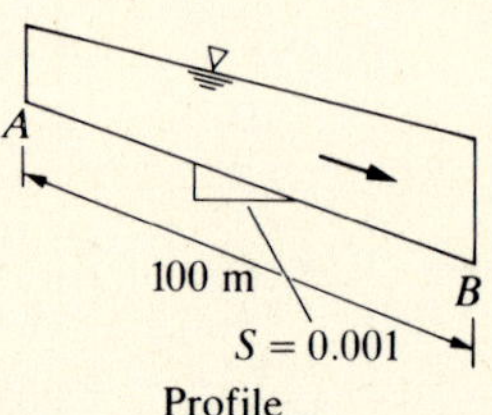

a) Develop a rating curve for water depths up to 5 m, based on the Manning equation with $n = 0.04$. The bottom slope is 1/1000.

b) Use this to plot the hydrograph, given the following stage (depth) measurements at point *A* (see profile).

TIME	STAGE AT *A*	STAGE AT *B* (RELATIVE TO DATUM AT *A*)
0	0.81	0.68
1	1.12	0.98
2	1.51	1.39
3	2.00	1.92
4	1.80	1.73
5	1.64	1.57
6	1.42	1.36

c) Now use the stage measurements obtained at point *B* downstream to "correct" the hydrograph obtained above.

(Contributed by David Tarboton.)

9. You are asked to make a preliminary design of a flood-control channel in New Orleans. The channel is to be built as a semicircle of radius *r*. The material is concrete with a Manning *n* of 0.01 and must have a slope of 0.004. The channel is to carry the outflow resulting from a historic rainfall event of cyclonic tropical origin. The maximum total precipitation recorded for the 18-hr long event was 12.8 in. with maximum intensity in the middle 6 hr, when the rainfall accumulation was twice that of any other 6-hr period. Total infiltration during the storm was 3.6 in. You are given the following 6-hr unit hydrograph corresponding to the 18.5-mi^2 area that the channel will be draining:

TIME (hr)	UNIT HYDROGRAPH ($ft^3 s^{-1} in.^{-1}$)
6	80
12	150
18	100
24	90
30	0

What is the required radius of the channel so that no overflow occurs during the storm?

10. Assume that the instantaneous unit hydrograph of given area takes a linear reservoir form with parameter $1/k$. Given the following set of rainfall and corresponding outflow data, calculate and obtain a formula for the hydrograph of an input that takes the form $2 \cdot \exp(-0.23t)$ in. hr^{-1}. What is the discharge at time $t = 8$ hr for that rainfall event?

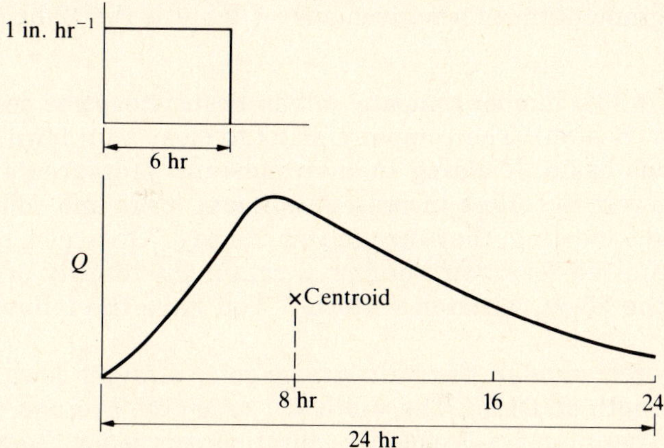

11. Over a 525-mi^2 area, with 10 rain gages, we have the following information about a storm with a 6-hr duration:

STATION	THIESSEN AREAS (mi^2)	PRECIPITATION TOTAL (in.)
A	72	3.50
B	34	4.46
C	36	4.28
D	40	5.29
E	76	6.34
F	92	5.62
G	46	5.20
H	40	5.26
I	86	3.83
J	6	3.30

The 6-hr unit hydrograph for the basin at hand is

TIME (hr)	UNIT HYDROGRAPH ($ft^3 s^{-1} in.^{-1}$)	TIME (hr)	UNIT HYDROGRAPH ($ft^3 s^{-1} in.^{-1}$)
0	0	24	212
3	343	27	149
6	830	30	99
9	1145	33	59
12	1023	36	27
15	744	39	9
18	455	42	0
21	302		

The Φ index is given by 0.1 in. hr^{-1}. Evaporation throughout the storm is a constant 0.05 in. hr^{-1}. Surface ponding is a total of 0.1 in.

What is the hydrograph of the storm given above? What is the 12-hr unit hydrograph?

12. You are a consultant to a lumber company in San Francisco (some people say that is better than Boston). The company plans to clear (cut down the trees of) a mountainous basin. Yielding to environmental pressures, they want to design a reservoir to collect increased sediment loads and control expected floods. During a meeting, they turn to you and say, "Give me a ball-park figure for the required reservoir volume to retain the totality of the flood resulting from the 25-yr recurrence storm." You have the following information.

The basin is nearly rectangular: the main stream going straight down the middle and having a length of 0.9 mi. The width is 1 mi on the average. The slope is a steep 0.01. The intensity–frequency–duration curves for the area are provided below [data from technical paper no. 40 of the *Rainfall Frequency Atlas of the United States,* U.S. Weather Bureau]. For a small consulting fee, you get me to suggest using a runoff coefficient for the stripped mountainside of about 0.7.

Suggest a quick and reasonable answer and clearly state your assumptions and their justification given the available data.

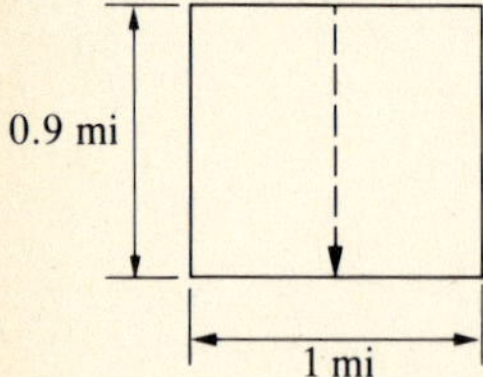

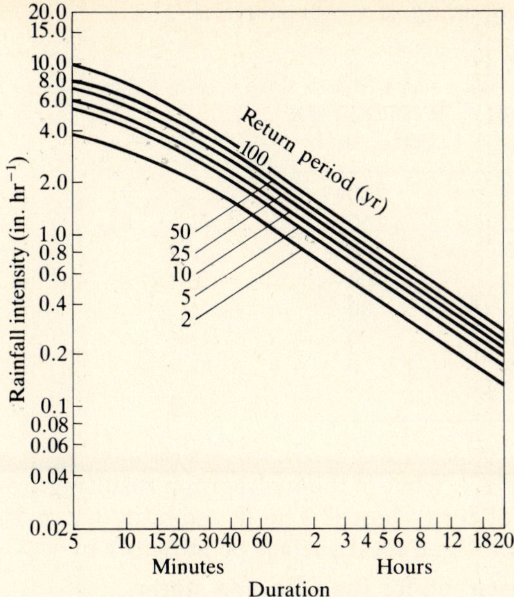

13. A rainfall event can be described by

$$I(t) = \frac{1}{3}\left(\frac{t}{3}\right)\frac{1}{\Gamma(2)}e^{-t/3}.$$

Assuming that the basin is linear with response function

$$h(t) = \frac{1}{3}\left(\frac{t}{3}\right)^4\frac{1}{\Gamma(5)}e^{-t/3},$$

what is the expression for the discharge from the basin as a function of time? What is the mean and second moment about the mean of the output?

14. Somebody suggests a conceptual instantaneous unit hydrograph consisting of two systems in series. The first has a response function that is a triangle with time to peak equal to a and base equal to $2a$. The consecutive system is a Nash model. What is the first moment about the origin of the joint system? What are the second and third moments around the center of the joint system? The "triangular" response function is many times used as a time–area curve.

15. Show the derivation of the Nash model.

16. Using Figure 9.20 and/or Table 9.5, determine the unit hydrograph for a rainfall duration of 2 hr if the drainage area is 85 mi^2 and $t_\ell = 1.44A^{0.6}$.

17. The unit hydrograph corresponding to a 6-hr storm is

TIME (hr)	UNIT HYDROGRAPH ($ft^3\,s^{-1}\,in.^{-1}$)	TIME (hr)	UNIT HYDROGRAPH ($ft^3\,s^{-1}\,in.^{-1}$)
0	0	24	210
3	340	27	150
6	830	30	100
9	1150	33	60
12	1020	36	30
15	750	39	0
18	450	42	0
21	300		

a) If the base flow is 500 $ft^3\,s^{-1}$, what is the hydrograph resulting from a 6-hr storm with 3 in. of total depth? The Φ index is 0.1 in. hr^{-1}. Evaporation is 0.05 in. hr^{-1} and total surface ponding 0.1 in.

b) Find the discharge hydrograph for the following storm:

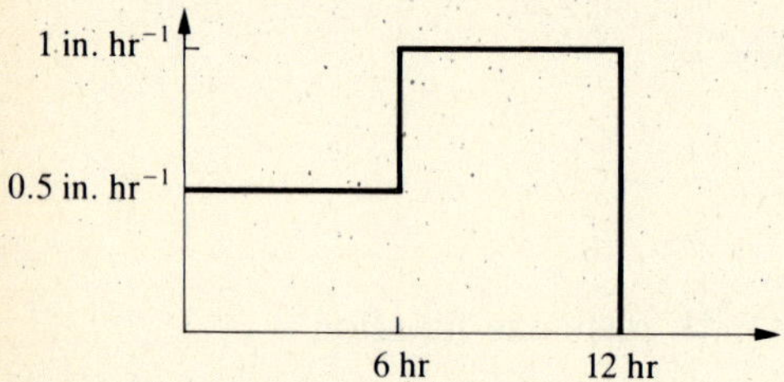

c) Compute the S hydrograph.

18. You have the following rainfall information about storms of 10-yr recurrence.

t (Duration in minutes)	i (in. hr^{-1})
20	5.04
25	4.57
30	4.19
40	3.58
45	3.34
60	2.78
120	1.66

You are to design the storm drain system for the area below:

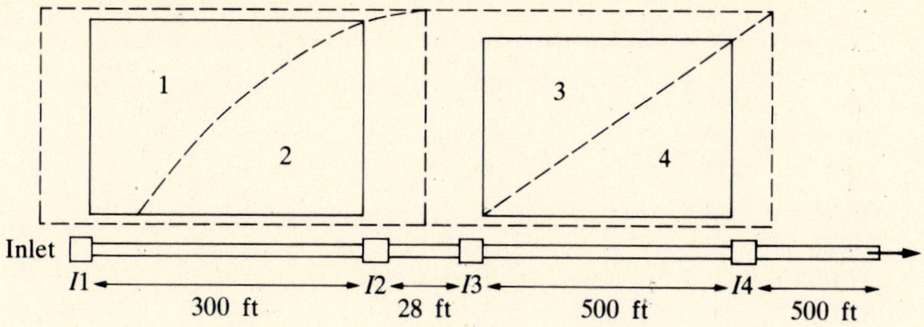

		TIME OF CONCENTRATION
Area 1 = 3.5 acres	$c_1 = 0.65$	$t_{c1} = 10$ min
Area 2 = 4 acres	$c_2 = 0.65$	$t_{c2} = 10$ min
Area 3 = 4 acres	$c_3 = 0.50$	$t_{c3} = 15$ min
Area 4 = 3.5 acres	$c_4 = 0.65$	$t_{c4} = 10$ min

Manning's $n = 0.013$ for all pipes, area 1 corresponds to inlet $I1$, and so on. Design the four pipes shown. Do not have more than a 10-ft drop in depth from the beginning to the end of the pipe system.

19. Assume you have a flat surface as shown in the figure below. There is uniform rainfall of constant intensity i and uniform constant infiltration f over the surface. Any surface runoff moves from left to right (direction of arrows) at a constant velocity V. The surface has a ponding (storage) capacity S. At any time, what would be the discharge per unit width at the rightmost edge of the area? Assume that each point in the area responds exactly the same as any other point. What is the shape of the "discharge" curve (total runoff hydrograph)?

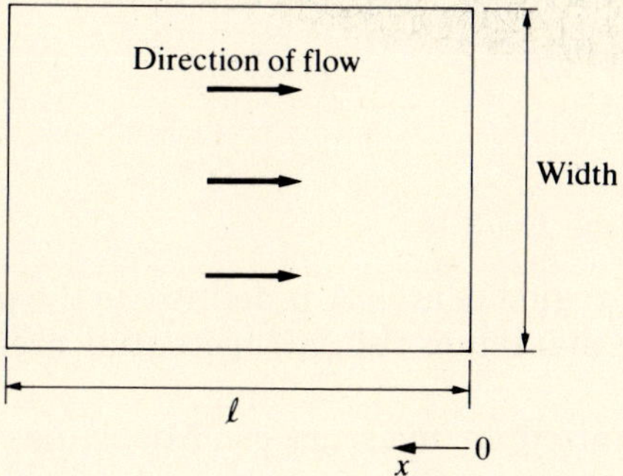

20. The time–area curve of a given watershed is given by

TIME (hr)	AREA (acres)
1	50
2	50
3	62.5
4	18.75
5	6.25
6	7.6875
7	32.09375
8	17.25

The following effective rainfall occurs in the area.

TIME (hr)	RAINFALL (in. hr^{-1})
1	4
2	2
3	3
4	4
5	1

Find the resulting hydrograph.

21. A steady rain of intensity i and duration t_r falls over a square parking lot, which drains uniformly to one of its sides. Compute the parking lot t_r-hour unit hydrograph if the velocity of travel of water on the lot is constant. Comment on the validity of your answer.

22. A steady rainfall of intensity 1 mm hr^{-1} falls over an area for 4 hr. Infiltration on the area can be characterized by the Horton infiltration equation with $f_c = 0$, $f_0 = 1$ mm hr^{-1}, and $\alpha = 8$ hr^{-1}. If the instantaneous unit hydrograph for the area is

$$h(t) = \lambda e^{-\lambda t} \qquad \lambda = 0.25 \text{ hr}^{-1},$$

compute the runoff from the area at time $t = 3$ hr.

23. The probable maximum flood for a region of area A is defined as the peak discharge that results if the probable maximum rainfall (potential precipitable depth) falls over the region.

A basin's atmosphere, when maximized for moisture conditions, has the following vertical pressure and temperature profiles.

Pressure (mb)	1000	900	800
Temperature (°C)	18	13	8

Such column is replenished on the average every hour, and storms are 30% efficient in converting atmospheric moisture to precipitation.

Find the probable maximum flood if the basin's rainfall–discharge transformation is given by a Nash model with parameters $n = 1$, $K = 2$ hr, and if the rain lasts for 1 hr. Ignore infiltration losses.

Chapter 10

Flood Routing

10.1 ROUTING (From Georgakakos and Bras [1980])

Routing refers to the movement of water across a given hydraulic component like overland flow segments or stream channels. Naturally, the movement of water can be described by the continuity and momentum equations as applied to an incompressible fluid. The solution of these equations is the ultimate, hydraulic, routing procedure. Although commonly done with sophisticated computer codes, the solution of these equations is not simple. Some particular cases will be discussed later.

In this chapter, the main concern will be routing an inflow hydrograph or wave through a channel or reservoir by simulating the delay (lag) and storage capacity of the hydraulic unit. Flood-routing models differ in the representation of the process physics, the ways of estimating model parameters, and the input data requirements. It is customary, in order to facilitate discussion of the subject, to group the different routing models into categories. In the past, division of the models has been based on the number of their parameters (Dooge [1973]) or on the aspects of the physical phenomenon (flood wave movement) that they best simulate (Weinmann and Laurenson [1977, 1979]). Here, flood-routing models will be divided in conceptual (or phenomenological), physically based, and regression black-box models.

All models that are based on the conservation of mass and momentum equations in channels (usually referred to as the St. Venant equations) are grouped under the heading "physically based" models. Representatives of this group of models are those utilizing the full equations for gradually varied, unsteady channel flow; those that are based on equations derived from a linearization of the complete nonlinear St. Venant equations; and an approximation of the full equations—the kinematic wave model.

Conceptual hydrologic models are those that retain some of the physical laws (e.g., conservation of mass) in their mathematical formulation, without trying to be exact representations of the physical reality. They are commonly based on analogies of river channels and basins as a set of storage reservoirs of varied properties.

The last category includes the regression type of models. They rely heavily on an input–output description of the phenomenon, without simulating any of the physical processes involved.

Admittedly, there is no exact representation of the physical reality given the assumptions inherent in the best available description of the flood-wave propagation in channel networks. The main assumptions commonly are one-dimensional flow and resistance laws similar to those of steady uniform flow in prismatic channels. Furthermore, general exact analytical solution to the partial differential equations of motion and continuity for a given set of initial and boundary conditions is not available. Consequently, a major abstraction is introduced in most cases through the discretization schemes used for the numerical integration of these equations. Therefore one is tempted to treat physically based models as conceptual ones. However, the distinction is made since it is believed that physical models are superior to the others, in that they are theoretically capable of good performance even with little data for calibration.

Concerning the so-called conceptual models, the common procedure to evaluate their parameters is to use input–output data; as a consequence, little or no physical meaning can be attributed to the obtained values. In this way, they resemble very much the models of the third category. Nevertheless, conceptual models are expected to have a better performance than the "black-box" models in forecasting under conditions dissimilar to those of their calibration period (Kitanidis and Bras [1978]).

10.2 CONCEPTUAL MODELS

10.2.1 Channel Routing: The Muskingum Method

Figure 10.1 represents an input going through a channel that modifies it and results in a hydrograph. In Chapter 9, the concept of simple delays as a linear channel was suggested. In this model there were no storage effects and the

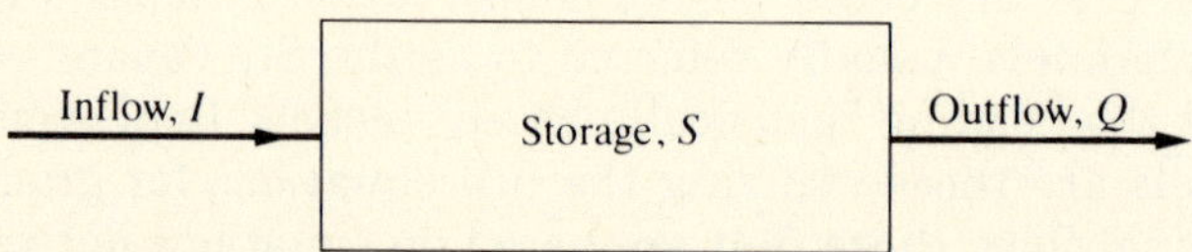

FIGURE 10.1 Conceptualization of a river reach. Source: Gray [1973]. Reproduced by permission of the National Research Council of Canada.

output was simply a delayed input,

$$Q(t) = I(t - T), \tag{10.1}$$

where T equals lag or delay time.

In fact, channels and reservoirs do have the ability to store water and thus alter the shape of the inflow hydrograph. Considering this storage effect and writing a mass balance for the channel results in

$$I_{\text{avg}}\Delta t - Q_{\text{avg}}\Delta t = \Delta S \tag{10.2}$$

or

$$\frac{I_n + I_{n+1}}{2} - \frac{Q_n + Q_{n+1}}{2} = \frac{S_{n+1} - S_n}{\Delta t}, \tag{10.3}$$

where I_n and Q_n are input and output at the beginning of period Δt; I_{n+1} and Q_{n+1} are input and output at the end of period Δt; S_n and S_{n+1} are storage at the beginning and end of periods Δt; and Δt is the time interval over which the balance is performed. The time interval should be small enough so that the input and the output change slowly enough to make the averaging procedure reasonable.

All variables in Eq. (10.3) are known except for Q_{n+1} and S_{n+1}. Since the hydrologist's goal is to obtain the routed flow Q_{n+1} another relation between discharge and storage must be suggested.

Generally, output is a function of storage and inputs or, vice versa, storage is a function of inputs and outputs. Assume then that total storage is

$$S = xS_I + (1 - x)S_Q \tag{10.4}$$

or storage is a weighted average (weight $x \leq 1$) of storage due to inflow S_I and storage due to outflow S_Q.

Express inflows and outflows as a power function of depth, which is a reasonable assumption (the Chezy and Manning equations are such relations):

$$\begin{aligned} I &= ad^c, \\ Q &= ad^c. \end{aligned} \tag{10.5}$$

Similarly, express storage due to inflow and outflow as power functions of depth,

$$\begin{aligned} S_I &= bd^m, \\ S_Q &= bd^m. \end{aligned} \tag{10.6}$$

Combining Eqs. (10.4), (10.5), and (10.6) results in

$$S = xb\left(\frac{I}{a}\right)^{m/c} + (1 - x)b\left(\frac{Q}{a}\right)^{m/c}, \tag{10.7}$$

where x, m, c, a, and b are coefficients.

Equation (10.7) is the necessary relation that jointly with Eq. (10.3) will permit the solution for Q_{n+1} and S_{n+1}. Note that for $x = 0$ (no inflow storage effect) and $m/c = 1$, Eq. (10.7) reverts to the linear reservoir relation ($b/a = K$ time constant) seen in the discussion of the conceptual instantaneous unit hydrograph (Chapter 9).

In uniform prismatic channels $m = 1$ and $c = 5/3$, resulting in

$$S = xb\left(\frac{I}{a}\right)^{0.6} + (1 - x)b\left(\frac{Q}{a}\right)^{0.6}. \tag{10.8}$$

In natural channels, m/c can be greater than 1. The value of x in natural channels varies with the extent of its flood plain and the length of the reach. For a given length, a channel with large flood plains and wide berm will behave closer to a linear reservoir and x will be relatively small. The same reach with deep narrow channels will be relatively more dependent on inflow and have larger x values.

When assuming a linear response of storage to inflow and outflow, $m/c = 1$, Eq. (10.7) becomes

$$S = x\frac{b}{a}I + (1 - x)\frac{b}{a}Q = K[xI + (1 - x)Q], \tag{10.9}$$

where $K = b/a$ is a storage factor with units of time.

Substituting Eq. (10.9) in the continuity expression, Eq. (10.3), results in

$$\frac{I_n + I_{n+1}}{2} - \frac{Q_n + Q_{n+1}}{2} = \frac{K[xI_{n+1} + (1 - x)Q_{n+1} - xI_n - (1 - x)Q_n]}{\Delta t}, \tag{10.10}$$

which after manipulations becomes

$$Q_{n+1} = C_0 I_{n+1} + C_1 I_n + C_2 Q_n, \tag{10.11}$$

where

$$C_0 = -\left(\frac{Kx - 0.5\Delta t}{K - Kx + 0.5\Delta t}\right),$$

$$C_1 = \frac{Kx + 0.5\Delta t}{K - Kx + 0.5\Delta t},$$

$$C_2 = \frac{K - Kx - 0.5\Delta t}{K - Kx + 0.5\Delta t},$$

$$C_0 + C_1 + C_2 = 1.$$

Equation (10.11) is the so-called Muskingum routing procedure.

Parameters K and x in Eq. (10.11) can be obtained graphically. Given an input and output as in Figure 10.2, the storage at any one time and the

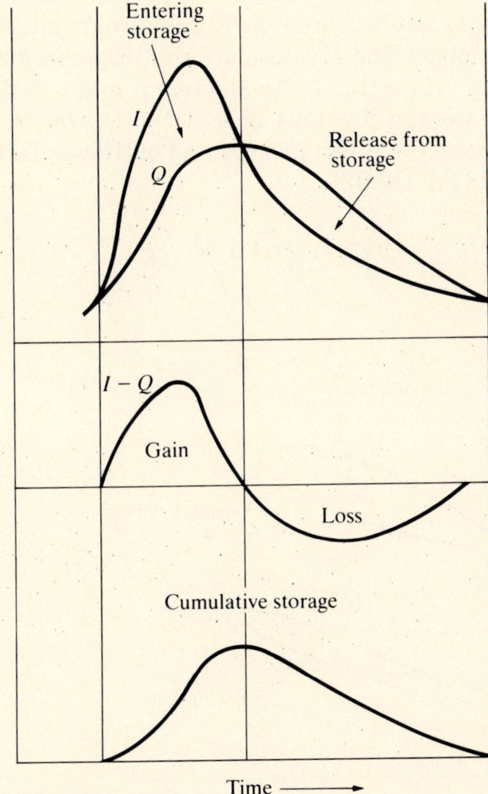

FIGURE 10.2 Computation of storage from observed hydrographs. Source: Gray [1973]. Reproduced by permission of the National Research Council of Canada.

cumulative storage can be obtained (solving for S_{n+1} in Eq. 10.3). Note that in the rising portion of the hydrograph there is a gain in storage with a corresponding loss in the receding phase. The rate of gain and loss are not the same, though, leading to a nonsymmetrical cumulative storage curve. If the cumulative storage is plotted against the weighted input and output $[xI + (1 - x)Q]$, the plot must result in a straight line for Eq. (10.9) to hold. In fact, a hysteresis loop is observed because of the higher storage in the rising phase. The value of x that results in the narrowest loop, most resembling a line, is the correct parameter. The slope of the line fit gives K. This procedure is illustrated in Figure 10.3. Given the parameters K and x, Eq. (10.11) is sequentially used to compute outflow from a given stream channel.

EXAMPLE 10.1

The Muskingum Method

You are given a history (Table 10.1) of inflow and outflow hydrographs for a river reach. As part of a new bridge design, you are asked to compute the outflow that would result from a new inflow scenario, as shown in Table 10.2.

The problem is to route a new inflow hydrograph, hence a resulting model must be developed. The Muskingum routing procedure is a good alternative. The first step is to estimate parameters K and x in Eq. (10.9). To do so, we plot cumulative storage S versus $xI + (1 - x)Q$, where x is set at various values, and I and Q are the given inflows and outflows (Table 10.1). The storage is obtained from Eq. (10.3):

$$S_{n+1} = S_n + \Delta t\left(\frac{I_n + I_{n+1}}{2}\right) - \Delta t\left(\frac{Q_n + Q_{n+1}}{2}\right).$$

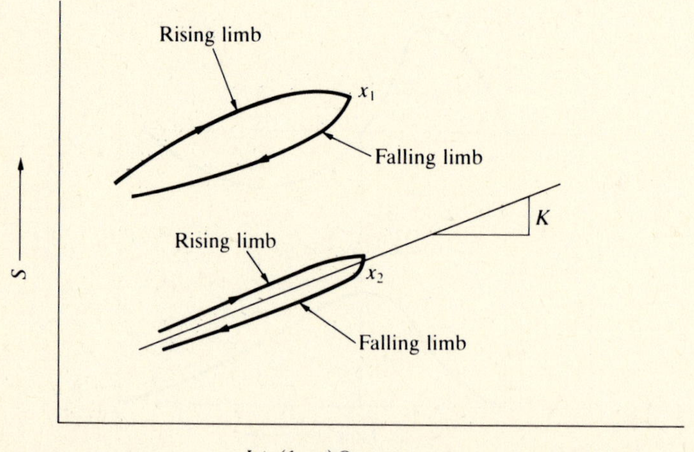

FIGURE 10.3 Determination of x and K for the Muskingum routing method. Source: Gray [1973]. Reproduced by permission of the National Research Council of Canada.

TABLE 10.1 The Inflow and Outflow Hydrographs for a Reach of a River

DATE	TIME (hr)	INFLOW ($m^3 s^{-1}$)	OUTFLOW ($m^3 s^{-1}$)
1	0600	30	30
	1200	60	32
	1800	120	54
	2400	210	101
2	0600	330	181
	1200	420	278
	1800	480	370
	2400	510	440
3	0600	480	480
	1200	420	475
	1800	330	434
	2400	210	360
4	0600	120	261
	1200	60	169
	1800	30	99
	2400	30	56

TABLE 10.2 New Inflow and Resulting Outflow

DATE	TIME (hr)	TIME STEP	INFLOW ($m^3 s^{-1}$)	PREDICTED OUTFLOW ($m^3 s^{-1}$)
1	0600	1	40	40
	1200	2	180	54
	1800	3	250	142
	2400	4	400	226
2	0600	5	370	335
	1200	6	320	353
	1800	7	250	325
	2400	8	175	269
3	0600	9	120	203
	1200	10	80	146
	1800	11	75	103
	2400	12	60	83
4	0600	13	50	67
	1200	14	45	56
	1800	15	40	48
	2400	16	40	43

TABLE 10.3 Storage and Parameter Computations

DATE	TIME (hr)	TIME STEP	INFLOW ($m^3 s^{-1}$)	OUTFLOW ($m^3 s^{-1}$)	STORAGE[a]	$xI + (1 - x)Q$, $m^3 s^{-1}$		
						$x = 0.15$	$x = 0.25$	$x = 0.40$
1	0600	1	30	30	0	30	30	30
	1200	2	60	32	14	36	39	43
	1800	3	120	54	61	64	71	80
	2400	4	210	101	149	117	128	145
2	0600	5	330	181	278	203	218	241
	1200	6	420	278	423	299	314	335
	1800	7	480	370	549	387	398	414
	2400	8	510	440	639	451	458	468
3	0600	9	480	480	674	480	480	480
	1200	10	420	475	647	467	461	453
	1800	11	330	434	567	418	408	392
	2400	12	210	360	440	338	323	300
4	0600	13	120	261	295	240	226	205
	1200	14	60	169	170	153	142	125
	1800	15	30	99	81	89	82	71
	2400	16	30	56	33	52	50	46

[a]Computed using $\Delta t = 1$, hence units are cubic meters per second per time step.

Table 10.3 gives the obtained values of storage as well as corresponding values of $xI + (1 - x)Q$ for $x = 0.15$, 0.25, and 0.40. Figure 10.4 shows the corresponding graphs. The loop for $x = 0.25$ shows the least hysteresis and most resemblance to a straight line. The slope of that line yields

$$K = \frac{675}{480} = 1.4 \text{ (6-hr units)}$$

or $K = 8.4$ hr. Note that Table 10.3 took Δt as 1 unit, and from the data we know that $\Delta t = 6$ hr.

The formula for Muskingum routing (Eq. 10.11) requires constants C_0, C_1, and C_2, which, using $K = 1.40$ (in 6-hr units), $x = 0.25$ and $\Delta t = 1$ (6-hr unit) result in:

$$C_0 = 0.097\,, \qquad C_1 = 0.548\,, \qquad C_2 = 0.355\,.$$

Be aware that the constants C_0, C_1, and C_2 should be positive. Negative values would most probably imply the need for adjustments in the length of the reach or in the interval Δt, or both. Using the above constants, the discharge corresponding to any inflow can be computed. As a test, Figure 10.5 shows

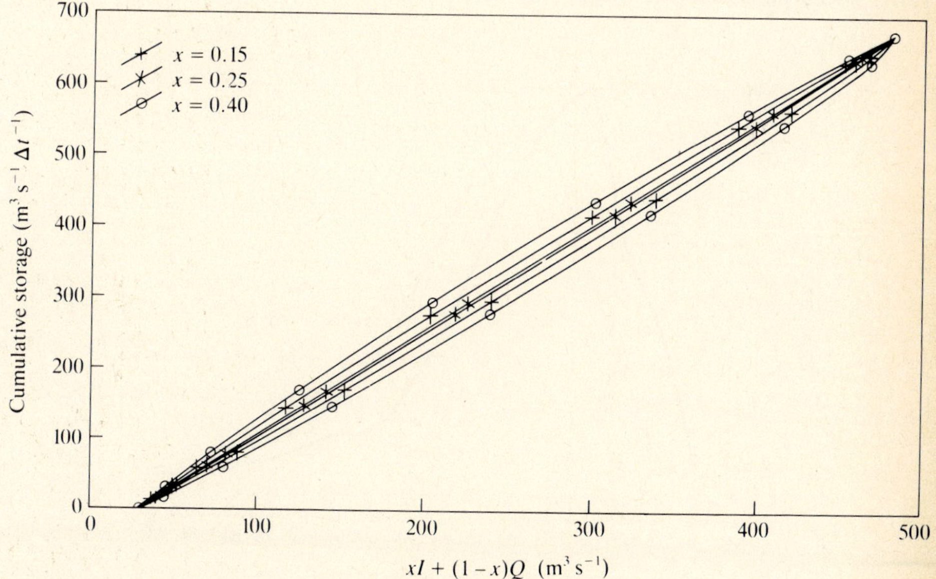

FIGURE 10.4 Muskingum fitting procedure for Example 10.1.

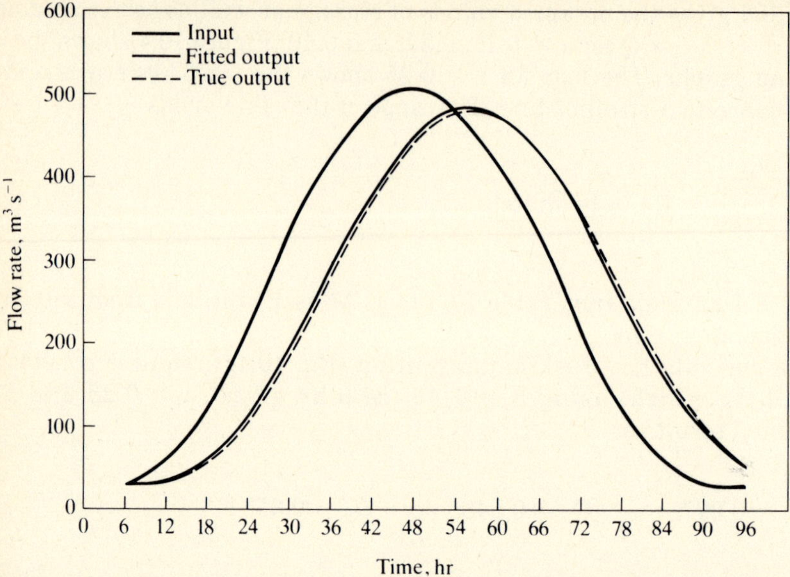

FIGURE 10.5 Comparative plot of true and fitted outflow hydrographs for Example 10.1.

the original inflow and discharge (Table 10.1) together with the discharge produced by the fitted Muskingum method.

The outflow to the new inflow can be computed similarly. It is shown in Table 10.2 and plotted in Figure 10.6. ◆

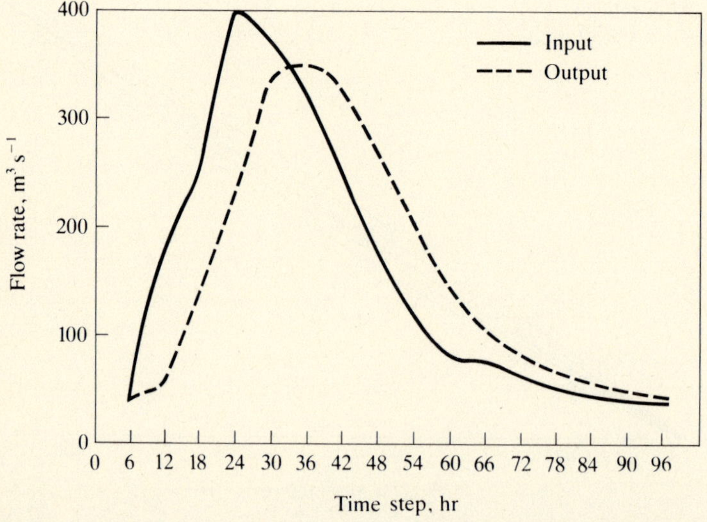

FIGURE 10.6 New inflow and predicted outflow hydrograph for Example 10.1.

10.2.2 Reservoir Routing

Routing through reservoirs is slightly different from channel routing because in the case of reservoirs storage is generally related to outflow, and outflow is either regulated or functionally related to head (storage) due to spillway behavior.

The continuity equation given by Eq. (10.3) can be rewritten as

$$I_n + I_{n+1} + \left(\frac{2S_n}{\Delta t} - Q_n\right) = \frac{2S_{n+1}}{\Delta t} + Q_{n+1}, \tag{10.12}$$

where n indexes the time and the right-hand term is completely unknown. To solve Eq. (10.12), a relation between S and Q must be used. It is generally convenient to relate $(2S/\Delta t) + Q$ and Q.

In a reservoir storage S is geometrically related to water-surface elevation (or surface area). Similarly, spillway discharge is generally given by

$$Q = CLH^a, \tag{10.13}$$

where C and a are coefficients, L is the length of the spillway, and H equals the total head or water-surface elevation. If for a given set of elevations the discharge Q and storage S are evaluated, then a function or plot of $(2S/\Delta t) + Q$ versus Q can be developed as shown in Figure 10.7. With that relationship, Eq. (10.12) can be solved.

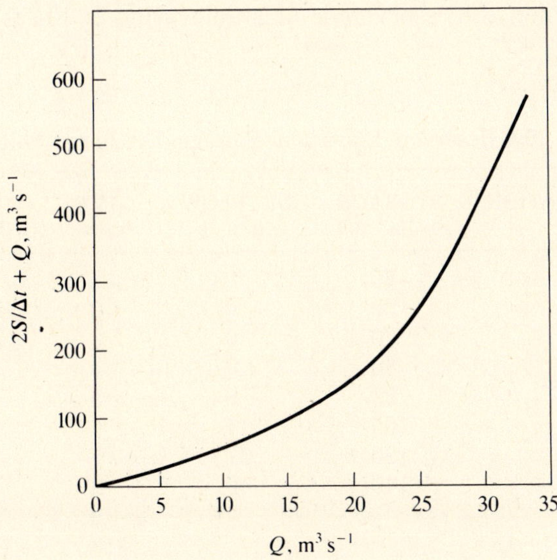

FIGURE 10.7 Storage–discharge relationship for reservoir routing.

To obtain a given discharge Q_{n+1}, the steps to follow are the following:

1. Find $I_n + I_{n+1}$.
2. Add

$$(I_n + I_{n+1}) + \left(\frac{2S_n}{\Delta t} - Q_n\right),$$

resulting in

$$\frac{2S_{n+1}}{\Delta t} + Q_{n+1}.$$

3. Use the $(2S/\Delta t) + Q$ versus Q relation to find the corresponding Q_{n+1}.
4. Find $(2S_{n+1}/\Delta t) - Q_{n+1}$ by subtracting $2Q_{n+1}$ from the obtained $(2S_{n+1}/\Delta t) + Q_{n+1}$.
5. Repeat calculations for next discharge value.

The inconvenience of this approach is that for a different Δt a new $(2S/\Delta t) + Q$ versus Q relationship must be constructed. Koussis and Osborne [1986] provide an alternative approach.

EXAMPLE 10.2

Reservoir Routing

A reservoir has an uncontrolled spillway such that the relation between elevation, storage, and spillway discharge is as given in Table 10.4. Note that outflow starts when the reservoir is at an elevation of 104 ft. The table also

TABLE 10.4 Reservoir Elevation–Storage–Discharge Relationship

ELEVATION (ft)	STORAGE ($\times 10^{-5}$ ft^3)	OUTFLOW (ft^3 s^{-1})	TIME (days)	INFLOW (ft^3 s^{-1})
100	50	0	1	10
101	60	0	2	20
102	70	0	3	30
103	80	0	4	35
104	92	8	5	30
105	105	17	6	22
106	120	27	7	15
107	140	40	8	10
			9	10
			10	10

gives an inflow into the reservoir over a 10-day period. What will the spillway discharge be if the reservoir is at an elevation of 100 ft at the beginning of the inflow hydrograph?

The reservoir routing will be done using Eq. (10.12). We first use the information in Table 10.4 to compute a relationship of Q versus $(2S/\Delta t) + Q$. The result is (starting at an elevation of 103 ft):

Q, $ft^3 s^{-1}$	$\frac{2S}{\Delta t} + Q$, $ft^3 s^{-1}$
0.0	185
8.0	221
16.6	260
26.6	305
40.0	364

The above relationship is plotted in Figure 10.8. The relationship is linear, of the form:

$$Q \approx \frac{1}{4.5}\left(\frac{2S}{\Delta t} + Q - 185\right) \qquad \text{for } \frac{2S}{\Delta t} + Q > 185 .$$

Figure 10.8 or the above equation is then used in Eq. (10.12) to obtain the spillway discharge. Table 10.5 shows the computation. Note the attenuation and delay of the incoming inflow. ◆

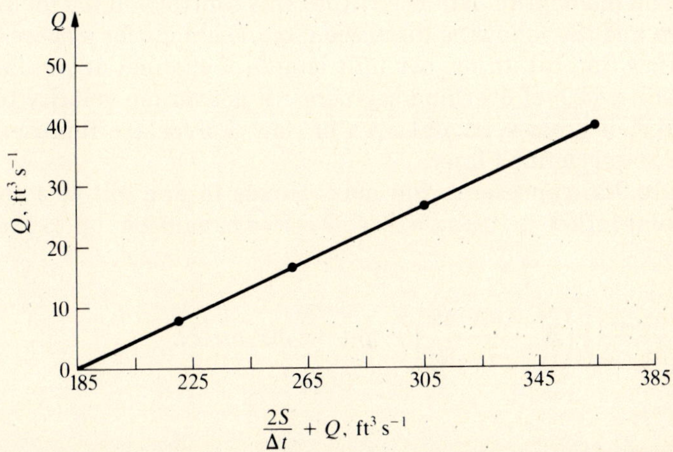

FIGURE 10.8 Discharge–storage relationship of reservoir.

TABLE 10.5 Reservoir Routing Computation

TIME (days)	I_i ($ft^3 s^{-1}$)	$\frac{2S_i}{\Delta t} - Q_i$ ($ft^3 s^{-1}$)	$\frac{2S_i}{\Delta t} + Q_i$ ($ft^3 s^{-1}$)	Q_i ($ft^3 s^{-1}$)
1	10	116	116	0
2	20	146	146	0
3	30	191	196	2.4
4	35	224	256	15.8
5	30	243	289	23.1
6	22	246	295	24.4
7	15	239	283	21.8
8	10	229	264	17.6
9	10	221	249	14.2
10	10		241	12.4

10.3 HYDRAULIC ROUTING: THE ST. VENANT EQUATIONS

As this section will show, the hydrologic routing schemes just discussed have a solid theoretical basis. Their apparent empiricism may be misleading. In order to understand the reasoning behind such schemes, it is now necessary to study the full equations of motion in an open channel. They are usually called the St. Venant equations.

Figure 10.9 delineates a control profile of unit length Δx, corresponding to one-dimensional open channel flow. The equation of motion can be derived from mass and momentum balance within this control volume (of unit width). In the figure and the following discussion, $q(x, t)$ stands for a space- and time-varying inflow (lateral inflow per unit length of channel or rainfall per unit area for a unit width of overland segment). V is average velocity in the cross section, and A is cross-sectional area of flow. Forces are represented by Fs and will be better defined later.

Figure 10.9(a) represents the mass fluxes in and out of a segment Δx long. The total inflow is (using a Taylor series expansion, up to linear terms, of V and A):

$$\rho\left(V - \frac{\partial V}{\partial x}\frac{\Delta x}{2}\right)\left(A - \frac{\partial A}{\partial x}\frac{\Delta x}{2}\right) + \rho\int_x^{x+\Delta x} q(x, t)\,dx\,. \tag{10.14}$$

The outflow is

$$\rho\left(V + \frac{\partial V}{\partial x}\frac{\Delta x}{2}\right)\left(A + \frac{\partial A}{\partial x}\frac{\Delta x}{2}\right) \tag{10.15}$$

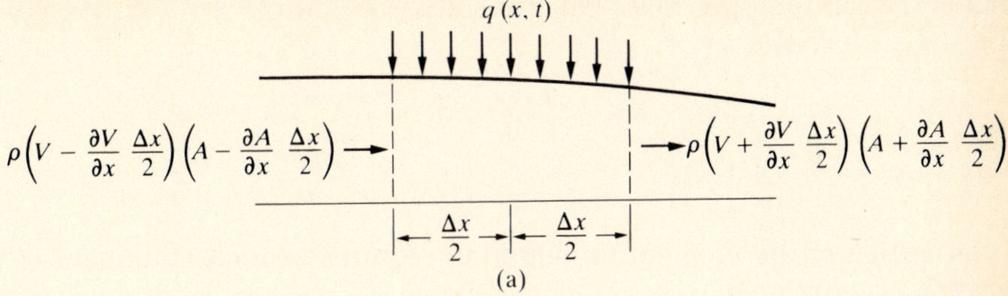

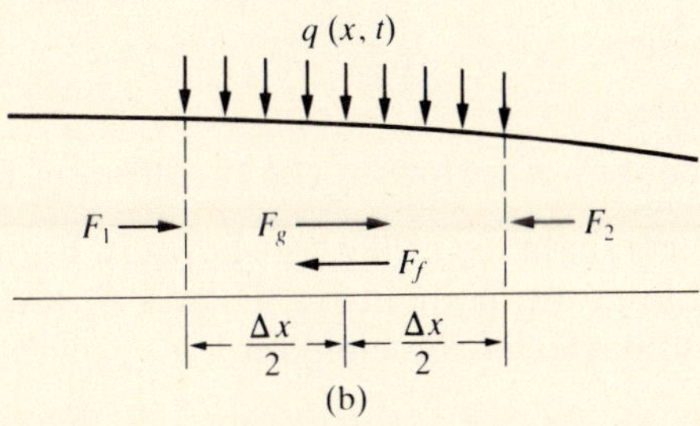

FIGURE 10.9 Unit volumes for the derivation of continuity and momentum equations.

and the change in storage is given by

$$\rho \frac{\partial A}{\partial t} \Delta x , \tag{10.16}$$

where ρ is the density of water.

Since

$$I - Q = \frac{dS}{dt} ,$$

combining the above equations results in

$$-\rho\left(A \frac{\partial V}{\partial x} \Delta x + V \frac{\partial A}{\partial x} \Delta x\right) + \rho \overline{q} \Delta x - \rho \frac{\partial A}{\partial t} \Delta x = 0 , \tag{10.17}$$

where $\overline{q}$ is the spatially averaged lateral inflow rate in units of volume per time per unit length. Dividing by $\rho \Delta x$ and reorganizing terms,

$$A \frac{\partial V}{\partial x} + V \frac{\partial A}{\partial x} + \frac{\partial A}{\partial t} = \overline{q} , \tag{10.18}$$

which is the continuity or mass balance equation. Since $Q = A \cdot V$, Eq. (10.18) can also be expressed as

$$\frac{\partial Q}{\partial x} + \frac{\partial A}{\partial t} = \overline{q}\,. \tag{10.19}$$

Derivation of the momentum equation requires explicit statement of the following assumptions:

1. unidirectional flow and uniform velocity distribution,
2. hydrostatic pressure,
3. small bottom slope, and
4. no velocity component of inflow in the direction of flow.

Given the above, we can balance all forces shown in Figure 10.9(b) and equate them to the total change in momentum. F_1 and F_2 are hydrostatic pressure forces at the two boundaries of the element,

$$F_1 = \gamma\left(\overline{y}A - \frac{\partial(\overline{y}A)}{\partial x}\frac{\Delta x}{2}\right) \tag{10.20}$$

$$F_2 = \gamma\left(\overline{y}A + \frac{\partial(\overline{y}A)}{\partial x}\frac{\Delta x}{2}\right), \tag{10.21}$$

where $\overline{y}$ is the depth to the centroid of the flow cross section. The resultant pressure force is

$$F_1 - F_2 = -\gamma\frac{\partial(\overline{y}A)}{\partial x}\Delta x\,. \tag{10.22}$$

Gravitational force assuming small angle of slope is

$$F_g = \gamma A\,\Delta x\,\sin\alpha = \gamma AS\,\Delta x\,, \tag{10.23}$$

where α is the angle of the channel bed with the horizontal and S is the channel slope.

Friction force in the channel bottom can be expressed in a similar manner by introducing the concept of friction slope S_f,

$$F_f = \gamma AS_f\,\Delta x\,. \tag{10.24}$$

The form of S_f will be discussed in detail later in this chapter.

The addition of all the above forces must be equal to the change of momentum,

$$\frac{d(mV)}{dt} = \frac{m\,dV}{dt} + V\frac{dm}{dt} = \rho A\,\Delta x\frac{dV}{dt} + \rho V\overline{q}\,\Delta x, \tag{10.25}$$

where $\overline{q}$ is the average over Δx of lateral inflow. Since

$$\frac{dV}{dt} = \frac{\partial V}{\partial t} + V\frac{\partial V}{\partial x}, \tag{10.26}$$

the change of momentum is

$$\frac{d(mV)}{dt} = \rho A\,\Delta x\left(\frac{\partial V}{\partial t} + V\frac{\partial V}{\partial x}\right) + \rho V\overline{q}\,\Delta x. \tag{10.27}$$

Equating forces to momentum change and dividing by $\rho A\,\Delta x$ results in

$$\frac{\partial V}{\partial t} + V\frac{\partial V}{\partial x} + \frac{g}{A}\frac{\partial(\overline{y}A)}{\partial x} + \frac{V\overline{q}}{A} = g(S - S_f), \tag{10.28}$$

which is the momentum equation. The terms of Eq. (10.28) correspond to particular elements contributing to momentum change. The first term in the left-hand side of the equation $\partial V/\partial t$ corresponds to local acceleration or velocity change. The second term $V(\partial V/\partial x)$ is the "advective" acceleration, corresponding to velocity change in space. The third term is momentum change induced by pressure differentials related to large water-surface changes. The fourth term is the momentum change caused by the incoming mass of lateral inflow. The terms on the right are the gravity force and the friction force related to channel slope and roughness.

Following Georgakakos and Bras [1980], one can rewrite the governing equations (Eqs. 10.18 and 10.28) as

$$\frac{\partial A}{\partial t} + V\frac{\partial A}{\partial x} + A\frac{\partial V}{\partial x} = 0 \tag{10.29}$$

$$S_f = S - \frac{1}{A}\frac{\partial(\overline{y}A)}{\partial x} - \frac{1}{g}\frac{\partial V}{\partial t} - \frac{V}{g}\frac{\partial V}{\partial x}, \tag{10.30}$$

where lateral flows are ignored. If the assumption is made that the friction slope can be determined as in steady, uniform flow, then,

$$S_f = V^2\frac{1}{C^2R}, \tag{10.31}$$

where C is a constant (Chezy) coefficient, R is the hydraulic radius A/P, and P is the wetted perimeter.

Substitution of Eq. (10.31) in the momentum Eq. (10.30) yields

$$V = CR^{1/2}\left(S - \frac{1}{A}\frac{\partial(\bar{y}A)}{\partial x} - \frac{1}{g}\frac{\partial V}{\partial t} - \frac{V}{g}\frac{\partial V}{\partial x}\right)^{1/2}. \tag{10.32}$$

It should be noted that Eq. (10.32) implies that the velocity (and discharge) are not single-valued functions of the depth. Thus it is common to refer to Eq. (10.32) as looped-rating curves. The second term in parentheses can be written as $\partial y/\partial x$, explicitly showing the influence of the change in water depth.

10.3.1 Solutions to St. Venant Equations
(Adapted from Georgakakos and Bras [1980])

Equations (10.29) and (10.30) have been used for the simulation of the flood-wave propagation in rivers. They are coupled, nonlinear, first-order partial differential equations of the hyperbolic type and require one initial and two boundary conditions for their solution. For subcritical flow (Froude numbers, $F = V/(gy)^{1/2}$, less than 1), the boundary conditions are specified at both ends of the river reach under consideration; while for supercritical flow (Froude numbers greater than 1), both boundary conditions are specified upstream (since disturbances propagate only downstream). Discharge or depth hydrographs and stage versus discharge (rating) curves are utilized as boundary conditions.

The objective in solving these equations is to determine the water depths and velocities at all points in the reach, for all times. Unfortunately, there is no known general analytical solution. Numerical schemes are required to discretize the space and time axes and convert the differential equations to difference ones. Different numerical schemes have been suggested.

Because of the computational burden associated with the numerical solution of the "complete" equations (Eqs. 10.29 and 10.30), as well as due to the quantity and quality of the input data required for their solution, several approximations have been proposed.

Dooge and Harley [1967] presented a linearized version of Eqs. (10.29) and (10.30). Expressed in terms of discharge and linearizing about a reference discharge per unit width of a wide channel $q_\circ$ results in

$$(gy_\circ - V_\circ^2)\frac{\partial^2 \delta q}{\partial x^2} - 2V_\circ \frac{\partial^2 \delta q}{\partial x\, \partial t} - \frac{\partial^2 \delta q}{\partial t^2} = 3gS_\circ \frac{\partial\, \delta q}{\partial x} + \frac{2gS_\circ}{V_\circ}\frac{\partial\, \delta q}{\partial t}, \tag{10.33}$$

where δq represents a deviation from a nominal discharge $q_\circ$, and where $y_\circ$ and $V_\circ$ are the reference depth and velocity, respectively, corresponding to the reference discharge and slope $S_\circ$. Note that the lateral inflow has been assumed insignificant, and it does not enter Eq. (10.33). Since the perturbation analysis was based on small deviations from the reference trajectories, these investigators performed sensitivity analysis with respect to the values of $q_\circ$. They reported that (a) an increase in $q_\circ$ results in a decrease in the lag time for the channel reach under study; (b) the second moment of the outflow hydrograph was not sensitive to changes in the value of $q_\circ$; and (c) for floods of long duration, changes in $q_\circ$ have a small effect on the shape of the hydrograph. Based on these observations, they subsequently suggested that the operation of the linearized channel on the inflow appears to consist of a translation dependent on the reference discharge $q_\circ$, and an attenuation effect practically independent of the reference discharge. That is, the system has been replaced by two subsystems, one causing a nonlinear translation of the inflow and the other causing a linear attenuation to the translated inflow.

With initial conditions

$$\delta q(x, 0) = 0$$

and

$$\left.\frac{\partial\, \delta q(x, t)}{\partial t}\right|_{t=0} = 0,$$

the solution of Eq. (10.33) is

$$\delta q(x, t) = \delta(t - x/c_1)\exp(-px) + \exp(-rt + zx) \cdot (x/c_1 - x/c_2)h\,\frac{I_1\{2h[(t - x/c_1)\,(t - x/c_2)]^{1/2}\}}{[(t - x/c_1)\,(t - x/c_2)]^{1/2}}\,u(t - x/c_1), \quad (10.34)$$

where $\delta q(x, t)$ is the response of a channel to an instantaneous input at the channel's most upstream point, with parameters

$$c_1 = V_\circ + (gy_\circ)^{1/2}; \qquad c_2 = V_\circ - (gy_\circ)^{1/2}$$

$$p = \frac{S_\circ}{2y_\circ}\,\frac{2 - F}{F(1 + F)}; \qquad r = \frac{S_\circ V_\circ}{2y_\circ}\,\frac{2 + F^2}{F^2}$$

$$z = \frac{S_\circ}{2y_\circ}; \qquad h = \frac{S_\circ V_\circ}{2y_\circ}\,\frac{[(4 - F^2)\,(1 - F^2)]^{1/2}}{2F^2}$$

$$F = \frac{V_\circ}{(gy_\circ)^{1/2}}.$$

$I_1\,[\cdot]$ is a first-order modified Bessel function; $u(\cdot)$ is the unit step function taking a value of 1 for positive arguments and 0 otherwise; and $\delta(t - x/c_1)$ is a Dirac delta function with units of inverse time.

Hence, analogously to the discussion of the instantaneous unit hydrograph of Chapter 9, the response to an upstream input $I(t)$ at a point x in the channel is given by the convolution integral,

$$q(x, t) = \int_0^t \delta q(x, \tau) I(t - \tau)\, d\tau + q_\circ . \tag{10.35}$$

Figure 10.10 shows $\delta q(x, t)$ for channels of various length. Note the translation effect and the dispersion of the wave, both increasing with length. Kirshen and Bras [1983] and Diaz-Granados et al. [1986] extend the above concepts to the case where there is uniform lateral inflow into the channel and where there are infiltration losses in the channel, respectively.

Simplification of Eq. (10.33) leads to the so-called diffusion analogy models whose describing equation is of the type:

$$\frac{\partial Q}{\partial t} + c\,\frac{\partial Q}{\partial x} = D\,\frac{\partial^2 Q}{\partial x^2}, \tag{10.36}$$

where c is a coefficient related to the translation of the flood wave and D is a diffusion coefficient that introduces attenuation of the flood wave. Price [1973] proposed analytical expressions for the determination of the coefficients D and c in natural channels. This type of model is well suited for very mild slopes of the channel bed, where pressure terms become significant. The diffusion analogy models can commonly be solved analytically, since Eq. (10.36) is the well-studied heat-flow equation (Carslaw and Jaeger [1959]).

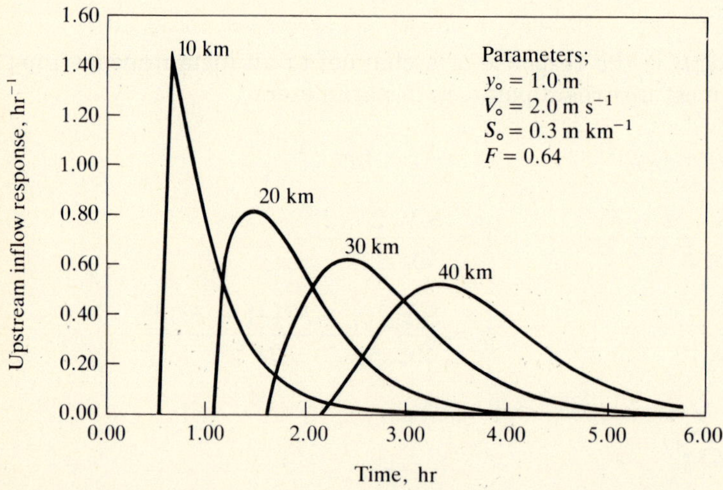

FIGURE 10.10 Upstream input response for different channel lengths. Source: D. M. Kirshen and R. L. Bras, "The Linear Channel and Its Effect on the Geomorphologic IUH," *J. Hydrol*. 65:175–208, 1983.

Perhaps the physically based approximation used most often is the kinematic wave model (Lighthill and Whitham [1955]). In this case, the momentum equation is reduced to the following:

$$S_f = S\,. \tag{10.37}$$

Consequently, there is a one-to-one correspondence between stage and discharge (through Eq. 10.31). In this case, the describing equations can be determined to be (Eagleson [1970])

$$\frac{\partial A}{\partial t} + \frac{\partial Q}{\partial x} = \bar{q}\,, \tag{10.38}$$

$$Q = \alpha A^m\,, \tag{10.39}$$

where α and m are constant coefficients, or in a combined form (Lighthill and Whitham [1955]),

$$\frac{1}{c} \cdot \frac{\partial Q}{\partial t} + \frac{\partial Q}{\partial x} = \bar{q}\,, \tag{10.40}$$

where c is the celerity of the kinematic wave and can be determined as

$$c = \left.\frac{dQ}{dA}\right|_{x=x_c}, \tag{10.41}$$

where x_c is any value of the position coordinate.

The kinematic wave model is a good representation of the physical process of the flood-wave movement when inflow, free-surface slope, and inertia terms are negligible in comparison with those of bottom slope and friction (Eagleson [1970]). Henderson [1966] suggests that the kinematic wave behavior is very close to the one observed for natural floods, in steep rivers whose slopes are of the order of 0.002 or more. It is also commonly used for overland flow in steep hillslopes.

A kinematic wave does not subside or disperse, but it will change shape (it steepens) due to the dependence of the velocity on the depth. If the steepening process ceases, the result will be a steady-state formation called the kinematic shock (or monoclinal wave) (Henderson [1966]). This kind of behavior is confirmed by observed natural floods, but the waves in nature do not steepen as much and do exhibit attenuation (Weinmann [1977]). Attenuation effects can be introduced by numerical procedures used in the discretization of the kinematic wave equations. One such procedure follows.

Discretizing (Eq. 10.40) in space only and solving the resultant differential equation for the unknown discharge leads to

$$Q_{n+1} = k_1 I_n + k_2 I_{n+1} + k_3 Q_n\,, \tag{10.42}$$

where

$$k_1 = \frac{\Delta x}{\bar{c}\,\Delta t}(1 - k_3) - k_3 \tag{10.43}$$

$$k_2 = 1 - \frac{\Delta x}{\bar{c}\,\Delta t}(1 - k_3) \tag{10.44}$$

$$k_3 = e^{-[\bar{c}\,\Delta t/\Delta x(1 - \theta)]} \tag{10.45}$$

and $\bar{c}$ is the mean celerity of the kinematic wave in the reach; Q_n is the discharge at the end of the reach, at the beginning of interval Δt; I_n and I_{n+1} are the upstream inflows into the reach at the beginning and end of Δt; and θ is an empirical parameter, generally taking values between 0 and 0.5.

It can be shown (Weinmann and Laurenson [1977]) that the model described by Eqs. (10.42) through (10.45) can be reduced to the model proposed by Nash [1959] and to the successful routing method of Kalinin and Milyukov (Dooge [1973]). The above formulation is attributed to Koussis [1976].

10.3.2 Numerical Solutions

Kinematic-Wave Equations

Numerical solutions to the kinematic-wave approximation, Eqs. (10.38) and (10.39), are probably the most popular method for routing through a channel network.

Combining Eqs. (10.38) and (10.39) leads to

$$\frac{\partial A}{\partial t} + \alpha \frac{\partial A^m}{\partial x} = \bar{q}\,. \tag{10.46}$$

Explicit differences are used as an approximation to obtain a numerical solution to Eq. (10.46). This is best explained in terms of the two-dimensional grid in Figure 10.11. This grid has x (length in the direction of flow) as the horizontal axis and t (time) as the vertical axis. The maximum value of x is the length of the channel segment in question. Each point in the grid is a discrete point in time and space. The values of x and t increase in finite steps Δx and Δt, respectively. Therefore the grid can be made as fine as necessary by reducing Δx and Δt; in the limit, such a reduction will form a continuous space in x and t in which the exact solution to the equation will lie.

A finite-difference analogy to Eq. (10.46) is

$$\frac{\Delta A}{\Delta t} + \alpha \frac{\Delta A^m}{\Delta x} = \bar{q}\,. \tag{10.47}$$

The solution then depends on the definition of the terms $\Delta A/\Delta t$ and $\Delta A^m/\Delta x$. Different definitions yield different numerical schemes with their own characteristic behavior with respect to stability and convergence.

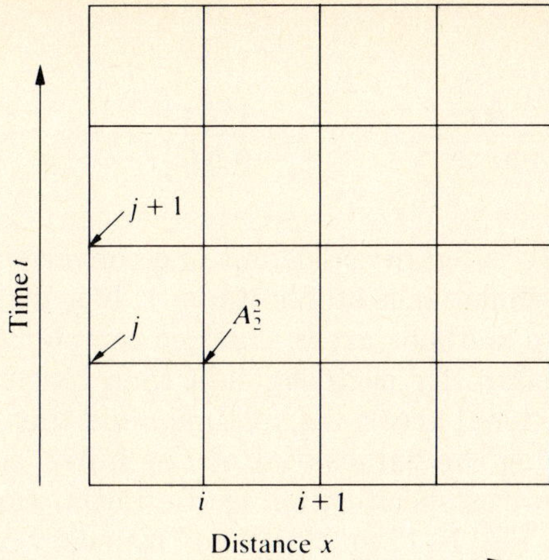

FIGURE 10.11 Finite-difference grid.

Two complementary definitions are commonly used. The first one (refer to Figure 10.11 for definition of indices) redefines Eq. (10.47) as

$$\frac{A_{i+1}^{j+1} - A_{i+1}^{j}}{\Delta t} + \alpha\left[\frac{(A_{i+1}^{j})^m - (A_i^j)^m}{\Delta x}\right] = \overline{q}\,. \tag{10.48}$$

In this case, solving for the unknown yields

$$A_{i+1}^{j+1} = \overline{q}\,\Delta t + A_{i+1}^{j}\left[1 - \frac{\alpha\,\Delta t}{\Delta x}(A_{i+1}^{j})^{m-1}\right] + \frac{\alpha\Delta t}{\Delta x}(A_i^j)^m\,. \tag{10.49}$$

This scheme is consistent and convergent, and it is stable if the following condition is satisfied:

$$\theta = \alpha m A^{m-1}\frac{\Delta t}{\Delta x} \le 1\,. \tag{10.50}$$

Equation (10.50) depends on the unknown A. An operational solution is to define a reference cross-sectional area, for example,

$$A_* = (A_{i+1}^{j} + A_i^{j+1})/2.0\,. \tag{10.51}$$

(*Note:* the definition of A_* may have an effect on the behavior of the scheme; see Bras [1972] for a discussion of this.)

The second approximation is

$$\frac{A_i^{j+1} - A_i^{j}}{\Delta t} + \alpha\left[\frac{(A_{i+1}^{j+1})^m - (A_i^{j+1})^m}{\Delta x}\right] = \overline{q}\,, \tag{10.52}$$

which gives

$$A_{i+1}^{j+1} = \left[\frac{\overline{q}\,\Delta x}{\alpha\,\Delta t} - \frac{A_i^{j+1}\,\Delta x}{\alpha\,\Delta t} + (A_i^{j+1})^m + \frac{A_i^j\,\Delta x}{\alpha\,\Delta t}\right]^{1/m}. \tag{10.53}$$

This second scheme is again consistent and convergent. However, in contrast to the previous scheme it is stable if $\theta \geq 1$. It is for this reason that the pair of finite-difference methods are considered complementary.

The solution procedure for each segment then consists of solving for the values of the cross-sectional areas, As, at time t and then advancing the solution to time $t + \Delta t$ using the values of A at t as initial conditions. Obviously then, the solution requires specification of the initial conditions (i.e., values of A along the x axis in Fig. 10.11 for $t = 0$) and boundary conditions (i.e., values of A for all time at $x = 0$) as starting points for the iterative-solution technique. In applications the boundary conditions are given by the discharge from upstream segments to which the present segment is connected.

For example (refer to Fig. 10.11), the solution for a segment will proceed as follows: At $t = 0$, $x = 0$, the values of j and i are 1. So, using the given boundary and initial conditions A_1^1, A_1^2, and A_2^1, Eq. (10.49) or Eq. (10.53) can be used to find A_2^2. From that result, in the same fashion, move to the right in the grid diagram, incrementing the value of x by Δx at each step. When the value of x is equal to the length of the segment, increment the time by Δt, go back to $x = 0$ and proceed with the solution in the same fashion, this time using the solution values at the previous time step as initial conditions. At the end of this iterative solution, after reaching the specified finishing time, the values for A at x equal to the length for all times are known. Using Eq. (10.39) the output hydrograph of the segment can then be found.

The kinematic-wave approximation has two parameters—α and m. Although operationally these should ultimately be obtained by calibration, it is possible to relate them to geometry and slope of prismatic channels. Equation (10.37) basically states the uniform flow assumption. Hence, using the Manning (or alternatively, the Chezy) equation, we have

$$Q = VA = \frac{1}{n} R^{2/3} S^{1/2} A = \frac{1}{n} \frac{A^{5/3}}{P^{2/3}} S^{1/2}, \tag{10.54}$$

where P is the wetted perimeter. For prismatic, simple geometry channels, the wetted perimeter is a function of A; hence, Eq. (10.54) is of the form $Q = \alpha A^m$. For example, for a wide rectangular channel of width B and depth of flow h, we obtain

$$Q = \frac{1}{n} \frac{(Bh)^{5/3}}{(B + 2h)^{2/3}} S^{1/2},$$

but given that $B \gg h$, the above reduces to

$$Q = \frac{1}{n}\frac{S^{1/2}}{B^{2/3}}A^{5/3}.$$

Hence, $\alpha = (1/n)(S^{1/2}/B^{2/3})$; $m = 5/3$. Table 10.6 gives α and m for a variety of shapes useful in approximating natural and man-made channels.

The Full Equations

The National Weather Service model known as DWOPER (Dynamic Wave Operational Model) developed by Fread [1973, 1974, 1975, 1976, 1978] is an example of a numerical solution of the complete St. Venant equation. The model is perhaps the most efficient, stable, and well-tested technique available for solving the full dynamic equations of motion. It uses the implicit finite-difference scheme recognized by Amein and Fang [1969] to take greatest advantage of the simultaneous algebraic equations that need to be solved at each iteration.

The mathematical basis for the DWOPER model is a finite-difference solution for a version of the full dynamic equations; in terms of discharge and cross-sectional area,

$$\frac{\partial Q}{\partial x} + \frac{\partial(A + A_o)}{\partial t} - \bar{q} = 0 \tag{10.55}$$

$$\frac{\partial Q}{\partial t} + \frac{\partial(\beta Q^2/A)}{\partial x} + gA\frac{\partial y}{\partial x} + gA(S_f + S_e) - \beta\bar{q}v_x + W_f B = 0, \tag{10.56}$$

TABLE 10.6 Kinematic-Parameter Definitions for Various Hydraulic Elements

TYPE OF SEGMENT	ALPHA (α)	m	TERM DEFINITIONS
Rectangular channel	$\frac{1.49}{nB^{2/3}}S^{1/2}$	1.67	S = slope
Triangular channel (isosceles)	$\frac{0.94}{n}S^{1/2}\left(\frac{z}{1+z^2}\right)^{1/3}$	1.33	n = Manning's roughness coefficient
Gutter flow	$\frac{1.182}{n}S^{1/2}\left[\frac{z^{1/2}}{1+(1+z^2)^{1/2}}\right]^{2/3}$	1.33	z = horizontal component of lateral slope
Curb; Street width; Gutter max. width			
Overland flow	$\frac{1.49}{n}S^{1/2}$	1.67	B = width of channel (ft)

where x is the downstream distance along the channel in meters, A is the cross-sectional area of flow in square meters, A_o is the off-channel flow area in square meters, Q is the flow rate in cubic meters per second, t is time in seconds, $\bar{q}$ is lateral inflow per unit length along channel in cubic meters per second per meter, β is a momentum correction factor, g is the gravitational constant in meters per square second, y is the water-surface elevation in meters, S_f is the friction slope, S_e is the eddy loss slope (to account for turbulent losses due to cross-sectional changes or obstructions in the flow), v_x is the velocity of lateral inflow in the direction of channel flow in feet per second, W_f is a wind resistance effect in square meters per square second, and B is top width of channel at surface in meters.

The friction slope S_f is evaluated through an empirical resistance expression such as Manning's equation as

$$S_f = \frac{n^2|Q|Q}{A^2R^{1.33}}, \tag{10.57}$$

where R is the hydraulic radius and n is Manning's coefficient.

The eddy loss slope is evaluated through

$$S_e = \frac{K_e}{2g}\frac{\partial V^2}{\partial x}, \tag{10.58}$$

where V is the mean velocity of flow across a cross section (in feet per second) and K_e is a loss coefficient.

A weighted four-point implicit scheme (with unequal distances Δx if needed) is used to solve these equations. Figure 10.12 gives the characteristic space–time discretization used in the finite-difference scheme. The length of the channel is divided into intervals of variable size Δx_m. The time is divided in equal intervals of equal size Δt. In this scheme the space and time derivatives are given by (Fig. 10.12):

$$\frac{\partial K}{\partial x} \simeq \theta\left(\frac{K_{m+1}^{n+1} - K_m^{n+1}}{\Delta x_m}\right) + (1-\theta)\left(\frac{K_{m+1}^{n} - K_m^{n}}{\Delta x_m}\right), \tag{10.59a}$$

$$\frac{\partial K}{\partial t} \approx \frac{K_m^{n+1} + K_{m+1}^{n+1} - K_m^n - K_{m+1}^n}{2\Delta t}, \tag{10.59b}$$

where K is a generic variable (i.e., cross-sectional area). Variables or functions other than derivatives are approximated by using weighting factors

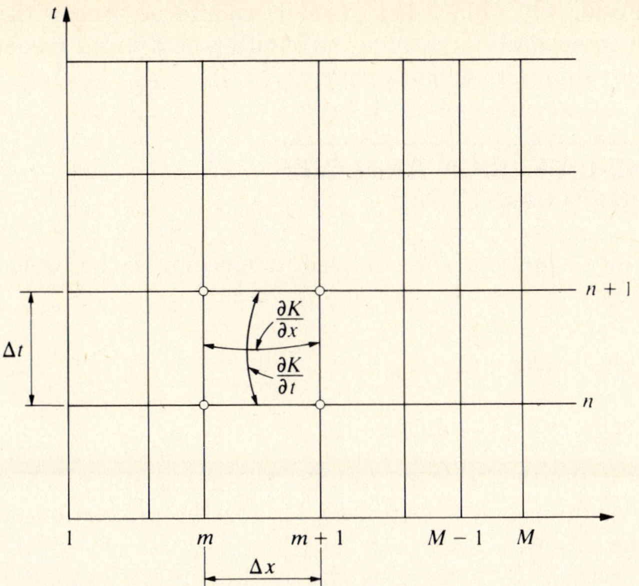

FIGURE 10.12 Definition sketch for program DWOPER: The *x*–*t* solution region.

similar to those of Eq. (10.59). Thus a function F will be given a value equivalent to

$$F \simeq \theta\left(\frac{F_m^{n+1} + F_{m+1}^{n+1}}{2}\right) + (1 - \theta)\left(\frac{F_m^n + F_{m+1}^n}{2}\right). \qquad (10.60)$$

For initial conditions, values of y and Q for $t = 0$ must be specified for all locations.

10.4 BLACK-BOX MODELS (From Georgakakos and Bras [1980])

The models of this category depart completely from any physical analogy to the flood-routing phenomenon. They describe the discharge (or depth) at a given position and time as a (usually linear) function of the discharges at n previous time steps in all m positions, as well as a function of the inflows at ℓ previous time steps in p inflow locations. The major problem in those models is to estimate the parameters involved. For example, if the model is linear, then estimation of the regression coefficients and the integers n, m, ℓ, and p is sought. This is done exclusively using observed data on inflow–outflow and statistical techniques (Dooge [1973]).

Concern regarding the predictive capability of black-box–type models for periods not similar to the calibration period suggests their use in short-term prediction. Preferably, "black-box" models should be used with observation processors to incorporate additional streamflow and input measurements in the estimated value of their parameters.

10.5 THE DIFFUSION ANALOGY (After Koussis [1978])

If the continuity equation is discretized in space only, the basic Muskingum model becomes

$$\frac{dS}{dt} = Q(x,t) - Q(x + \Delta x, t), \tag{10.61}$$

$$S = K[\theta Q(x,t) + (1-\theta)Q(x + \Delta x, t)], \tag{10.62}$$

where $Q(x,t)$ is the flow at time t at the beginning of a reach of length Δx. The outflow of the reach is $Q(x + \Delta x, t)$. θ is the proportion of storage attributed to the inflow.

Koussis proceeds by expanding the outflow in a truncated Taylor series around the inflow value,

$$Q(x + \Delta x, t) \approx Q(x,t) + \frac{\partial Q}{\partial x}\Delta x + \frac{1}{2}\frac{\partial^2 Q}{\partial x^2}(\Delta x)^2. \tag{10.63}$$

Using the above expression in Eqs. (10.61) and (10.62) results in

$$\frac{dS}{dt} = -\frac{\partial Q}{\partial x}\Delta x - \frac{1}{2}\frac{\partial^2 Q}{\partial x^2}(\Delta x)^2, \tag{10.64}$$

$$S = K\left[Q + \frac{\partial Q}{\partial x}\Delta x + \frac{1}{2}\frac{\partial^2 Q}{\partial x^2}(\Delta x)^2 - \theta\frac{\partial Q}{\partial x}\Delta x - \frac{\theta}{2}\frac{\partial^2 Q}{\partial x^2}\Delta x^2\right]. \tag{10.65}$$

Taking parameters K and θ as constants, we can take the derivative of Eq. (10.65) with respect to time (ignore third-order terms) and equate to Eq. (10.64). The result is

$$-\frac{\partial Q}{\partial x}\Delta x - \frac{1}{2}\frac{\partial^2 Q}{\partial x^2}(\Delta x)^2 = K\frac{\partial Q}{\partial t} + K\Delta x(1-\theta)\frac{\partial}{\partial x}\left(\frac{\partial Q}{\partial t}\right). \tag{10.66}$$

Equation (10.66) is complicated by the mixed space and time derivative appearing in the right-hand side. To avoid this problem, an attempt is made to express $\partial Q/\partial t$ in terms of spatial derivatives.

Recalling the continuity equation (Eq. 10.19),

$$\frac{\partial Q}{\partial x} + \frac{\partial A}{\partial t} = 0, \tag{10.67}$$

assuming no lateral inflow. The above can be further expressed as

$$\frac{\partial Q}{\partial x} + \left(\frac{dA}{dQ}\right)\frac{\partial Q}{\partial t}\bigg|_{x=\text{constant}} = 0, \tag{10.68}$$

from which the time derivative becomes

$$\frac{\partial Q}{\partial t} = -\left(\frac{dQ}{dA}\right)\bigg|_{x=\text{constant}} \frac{\partial Q}{\partial x}. \tag{10.69}$$

The term $-\frac{dQ}{dA}\big|_{x=\text{constant}}$ has units of velocity and in fact represents the slope of the stage–discharge relationship at a given x. Assuming that the stage–discharge relationship is single-valued, dQ/dA becomes the velocity c of propagation of "kinematic waves," a term that will be better defined later. Therefore Eq. (10.69) is now

$$\frac{\partial Q}{\partial t} = -c\frac{\partial Q}{\partial x}. \tag{10.70}$$

Substitution of Eq. (10.70) in the last term of Eq. (10.66) and rearranging yields

$$\frac{\partial Q}{\partial t} + \frac{\Delta x}{K}\frac{\partial Q}{\partial x} = \left[\Delta x(1-\theta)c - \frac{1}{2}\frac{(\Delta x)^2}{K}\right]\frac{\partial^2 Q}{\partial x^2}. \tag{10.71}$$

Students familiar with heat flow or the dynamics of conservative pollutants in rivers will recognize Eq. (10.71) as the diffusion equation, previously alluded to in Eq. (10.36).

Cunge [1969] also obtained the diffusion equation by linearizing the full dynamic equation around a perturbation and ignoring inertial terms. His result led to

$$\frac{\partial Q}{\partial t} + C_1\frac{\partial Q}{\partial x} = C_2\frac{\partial^2 Q}{\partial x^2}, \tag{10.72}$$

where C_1 is the discharge kinematic velocity and C_2 is the diffusion coefficient:

$$C_1 = c \qquad \text{and} \qquad C_2 = \frac{Q_\circ}{2B_\circ S_\circ}$$

In the above, $Q_\circ$, $B_\circ$, and $S_\circ$ are base (for linearization) discharge, channel top width, and slope, respectively. The obvious similarity of Eqs. (10.71) and (10.72) led Koussis [1978] to equate coefficients; therefore, after solving:

$$K = \frac{\Delta x}{c} \qquad \text{and} \qquad \theta = \frac{1}{2} - \frac{KQ_\circ}{2B_\circ S_\circ(\Delta x)^2} \tag{10.73}$$

or

$$K = \Delta x \Big/ \left(\frac{dQ}{dA}\right)\Bigg|_{x=\text{constant}}$$

Theoretically, Eq. (10.63) leads to constant parameters, since a single linearization was performed. The possibility of linearizing piecewise to obtain variable parameters immediately suggests itself. All that is required is a single-valued stage–discharge relationship to obtain dQ/dA at different discharge values. This also implies a fixed geometry that allows finding a linearization width $B_{\circ}$ for a given discharge $Q_{\circ}$. The slope $S_{\circ}$ can be taken as the average slope of the water surface at the linearization time.

10.6 SUMMARY

Once runoff gets into stream channels, its movement, as a flood wave, is described by the concepts presented in this chapter. In large basins where the hillslope flow time may be small relative to the time in channels (where overland or Hortonian runoff may be dominant), routing essentially accounts for the shape of the outflow hydrograph. Routing is also important in the design of channels to carry flood flows or in the design of other flood-prevention structures such as levees, diversions, and dams.

With modern-day computers the numerical solution of the full (or subset) equations of motions in the channels is very common. Nevertheless, it remains a difficult task requiring a lot of data and experience. In nonprismatic and natural channels, detailed and expensive information on channel cross sections would be required. Determination of roughness (i.e., Manning's n) or loss coefficients (i.e., K_e in Eq. 10.58) is hard. Ultimately, calibration against flow data is the best way of guaranteeing the reasonableness of the solution.

Simple conceptual, or black-box, procedures for routing are still used because of data demands of the numerical solutions. As we have seen in Sections 10.3.1 and 10.5, these conceptualizations can be justified from a solid theoretical argument and, if used correctly, can yield good results. This is particularly true when good calibration data are available.

REFERENCES

Amein, M., and C. S. Fang [1969]. "Stream Flow Routing (With Applications to North Carolina Rivers)." Raleigh, N. C.: Water Resources Research Institute of the University of North Carolina. (Report no. 17.)

Bras, R. L. [1972]. "Effects of Urbanization on the Runoff Characteristics of Small Basins in Puerto Rico." Cambridge, Mass.: MIT Department of Civil Engineering. (Bachelor's thesis.)

Carslaw, H. S., and C. Jaeger [1959]. *Conduction of Heat in Solids*. 2nd ed. Fair Lawn, N. J.: Oxford University Press.

Cunge, J. A. [1969]. "On the Subject of a Flood Propagation Method (Muskingum Method)." *J. Hydrol. Res.* 7(2):205–230.

Diaz-Granados, M., R. L. Bras, and J. B. Valdes [1986]. "Incorporation of Channel Losses in the Geomorphologic IUH." In: V. K. Gupta, I. Rodriguez-Iturbe, and E. F. Wood, eds. *Scale Problems in Hydrology*. Dorcrecht, Holland: D. Reidel, p. 246.

Dooge, J. C. I. [1973]. "Linear Theory of Hydrologic Systems." U.S. Department of Agriculture technical bulletin no. 1468, pp. 50–53, 148–183, 232–260.

Dooge, J. C. I., and B. M. Harley [1967]. "Linear Routing in Uniform Channels." *Proceedings of the International Hydrology Symposium*. Vol. 1. Fort Collins, Colo.: Colorado State University, pp. 57–63.

Dooge, J. C. I., et al. [1980]. Lecture notes. Caracas, Venezuela: Simon Bolivar University.

Eagleson, P. S. [1970]. *Dynamic Hydrology*. New York: McGraw-Hill.

Fread, D. L. [1973]. "A Dynamic Model of Stage Discharge Relations Affected by Changing Discharge." National Oceanographic and Atmospheric Administration technical memorandum NWS HYDRO-16. (Revised, 1976.)

Idem. [1974]. "Implicit Dynamic Routing of Floods and Surges in the Lower Mississippi." Presented at the American Geophysical Union Spring National Meeting. April 8–12, 1974, Washington, D. C.

Idem. [1975]. "Numerical Properties of Implicit Four-Point Finite Difference Equations of Unsteady Flow." National Oceanographic and Atmospheric Administration technical memorandum NWS HYDRO-18.

Idem. [1976]. "Flood Routing in Meandering Rivers with Flood Plains." In: *Rivers 1976*. Vol. I. Proceedings of the Symposium on Inland Waterways for Navigation, Flood Control and Water Diversions. Fort Collins, Colo.: Colorado State University, August 10–12, 1976.

Idem. [1978]. "National Weather Service Operational Dynamic Wave Model." Silver Spring, Md.: Hydrologic Research Laboratory.

Georgakakos, K. P., and R. L. Bras [1980]. "A Statistical Linearization Approach to Real Time Nonlinear Flood Routing." Cambridge, Mass.: MIT Department of Civil Engineering, Ralph M. Parsons Laboratory. (Technical report no. 256.)

Idem. [1982]. "Real Time, Statistically Linearized, Adaptive Flood Routing." *Water Resources Res.* 18(3):513–524.

Gray, D. M., ed. [1973]. *Handbook on the Principles of Hydrology*. Port Washington, N. Y.: Water Information Center.

Henderson, F. M. [1966]. *Open Channel Flow*. New York: MacMillan, pp. 125–164, 285–398.

Kirshen, D. M., and R. L. Bras [1983]. "The Linear Channel and Its Effect on the Geomorphologic IUH." *J. Hydrol*. 65:175–208.

Kitanidis, P. K., and R. L. Bras [1978]. "Real Time Forecasting of River Flows." Cambridge, Mass.: MIT Department of Civil Engineering, Ralph M. Parsons Laboratory. (Technical report no. 235.)

Koussis, A. D. [1976]. "An Approximate Dynamic Flood Routing Method." Presented at the International Symposium on Unsteady Open-Channel Flow, Newcastle-Upon-Tyne, England, April 12–15, 1976.

Idem. [1978]. "Theoretical Estimation of Flood Routing Parameters." *J. Hydraul. Div. A.S.C.E.* 104(HY1):109–115.

Koussis, A. D., and B. J. Osborne [1986]. "A Note on Nonlinear Storage Routing." *Water Resource Res.* 22(13):2111–2113.

Lighthill, M. J., and G. B. Whitham [1955]. "On Kinematic Waves I. Flood Movement in Long Rivers." *Proc. R. Soc. London* [*A*]. 229:281–316. Series 22g.

Nash, J. E. [1959]. "A Note on the Muskingum Flood-Routing Method." *J. Geophys. Res.* 64:1053–1056.

Price, R. K. [1973]. "Flood Routing Methods for British Rivers." Wallingford, England: Hydraulic Research Station. (Technical report INT 111.)

Viessman, W., Jr., J. W. Knapp, G. L. Lewis, and T. E. Harbaugh [1977]. *Introduction to Hydrology*. 2nd ed. New York: Harper & Row.

Weinmann, P. E. [1977]. "Comparison of Flood Routing Methods for Natural Rivers." Clayton, Victoria, Australia: Monash University. (M. Eng. Sci. thesis.)

Weinmann, P. E., and E. M. Laurenson [1977]. "Modern Methods of Flood Routing." Presented at the Hydrology Symposium at Brisbane, Australia. June 28–30, 1977.

Idem. [1979]. "Approximate Flood Routing Methods: A Review." *J. Hydraul. Div. A.S.C.E.* 105(HY12):1521–1536.

PROBLEMS

1. Given below is the "design" input hydrograph of a reservoir.

TIME (days)	INFLOW ($ft^3 s^{-1}$)	TIME (days)	INFLOW ($ft^3 s^{-1}$)
0	0	6	22
1	10	7	15
2	20	8	10
3	30	9	10
4	35	10	10
5	30		

You are asked to choose between two designs that have the following storage–outflow characteristics:

Design a

$$Q = \begin{cases} \dfrac{1}{5}\left(\dfrac{2S}{\Delta t} + Q - 150\right) & \text{for } 2S + Q > 150 \\ 0 & \text{otherwise} \end{cases}$$

Design b

$$Q = \begin{cases} \dfrac{1}{3}\left(\dfrac{2S}{\Delta t} + Q - 170\right) & \text{for } 2S + Q > 170 \\ 0 & \text{otherwise} \end{cases}$$

If the inflow hydrograph peak should be attenuated at least 25%, which design should be built? At time zero the storage is 50 $ft^3 s^{-1} day^{-1}$, and the outflow is 0, $\Delta t = 1$ day.

2. The dynamics of a river can be approximated by the kinematic-wave equation. The cross-sectional area versus time relationship at the river's outlet looks like a Nash model, i.e.,

$$A(t) = 5f(t)$$

in square meters, time in seconds, with $f(t)$ given by

$$f(t) = \frac{1}{k}\left(\frac{t}{k}\right)^{n-1}\frac{1}{\Gamma(n)}\,e^{-t/k},$$

and $k = 3$, $n = 2$. If the kinematic-wave parameters are $\alpha = 1.5$ ($m^{-1} s^{-1}$), $m = 2$, compute the peak velocity at the river's outlet.

3. Derive the expressions for the kinematic-wave parameters in an isosceles triangular channel and for gutter flow as shown in Table 10.6.

4. The time–area curve of a given watershed is given by

TIME (hr)	INCREMENTAL AREA (acres)
1	50
2	50
3	62.5
4	18.75
5	6.25
6	7.6875
7	32.09375
8	18.25

The following storm occurs in the area:

TIME (hr)	RAINFALL (in. hr^{-1})
1	4
2	2
3	3
4	4
5	1

If $\Delta t = 1$ hr, $C_0 = 0.1$, $C_1 = 0.4$ in the river reach transversing the area, what is the outflow hydrograph for the given storm? Use the Muskingum method to calculate it.

5. In continuous time, the Muskingum routing equations were seen to be

$$\frac{dS}{dt} = I - Q, \tag{1}$$

$$S = K[xI + (1 - x)Q]. \tag{2}$$

Assume you have an empty ($Q = 0$ at $t = 0$) channel below a dam. The dam breaks and the release (input to the channel) is very high at the beginning and decreases with time. It can be described by a function

$$I(t) = I_o e^{-\lambda t}.$$

Derive the expression for discharge in the channel. The solution to

$$\frac{dY}{dt} + P(t)Y = g(t)$$

is

$$Y(t) = \frac{1}{\mu(t)}\left[\int \mu(t)g(t)\,dt + c\right],$$

where

$$\mu(t) = \exp\left[\int P(t)\,dt\right].$$

Show that the answer is

$$Q(t) = \frac{b_3}{(1/b_2 - \lambda)}(e^{-\lambda t} - e^{-t/b_2}),$$

where

$$b_3 = \frac{I_o + b_1\lambda}{b_2}, \qquad b_1 = KAx, \qquad b_2 = K(1 - x).$$

6. It was shown that following the "diffusion analogy" for flow in open channels, the Muskingum parameters can be expressed as

$$K = \frac{\Delta x}{c} \qquad X = \frac{1}{2} - \frac{KQ_o}{2B_o S_o \Delta x^2},$$

where $Q_\circ$ is a reference discharge, $B_\circ$ is a reference channel width, $S_\circ$ is a reference slope, Δx is the channel segment length, and c is the kinematic velocity.

Generally, $c \approx 1.5U_\circ$, where $U_\circ$ is the normal velocity, i.e., that corresponding to reference discharge $Q_\circ$.

Assume the channel in Problem 5 is rectangular, with width $B_\circ = 50$ m. Assume $Q_\circ = 1000\ \text{m}^3\,\text{s}^{-1}$ and $S_\circ = 0.01$ (bottom channel slope). If $I_\circ = 5000\ \text{m}^3\,\text{s}^{-1}$ and $\lambda = 2\ \text{hr}^{-1}$, estimate how long it will take the peak of the hydrograph to reach a town 5 km downstream of the dam (let $\Delta x = 5$ km).

7. The U.S. Army Corps of Engineers is worried about the safety implications of one of their reservoirs. They figure that if the dam fails, a sharp-crested hydrograph of the following form will occur.

TIME (hr)	Q ($\text{m}^3\,\text{s}^{-1}$)
6	50
12	500
18	300
24	200
30	100

Below the dam, there is a channel with routing parameters $C_0 = 0.1$ and $C_1 = 0.4$. The channel flows into another reservoir, which is shaped like a cube with bottom area of $20 \times 10^6\ \text{m}^2$. A cross section of the reservoir is shown below. The spillway discharge is given by

$$Q = CLH^b,$$

where Q = discharge in cubic meters per second, $L = 200$ m, $C = 1.0$, $b = 1.5$, and H = head above spillway crest. The maximum capacity of the spillway is $200\ \text{m}^3\,\text{s}^{-1}$. Will the spillway be able to handle the flow resulting from the upstream dam failure? Assume that the second reservoir is full when the upstream flow arrives.

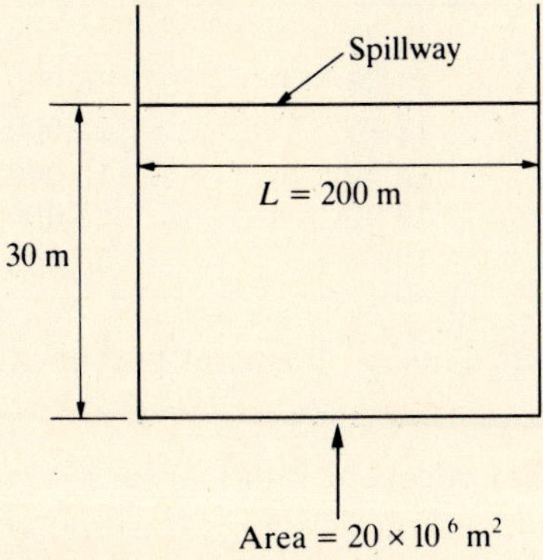

8. Below is the profile of a circular pipe of radius 1 ft, with water flowing under pressure. The net effect is the indicated water accumulation, to the depth of 1 ft, in the upstream manhole. Using the kinematic-wave approximation, give an estimate of the flow through the pipe and of the effective kinematic parameter α.

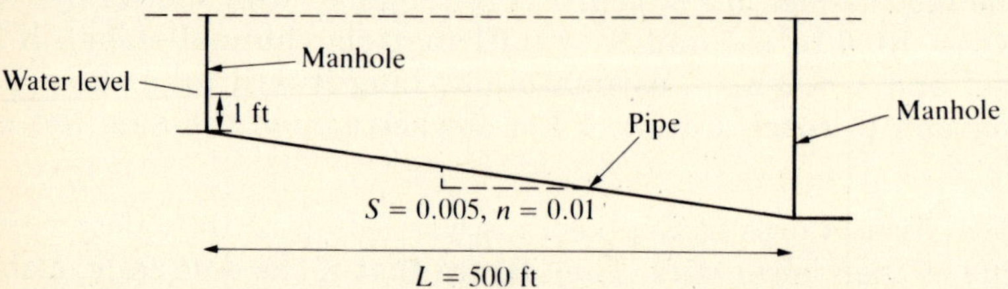

9. A lake having steep banks and a surface area of 500 acres discharges into a steep channel that is approximately rectangular in shape, with a width of 25 ft. Initially, conditions are steady, with a flow of 1000 $ft^3 s^{-1}$ passing through the lake; then a flood comes down the river feeding the lake, giving rise to the following inflow hydrograph:

TIME FROM START (hr)	INFLOW ($ft^3 s^{-1}$)
0	1000
6	1600
12	2630
18	3050
24	2840
30	2300
36	1700
42	1200
48	1000

Calculate the outflow hydrograph for the 48-hr period.

LAKE LEVEL (ft)	CHANNEL DEPTH (ft)	VELOCITY ($ft\,s^{-1}$)	LAKE VOLUME ($ft^3 s^{-1} hr$)
5.0	3.33	10.35	0.0
6.0	4.00	11.35	6,000.0
7.0	4.67	12.25	12,000.0
8.0	5.33	13.10	18,000.0
9.0	6.00	13.90	24,000.0
10.0	6.67	14.63	30,000.0

10. Derive Eq. (10.33) by expressing all quantities in terms of perturbations around a nominal value, i.e., $q = q_\circ + \delta q$.

11. Equation (10.34) assumes that a channel receives input only at the upstream end. Suggest how to obtain the response function for a channel with a uniform lateral inflow. (*Hint:* Try to build the solution from the available upstream inflow case.)

12. Derive Eqs. (10.66) and (10.71). Review the assumptions made.

13. Assuming typical values for the variables involved, perform an order-of-magnitude analysis of the terms in the St. Venant momentum equation (Eq. 10.28).

14. Derive a stage–discharge relationship for (a) a triangular channel and (b) a semicircular channel.

15. If the channel in Example 10.1 is rectangular, with a bottom width of 50 m, slope of 0.001, and Manning's n of 0.01, estimate what its length might be. (*Hint:* Define and assume a normal discharge around which to linearize and use the diffusion analogy of Section 10.5.)

16. A wide rectangular channel 41 km long and 25 m wide has a normal discharge (Q_o) of 50 $m^3 s^{-1}$. The slope is 0.0001 and Manning's n is 0.013. Route the inflow given in Table 10.2 using the Muskingum–Cunge method if $\Delta t = 6$ hr. (See Problem 20 for definition of the Muskingum–Cunge method.)

17. Repeat Problem 16, using Eq. (10.42) and if $\Delta t = 4$ hr.

18. A 12-acre site (1 acre $\approx$ 4000 m^2) is to be developed for an industrial park. About 8 acres are pavement, roofs or other impervious areas. The other 4 acres are undisturbed and wooded. Developed and wooded areas are fairly uniformly distributed throughout the site. A plan view of the site is shown below, indicating the preferred flow directions.

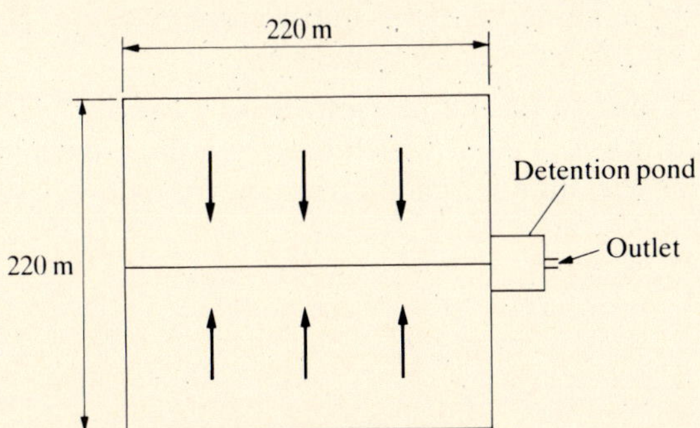

It has been determined that the response function (instantaneous unit hydrograph) of the site is well approximated by an exponential (linear reservoir) model. For a 24-hr long, uniform-duration storm, the mean of the resulting hydrograph occurs at 12.5 hr.

The outflow from the site goes into a detention pond, intended to reduce the peak discharge. The pond has vertical sides and is a square 1000 m^2 in bottom area (i.e., 31.6 m to the side). The pond is 5 m deep. At the bottom of the pond there is a 6-in.-square drain culvert that serves as an outlet for the detention basin. A side view of the pond is given below.

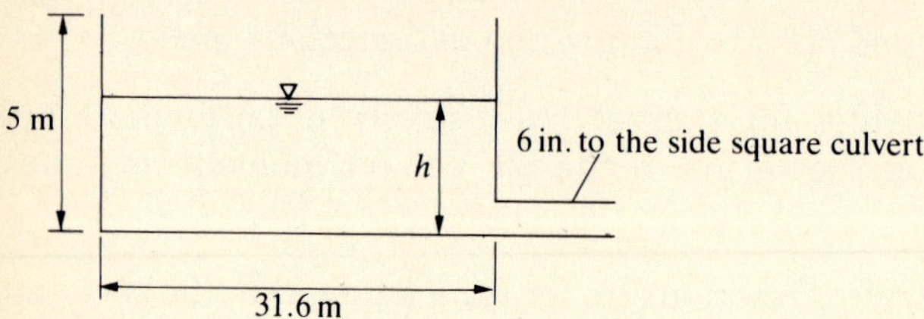

From a hydraulics handbook you find that the pipe discharges proportionally to the square root of head above it as

$$Q = 0.3a\sqrt{2gh},$$

where a is the cross-sectional area of the culvert and $g = 9.8\ \mathrm{m\,s^{-2}}$.

What would be (approximately) the maximum discharge and elevation in the detention pond given a 175-mm storm of 24 hours' duration on the site?

19. Using the dynamic-wave approximation, it can be shown that the coefficients for the diffusion equation (Eq. 10.36) can be expressed as

$$c = \frac{dQ}{dA}$$

$$D = \frac{Q_p}{2S_\circ B_\circ},$$

where Q_p = peak discharge, c = celerity, $S_\circ$ = main stream slope, $B_\circ$ = channel width at the surface, and A = cross-sectional area.

a) Determine the coefficients c, D for a channel reach of slope $S_\circ = 0.0001$ and Manning's $n = 0.04$, which carries a peak discharge $Q_p = 30\ \mathrm{m^3\,s^{-1}}$. Do it for the channel geometries shown below.

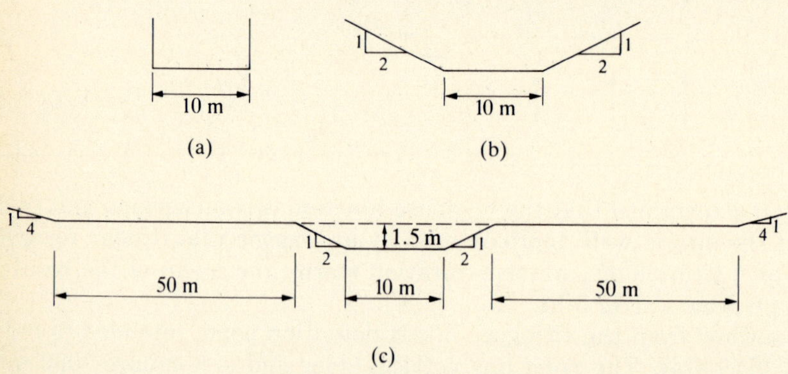

b) For channel geometry (c), what happens to these coefficients when $Q = 17.1\ \text{m}^3\,\text{s}^{-1}$? (*Hint:* The flow depth = 1.5 m.) What is the reason for this behavior? Do you think this behavior would occur in nature? What are the routing implications of the behavior?

(Contributed by Garry Willgoose.)

20. *Muskingum–Cunge Routing:* Cunge [1967] showed that the coefficients of the Muskingum method K and x of Eq. (10.9) can be related to the diffusion constants c, D in the diffusion equation (Eq. 10.36). These relationships are

$$K = \frac{\Delta x}{c}$$

$$x = \frac{1}{2} = \frac{D}{c\,\Delta x},$$

where Δx = length of the channel reach.

a) For the channel geometries of Problem 19a and a stream reach length of 2000 m, determine K and x.

b) Using the K, x determined in (a), route the hydrograph shown below through channel geometries (b) and (c) of Problem 19.

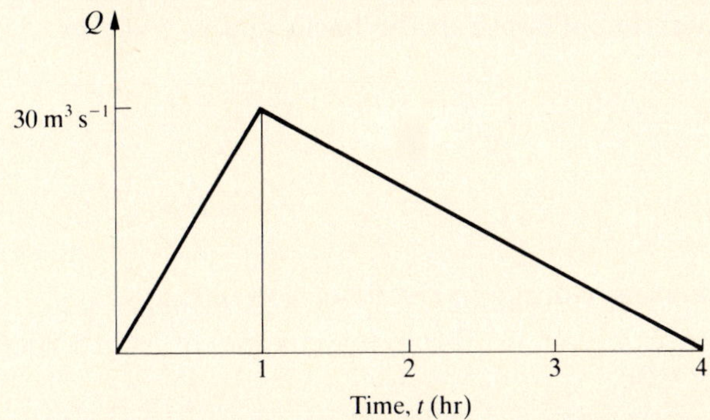

c) For channel geometry (c) of Problem 19 and using values of depth y = 1.40 and 1.6, determine dQ/dA by finite differences. Determine c, D, K, and x for a channel length of 2000 m. Using these values of K and x, route the storm shown above. Compare the results with those of part (b).

d) Do you think it is reasonable to use a constant K, x in part (b) for geometry (c) of Problem 19? Consider the cases where you are interested in peak discharge only and total hydrograph volume.

e) Comment on the relation between K, x and overbank storage. What is the effect of overbank storage on the routing of the storm?

f) In the Muskingum–Cunge relationships above, it is possible to obtain a negative value for x if Δx is chosen small enough. Comment on this anomalous result. On the other hand, verify that for a fixed-channel cross section (i.e., fixed c, D) a single channel of length L provides the same routing effect as does two channels, end to end of length $L/2$, provided only that the discharge at the midpoint is related as

$$Q_{\text{midpoint}} = \frac{1}{2}(Q_{\text{in}} + Q_{\text{out}}).$$

(*Hint:* This apparent contradiction may be clarified by viewing x as an interpolation coefficient between the storage due to the outlet discharge and storage due to the inflow discharge. Negative x corresponds to an extrapolation to a discharge downstream of the reach. For numerical stability, though, it is recommended that x be between 0 and 0.5.)

(Contributed by Garry Willgoose.)

21. a) Route the hydrograph in Problem 20b through a detention basin with a box culvert outlet. The culvert has a square cross section of sides 1.25 m. Storage (in cubic meters) in the detention basin can be expressed as a function of depth $S = 1720y^2$. The discharge through the culvert is also a function of depth in the basin and is given by

$$Q_{\text{out}} = \begin{cases} 0.6B^2[2g(y - 0.6B)]^{1/2} & B = 1.25 \qquad y > 1.25B \\ \dfrac{2}{3}By\sqrt{\dfrac{2}{3}gy} & \qquad\qquad\quad y < 1.25B \end{cases}$$

Calculate the maximum storage in the detention basin.

b) Route the hydrograph of part (a) through a channel of length 2000 m with $K = 320$, $x = 0.05$.

c) Route the hydrograph of Figure 10.2 through a channel of length 2000 m with $K = 12{,}800$, $x = 0.1$.

d) You are told that the K, x parameters of part (c) correspond to a 2000-m channel with an overbank region 100 m wide and that those of part (b) correspond to the same channel without the overbank region. Compare the results of (a), (b), and (c) and consider the implications they have for flood-mitigation strategies based on levees and detention basins.

(Contributed by Garry Willgoose.)

Chapter **11**

Concepts of Probability in Hydrology

11.1 INTRODUCTION

There are many hydrologic phenomena in which the variable of interest cannot be uniquely specified as a function of known related variables and conditions. For example, this is the case for the highest discharge that a certain river will attain in the next five years. The previous example is clearly the case of a random variable where the outcome can never be uniquely predicted with the help of physical laws, no matter how much information we gather.

The random aspect of a hydrologic phenomenon may also arise due to our inability to understand all the details of a causal relationship between an input (i.e., rainfall) and an output (i.e., discharge). This lack of knowledge may be inherent in the physical description of the processes or in limitations on availability of data. Summarizing, uncertainty is introduced into hydrologic problems through

1. inherent unexplainable variability of nature;
2. lack of understanding of all causes and effects in physical systems; and
3. lack of sufficient data.

The future of a random variable is not subject to precise prediction and must be described within the domain of the set of possible values it may take (its sample space). The description of the random variable is then accomplished through the concept of probability distributions. The hydrologist or analyst must consider the possibility of occurrence of particular events and then determine the likelihood of their occurrence.

The collection of all possible outcomes of an experiment or random phenomenon is called its sample space. This space may be discrete or continuous and consists of a set of points—sample points—each of which is associated with one and only one distinguishable outcome. An "event" is a collection of sample points, and it may be simple if the event consists only of one sample point and compound if consisting of two or more sample points.

11.2 REVIEW OF PROBABILITY

Some parts of this section have been adapted from notes by I. Rodriguez-Iturbe.

The classical and simplest interpretation of probability is one of frequency. If A is a given event that may occur from an experiment (i.e., a given streamflow in nature), then the probability of A, $P(A)$, is given by

$$P(A) = \lim_{m\to\infty} \frac{n_A}{m}, \tag{11.1}$$

where n_A is the number of times that event A occurs in m repeated experiments. The true probability measure will only be obtained as the number of experiments m goes to infinity.

Mathematically the probability $P(A)$ must satisfy a series of axioms:

1. The probability of an event is a number between 0 and 1.
2. The probability of a *certain* event is 1. A certain event is one that covers all possible outcomes of an experiment. For example, the probability of a streamflow being greater than or equal to 0, but less than infinity, is 1.
3. The probability of an event that is the sum of two mutually exclusive events is the sum of the probabilities of those two events. Mutually exclusive events are those that by definition preclude the occurrence of each other. For example, a streamflow cannot be less than 300 $m^3 s^{-1}$ and greater than 500 $m^3 s^{-1}$ at the same time. Those are mutually exclusive events. A set of exhaustive events covers all possible outcomes of an experiment.

Union and Intersection

The above axioms allow the definition of some important probabilistic concepts. For example, if two events A and B are not exclusive, they have overlapping components; the set of common points is called the intersection of A and B, $A \cap B$. The union of two events A and B is the collection of all sample points occurring in either A or B. It is represented by $A \cup B$.

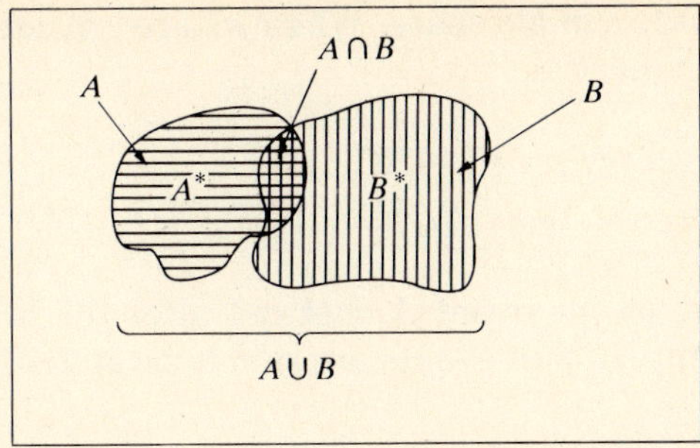

FIGURE 11.1 Illustration of the concept of intersection and union of two events *A* and *B*.

Figure 11.1 illustrates an intersection and a union. The probability of event A may be expressed as

$$P[A] = P[A \cap B] + P[A^*], \tag{11.2}$$

since $A \cap B$ and A^* are mutually exclusive (see Fig. 11.1). Similarly,

$$P[B] = P[A \cap B] + P[B^*] \tag{11.3}$$

and

$$P[A \cup B] = P[A^*] + P[B^*] + P[A \cap B]. \tag{11.4}$$

Combining the above leads to

$$P[A \cup B] = P[A] + P[B] - P[A \cap B]. \tag{11.5}$$

Conditional Probability and Probabilistic Independence

The conditional probability of the event A given that the event B has occurred, denoted $P[A \mid B]$, is defined as the ratio

$$P[A \mid B] = \frac{P[A \cap B]}{P[B]}. \tag{11.6}$$

The condition that "B has occurred" restricts the sample space to the set of sample points in B but should not change the relative likelihood of the simple events in B; we then renormalize the probability measure of those points in B that are also in A, $P[A \cap B]$, dividing by $P[B]$ in order to account for the reduction in the sample space. When we are considering many events, Eq. (11.6) can be expanded to

$$P[A \cap B \cap C \ldots \cap N] = P[A|BC\ldots N] \cdot P[B|C\ldots N]P[C|D\ldots N]\ldots P[N]. \tag{11.7}$$

If the occurrence of one event A does not alter the probability of occurrence of another event B, both events are said to be statistically independent,

$$P[A|B] = P[A], \tag{11.8}$$

which implies

$$P[A \cap B] = P[A]P[B] \tag{11.9}$$

and

$$P[B|A] = P[B]. \tag{11.10}$$

N events are independent if it holds that

$$P[A \cap B \cap C \ldots \cap N] = P[A]P[B]\ldots P[N]. \tag{11.11}$$

The hydrologist usually relies on his/her knowledge of the physical situation at hand in order to assume independence or dependence among events. Thus, high flows in a river from year to year may be assumed independent with more confidence than low flows, which are very much influenced by the carryover between years.

Total Probability Theorem

This theorem concerns the probability of a compound event A in a random experiment. Given a set of mutually exclusive, collectively exhaustive events $B_1, B_2, \ldots, B_n$, it is always possible to express the probability of any event A as

$$P[A] = P[A \cap B_1] + P[A \cap B_2] + \cdots + P[A \cap B_n]. \tag{11.12}$$

Every term of Eq. (11.12) can be expressed in the form of a conditional probability to yield

$$P[A] = \sum_{i=1}^{n} P[A|B_i]P[B_i], \tag{11.13}$$

which is the most common version of the total probability theorem.

Bayes Theorem

The Bayes theorem is fundamental to engineering and hydrologic analysis and is very important when considering the conditional probability of an event B_j given another event A. We know that

$$P[B_j|A] = \frac{P[B_j \cap A]}{P[A]} = \frac{P[A \cap B_j]}{P[A]}, \tag{11.14}$$

but we have

$$P[A \cap B_j] = P[A|B_j]P[B_j]$$

and

$$P[A] = \sum_{i=1}^{n} P[A|B_i]P[B_i],$$

which substituted in Eq. (11.14) yields

$$P[B_j|A] = \frac{P[A|B_j]P[B_j]}{\sum_{i=1}^{n} P[A|B_i]P[B_i]}. \tag{11.15}$$

The importance of Eq. (11.15) is that it allows the hydrologist to express his/her experience and judgment—which is valuable information—in the form of $P[B_j]$ or probabilities of a certain state of nature, before any sample has been taken. The information obtained from sample A is incorporated through conditional probabilities of the sample given a certain state of nature. In this manner Eq. (11.15) can also be expressed as

$$P[\text{state}|\text{sample}] = \frac{P[\text{sample}|\text{state}]\,P[\text{state}]}{\sum_{\text{all states}} P[\text{sample}|\text{state}]\,P[\text{state}]}, \tag{11.16}$$

where the P[state] in the right-hand side of Eq. (11.16) represents prior probabilities and the left-hand side of the equation is the posterior probability, which incorporates both the information available before the sample was taken and the information yielded by the sample about the state of nature.

Random Variables, Ensembles, and Distributions

A random variable may be defined as a numerical variable not subject to precise prediction. It must be described in the domain of all its possible values through the aid of probability distributions.

Discrete Random Variables

A random variable is discrete when the value it can take is restricted to countable numbers. An example may be the number of rainy days during one year in Boston.

The function $P_X(x_i)$, which gives the probability of the discrete random variable X taking any possible value x_i, is called the probability mass function (pmf) of X,

$$P_X(x_i) = P[X = x_i].$$

$P_X(x_i)$ is a nonnegative function in accordance with the definition of probability and

$$\sum_{\text{all } i} P_X(x_i) = 1.$$

Clearly, the probability of X being between any two values x_j and $x_k (x_j < x_k)$ is

$$P[x_j \le X \le x_k] = \sum_{i=j}^{i=k} P_X(x_i). \tag{11.17}$$

Any random variable can also be described through its cumulative distribution function, which simply gives the probability of the event that the random variable takes a value equal to or less than the argument

$$F_X(x) = P[X \le x].$$

For discrete random variables we have

$$F_X(x) = \sum_{\text{all } x_i \le x} P_X(x_i). \tag{11.18}$$

Continuous Random Variables

When the range of variation of a random variable is continuous, the variable is called a continuous random variable. Unlike the discrete random variable, the continuous one is free to take any value on the real line, although this does not mean it must take on values over the entire domain of real numbers.

Separating the real line into a number of intervals of infinitesimal length dx, the probability that a continuous random variable X falls between x and $x + dx$ is given by $f_X(x)\,dx$, where the function $f_X(x)$ is called the probability density function (pdf) of X.

The occurrence of x in different intervals dx constitutes mutually exclusive events and thus, according to the theorem of total probability, we can compute the probability that X takes a value in an interval of finite length,

$$P[x_1 \leq X \leq x_2] = \int_{x_1}^{x_2} f_X(x)\, dx\,. \tag{11.19}$$

The value $f_X(x)$ is not itself a probability; it is, rather, a measure of the probability density. The meaning of Eq. (11.19) is illustrated in Figure 11.2.

From the axioms of probability theory, we have

$$f_X(x) \geq 0$$

$$\int_{-\infty}^{\infty} f_X(x)\, dx = 1\,.$$

Similarly to the case of discrete random variables, we define the cumulative probability distribution or cumulative density function (cdf) of a continuous random variable by

$$F_X(x) = P[X \leq x] = \int_{-\infty}^{x} f_X(u)\, du \tag{11.20}$$

and we have the relationship

$$\frac{dF_X(x)}{dx} = f_X(x)\,. \tag{11.21}$$

In hydrology there are many cases where the distribution of the random variable is composed of two parts: a discontinuous part, or probability mass; and a continuous part, or probability density. These are the so-called mixed distributions. An example may be the distribution of daily flows for an ephemeral

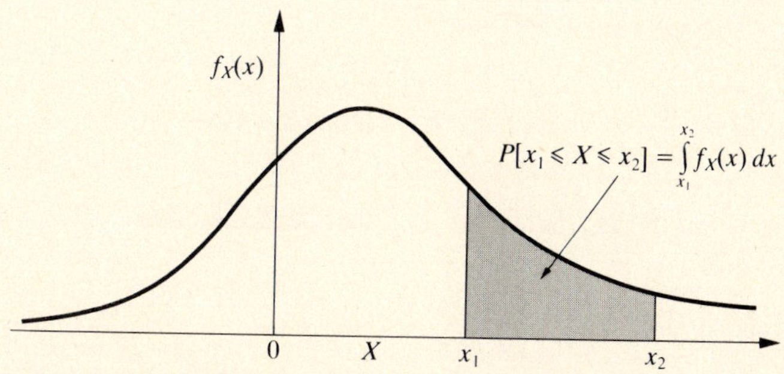

FIGURE 11.2 A probability density function and the probability of a random variable X taking values between x_1 and x_2.

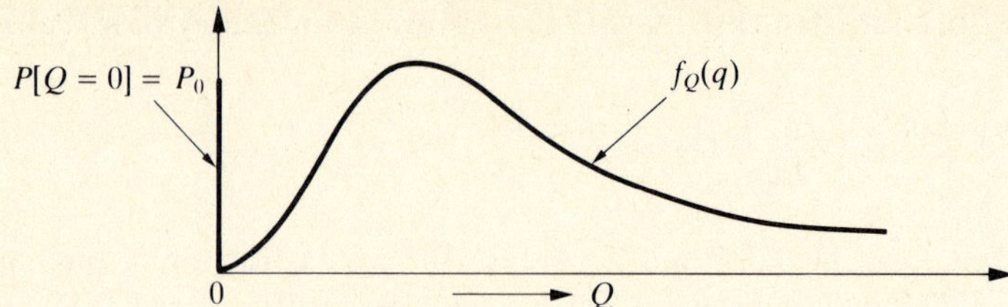

FIGURE 11.3 Example of a mixed distribution with probability P_0 of random variable Q (i.e., discharge) taking a value of 0.

stream. The stream may be dry during certain times of the year and thus we have a probability distribution that looks like Figure 11.3.

Moments and Expectation

Simple numbers are sometimes used to describe the dominant features of the behavior of a random variable, in other words, to describe the general shape of a probability density function. "These numbers usually take the form of weighted averages of certain functions of the random variable. The weights used are the pmf or pdf of the variable, and the average is called the expectation of the function" (Benjamin and Cornell [1970]).

We define the mean or the expected value $E[X]$ of a random variable X as

$$\mu_X = E[X] = \sum_{\text{all } x_i} x_i P_X(x_i) \quad \text{Discrete case}$$
$$\mu_X = E[X] = \int_{-\infty}^{\infty} x f_X(x)\, dx \quad \text{Continuous case} \tag{11.22}$$

The mean is a measure of central tendency.

In many problems we are most interested in the dispersion that the random variable X can have around its expected value. The most useful measure of dispersion is the variance defined as

$$\sigma_X^2 = \text{Var}[X] = \sum_{\text{all } x_i} (x_i - \mu_X)^2 P_X(x_i) \quad \text{Discrete case}$$
$$\sigma_X^2 = \int_{-\infty}^{\infty} (x - \mu_X)^2 f_X(x)\, dx. \quad \text{Continuous case} \tag{11.23}$$

The standard deviation is

$$\sigma_X = \sqrt{\sigma_X^2} = \left[\int_{-\infty}^{\infty} (x - \mu_X)^2 f_X(x)\right]^{1/2}.$$

From the definition of expectation we can write

$$\begin{aligned}\operatorname{Var}[X] &= E[(X - \mu_X)^2] \\ &= E[X^2 - 2\mu_X X + \mu_X^2] \\ &= E[X^2] - 2\mu_X E[X] + \mu_X^2 \\ &= E[X^2] - E^2[X].\end{aligned}$$

The third moment around the mean is called the skewness,

$$\begin{aligned}G &= \sum_{\text{all } x_i} (x_i - \mu_X)^3 P_X(x_i) \quad \text{Discrete case} \\ G &= \int_{-\infty}^{\infty} (x - \mu_X)^3 f_X(x)\, dx \quad \text{Continuous case}\end{aligned} \tag{11.24}$$

The commonly used skewness coefficient is defined as

$$\gamma = \frac{G}{\sigma^3} \tag{11.25}$$

and takes negative and positive values. Streamflows usually have positive skewness, implying that their probability density functions have tails that extend to the right (high streamflow values). This is shown in Figure 11.4(a). Figure 11.4(b) shows a negatively skewed pdf. A symmetrical probability density function would have a skewness value of 0.

The fourth moment around the mean is called the kurtosis. It is defined analogously to the skewness.

Joint and Conditional Probability Distribution Functions

The joint probability mass function of two discrete random variables is defined as

$$P_{X,Y}(x_i, y_j) = P[X = x_i;\ Y = y_j].$$

The joint cumulative distribution function is then defined as

$$P[X \le x;\ Y \le y] = F_{X,Y}(x, y) = \sum_{x_i \le x} \sum_{y_j \le y} P_{X,Y}(x_i, y_j), \tag{11.26}$$

which has the property

$$\sum_{\text{all } x_i} \sum_{\text{all } y_j} P_{X,Y}(x_i, y_j) = 1.$$

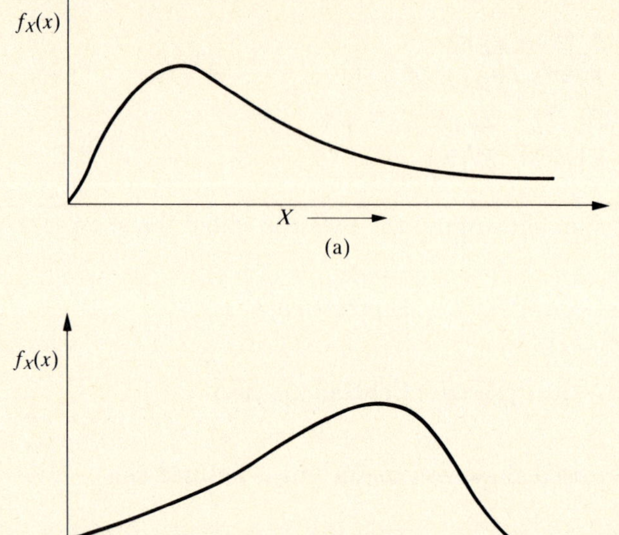

FIGURE 11.4 (a) Probability density function with positive skew coefficient. (b) Probability density function with negative skew coefficient.

The behavior of the random variable X, irrespective of the other random variable Y, is described by the marginal pmf

$$P_X(x_i) = P[X = x_i] = \sum_{\text{all } y_j} P_{X,Y}(x_i, y_j) .$$

A different type of distribution is the one that describes the behavior of the random variable X given the random variable Y. It is called the conditional pmf of X given Y:

$$\begin{aligned} P_{X|Y}(x_i, y_j) &= P[X = x_i \,|\, Y = y_j] \\ &= \frac{P[X = x_i;\, Y = y_j]}{P[Y = y_j]} \\ &= \frac{P_{X,Y}(x_i, y_j)}{P_Y(y_j)} . \end{aligned} \qquad (11.27)$$

"It is the relationship between the conditional distribution and the marginal distribution that determines how much an observation of one variable helps in the prediction of the other" (Benjamin and Cornell [1970]).

As for discrete random variables, we can similarly define joint probability functions for continuous random variables. Thus the probability that X lies in the interval $[x, x + dx]$ and Y lies in the interval $[y, y + dy]$ is given by $f_{X,Y}(x,y)\,dx\,dy$, where the function $f_{X,Y}(x,y)$ is called the joint pdf of X and Y.

We then have

$$P[x_1 \leq X \leq x_2;\, y_1 \leq Y \leq y_2] = \int_{x_1}^{x_2} \int_{y_1}^{y_2} f_{X,Y}(x,y)\,dx\,dy\,. \tag{11.28}$$

Clearly, $f_{X,Y}(x,y)$ must satisfy

$$f_{X,Y}(x,y) \geq 0$$

$$\int_{-\infty}^{\infty} \int_{-\infty}^{\infty} f_{X,Y}(x,y)\,dx\,dy = 1\,.$$

The joint cumulative distribution function is now

$$F_{X,Y}(x,y) = P[X \leq x;\, Y \leq y] = \int_{-\infty}^{x} \int_{-\infty}^{y} f_{X,Y}(u,v)\,du\,dv\,, \tag{11.29}$$

with the property that

$$\frac{\partial^2}{\partial x\,\partial y} F_{X,Y}(x,y) = f_{X,Y}(x,y)\,. \tag{11.30}$$

The behavior of X, irrespective of Y, is given by the marginal pdf of X:

$$f_X(x) = \int_{-\infty}^{\infty} f_{X,Y}(x,y)\,dy\,. \tag{11.31}$$

11.3 MODELS OF PROBABILITY

There are innumerable functions that satisfy the probability axioms and hence are adequate models of probability. Next, we will review some of the simplest and most commonly used models to represent hydrologic variables.

11.3.1 Models of Discrete Random Variables

Bernoulli Trials

Imagine a river discharge of a given magnitude Q^*. From many years of observations it is determined that in any one year the probability of a flood larger or equal to Q^* is P. We can reasonably assume that the flood of any one year is independent of that of another year. We can define a discrete random variable X that takes a value of 1 if a flood greater than or equal to Q^* occurs in a year; the variable X takes a value of 0 otherwise. The probability mass function of X is then

$$P_X(x) = \begin{cases} P & \text{if } x = 1 \\ 1 - P & \text{if } x = 0 \end{cases} \tag{11.32}$$

The mean or expected value of X is

$$\mu_X = E[X] = 1P + 0(1 - P) = P. \tag{11.33}$$

We have defined expectation as the operation of taking the mean of a function.

The variance of X is

$$E[(X - \mu_X)^2] = (1 - P)P. \tag{11.34}$$

Binomial Distribution

The Bernoulli trial has but two outcomes and is repeated only once. Nevertheless, every year we have the same probability P that a flood greater or equal to Q^* will occur. In other words, the independent Bernoulli trial is repeated year after year. A valid question is then: What is the probability that our flood will occur two times in the next three years? In three trials (three years) two floods can occur in three different ways: in the first and second years; in the second and third; or in the first and third. Since every year is independent of the others we can find the joint probability of the above three-year sequences (after Eq. 11.11):

$$P[1,0,1] = P(1 - P)P = P^2(1 - P)$$

$$P[0,1,1] = (1 - P)PP = P^2(1 - P)$$

$$P[1,0,1] = P(1 - P)P = P^2(1 - P)$$

where 1s are used for indicating a flood, and 0s are years of no flood. Since we have three independent ways of achieving our two years of floods, the desired probability is

$$P[2 \text{ floods in 3 years}] = 3P^2(1 - P).$$

The above result can be generalized. The probability of k floods in n years is

$$P[K = k] = B(n, P) = \binom{n}{k} P^k(1 - P)^{n-k}, \tag{11.35}$$

where K is the random variable representing the number of floods in n years and

$$\binom{n}{k} = \frac{n!}{(n-k)!\,k!}$$

represents the number of ways that k events (i.e., floods) can occur in n trials (i.e., years).

The above is the binomial distribution. Its first two moments are

$$\mu_K = E[K] = nP \tag{11.36}$$

$$\text{Var}[K] = nP(1-P). \tag{11.37}$$

The results should be clear in the context of the independence of the n Bernoulli trials that lead to the binomial distribution.

If the probability of discharge Q^* being exceeded in any one year is 0.02, then the mean number of times that the flood will be exceeded in 50 years is 1, with a variance of 0.98. Figure 11.5 shows various forms of the binomial for various values of parameters P and n.

Geometric Distribution

Another relevant question in hydrology is: How many years will pass before discharge Q^* is equalled or exceeded? Using the binomial probability concept,

$$P[\text{flood in the } n\text{th year}] = (1-P)^{n-1}P, \tag{11.38}$$

or $n-1$ consecutive failures followed by a flood in the nth year.

The above is called the geometric distribution and represents the probability mass function that a random variable N representing the number of years to the first flood is n. Figure 11.6 shows the geometric distribution.

We can now ask: What is the probability that the next flood will occur in n or less years? The answer is given by the geometric cumulative mass function. A simple way to obtain it is to ask an equivalent question: What is the probability that at least one flood will occur in the next n years? This is the complement of the probability of no floods in n years, which is $(1-P)^n$. Hence,

$$P[N \le n] = 1 - (1-P)^n \qquad n = 1, 2, \ldots. \tag{11.39}$$

The mean of the geometric distribution is

$$\mu_N = 1/P. \tag{11.40}$$

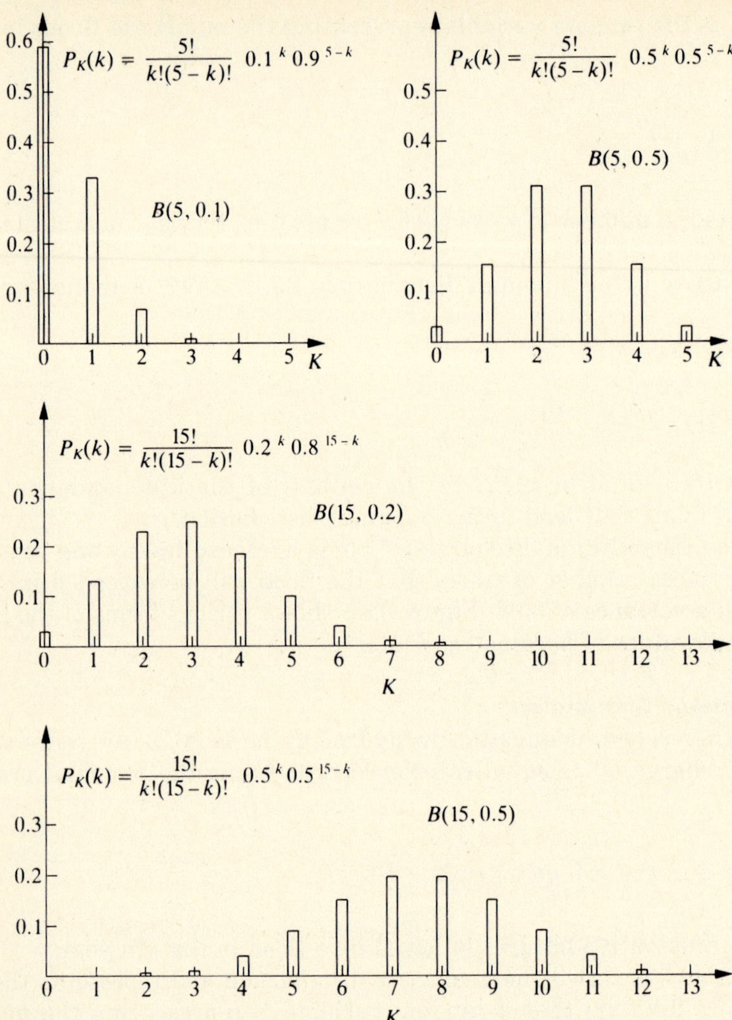

FIGURE 11.5 Binomial distribution $B(n, P)$.

The variance is

$$\sigma_N^2 = (1 - P)/P^2. \tag{11.41}$$

Note that since P is the probability of exceedance in any one year, the mean of the geometric distribution is the recurrence interval or the average number of years that will pass before a flood of magnitude Q^* or greater occurs.

Equation (11.39) is commonly called the risk of flood, since it answers the question: What is the risk of having at least one event (i.e., flood) of recur-

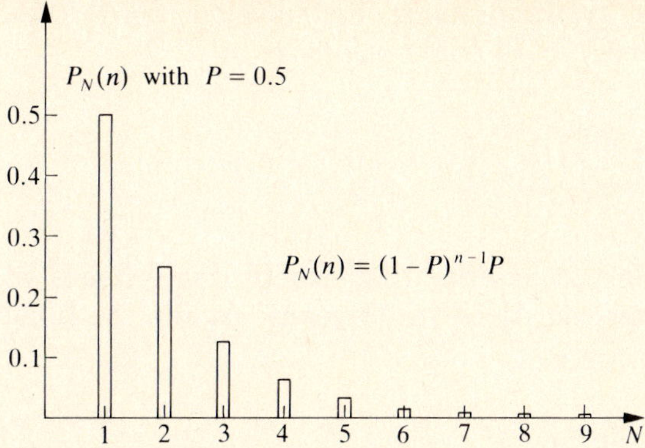

FIGURE 11.6 Geometric distribution $G(P)$.

rence $1/P$ in n years? Table 11.1 evaluates risk for several values of $T = 1/P$ and n, where P is the probability of exceedance in a year.

EXAMPLE 11.1

Uses of the Binomial Distribution

The magnitude of the T-year flood has been defined as that which is exceeded with probability $1/T$ in any given year. If we assume that successive annual floods are independent, several interesting questions can be answered.

A. What is the probability that exactly one flood equal to or in excess of the 50-year flood will occur in a 50-year period? The answer to this type of

TABLE 11.1 Risk of Event Occurring in Specified Number of Years

	RECURRENCE INTERVAL			
NUMBER OF YEARS	10 yr (%)	50 yr (%)	100 yr (%)	500 yr (%)
1	10	2	1	0.2
2	19	4	2	0.4
5	41	10	5	1
10	65	18	10	2
20	88	33	18	4
30	96	45	26	6
50	99	64	40	10
75	99.9	78	53	14
100	99.99	87	63	18
200	99.999	98	87	33
500	99.999	99.99	99	63

question is given by the binomial distribution, Eq. (11.35). The probability of one success (flood) in 50 trials (years) is

$$P[K = 1] = \binom{50}{1}\left(\frac{1}{50}\right)^1\left(1 - \frac{1}{50}\right)^{49} = 0.37\,.$$

B. What is the probability that exactly three floods will be equal to or will exceed the 50-year flood in 50 years? Again, the binomial distribution gives the answer.

$$P[K = 3] = \binom{50}{3}\left(\frac{1}{50}\right)^3\left(1 - \frac{1}{50}\right)^{47} = 0.06\,.$$

Note that the probability of three floods of 50-year recurrence in 50 years is much less than the probability of one such flood occurring.

C. What is the probability that one or more floods will equal or exceed the 50-year flood in 50 years? The key to this question is in the words "one or more." Exploiting the properties of mutually exclusive and collectively exhaustive events we can say

$$P[\text{one or more floods in 50 years}] = 1 - P[\text{no floods in 50 years}]$$

or

$$P[\text{one or more floods in 50 years}] = 1 - \binom{50}{0}\left(\frac{1}{50}\right)^0\left(1 - \frac{1}{50}\right)^{50} = 0.64\,.$$

Note that the probability of one or more floods is nearly twice the probability of a single flood, which was obtained in part A above.

D. If an agency designs each of 20 independent flood-control systems (i.e., systems in widely scattered locations with independent hydrology) for a particular 500-year flood, what is the distribution of the number of systems that will fail, because of the occurrence of floods with 500-year return periods or larger, at least once within the first 50 years after their construction? If each system is independent, the answer to this question is given by a binomial distribution. Define a random variable K as the number of systems that fail. Then

$$P[K = k] = \binom{20}{k} P_1^k(1 - P_1)^{20-k},$$

where P_1 is the probability that any one system fails at least once in 50 years given that it has a probability of failure of 1/500 in any one year. P_1 is obtained as in part C above.

$$\begin{aligned} P_1 &= P[\text{one or more failures of any one system}] \\ &= 1 - P[\text{no failure by any one system}] \\ &= 1 - \binom{50}{0}\left(\frac{1}{500}\right)^0\left(1 - \frac{1}{500}\right)^{50} \\ &= 1 - \left(\frac{499}{500}\right)^{50} = 0.095\,. \end{aligned}$$

P_1 can be used in the previous equation to find the distribution of the number of systems that fail:

$$P[K = k] = \binom{20}{k}(0.095)^k\,(1 - 0.095)^{20-k}\,.$$

For example,

$$P[K = 0] = 0.136\,,$$

$$P[K = 1] = 0.285\,.\ \blacklozenge$$

11.3.2 Models of Continuous Random Variables

Throughout the previous section we have assumed that a probability P of exceeding a given event (i.e., flood) in a given unit time period is known. For example, P could be the probability that a flood of magnitude Q^* is equaled or exceeded in a year or the probability that total rainfall depth in any given day exceeds a magnitude D^*. Both discharge and rainfall depth are continuous variables that, during any time period, can take values between 0 and infinity. The probability of them taking any value within a range must be described by a probability density function. Only through their pdf can we define P.

There are innumerable functions that are valid probability density functions. Following is a very limited collection of some of the most common, particularly in hydrology.

Gaussian or Normal Distribution

The Gaussian probability density curve is probably the most common model of probability. It arises from the argument of the central limit theorem. Briefly, this theorem states that the normal distribution is asymptotically the model for a sum of a large (infinite) number of identically distributed random variables. Given that many natural and man-made processes result from additive mechanisms, the commonality of this distribution is not surprising.

The Gaussian pdf is shown in Figure 11.7. It is characteristically bell-shaped, symmetrical, and extends from minus infinity to infinity. Since natural processes, like river discharges, are rarely defined as negative, the Gaussian pdf is obviously limited in its relevance to hydrology. It can nevertheless be useful. The curves of Figure 11.7 are described by a two-parameter function,

$$f_X(x) = \frac{1}{\sigma\sqrt{2\pi}} \exp\left[-\frac{1}{2}\left(\frac{x-\mu}{\sigma}\right)^2\right], \qquad -\infty < x \leq \infty. \tag{11.42}$$

The two parameters are the mean μ, which centers the distribution around a preferred value and σ, the standard deviation, which tells us how dispersed are occurrences of the random variable around its mean. The cumulative Gaussian distribution is commonly tabulated as in Table 11.2.

The Gaussian pdf, like any of the other functions we will see in this section, could be used to analyze data sets. The following example defines two common types of streamflow data sets and analyzes them using the Gaussian pdf.

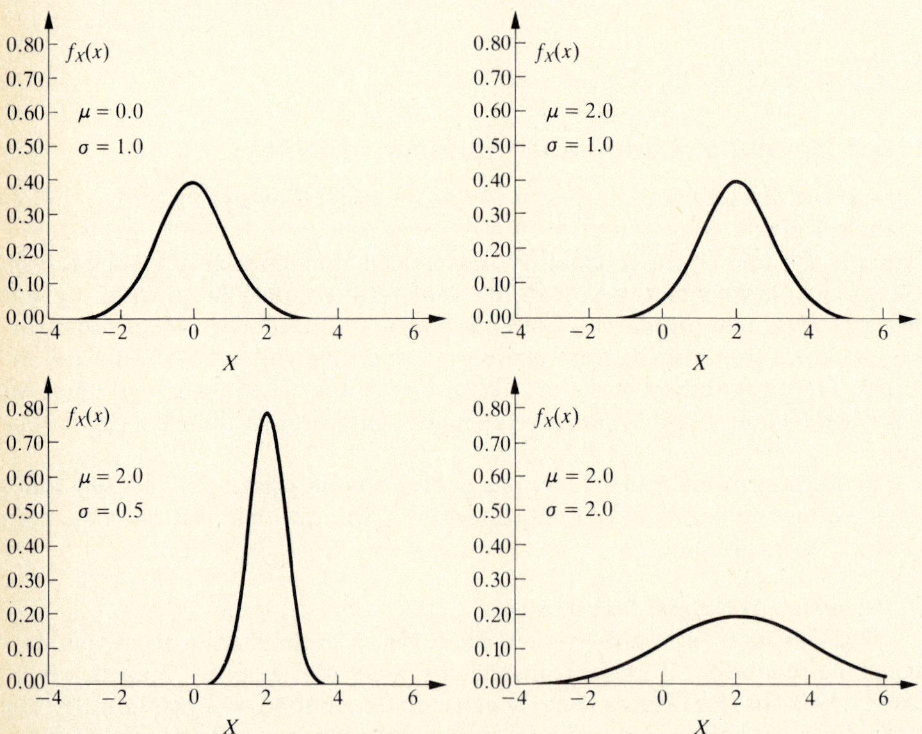

FIGURE 11.7 Gaussian probability density function.

TABLE 11.2 Values of the Standardized Normal Distribution

THE CUMULATIVE DISTRIBUTION FUNCTION, $F_U(u) = \int_{-\infty}^{u} f_U(u)\,du$

u	0.00	0.01	0.02	0.03	0.04	0.05	0.06	0.07	0.08	0.09
0.0	0.5000	0.5040	0.5080	0.5120	0.5160	0.5199	0.5239	0.5279	0.5319	0.5359
0.1	0.5398	0.5438	0.5478	0.5517	0.5557	0.5596	0.5636	0.5675	0.5714	0.5753
0.2	0.5793	0.5832	0.5871	0.5910	0.5948	0.5987	0.6026	0.6064	0.6103	0.6141
0.3	0.6179	0.6217	0.6255	0.6293	0.6331	0.6368	0.6406	0.6443	0.6480	0.6517
0.4	0.6554	0.6591	0.6628	0.6664	0.6700	0.6736	0.6772	0.6808	0.6844	0.6879
0.5	0.6915	0.6950	0.6985	0.7019	0.7054	0.7088	0.7123	0.7157	0.7190	0.7224
0.6	0.7257	0.7291	0.7324	0.7357	0.7389	0.7422	0.7454	0.7486	0.7517	0.7549
0.7	0.7580	0.7611	0.7642	0.7673	0.7703	0.7734	0.7764	0.7794	0.7823	0.7852
0.8	0.7881	0.7910	0.7939	0.7967	0.7995	0.8023	0.8051	0.8078	0.8106	0.8133
0.9	0.8159	0.8186	0.8212	0.8238	0.8264	0.8289	0.8315	0.8340	0.8365	0.8389
1.0	0.8413	0.8438	0.8461	0.8485	0.8508	0.8531	0.8554	0.8577	0.8599	0.8621
1.1	0.8643	0.8665	0.8686	0.8708	0.8729	0.8749	0.8770	0.8790	0.8810	0.8830
1.2	0.8849	0.8869	0.8888	0.8907	0.8925	0.8944	0.8962	0.8980	0.8997	0.90147
1.3	0.90320	0.90490	0.90658	0.90824	0.90988	0.91149	0.91309	0.91466	0.91621	0.91774
1.4	0.91924	0.92073	0.92220	0.92364	0.92507	0.92647	0.92785	0.92922	0.93056	0.93189
1.5	0.93319	0.93448	0.93574	0.93699	0.93822	0.93943	0.94062	0.94179	0.94295	0.94408
1.6	0.94520	0.94630	0.94738	0.94845	0.94950	0.95053	0.95154	0.95254	0.95352	0.95449
1.7	0.95543	0.95637	0.95728	0.95818	0.95907	0.95994	0.96080	0.96164	0.96246	0.96327
1.8	0.96407	0.96485	0.96562	0.96638	0.96712	0.96784	0.96856	0.96926	0.96995	0.97062
1.9	0.97128	0.97193	0.97257	0.97320	0.97381	0.97441	0.97500	0.97558	0.97615	0.97670
2.0	0.97725									
2.1	0.98214									
2.2	0.98610									
2.3	0.98928									
2.4	0.99180									
2.5	0.99379									
3.0	0.99865									
3.5	0.999767									
4.0	0.9999683									
4.5	0.9999966									
5.0	0.99999971									
5.5	0.999999981									

u	2.32	3.09	3.72	4.27	4.75	5.20	5.61	6.00	6.36	6.71
$1 - F_U(u)$	10^{-2}	10^{-3}	10^{-4}	10^{-5}	10^{-6}	10^{-7}	10^{-8}	10^{-9}	10^{-10}	10^{-11}

Sources: A. Hald, *Statistical Tables and Formulas*. Copyright © 1952 by Wiley. Reprinted by permission of John Wiley & Sons, Inc. National Bureau of Standards [1953].

EXAMPLE 11.2

Annual Exceedance and Annual Maxima Series

Hydrologists are commonly interested in extremes, particularly high (floods) and low (droughts) streamflows. Assume you have a daily record of streamflow over N years. That is called the complete duration series because it includes all available information. If you are interested in the very high flows (floods), it is important to study separately that portion of the record that includes the desired extremes. To do that we commonly form a partial duration series. These are series that include only some of the most extreme events of the complete set, regardless of chronological order. Hence, it is assumed that extremes occur independently of each other. There are two types of partial duration series. An annual exceedance series is composed of the N highest (lowest if you are studying droughts) observed values of the process, where N is the total number of years of observation. An annual maxima (or minima) extreme value series is composed of the largest value in each year of observation; therefore it also has N points. Figure 11.8 illustrates these two types of partial duration series.

Although similar, annual exceedance and maxima series are clearly not the same; the difference being mainly at the lower-valued end, where the annual exceedance series will tend to contain larger values.

If, for example, the annual maxima series of a given river is assumed to follow a Gaussian probability distribution with known parameters, it would then be possible to find the annual flood of any desired recurrence. Imagine that indeed you have such a flood series, with mean of 300 $m^3 s^{-1}$ and standard deviation of 100 $m^3 s^{-1}$ and you desire to find the magnitude of the 100-year flood. The object is then to solve the following equation for Q_{100}.

$$P[Q > Q_{100}] = 1 - F(Q_{100}) = \frac{1}{100} = 0.01,$$

where $F(Q)$ is in this case the cumulative density function of the Gaussian distribution. The above is the same as

$$\int_{-\infty}^{Q_{100}} f_Q(q)\, dq = F(Q_{100}) = 0.99.$$

From Table 11.2, for $F_U(U_{100}) = 0.99$, we find that $U_{100} = 2.32$. The variable U in such tables is a standardized variate of mean 0 and variance 1 (standard normal deviate); therefore

$$U_{100} = \frac{Q_{100} - \mu}{\sigma} = 2.32$$

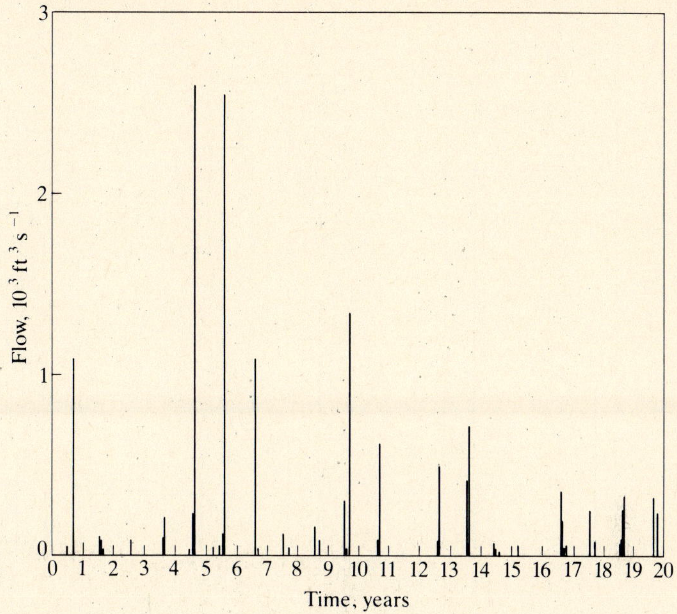

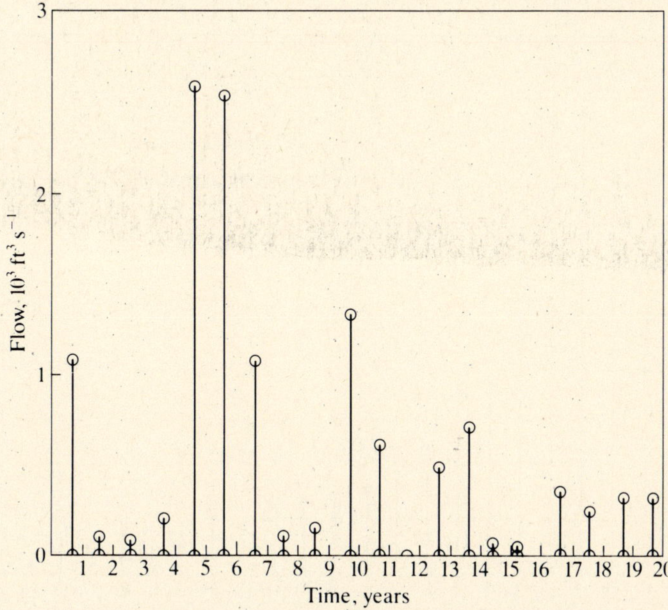

FIGURE 11.8 (a) Definition of annual maxima and annual exceedance series. (Continued on next page.) (b) Difference between the ordered annual maxima and annual exceedance series.

III. Exceedances (20 largest flows)

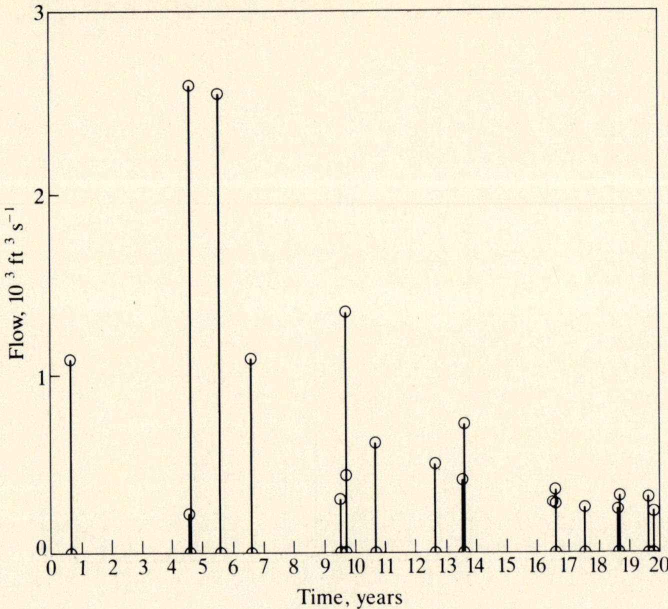

(b) Annual exceedance and maximum values

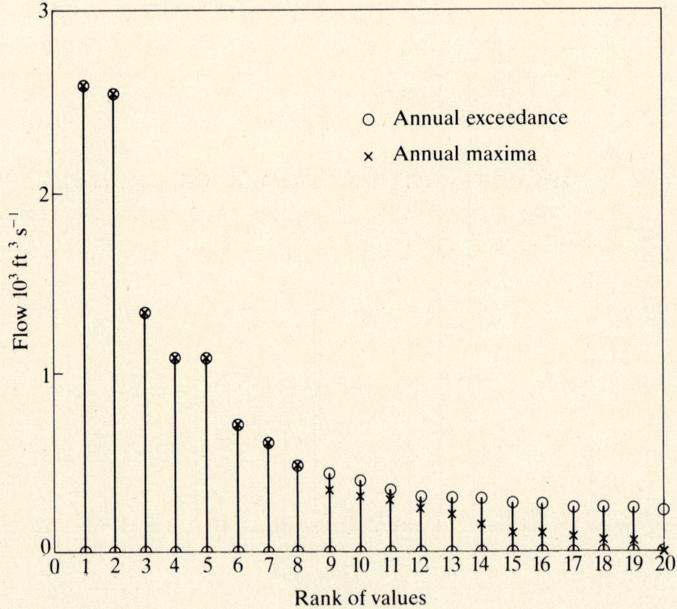

FIGURE 11.8 (*Continued*)

or

$$Q_{100} = \sigma U_{100} + \mu .$$

From the given values of σ and μ, we get $Q_{100} = 532 \text{ m}^3 \text{s}^{-1}$.

Note that the desired discharge was the result of an equation of the form

$$Q_T = K_T \sigma + \mu . \tag{11.43}$$

The coefficient K_T is called the frequency factor and is dependent on the probability model assumed. For the normal distribution, K_T is simply the standard normal deviate. ◆

Log-Normal Distribution

Since processes like streamflow or rainfall do not take negative values, the use of the Gaussian pdf is an obvious approximation. If the mean and variance of the modeled process are such that significant probabilities of negative values exist, then the Gaussian approximation would be very bad.

One distribution which is only defined for positive values is the log-normal distribution. It arises naturally from processes that are multiplicative in nature. Let

$$Y = Y_1 \times Y_2 \times \cdots \times Y_n ,$$

where Y_i are independent, identically distributed random variables. Then

$$X = \ln Y = \ln Y_1 + \ln Y_2 + \cdots + \ln Y_n . \tag{11.44}$$

If the $\ln Y_i$ are independent and identically distributed, then for large n the central limit theorem would imply that $X = \ln Y$ is Gaussian or normally distributed. In such cases Y is said to be log-normally distributed.

The log-normal pdf takes the form

$$f_Y(y) = \frac{1}{y\sigma_X\sqrt{2\pi}} \exp\left[-\frac{1}{2}\left(\frac{\ln y - m_X}{\sigma_X}\right)^2\right] \qquad y \geq 0 , \tag{11.45}$$

where m_X and σ_X are the mean and variance of the transformed variable, $X = \ln Y$. Figure 11.9 illustrates the above equation.

Note that Eq. (11.45) is parameterized in terms of moments of the transformed variable X. It can be shown that the median (value with a 0.5 probability of being exceeded) of Y, $\breve{m}_Y$, is related to the mean of $\ln Y$, m_X, by

$$\ln \breve{m}_Y = \breve{m}_X = m_X . \tag{11.46}$$

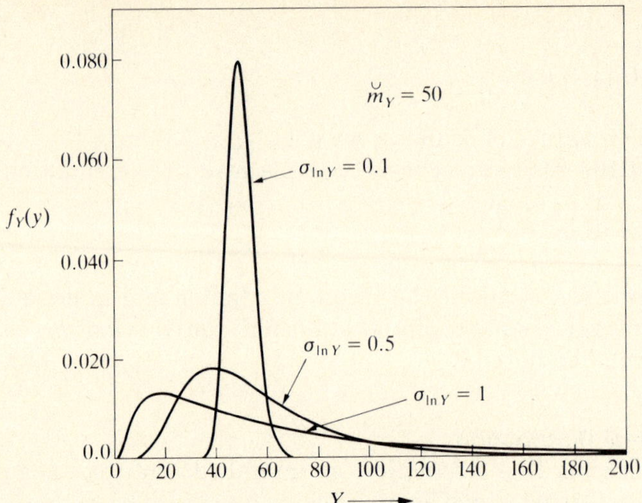

FIGURE 11.9 Log-normal distributions showing influence of $\sigma_{\ln y}$.

Hence, Eq. (11.45) can be expressed as

$$f_Y(y) = \frac{1}{y\sigma_X\sqrt{2\pi}} \exp\left\{-\frac{1}{2}\left[\frac{1}{\sigma_X}\ln\left(\frac{y}{\breve{m}_Y}\right)\right]^2\right\} \qquad y \geq 0. \tag{11.47}$$

Since the logarithm is a one-to-one monotonically increasing transformation, then it should be clear that

$$F_Y(y) = F_X(\ln y), \tag{11.48}$$

where $F_X(x)$ is the cumulative Gaussian distribution in this case.

In essence, to use the log-normal distribution, the hydrologist should simply take the logarithms of the variable of interest (i.e., peak discharge) and define a standard normal deviate

$$U = \frac{\ln Y - m_{\ln Y}}{\sigma_{\ln Y}} = \frac{\ln(Y/\breve{m}_Y)}{\sigma_{\ln Y}},$$

which can be treated as illustrated in the previous section. An exponentiation of the logarithms is required to revert back to original variables after all operations are finished.

There are some useful relations between moments of log transformed and untransformed variables. The median of the untransformed log-normal variable Y is related to its mean by

$$\breve{m}_Y = m_Y \exp\left(-\frac{1}{2}\sigma^2_{\ln Y}\right). \tag{11.49}$$

The variance of Y is given by

$$\sigma^2_Y = m^2_Y(\exp \sigma^2_X - 1). \tag{11.50}$$

Also,

$$m_X = \ln m_Y - \frac{1}{2}\sigma^2_X. \tag{11.51}$$

EXAMPLE 11.3

Frequency Analysis with Log-Normal Distribution

Let us now assume that the annual maxima streamflows of Example 11.2 are log-normally distributed. What is the 100-year flood in that case?

From Eq. (11.50), we have

$$\sigma^2_X = \ln\left(\frac{\sigma^2_Y}{m^2_Y} + 1\right).$$

Using $m_Y = 300 \text{ m}^3\text{s}^{-1}$ and $\sigma_Y = 100 \text{ m}^3\text{s}^{-1}$ as given in Example 11.2, we get $\sigma^2_X = 0.105$. Using this result in Eq. (11.51),

$$m_X = \ln(300) - \frac{1}{2}(0.105)$$
$$= 5.65.$$

In Example 11.2 we saw that the standard normal deviate with a probability of exceedance of 0.01 (100-year recurrence) is 2.32. Using Eq (11.48),

$$\ln Y_{100} = \sqrt{0.105}\ 2.32 + 5.65$$

or

$$Y_{100} = e^{6.4} = 602 \text{ m}^3\text{s}^{-1}.$$

Note that this answer is different from that obtained under the normal-distribution assumption. Different models will yield different results, sometimes very different! This highlights the importance of model selection. This is a subject of tremendous controversy in hydrology as well as in statistics and probability. It is a difficult issue and one that remains largely unresolved.

The log-normal distribution is very common in hydrology. Annual maxima of discharges, storm depths, and hydraulic conductivities are a few of the hydrologic variables that have been modeled as log-normally distributed. ◆

Extreme-Value Distributions— The Gumbel Distribution

Mathematically the annual maxima series can be expressed as

$$Y = \max\{X_1, X_2, \ldots , X_m\}, \tag{11.52}$$

where X_i; $i = 1, \ldots , m$ is the set of flows from which the maximum is chosen. If the X_is are independent and identically distributed, then the cumulative probability function of Y is

$$F_Y(y) = P[Y \le y] = P[X_1 < y]P[X_2 < y] \ldots P[X_m \le y] = F_X^m(y), \tag{11.53}$$

where $F_X(x)$ is the cdf (cumulative density function) of the Xs. The pdf of Y is then (taking derivatives of Eq. 11.53)

$$f_Y(y) = mF_X^{m-1}(y)f_X(y). \tag{11.54}$$

A powerful asymptotic result is that as m becomes large, and if $F_X(x)$ is such that it goes as $1 - e^{-g(x)}$ for large values of x, then,

$$\begin{aligned} F_Y(y) &= \exp[-e^{-\alpha(y-u)}] \qquad -\infty \le Y \le \infty \\ f_Y(y) &= \alpha \exp[-\alpha(y - u) - e^{-\alpha(y-u)}]. \end{aligned} \tag{11.55}$$

The above is the Type I extreme, large-value, distribution. In hydrology it is also called the Gumbel distribution. The Gumbel is a two-parameter distribution. The parameter u is the mode (most probable) value of the distribution and α is a measure of dispersion.

The two parameters are related by

$$\mu = u + \frac{0.577}{\alpha} \tag{11.56}$$

$$\sigma^2 = \frac{1.645}{\alpha^2}. \tag{11.57}$$

The distribution has a skewness of 0.577. The previous equations can be used with sample estimates of μ and σ^2 to find parameters u and α. This exercise is called the method of moments. The general topic of parameter estimation will be briefly treated later.

Figure 11.10 shows the Gumbel distribution and Table 11.3 gives the cdf for a standardized variable of the form $\alpha(y - u)$. The frequency factor of Eq. (11.43) for the Gumbel distribution is given by Chow [1964] as

$$K_T = -\frac{\sqrt{6}}{\pi}\{0.577 + \ln[\ln T - \ln(T - 1)]\}, \tag{11.58}$$

where T is the desired recurrence.

The theoretical underpinning of the Gumbel distribution has made it very popular in the analysis of floods. Nevertheless, it should be viewed as another alternative whose performance must be judged relative to its ability to reproduce observed data.

There are other types of asymptotic extreme-value distributions, including several applicable to small values (droughts). The reader is referred to Gumbel [1958] for some of the original work on extremes and Benjamin and Cornell [1970] for a good summary of extreme-value functions.

EXAMPLE 11.4

Frequency Analysis with Gumbel Distribution

The Gumbel distribution can be used to find the 100-year flood of Example 11.2. One way would be to use Eq. (11.58) to find the approximate

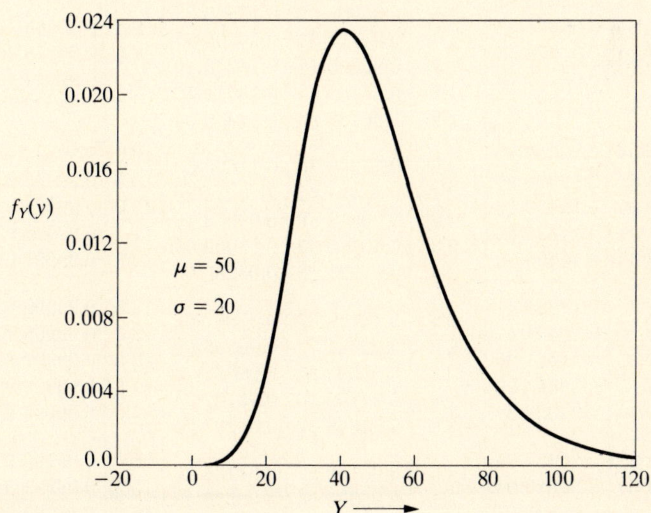

FIGURE 11.10 Type I extreme-value distribution for largest value: $\mu = 50$, $\sigma = 20$.

TABLE 11.3 Values of the Standardized Type I Extreme-Value Distribution (Largest Value) $F_W(w) = e^{-e^{-w}}$

w	cdf	pdf
−3.0		0.00000 00
−2.9	0.00000 00	0.00000 02
−2.8	0.00000 01	0.00000 12
−2.7	0.00000 03	0.00000 51
−2.6	0.00000 14	0.00001 91
−2.5	0.00000 51	0.00006 24
−2.40	0.00001 63	0.00017 99
−2.35	0.00002 79	0.00029 29
−2.30	0.00004 66	0.00046 47
−2.25	0.00007 58	0.00071 89
−2.20	0.00012 04	0.00108 63
−2.15	0.00018 69	0.00160 46
−2.10	0.00028 41	0.00232 00
−2.05	0.00042 31	0.00328 66
−2.00	0.00061 80	0.00456 63
−1.95	0.00088 61	0.00622 81
−1.90	0.00124 84	0.00834 67
−1.85	0.00172 97	0.01100 04
−1.80	0.00235 87	0.01426 93
−1.75	0.00316 82	0.01823 15
−1.70	0.00419 46	0.02296 12
−1.65	0.00547 82	0.02852 48
−1.60	0.00706 20	0.03497 81
−1.55	0.00899 15	0.04236 34
−1.50	0.01131 43	0.05070 71
−1.45	0.01407 84	0.06001 78
−1.40	0.01733 20	0.07028 48
−1.35	0.02112 23	0.08147 77
−1.30	0.02549 44	0.09354 65
−1.25	0.03049 04	0.10642 20
−1.20	0.03614 86	0.12001 76
−1.15	0.04250 25	0.13423 10
−1.10	0.04958 01	0.14894 68
−1.05	0.05740 34	0.16403 90
−1.00	0.06598 80	0.17937 41
−0.95	0.07534 26	0.19481 41
−0.90	0.08546 89	0.21021 95
−0.85	0.09636 17	0.22545 23
−0.80	0.10800 90	0.24037 84
−0.75	0.12039 23	0.25487 04
−0.70	0.13348 68	0.26880 94
−0.65	0.14726 22	0.28208 67
−0.60	0.16168 28	0.29460 53
−0.55	0.17670 86	0.30628 08

w	cdf	pdf
−0.50	0.19229 56	0.31704 19
−0.45	0.20839 66	0.32638 10
−0.40	0.22496 18	0.33560 30
−0.35	0.24193 95	0.34332 85
−0.30	0.25927 69	0.34998 72
−0.25	0.27692 03	0.35557 27
−0.20	0.29481 63	0.36008 95
−0.15	0.31291 17	0.36355 15
−0.10	0.33115 43	0.36598 21
−0.05	0.34949 32	0.36741 21
0.0	0.36787 94	0.36787 94
0.1	0.40460 77	0.36610 42
0.2	0.44099 10	0.36105 29
0.3	0.47672 37	0.35316 56
0.4	0.51154 48	0.34289 88
0.5	0.54523 92	0.33070 43
0.6	0.57763 58	0.31701 33
0.7	0.60860 53	0.30222 45
0.8	0.63805 62	0.28669 71
0.9	0.66593 07	0.27074 72
1.0	0.69220 06	0.25464 64
1.1	0.71686 26	0.23862 28
1.2	0.73993 41	0.22286 39
1.3	0.76144 92	0.20751 91
1.4	0.78145 56	0.19270 46
1.5	0.80001 07	0.17850 65
1.6	0.81717 95	0.16498 57
1.7	0.83303 17	0.15218 12
1.8	0.84764 03	0.14011 40
1.9	0.86107 93	0.12879 04
2.0	0.87342 30	0.11820 50
2.1	0.88474 45	0.10834 26
2.2	0.89511 49	0.09918 16
2.3	0.90460 32	0.09069 45
2.4	0.91327 53	0.08285 05
2.5	0.92119 37	0.07561 62
2.6	0.92841 77	0.06895 69
2.7	0.93500 30	0.06283 74
2.8	0.94100 20	0.05722 24
2.9	0.94646 32	0.05207 75
3.0	0.95143 20	0.04736 90
3.1	0.95595 04	0.04306 48
3.2	0.96005 74	0.03913 41
3.3	0.96378 87	0.03554 76
3.4	0.96717 75	0.03227 79

w	cdf	pdf
3.5	0.97025 40	0.02929 91
3.6	0.97304 62	0.02658 72
3.7	0.97557 96	0.02411 98
3.8	0.97787 76	0.02187 59
3.9	0.97996 16	0.01983 63
4.0	0.98185 11	0.01798 32
4.2	0.98511 63	0.01477 24
4.4	0.98779 77	0.01212 75
4.6	0.98999 85	0.00995 13
4.8	0.99180 40	0.00816 23
5.0	0.99328 47	0.00669 27
5.2	0.99449 86	0.00548 62
5.4	0.99549 36	0.00449 62
5.6	0.99630 90	0.00368 42
5.8	0.99697 70	0.00301 84
6.0	0.99752 43	0.00247 26
6.2	0.99797 26	0.00202 53
6.4	0.99833 98	0.00165 88
6.6	0.99864 06	0.00135 85
6.8	0.99888 68	0.00111 25
7.0	0.99908 85	0.00091 11
7.2	0.99925 37	0.00074 60
7.4	0.99938 89	0.00061 09
7.6	0.99949 97	0.00050 02
7.8	0.99959 03	0.00040 96
8.0	0.99966 46	0.00033 54
8.5	0.99979 66	0.00020 34
9.0	0.99987 66	0.00012 34
9.5	0.99992 51	0.00007 48
10.0	0.99995 46	0.00004 54
10.5	0.99997 25	0.00002 75
11.0	0.99998 33	0.00001 67
11.5	0.99998 99	0.00001 01
12.0	0.99999 39	0.00000 61
12.5	0.99999 63	0.00000 37
13.0	0.99999 77	0.00000 23
13.5	0.99999 86	0.00000 14
14.0	0.99999 92	0.00000 08
14.5	0.99999 95	0.00000 05
15.0	0.99999 97	0.00000 03
15.5	0.99999 98	0.00000 02
16.0	0.99999 99	0.00000 01
16.5	0.99999 99	0.00000 01
17.0	1.00000 00	0.00000 00

Source: National Bureau of Standards [1953].

frequency factor for the 100-year event.

$$K_{100} = -\frac{\sqrt{6}}{\pi}\{0.577 + \ln[\ln(100) - \ln(99)]\}$$
$$= 3.14.$$

The 100-year flood is then (from Eq. 11.43)

$$Q_{100} = \mu + K_{100}\sigma$$
$$= 300 + 3.14 \times 100 = 614 \text{ m}^3\text{s}^{-1},$$

which is the largest of the estimates we have obtained for the 100-year flood in this example.

Rather than using Eq. (11.58), we could have used Eqs. (11.56) and (11.57) to estimate Gumbel distribution parameters and Table 11.3 to get the 100-year flood.

From Eq. (11.57)

$$\alpha = \left(\frac{1.645}{100^2}\right)^{1/2} = 0.0128\,.$$

From Eq. (11.56)

$$u = 300 - \frac{0.577}{0.013} = 255.0\,.$$

From Table 11.3 the standardized variable with probability of exceedance of 0.01 (or cdf = 0.99) is about 4.6 or

$$\alpha(Q_{100} - u) = 4.6\,.$$

Therefore,

$$Q_{100} = \frac{4.6}{\alpha} + u = 614 \text{ m}^3\text{s}^{-1}.$$ ◆

The Log-Pearson Type III Distribution

Motivated by a desire to minimize coordination efforts among U.S. government agencies making flood frequency estimates and to provide a consistent approach for defining flood risk, losses, and insurance provisions, the U.S. Water Resources Council [1967] recommended the use of the log-Pearson Type III distribution to model the annual maxima streamflow series. (*Note:* the Water Resources Council is no longer in existence. Its duties in establishing guidelines for flood frequency determinations have been taken over by the Interagency Advisory Committee on Water Data, coordinated by the U.S. Geological Survey.) The selection was based on a comparison with five other probabilistic models and a nonparametric procedure of estimating flood frequency. The guidelines on how to use the log-Pearson Type III distribution to compute flood frequencies have been revised several times (Water Resources Council [1967, 1976, 1977]; U.S. Geological Survey [1982]). These guidelines specify recommended procedures on how to fit parameters to the distribution and handle several other statistical questions. Some of these will be discussed below at some length since the concepts are general in nature.

The Pearson Type III distribution is

$$f_X(x) = P_0\left(1 + \frac{x}{\alpha}\right)^{\alpha/\delta} e^{-x/\delta}, \tag{11.59}$$

where the most probable point (highest point) or mode is at $X = 0$. The difference between the mean and the mode is δ. The distribution extends from $X = -\alpha$ to infinity. P_0 is the value of the distribution at the mode. The above distribution has three parameters to be estimated. A variable Y is log-Pearson Type III distributed if its logarithm, $X = \log Y$, has Eq. (11.59) as a model.

The Pearson Type III distribution can be shown to be equivalent to a three-parameter gamma distribution, which takes the form (Wallis et al. [1974])

$$f_X(x) = \frac{1}{a\Gamma(b)}\left(\frac{y - m}{a}\right)^{b-1} \exp\left[-\left(\frac{y - m}{a}\right)\right], \tag{11.60}$$

where the first four moments are related to the parameters m, a, and b by

Mean

$$\mu = m + ab$$

Standard deviation

$$\sigma = a(b)^{1/2}$$

Skewness coefficient

$$\gamma = \frac{2}{b^{1/2}}$$

Kurtosis

$$\lambda = 3 + \frac{6}{b}$$

The Water Resources Council recommended that the base 10 logarithms of the annual maxima flood series, $X = \log_{10} Y$, be fitted to a Pearson Type III distribution.

Fitting distributions to data is a major and difficult problem. Many possible procedures exist. Two of the most common are the method of moments and maximum-likelihood techniques. The many issues of statistical inference, as important as they are, are beyond the scope of this book. Nevertheless, the easiest procedure to understand is the method of moments, which we have already utilized in Examples 11.1 through 11.3. In that procedure, parameters are obtained by relating them to the sample moments (i.e., estimates of the true moments obtained from finite-sized records).

The Water Resources Council recommended the method of moments to fit the Pearson Type III distribution to the base 10 logarithms of the annual floods. The procedure is as follows. First, find the logarithm of all records and define a transformed series

$$x_i = \log_{10} y_i . \tag{11.61}$$

Then, compute the first three sample moments using

Mean

$$\overline{X} = \frac{1}{N}\sum_{i=1}^{N} x_i \tag{11.62}$$

Standard deviation

$$S = \left[\frac{1}{n-1}\sum_{i=1}^{N}(x_i - \overline{X})^2\right]^{0.5} = \left\{\frac{1}{N-1}\left[\sum_{i=1}^{N} x_i^2 - \frac{\left(\sum x_i\right)^2}{N}\right]\right\}^{0.5} \tag{11.63}$$

Skewness coefficient

$$G = \frac{N\sum_{i=1}^{N}(x_i - \overline{X})^3}{(N-1)(N-2)S^3} = \frac{N^2\left(\sum_{i=1}^{N} x^3\right) - 3N\left(\sum_{i=1}^{N} x_i\right)\left(\sum_{i=1}^{N} x_i^2\right) + 2\left(\sum_{i=1}^{N} x_i\right)^3}{N(N-1)(N-2)S^2} \tag{11.64}$$

where N is the number of data points available. A reasonable number of data points should be used. More than 25 years (points) of data is a good guideline.

The above estimates of the first three moments are based on a finite number of data points and hence have errors of estimation. Wallis et al. [1974] give the distribution functions and moments for the three statistics, based on small samples ($N = 10$ through 90 data points) from several parent distributions including the Pearson Type III. Generally, the smaller N is, the greater is the variance of the estimate and the worse is its bias.

The skewness coefficient estimate is particularly sensitive to large values when a small number of data points are available. For values of N less than 100 the Water Resources Council recommended the use of the so-called generalized skew. The generalized skew is a regional estimate of skew based on data from surrounding stations. The hypothesis is that regions of similar climatologic, topographic, and hydrologic characteristics should exhibit similar relations between moments of flood flows and variables such as area, basin shape, slope, channel length, annual precipitation, etc. It is recommended that skew values from surrounding stations (i.e., 40) with at least 25 years of record be related via regression to some of the possible explanatory variables. Efforts should also be made to determine regional parameter patterns, if any. Skew coefficients estimated this way are called generalized skews. The guidelines also provide estimates of generalized skews in the form of the map in Figure 11.11, which should be used except when the analyst feels that he or she has a more accurate procedure to find the generalized skew.

The skew coefficient of the logarithms actually used in fitting the log-Pearson Type III distribution is a weighted average of that computed for the station at hand and the generalized skew,

$$G_w = \frac{\text{MSE}_{\overline{G}} \cdot G + \text{MSE}_G \cdot \overline{G}}{\text{MSE}_{\overline{G}} + \text{MSE}_G}. \tag{11.65}$$

The weighting procedure is adapted from Tasker [1978]. In the above $\text{MSE}_{\overline{G}}$ is the mean square error of the generalized skew and MSE_G is the same for the station skew. The former is a function of the regression equations used. If the generalized skew is obtained from Figure 11.11, the $\text{MSE}_{\overline{G}}$ is taken as the estimated error of that map, or 0.302. The station mean square error is obtained from results of bias and variance of skew coefficients obtained from Pearson Type III distributed variables and given by Wallis et al. [1974]. An approximation to their numerical results is

$$\text{MSE}_G \simeq 10^{\{A-B[\log_{10}(N/10)]\}}, \tag{11.66}$$

where

$$\begin{aligned} A &= -0.33 + 0.08|G| \quad \text{if } |G| \le 0.90 \\ &= -0.52 + 0.30|G| \quad \text{if } |G| > 0.90 \\ B &= 0.94 \quad - 0.26|G| \quad \text{if } |G| \le 1.5 \\ &= 0.55 \qquad\qquad\quad \text{if } |G| > 1.5 \end{aligned}$$

in which $|G|$ is the absolute value of the station skew (used as an estimate of the population skew) and N is the record length in years.

Table 11.4 gives mean square error values for station skew as a function of N and G, according to Eq. (11.66).

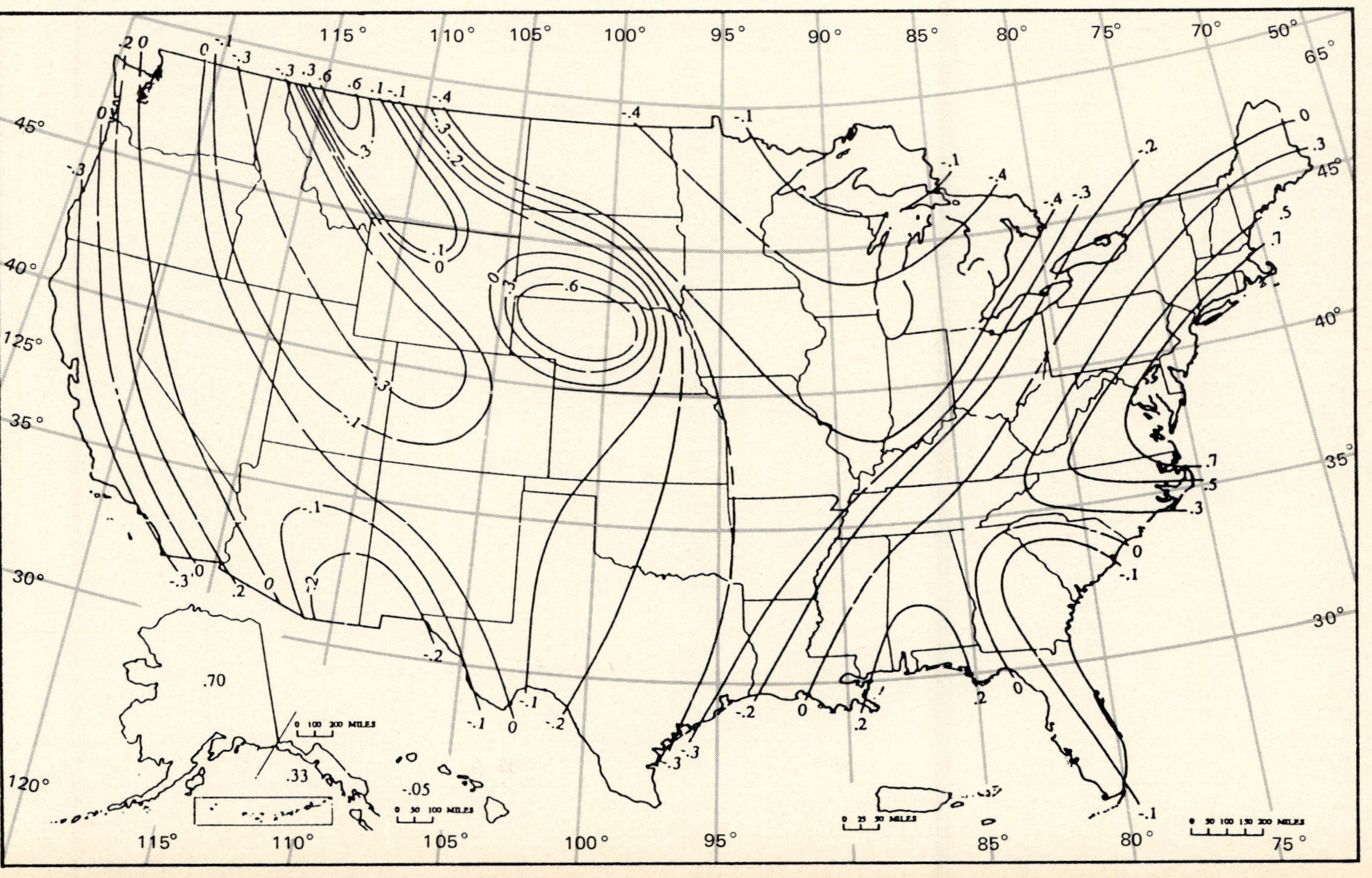

FIGURE 11.11 Generalized skew coefficients of annual maximum streamflow. Source: U.S. Geological Survey [1982].

TABLE 11.4 Summary of Mean Square Error of Station Skew as a Function of Record Length and Station Skew

STATION SKEW (*G*)	RECORD LENGTH, YEARS (*N*)									
	10	20	30	40	50	60	70	80	90	100
0.0	0.468	0.244	0.167	0.127	0.103	0.087	0.075	0.066	0.059	0.054
0.1	0.476	0.253	0.175	0.134	0.109	0.093	0.080	0.071	0.064	0.058
0.2	0.485	0.262	0.183	0.142	0.116	0.099	0.086	0.077	0.069	0.063
0.3	0.494	0.272	0.192	0.150	0.123	0.105	0.092	0.082	0.074	0.068
0.4	0.504	0.282	0.201	0.158	0.131	0.113	0.099	0.089	0.080	0.073
0.5	0.513	0.293	0.211	0.167	0.139	0.120	0.106	0.095	0.087	0.079
0.6	0.522	0.303	0.221	0.176	0.148	0.128	0.114	0.102	0.093	0.086
0.7	0.532	0.315	0.231	0.186	0.157	0.137	0.122	0.110	0.101	0.093
0.8	0.542	0.326	0.243	0.196	0.167	0.146	0.130	0.118	0.109	0.100
0.9	0.562	0.345	0.259	0.211	0.181	0.159	0.142	0.130	0.119	0.111
1.0	0.603	0.376	0.285	0.235	0.202	0.178	0.160	0.147	0.135	0.126
1.1	0.646	0.410	0.315	0.261	0.225	0.200	0.181	0.166	0.153	0.143
1.2	0.692	0.448	0.347	0.290	0.252	0.225	0.204	0.187	0.174	0.163
1.3	0.741	0.488	0.383	0.322	0.281	0.252	0.230	0.212	0.197	0.185
1.4	0.794	0.533	0.422	0.357	0.314	0.283	0.259	0.240	0.224	0.211
1.5	0.851	0.581	0.465	0.397	0.351	0.318	0.292	0.271	0.254	0.240
1.6	0.912	0.623	0.498	0.425	0.376	0.340	0.313	0.291	0.272	0.257
1.7	0.977	0.667	0.534	0.456	0.403	0.365	0.335	0.311	0.292	0.275
1.8	1.047	0.715	0.572	0.489	0.432	0.391	0.359	0.334	0.313	0.295
1.9	1.122	0.766	0.613	0.523	0.463	0.419	0.385	0.358	0.335	0.316
2.0	1.202	0.821	0.657	0.561	0.496	0.449	0.412	0.383	0.359	0.339
2.1	1.288	0.880	0.704	0.601	0.532	0.481	0.442	0.410	0.385	0.363
2.2	1.380	0.943	0.754	0.644	0.570	0.515	0.473	0.440	0.412	0.389
2.3	1.479	1.010	0.808	0.690	0.610	0.552	0.507	0.471	0.442	0.417
2.4	1.585	1.083	0.866	0.739	0.654	0.592	0.543	0.505	0.473	0.447
2.5	1.698	1.160	0.928	0.792	0.701	0.634	0.582	0.541	0.507	0.479
2.6	1.820	1.243	0.994	0.849	0.751	0.679	0.624	0.580	0.543	0.513
2.7	1.950	1.332	1.066	0.910	0.805	0.728	0.669	0.621	0.582	0.550
2.8	2.089	1.427	1.142	0.975	0.862	0.780	0.716	0.666	0.624	0.589
2.9	2.239	1.529	1.223	1.044	0.924	0.836	0.768	0.713	0.669	0.631
3.0	2.399	1.638	1.311	1.119	0.990	0.895	0.823	0.764	0.716	0.676

Source: U.S. Geological Survey [1982].

Once the first three sample moments are available, the logarithm of the base 10 discharge of a given recurrence T is

$$X_T = \overline{X} + K_T S\,, \tag{11.67}$$

where the frequency factor K_T is given in Table 11.5 as a function of the skewness coefficient G_w. Approximate values for K_T can be obtained, when the skewness is between -1.0 and 1.0, using

$$K_T = \frac{2}{G_w}\left\{\left[\left(K_T^n - \frac{G_w}{6}\right)\frac{G_w}{6} + 1\right]^3 - 1\right\}, \tag{11.68}$$

where K_T^n is the standard normal deviate corresponding to recurrence T.

The Water Resources Council [1977] and the U.S. Geological Survey [1982] deal with many other aspects of how to process data and use the log-Pearson Type III distribution. They describe how to handle incomplete records, zero flood years, mix historic and systematic data records (this will be touched on in a more general way later), estimate confidence limits, adjust for limited data, etc. The reader is urged to study the details in the above references. A very brief discussion of the important topic of confidence limits follows.

We have seen that moments estimates are themselves random variables because of the limited size of the data sample used to estimate them. Since these moments are used in the fitting of the distribution, the estimates of exceedance probability for a given discharge or the discharge for a given exceedance probability are themselves uncertain. It is theoretically possible to put bounds, confidence limits, within which we can state that the true answer lies with a given level of certainty. The confidence limits are dependent on the distributions being used and, for nonnormal models, are different for the exceedance probability (for a given discharge) and for the discharge (for a given probability).

A two-sided confidence interval on the logarithm of the discharge with T years recurrence X_T^* is given by

$$P[L_{T,c}(X) \le X_T^* \le U_{T,c}(X)] = 2c - 1\,, \tag{11.69}$$

where c is the "confidence level" such that

$$P[X_T^* > U_{T,c}(X)] = 1 - c \quad \text{and} \quad P[X_T^* \le L_{T,c}(X)] = 1 - c\,.$$

$L_{T,c}(X)$ is a lower limit on X_T^* with a probability of exceedance of c. $U_{T,c}(X)$ is an upper limit on X_T^* with a probability of exceedance of $1 - c$. These limits are given by the U.S. Geological Survey [1982] as

$$U_{T,c}(X) = \overline{X} + K_{T,c}^U S \tag{11.70}$$

$$L_{T,c}(\overline{X}) = \overline{X} + K_{T,c}^L S\,, \tag{11.71}$$

TABLE 11.5 Frequency Factors for the Pearson Type III Distribution

	RETURN PERIOD IN YEARS						
	2	5	10	25	50	100	200
SKEW COEFFICIENT	EXCEEDANCE PROBABILITY						
G_w	0.50	0.20	0.10	0.04	0.02	0.01	0.005
3.0	−0.396	0.420	1.180	2.278	3.152	4.051	4.970
2.9	−0.390	0.440	1.195	2.277	3.134	4.013	4.909
2.8	−0.384	0.460	1.210	2.275	3.114	3.973	4.847
2.7	−0.376	0.479	1.224	2.272	3.093	3.932	4.783
2.6	−0.368	0.499	1.238	2.267	3.071	3.889	4.718
2.5	−0.360	0.518	1.250	2.262	3.048	3.845	4.652
2.4	−0.351	0.537	1.262	2.256	3.023	3.800	4.584
2.3	−0.341	0.555	1.274	2.248	2.997	3.753	4.515
2.2	−0.330	0.574	1.284	2.240	2.970	3.705	4.444
2.1	−0.319	0.592	1.294	2.230	2.942	3.656	4.372
2.0	−0.307	0.609	1.302	2.219	2.912	3.605	4.298
1.9	−0.294	0.627	1.310	2.207	2.881	3.553	4.223
1.8	−0.282	0.643	1.318	2.193	2.848	3.499	4.147
1.7	−0.268	0.660	1.324	2.179	2.815	3.444	4.069
1.6	−0.254	0.675	1.329	2.163	2.780	3.388	3.990
1.5	−0.240	0.690	1.333	2.146	2.743	3.330	3.910
1.4	−0.225	0.705	1.337	2.128	2.706	3.271	3.828
1.3	−0.210	0.719	1.339	2.108	2.666	3.211	3.745
1.2	−0.195	0.732	1.340	2.087	2.626	3.149	3.661
1.1	−0.180	0.745	1.341	2.066	2.585	3.087	3.575
1.0	−0.164	0.758	1.340	2.043	2.542	3.022	3.489
0.9	−0.148	0.769	1.339	2.018	2.498	2.957	3.401
0.8	−0.132	0.780	1.336	1.993	2.453	2.891	3.312
0.7	−0.116	0.790	1.333	1.967	2.407	2.824	3.223
0.6	−0.099	0.800	1.328	1.939	2.359	2.755	3.132
0.5	−0.083	0.808	1.323	1.910	2.311	2.686	3.041
0.4	−0.066	0.816	1.317	1.880	2.261	2.615	2.949
0.3	−0.050	0.824	1.309	1.849	2.211	2.544	2.856
0.2	−0.033	0.830	1.301	1.818	2.159	2.472	2.763
0.1	−0.017	0.836	1.292	1.785	2.107	2.400	2.670
0.0	0	0.842	1.282	1.751	2.054	2.326	2.576
−0.1	0.017	0.846	1.270	1.716	2.000	2.252	2.482
−0.2	0.033	0.850	1.258	1.680	1.945	2.178	2.388
−0.3	0.050	0.853	1.245	1.643	1.890	2.104	2.294
−0.4	0.066	0.855	1.231	1.606	1.834	2.029	2.201
−0.5	0.083	0.856	1.216	1.567	1.777	1.955	2.108
−0.6	0.099	0.857	1.200	1.528	1.720	1.880	2.016
−0.7	0.116	0.857	1.183	1.488	1.663	1.806	1.926
−0.8	0.132	0.856	1.166	1.448	1.606	1.733	1.837
−0.9	0.148	0.854	1.147	1.407	1.549	1.660	1.749

(continued)

TABLE 11.5 (*Continued*)

SKEW COEFFICIENT G_w	RETURN PERIOD IN YEARS 2	5	10	25	50	100	200
	EXCEEDANCE PROBABILITY 0.50	0.20	0.10	0.04	0.02	0.01	0.005
− 1.0	0.164	0.852	1.128	1.366	1.492	1.588	1.664
− 1.1	0.180	0.848	1.107	1.324	1.435	1.518	1.581
− 1.2	0.195	0.844	1.086	1.282	1.379	1.449	1.501
− 1.3	0.210	0.838	1.064	1.240	1.324	1.383	1.424
− 1.4	0.225	0.832	1.041	1.198	1.270	1.318	1.351
− 1.5	0.240	0.825	1.018	1.157	1.217	1.256	1.282
− 1.6	0.254	0.817	0.994	1.116	1.166	1.197	1.216
− 1.7	0.268	0.808	0.970	1.075	1.116	1.140	1.155
− 1.8	0.282	0.799	0.945	1.035	1.069	1.087	1.097
− 1.9	0.294	0.788	0.920	0.996	1.023	1.037	1.044
− 2.0	0.307	0.777	0.895	0.959	0.980	0.990	0.995
− 2.1	0.319	0.765	0.869	0.923	0.939	0.946	0.949
− 2.2	0.330	0.752	0.844	0.888	0.900	0.905	0.907
− 2.3	0.341	0.739	0.819	0.855	0.864	0.867	0.869
− 2.4	0.351	0.725	0.795	0.823	0.830	0.832	0.833
− 2.5	0.360	0.711	0.771	0.793	0.798	0.799	0.800
− 2.6	0.368	0.696	0.747	0.764	0.768	0.769	0.769
− 2.7	0.376	0.681	0.724	0.738	0.740	0.740	0.741
− 2.8	0.384	0.666	0.702	0.712	0.714	0.714	0.714
− 2.9	0.390	0.651	0.681	0.683	0.689	0.690	0.690
− 3.0	0.396	0.636	0.666	0.666	0.666	0.667	0.667

Source: U.S. Geological Survey [1982].

where $K^U_{T,c}$ and $K^L_{T,c}$ are new frequency factors (as in Eq. 11.67), approximated as

$$K^U_{T,c} = \frac{K_T + \sqrt{K_T^2 - ab}}{a} \tag{11.72}$$

$$K^L_{T,c} = \frac{K_T - \sqrt{K_T^2 - ab}}{a} \tag{11.73}$$

in which

$$a = 1 - \frac{U_c^2}{2(N-1)}$$

$$b = K_T^2 - \frac{U_c^2}{N}.$$

In Eqs. (11.72) and (11.73), K_T is the frequency factor with recurrence T (probability of exceedance $P = 1/T$) appearing in Eq. (11.67). It is a function of skewness G_w. U_c is a standard normal deviate with exceedance probability $1 - c$. The length of the record is N.

EXAMPLE 11.5

Calculation of Confidence Limits with the Log-Pearson Type III Distribution

Imagine that from 50 years of records (i.e., 50 data points), we had computed a mean of the base 10 logarithms of the data to be $\overline{X} = 1.5$, the standard deviation of the logarithms to be $S = 1.0$, and the skewness to be $G_w = 0.5$. The logarithm of the 100-year flood would be given by Eq. (11.67) with K_T from Table 11.5. For an exceedance probability of 0.01 and $G_w = 0.5$ the table gives $K_{100} = 2.686$ or

$$X_{100} = 1.5 + 2.686(1.0) = 4.2$$

or $Y = 10^{4.2} = 15{,}850$ in units of volume per unit time.

With 50 years of data we can compute the range in which the true X_{100} must lie, with probability of 0.9. That would imply that there must be a 0.05 probability that it is larger than the upper limit and a 0.05 probability that it is smaller than the lower limit, $c = 0.95$. (See Eq. 11.69.) We can use Eqs. (11.70) to (11.73) to answer the question.

From Table 11.2, the standard normal deviate with cumulative probability of 0.95 ($c = 0.95$) is about 1.645. Hence,

$$a = 1 - \frac{(1.645)^2}{2(49)} = 0.972$$

$$b = (2.686)^2 - \frac{(1.645)^2}{50} = 7.160\,.$$

From Eq. (11.72),

$$K^U_{T,c} = \frac{2.686 + \sqrt{(2.686)^2 - (0.972)(7.160)}}{0.972}$$

$$= 3.283\,.$$

Similarly, Eq. (11.73) yields

$$K^L_{T,c} = 2.244\,.$$

Using Eqs. (11.70) and (11.71)

$$U_{T,c}(X) = 1.5 + 3.283(1) = 4.783$$

$$L_{T,c}(X) = 1.5 + 2.244(1) = 3.744\,.$$

The 90% confidence limits in discharge units are obtained by raising 10 to the above powers. The results are $\{5546, 60674\}$. ◆

11.4 NONPARAMETRIC ESTIMATES OF EXCEEDANCE PROBABILITY

Exceedance probabilities can be estimated using the concept of plotting positions. If we are interested in floods, the N values of the annual maxima series are ordered in descending magnitude. The order or position is called m, giving the value $m = 1$ to the largest value in the set and $m = N$ to the smallest. The plotting position, recurrence ($T = 1/P$), is obtained by using one of the many formulas available, some of which are shown in Table 11.6 (Chow [1964]). The third equation in the table seems to have the best theoretical justification (Chow [1953], Thomas [1948]). Using that equation then,

$$T = \frac{N+1}{m} \quad \text{or} \quad P = \frac{1}{T} = \frac{m}{N+1}. \tag{11.74}$$

Most formulas give essentially the same value for middle values of m but differ in the extremes.

Thomas [1948] has shown that the formula

$$T = \frac{N+1}{m}$$

is in fact the mean recurrence (or probability) obtained by taking the expected value of the distribution of T for a given value of N and m.

TABLE 11.6 Plotting-Position Formulas

FORMULA* FOR T OR $1/P(X \geq x)$	
$T = \dfrac{N}{m}$	$T = \dfrac{N+0.4}{m-0.3}$
$T = \dfrac{2N}{2m-1}$	$T = \dfrac{N+1/3}{m-3/8}$
$T = \dfrac{N+1}{m}$	$T = \dfrac{3N+1}{3m-1}$
$T = \dfrac{1}{1-0.5^{1/N}}$	$T = \dfrac{N+0.12}{m-0.44}$

*N = total number of items and m = order number of the items arranged in descending magnitude; thus $m = 1$ for the largest item.

Source: Adapted from Chow [1964].

The distribution of $\overline{P}^c$ (probability of $X \le x$) is an incomplete beta function (Thomas [1948]), so

$$\text{Prob}[\overline{P}^c \le P_\circ] = \theta = \binom{N}{m} m \int_0^{P_\circ} P^{N-m}(1-P)^{m-1}\,dP\,. \tag{11.75}$$

For the largest flood $m = 1$, so

$$\theta = N \int_0^{P_\circ} P^{N-1}\,dP = P_\circ^N\,. \tag{11.76}$$

For example, assume that in 25 years of data, the largest flow is 4720 $\text{ft}^3\,\text{s}^{-1}$. What is the mean recurrence? What is the probability that in fact the recurrence is between 20 and 100 years? The mean recurrence is

$$\frac{N+1}{1} = 26 \text{ years.}$$

The probability of exceedance corresponding to the two recurrence limits are $1/100 = 0.01$ and $1/20 = 0.05$.

Therefore, $P[Q \le 4720] = 0.99$ or 0.95. Using Eq. (11.76) for θ

$$\theta_{0.99} - \theta_{0.95} = 0.99^{25} - 0.95^{25} = 0.5004\,.$$

So there is only a 50% chance that the actual recurrence is within these limits and 50% that it is outside them.

Using a similar approach we can compute the 50% confidence limits on the recurrence of the five largest floods in 25 years of data:

m (RANK)	LOWER LIMIT (YEARS)	UPPER LIMIT (YEARS)
1	18	87
2	10	26
3	6.6	14
4	5.1	9.8
5	4.2	7.3

Note that confidence limits are narrower for smaller floods, relative to N. To be 95% confident of the largest ($m = 1$) of 25 years, you must state the recurrence may be between 7.3 and 987 years!

Thomas [1948] weighted the binomial distribution by possible values of the probability of exceedance P of the annual flood to obtain a nonparametric statement of risk. (See Eqs. 11.35 and 11.39 for the definition of risk.)

$$\phi_k = \frac{m\binom{t}{k}\binom{N}{m}}{(m+k)\binom{t+N}{m+k}}, \tag{11.77}$$

where ϕ_k is the probability that in t future years the mth of N past floods will be exceeded in exactly k years. As before the notation $\binom{N}{m}$ is the number of combinations of m in N,

$$\binom{N}{m} = \frac{N!}{m!(N-m)!}.$$

When $k = 0$,

$$\phi_0 = \frac{\binom{N}{m}}{\binom{t+N}{m}},$$

if $m = 1$,

$$\phi_0 = \frac{\binom{N}{1}}{\binom{t+N}{1}} = \frac{N}{t+N},$$

so if $t = N$ the probability of no exceedance of the largest flood in past N years in a future period of N years is 50%.

An illustrative example is to design a cofferdam to protect against the sixth largest flow in the past 25 years during the five-year construction period of a bigger dam. The probability of no exceedance ($k = 0$) in the next five years is 0.298. So the probability of exceedance at least once in five years is $1 - 0.298 = 0.702$.

The probability of the design flow being exceeded more than twice can be obtained as one minus the probability of zero, one, or two exceedances:

$$1 - \phi_0 - \phi_1 - \phi_2 = 1 - 0.898 = 0.102.$$

*11.5 NOVEL APPROACHES AND FUTURE DIRECTIONS

11.5.1 Derived Distributions

Many times, data may be lacking or completely nonexistent. It may be possible, though, to deterministically relate the variable of interest to a better-defined random variable. Typical examples are soil moisture, basin total yield, and discharge volumes, all related to accessible rainfall data. The derived distribution approach augments information by introducing the engineer's/scientist's knowledge of the physical processes at hand into the probabilistic analysis.

In 1972, Eagleson utilized the concept to derive the distribution of flood peaks as a function of rainfall probabilistic properties and basin characteristics. Eagleson [1978] followed the same ideas in obtaining the probability density function of annual basin yield (surface plus groundwater) as a function of climate and basin-soil properties. Howard [1976] uses derived distributions in obtaining a procedure for the design of combined storage and water-treatment facilities.

The mechanics of derived distributions are well established in probability theory (Benjamin and Cornell [1970]). The conceptual framework is the following. Assume that a variable Q is functionally related to a vector of parameters $\boldsymbol{\theta}$ by

$$q = Q(\boldsymbol{\theta}). \tag{11.78}$$

The elements of the vector $\boldsymbol{\theta}$ are random variables with a given joint probability density function $f_{\theta}(\boldsymbol{\theta})$ and corresponding cumulative distribution $F_{\theta}(\boldsymbol{\theta})$.

Due to the randomness of $\boldsymbol{\theta}$, the variable Q is also a random variable with cumulative distribution,

$$F_Q(q) = \int_{R_q} f_{\theta}(\boldsymbol{\theta})\, d\boldsymbol{\theta}, \tag{11.79}$$

where R_q is the region within the possible values of $\boldsymbol{\theta}$ for which $Q \le q$.

Chan and Bras [1979] concentrated on obtaining the theoretical distribution of the volume of the hydrograph above a given threshold. The variable in question is illustrated in Figure 11.12. The volume above a threshold discharge is required for the design of flood storage devices in urban areas. For example, the threshold discharge shown in the figure may be the maximum capacity of treatment of water from a combined sewer system. The exceedance volume must be stored or spilled, possibly contaminating receiving water bodies.

Since hydrograph volume data is scarce or nonexistent, the approach taken was to relate the available probabilistic description of the rainfall process to a deterministic model of flood volumes in order to derive the distribution of the latter. The typical behavior of the cumulative density function of volume derived by Chan and Bras [1979] is shown in Figure 11.13. There, only the parameter corresponding to the length of overland flow L is varied while all others remain constant.

The important features are that the probabilistic distribution is mixed. There is a finite probability of attaining zero volume above a given threshold and a continuous probability density for volumes greater than zero. As length of overland flow decreases, the peak discharges relative to qth decrease, leading to an increasing probability of zero volume above a fixed threshold discharge. Any parameter influencing peak discharge in a similar manner will have the same effect on the cumulative distribution function. For example,

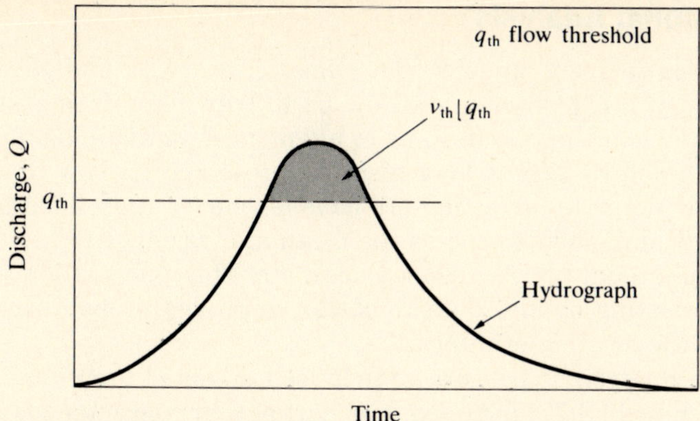

FIGURE 11.12 Volume above a given threshold discharge. Source: S.-O. Chan and R. L. Bras, "Urban Storm Water Management: Distribution of Flood Volumes," *Water Resources Res.* 15(2):371–382, 1979. Copyright by the American Geophysical Union.

increasing qth has the same qualitative effect on the distribution as decreasing length of overland flow.

Hebson and Wood [1982] and Diaz-Granados et al. [1984] used the derived distribution method to obtain the distribution of annual maxima from simple assumptions for the distributions for rainfall intensity and duration. They related rainfall properties to floods via river basin geomorphology (tree network) and models of runoff production (see Chapter 12).

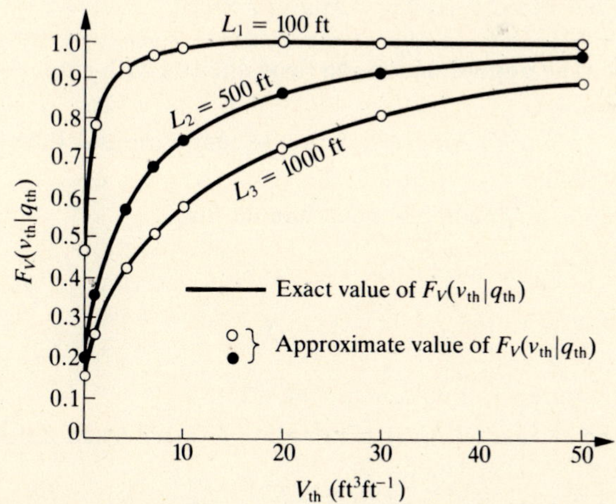

FIGURE 11.13 Cumulative density function of volume above a given threshold discharge. Source: S.-O. Chan and R. L. Bras, "Urban Storm Water Management: Distribution of Flood Volumes," *Water Resources Res.* 15(2):371–382, 1979. Copyright by the American Geophysical Union.

11.5.2 Regional Analysis

It should be intuitively obvious that climatic, geologic, and geomorphologic homogeneity of a region should allow us to transfer information from one basin to another in the region so as to augment data available for statistical inference. There has been a long history of research on how to achieve this objective. For example, investigators have developed regression equations to relate statistical moments such as the mean and variance to basin properties such as area, annual precipitation, slopes, etc. This information is then combined with existing on-site data to obtain improved probabilistic models of extremes at the location of interest.

Regional analysis is by now a fairly well accepted procedure. The Water Resources Council [1977] included it as part of their recommendation for estimating parameters of the log-Pearson Type III distribution of annual streamflow maxima as discussed in Section 11.3.2.

Most recently, Hosking and Wallis [1986a, 1988] reported very encouraging results using one of the simplest of regionalizing procedures, the flood index method. In the flood index method, data from several basins in a region are scaled by dividing by a characteristic quantity, commonly taken as the mean annual flow. The scaled data sets are then lumped and fitted with a common distribution, which becomes the regional distribution. The flow of a given recurrence for any basin in the region is then obtained from the regional distribution multiplied by the sample mean annual flow in that basin. This procedure is summarized in Table 11.7.

TABLE 11.7 Regional Flood Frequency Analysis Steps

1. Define $Q_i(F)$ as the annual maximum flood at site i with cumulative probability F (growth curve).
2. Assume $Q_i(F) = \mu_i q(F)$, where μ_i is a site mean annual flow and $q(F)$ is a regional growth curve.
3. Estimate μ_i by $\hat{\mu}_i = \overline{Q}_i$ (sample mean annual flow).
4. Fit distribution to

   $$q_{ij} \equiv Q_{ij} | \overline{Q}_{ij} \qquad j = 1, \ldots n_i; \qquad i = 1, \ldots N,$$

   where n_i is the number of data points at site i.
5. Let $\hat{q}(F)$ be the estimated inverse cumulative distribution function of the scaled streamflows.
6. The quantile estimator for site i is $\hat{Q}_i(F) = \hat{\mu}_i \hat{q}(F)$.

Source: Adapted from J. R. M. Hosking and J. R. Wallis, "Regional Flood Frequency Analysis Using the Log Normal Distribution," *EOS* 67(44), 1986.

Hosking and Wallis [1986a] suggest that if a three-parameter log-normal distribution is fitted to the regional flows (using a method called probability weighted moments), the results of the index method are surprisingly good and accurate. Using numerical experiments, they show consistently low mean square errors and bias of estimation of floods of various probabilities of exceedance.

Figure 11.14 is one example where real annual maxima are assumed to be distributed with an extreme-value distribution, Type I. Root mean square errors and bias of estimating the 1000-year flood are shown for 21 sites. The coefficients of variation (σ/μ) were 0.5, the skewness 1.14, and the number of data points in each site varied from 10 to 30. Sites were uncorrelated. Tested were assumptions for generalized extreme value distributions at each site (GEV/AS), log-normal distribution at each site (LN3/AS), generalized extreme value as a regional distribution (GEV/R), a Wakeby distribution for the region (WAK/R), and a log-normal distribution for the region (LN3/R). It is surprising how well the latter technique performs even though the assumed distribution, log-normal, is different than the true underlying distribution, which was extreme-value, Type I. Figure 11.15 is even more striking. Discharges of various recurrences Q_T are computed under the assumption that the underly-

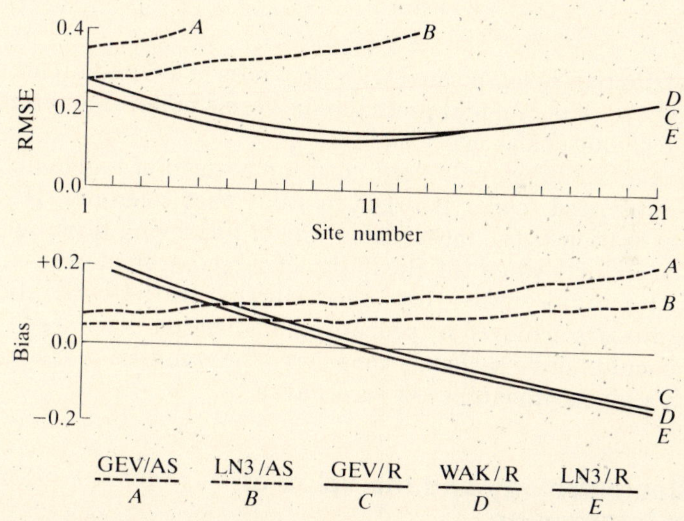

FIGURE 11.14 Estimating the 1000-year flood in 21 sites by different on-site and regional procedures when the population follows an extreme-value distribution, Type I. RMSE denotes root mean square error, GEV generalized extreme value, AS at site, LN3 log-normal distribution, R regional distribution, and WAK Wakeby distribution. Source: J. R. M. Hosking and J. R. Wallis, "Regional Flood Frequency Analysis Using the Log Normal Distribution," *EOS* 67(44), 1986. Copyright by the American Geophysical Union.

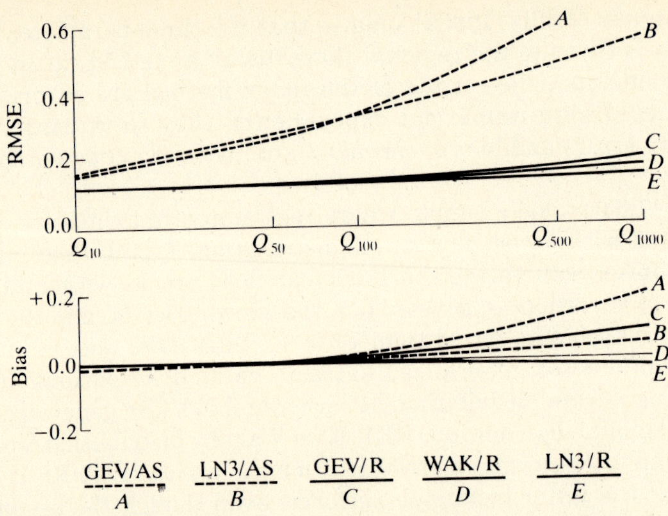

FIGURE 11.15 Estimation of different quantiles when population is homogeneous and log-normal. Different on-site and regional methods are compared. Abbreviations are explained in the legend to Figure 11.14. Source: J. R. M. Hosking and J. R. Wallis, "Regional Flood Frequency Analysis Using the Log Normal Distribution," *EOS* 67(44), 1986. Copyright by the American Geophysical Union.

ing true distribution is log-normal with skewness of 2.5 and all other characteristics as above. The regional procedure performs better than attempts to fit individual distributions at every site.

Lettenmaier [1988] gives a very good comparison of regionalizing procedures. The interested reader is urged to study that reference. He concludes that "regionalization is the most viable way of improving flood quantile estimation." Particularly when the flexibility of three-parameter distributions is required, Lettenmaier states that "the reduction in the variability of flood quantile estimators achieved by proper regionalization is so large that at site estimators should not be seriously considered." A quantile is the value of discharge with a given probability of exceedance.

11.5.3 Paleohydrology and the Value of Historical Information

The Water Resources Council also recognized the value of using historical (vs. systematic record) data. The historical record can consist of written or unwritten accounts of past floods or fragmented records of past civilizations. Information is also available in the geology or related time series like varves, tree rings, etc.

Recently there have been considerable advances in methodology and evaluation of methods to include historical information. Hosking and Wallis [1986b, 1986c] address the information of both paleohydrology and historical data in relation to systematic records. In their work, paleologic information is measured in thousands of years and historical information in hundreds of years. Otherwise there is no significant difference in treatment. Their approach is a [Monte Carlo] simulation where they assume that they have only one historical or paleological maximum event in a period of m years. They process the historical data using the incomplete data likelihood (a maximum-likelihood procedure) and account for errors in estimating the magnitude of the historical maximum event. Typical of their results are Figures 11.16 and 11.17. Their conclusions can be summarized as

1. Historical and paleohydrologic information can be valuable in estimating three-parameter distributions at a single site. It is much less effective when dealing with two-parameter distributions.
2. The effectiveness of historical information reduces with increased sample length.
3. The effectiveness of historical information reduces with increased error in estimating the magnitude of the event. Yet improvements are sometimes observed even in cases when the paleologic maximum event is subject to an error of ±50%.

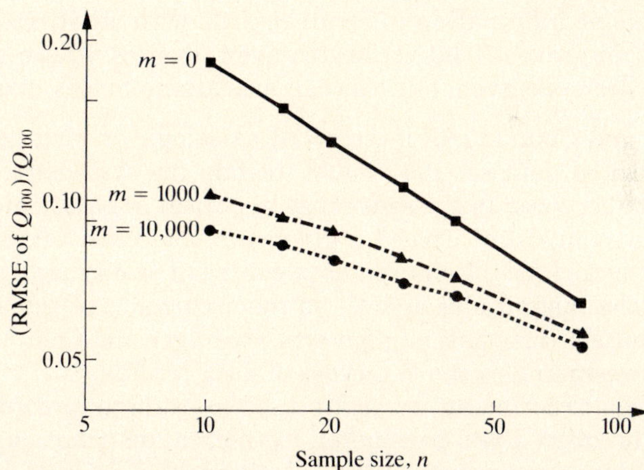

FIGURE 11.16 (Root mean square error of Q_{100})/Q_{100} as a function of gaged record length n and historic period m of paleologic maximum event. Single-site analysis, parent distribution is an extreme-value, Type I (EVI) with coefficient of variation of 0.4 and fitted distribution EVI. Source: J. R. M. Hosking and J. R. Wallis, "Paleoflood Hydrology and Flood Frequency Analysis," *Water Resources Res.* 22(4):543–550, 1986. Copyright by the American Geophysical Union.

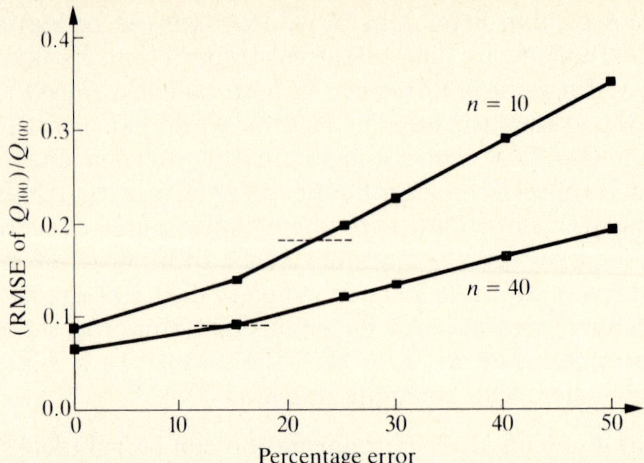

FIGURE 11.17 (Root mean square error of Q_{100})/Q_{100} as a function of measurement error of 10,000-year paleologic maximum event. Dashed lines denote (root mean square error of Q_{100})/Q_{100} when there are no paleologic data; n is length of gaged record. Single-site analysis, parent distribution EVI with coefficient of variation of 0.4 and fitted distribution EVI. Source: J. R. M. Hosking and J. R. Wallis, "Paleoflood Hydrology and Flood Frequency Analysis," *Water Resources Res.* 22(4):543–550, 1986. Copyright by the American Geophysical Union.

4. In a regional analysis using a large number of sites the inclusion of a realistic amount of historical information is unlikely to be useful in practice. Paleologic information in small regions with short records is worthwhile and improves flood estimates even at sites where no paleologic event has been observed, but this can exacerbate biases in estimates.

Stedinger and Cohn [1986] interpreted paleologic or historical records as parts of a censored data set that would include the systematic information. The records are censored in the sense that historical or paleologic information corresponds to events that exceed a given threshold discharge of the given probability of exceedance. Censored data can be of two types, in one there is knowledge of the magnitudes as well as the occurrence of the events, in the other no magnitude information is known. Stedinger and Cohn use likelihood functions corresponding to the two types of data records.

A simulation experiment was again used to evaluate procedures. Reality was assumed to follow a two-parameter, log-normal distribution, and samples were fitted with the same distribution. Their results are well represented in Figures 11.18 and 11.19. The figures give the equivalent number of systematic years of data achieved by using a given historical record length (over and above 20 years of systematic record) for two threshold levels (90% for 10-year recurrence and 99% for 100-year recurrence). It is concluded that

1. Historical information can be very effective in augmenting systematic records.

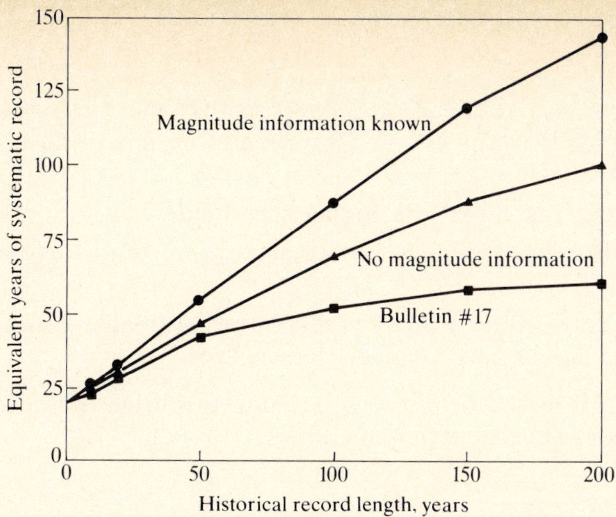

FIGURE 11.18 Effective record length of the two maximum-likelihood estimators and the Bulletin 17 procedures when estimating the 100-year flood. The cases have a 20-year systematic record, a censoring threshold at the 90th percentile of the flow distribution (i.e., 10-year recurrence flood) and between 0 and 200 years of historical information. Source: J. R. Stedinger and T. A. Cohn, "Flood Frequency Analysis with Historical and Paleoflood Information," *Water Resources Res.* 22(5):785–793, 1986. Copyright by the American Geophysical Union.

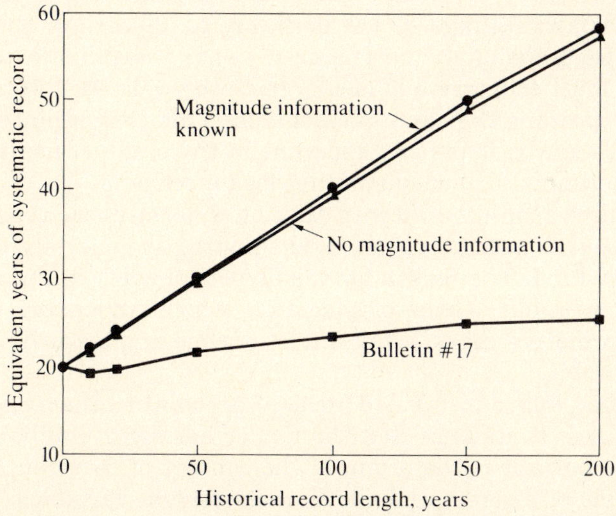

FIGURE 11.19 Effective record length of the two maximum-likelihood estimators and the Bulletin 17 estimator when estimating the 100-year flood. The cases have a 20-year systematic record, a censoring threshold at the 99th percentile of the flow distribution (i.e., 100-year recurrence flood) and between 0 and 200 years of historical information. Source: J. R. Stedinger and T. A. Cohn, "Flood Frequency Analysis with Historical and Paleoflood Information," *Water Resources Res.* 22(5):785–793, 1986. Copyright by the American Geophysical Union.

2. The maximum-likelihood procedures are far more effective than those recommended by the Water Resources Council to handle historical information.
3. Effectiveness increases with higher thresholds.
4. Magnitude information is less important as the censoring threshold increases.
5. The procedures showed robustness, even when reality was assumed to be other than log-normal (i.e., log-Pearson Type III).

Readers interested in plotting position formulas including historical information are referred to Hirsch [1987].

11.6 SUMMARY

Probability and statistics are integral tools of hydrology. As in many natural systems, hydrologic processes are never predictable in exact, deterministic, ways; hence, the need for probabilistic models. The hydrologist also depends on data collection and handling; hence, the need for statistics. You may ask why it took 10 chapters of deterministic conceptualizations to get to the important probability concept. There are two reasons. One is the philosophical belief that a good understanding of physical, deterministic, mechanisms, and their strengths and failings, is the best way to appreciate and ultimately intelligently use probabilistic models. In the end, probabilistic and deterministic thinking must converge and ideally be closely related. After all, they are intended to represent the same phenomena. This convergence of concepts is occurring. Recently, it is most apparent in the study of nonlinear chaotic systems—deterministic phenomena that look stochastic.

The second reason for concentrating on deterministic concepts in this introductory course is that probabilistic thinking is generally harder to accept and understand. The reason for this is not inherent in the mathematics or the ideas but in our system of education, which emphasizes determinism during most formative years. It takes a lot of time and maturity to break the deterministic habit.

This chapter introduced basic ideas of probability theory as applied to random variables. It assumes that the reader has some familiarity with the topics. Most important is the origin of the concept of recurrence, which ties together models of discrete and continuous random variables. The chapter does provide a lot of useful and pragmatic tools for everyday hydrology.

The study of random functions, i.e., random processes in time and space, is not covered, except for a brief hint in the problem set. Most hydrologic variables are indeed random processes. Examples are rainfall in time, daily streamflow, evaporation in time, rainfall distribution in space, and soil hydraulic properties in space. This topic is the next level of study for the serious hydrologist. A general, good, random processes book is Parzen [1962]. For hydrologic emphasis at an advanced level see Bras and Rodriguez-Iturbe [1985]. Other useful books are Haan [1977], McCuen and Snyder [1986], and Kottegoda [1980].

Statistics deals with the tools of data handling and parameter estimation in particular. It is barely touched in this chapter. The field is extensive and specialized. The references are innumerable. Some of the references given above present the most common statistical concepts in hydrology.

REFERENCES

Benjamin, J. R., and C. A. Cornell [1970]. *Probability, Statistics and Decision for Civil Engineers*. New York: McGraw-Hill.

Bobée, B. B. [1975]. "The Log-Pearson Type III Distribution and Its Application in Hydrology." *Water Resources Res.* 11(5):681–689.

Bobée, B. B., and R. Robitaille [1977]. "The Use of the Pearson Type III and Log-Pearson Type III Distribution Revisited." *Water Resources Res.* 13(2):427–443.

Bras, R. L., and I. Rodriguez-Iturbe [1985]. *Random Functions and Hydrology.* Reading, Mass.: Addison–Wesley.

Burges, S. J., D. P. Lettenmaier, and C. L. Bates [1975]. "Properties of the Three-Parameter Log Normal Probability Distribution." *Water Resources Res.* 11(2):229–235.

Chan, S.-O., and R. L. Bras [1979]. "Urban Storm Water Management: Distribution of Flood Volumes." *Water Resources Res.* 15(2):371–382.

Chow, V. T. [1953]. "Frequency Analysis of Hydrologic Data with Special Application to Rainfall Intensities." Urbana, Ill.: University of Illinois Bulletin. 5(31).

Idem. [1964]. *Handbook of Applied Hydrology.* New York: McGraw-Hill.

Diaz-Granados, M. S., J. B. Valdes, and R. L. Bras [1984]. "A Physically Based Flood Frequency Distribution." *Water Resources Res.* 20(7):995–1002.

Eagleson, P. S. [1972]. "Dynamics of Flood Frequency." *Water Resources Res.* 8(4):878–898.

Idem. [1978]. "Climate, Soil, and Vegetation: 1. Introduction to Water Balance Dynamics." *Water Resources Res.* 14(5):705–712.

Gumbel, E. J. [1958]. *Statistics of Extremes*. New York: Columbia University Press.

Haan, C. T. [1977]. *Statistical Methods in Hydrology.* Ames, Iowa: Iowa State University Press.

Hald, A. [1952]. *Statistical Tables and Formulas*. New York: Wiley.

Hebson, C., and E. F. Wood [1982]. "A Derived Flood Frequency Distribution Using Horton Order Ratios." *Water Resources Res.* 18(5):1509–1518.

Hirsch, R. M. [1987]. "Probability Plotting Position Formulas for Flood Records with Historical Information." In: *Analysis of Extraordinary Flood Events,* Special Issue of the *Journal of Hydrology,* 96(1–4):185–199.

Hosking, J. R. M., and J. R. Wallis [1986a]. "Regional Flood Frequency Analysis Using the Log Normal Distribution." *EOS*. 67(44).

Idem. [1986b]. "Paleoflood Hydrology and Flood Frequency Analysis." *Water Resources Res.* 22(4):543–550.

Idem. [1986c]. "The Value of Historical Data in Flood Frequency Analysis." *Water Resources Res.* 22(11):1606–1612.

Idem. [1988]. "The Effect of Intersite Dependence on Regional Flood Frequency Analysis." *Water Resources Res.* 24(4):588–600.

Howard, C. [1976]. "Theory of Storage and Treatment-Plant Overflows." *J. Environ. Engin. Div., A.S.C.E.* 102(EE4):709–722.

Keeney, R.L., and E.F. Wood [1977]. "An Illustrative Example of the Use of Multiattribute Utility Theory for Water Resources Planning." *Water Resources Res.* 13(4):705–712.

Kottegoda, N.T. [1980]. *Stochastic Water Resources Technology*. New York: Wiley.

Landwehr, J.M., N.C. Matalas, and J.R. Wallis [1978]. "Some Comparisons of Flood Statistics in Real and Log Space." *Water Resources Res.* 14(5):902–920.

Lenton, R.L., I. Rodriguez-Iturbe, and J.C. Schaake, Jr. [1974]. "The Estimation of ρ in the First-Order Autoregressive Model: A Bayesian Approach." *Water Resources Res.* 10(2):227–241.

Lettenmaier, D.P. [1988]. "Evaluation and Testing of Flood Frequency Estimation Methods," Proceedings of the International Workshop on Natural Disasters in European Mediterranean Countries, Villa Colombella, Perugia, Italy, June 27–July 1, 1988.

McCuen, R.H., and W.M. Snyder [1986]. *Hydrologic Modeling: Statistical Methods and Applications*. Englewood Cliffs, N.J.: Prentice-Hall.

Moughamian, M.S. [1986]. "Flood Frequency Estimation: A Testing and Analysis of Physically Based Models." Cambridge, Mass.: MIT Department of Civil Engineering. M.S. thesis.

National Bureau of Standards [1953]. "Probability Tables for the Analysis of Extreme-Value Data." *Appl. Math Series* 22. Washington, D.C.

Parzen, E. [1962]. *Stochastic Processes*. San Francisco: Holden Day.

Rao, D.U. [1980]. "Log-Pearson Type III Distribution—Method of Mixed Moments." *J. Hydraul. Div. A.S.C.E.* 106(HY6):999–1019.

Russell, S.O. [1982]. "Flood Probability Estimation." *J. Hydraul. Div. A.S.C.E.* 108(HY1):63–72.

Stedinger, J.R. [1980]. "Fitting Lognormal Distributions to Hydrologic Data." *Water Resources Res.* 16(2):481–490.

Stedinger, J.R., and T.A. Cohn [1986]. "Flood Frequency Analysis with Historical and Paleoflood Information." *Water Resources Res.* 22(5):785–793.

Tasker, G.D. [1978]. "Flood Frequency Analysis with a Generalized Skew Coefficient." *Water Resources Res.* 14(2):373–376.

Thomas, H.R., Jr. [1948]. "Frequency of Minor Floods." *Boston Soc. Civil Eng.* 34:425–442.

U.S. Water Resources Council [1967]. "A Uniform Technique for Determining Flood Flow Frequencies." *Water Resources Council Bull.* 15. Washington, D.C.

Idem. [1976]. "Guidelines for Determining Flood Flow Frequency." *Water Resources Council Bull.* 17. Washington, D.C.

Idem. [1977]. "Guidelines for Determining Flood Flow Frequency." *Water Resources Council Bull.* 17A. Washington, D.C.

U.S. Geological Survey [1982]. "Guidelines for Determining Flood Flow Frequency." *Water Resources Council Bull.* 17B. Washington, D.C.

Wallis, J.R., N. Matalas, and J.R. Slack [1974]. "Just a Moment!" *Water Resources Res.* 10(2):211–219.

Wood, E. F. [1978]. "Analyzing Hydrologic Uncertainty and Its Impact Upon Decision Making in Water Resources." *Adv. Water Resources.* 1(5):299–306.
Wood, E. F., and I. Rodriguez-Iturbe [1975a]. "Bayesian Inference and Decision Making for Extreme Hydrologic Events." *Water Resources Res.* 11(4):533–542.
Idem. [1975b]. "A Bayesian Approach to Analyzing Uncertainty Among Flood Frequency Models." *Water Resources Res.* 11(6):839–843.

PROBLEMS

1. Chapter 4 briefly mentioned intensity–frequency–duration curves in the analysis of precipitation. These are curves that show the relationship between rainfall intensity and recurrence for a given duration. The intensity is higher for storms of larger recurrence intervals (rare, low-probability storms) for a fixed duration. For a fixed recurrence the intensity decreases with duration. In order to obtain intensity–frequency–duration (IFD) curves, the hydrologist groups storms by duration and analyzes each duration group as a realization (i.e., a group of possible values) of a random variable. The recurrence or the probability of exceedance is estimated for each value in the set of equal-duration storms by fitting a distribution to the set or by using nonparametric plotting position concepts as discussed in this chapter.

Given the following data, compute and draw the intensity–frequency–duration curves:

AVERAGE INTENSITY (in. hr^{-1})	DURATION (hr)	AVERAGE INTENSITY (in. hr^{-1})	DURATION (hr)
0.26	12	0.68	6
0.48	6	0.75	6
0.13	24	1.10	6
0.85	3	1.40	6
1.00	3	2.10	3
1.80	3	2.40	2
0.34	24	3.60	1
0.28	24	4.20	1
0.40	24	3.40	1
0.21	24	2.40	1
0.32	12	2.30	1
0.58	6	1.10	3
1.50	2	1.50	2
2.10	1	1.60	2
1.80	1	1.15	3
1.15	2	0.18	24
0.11	24	0.19	24
0.55	12	0.30	24
0.36	12	0.27	24

(continued)

AVERAGE INTENSITY (in. hr^{-1})	DURATION (hr)	AVERAGE INTENSITY (in. hr^{-1})	DURATION (hr)
0.34	12	0.48	12
0.64	6	0.60	12
0.52	12	0.42	12
1.00	6	0.40	12
0.93	6	0.25	24
0.78	6	1.60	3
1.25	3	1.40	3
2.10	2	1.30	3
1.70	2	3.20	1
2.00	2	3.05	1
3.00	1	3.40	1
2.60	1	2.30	2
0.90	6	2.00	2
0.85	6	1.85	2
0.80	6	2.70	1
0.44	12	2.90	1
1.70	2	1.50	3
1.40	6	1.45	3
0.23	24	0.24	24
0.67	12	0.47	2
2.80	2		

Fit a function to the intensity–frequency–duration curves.

2. You are building a dam in a U.S. river with 50 years of record. The base 10 logarithms of the annual maxima series have a mean of 0.6, a standard deviation of 0.3, and a skewness coefficient of 0.6. The original data is in cubic meters per second. During construction you want to build a buffer dam to divert the river. You want to design the buffer dam so that the probability that the dam will fail in any of the five years is 0.1. What is the necessary design recurrence and the magnitude of the design flood? (*Hint:* The exceedance probability in any one year is between 0.025 and 0.015.)

3. The annual minimum rate of discharge of a particular river is thought to have the Type III extreme-value distribution for the smallest value:

$$F_Z(z) = 1 - \exp\left[-\left(\frac{z-\varepsilon}{u-\varepsilon}\right)^2\right] \qquad z \geq \varepsilon \geq 0\,.$$

a) If the observed first and second moments of annual minimum discharge for the river are 350 $\text{ft}^3\,\text{s}^{-1}$ and $(160)^2\ \text{ft}^6\,\text{s}^{-2}$, find the probability that the annual minimum runoff is below 100 $\text{ft}^3\,\text{s}^{-1}$.

b) What is the probability of having less than two droughts (annual flows less than 100 $\text{ft}^3\,\text{s}^{-1}$) in 50 years?

c) If 100 $ft^3 s^{-1}$ is the magnitude of the lowest runoff value in a series of 30 years, give a nonparametric answer to the question in part b. (*Hint:* The first and second moments of the Type III distribution are $m_z = \varepsilon + (u - \varepsilon)(\sqrt{\pi}/2)$ and $\sigma_z^2 = (u - \varepsilon)^2 (1 - \pi/4)$.)

4. The 24-hour-duration storm depth at a site follows a Gumbel distribution with mean of 83 mm and a standard deviation of 30 mm. What is the 100-year 24-hour-duration storm at the site? (See Problem 1 and Chapter 4 for a description of intensity–frequency–duration curves.)

5. An impounding reservoir is designed so as to have sufficient capacity to meet the water requirements of a city during a drought year in which the mean rate of streamflow during the filling period is 4.5 $ft^3 s^{-1}$. During 30 years of records, the lowest flow observed during the filling period was 3.8 $ft^3 s^{-1}$ and the second lowest 4.5 $ft^3 s^{-1}$. Find the probability that during the next 20 years a flow less than 4.5 $ft^3 s^{-1}$ will occur two or more times. (*Hint:* In nonparametric analysis, floods and droughts are treated the same way.)

6. In 1958, the 50-year flood was estimated to be of a particular size. In the next 10 years, two floods were observed in excess of that size. If the original estimate was correct, what is the probability of such an observation?

7. Exceedance series—flows greater than some arbitrary base value—have been treated as random variables with a shifted exponential distribution. For 23 years of records during which 56 flows in excess of 5500 $ft^3 s^{-1}$ were observed, the following density function was found adequate:

$$f_X(x) = \left(\frac{1}{6500}\right) e^{-(x-5500)/6500} \qquad x \geq 5500\,,$$

where X is the peak-flow magnitude given a flood in excess of 5500 occurred.

a) Sketch the density function.

b) Compute the probability that a flow in excess of 20,000 $ft^3 s^{-1}$ will be observed given that the flow exceeds the base value of 5500 $ft^3 s^{-1}$.

c) Compute the flow above 5500 $ft^3 s^{-1}$ such that the probability of being exceeded is 1/50. Is this the 50-year flow?

8. A temporary cofferdam is to be built to protect the five-year construction activity for a major cross-valley dam. If the cofferdam is designed to withstand the 20-year flood, what is the risk that the structure will be overtopped (a) in the first year; (b) in the third year exactly; (c) at least once in the five-year construction period; and (d) not at all during the five-year period?

9. Complete the following mathematical statements about the properties of a probability density function by matching the statements on the left with the correct item number on the right. Assume x is a series of annual occurrences from a normal distribution.

a) $\int_{-\infty}^{\infty} F(x)\,dx =$

b) $\int_{-\infty}^{m_1} f(x)\,dx =$

c) $\int_{m_1}^{m_2} f(x)\,dx =$

d) $\int_{-\infty}^{\square} f(x)\,dx = 0.5$

e) $1 - \int_{m_1}^{m_2} f(x)\,dx =$

1. $P[m_1 \geq x \cup x \geq m_2]$
2. unity
3. median
4. $P[x \leq m_1]$
5. standard deviation
6. $p(m_1 \leq x \leq m_2)$

10. A temporary flood wall has been constructed to protect several homes in the flood plain. The wall was built to withstand any discharge up to the 20-year flood magnitude. The wall will be removed at the end of the three-year period after all the homes have been relocated. Determine the probability that (a) the wall will be overtopped in any year, (b) the wall will not be overtopped during the relocation operation, (c) the wall will be overtopped at least once before all of the homes are relocated, (d) the wall will be overtopped exactly once before all the homes are relocated, and (e) the wall will be adequate for the first two years and overtopped in the third year.

11. A cofferdam designed to withstand flows up to and including the 1944 flow of 1870 $ft^3 s^{-1}$ is planned for the Middle Branch Westfield River. Investigate the degree of protection afforded by the dam of this size during a five-year program of channel improvement and dam construction. (The 1944 flood was the sixth largest during a 25-year record from 1921 to 1945.)

12. The annual maximum rate of discharge of a particular river is thought to have the Type I extreme-value distribution, i.e., Gumbel, with mean 10,000 $ft^3 s^{-1}$ and standard deviation 300 $ft^3 s^{-1}$.

a) Compute P[annual maximum discharge $\geq$ 15,000 $ft^3 s^{-1}$].

b) Find an expression for the cumulative density function of the river's maximum discharge during the 20-year lifetime of an anticipated flood-control project. Assume that the individual annual maxima are mutually independent random variables.

c) What is the probability that the maximum of 20 years will exceed 15,000 $ft^3 s^{-1}$? *Hint:* The Gumbel cumulative density function is

$$F_Y(y) = \exp[-e^{-\alpha(y-u)}] \quad -\infty \leq y \leq \infty$$

$$m_Y = u + \frac{0.577}{\alpha} \qquad \sigma_Y^2 = \frac{1.645}{\alpha^2}.$$

13. Using the rainfall data and the sketch of an urban development given below, what is the discharge corresponding to a recurrence interval of about seven years? In the analysis you can ignore the road as an area contributing to runoff. See Problem 1 and Chapter 4 for a discussion of intensity–frequency–duration curves.

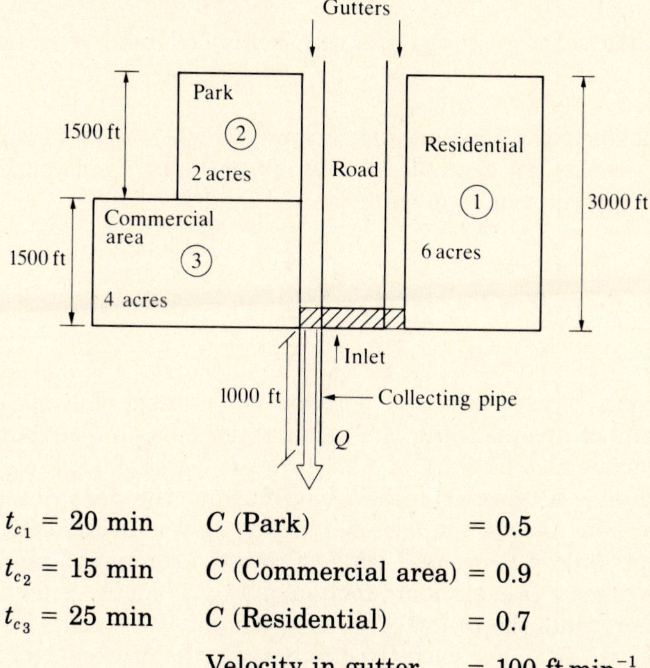

$t_{c_1} = 20$ min $\quad C$ (Park) $= 0.5$

$t_{c_2} = 15$ min $\quad C$ (Commercial area) $= 0.9$

$t_{c_3} = 25$ min $\quad C$ (Residential) $= 0.7$

Velocity in gutter $= 100 \text{ ft min}^{-1}$

Velocity in pipes $= 200 \text{ ft min}^{-1}$

RAINFALL DATA

Intensity (in. hr^{-1})	Duration (min)	Intensity (in. hr^{-1})	Duration (min)
1.19	45	0.28	55
0.65	50	0.40	55
1.12	50	1.05	55
0.47	55	0.37	50
0.61	55	0.30	50
0.32	55	0.32	45
0.31	50	0.29	45
0.36	45	0.69	45
0.45	45	0.40	45
0.53	45	0.34	50
0.42	50	0.27	50
0.50	50	0.29	55
0.24	55	0.26	55
0.35	55	0.33	45
0.27	45	0.26	50

14. In general, does the discharge of a given recurrence correspond to a rainfall event of the same recurrence? Explain and discuss.

15. Show that the mean of the binomial distribution is $E[K] = nP$ and the variance, $\text{Var}[K] = nP(1 - P)$.

16. Show that the mean of the geometric distribution is $1/P$ and the variance is $(1 - P)/P^2$.

17. A very useful model is the Poisson distribution, which is used to represent the number of "events," for example rainstorms, arrivals in a given time period t. The Poisson distribution is given by

$$P_X(x) = \frac{(\lambda t)^x e^{-\lambda t}}{x!} \qquad x = 0, 1, 2, \ldots,$$

where λ is rate of arrival of events in time period t (i.e., number of events per unit time). It has units of inverse time. The mean of the Poisson distribution is $\mu = \lambda t$. The variance is also $\sigma^2 = \lambda t$.

The Poisson is hence a one-parameter distribution; the parameter is $\nu = \lambda t$, or for a fixed time period the parameter is simply ν. In introducing time, the Poisson distribution is a small window into the concept of random processes. A random process is a random function of an argument, commonly time. The Poisson represents the process $X(t)$ of the number of events occurring in the time interval $\{0, t\}$. For each fixed value of t, the process $X(t)$ is a random variable, Poisson distributed, with parameter λt.

The Poisson assumption requires that the process in question satisfy the following conditions: (1) The probability of an event occurring in an arbitrary and small time interval Δt is always $\lambda \Delta t$. This is called stationarity, i.e., the distribution does not change with time. (2) Only one event can occur in a short interval Δt, with probability $\lambda \Delta t$. (3) The occurrence of an event in an interval Δt is independent of the occurrence of an event in any other Δt. This is independence as defined at the beginning of this chapter.

Having introduced the Poisson distribution, here are some useful questions. Assume that daily rainfall in Boston follows a Poisson process, hence only one storm per day is allowed per condition 2 above. Storms arrive in Boston every three days in the average summer month.

a) Draw the Poisson distribution for the 30 days of the month of June.

b) What is the probability that four or less storms occur in June?

c) Call the random variable T the time to the first storm arrival in June. Derive the distribution of T. (*Hint:* The time T is exponentially distributed as $f_T(t) = \lambda e^{-\lambda t}$.)

d) What is the distribution of the time between storm arrivals? (*Hint:* Use the independence property of Poisson events and part c.)

e) What are the mean and variance of the exponential distribution?

f) Show that the conditional distribution of the time T to the next event given that no event has occurred up to time t_0 is exponential.

$$f_{T|T>t_0}(\tau) = \lambda e^{-\lambda\tau} \qquad t \geq 0,$$

where $\tau = t - t_0$.

g) What is the distribution of the time to the kth storm? (*Hint:* The time to the kth storm Γ_k is the sum of the times to each of the storms prior to that time. That is, $\Gamma_k = T_1 + T_2 + \cdots + T_k$. Each of the T_i are independent and exponentially distributed variables with common parameter λ. The answer is the gamma distribution

$$f_\Gamma(t) = \frac{\lambda(\lambda t)^{k-1}e^{-\lambda x}}{(k-1)!} \qquad t \geq 0.$$

This is analogous to the derivation of the Nash model of the instantaneous unit hydrograph. It involves a convolution of distributions.

18. Assume that coastal flooding in an estuary is due to the combination of wind velocity and river discharge in the estuary. The annual maximum wind velocity is taken to obey an exponential distribution:

$$f_W(w) = \lambda_1 e^{-\lambda_1 w}.$$

The annual maximum discharge is also assumed exponential:

$$f_Q(q) = \lambda_2 e^{-\lambda_2 q}.$$

The level of flooding H is proportional to the sum of W and Q. If winds and river discharges are independent, what is the distribution of Q? What is the recurrence of the coastal flooding resulting from the 50-year discharge and the 50-year wind?

19. For a river basin near you, find 20 or more years of streamflow records and fit the log-Pearson Type III distribution. Estimate the 50-year flood. Draw confidence limits on your distribution. In the United States, streamflow records are available from the U.S. Geological Survey or its publications. The U.S. Army Corps of Engineers and state agencies also have streamflow records.

20. Simulation is a powerful tool in hydrology and other earth sciences. Monte Carlo simulation refers to the generation of sequences of numbers following a particular probabilistic distribution. Given a probabilistic model, simulation allows the study of possible sequences (and effects) of hydrologic processes like rainfall and discharges. Simulation exercises are a day-to-day tool in practice and research.

If a random variable obeys a cumulative density function

$$P[X \leq x] = F_X(x)\,. \tag{1}$$

The goal of Monte Carlo simulation is to invert the above equation to obtain values of X. Theoretically, we want to solve

$$x = F^{-1}(P)\,, \tag{2}$$

where P is the probability of X being less than or equal to x. Hence, P is uniformly distributed between 0 and 1. Given Eq. (2), all that needs to be done is to introduce values uniformly distributed between 0 and 1 for P, and solve for x. Uniformly distributed values between 0 and 1 are available from most computers or scientific calculators. Alternatively, look at your local telephone directory and add a period before the last three digits of telephone numbers randomly selected!

Analytical solutions of Eq. (2) are sometimes possible. Most times numerical solutions are required.

The concept of Monte Carlo simulation can be used to create models of hydrologic phenomena. For example, assume that storms can be represented as instantaneous pulses of depth D arriving as Poisson events on the average once every 10 days. The depth D is exponentially distributed.

$$f_D(d) = \alpha e^{-\alpha d}\,.$$

Generate a sequence of five years of storms, showing time of arrival and depth of each event.

21. In Chapter 4 we saw that large-duration storms have generally small intensities. Let a storm be represented by rectangular pulses of duration t and intensity i (i.e., depth $= it$). Assume the conditional distribution of intensity i on duration t is of the form

$$f_{I|T}(i|t) = \alpha t e^{-\alpha t i}$$

and that the storm duration follows

$$g_T(t) = \beta e^{-\beta t}.$$

What is the joint distribution of I and T? If $\beta = 0.25\ \text{hr}^{-1}$ and $\alpha = 2\ \text{cm}^{-1}$, what is the probability of obtaining a storm of mean intensity $0.5\ \text{cm}\,\text{hr}^{-1}$ and three-hour duration? What is the probability of obtaining a storm of total depth equal to 2 cm?

22. A hydrologist is trying to determine the hydraulic conductivity of a formation that he knows must have one of four geologic origins. From his or her hydrogeologic knowledge he or she can assign the following probabilities to each possibility.

GEOLOGIC ORIGIN OR STATE	ASSOCIATED HYDRAULIC CONDUCTIVITY ($\text{cm}\,\text{s}^{-1}$)	PROBABILITY
1	10^{-4}	0.1
2	5×10^{-4}	0.2
3	10^{-3}	0.5
4	10^{-2}	0.2

To determine hydraulic conductivity a well test is performed. Such procedures are not completely accurate. From experience the hydrologist can assign the probability that the test indicates geologic state i given that the true state is j. This probability table is

WELL TEST RESULT	TRUE STATE 1	2	3	4
1	0.5	0.2	0.0	0.0
2	0.3	0.6	0.1	0.0
3	0.1	0.1	0.7	0.2
4	0.0	0.1	0.2	0.8

What is the probability mass function of hydraulic conductivity if the well test indicates that the formation is of type 2? If another test is repeated and the test indicates type 1, what is the new probability mass function of hydraulic conductivity?

23. Using concepts and introductory ideas presented in this chapter to use historical information for frequency analysis, how many years of historical record would you need to augment a 20-year systematic record to an equivalent of 40 years if you are estimating the 100-year flood? Assume your historical record contains (a) only the events (no magnitude) larger than the 10-year flood and (b) only the events (no magnitude) larger than the 100-year flood.

24. Probability paper is one on which a graph of a particular cumulative density function (cdf) plots as a straight line. This is achieved by distorting the ordinate scale in a plot of $F_X(x)$ versus x. The goal is to obtain a relationship of the form

$$y = ax + b,$$

where y is a function of $F_X(x)$. What would y (the necessary scale transformation) be for the case of the exponential distribution

$$F_X(x) = 1 - e^{-\lambda x}?$$

Construct probability paper for the normal distribution. (*Hint:* Do it graphically, if necessary, by first drawing the cdf of the normal on arithmetic scales and seeing how it needs to be distorted to produce a straight line.)

Chapter 12

Concepts of Fluvial Geomorphology

12.1 INTRODUCTION

The two most obvious features of a river basin are probably the channel network and the surrounding landscape, which is made up of interconnected hillslopes. The role of both hillslope and channels is to deliver and transport sediment and water to the basin outlet. The two features are very much interrelated, each one being molded, shaped, and maintained by the other in a not fully understood synergistic relationship.

The hillslope receives precipitation and supplies runoff. The channels grow and develop to drain this runoff in some efficient manner. The principles of efficiency leading to the channel network growth are not yet known. It is nevertheless true that growing channels dissect the landscape and hence fix the hillslope pattern. In turn, for a channel to exist there must be sufficient water delivered and sediment removed from the surrounding hillslope.

The precarious balance between hillslope and channel is evidenced in hollows and hillslope regions near channels. These are preferred sites for runoff production by saturation of the soil column from below (see Chapter 8). Hollows are also preferred patterns for upward-growing channels. When and where does a hillslope section develop a channel? What is hillslope and what is channel? The above are nontrivial questions with unclear answers. For example, a model could be an infinitely long, fine, filigreed pattern of channels extending everywhere on the basin. In such a conceptualization a "channel" would drain every point; there would be no hillslopes. The validity or failure of this or any conceptualization depends on the definition of channels and on the related issue of the resolution at which the basin is being viewed. After all, an ant may see "channels" where we do not!

Channels in a river basin generally develop in a treelike structure with the outlet stream as the main trunk and upstream having a bifurcating pattern of smaller and smaller branches. These networks exhibit tremendous regularity and organization. This regularity is expressed not only in the geometry but in the size and relationship between the parts. A lot of results exist describing, at least in two planar dimensions, the structure of the channel network.

The references give an extensive list of publications on the subjects of network growth, channel network organization, channel geometry studies, landscape morphology and patterns, and the relationship between hydrologic response and fluvial geomorphology. The following sections will, nevertheless, be limited in scope. First we will talk about the channel networks and potentially useful descriptors of the patterns they form. Then we will concentrate on reviewing recent suggested links between geomorphology and hydrology. A lot will be left untouched, particularly issues of the regularities in the hillslopes and possible relationships to runoff production. Nevertheless, this chapter should introduce you to one of the most fascinating frontiers in hydrology.

12.2 DESCRIPTIONS OF DRAINAGE BASIN COMPOSITION

12.2.1 Two-Dimensional Planar Descriptors

Horton's Law

The quantitative analysis of channel networks begins with Horton's [1945] method of classifying streams by order. Strahler [1957] revised Horton's classification scheme such that the ordering scheme is, unlike Horton's, purely topological, for it refers to only the interconnections and not to the lengths, shapes, or orientation of the links comprising a network. Figure 12.1 shows a hypothetical river network. The network has a single outlet (root or trunk). Inner nodes are points where lines or river segments join. Outer nodes are sources with one line or stream segment originating from them. Links are segments between nodes. Interior links connect interior nodes. Exterior links are between a source (outer node) and a downstream interior node. Mock [1971] refines the classification of links. Based on the above representation Strahler classified streams according to the following procedure (see Fig. 12.1).

1. Channels that originate at a source are defined to be first-order streams;
2. When two streams of order ω join, a stream of order $(\omega + 1)$ is created;

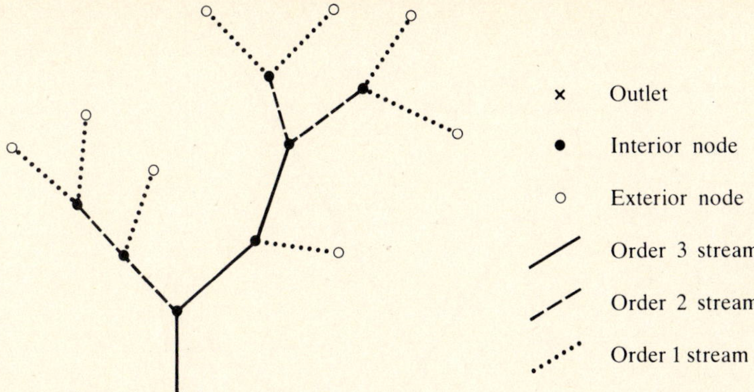

FIGURE 12.1 Hypothetical river network illustrating Strahler's stream-ordering system.

3. When two streams of different order merge, the channel segment immediately downstream is taken to be the continuation of the higher order stream; and
4. The order of the basin is the highest stream order Ω.

Note that Strahler streams of order higher than 1 may be composed of several links.

Horton [1945] first developed a somewhat different stream-ordering scheme and suggested several empirical laws: the law of stream numbers and the law of stream lengths. These results have been confirmed many times using Strahler's ordering system. Schumm [1956] proposed a Horton-type law of stream areas.

The law of stream numbers states that the number of streams of a given order follows an inverse geometric relationship with stream order:

$$N_\omega = R_B^{\Omega-\omega}, \tag{12.1}$$

where Ω is the order of the highest-order stream in the network, ω is the order of interest, and R_B is a constant for a given network. R_B is called the bifurcation ratio. A plot of the logarithm of N_ω versus order ω will approximately yield a straight line with negative slope. The magnitude of that slope is the logarithm of R_B. Figure 12.2 illustrates that result.

Note that Eq. (12.1) leads to the conclusion that the total number of streams in the network is

$$\sum_{\omega=1}^{\Omega} N_\omega = 1 + R_B + R_B^2 + \cdots + R_B^{\Omega-1} = \sum_{\omega=1}^{\Omega} R_B^{\Omega-\omega} = \frac{R_B^\Omega - 1}{R_B - 1}. \tag{12.2}$$

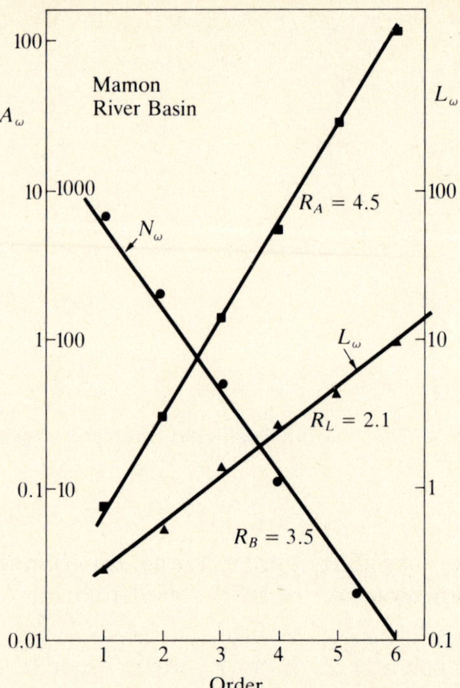

FIGURE 12.2 Stream numbers, lengths, and areas versus stream order illustrating Horton's laws for the Mamon Basin in Venezuela. Source: J. B. Valdes, Y. Fiallo, and I. Rodriguez-Iturbe, "A Rainfall Runoff Analysis of the Geomorphologic IUH," *Water Resources Res.* 15(6):1421–1434, 1979. Copyright by the American Geophysical Union.

Also note that

$$R_B = \frac{N_{\omega-1}}{N_\omega}, \tag{12.3}$$

which implies that on the average there are R_B streams of order $\Omega - 1$.

The concept of the laws of stream lengths and stream areas is the same as the law of stream numbers, the ratios of the series being the length ratio R_L and the area ratio R_A, respectively. R_L and R_A are calculated using the following quantities: the average stream lengths of each order $\overline{L}_\omega$ are given by

$$\overline{L}_\omega = \frac{1}{N_\omega} \sum_{i=1}^{N_\omega} L_{\omega_i}, \tag{12.4}$$

where L_{ω_i} is the length of a stream of order ω, and the average stream area of each order $\overline{A}_\omega$ is given by

$$\overline{A}_\omega = \frac{1}{N_\omega} \sum_{i=1}^{N_\omega} A_{\omega_i}, \tag{12.5}$$

where A_{ω_i} is the area contributing runoff to a stream of order ω and its tributaries. For example, $\overline{A}_\Omega$ is the total area of the basin. The quantitative expressions of Horton's laws are summarized below:

Law of stream numbers $$\frac{N_{\omega-1}}{N_\omega} = R_B \tag{12.6}$$

Law of stream lengths $$\frac{\overline{L}_\omega}{\overline{L}_{\omega-1}} = R_L \tag{12.7}$$

Law of stream areas $$\frac{\overline{A}_\omega}{\overline{A}_{\omega-1}} = R_A \tag{12.8}$$

Empirical results indicate that for natural basins the value for R_B normally ranges from 3 to 5, for R_L from 1.5 to 3.5, and for R_A from 3 to 6 (Smart [1972]). Figure 12.2 also shows plots of $\overline{L}_\omega$ and $\overline{A}_\omega$ versus ω.

The law of stream numbers and lengths can be combined into an equation for the total length of channels of order ω:

$$\sum_{i=1}^{N_\omega} L_{\omega_i} = \overline{L}_1 R_B^{\Omega-\omega} R_L^{\omega-1}. \tag{12.9}$$

The total length of channels is also shown (Horton [1945]) to be

$$\sum_{\omega=1}^{\Omega} \sum_{i=1}^{N_\omega} L_{\omega_i} = \overline{L}_1 R_B^{\Omega-1} \frac{R_{LB}^{\Omega} - 1}{R_{LB} - 1}, \tag{12.10}$$

where

$$R_{LB} = \frac{R_L}{R_B}$$

and $\overline{L}_1$ is the mean length of first-order streams.

Shreve [1969] reports, based on studies by Strahler [1952] and Broscoe [1959], that the law of stream lengths often does not fit data very well, in terms of Strahler ordering.

Horton [1945], in formulating the laws of drainage composition, viewed them as reflecting the orderly evolution of channel networks and their drainage basins. However, Shreve [1966, 1967] effectively showed that they were derivable from a link-based random model. Shreve [1966, p. 27] stated the fundamental geomorphologic hypothesis that "in the absence of geologic controls a natural population of channel networks will be topologically random." A second postulate introduced by Smart [1968] suggests that the exterior and interior link lengths are independent random variables. Based on these assumptions, Shreve [1969] showed that Horton's laws were no more than the outcome of applying ordering to a set of random channel networks.

Tokunaga's Characterization

An alternative to the Horton formulation of geometric similarity laws is given by Tokunaga [1978]. This is also based on Strahler ordering. Let ${}_{\omega}E_{\lambda}$ denote the average number of streams of order λ entering a stream of order ω from the side. Then Tokunaga suggests that ${}_{\omega}E_{\omega-\ell}$ is on average constant for all ω, denoted $E_{\ell} = {}_{\omega}E_{\omega-\ell}$ and that furthermore

$$K \triangleq \frac{E_{\ell}}{E_{\ell-1}} \tag{12.11}$$

is constant. Two parameters E_1 and K are therefore required to describe the network composition as far as bifurcation and stream numbers are concerned. Tokunaga gives

$$N(\Omega, \omega) = \frac{Q^{\Omega-\omega-1} - P^{\Omega-\omega-1}}{Q - P} Q(2 + E_1 - P) + P^{\Omega-\omega-1}(2 + E_1) \tag{12.12}$$

for the number of streams of order ω within a basin of order Ω. Here, P and Q are parameters given by

$$P = \frac{2 + E_1 + K - \sqrt{(2 + E_1 + K)^2 - 8K}}{2}$$

$$Q = \frac{2 + E_1 + K + \sqrt{(2 + E_1 + K)^2 - 8K}}{2}.$$

Equation (12.12) gives a law of stream numbers that when plotted on a Horton diagram gives a curve slightly concave upward, thus agreeing qualitatively with this tendency reported by Shreve [1966] (Fig. 12.3). Tokunaga [1978] also introduces the notion of infinite networks in the following sense: "Assume a basin of finite size consisting of infinitesimally small subbasins of the lowest order and interbasin areas, instead of an infinite basin consisting of subbasins of the lowest order and interbasin areas of finite size."

With assumptions that the interbasin areas, i.e., areas draining directly into internal links, are less than lowest-order stream areas on average, and $K < 2 + E_1$ as usually borne out in practice, Tokunaga derives

$$A_{\omega} = Q^{\omega-\lambda} A_{\lambda}, \tag{12.13}$$

where A_{ω} denotes the area draining a basin of order ω. This is Horton's area law (Eq. 12.8), with area ratio Q.

Tokunaga [1978] relates his formulation to the random model of Shreve [1966, 1967, 1969] by showing that the expected values of E_1 and K in an

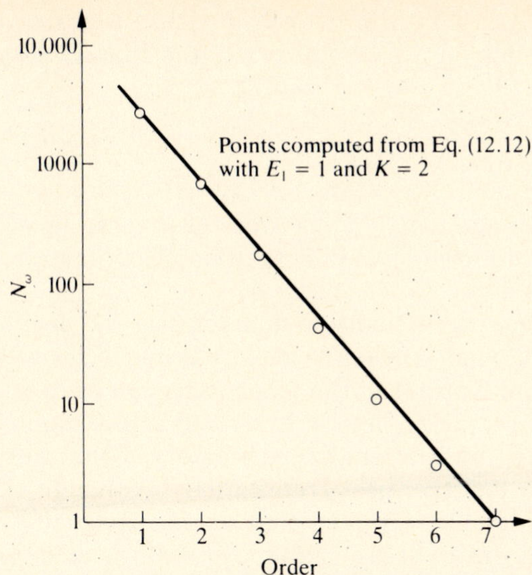

FIGURE 12.3 Stream numbers versus order showing concavity and Tokunaga's parameterization.

infinite topologically random channel network are 1 and 2, respectively. This results in $Q = 4$, implying $R_A = 4$ as obtained by Shreve [1967]. Also, Eq. (12.12) with $\omega = 1$ and the expected values for E_1 and K gives

$$N(\Omega, 1) = \frac{2^{2\Omega - 1} + 1}{3}, \tag{12.14}$$

which is the expected number of first-order streams in a network of order Ω given by Shreve [1969].

While the theoretical framework suggested by Tokunaga [1978] is appealing, it has not been widely tested against data. Estimates for E_1 and K seem to show considerable scatter around their expected values, and tests of whether these differences are significant may provide more evidence for the acceptance or rejection of the random model.

Drainage Density and Scale

The notion that there is a fundamental length scale in drainage networks has existed at least since Horton [1945]. Horton defined drainage density as

$$D = \frac{L_T}{A_\Omega} = \sum_{\omega=1}^{\Omega} \sum_{i=1}^{N_\omega} \frac{L_{\omega_i}}{A_\Omega} = \sum_{\omega=1}^{\Omega} \frac{N_\omega \overline{L}_\omega}{A_\Omega}, \tag{12.15}$$

where L_T is the total length of channels within a basin with area A_Ω and order Ω. Horton relates drainage density to the Horton numbers as

$$D = \frac{\overline{L}_1 R_B^{\Omega-1}}{A_\Omega} \frac{R_{LB}^{\Omega} - 1}{R_{LB} - 1}. \tag{12.16}$$

Note that the drainage density has units of inverse length and hence is a quantity dependent on the level of resolution of the map from which lengths are measured.

Horton used drainage density to characterize the degree of drainage development within a basin. Drainage density is one of several linear measures by which the scale of features of the topography can be compared. Other measures include texture, defined by Smith [1950] as the number of crenulations of a contour divided by the length (perimeter) of the contour. The contour used is that which has the most crenulations. Although these notions may sometimes be ambiguous, the idea that drainage density or texture represents fundamental differences in landform scale is clear from Figure 12.4, taken from Smith [1950].

The drainage density has been suggested as a basic length scale. If drainage density is constant everywhere in the basin, then the average length of a contributing hillslope is approximately half the average distance between stream channels and hence

$$L_c = \frac{1}{2D}. \tag{12.17}$$

The issue of scale in the river basin is fundamental to hydrology. The previous section quoted Tokunaga's [1978] notion of a river network growing infinitely in space, so that smaller and smaller streams reach every point in the basin. If this growth occurs so that streams and the patterns they form have properties that are invariant (except for a scaling factor) at all levels of resolution, the system is called *scaling*. An object invariant under ordinary geometric similarity is called self-similar (Mandelbrot [1983]). The concept is extended to statistical self-similarity when the rescaled objects are random with identical distributions.

Indeed Horton's and Tokunaga's characterizations imply self-similarity. This is discussed and shown in Tarboton et al. [1988]. The crucial question is whether there is a dominating scale that limits the resolution of the channel network. Figure 12.5 shows a river basin inferred from digitized elevation information as viewed with increased resolution. Which of the channel representations is the hydrogeomorphologic reality? The answer is tied to the drainage density, the scale that controls the boundary between channel and hillslope. Although this scale could be accurately inferred by extensive field observations at a site, predicting it or even inferring it from elevation data, remote sensing techniques, or other less direct observations remains a very difficult problem.

Stream Frequency

Stream frequency is defined as the number of stream segments per unit area

$$F = \frac{\sum_{\omega=1}^{\Omega} N_\omega}{A_\Omega}, \tag{12.18}$$

where N_ω is the number of streams of order ω and A_Ω is the total area of the basin of order Ω. Like drainage density, stream frequency is a scale of resolution dependent quantity. Although stream frequency and drainage density measure different properties, Melton [1958] found a very good relationship between the two:

$$F = 0.694D^2. \tag{12.19}$$

The implication is that the dimensionless ratio F/D^2 approaches a constant value of 0.694, independently of scale. Since stream frequency and drainage density can be expressed as functions of Horton numbers, then Eq. (12.19) hints that the Horton numbers may not be independent characterizations of the basin. That is, the three quantities R_B, R_L, and R_A may be redundant descriptors, with probably R_B being the one fundamental quantity.

Relationship between Length and Area

Figure 12.6 shows the relationship between catchment area and catchment length. Length is usually taken as the "mainstream" length defined as the distance from the catchment outlet to the projected intersection with the upstream catchment boundary. The "mainstream" follows the highest-order stream. At intersections where two streams of equal order join, the "mainstream" follows the dominant branch that drains the largest catchment area.

The line of best fit in Figure 12.6 is

$$L = 1.40A^{0.568}. \tag{12.20}$$

The fact that length does not go as the square root of area implies that river basins are not fully geometrically similar, i.e., the ratio of area to the square of length is not constant for all basins. The figure gives the best geometrically similar fit as

$$L = 1.73A^{0.5}$$

or

$$\frac{A}{L^2} = \frac{1}{3}. \tag{12.21}$$

Topographic map, coarse texture

Drainage map, coarse texture

FIGURE 12.4 Coarse and medium texture topography and drainage maps. Source: K. G. Smith, "Standards for Grading Texture of Erosional Topography," *Am. J. Sci.* 248:655–658, 1950. Reprinted by permission of American Journal of Science.

Topographic map, medium texture

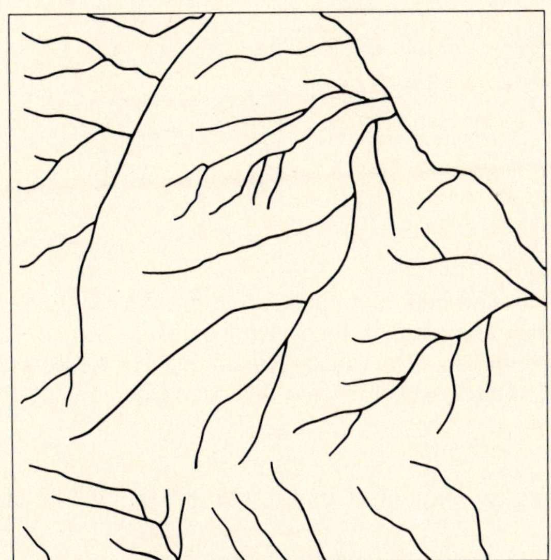

Drainage map, medium texture

FIGURE 12.4 (*Continued*)

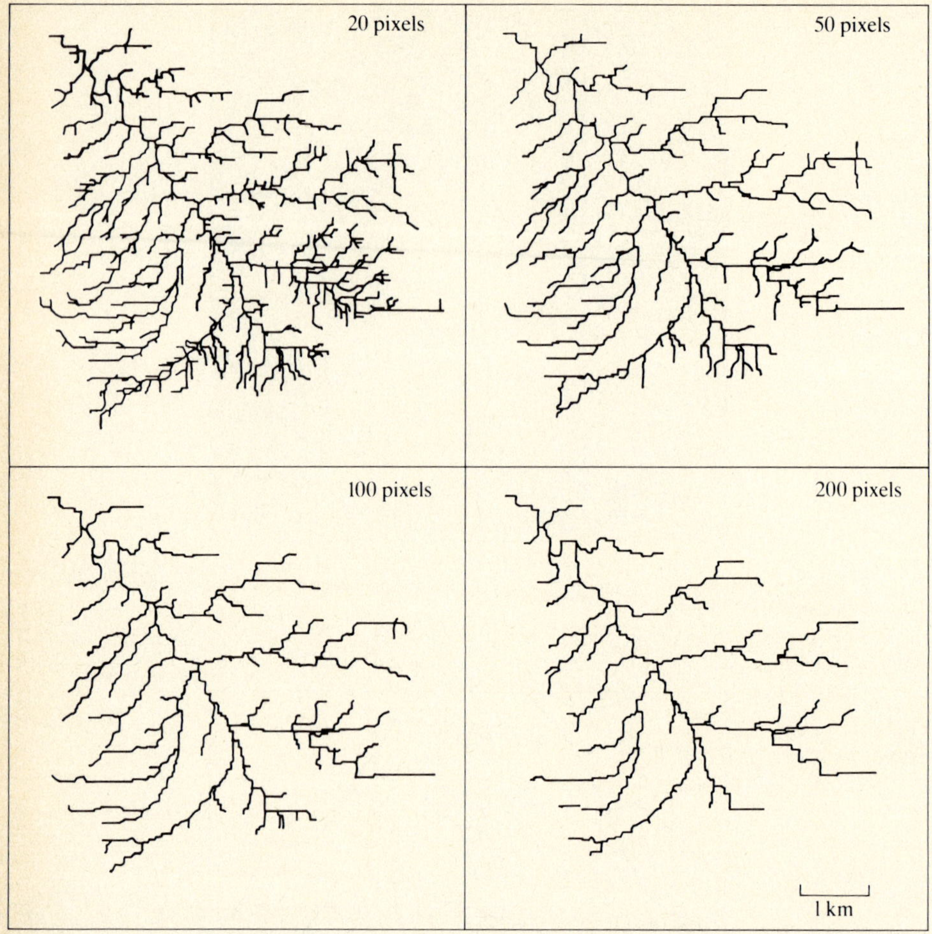

FIGURE 12.5 Channel networks of a subbasin of the Walnut Gulch in Arizona inferred from digitized elevation maps as the resolution increases. Source: D. G. Tarboton, R. L. Bras, and I. Rodriguez-Iturbe, "The Fractal Nature of River Networks," *Water Resources Res.* 24(8):317–322, 1988. Copyright by the American Geophysical Union.

The fact is that the exponent in Eq. (12.20) is not 0.5; there are two possible explanations:

1. Larger basins are more elongated than smaller basins since the ratio A/L^2 falls as area increases.
2. Are large basins in Figure 12.6 measured with a coarser resolution? If they are, is stream length a function of the resolution of the ruler measuring it? If that is the case, then the stream is then said to have a "fractal" dimension, which can lead to Eq. (12.20). Mandelbrot [1983] and Tarboton et al. [1988] discuss this issue.

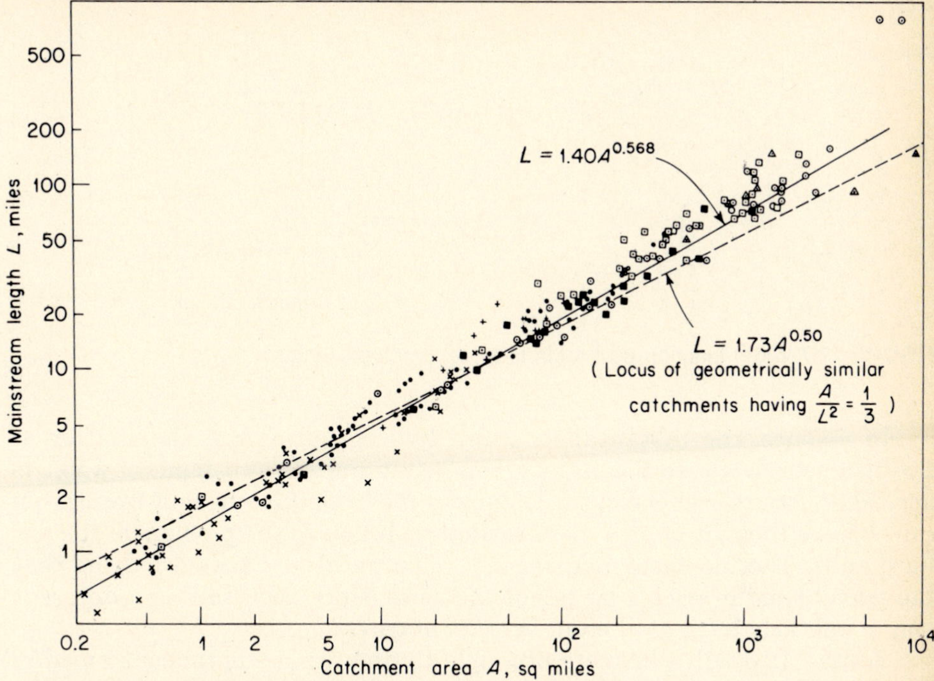

FIGURE 12.6 Relationship between catchment area and catchment length. Source: P. S. Eagleson, *Dynamic Hydrology,* McGraw-Hill, 1970.

Width Function

A basin descriptor shown to have important effects on hydrograph peak and shape is the width function (Surkan [1968]; Calver et al. [1972]; Kirkby [1976]; Gupta et al. [1986]; Troutman and Karlinger [1986]). The width function $N(x)$ measures the number of links at a given distance x from the outlet. The distance x may be the geometric length or straight-line distance between nodes, the actual length, or the topologic length. Topologic length is measured in terms of the number of links from the outlet. Figure 12.7 illustrates the derivation of the width function based on topologic length. Figure 12.8 shows the width function based on actual lengths of a river basin.

Using Figure 12.7 it is easy to see why the width function is closely related to hydrologic response. Imagine a basin, like that in Figure 12.7, with the property that the travel time of water in each link is the same. Hence if you are two links away from the outlet it would take two units of time to reach the outlet. Think of available runoff as billiard balls released instantaneously at each node, including the sources, hence 19 balls are used. The number of balls reaching the outlet in each time interval is given by the width function. The width function is the instantaneous response function of the billiard balls. In this simplified example, it is the instantaneous unit hydrograph.

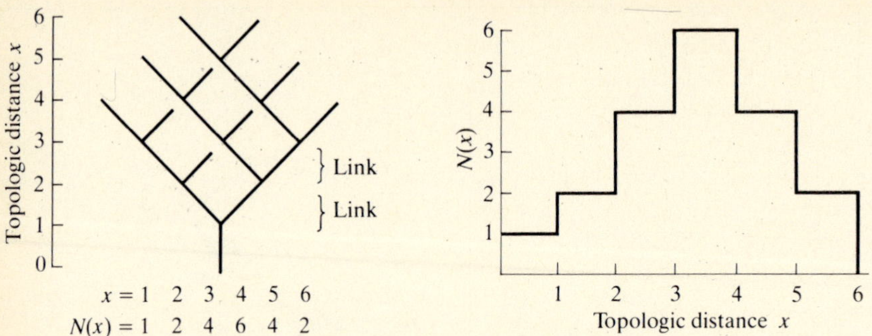

FIGURE 12.7 Width function defined in terms of topologic length.

In essence, the instantaneous unit hydrograph is the distribution of travel time of water to the outlet when instantaneous unit input of water is applied uniformly over the basin. Obtaining the instantaneous unit hydrograph can then be thought of as a more realistic solution to the problem of the billiard balls. The increased realism will come mostly by acknowledging that the travel time in each link is not the same since each link is a different length and may have a different velocity of travel.

Section 12.3 will illustrate attempts to find the instantaneous unit hydrograph by relating travel time distribution to geomorphologic measures.

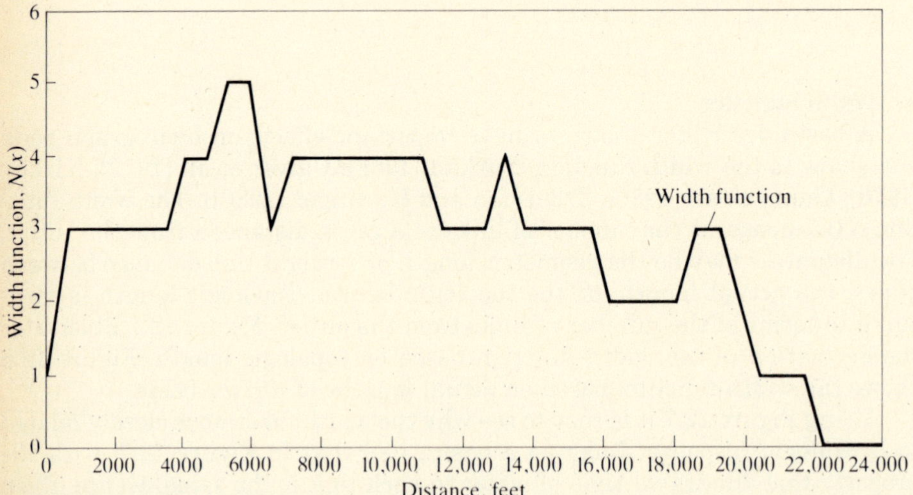

FIGURE 12.8 Width function of the Blue Creek Basin in Alabama. Source: M. R. Karlinger and B. M. Troutman, "Assessment of the Instantaneous Unit Hydrograph Derived from the Theory of Topologically Random Networks," *Water Resources Res.* 21(11):1693–1702, 1985. Copyright by the American Geophysical Union.

Magnitude and Diameter

The magnitude M of a basin (or subbasin) is the number of exterior links in the network, or equivalently the number of first-order streams. In any network the number of internal links is $M - 1$, and hence the total number of links is $2M - 1$.

An important observation is that there is a very good linear relationship between total area and the number of links, or the magnitude. Figure 12.9 shows one such relationship. The implication is that each link has an associated area that does not seem to vary much.

Diameter D refers to the longest topologic length of the network. That is, the diameter is the path from the outlet to a source with the largest number of links.

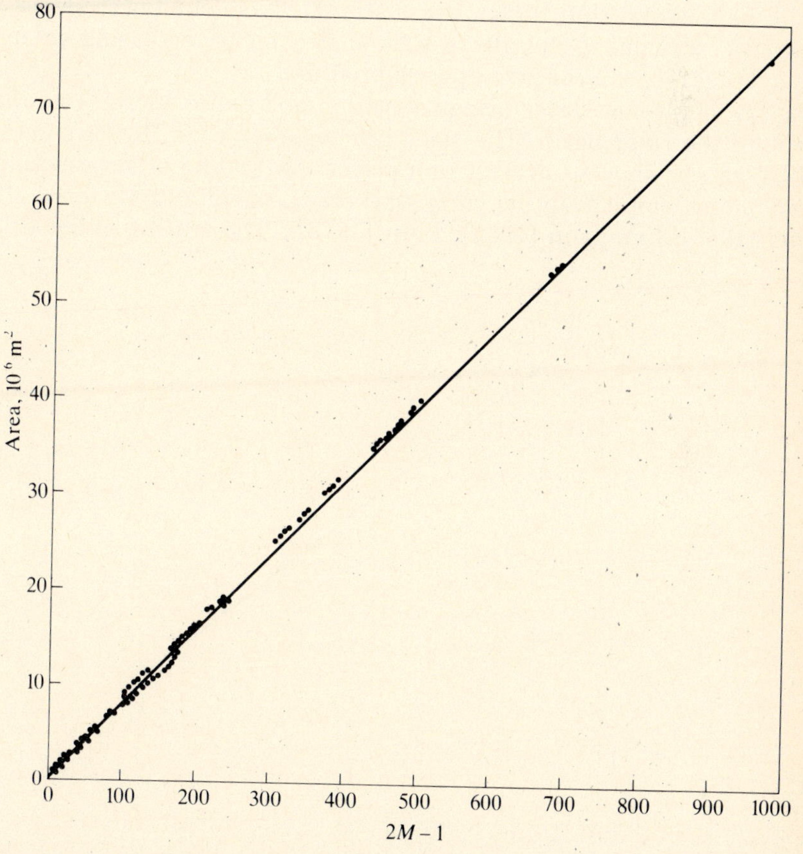

FIGURE 12.9 Area versus magnitude relationship obtained from the Walnut Gulch River Basin in Arizona.

12.2.2 Descriptors of Relief

Law of Stream Slopes

Horton [1945] also suggested that stream slopes follow a geometric series. The slope ratio R_s is defined by

$$\overline{S}_\omega = \overline{S}_\Omega R_s^{\Omega-\omega}, \tag{12.22}$$

where $\overline{S}$ is the average slope of a channel of order ω. This relationship is illustrated in Figure 12.10.

The slope ratio is

$$R_s = \frac{\overline{S}_{\omega-1}}{\overline{S}_\omega}, \tag{12.23}$$

with the implication that higher-order streams have shallower slopes than smaller-order streams. Generally, a path of streams of increasing order will, on the average, show a concave-upward profile.

The law of stream slopes brings elevation and hence energy into the description of the river basin. The slope law together with Horton's area law (Eq. 12.8) can be shown to result in a power law scaling of slope with area. This was noted empirically by early workers (Leopold and Maddock [1953]; Wolman [1955]; Langbein [1964]; Flint [1974]). The scaling of slope is be-

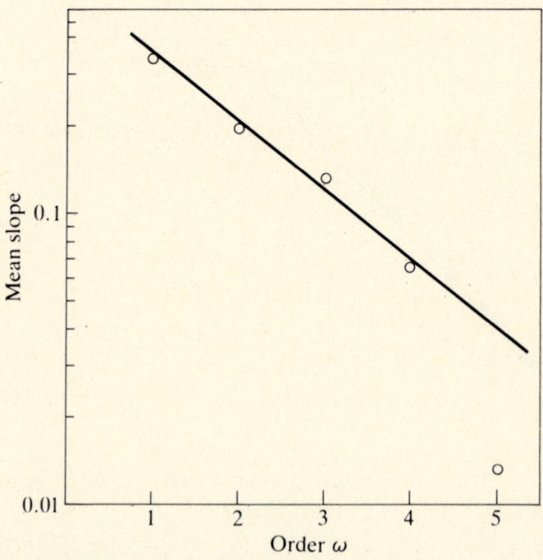

FIGURE 12.10 Slope versus order for the Squannacook Basin in New Hampshire. Illustration of the law of stream slopes.

lieved to be due to notions of balance or dynamic equilibrium in the expenditure of potential energy during the movement of river sediments and water, a topic of current research. The recent work of Gupta and Waymire [1989] and Tarboton et al. [1989] has related the scaling of slopes to notions of self-similarity and multi-scaling.

Relationship between Hillslope and Stream Slopes

Section 12.1 alluded to the close relationship that must exist between hillslopes and streams. Strahler [1950] suggested that the average maximum valleyside slope is related to the slope of the immediately adjacent stream channel by

$$\log G = 0.6 + 0.8 \log S\,, \tag{12.24}$$

where G is the average maximum valleyside slope and S is the adjacent stream slope. The above was empirically derived from nine regions of variable relief, climate, and rock type. No theoretical equivalent to Eq. (12.24) exists. Figure 12.11 illustrates a topography (elevation contours) generated from stream-slope data using Eq. (12.23).

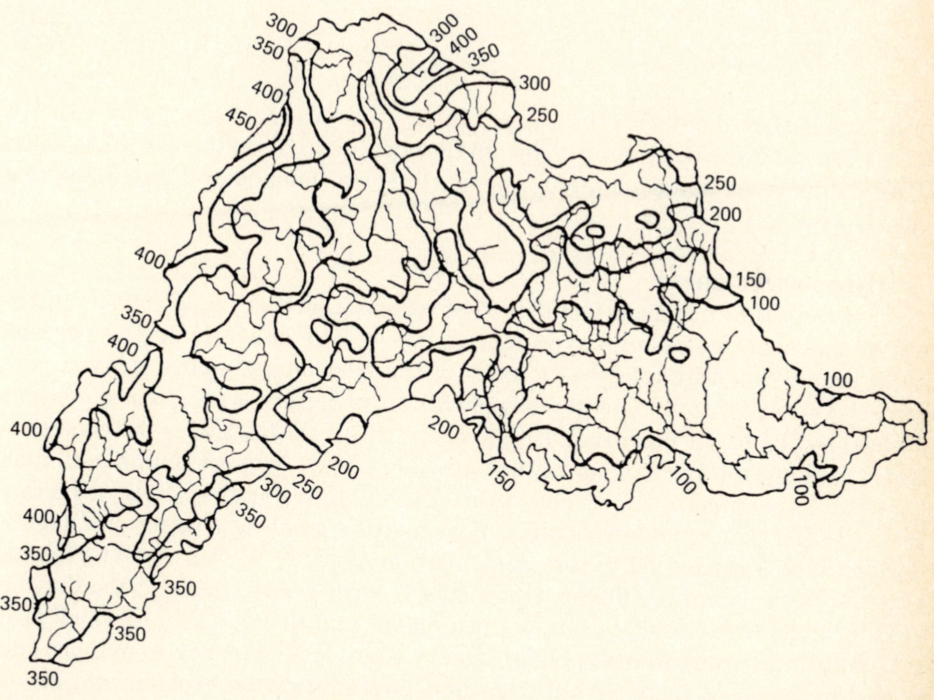

FIGURE 12.11 Artificial elevation contours generated for the Souhegan River Basin in New Hampshire. Source: Wyss [1988].

Measures of Relief

Relief measures elevation (or elevation differences) in the basin. It is an indicator of the potential energy of the water being drained by the system. Maximum basin relief is defined as the difference in elevations between the basin mouth and the highest point in the basin perimeter. Another definition may use the average elevation along the basin divide or perimeter in order to avoid the effect of local, unrepresentative peaks. Other definitions of where elevations are measured are possible.

A relief ratio is defined as the division of the relief measure (elevation) by the horizontal length over which it was calculated. Strahler [1964] suggests the use of a ruggedness number defined as the product of relief H and drainage density D. This dimensionless measure is like a slope. Note that if a local relief is defined as the difference between the highest point on a hillslope and the adjacent stream, along a line perpendicular to the contours, and Eq. (12.17) is used to represent the mean hillslope length, then a local slope would be

$$G = H \times 2D . \tag{12.25}$$

Strahler [1958] defines a geometry number as

$$K = \frac{HD}{G} \tag{12.26}$$

and argues that it falls in the range 0.4 to 1 in various regions, not exactly 0.5 as Eq. (12.25) would imply. The small variability nevertheless hints, once again, that river basin elements, particularly hillslopes and channels, are closely related.

Hypsometric Curve

To study the topography of drainage basins, Langbein et al. [1947] introduced the nondimensional hypsometric curve. The curve plots the percent area (area divided by total basin area) of the basin found above a given percent elevation contour. The percent elevation is defined as a given elevation divided by the basin maximum relief H. Figure 12.12 illustrates the construction of a hypsometric curve. Figure 12.12(c) shows several possible shapes that the hypsometric curve may take. Davis [1899] suggested that basins evolved after some sudden tectonic uplifting and ensuing erosion and degradation. Such a model led to the identification of curve A with a young basin, curve B with a mature basin, and curve C with an old basin. Scheidegger [1987] rebuts this classification by arguing that uplifting is a continuous process and that throughout the basin history there is a tendency to balance the opposing forces of tectonic build-up and degradation by erosion and other mechanisms. He argues that "if a landscape shows any sort of permanent character, these two antagonistic processes are in dynamic equilibrium."

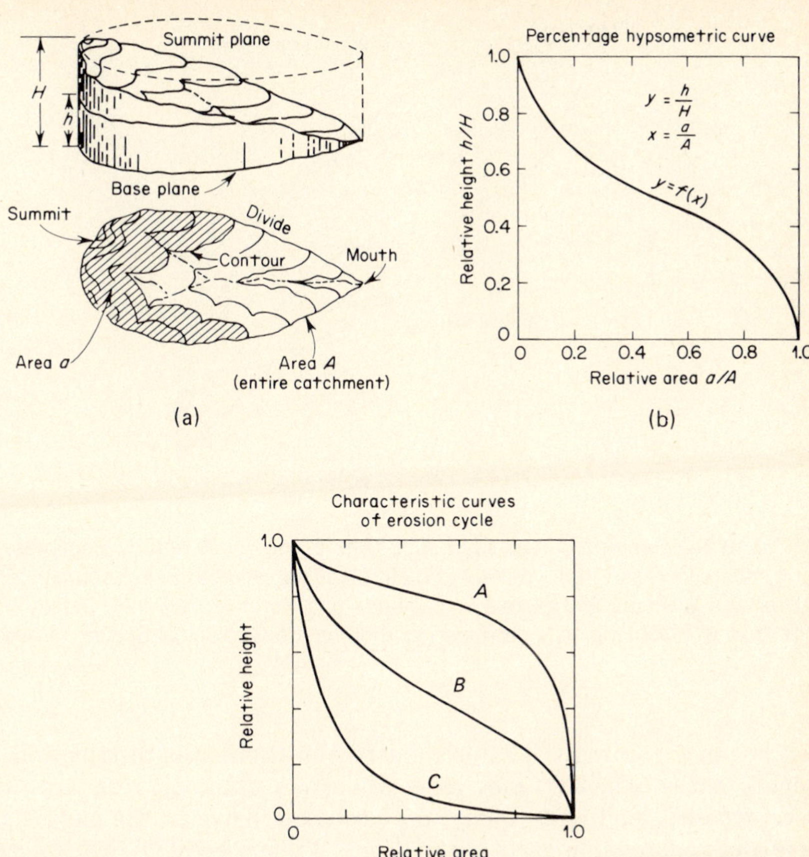

FIGURE 12.12 Construction of the hypsometric curve. Source: After A. N. Strahler, "Quantitative Analysis of Watershed Geomorphology," *Trans. Am. Geophys. Union* 63:1117–1142, 1957. Copyright by the American Geophysical Union.

Scheidegger then attributes the various shapes of the hypsometric curve to the levels of activity of the antagonistic processes. Curve A corresponds to high activity, curve B to medium activity, and curve C to low activity. The level of activity is not necessarily related to basin age.

A third interpretation is that the shape of the hypsometric curve is a measure of the relative equilibrium, in terms of tectonic uplift and erosional activity, of the basin. This may or may not be a measure of basin age. Curve A shows inequilibrium between opposing processes. The "mature" curve B corresponds to dynamic equilibrium of aggradation and degradation. Erosion continues during the equilibrium phase, leading to a general reduction

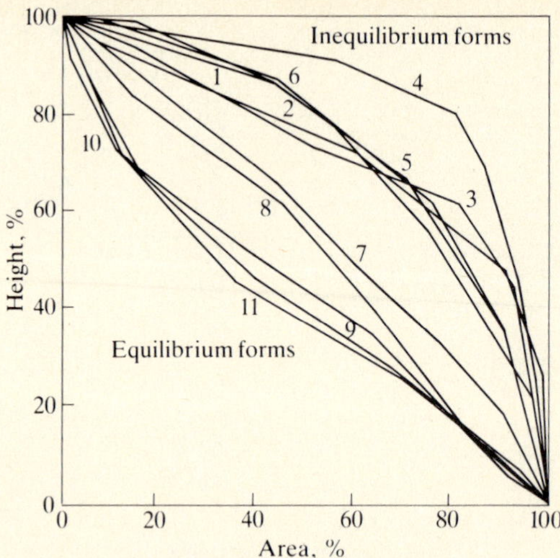

FIGURE 12.13 Hypsometric curves from subbasins in the Perth Amboy Badlands, New Jersey. Numbers increase from youthful to mature basins. Source: S. A. Schumm, "Evolution of Drainage Systems and Slopes in Badlands and Perth Amboy, New Jersey," *Geol. Soc. Am. Bull.* 67:615. Originally published in 1956 by the Geological Society of America.

in relief but maintaining the same relative elevation–area distribution. The Monadnock phase (curve C) may arise when rock outcrops, resistant to erosion, are exposed, leading to fairly large contrast between the elevations of erodable and nonerodable parts of the basin. Willgoose [1989], using a model of network growth, argues that the hypsometric curve shapes are as much a reflection of the tectonic history of the basin and the various erosive processes operating in the basin as they are a result of dynamic equilibrium or age. He disagrees with Scheidegger's interpretation. Figure 12.13 shows the hypsometric curves of basins of different ages, which according to Schumm [1956] illustrates the progression of shapes.

Link Concentration Function

Motivated by the search for the possible connection between runoff sediment production and basin elevation, Mesa [1986] and Gupta et al. [1986] introduced the link concentration function. In analogy to the width function, the link concentration function $N(h)$, $0 < h < H$, measures the number of links in the network at elevation h; in other words, the number of links cut by a particular contour line. Figure 12.14 shows the link concentration function for a particular basin.

The significance of the link concentration function is that, like the hypsometric curve, it addresses the elevation (energy) distribution within the network. The difference between the two is that the hypsometric curve

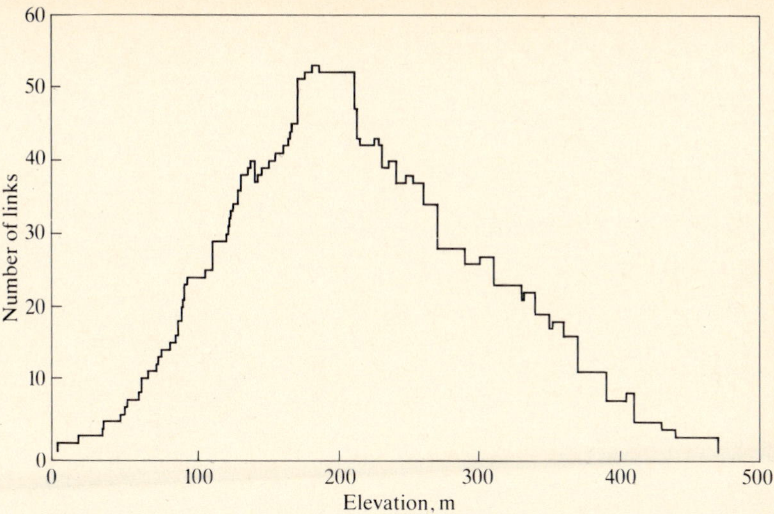

FIGURE 12.14 Link concentration function for the Agua Fria Basin in Venezuela. Source: V. K. Gupta and O. J. Mesa, "Runoff Generation and Hydrologic Response via Channel Network Geomorphology: Recent Progress and Open Problems," *J. Hydrol.* 102(1-4):3–28, 1988.

describes elevation relative to area while the link concentration function describes elevation relative to the channel network.

12.2.3 Stream Channel Geometry

Water and erosion not only shape the geometric arrangement of streams but also their cross sections. We have argued that the network develops to drain the basin in some efficient way. Similarly, the water and sediment conveyance properties of the channels must develop in a manner consistent with their ordered interconnections. Since from Manning's equation we know that conveyance is a function of slope and cross-sectional areas, we expect an interrelationship of these properties throughout the river network.

Leopold and Maddock [1953] performed the key studies of the hydraulic properties of streams. They studied how velocity, depth of flow in the channel, width of flow in the channel, sediment load, channel roughness (apparent Manning's roughness coefficient), and slope varied for different discharges. They studied discharges of the same frequency (probability of occurrence) at different points moving downstream in the basin. This is not necessarily equivalent to discharge resulting from the same event at different points; the results are only a reasonable indicator of what may occur in that situation. They also studied discharges of various frequencies at a point; hence how hydraulic properties at a point changed with varying discharges.

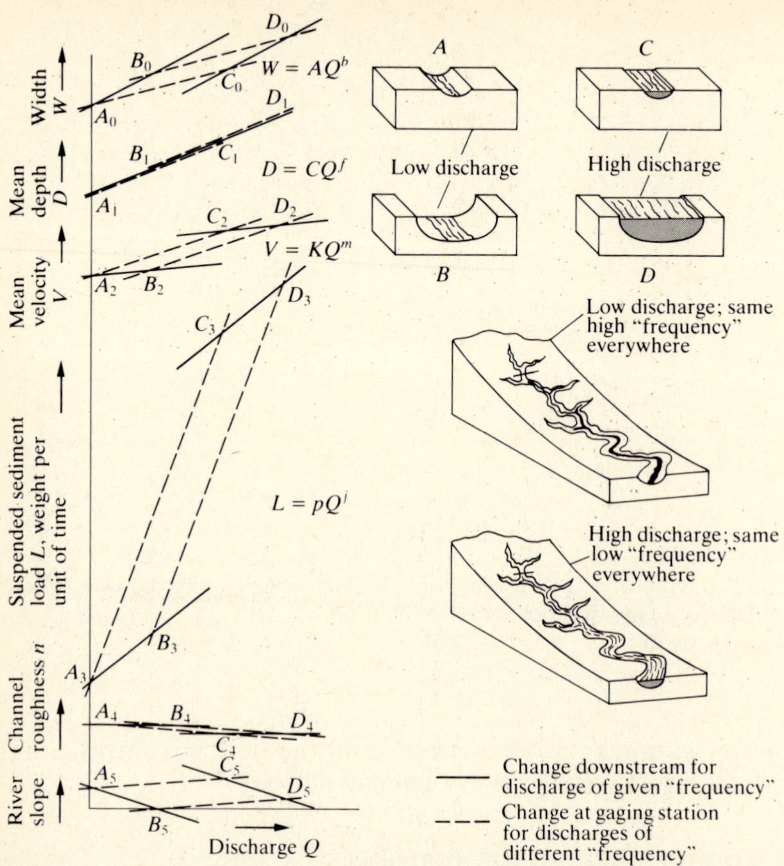

FIGURE 12.15 Variation of velocity, depth, width, sediment load, roughness, and slope in channels. All scales are logarithmic. Discharges of the same "frequency" are equaled or exceeded the same percentage of time. Source: Leopold and Maddock [1953].

Figure 12.15 gives their results. All scales are logarithmic, so the straight-line relationships imply power functions of discharge for all hydraulic variables studied. The velocity, width, and depth of flow are given by

$$V = KQ^m \qquad W = AQ^b \qquad D = CQ^f, \tag{12.27}$$

where Q is discharge; K, A, and C are proportionality constants; and m, b, and f are exponents from the slopes in Figure 12.15. Now the discharge is $Q \approx V \cdot W \cdot D$ (assuming a rectangular cross section). Hence Eq. (12.27) implies

$$m + b + f = 1$$

and

$$K \cdot A \cdot D = 1 .$$

For the discharges of equal frequencies, Leopold and Maddock [1953] report average values for semi-arid regions in the United States as

$$m = 0.1 \qquad b = 0.5 \qquad f = 0.4 .$$

For discharges at a point of different frequencies, they report

$$m = 0.34 \qquad b = 0.26 \qquad f = 0.4 .$$

It is interesting to note that the exponent for the velocity relationship with a given frequency discharge is very small (~0.1). This has been extrapolated to imply that for a given instant in time the velocity of flow anywhere in the basin is nearly constant. Other investigators (Pilgrim [1977]) have argued that to be approximately the case, based on simultaneous discharge observations in basins. This fact will be important in the development of the geomorphologic unit hydrograph below.

*12.3 FLUVIAL GEOMORPHOLOGY AND HYDROLOGY

Hydrologists have always tried to relate morphologic or topographic features of the basin to its response. Some of the synthetic unit hydrograph concepts of Chapter 9 as well as regionalization procedures of Chapter 11 are illustrations of these efforts. These generally statistical approaches have had varied success. They have always been geared to find the response of a particular basin with particular properties.

The latest hydrogeomorphologic efforts differ from past results in favoring a theoretical link between hydrology and geomorphology and in concentrating in finding the "ensemble" average response of a basin with given geomorphologic features.

The latter point has its origins in Shreve's [1967] view of river basins as particular outcomes from a set, the ensemble, of topologically distinct tree patterns with the given properties. For example, Shreve talks of the population of networks having a given magnitude (number of first-order streams) and a given Strahler order. The random model assumes that topologically distinct channel networks of the same magnitude occur with the same probability. When these networks are classified into classes of Strahler stream number $\{N_\omega\}$, the sets that have highest probability of occurrence are found to be those that correspond closely to Horton's laws. However, several workers (see Abrahams [1984] for a review) have noted small but consistent deviations from the random model.

Using the random topology model as a philosophical and theoretical cornerstone, some hydrologists argue that basin response functions in nature must generally correspond to the most probable basins. Furthermore, the response must be related to geomorphologic measures like the width function, magnitude, diameter, and the Horton numbers.

The theoretical link between basin response and geomorphology is usually based on the assumption of linearity and the existence of an instantaneous unit hydrograph as the ensemble mean response. The instantaneous unit hydrograph is interpreted as the probability distribution of water travel time in the basin. This interpretation was motivated in Section 12.2 when the width function was discussed. That this interpretation is correct is shown by Gupta et al. [1980].

Before proceeding to some of the existing results, it is useful to emphasize that the topologically random model of geomorphology and hence the hydrologic results derived from it is based on only two postulates given by Shreve [1967] and modified by Gupta and Waymire [1983] as

1. In the absence of environmental controls, natural networks are topologically random in the sense that sources and junctions occur independently of one another and each has a probability of occurrence equal to one-half. This can be shown to imply that all networks of a given magnitude are equally likely.
2. In drainage basins with uniform environments, interior and exterior link lengths, and the associated areas, are independent random variables and independent of location within a basin, with separate probability density functions.

Just the above two postulates allow the development of a consistent and beautiful theory of river basins. Details require far more than this book can afford at this stage. Nevertheless, the reader is urged to study this rich literature.

12.3.1 Geomorphologic Instantaneous Unit Hydrograph

Rodriguez-Iturbe and Valdes [1979] first introduced the geomorphologic instantaneous unit hydrograph, which led to the renewal of research in hydrogeomorphology. The concept was restated by Gupta et al. [1980]. The following is based on their presentation.

In order to determine the basin instantaneous unit hydrograph, let us consider the input as a unit volume composed of an infinite number of water drops. The following analysis will focus on the travel of one drop, chosen at random, through the basin. The drop travels through the basin making transitions from streams of lower order to those of higher order. A transition can be referred to as a change of state where the state is the order of the channel where the drop is travelling. The states of the process are defined to be the hillslope region a_ω or stream r_ω of order ω where the drop is located at time t. The travel of a drop is governed by the following rules:

Rule 1: When the drop is still in the hillslope phase, the state a_ω is of the order of the stream to which the land drains directly.

Rule 2: The only possible transitions out of state a_ω are into the corresponding r_ω. From r_ω transitions of the form $\omega \rightarrow j$ for some $j > \omega$ $(j = \omega + 1, \ldots, \Omega)$ are possible.

Rule 3: Defining the outlet as a trapping state $\Omega + 1$ the final state of the drop is $\Omega + 1$ from which transitions are impossible.

The above set of rules defines a finite set of possible paths that a drop falling randomly on the basin may follow to reach the outlet. For example, suppose that the basin of interest is of order 3 (Fig. 12.16), then the path space $S = \{s_1, s_2, s_3, s_4\}$ is given by

path s_1: $a_1 \rightarrow r_1 \rightarrow r_2 \rightarrow r_3 \rightarrow$ outlet

path s_2: $a_1 \rightarrow r_1 \rightarrow r_3 \rightarrow$ outlet

path s_3: $a_2 \rightarrow r_2 \rightarrow r_3 \rightarrow$ outlet

path s_4: $a_3 \rightarrow r_3 \rightarrow$ outlet

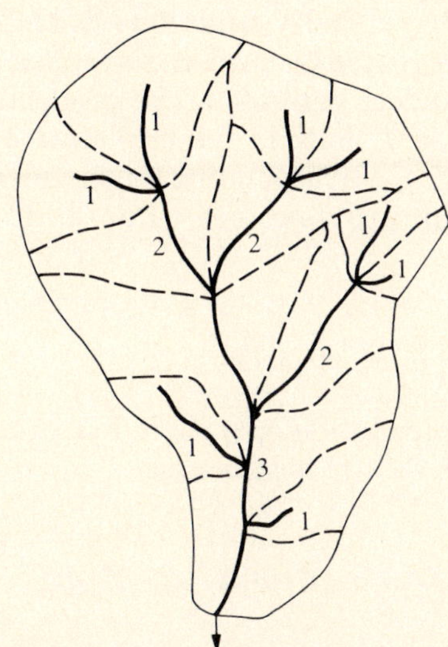

FIGURE 12.16 Basin of order 3 according to Strahler's ordering system. Source: I. Rodriguez-Iturbe and J. B. Valdes, "The Geomorphologic Structure of Hydrologic Response," *Water Resources Res.* 15(6):1409–1420, 1979. Copyright by the American Geophysical Union.

A drop of water is assumed to always fall on the hillslope, i.e., one of the initial states a_ω in a path. There is always an implied transition from the hillslope of order ω to the stream of order ω.

Following Gupta et al. [1980], the cumulative density function (Chapter 11) of the time a drop takes to travel to the basin outlet is given by

$$P(T_B \leq t) = \sum_{s \varepsilon S} P(T_s \leq t) P(s)\,, \tag{12.28}$$

where $P(\cdot)$ stands for the probability of the set given in parenthesis, T_B is the time of travel to the basin outlet, T_s is the travel time in a particular path s, $P(s)$ is the probability of a drop taking path s, and S is the set of all possible paths that a drop can take upon falling in the basin.

The travel time T_s in a particular path $a_i \rightarrow r_i \rightarrow r_j \rightarrow \cdots \rightarrow r_\Omega$, $i < j < \Omega$, must be equal to the sum of travel times in the elements of that path:

$$T_s = T_{a_i} + T_{r_i} + T_{r_j} + \cdots + T_{r_\Omega}\,, \tag{12.29}$$

where T_{a_i} is the travel time in a hillslope region and T_{r_i} is the travel time in a stream of order i, $i \varepsilon \{1, \ldots, \Omega\}$. Given the many hillslopes and streams of given orders and their various properties, the various times are taken as random variables with probability density functions $f_{T_{a_i}}(t)$ or $f_{T_{r_i}}(t)$, respectively. Furthermore, there is no reason to suspect the times are anything but independently distributed random variables. The probability density function of the total path travel time T_s is then the convolution of probability density functions $f_{T_{a_i}}(t)$ and $f_{T_{r_i}}(t)$, corresponding to the elements of path s. (This is a derived distribution problem; see Chapter 11, and Problem 17g in Chapter 11.)

$$f_{T_s}(t) = f_{T_{a_i}}(t) * f_{T_{r_i}}(t) * \cdots * f_{T_{r_\Omega}}(t)\,, \tag{12.30}$$

where $*$ stands for the convolution operation. For example, for path s_2 above, $a_1 \rightarrow r_1 \rightarrow r_3 \rightarrow$ outlet, $P(T_s \leq t)$ is given by

$$f_{T_{s_2}}(t) = \int_0^t \int_0^{t''} f_{T_{a_1}}(t') f_{T_{r_1}}(t'' - t')\, dt' f_{T_{r_3}}(t - t'')\, dt'' \tag{12.31}$$

and

$$P[T_{s_2} \leq t] = \int_0^t f_{T_{s_2}}(t)\, dt\,.$$

In their original development Rodriguez-Iturbe and Valdes [1979] ignored the travel time in the hillslope relative to the overall time that the drop spends in the basin. This will be addressed later. Equations (12.29) and (12.30) then simplify to

$$T_s = T_{r_i} + T_{r_j} + \cdots + T_{r\Omega} \tag{12.32}$$

$$f_{T_s}(t) = f_{T_{r_i}}(t) * f_{T_{r_j}}(t) * \cdots * f_{T_{r\Omega}}(t)\,. \tag{12.33}$$

The probability of following a given path s, $P(s)$, is given by

$$P(s) = \theta_i P_{ij} P_{jk} \cdots P_{\ell\Omega}\,, \tag{12.34}$$

where θ_i is the probability that the drop starts its travel in a hillslope segment draining into a stream of order i, P_{ij} is the transition probability from streams of orders i to streams of order j. Note that due to Rule 1, a drop initially falling in an area which drains to a stream of order i, goes to a stream of order i with probability 1.

Rodriguez-Iturbe and Valdes [1979] show that the initial state probabilities θ_i and the transition probabilities P_{ij} are functions only of the geomorphology and geometry of the river basin. The physical interpretation of the probabilities is as follows:

$$\theta_i = \frac{(\text{total area draining directly into streams of order } i)}{(\text{total basin area})} \tag{12.35}$$

$$P_{ij} = \frac{(\text{number of streams of order } i \text{ draining into streams of order } j)}{(\text{total number of streams of order } i)}\,. \tag{12.36}$$

The transition probabilities can be approximated as a function of the number of Strahler streams of each order N_i using the following general expression given by Gupta et al. [1980]:

$$P_{ij} = \frac{(N_i - 2N_{i+1})E(j,\Omega)}{\sum_{k=i+1}^{\Omega} E(k,\Omega)N_i} + \frac{2N_{i+1}}{N_i}\delta_{i+1,j} \qquad 1 \le i \le j \le \Omega\,, \tag{12.37}$$

where $\delta_{i+1,j} = 1$ if $j = i + 1$ and 0 otherwise. $E(i,\Omega)$ denotes the mean number of interior links of order i in a finite network of order Ω. The expression is given by Smart [1972]:

$$E(i,\Omega) = N_i \prod_{j=2}^{i} \frac{(N_{j-1} - 1)}{2N_j - 1}\,, \qquad i = 2,\ldots,\Omega\,. \tag{12.38}$$

An interior link is a segment of channel network between two successive junctions or between the outlet and the first junction upstream.

Similarly, the probability that a drop falls in an area of order ω can be approximated using the following general expression:

$$\theta_1 = \frac{N_1 \overline{A}_1}{A_\Omega}$$

$$\theta_\omega = \frac{N_\omega}{A_\Omega}\left[\overline{A}_\omega - \sum_{j=1}^{\omega-1} \overline{A}_j\left(\frac{N_j P_{j\omega}}{N_\omega}\right)\right], \qquad \omega = 2,\ldots,\Omega. \tag{12.39}$$

Equations (12.37) and (12.39) refer to ensemble properties of basins and hence will differ from statistics of any particular basin. van der Tak [1988] shows that Eqs. (12.37) and (12.39) are also slightly inaccurate. The problems arise because:

1. Equation (12.37) assumes that stream transitions from streams of order i to streams of order j are uniformly distributed between different drainage pathways $i \to j$, $j = 1,\ldots,\Omega$, in direct proportion to the number of streams of each order j and i.
2. Equation (12.39) is also inaccurate because it assumes that the area contributing to order j streams is uniformly distributed between different types of stream drainage paths.

Fortunately, the effects of the above approximations are relatively minor. Note that for a particular basin, Eqs. (12.35) and (12.36) could be evaluated exactly. The most serious difficulty with Eqs. (12.37) and (12.39) is that small negative probabilities may arise for θ_ω, particularly for the higher orders. Obviously this does not make sense and the distribution of θ_ω, $\omega = 1,\ldots,\Omega$ should be adjusted to eliminate the aberrant behavior.

Table 12.1 presents a complete list of the initial and transition probabilities for a third-order basin. By using Horton's bifurcation and area laws (Eqs. 12.6 and 12.8) to substitute for each N_ω and $\overline{A}_\omega$ in Eqs. (12.37) and (12.39), the initial and transition probabilities can be derived as functions only of R_A and R_B.

The probability function for a drop's travel time in a basin, $P(T_B \leq t)$, is now fully defined in terms of the geomorphologic basin properties and the probability functions $f_{T_{r_i}}(t)$ corresponding to the travel time of a drop in a given channel T_{r_i}. As previously stated, the instantaneous unit hydrograph is defined to be the probability density function of T_B and therefore

$$h_B(t) = dP(T_B \leq t)/dt = \sum_{s \varepsilon S} f_{T_{r_i}}(t) * \cdots * f_{T_{r_\Omega}}(t)P(s), \tag{12.40}$$

where $f_{T_{r_i}}(t)$ is the probability density function of T_{r_i}.

In summary, the instantaneous unit hydrograph is a function of the probability that a drop initially falls in an area that drains to a channel of a

TABLE 12.1 Initial and Transition Probabilities for a Third-Order Basin

$\Theta_1 = \dfrac{R_B^2}{R_A^2}$	$P_{12} = \dfrac{R_B^2 + 2R_B^2 - 2}{2R_B^2 - R_B}$
$\Theta_2 = \dfrac{R_B}{R_A} - \dfrac{R_B^3 + 2R_B^2 - 2R_B}{R_A^2(2R_B - 1)}$	$P_{13} = \dfrac{R_B^2 - 3R_B + 2}{2R_B^2 - R_B}$
$\Theta_3 = 1 - \dfrac{R_B}{R_A} - \dfrac{R_B^3 - 3R_B^2 + 2R_B}{R_A^2(2R_B - 1)}$	$P_{23} = 1$

Source: I. Rodriguez-Iturbe and J.B. Valdes, "The Geomorphologic Structure of Hydrologic Response," *Water Resources Res.* 15(6):1409–1420, 1979. Copyright by the American Geophysical Union.

given order, the transition probabilities from a stream of one order to another that are functions of the basin's geomorphologic characteristics R_A and R_B, and the travel time distribution in streams of a given order. Initial and transition probabilities provide a probabilistic description of the drainage network and a link between quantitative geomorphology and hydrology.

Critical to this work is the assumption usually made for each $f_{T_{r_i}}(t)$ in Eq. (12.40). Rodriguez-Iturbe and Valdes [1979] introduced the idea that the travel time in a channel or order ω obeys an exponential probability density function:

$$f_{T_{r_\omega}}(t) = K_\omega \exp(-K_\omega t), \tag{12.41}$$

where K_ω is a parameter of units t^{-1}, characteristic of channels or order ω. Since $1/K_\omega$ is also the mean travel time of the above distribution, Rodriguez-Iturbe and Valdes [1979] suggested estimating K_ω as

$$K_\omega = \frac{V}{\overline{L}_\omega}, \tag{12.42}$$

where V is a characteristic velocity, their "dynamic parameter." It was assumed that V was the same anywhere in the basin at any given time (see Section 12.2.3). Valdes et al. [1979] suggested that using an estimated peak velocity for any event in the basin for V is an adequate parameterization. Figures 12.17 and 12.18 show how the geomorphologic instantaneous unit hydrograph changes with parameters. Clearly there is extensive variety in possible geomorphologic instantaneous unit hydrograph shapes.

Equation (12.40) yields full analytical, but complicated, expressions for the instantaneous unit hydrograph $h(t)$. Rodriguez-Iturbe and Valdes [1979] suggested that it is adequate to assume a triangular instantaneous unit hy-

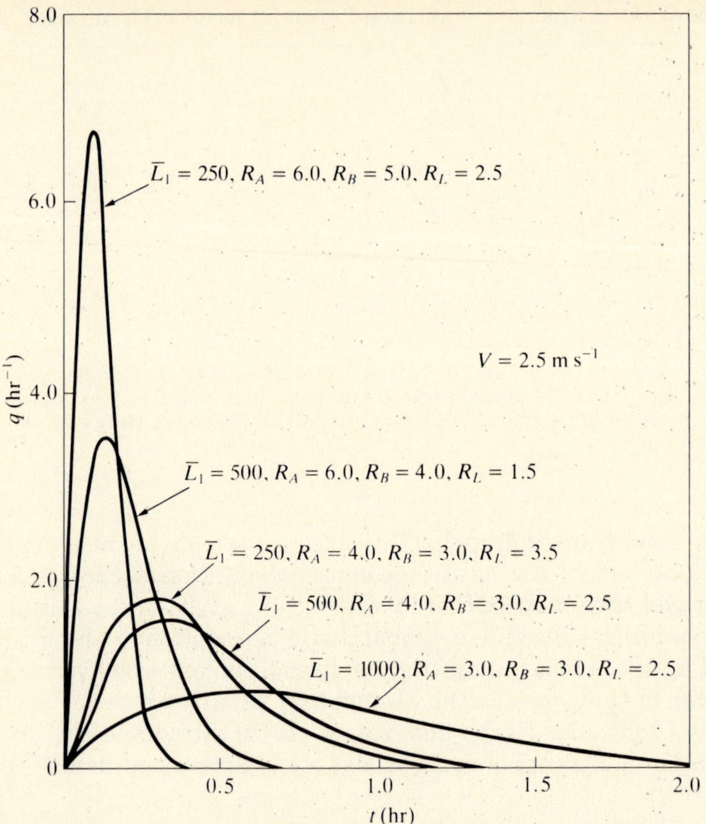

FIGURE 12.17 Changes in the geomorphologic instantaneous unit hydrograph when the velocity is kept constant and geomorphologic properties vary. Source: I. Rodriguez-Iturbe and J. B. Valdes, "The Geomorphologic Structure of Hydrologic Response," *Water Resources Res.* 15(6):1409–1420, 1979. Copyright by the American Geophysical Union.

drograph and only specify the time to peak and peak of $h(t)$. These characteristics have simple expressions obtained by regression of the peak and time to peak of the analytic solution to Eq. (12.40) for a wide range of parameters.

$$q_p = \frac{1.31}{L_\Omega} R_L^{0.43} V$$
$$t_p = \frac{0.44 L_\Omega}{V} \left(\frac{R_B}{R_A}\right)^{0.55} R_L^{-0.38}, \tag{12.43}$$

where L_Ω is the length in kilometers of the highest-order stream and V, as previously stated, is the expected peak velocity in meters per second. The peak q_p is given in units of inverse hours and t_p in hours.

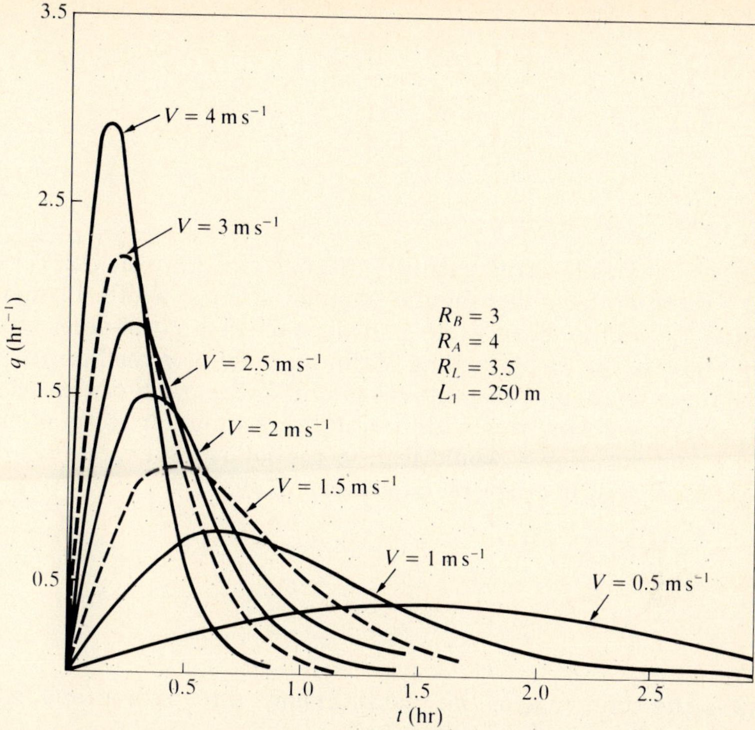

FIGURE 12.18 Changes in the geomorphologic instantaneous unit hydrograph when the geomorphologic characteristics are kept constant and the velocity varies. Source: I. Rodriguez-Iturbe and J. B. Valdes, "The Geomorphologic Structure of Hydrologic Response," *Water Resources Res.* 15(6):1409–1420, 1979. Copyright by the American Geophysical Union.

12.3.2 Geomorphoclimatic Instantaneous Unit Hydrograph

The difficulty with the geomorphologic instantaneous unit hydrograph is the dependence on the peak velocity V. This is a parameter that must be subjectively evaluated. Rodriguez-Iturbe et al. [1982a] rationalized that V must be a function of the effective rainfall intensity and duration and proceeded to eliminate V from the results. The details of how this is achieved are beyond the scope of this book. Nevertheless, the results are simple and easy to understand. Most useful is the restatement of q_p and t_p as

$$q_p = \frac{0.871}{\Pi_i^{0.4}}, \tag{12.44}$$

$$t_p = 0.585\Pi_i^{0.4}, \tag{12.45}$$

where

$$\Pi_i = \frac{L_\Omega^{2.5}}{i_r A_\Omega R_L \alpha_\Omega^{1.5}}, \tag{12.46}$$

$$\alpha_\Omega = \frac{1}{n b_\Omega^{2/3}} S_\Omega^{1/2}, \tag{12.47}$$

and i_r is the mean effective rainfall intensity. Equation (12.47) is just Manning's equation for a rectangular channel of large width. In that equation, b_Ω and S_Ω are the mean width and slope of the highest-order stream in the basin and n is the corresponding Manning roughness coefficient.

Assuming a triangular instantaneous unit hydrograph of characteristics q_p and t_p and convoluting this with a uniform rectangular storm of effective intensity i_r and duration t_r would lead to the following expressions for the peak of the discharge hydrograph (Henderson [1963]):

$$\frac{Q_p}{Q_e} = t_r q_p \left(1 - \frac{t_r q_p}{4}\right) \qquad t_r \leq t_c = t_B \tag{12.48}$$

$$Q_p = Q_e = i_r A \qquad t_r = t_c\,, \tag{12.49}$$

where t_B is the time base of the instantaneous unit hydrograph, which is equal to the basin time of concentration.

Using Eqs. (12.44) and (12.45) in the previous two equations leads to

$$Q_p = 2.42 \frac{i_r A_\Omega t_r}{\Pi_i^{0.4}} \left(1 - \frac{0.218 t_r}{\Pi_i^{0.4}}\right). \tag{12.50}$$

The time to this peak discharge is given by Rodriguez-Iturbe et al. [1982b] as

$$T_p = 0.585 \Pi_i^{0.4} + 0.75 t_r\,. \tag{12.51}$$

Equations (12.50) and (12.51) are potentially extremely useful results. They imply a rainfall–runoff relationship that is not dependent on calibration with input–output data and is theoretically only dependent on geomorphologic and climatic data, hence the name geomorphoclimatic instantaneous unit hydrograph. In fact, since the geomorphoclimatic instantaneous unit hydrograph is now dependent on the input i_r, it departs from the linearity assumptions of the traditional theory.

The concepts presented in this section are certainly speculative and on the borderline of research. Nevertheless, results up to now have been promising. For example, Figure 12.19 plots the peak discharge for a real basin obtained with a sophisticated, nonlinear, spatially distributed, model of the

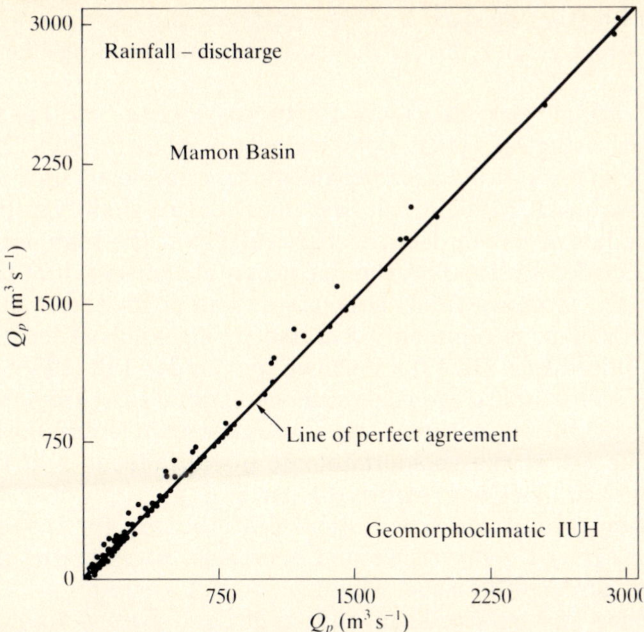

FIGURE 12.19 Comparison of peak discharge predictions from the geomorphoclimatic instantaneous unit hydrograph and a nonlinear, distributed, rainfall–runoff model. Source: I. Rodriguez-Iturbe, M. Gonzalez-Sanabria, and G. Caamaño, "On the Climatic Dependence of the IUH: A Rainfall–Runoff Analysis of the Nash Model and the Geomorphoclimatic Theory," *Water Resources Res.* 18(4):887–903, 1982. Copyright by the American Geophysical Union.

runoff–discharge process versus the results of the lumped geomorphoclimatic instantaneous unit hydrograph model with no calibration. Perfect agreement would imply points on a 45-degree line. The results of the many experiments in the figure are certainly very good. Predictions of time to peak discharge are similarly good. Extensions to the hydrogeomorphology theory attempt to account for infiltration in overland and stream segments, avoiding the need to estimate a priori effective rainfall (Diaz-Granados et al. [1984]; Hebson and Wood [1982]).

12.3.3 Comments and Further Developments of the Geomorphologic Instantaneous Unit Hydrograph

The geomorphologic instantaneous unit hydrograph described in the past sections is not without problems. Gupta and Waymire [1983] argue that the use of Strahler ordering and Horton numbers as a way of parameterizing the geomorphologic instantaneous unit hydrograph can be problematic. Roughly, the argument is that these characterizations are too rough and coarse to specify the response function adequately. They suggest that an approach based on the network links, rather than streams, as a basic building block would be more natural.

In their original work, Rodriguez-Iturbe and Valdes [1979] parameterized stream lengths using R_L. This is particularly obvious in Eq. (12.43) for the peak and time to peak of the geomorphologic instantaneous unit hydrograph. Bowden and Wallis [1964] point out that many basins show significant deviations from the law of stream lengths, particularly at the high-order streams. This is compounded by the fact the definition of the length of the highest-order stream L_Ω is many times subjective. The downstream cutoff of the highest-order stream is commonly dictated by streamgage location or map availability. Indeed Eqs. (12.43) are shown by van der Tak [1988] to have the tendency to underestimate the peak and overestimate the time to peak relative to parameterizations not invoking Horton's stream length law. In fact the generic description of the geomorphologic instantaneous unit hydrograph given in Eq. (12.40) does not require the use of R_L.

The assumption of an exponential distribution for travel times in Eq. (12.41) has raised many issues and questions. Kirshen and Bras [1983], among others, discuss the issue. The exponential assumption is convenient because it allows an analytical convolution of Eq. (12.40). Gupta et al. [1980] cite the solution from Feller [1971]. The exponential assumption also makes the geomorphologic instantaneous unit hydrograph mathematically equivalent, not physically, to the response of a conceptual instantaneous unit hydrograph (Chapter 9) consisting of linear storage elements in parallel and in series, much like the model of Diskin in Chapter 9. Since the travel time is effectively the stream length divided by a velocity, the basic question is what is the distribution of lengths in a Strahler ordering system. Gupta and Mesa [1988] point out that the theoretical answer to this remains to be found. van der Tak [1988] argues that the gamma distribution best represents Strahler's lengths and hence should be used for the travel time distribution. In a comparison (Fig. 12.20) of geomorphologic instantaneous unit hydrographs derived with exponential and gamma (shape parameter 2) to a data-based instantaneous unit hydrograph for the Souhegan Basin in New Hampshire, the latter gave better results.

The assumption that the hillslope travel time can be ignored is tenuous at best. The water velocity in the hillslope is so small that even though travel lengths are shorter, the travel time can be significant. Quoting from Troendle [1985], "velocity of flow increases by an order of magnitude or more once water is in the channel regardless of the pathway by which it arrived. A key to hydrological response, then, is how far water has to travel (slope length) to get to or influence channel flow and the mechanics by which it is transferred." The latter point in the quote is crucial to the introduction of a hillslope component in the geomorphologic instantaneous unit hydrograph. Conceptually, all that is needed is the introduction of a form for the hillslope travel time $f_{T_{a_i}}(t)$ (or length if a fixed velocity is assumed) in Eq. (12.30).

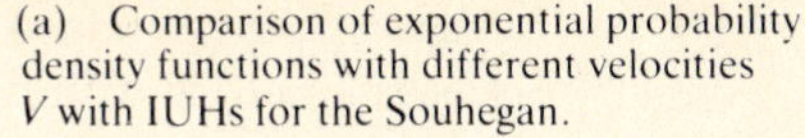

(a) Comparison of exponential probability density functions with different velocities V with IUHs for the Souhegan.

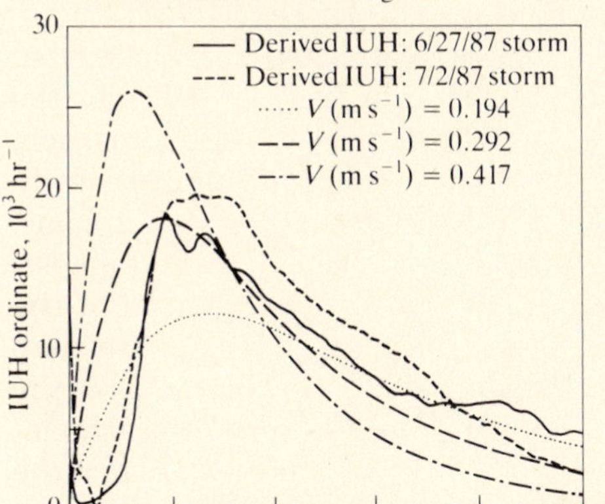

(b) Comparison of gamma probability density functions with shape parameter 2 and different velocities V with IUHs for the Souhegan.

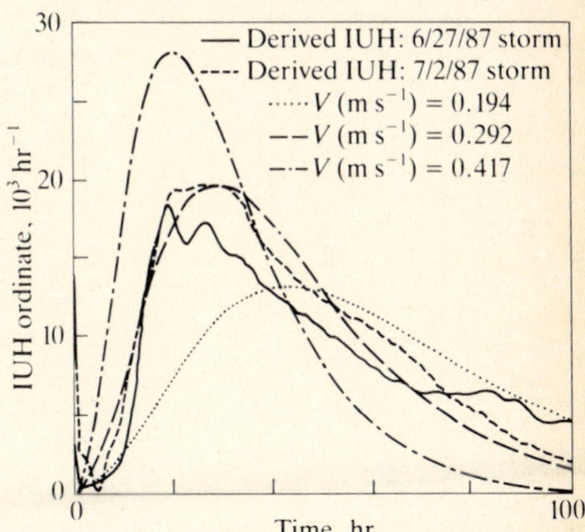

FIGURE 12.20 Comparison of the geomorphologic instantaneous unit hydrograph with stream travel times distributed exponentially or as gamma functions. Also shown is the instantaneous unit hydrograph derived from data for the Souhegan River Basin in New Hampshire. IUH denotes instantaneous unit hydrograph. Source: van der Tak [1988].

Based on studies of two basins in New England and on some theoretical arguments, Wyss [1988] concludes that the travel distance of any point in a hillslope to the closest stream is exponentially distributed with the mean distance given by Eq. (12.17), which says that the mean hillslope has a length equal to the inverse of twice the local drainage density. In other words, Wyss [1988] results argue for

$$f_{T_{a_i}}(t) = K_{a_i} e^{-K_{a_i} t}, \tag{12.52}$$

with $K_{a_i} = 2D_i V_h$ and where D_i is the local drainage density (it could vary with order) and V_h is a parameter representing velocity in the hillslopes.

van der Tak [1988] used Eq. (12.52) in the general geomorphologic instantaneous unit hydrograph formulation (Eqs. 12.30 and 12.40). He also compared this approach to assuming that the hillslope travel time $f_{T_{a_i}}(t)$ is gamma distributed with shape parameter $a = 2$ and scale parameter $b = a2D_i V_h$. Figure 12.21 gives the results together with a data-based instantaneous unit hydrograph derived for the Souhegan River Basin in New Hampshire. van der Tak [1988] suggests a method-of-moments approach for the estimation of channel V and hillslope V_h velocities. More work is needed before a conclusive statement on hillslope travel time is made. Nevertheless, it is encouraging that Figure 12.21 shows a very good fit of data with channel and hillslope velocities that agree with typical values observed in nature. Wyss [1988] quotes an estimated channel velocity (via measurement) in the

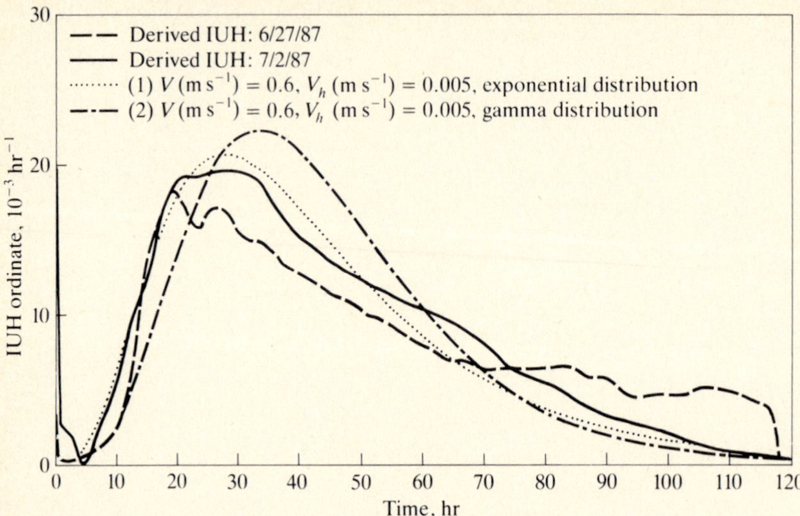

FIGURE 12.21 Geomorphologic instantaneous unit hydrograph with hillslope component compared to instantaneous unit hydrograph of the Souhegan River Basin in New Hampshire. IUH denotes instantaneous unit hydrograph. Source: van der Tak [1988].

Souhegan of 0.6 m s^{-1}. The representative hillslope velocity of 0.005 m s^{-1} is consistent with observed soil macropore velocities in hillslopes by Beven and Germann [1982]. For basins where overland flow (associated with Hortonian runoff production) is significant, Mesa and Mifflin [1986] suggest to split the overland response function into slow and fast components, each weighted by a proportion of the precipitation following each path. The sum of the two hillslope responses is then convoluted with the stream response.

12.3.4 Link-Based Derivations of the Geomorphologic Instantaneous Unit Hydrograph

In introducing the width function $N(x)$, Section 12.2.1 hinted at how it can be related to the instantaneous unit hydrograph. Gupta et al. [1986] and Mesa and Mifflin [1986] formulate a generic expression for the geomorphologic instantaneous unit hydrograph as a function of the width function:

$$h(t) = \int_0^{\infty} g(x,t)N(x)Z^{-1}\,dx\,, \tag{12.53}$$

where $g(x,t)$ is defined as the hydraulic response function of a single channel a distance x from the basin outlet. In other words, a unit impulse a distance x away from the outlet will result in an outflow given by $g(x,t)$. The area under $N(x)$ is represented by Z. Equation (12.53) can then be interpreted as the average of $g(x,t)$, where the probability density of channels of length x is given by $N(x)Z^{-1}$.

Equation (12.53) can be used given $N(x)$ for a particular basin and a function $g(x,t)$. The function $g(x,t)$ can be any linear routing procedures presented in Chapter 10 or any other. Nevertheless, in the spirit of the topologically random model of channel networks, Troutman and Karlinger [1984, 1985, 1986] and Karlinger and Troutman [1985] studied approximations to Eq. (12.53) that would require specification of a finite number of topologic features rather than the complete width function. Specifically, they suggested approximating $h(t)$ by its mean value conditional on a topological parameter vector $\boldsymbol{\lambda}$. That is, the approximation of h is

$$\hat{h}(t) = E[h(t)\,|\,\boldsymbol{\lambda}]. \tag{12.54}$$

To obtain $\hat{h}(t)$, Gupta and Mesa [1988] approximate the area under the link concentration function by

$$Z = (M - 1)\bar{\ell}_i + M\bar{\ell}_e \simeq \frac{2M(\bar{\ell}_i + \bar{\ell}_e)}{2} = 2M\bar{\ell}, \tag{12.55}$$

where $\bar{\ell}_i$ and $\bar{\ell}_e$ are the mean length of interior and exterior links, respectively. There are $M - 1$ interior links and M exterior links, where M is the basin magnitude. The overall mean link length is $\bar{\ell}$. Substituting in Eq. (12.53) and taking expected values yields

$$E[h(t)\,|\,\boldsymbol{\lambda}] \approx \int_0^\infty g(x,t)\frac{E[N(x)\,|\,\boldsymbol{\lambda}]}{2M\bar{\ell}}\,dx. \tag{12.56}$$

Troutman and Karlinger [1984, 1986] evaluate $E[N(x)\,|\,\boldsymbol{\lambda}]$ under the assumption of topologically random networks for a variety of parameter vectors $\boldsymbol{\lambda}$. These included M, $\{M, D\}$, $\{M, \Omega\}$, $\{\tilde{D}\}$, and $\{M, \tilde{D}\}$, where network magnitude M, diameter D, and Strahler order Ω have been previously defined. The mainstream length $\tilde{D}$ is the maximum actual length along the longest path from a source to the outlet. The deviation of the various conditional mean values of the link function is not easy. One particularly nice result corresponds to $\boldsymbol{\lambda} = M$. In that case, for large values of M the asymptotic result is

$$\frac{E[N(x)\,|\,M]}{2M\bar{\ell}} = \frac{X}{2M\bar{\ell}_i^2}\exp\left(\frac{-X^2}{4M\bar{\ell}_i^2}\right). \tag{12.57}$$

The above is a Weibull probability density function, that looks much like a gamma probability density function. Particularly illustrative is the case when the hydraulic response function is a simple translation,

$$g(x,t) = \delta\left(t - \frac{x}{V}\right), \tag{12.58}$$

where δ is a Dirac delta function and V is a velocity of translation. Substitution of Eqs. (12.57) and (12.58) into (12.56) and letting $x = Vt$ yields, for large M,

$$E[h(t)|M] = \frac{t}{2M(\overline{\ell}_i/V)^2}\exp\left(\frac{-V^2t^2}{4M\overline{\ell}_i^2}\right); \qquad t > 0. \tag{12.59}$$

Figure 12.22 shows Eq. (12.59) as evaluated by Karlinger and Troutman [1985] for the Blue Creek Basin in Alabama. Also shown is the nonasymptotic evaluation of Eq. (12.53) for that basin. Given that the magnitude of the Blue Creek Basin is only 13, the closeness of the two results is surprising. Troutman and Karlinger [1985] show that asymptotically, Eq. (12.59) is always the answer, irrelevant of the form of $g(x,t)$. The only difference is that V is interpreted as the mean celerity in an interior link, which will vary with the routing procedure. The result is also independent of the distribution of link lengths and the exterior link lengths. They point out that $(\overline{\ell}_i/V)$ is then a characteristic scaling parameter that controls the hydrologic response. All basins with the same parameter $\overline{\ell}_i/V$ will have the same response, a powerful concept reminiscent of the geomorphoclimatic parameter Π_i.

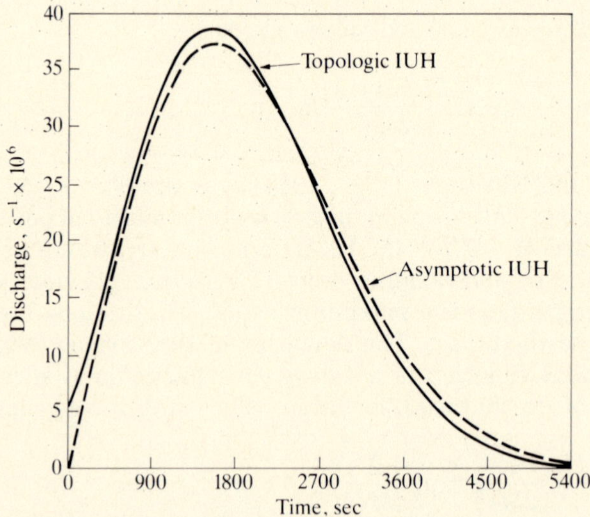

FIGURE 12.22 The asymptotic instantaneous unit hydrograph based on the mean width function conditional on basin magnitude, for the Blue Creek Basin in Alabama. IUH denotes instantaneous unit hydrograph. Source: M. R. Karlinger and B. M. Troutman, "Assessment of the Instantaneous Unit Hydrograph Derived from the Theory of Topologically Random Networks," *Water Resources Res.* 21(11):1693–1702, 1985. Copyright by the American Geophysical Union.

12.4 SUMMARY

Nature shapes river basins in an orderly and organized manner. The river basin must reflect the interdependence of geology, soils, vegetation, topography, and climate. Hydrologists and other earth scientists have but scratched the surface in trying to explain the way nature operates and why river basins come to be. This chapter has attempted to summarize a lot of fluvial geomorphology and its connection to hydrology. These are topics that require long study for deep understanding; nevertheless, this chapter provides a solid cornerstone to begin that study.

A lot of results of this chapter, particularly, in relation to the concept of the geomorphologic instantaneous unit hydrograph, are tempting to the practitioner. Keep in mind that these results are still actively developing research subjects and most of them refer to average responses over an ensemble of basins with given properties. Although many successful uses can be documented, it would be wrong to perceive these results as substitutions for more traditional tools.

It is fitting to end this book with this chapter because it embodies the excitement of all that is yet to be discovered in hydrology. Besides the obvious relationships between geomorphology, topography, geology, soil, climate, and hydrology, some specific questions that arise are

1. Is there organization in the production of runoff? This chapter always assumes that a runoff is available to the geomorphologic instantaneous unit hydrograph. Are there "laws" of runoff production closely intertwined with the hillslope properties and the channel network?
2. Do scaling properties in the geomorphology translate to scaling properties in the hydrology? What are the nondimensional numbers, analogous to those in fluid mechanics, that characterize hydrologic response? Even though that may be a futile search, it should shed a lot of light on the nature of the processes involved.
3. What are the links of hydrology at the basin scale (as discussed in this chapter) with reductionist hydrology? That is, what is the relationship between behavior and its quantification on large and small scales?

REFERENCES

Abrahams, A. D. [1984]. "Channel Networks: A Geomorphologic Perspective." *Water Resources Res.* 20(2):161–168.

Band, L. E. [1986]. "Topographic Partition of Watersheds with Digital Elevation Models." *Water Resources Res.* 22(1):15–24.

Beven, K., and P. Germann [1982]. "Macropores and Water Flow in Soils." *Water Resources Res.* 18(5):1311–1325.

Beven, K. J., and E. F. Wood [1983]. "Catchment Geomorphology and the Dynamics of Runoff Contributing Areas." *J. Hydrol.* 65:139–158.

Bowden, K. L., and J. R. Wallis [1964]. "Effect of Stream Ordering in Horton's Law of Drainage Composition." *Geol. Soc. Am. Bull.* 75:767–774.

Broscoe, A. J. [1959]. "Quantitative Analysis of Longitudinal Stream Profiles of Small Watersheds." Office of Naval Research, Project NR 389-042. New York: Columbia University Department of Geology. (Technical report no. 18.)

Brunsden, D. [1980]. "Applicable Models of Long Term Landform Evolution." *Geomorph. Supp.* 36:16–26.

Calver, A., M. J. Kirkby, and D. R. Weyman [1972]. "Modelling Hillslope and Channel Flows." In: R. J. Chorley, ed. *Spatial Analysis in Geomorphology.* London: Methuen, pp. 197–218.

Chorley, R. J., and P. F. Dale [1972]. "Cartographic Problems in Stream Channel Delineation." *Cartography.* 7:150–162.

Culling, W. E. H. [1986]. "On Hurst Phenomena in the Landscape." *Trans. Japanese Geomorphol. Union.* 7(4):221–243.

Culling, W. E. H., and M. Datko [1987]. "The Fractal Geometry of the Soil-Covered Landscape." *Earth Surfaces Processes and Landforms.* 12:369–385.

Dacey, M. F., and W. C. Krunbein [1976]. "Three Growth Models for Stream Channel Networks." *J. Geol.* 84:153–163.

Davis, W. M. [1899]. "The Geographical Cycle." *Geogr. J.* 14:481–504. In: W. M. Davis. *Geographical Essays.* Boston: Ginn [1909].

Diaz-Granados, M. A., J. B. Valdes, and R. L. Bras [1984]. "A Physically Based Flood Frequency Distribution." *Water Resources Res.* 20(7):995–1002.

Dietrich, W. E., D. R. Montgomery, S. L. Reneau, and P. Jordan [1988]. "The Use of Hillslope Convexity to Calculate Diffusion Coefficients for a Slope Dependent Transport Law." *EOS.* 69(16).

Eagleson, P. S. [1970]. *Dynamic Hydrology.* New York: McGraw-Hill.

Feller, W. [1971]. *Introduction to Probability Theory and Its Applications.* New York: Wiley.

Flint, J. J. [1974]. "Stream Gradient as a Function of Order, Magnitude and Discharge." *Water Resources Res.* 10:969–973.

Glock, W. S. [1931]. "The Development of Drainage Systems: A Synoptic View." *Geogr. Rev.* 21:475–482.

Gregory, K. J., and D. E. Walling [1968]. "The Variation of Drainage Density within a Catchment." *Int. Assoc. Sci. Hydrol. Bull.* 13:61–68.

Gupta, V. K., and O. J. Mesa [1988]. "Runoff Generation and Hydrologic Response via Channel Network Geomorphology: Recent Progress and Open Problems." *J. Hydrol.* 102(1-4):3–28.

Gupta, V. K., and E. Waymire [1983]. "On the Formulation of an Analytical Approach to Hydrologic Response and Similarity at the Basin Scale." *J. Hydrol.* 65:95–123.

Idem. [1989]. "Statistical Self-Similarity in River Networks Parameterized by Elevation." *Water Resources Res.* 25(3):463–476.

Gupta, V. K., E. Waymire, and I. Rodriguez-Iturbe [1986]. "On Scales, Gravity and Network Structure in Basin Runoff." In: V. Gupta, I. Rodriguez-Iturbe, and E. Wood, eds. *Scale Problems in Hydrology.* Dordrecht, Holland: D. Reidel, pp. 159–184.

Gupta, V. K., E. Waymire, and C. T. Wang [1980]. "Representation of an Instantaneous Unit Hydrograph from Geomorphology." *Water Resources Res.* 16(5):855–862.

Hebson, C., and E. F. Wood [1982]. "A Derived Flood Frequency Distribution Using Horton Order Ratios." *Water Resources Res.* 18(5):1509–1518.

Henderson, F. M. [1963]. "Some Properties of the Unit Hydrograph." *J. Geophys. Res.* 68(16):4785–4793.

Horton, R. E. [1932]. "Drainage Basin Characteristics." *Trans. Amer. Geophys. Union.* 13:350–361.

Idem. [1945]. "Erosional Development of Streams and Their Drainage Basins: Hydrophysical Approach to Quantitative Morphology." *Bull. Geol. Soc. Am.* 56:275–370.

Hugget, R. J. [1985]. *Earth Surface Systems.* New York: Springer-Verlag.

Idem. [1988]. "Dissipative Systems: Implications for Geomorphology." *Earth Surface Processes and Landforms.* 13:45–49.

Jarvis, R. S., and M. J. Woldenberg [1984]. *River Networks.* Stroudsburg, Pa.: Hutchinson Ross.

Karlinger, M. R., and B. M. Troutman [1985]. "Assessment of the Instantaneous Unit Hydrograph Derived from the Theory of Topologically Random Networks." *Water Resources Res.* 21(11):1693–1702.

Kirkby, M. J. [1976]. "Tests of the Random Network Model, and Its Application to Basin Hydrology." *Earth Surface Processes.* 1:197–212.

Kirshen, D. M., and R. L. Bras [1983]. "The Linear Channel and Its Effect on the Geomorphologic IUH." *J. Hydrol.* 65:175–208.

LaBarbera, P., and R. Rosso [1987]. "Fractal Geometry of River Networks." *EOS Trans. Am. Geophys. Union.* 68(44):1276.

Langbein, W. B. [1964]. "Geometry of River Channels." *J. Hydraul. Div. A.S.C.E.* HY2:301–312.

Langbein, W. B., et al. [1947]. "Topographic Characteristics of Drainage Basins." U.S. Geological Survey water supply paper no. 968-C. Washington, D.C.

Leopold, L. B., and W. B. Langbein [1962]. "The Concept of Entropy in Landscape Evolution." U.S. Geological Survey professional paper no. 500-A.

Leopold, L. B., and T. Maddock, Jr. [1953]. "The Hydraulic Geometry of Stream Channels and Some Physiographic Implications." U.S. Geological Survey professional paper no. 252. Washington, D.C.

Leopold, L. B., and J. P. Miller [1956]. "Ephemeral Streams—Hydraulic Factor and Their Relation to the Drainage Net." U.S. Geological Survey professional paper no. 282-A, pp. 16–24.

Liao, K. H., and A. E. Scheidegger [1968]. "A Computer Model for Some Branching-Type Phenomena in Hydrology," *Int. Assoc. Sci. Hydrol. Bull.* 13:5–13.

Mandelbrot, B. B. [1975]. "Stochastic Models for the Earth's Relief, the Shape and the Fractal Dimensions of the Coastlines, and the Number-Area Rule for Islands." *Proc. Natl. Acad. Sci. USA.* 72(10):3825–3828.

Idem. [1983]. *The Fractal Geometry of Nature.* San Francisco: Freeman.

Mark, D. M., and P. B. Aronson [1984]. "Scale-Dependent Fractal Dimensions of Topographic Surfaces: An Empirical Investigation with Applications in Geomorphology and Computer Mapping." *Math Geol.* 16(7):671–683.

Melton, M. A. [1958]. "Geometric Properties of Mature Drainage Systems and Their Representation in an E_4 Phase Space." *J. Geol.* 66:35–54.

Idem. [1959]. "A Derivation of Strahler's Channel-Ordering System." *J. Geol.* 67:345–346.

Mesa, O. J. [1986]. "Analysis of Channel Networks Parameterized by Elevation." University, Miss.: University of Mississippi. Ph.D. dissertation.

Mesa, O. J., and E. R. Mifflin [1986]. "On the Relative Role of Hillslope and Network Geometry in Hydrologic Response." In: V. Gupta, I. Rodriguez-Iturbe, and E. Wood, eds. *Scale Problems in Hydrology*. Dordrecht, Holland: D. Reidel.

Mock, S. J. [1971]. "A Classification of Channel Links in Stream Networks." *Water Resources Res.* 7:1558–1566.

Montgomery, D. R., and W. E. Dietrich [1988]. "Where Do Channels Begin?" *Nature.* 336:232–234.

Morisawa, M. [1964]. "Development of Drainage Systems on an Upraised Lake Floor." *Am. J. Sci.* 262:340–354.

O'Loughlin, E. M. [1986]. "Prediction of Surface Saturation Zones in Natural Catchments by Topographic Analysis." *Water Resources Res.* 22(5):794–804.

Pilgrim, D. H. [1977]. "Isochrones of Travel Time and Distribution of Flood Storage From a Tracer Study on a Small Watershed." *Water Resources Res.* 13(3):587–595.

Rodriguez-Iturbe, I., G. Devoto, and J. B. Valdes [1979]. "Discharge Response Analysis and Hydrologic Similarity: The Interrelation between the Geomorphologic IUH and Storm Characteristics." *Water Resources Res.* 5(6):1435–1444.

Rodriguez-Iturbe, I., M. Gonzalez-Sanabria, and R. L. Bras [1982a]. "A Geomorphoclimatic Theory of the Instantaneous Unit Hydrograph." *Water Resources Res.* 18(4):877–886.

Rodriguez-Iturbe, I., M. Gonzalez-Sanabria, and G. Caamaño [1982b]. "On the Climatic Dependence of the IUH: A Rainfall–Runoff Analysis of the Nash Model and the Geomorphoclimatic Theory." *Water Resources Res.* 18(4):887–903.

Rodriguez-Iturbe, I., and J. B. Valdes [1979]. "The Geomorphologic Structure of Hydrologic Response." *Water Resources Res.* 15(6):1409–1420.

Scheidegger, A. E. [1964]. "Some Implications of Statistical Mechanics in Geomorphology." *Int. Assoc. Hydrol. Sci. Bull.* 9(1):12–16.

Idem. [1987]. *Systematic Geomorphology.* Vienna: Springer-Verlag.

Schumm, S. A. [1956]. "Evolution of Drainage Systems and Slopes in Badlands and Perth Amboy, New Jersey." *Geol. Soc. Am. Bull.* 67:597–646.

Shreve, R. L. [1966]. "Statistical Law of Stream Numbers." *J. Geol.* 74:17–37.

Idem. [1967]. "Infinite Topologically Random Channel Networks." *J. Geol.* 75:178–186.

Idem. [1969]. "Stream Lengths and Basin Areas in Topologically Random Channel Networks." *J. Geol.* 77:397–414.

Smart, J. S. [1968]. "Statistical Properties of Stream Lengths." *Water Resources Res.* 4(5):1001–1014.

Idem. [1969]. "Topological Properties of Channel Networks." *Geol. Soc. Am. Bull.* 80:1757–1774.

Idem. [1972]. "Quantitative Characterization of Channel Network Structure." *Water Resources Res.* 8(6):1487–1496.

Smith, K. G. [1950]. "Standards for Grading Texture of Erosional Topography." *Am. J. Sci.* 248:655–668.

Strahler, A. N. [1950]. "Equilibrium Theory of Erosional Slopes Approached by Frequency Distribution Analysis." *Am. J. Sci.* 248:673–696.

Idem. [1952]. "Hypsometric (Area Altitude) Analysis of Erosional Topography." *Geol. Soc. Am. Bull.* 63:1117–1142.

Idem. [1957]. "Quantitative Analysis of Watershed Geomorphology." *Trans. Am. Geophys. Union.* 38(6):913–920.

Idem. [1958]. "Dimensional Analysis Applied to Fluvially Eroded Landforms." *Bull. Geol. Soc. Am.* 69:279–300.

Idem. [1964]. "Quantitative Geomorphology of Drainage Basins and Channel Networks." pp. 4.39–4.76. In: V. T. Chow, ed. *Handbook of Applied Hydrology*. New York: McGraw-Hill.

Surkan, A. J. [1968]. "Synthetic Hydrographs: Effects of Network Geometry." *Water Resources Res.* 5(1):112–128.

Tarboton, D. G., R. L. Bras, and I. Rodriguez-Iturbe [1988]. "The Fractal Nature of River Networks." *Water Resources Res.* 24(8):1317–1322.

Idem. [1989]. "Scaling and Elevation in River Networks." *Water Resources Res.* (In press.)

Thomas, J. B. [1983]. "Evolutionary Geomorphology." *Geography*. 68:225–235.

Tokunaga, E. [1978]. "Consideration on the Composition of Drainage Networks and Their Evolution." *Geographical Report No. 13*. Tokyo Metropolitan University.

Idem. [1984]. "Ordering of Divide Segments and Law of Divide Segment Numbers." *Trans. Jpn. Geomorphol. Union*. 5-2:71–77.

Troendle, C. A. [1985]. "Variable Source Area Models." In: M. G. Anderson and T. P. Burt, eds. *Hydrological Forecasting*. New York: Wiley, pp. 347–403.

Troutman, B. M., and M. R. Karlinger [1984]. "On the Expected Width Function for Topologically Random Channel Networks." *J. Appl. Prob.* 21:836–884.

Idem. [1985]. "Unit Hydrograph Approximation Assuming Linear Flow through Topologically Random Channel Networks." *Water Resources Res.* 21(5):743–754.

Idem. [1986]. "Averaging Properties of Channel Networks Using Methods in Stochastic Branching Theory." In: V. K. Gupta, I. Rodriguez-Iturbe, and E. F. Wood, eds. *Scale Problems in Hydrology*. Dordrecht, Holland: D. Reidel.

Valdes, J. B., Y. Fiallo, and I. Rodriguez-Iturbe [1979]. "A Rainfall Runoff Analysis of the Geomorphologic IUH." *Water Resources Res.* 15(6):1421–1434.

van der Tak, L. D. [1988]. "Part I: Stream Length Distributions, Hillslope Effects and Other Refinements of the Geomorphologic IUH; Part II: Topologically Random Channel Networks and Horton's Laws: The Howard Network Simulation Model Revisited." Cambridge, Mass.: MIT Department of Civil Engineering. Civil Engineer thesis.

Warntz, W. [1975]. "Stream Ordering and Contour Mapping." *J. Hydrol.* 25:209–227.

Werner, C., and J. S. Smart [1973]. "Some New Methods of Topologic Classification of Channel Networks." *Geogr. Anal.* 5:271–295.

Willgoose, G. [1989]. "A Physically Based Channel Network and Catchment Evolution Model." Cambridge, Mass.: MIT Department of Civil Engineering. Ph.D. thesis.

Wolman, M. G. [1955]. "The Natural Channel of Brandywine Creek, Pennsylvania." U.S. Geological Survey professional paper no. 271. Washington, D.C.

Idem. [1988]. "Magnitude and Frequency of Geomorphic Events: Matching Geography, Process and Form." *EOS*. 69(16).

Wyss, J. [1988]. "Hydrologic Modelling of New England River Basins Using Radar Rainfall Data." Cambridge, Mass.: MIT Department of Earth, Atmospheric, and Planetary Sciences. M.S. thesis.

Yang, C. T. [1971]. "Potential Energy and Stream Morphology." *Water Resources Res.* 7(2):311–322.

PROBLEMS

1. Show how the estimate of the total length of channels in a network (Eq. 12.9) is obtained from Horton's laws. What would the expression be if the length of the highest-order stream L_Ω is used instead of $\overline{L}_1$?

2. The following data is for a tributary of the Walnut Gulch Creek in Arizona. The Creek is classified as fifth order.

ORDER	NUMBER OF STREAMS	AVERAGE STREAM LENGTH (m)
1	107	404
2	25	758
3	7	1134
4	2	807
5	1	5989

a) Using semilog paper, plot number of streams versus order, and average length of streams versus order. What are your estimates of R_B and R_L?

b) Compute the number of streams and mean length per order implied by your Horton numbers estimates.

3. Using the concept of drainage density, develop a relationship between area of a basin and its Strahler order. Can you use this equation to explain why it is not possible to obtain basins of too high an order?

4. Derive an estimate for the total area of a basin of order Ω based on R_L, R_B, $\overline{A}_1$, and Ω.

5. In any area of interest select a river basin with available topographic map at reasonable resolution and little or no development (i.e., no urbanization, large man-made lakes, etc.).

a) Number all streams using Strahler's ordering system.

b) Prepare plots of stream numbers, mean lengths, and mean areas versus order.

c) Estimate K and E_1 of Tokunaga's characterization of the basin.

d) Compare results of Eq. (12.12) against the stream number plot in (b) above.

6. For a basin of choice, as in Problem 5, plot number of links versus area, as in Figure 12.9. Estimate Horton's slope ratio R_S.

7. Assuming that the random topology model governs basin geomorphology, Smart [1972] derived expressions for stream-length ratios

$$\frac{\overline{L}_2}{\overline{L}_1} = \frac{(N_1 - 1)}{(2N_2 - 1)} \frac{\overline{\ell}_i}{\overline{\ell}_e}$$

$$\frac{\overline{L}_\omega}{\overline{L}_{\omega-1}} = \frac{N_{\omega-1} - 1}{(2N_\omega - 1)} \qquad \omega = 3, 4, \ldots, \Omega,$$

where N_ω is the number of streams of order ω and $\overline{\ell}_i$ and $\overline{\ell}_e$ are the mean length of interior and exterior links, respectively. Generally, $\overline{\ell}_e/\overline{\ell}_i > 1$, but a variety of values have been reported. Shreve [1969] summarizes results of other investigators giving a value of 2. The Souhegan and Squannacook River basins have link ratios of 1.09 and 1.26, respectively.

Compare the mean stream-length ratios given by the above equations to the data given below.

SQUANNACOOK

Order ω	Number of Streams N_ω	Number of Links	Mean Stream Length $\overline{L}_\omega$	Mean Link Length $\overline{\ell}$ (km)	$\frac{\overline{L}_\omega}{\overline{L}_{\omega-1}}$	$\frac{N_{\omega-1}}{N_\omega}$	Mean Slope
1	133	133	1.012	1.012			0.3461
2	32	66	1.512	0.733	1.494	4.156	0.1979
3	7	42	5.472	0.912	3.619	4.571	0.1323
4	2	4	2.046	1.023	0.374	3.500	0.0665
5	1	20	15.113	0.756	7.387	2.000	0.0132

SOUHEGAN

Order ω	Number of Streams N_ω	Number of Links	Mean Stream Length $\overline{L}_\omega$	Mean Link Length $\overline{\ell}$ (km)	$\frac{\overline{L}_\omega}{\overline{L}_{\omega-1}}$	$\frac{N_{\omega-1}}{N_\omega}$	Mean Slope
1	177	177	1.352	1.352			0.3552
2	40	80	2.362	1.181	1.747	4.425	0.1667
3	12	48	4.845	1.211	2.051	3.333	0.1026
4	4	23	7.683	1.336	1.586	3.000	0.0922
5	1	25	35.645	1.426	4.639	4.000	0.0247

8. It is recommended by some to use the geometric mean, rather than arithmetic means, to estimate Horton's numbers. For example, the arithmetic mean length ratio is

$$R_L = \frac{1}{\Omega - 1} \sum_{\omega=2}^{\Omega} \frac{\overline{L}_\omega}{\overline{L}_{\omega-1}}.$$

The geometric mean is

$$R_L = \left(\prod_{\omega=2}^{\Omega} \frac{\overline{L}_\omega}{\overline{L}_{\omega-1}}\right)^{1/(\Omega-1)} = \left(\frac{\overline{L}_\Omega}{\overline{L}_1}\right)^{1/(\Omega-1)}.$$

Find R_B and R_L using both arithmetic and geometric means for the two river basins of Problem 7. Using semilog paper plot the number of streams versus order and average stream length versus order. Estimate R_B and R_L from the "best fit" lines and draw on these figures the lines representing R_B and R_L estimated by arithmetic and geometric means.

9. The probability of precipitation falling in an area draining into a stream of order ω is given in Eq. (12.35). It can be written as

$$\theta_\omega = \frac{N_\omega A'_\omega}{A_\Omega},$$

where A'_ω is the mean area draining directly into streams of order ω. If we assume that drainage density is constant, find an expression for θ_ω in terms of mean stream lengths. (a) Express the result using Horton's ratios and compute θ_ω, $\omega = 1, \ldots, 5$ for the Souhegan Basin. (b) Do the same using Smart's expressions for mean stream lengths given in Problem 7.

10. The following figure (from van der Tak [1988]) shows two hypothetical basins with identical stream numbers but with diferent distribution of links. The Strahler orders of the links are shown. Compute the exact transition probabilities P_{13} and those given by Eq. (12.37) for both basins.

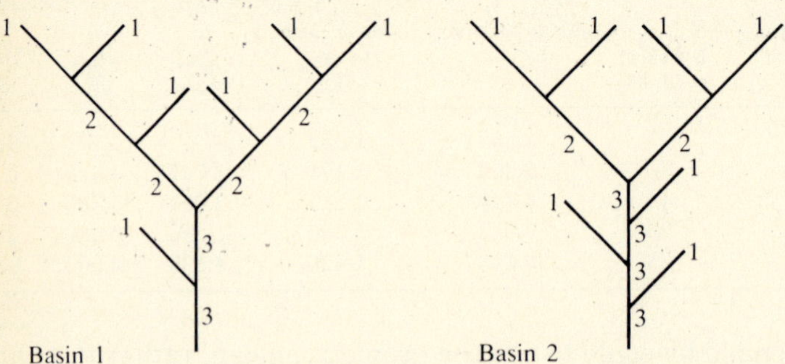

11. Assume that each link in the following schematic basin is of the same length, and that the velocity of travel in each link is 1 m per second. Compute the unit response function implied by the width function and that resulting from the geomorphologic instantaneous unit hydrograph. Assume each link drains a unit area. (Based on Gupta and Waymire [1983].)

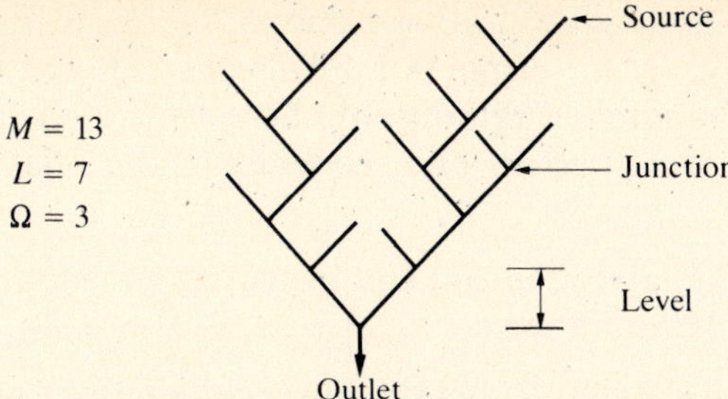

12. For a basin of your choice, as in Problem 5, estimate (a) the width function, (b) the link concentration function, and (c) compare the width function to Eq. (12.57).

13. The concept of fractal dimensions was introduced by Mandelbrot [1983]. One of the properties of a fractal is that measures like length or area are a function of the resolution they are measured with. This is best illustrated by example. A perfectly straight line will have a dimension of 1. No matter how small a ruler is used, the length of the line will always come out the same, assuming accurate experimental techniques. On the other hand imagine measuring a coast. A large "ruler" will miss some inlets and bays, a smaller ruler will capture more and more of these details. The measured length increases with the resolution of the ruler. In that case the length is often proportional to the ruler resolution as

$$L \sim s^{1-D},$$

where D is a fractal dimension greater than or equal to 1, and s is the size of the ruler. Hence, for $D > 1$, L increases as s gets smaller.

The total length of channels in a river basin obeying Horton's laws can be shown to have a fractal dimension (Tarboton et al. [1988]). Let the resolution of choice be the average length of first-order streams

$$s = \left(\frac{1}{R_L}\right)^{\Omega-1} L_\Omega .$$

Note that for fixed area as s gets smaller we see more streams and order Ω increases. The main stream remains the same and hence L_Ω does not change. The implication is that order is a function of map resolution. Now assume that Horton's laws describe the network at all levels of resolution. Show that the fractal dimension of the river network, when s is defined as above, is

$$D = \max\left\{\frac{\log R_B}{\log R_L}, 1\right\}.$$

(*Hint:* Find an estimate for Ω in terms of s and use this in Eq. (12.10) to express total length as a power function of s.)

14. If the fractal dimension of the length of river network, given in Problem 13, is 2 (use values of R_B and R_L for common river basins and you will find that it is near 2), what is the implied relationship between R_B and R_L? Search the literature for values of R_B and R_L and see whether the relationship is reasonable in practice.

15. Derive the results of Table 12.1, giving initial and transition probabilities for a third-order basin following Horton's laws.

16. Horton [1932, 1945] modified Eq. (12.17) for the average length of overland flow to

$$L_c = \frac{1}{2D\left[1 - \left(\frac{S_c}{S_g}\right)^2\right]^{1/2}},$$

where S_c and S_g are the average channel and ground slope in the area, respectively.

Discuss the reasoning behind this modification. Do you expect L_c to change with the age of the basin?

17. Assume that effective rainfall is uniform over an area with constant intensity i_r and duration longer than the time of concentration; peak stream velocity is constant throughout the basin for any given storm; and velocity is given by $V_\omega = K_\omega Q_\omega^m$ as in Eq. (12.27). Find a relationship among K_ωs for different-order streams as a function of R_A and m. (*Hint:* Peak discharge in any stream must be $Q = i_r A$, where A is the contributing area, assuming the storm is sufficiently long.)

18. The Mamon River Basin in Venezuela has a area of 103 km^2 and is of order 6. The highest-order stream has a length of 12.25 km and a kinematic wave parameter α_Ω of 0.642 m$^{-1/3}$s^{-1}. A geomorphologic analysis yields $R_A = 4.5$, $R_B = 3.5$, and $R_L = 2.1$. Estimate the peak discharge and time to the peak resulting from a 3 cm hr^{-1} storm of 2.5-hr duration.

19. Rodriguez-Iturbe et al. [1979] defined hydrologic similarity between basins when two basins have identical peak and time to peak of the geomorphologic instantaneous unit hydrograph (see Eq. 12.43) in which case the product

$$q_p t_p \approx 0.57\left(\frac{R_B}{R_A}\right)^{0.55}.$$

Using geomorphoclimatic instantaneous unit hydrograph results, what is the implied value of R_B/R_A? Compare to values in the literature.

20. In the theory of topologically random networks of Shreve [1967], any topologically distinct basin with a given number of sources (magnitude) is equally likely. Topologically distinct means networks that cannot be made identical by simple stretching of links, rotation, or bending of junctions within a plane. For example there is one topologically distinct network of magnitude 2, two of magnitude 3, and five of magnitude 4. These last are:

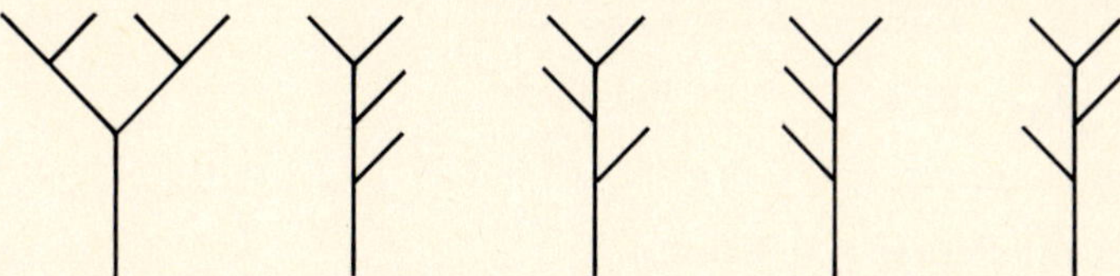

The number of topologically distinct configurations for a network of magnitude M is given by

$$N(M) = \frac{1}{2M - 1}\binom{2M - 1}{M},$$

where

$$\binom{a}{b} = \frac{a!}{(a - b)!\,b!}.$$

Shreve [1966] also gives the number of topologically distinct channel networks corresponding to a given set of Strahler's stream numbers N_1, $N_2, \ldots, N_{\Omega-1}$, 1, as

$$N(N_1 = M, N_2, \ldots, N_{\Omega-1}, 1) = \prod_{\omega=1}^{\Omega-1} 2^{N_\omega - 2N_{\omega+1}} \binom{N_\omega - 2}{N_\omega - 2N_{\omega+1}}.$$

Also available is the number of topologically distinct channel networks of magnitude M for a given-order Ω.

$$N(M, \Omega) = \sum_{i=1}^{M-1} \left[N(i; \Omega - 1)N(M - i; \Omega - 1) + 2N(i; \Omega) \sum_{\omega=1}^{\Omega-1} N(M - i; \omega) \right]$$

$$N(1; 1) = 1$$

$$N(1; \Omega) = 0 \qquad \Omega = 2, 3, \ldots$$

$$N(M; 1) = 0$$

It has been argued that networks in nature have configurations that are most probable. What is the probability of seeing basins with the properties of the Souhegan and Squannacook of Problem 7? Answer in terms of the observed magnitude as well as the observed set of Strahler's streams.

21. Find the geomorphologic instantaneous unit hydrograph of an order 3 basin with the following properties:

$$\Omega = 3; \quad R_B = 4.0; \quad R_A = 5.6; \quad R_L = 2.8;$$

$$\overline{L}_1 = 1.56; \quad \overline{L}_2 = 4.38; \quad \overline{L}_3 = 12.25 \text{ km};$$

$$\overline{A}_1 = 3.3; \quad \overline{A}_2 = 18.4; \quad \overline{A}_3 = 103.0 \text{ km}^2;$$

V_s (velocity in channels) $= 0.6 \text{ m s}^{-1}$;

V_c (velocity in hillslopes) $= 0.005 \text{ m s}^{-1}$.

(*Hints:* Approximate drainage density. The convolution of nonidentical exponential probability density functions can be obtained as

$$f_T^i(t) * \cdots * f_T^\Omega(t) = \sum_{j=1}^{\Omega} \frac{(\lambda_i \cdots \lambda_{\Omega-1}) \exp(-\lambda_j t)}{[(\lambda_i - \lambda_j) \cdots (\lambda_{j-1} - \lambda_j)(\lambda_{j+1} - \lambda_j) \cdots (\lambda_\Omega - \lambda_j)]}.$$

You need a good calculator or a small computer to do this quickly.)

22. (*Note:* This is a fairly difficult and challenging question. The dedicated student may want to study Tokunaga [1978] for the solution.)

Tokunaga gives an expression for the number of interbasin areas in contact with a stream of order λ as

$$I(\lambda, \ell) = 2 + E_1 + \frac{E_1 K(K^{\lambda-\ell-1} - 1)}{K - 1}.$$

Interbasin areas are defined according to the figure below. ℓ is the lowest order stream. Show how this expression is derived. Next express the area draining into the basin of order λ as a sum of areas of inflowing lower-order streams plus interbasin areas to get

$$A_\lambda = 2A_{\lambda-1} + \sum_{\eta=\ell}^{\lambda-k} E_1 K^{\lambda-\eta-1} A_\eta + \left[2 + E_1 + \frac{E_1 K(K^{\lambda-\ell-1} - 1)}{K - 1}\right] \beta_{\lambda,\ell},$$

where $\beta_{\lambda,\ell}$ denotes the average area of interbasin areas in contact with streams of order λ. Now if the resolution with which streams are measured is refined, we get more and more streams. Instead of keeping $\ell = 1$ and renumbering the stream order, we keep the highest order Ω constant and let ℓ decrease. This procedure is taken to the limit $\ell \to -\infty$. Show that under the assumptions: (1) $\beta_{\lambda,\ell} < A_\ell$ and (2) $K/(2 + E_1) < 1$, the series above converges under the limiting process $\ell \to -\infty$ and can be expressed

$$A_\lambda = 2A_{\lambda-1} + \sum_{\eta=-\infty}^{\lambda-1} E_1 K^{\lambda-\eta-1} A_\eta.$$

From this derive the recurrence relation

$$A_\lambda = (2 + K + E_1)A_{\lambda-1} - 2KA_{\lambda-2}.$$

Express this in terms of Q and P (Section 12.2.1) as

$$A_\lambda - PA_{\lambda-1} = Q(A_{\lambda-1} - PA_{\lambda-2})$$

and show that this leads to Eq. (12.13)

$$A_\lambda = QA_{\lambda-1}.$$

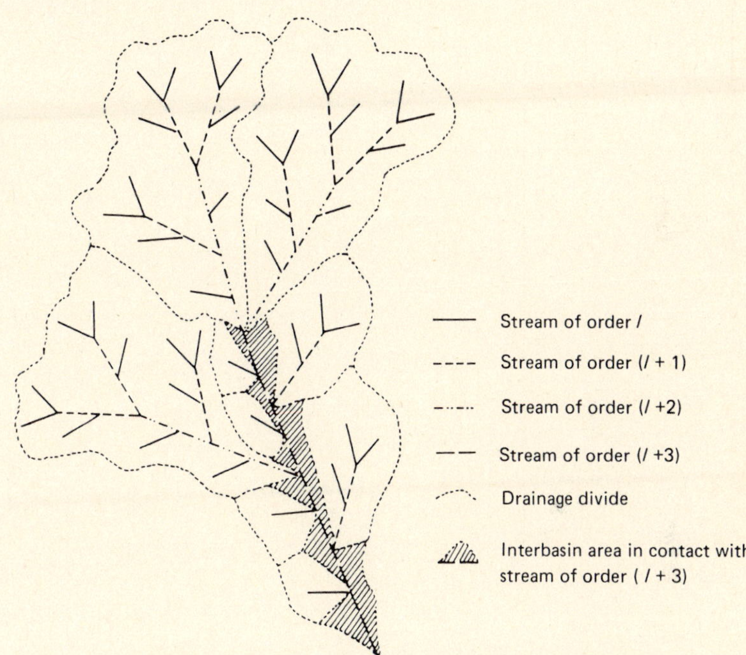

(Courtesy of David Tarboton. Figure courtesy of Eiji Tokunaga.)

Appendix ***A***

Tables of Water Properties

TABLE A.1 Properties of Water in Metric Units

TEMP., °C	SPECIFIC GRAVITY	DENSITY, $g\,cm^{-3}$	HEAT OF VAPORIZATION, $cal\,g^{-1}$	VISCOSITY		VAPOR PRESSURE		
				Dynamic, centipoise‡	Kinematic, centistokes§	mm Hg	Millibars (mb)	$g\,cm^{-2}$
0	0.99987	0.99984	597.3	1.79	1.79	4.58	6.11	6.23
5	0.99999	0.99996†	594.5	1.52	1.52	6.54	8.72	8.89
10	0.99973	0.99970	591.7	1.31	1.31	9.20	12.27	12.51
15	0.99913	0.99910	588.9	1.14	1.14	12.78	17.04	17.38
20	0.99824	0.99821	586.0	1.00	1.00	17.53	23.37	23.83
25	0.99708	0.99705	583.2	0.890	0.893	23.76	31.67	32.30
30	0.99568	0.99565	580.4	0.798	0.801	31.83	42.43	43.27
35	0.99407	0.99404	577.6	0.719	0.723	42.18	56.24	57.34
40	0.99225	0.99222	574.7	0.653	0.658	55.34	73.78	75.23
50	0.98807	0.98804	569.0	0.547	0.554	92.56	123.40	125.83
60	0.98323	0.98320	563.2	0.466	0.474	149.46	199.26	203.19
70	0.97780	0.97777	557.4	0.404	0.413	233.79	311.69	317.84
80	0.97182	0.97179	551.4	0.355	0.365	355.28	473.67	483.01
90	0.96534	0.96531	545.3	0.315	0.326	525.89	701.13	714.95
100	0.95839	0.95836	539.1	0.282	0.294	760.00	1013.25	1033.23

†Maximum density is 0.999973 $g\,cm^{-3}$ at 3.98°C.

‡Centipoise = $(g\,cm^{-1}\,s^{-1}) \times 10^2 = (Pa \cdot s) \times 10^3$.

§Centistokes = $(cm^2\,s^{-1}) \times 10^2 = (m^2\,s^{-1}) \times 10^6$.

(continued)

TABLE A.1 (*Continued*) Properties of Water in English Units

TEMP., °F	SPECIFIC GRAVITY	SPECIFIC WEIGHT, lb ft^{-3}	HEAT OF VAPORIZATION, Btu lb^{-1}	VISCOSITY		VAPOR PRESSURE		
				Dynamic, lb s ft^{-2}	Kinematic, ft^2 s^{-1}	in. Hg	Millibars (mb)	lb in.$^{-2}$
32	0.99986	62.418	1075.5	3.746×10^{-5}	1.931×10^{-5}	0.180	6.11	0.089
40	0.99998	62.426†	1071.0	3.229	1.664	0.248	8.39	0.122
50	0.99971	62.409	1065.3	2.735	1.410	0.362	12.27	0.178
60	0.99902	62.366	1059.7	2.359	1.217	0.522	17.66	0.256
70	0.99798	62.301	1054.0	2.050	1.058	0.739	25.03	0.363
80	0.99662	62.216	1048.4	1.799	0.930	1.032	34.96	0.507
90	0.99497	62.113	1042.7	1.595	0.826	1.422	48.15	0.698
100	0.99306	61.994	1037.1	1.424	0.739	1.933	65.47	0.950
120	0.98856	61.713	1025.6	1.168	0.609	3.448	116.75	1.693
140	0.98321	61.379	1014.0	0.981	0.514	5.884	199.26	2.890
160	0.97714	61.000	1002.2	0.838	0.442	9.656	326.98	4.742
180	0.97041	60.580	990.2	0.726	0.386	15.295	517.95	7.512
200	0.96306	60.121	977.9	0.637	0.341	23.468	794.72	11.526
212	0.95837	59.828	970.3	0.593	0.319	29.921	1013.25	14.696

†Maximum specific weight is 62.427 lb ft^{-3} at 39.2°F.

Source: R. K. Linsley, Jr., M. A. Kohler, and J. L. H. Paulhus, *Hydrology for Engineers*, 3rd ed., McGraw-Hill, 1982, pp. 477–478. Reproduced with permission.

TABLE A.2 Variation of Relative Humidity in Percent with Temperature and Wet-Bulb Depression on the Celsius Scale (Pressure = 990 mb = 29.24 in. = 99.0 kilopascals)

AIR TEMP., °C	WET-BULB DEPRESSION, DEGREES															
	0	1	2	3	4	5	6	7	8	9	10	11	12	13	14	15
−10	91	60	31	2												
−8	93	65	39	13												
−6	94	70	46	23	0											
−4	96	74	53	32	11											
−2	98	78	58	39	21	3										
0	100	81	63	46	29	13										
2	100	84	68	52	37	22	7									
4	100	85	71	57	43	29	16									
6	100	86	73	60	48	35	24	11								
8	100	87	75	63	51	40	29	19	8							
10	100	88	77	66	55	44	34	24	15	6						
12	100	89	78	68	58	48	39	29	21	12	4					
14	100	90	79	70	60	51	42	34	26	18	10	3				
16	100	90	81	71	63	54	46	38	30	23	15	8				
18	100	91	82	73	65	57	49	41	34	27	20	14	7			
20	100	91	83	74	66	59	51	44	37	31	24	18	12	6		
22	100	92	83	76	68	61	54	47	40	34	28	22	17	11	6	
24	100	92	84	77	69	62	56	49	43	37	31	26	20	15	10	5
26	100	92	85	78	71	64	58	51	46	40	34	29	24	19	14	10
28	100	93	85	78	72	65	59	53	48	42	37	32	27	22	18	13
30	100	93	86	79	73	67	61	55	50	44	39	35	30	25	21	17
32	100	93	86	80	74	68	62	57	51	46	41	37	32	28	24	20
34	100	93	87	81	75	69	63	58	53	48	43	39	35	30	26	23
36	100	94	87	81	75	70	64	59	54	50	45	41	37	33	29	25
38	100	94	88	82	76	71	66	61	56	51	47	43	39	35	31	27
40	100	94	88	82	77	72	67	62	57	53	48	44	40	36	33	29

Sources: *Radiosonde Observation Computation Tables,* Department of Commerce–Department of Defense, Washington, D.C., June 1972; R. K. Linsley, Jr., M. A. Kohler, and J. L. H. Paulhus, *Hydrology for Engineers,* 3rd ed., McGraw-Hill, 1982, p. 481.

TABLE A.2 (*Continued*) Variation of Relative Humidity in Percent with Temperature and Wet-Bulb Depression on the Fahrenheit Scale
(Pressure = 30.00 in. = 1015.9 mb = 101.59 kilopascals)

AIR TEMP., °F	WET-BULB DEPRESSION, DEGREES													
	0	1	2	3	4	6	8	10	12	14	16	18	20	25	30
0	84	56	27												
5	86	63	40	16											
10	89	69	50	30	11										
15	91	74	58	42	26										
20	94	79	65	51	37	10									
25	96	84	71	59	47	24	1								
30	99	88	77	66	56	35	15								
35	100	91	81	72	63	45	27	10							
40	100	92	84	76	68	52	37	22	7						
45	100	93	85	78	71	57	44	31	19	6					
50	100	93	87	80	74	61	49	38	27	16	5				
55	100	94	88	82	76	65	54	43	33	24	14	5			
60	100	94	89	83	78	68	58	48	39	30	21	13	5		
65	100	95	90	85	80	70	61	52	44	35	28	20	13		
70	100	95	90	86	81	72	64	55	48	40	33	26	19	3	
75	100	95	91	87	82	74	66	58	51	44	37	31	24	10	
80	100	96	91	87	83	75	68	61	54	47	41	35	29	15	3
85	100	96	92	88	84	77	70	63	56	50	44	38	33	20	8
90	100	96	92	89	85	78	71	65	58	53	47	41	36	24	13
95	100	96	93	89	86	79	72	66	60	55	49	44	39	28	17
100	100	96	93	89	86	80	74	68	62	57	51	46	42	31	21

Sources: U.S. Weather Bureau, Relative Humidity and Dew Point Table, TA 454-0-3E, September 1965; R. K. Linsley, Jr., M. A. Kohler, and J. L. H. Paulhus, *Hydrology for Engineers*, 3rd ed., McGraw-Hill, 1982, p. 480.

TABLE A.3 Variation of Dewpoint with Temperature and Wet-Bulb Depression and of Saturation Vapor Pressure over Water with Temperature on the Celsius Scale (Pressure = 1013.2 mb = 29.92 in. = 101.32 kilopascals)

AIR TEMP., °C	SATURATION VAPOR PRESSURE Milli-bars (mb)	SATURATION VAPOR PRESSURE in. Hg	WET-BULB DEPRESSION, DEGREES 0	1	2	3	4	5	6	7	8	9	10	11	12	13	14	15
−10	2.86	0.085	−11	−16	−24													
−8	3.35	0.099	−9	−13	−20	−33												
−6	3.91	0.115	−7	−11	−16	−24												
−4	4.55	0.134	−5	−8	−12	−19	−32											
−2	5.28	0.156	−2	−5	−9	−14	−22											
0	6.11	0.180	0	−3	−6	−11	−16	−27										
2	7.05	0.208	2	−1	−3	−7	−12	−19	−33									
4	8.13	0.240	4	2	−1	−4	−8	−13	−21	−47								
6	9.35	0.276	6	4	2	−1	−5	−9	−14	−23								
8	10.72	0.317	8	6	4	1	−2	−5	−9	−15	−26							
10	12.27	0.362	10	8	6	4	1	−2	−5	−10	−17	−29						
12	14.02	0.414	12	10	8	6	4	1	−2	−6	−11	−18	−34					
14	15.98	0.472	14	12	11	9	6	4	1	−2	−6	−11	−19					
16	18.17	0.532	16	14	13	11	9	7	4	1	−2	−6	−11					
18	20.63	0.609	18	16	15	13	11	9	7	4	2	−2	−6	−10				
20	23.37	0.690	20	19	17	15	14	12	10	7	5	2	−1	−4				
22	26.43	0.780	22	21	19	17	16	14	12	10	8	5	2	−1	−5			
24	29.83	0.881	24	23	21	20	18	16	15	13	11	8	6	3	−1	−5	−10	
26	33.61	0.992	26	25	23	22	20	19	18	15	13	11	9	6	4	0	−4	−9
28	37.80	1.116	28	27	25	24	22	21	19	18	16	14	12	10	7	4	1	−3
30	42.43	1.253	30	29	27	26	25	23	22	20	18	17	15	13	10	8	5	2
32	47.55	1.404	32	31	29	28	27	25	24	22	21	19	17	15	13	11	9	6
34	53.20	1.571	34	33	32	30	29	28	26	25	23	21	20	17	16	14	12	10
36	59.42	1.755	36	35	34	32	31	30	28	27	25	24	22	21	19	17	15	13
38	66.26	1.957	38	37	36	34	33	32	30	29	28	26	25	23	21	20	18	16
40	73.78	2.179	40	39	38	36	35	34	33	31	30	28	27	25	24	22	20	19

Sources: U.S. National Weather Service, Marine Surface Observations, *Weather Bureau Handbook* 1, 1969; R. K. Linsley, Jr., M. A. Kohler, and J. L. H. Paulhus, *Hydrology for Engineers*, 3rd ed., McGraw-Hill, 1982, p. 483.

TABLE A.3 *(Continued)* Variation of Dewpoint with Temperature and Wet-Bulb Depression and of Saturation Vapor Pressure over Water with Temperature on the Fahrenheit Scale (Pressure = 30.00 in. = 1015.9 mb = 101.59 kilopascals)

AIR TEMP., °F	SATURATION VAPOR PRESSURE Millibars (mb)	in. Hg	WET-BULB DEPRESSION, DEGREES 0	1	2	3	4	6	8	10	12	14	16	18	20	25	30
0	1.52	0.045	−4	−12	−26												
5	1.91	0.056	2	−5	−14	−31											
10	2.40	0.071	7	2	−5	−15	−34										
15	2.99	0.088	13	8	3	−4	−14										
20	3.71	0.110	18	15	10	5	−2	−27									
25	4.58	0.135	24	21	17	13	8	−7									
30	5.63	0.166	30	27	24	20	16	6	−12								
35	6.89	0.203	35	33	30	27	24	16	5	−16							
40	8.39	0.248	40	38	35	33	30	24	16	4	−18						
45	10.17	0.300	45	43	41	39	36	31	24	16	5	−18					
50	12.27	0.362	50	48	46	44	42	37	32	25	17	5	−17				
55	14.75	0.436	55	53	51	50	48	43	39	33	27	18	7	−15			
60	17.66	0.522	60	58	57	55	53	49	45	40	35	29	20	9	−11		
65	21.07	0.622	65	63	62	60	59	55	51	47	42	37	31	23	12		
70	25.03	0.739	70	69	67	66	64	61	57	53	49	45	39	33	26	−14	
75	29.63	0.875	75	74	72	71	69	66	63	59	56	52	47	42	36	14	
80	34.96	1.032	80	79	77	76	74	72	68	65	62	58	54	50	45	29	−9
85	41.10	1.214	85	84	82	81	80	77	74	71	68	64	61	57	53	39	18
90	48.15	1.422	90	89	87	86	85	82	79	76	73	70	67	63	60	48	33
95	56.24	1.661	95	94	93	91	90	87	85	82	79	76	73	70	66	56	44
100	65.47	1.933	100	99	98	96	95	93	90	87	85	82	79	76	73	64	53

Sources: U.S. Weather Bureau, Relative Humidity and Dew Point Table, TA 454-0-3E, September 1965; R. K. Linsley, Jr., M. A. Kohler, and J. L. H. Paulhus, *Hydrology for Engineers,* 3rd ed., McGraw-Hill, 1982, p. 482.

TABLE A.4 Variation of Pressure, Temperature, Density, and Boiling Point with Elevation (U.S. standard atmosphere)

ELEVATION FROM MEAN SEA LEVEL, *m*	PRESSURE			AIR TEMP., °C	AIR DENSITY, kg m^{-3}	BOILING POINT, °C
	mm Hg	Millibars (mb)	cm H_2O			
−500	806.15	1074.78	1096.0	18.2	1.285	101.7
0	760.00	1013.25	1033.2	15.0	1.225	100.0
500	716.02	954.61	973.4	11.8	1.167	98.3
1000	674.13	898.76	916.5	8.5	1.112	96.7
1500	634.25	845.60	862.3	5.3	1.058	95.0
2000	596.31	795.01	810.7	2.0	1.007	93.4
2500	560.23	746.92	761.6	−1.2	0.957	91.7
3000	525.95	701.21	715.0	−4.5	0.909	90.0
3500	493.39	657.80	670.8	−7.7	0.863	88.3
4000	462.49	616.60	628.8	−11.0	0.819	86.7
4500	433.18	577.52	588.9	−14.2	0.777	85.0
5000	405.40	540.48	551.1	−17.5	0.736	83.3

Sources: *U.S. Standard Atmosphere, 1962,* National Aeronautics and Space Administration, U.S. Air Force, and U.S. Weather Bureau; R.K. Linsley, Jr., M.A. Kohler, and J.L.H. Paulhus, *Hydrology for Engineers,* 3rd ed., McGraw-Hill, 1982, p. 480.

TABLE A.5 Ratio of Saturation Vapor Pressure over Ice to That over Water at the Same Temperature

DEG.	0	1	2	3	4	5	6	7	8	9
					CELSIUS					
DEG.	0.000	0.000	0.000	0.000	0.000	0.000	0.000	0.000	0.000	0.000
−0	1.000	0.990	0.981	0.971	0.962	0.953	0.943	0.934	0.925	0.916
−10	0.907	0.899	0.890	0.881	0.873	0.864	0.856	0.847	0.839	0.831
−20	0.823	0.815	0.807	0.799	0.791	0.784	0.776	0.769	0.761	0.754
−30	0.746	0.739	0.732	0.725	0.718	0.711	0.704	0.698	0.691	0.685
−40	0.678	0.672	0.666	0.660	0.654	0.648	0.642	0.636	0.630	0.625
					FAHRENHEIT					
30	0.989	0.994	1.000							
20	0.937	0.942	0.947	0.953	0.958	0.963	0.968	0.973	0.979	0.984
10	0.888	0.893	0.897	0.902	0.907	0.912	0.917	0.922	0.927	0.932
0	0.841	0.845	0.850	0.855	0.859	0.864	0.868	0.873	0.878	0.883
−0	0.841	0.836	0.832	0.827	0.823	0.818	0.814	0.809	0.805	0.801
−10	0.796	0.792	0.788	0.784	0.779	0.775	0.771	0.767	0.763	0.759
−20	0.755	0.750	0.746	0.742	0.739	0.734	0.731	0.727	0.723	0.719
−30	0.715	0.711	0.708	0.704	0.700	0.696	0.693	0.690	0.686	0.682
−40	0.678	0.675	0.672	0.668	0.665	0.661	0.658	0.654	0.651	0.648
−50	0.644	0.641	0.638	0.635	0.631	0.628	0.625	0.622	0.619	0.616

Sources: *Smithsonian Meterological Tables,* 6th ed., Smithsonian Institute, Washington, D.C., 1966, p. 370; R. K. Linsley, Jr., M. A. Kohler, and J. L. H. Paulhus, *Hydrology for Engineers,* 3rd ed., McGraw-Hill, 1982, p. 479.

Appendix **B**

Development of Unsteady Flow Equations for Saturated Media

The development of the unsteady saturated flow equation presented here is mainly due to Jacob [1949].*

The general continuity equation was previously derived (Eq. 7.16) as

$$-\left[\frac{\partial(\rho q_x)}{\partial x} + \frac{\partial(\rho q_y)}{\partial y} + \frac{\partial(\rho q_z)}{\partial z}\right] \Delta x\, \Delta y\, \Delta z = \frac{\partial}{\partial t}(\rho n\, \Delta x\, \Delta y\, \Delta z). \tag{B.1}$$

Only changes in water density ρ, porosity n, and vertical medium dimensions Δz will be allowed. Expanding the right-hand term of Eq. (B.1) leads to

$$\frac{\partial}{\partial t}(\rho n\, \Delta x\, \Delta y\, \Delta z) = \left\{n\frac{\partial \rho}{\partial t}\, \Delta z + \rho\frac{\partial n}{\partial t}\, \Delta z + \rho n\frac{\partial\, \Delta z}{\partial t}\right\} \Delta x\, \Delta y. \tag{B.2}$$

The height of the elemental volume Δz will vary with the vertical component of compressive stress σ_z;

$$\partial(\Delta z) = -\alpha\, \Delta z\, \partial\sigma_z\,, \tag{B.3}$$

where α is the vertical compressibility of the soil, or the inverse of its modulus of elasticity.

*C. E. Jacob, "Flow of Groundwater," in H. Rouse, ed., *Engineering Hydraulics*, Wiley, 1949.

Therefore from Eq. (B.3)

$$\frac{\partial(\Delta z)}{\partial t} = -\alpha\,\Delta z\frac{\partial \sigma_z}{\partial t}. \tag{B.4}$$

The volume of solid material in the elemental volume stays constant (particles do not compress relative to porosity change and water compressibility). Therefore

$$V_s = (1 - n)\,\Delta x\,\Delta y\,\Delta z = \text{constant}, \tag{B.5}$$

where V_s = volume of solids.

From Eq. (B.5),

$$\partial V_s = [(1 - n)\partial(\Delta z) - \Delta z\,\partial n]\,\Delta x\,\Delta y = 0$$

or

$$\frac{\partial n}{1 - n} = \frac{\partial(\Delta z)}{\Delta z}. \tag{B.6}$$

And from Eq. (B.6),

$$\frac{\partial n}{\partial t} = \frac{1 - n}{\Delta z}\frac{\partial(\Delta z)}{\partial t} = -\alpha(1 - n)\frac{\partial \sigma_z}{\partial t}. \tag{B.7}$$

Further assume that overburden and confining pressures are in balance with vertical compressive stress and pore pressures. Therefore

$$P + \sigma_z = \text{constant}$$

or

$$dP = -d\sigma_z\,. \tag{B.8}$$

Finally assume the water density increases proportionally to pressure according to a water compressibility factor β (reciprocal of modulus of elasticity):

$$\frac{\partial \rho}{\partial t} = \beta\rho\frac{\partial P}{\partial t}. \tag{B.9}$$

Using Eq. (B.8) with Eqs. (B.4), (B.7), and (B.9),

$$\frac{\partial \rho}{\partial t} = \beta \rho \frac{\partial P}{\partial t}$$

$$\frac{\partial \Delta z}{\partial t} = \alpha \Delta z \frac{\partial P}{\partial t} \tag{B.10}$$

$$\frac{\partial n}{\partial t} = \alpha(1 - n) \frac{\partial P}{\partial t}.$$

Substituting Eqs. (B.10) into Eq. (B.2),

$$\frac{\partial}{\partial t}(\rho n \Delta x \, \Delta y \, \Delta z) = \left[n \beta \rho \frac{\partial P}{\partial t} + \rho \alpha (1 - n) \frac{\partial P}{\partial t} + \rho n \alpha \frac{\partial P}{\partial t} \right] \Delta x \, \Delta y \, \Delta z$$

$$= (n \beta \rho + \rho \alpha) \frac{\partial P}{\partial t} \Delta x \, \Delta y \, \Delta z \,. \tag{B.11}$$

The continuity equation is then

$$-\left[\frac{\partial(\rho q_x)}{\partial x} + \frac{\partial(\rho q_y)}{\partial y} + \frac{\partial(\rho q_z)}{\partial z} \right] = n \rho \left(\beta + \frac{\alpha}{n} \right) \frac{\partial P}{\partial t}. \tag{B.12}$$

The left-hand side of Eq. (B.12) can be further expanded,

$$-\left[\rho \left(\frac{\partial q_x}{\partial x} + \frac{\partial q_y}{\partial y} + \frac{\partial q_z}{\partial z} \right) + \left(q_x \frac{\partial \rho}{\partial x} + q_y \frac{\partial \rho}{\partial y} + q_z \frac{\partial \rho}{\partial z} \right) \right] = n \rho \left(\beta + \frac{\alpha}{n} \right) \frac{\partial P}{\partial t}. \tag{B.13}$$

The second term in the parentheses in the left-hand side of Eq. (B.13) is of second-order magnitude and can usually be ignored. For example, taking the first element of that term,

$$\frac{\partial \rho}{\partial x} = \beta \rho \frac{\partial P}{\partial x} = \rho^2 \beta g \frac{\partial h}{\partial x}. \tag{B.14}$$

Therefore

$$q_x \frac{\partial \rho}{\partial x} = -K \rho^2 \beta g \left(\frac{\partial h}{\partial x} \right)^2. \tag{B.15}$$

Assuming small gradient, the term in Eq. (B.15) is very small.

Ignoring the second term in the left-hand side of Eq. (B.13), dividing by ρ, substituting the Darcy equation, and assuming isotropy, clearly results in

$$\frac{\partial^2 h}{\partial x^2} + \frac{\partial^2 h}{\partial y^2} + \frac{\partial^2 h}{\partial z^2} = \frac{n}{K}\left(\beta + \frac{\alpha}{n}\right)\frac{\partial P}{\partial t} = \frac{n}{K}\rho g\left(\beta + \frac{\alpha}{n}\right)\frac{\partial h}{\partial t}. \tag{B.16}$$

In an ideal confined aquifer, horizontal, and of uniform thickness b, Eq. (B.16) is expressed as

$$\nabla^2 h = \frac{n\rho g b}{Kb}\left(\beta + \frac{\alpha}{n}\right)\frac{\partial h}{\partial t} = \frac{S}{T}\frac{\partial h}{\partial t}, \tag{B.17}$$

where $T = \mathrm{Kb}$

$= \text{transmissivity}, L^2 t^{-1}$

$S = \text{coefficient of storage, storativity}$

$= n\rho g b\,(\beta + \alpha/n)$

$\beta = 3.3 \times 10^{-6}\ \text{in.}^2\,\text{lb}^{-1}$

$\alpha = 2 \times 10^{-6}\ \text{in.}^2\,\text{lb}^{-1}$

Equation (B.17) is the commonly used expression for unsteady saturated flow in a confined aquifer.

The interpretation of storativity is that volume of water released from a unit area column of aquifer when the head declines by one unit.

Index

A

Adiabatic rising, 93
Advection, 73
  oceans, 80
  winds, 73
Air,
  density, 85
  moist, 84
Albedo, 37
  of snow, 38, 262
Ambient lapse rate, 63
Annual exceedance series, 524–526
Annual maxima series, 524–526
Anomalous propagation, 138
Antecedent precipitation methods to compute infiltration, 365
Antecedent temperature index, 273
Anticyclones, 77
Aquiclude, 287
Aquifer, 285, 286
  phreatic, 286
  tests, 325
Aquifuge, 287
Areal average of precipitation, 161
Artesian aquifers/wells, 287
Atmosphere,
  characteristics, 54
  circulation, 73
  composition, 54
  density, 71
  humidity, 82, 88
  pressure, 68
  temperature, 58
  transport processes in, 56
Atmospheric emissivity, 42
Autumnal equinox, 24

B

Base flow, 395
Basin,
  profiles, 584
  shape, 584
Bayes theorem, 509
Bernoulli trials, 515
Binomial distribution, 515
  example, 519
Black body, 31
  radiation intensity, 32
Black-box models, 465, 491
Boltzmann constant, 32
Boltzmann transformation, 358
Boundary conditions, 300
  Cauchy, 300
  Dirichlet, 300
  Neumann, 300

Bowen ratio, 190, 193, 195
Bright-band, 137, 138
Budget equation, 3
  hydrologic, 3

C

Capillary forces, 350
Capillary potential, 350
Capillary rise, 382
Capillary soil zone, 285
Capillary water, 285
Cauchy boundary conditions, 300
Celerity, 485
Cells,
  of convective, 127
Centigrade, 20
Channel geometry, 587
Channel networks,
  ordering system, 568
  Strahler ordering system, 568
Chezy equation, 401
Cloud physics, 116
Coalescence, 118
Cold content, 259
Cold fronts, 110
Complete duration series, 524
Conceptual instantaneous unit hydrographs, 441
  Diskin's model, 447
  lag and route models, 445
  linear channel, 445
  linear reservoir, 441–444
  Nash model, 446
  time–area curve, 448
Condensation, 95, 114
  latent heat of, 83
Conditional probability mass function, 514
Conduction, 56
Conservation of mass, 2, 3
  in open channels, 467, 478
  in saturated porous media, 294
  in unsaturated porous media, 352
Convection,
  cooling, 114
  forced, 198
  free, 198
  turbulent, 57
Convective cells, 127
Convolution, 419, 431
Cooling mechanisms, 110. *See also* Lifting mechanisms
Cooling ponds, 217
Coriolis force, 74
Critical diameter,
  of a hydrometeor, 123
Critical point, 83
Cumulants, 439
Cumulative density function, 511
Cumulative distribution function, 510
Curve number, 367
Cyclones, 77, 110
Cyclonic storms, 110

D

Dalton analogy to compute evaporation, 197
Darcy's equation, 290
  in three-dimension, 292
Data analysis (of precipitation), 146
  consistency checks, 153
  mean areal precipitation, 154
  missing data, 149
Declination of the sun, 21, 25
Degree days, 67
Density,
  distribution in atmosphere, 71
  of snow and snowpack, 257
Depression storage, 234
Depth–area–duration curves, 157
Derived distributions, 545–547
Dew-point temperature, 83
Diameter,
  of a river basin, 581
  topological, 581
Diffusion analogy to flow in channels, 484, 492
Diffusivity of soil, 353
  effective, 360
  variable, 358, 378
Dirac-delta function, 431
Direct runoff, 401
Dirichlet boundary conditions, 300
Disdrometer, 136
Diskin's model, 447
Distance–drawdown curve, 322

Distribution of hydrometeors, 118
Double-mass analysis, 153
Drainage density, 573
Drop diameter, 136
Dry adiabatic lapse rate, 95
Dry adiabatic rising, 93
Duhamel's integral, 431. *See also* Convolution
Dupuit approximation, 296
DWOPER, 489

E

Easterlies, 76
Eddy loss slope, 490
Emissivity, 31
Energy balance to compute evaporation, 190
  advantages, 192
  disadvantages, 193
Energy budget of snowpack, 264
Equatorial convergence zone, 76
Equinox, 24
Equivalent potential temperature, 98
Error function, 356
Evaporation, 57, 183
  actual, 377
  annual evaporation over U.S., 186
  Central Sierra Snow Laboratory equation, 232
  Dalton analogy, 197
  direct measurement of, 210
  effects of forced convection on, 198
  effects of free convection, 199
  empirical equations to compute, 203
  energy balance method, 190
  equations, 203
  Fetch dependence, 200
  from free water surfaces, 188
  of hydrometeors, 120
  Kuzmin equation (for snow), 232
  from Lake Nasser, 188
  latent heat of, 83
  mass-transfer methods, 197
  mean annual over the world, 184–185
  pans, 210
  Penman equation, 201
  from snow, 231, 266
  Sverdrup equation (for snow), 231
  turbulent transport equation to compute, 195
  water balance method, 189
Evapotranspiration, 4, 219–231
  actual, 220, 225
  Blaney–Criddle equation, 223
  crop resistance factors, 221
  dependence on root development, 225
  potential, 220
  relationship to leaf area index, 225–226
  Thornthwaite equation, 224
  world distribution of, 185
Event, 506
  independent, 508
  mutually exclusive, 506
Exfiltration, 4, 355, 359, 360
Expectation, 512
Extreme-value distributions, 530
  example, 531
  frequency factor, 531
  tables of, 532

F

Fahrenheit, 20
Fetch dependence of evaporation, 200
Fick's law, 56
Field capacity, 285
Flood frequency analysis, 524, 529, 531
  with derived distributions, 545
  historical information in, 551
  with log-Pearson Type III distribution, 533–542
  nonparametric, 543–544
  paleohydrology in, 550
  regional analysis in, 548
Flood index method, 548
  for flood frequency analysis, 548
Flow in porous media,
  saturated, 283
Flow into a river bank, 310–313
Fluvial geomorphology, 567
  channel geometry, 587
  descriptors of relief, 582
  diameter, 581

drainage density, 573
geometry number, 584
geomorphoclimatic instantaneous unit hydrograph, 597
geomorphologic instantaneous unit hydrograph, 590
hillslope response, 600
Horton's laws, 568
Horton's numbers, 568
hypsometric curve, 584
issues of scale, 573, 574, 613
length–area relationships, 575
link concentration function, 586
magnitude, 581
planar descriptors, 568
random topology model, 571, 590, 611, 614
relationship to hydrology, 579, 589
ruggedness number, 584
stream frequency, 575
Tokunaga's characterization, 572
width function, 579, 602
Forced convection, effects on evaporation, 198
Fossil waters, 287
Fourier series for identifying IUH, 432
Fourier transforms for identifying IUH, 436
Fractal dimension, 613
Frequency,
for extreme-value Type I, 531
factor, 527
for Gumbel distribution, 531
for normal distribution, 527
streams, 575
Friction slope, 480
Friction velocity, 195
Frontal storms, 110

G

Gages, precipitation, 132
accuracy, 133
recording gages, 133
tipping bucket, 133
weighing type, 133
Gamma radiation devices, 133
Gas constant,
dry air, 84
universal, 84
water vapor, 121
Gaussian distribution, 521–527
example, 524
table of, 523
Geometric distribution, 517
Geomorphoclimatic instantaneous unit hydrograph, 597
Geomorphologic instantaneous unit hydrograph, 590
Geomorphology, 567. *See also* Fluvial geomorphology
Geostationary operational environmental satellites, 68, 140
Geostationary satellites, 68, 140
Geostrophic winds, 78
Generalized skew, 536
Global water balance, 3
GOES, 68, 140
Gravitational water, 285
Gray body, 33
Gray's synthetic unit hydrograph, 425
Green–Ampt infiltration model, 365
Greenhouse effect, 47
Ground clutter, 137
Groundwater, 283
Growth of hydrometeors, 118
Gumbel distribution, 530
example, 531
frequency factor, 531
tables of, 532

H

Hagen–Poiseville equation, 291
Heat conduction through snow, 267
Heat content of precipitation,
change of, 273
of precipitation, 265
Heat transport, 56
Hillslopes,
response of hillslopes in GIUH, 600
slopes, 582
Historical information,
in flood frequency analysis, 550–553

Homogeneity of hydraulic
conductivity, 293
Horton's equation for infiltration, 357, 362
Horton's numbers or laws, 568
area ratio, 571, 572, 616
bifurcation ratio, 569
law of stream areas, 571, 572, 616
law of stream lengths, 571
law of stream numbers, 569
law of stream slopes, 582
length ratio, 571
Hour angle of the sun, 25
Huggins–Monke infiltration equation, 364
Humidity, 82
absolute, 85
distribution, 88
estimation of, 86
measurement, 86
mixing ratio, 85
relative, 85
specific, 85
Hydraulic conductivity, 290, 352
homogeneity, 293
intrinsic, 292
Hydraulic properties of streams, 587
Hydraulic radius, 401
Hydraulics of wells, 313
in confined aquifers, steady state, 313
partially penetrating wells, 326
in unconfined aquifer, steady state, 316
unsteady flow in wells, 317, 328. *See also* Theis equation
Hydrograph, 395
design of a storm sewer system, 406
effect of basin shape on, 396
instantaneous unit, 430
mathematical formulation, 417
peak discharge, 405
rational formula, 405
receding limb, 396
recession, 396
response to rainfall, 395
S curve, 413
separation, 399
shape of, 404
synthetic, 419
unit, 409
Hydrologic budget equation, 5
groundwater, 6
surface, 5
Hydrologic cycle, 3
components of, 4
global, 4
Hydrology,
definition, 1
history, 1
Hydrometeors,
distribution of, 118
measurements of, 136
Hyetograph, 130
Hygrometer, 87
Hygroscopic water, 285
Hypsometric curve, 584
Hysteresis, 351

I

Ideal gas law, 70
Image wells, 323. *See also* Method of images
Independence,
probabilistic, 508
statistical, 508
Index flood method, 548
for flood frequency analysis, 548
Infiltration, 4, 355–368
antecedent precipitation methods, 365
constant infiltration indexes, 368
empirical equations, 362
examples, 378
Green–Ampt model, 365
Horton equation, 357, 362
Huggins–Monke, 364
indices, 368
Philip's equation, 361, 379
Soil Conservation Service, 366
Insolation, 23
Instantaneous unit hydrograph, 430
conceptual, 441
cumulatants to identify the, 439
Fourier series for identification, 432
Fourier transforms for identification, 436

identification of, 432
Laplace transforms for identification, 436
moments to identify the, 437
Intensity–frequency–duration curves, 157–158, 557
Intertropical convergence zone, 76
Interception, 232
Interflow, 4, 395
Intermediate soil water zone, 285
Internal energy, 93
Intersection, 506
probability of, 507
Intrinsic hydraulic conductivity, 292
Inversion, thermal, 63
upper air, 66
Island,
groundwater flow in, 300
Isohyetal method, 157
Isotropy of hydraulic conductivity, 293

J

Jet streams, 78
Joint cumulative distribution, 515
Joint probability density function, 515
Joint probability mass function, 513
Joule, 20

K

Kelvin, degrees, 20, 67
Kinematic wave equations, 485, 486
celerity, 485
numerical solutions of, 486
parameters, 488
Kirchhoff's law, 31
Kriging, 153
Kurtosis, 513

L

Lag and route models, 445
moments, 446
Lag time, 423
Langley, 21
Laplace equation, 296
Laplace transforms for identifying IUH, 436
Lapse rate,
ambient, 63
dry adiabatic, 95
Latent heat, 83
Leaf area index, 225
Leaky aquifers, 299, 306
wells in, 323
Length of channels in a basin, 571
Lifting mechanisms, 110
convective, 114
frontal, 110
nonfrontal, 110
orographic, 113
Linear channel, 445, 466
Linear reservoir, 441–444
IUH, 443
method of moments, 444
Linearity, 300, 409
Linearization of unconfined aquifer equations, 303
Link concentration function, 586
Links, 568
Liquid-water content, 261, 268
Log-normal distribution, 527–530
example, 529
Log-Pearson Type III distribution, 533–542
confidence limits, 539, 542
frequency factor, 539
moments, 534
skewness computation, 535–538
tables of, 540–541
Long-term mean areal precipitation, 154
Longwave radiation, 42
albedo, 44
atmospheric emissivity, 43
clear sky, 42
cloud effects, 44
forest cover effects, 45
Lysimeters, 220

M

Magnitude, 581
Manning equation, 401, 488

Marginal probability density function, 515
Mass balance, 5
Mass balance in saturated porous media, 294
  confined aquifers, 294
  cylindrical coordinates, 299
  unconfined aquifers, 296
Mass transport, 56
Matching method of solution for transmissivity and storativity, 318
Mean, 512, 535
Mean areal precipitation, 151
Melt factor, 270
Meltwater, 259
  routing, 275
Mesoscale, 127
Metamorphosis of snow, 256
Meteorology, 4, 53
Method of images, 323, 333
Method of moments, 437–439, 444
Microscale, 127
Mixed distributions, 511
Mixed ratio, 85
Models of probability,
  Bernoulli trials, 515
  binomial distribution, 516
  extreme-value distributions, 530–533
  Gaussian distribution, 521–527
  geometric distribution, 517
  Gumbel distribution, 530–531
  log-normal distribution, 527–530
  log-Pearson Type III distribution, 533–542
  normal distribution, 521–527
Modified Theis equation, 320, 329
Moist air density, 85
Moisture profiles,
  drying, 286
  wetting, 286
Molecular diffusivity, 56
Moments, 437, 512
  expectation, 512
  Kurtosis, 513
  mean, 512, 535
  skewness, 513, 535
    generalized, 536–538
    mean square error, 538
  standard deviation, 512, 535
  variance, 512, 535
Momentum transport, 56, 57
Monadnock Basin profile, 586
Monsoon, 77
Muskingum method, 466
  Example 10.1, 470
  parameters of, 470

N

Nash model, 446
  moments, 447
Network design (for monitoring precipitation), 158
Neumann boundary conditions, 300
Newton, 20
Nile River, 7, 189
Nonfrontal storms, 110
Nonparametric estimates of probability, 543
  confidence limits, 544
  risk, 544–545
Normal distribution, 521–527
  example, 524
  table of, 523
Normal ratio method, 149
Nucleation, 116
Number of streams in a basin, 569
Numerical solutions to flow in channels, 486
  kinematic wave equations, 486
  full equations, 489

O

Ocean currents, 80
Oceanography, 4, 53
Orographic lifting, 113
Overland flow, 235

P

Paleohydrology, 550–552
  in flood frequency analysis, 550–553
Pans, evaporation, 210
Partial area runoff, 374

Partial duration series, 524–526
Peak discharge, 421, 425
  formula, 405
Pearson Type III distribution, 533–542
  confidence limits, 539, 542
  frequency factor, 539
  moments, 534
  skewness computation, 535–538
  tables of, 540–541
Penman equation, 201
Percolation, 4, 382
Permanent wilting point, 285
Permeability, 290. *See also* Hydraulic conductivity
Philip equation for infiltration, 361, 379
Photosynthesis, 219
Planck's constant, 32
Planck's law, 32
Planetary motions, 21
Plan position indicator, 135
Plotting positions, 543
  confidence limits, 544
Poisson equation, 298
Ponding time, 372, 380
Pore disconnectedness, 351
Pore-size distribution, 351
Porosity, 290
  of snow, 261
Porous media,
  unsaturated, 349
Potential, 299
  equivalent, 98
  temperature, 94
Precipitable water, 90, 98
Precipitation, 109
  data analysis, 146
  forms, 126
  measurements of, 132
  requirements for, 109
  variability of, 148
Pressure,
  distribution, 68
  diurnal variations, 70
  global variations, 69–70
  versus elevation relationship, 270
  vertical distribution, 70
Probability, 505
  axioms of, 506
  conditional, 507
  density function, 510, 511, 515
  mass function, 510
  models of, 515
  nonparametric estimates, 543
  review, 506
  total probability theorem, 508
Pseudo-adiabatic, 95
Psychrometer, 87
Pumping at variable rates, 324
Pumping tests, 325

R

Radar, 134
  errors of, 138
Radiation, 19
  black-body, 31
  black-body radiation intensity, 32
  global distribution, 21
  longwave, 42
  physics, 31
  shortwave, 34
  shortwave clear sky, 34
Radiometers, 190
Rainfall–discharge, 395
Rainfall–runoff, 395, 404
Rainfall simulation, 130
Random topology model, 590, 614
Random variables, 509
  conditional, 507, 513, 514
  continuous, 510, 521
  discrete, 510, 515
Range height indicator, 135
Rankine, 67
Rating curves, 401, 482
Rational formula, 405
Receding limb of hydrograph, 396
Recession limb of hydrograph, 396
Recovery analysis, 332
Recurrence, 518
Recurrence interval, 158
Reflectivity factor, 136
Regional analysis, 548–550
  in flood frequency analysis, 548–550
Relative humidity, 85, 86
Reservoir routing, 475
  Example 10.2, 476

Respiration, 219
Reynolds number, 122, 291
Ripe (in reference in snowpack), 257
Risk, 519
  nonparametric, 544
River valley,
  groundwater flow in, 308
Rock–soil systems,
  water-bearing potential, 287
Roots,
  development of, 225
Roughness coefficient, 401
Routing, 465
  channels, 466
  conceptual models, 465, 466
  dynamic models, 465, 478
  hydraulic, 478
  models, 465
  physical models, 465
  regression models, 465
  reservoirs, 475
  St. Venant equations, 478
Routing meltwater, 275–277
  attenuation, 277
  time lag, 276
Ruggedness number, 584
Runoff, 366–377, 380, 382
  direct, 401
  mechanisms, 372–377
  partial area runoff, 374
  subsurface runoff, 376
  variable source runoff, 375

S

St. Venant equations, 478
  diffusion analogy, 484
  kinematic wave approximation, 486
  linearized solutions of, 482
  numerical solutions of, 486
  solutions of, 482
Sample space, 506
Satellites, 140
  precipitation estimates from, 140–146
Saturated porous media,
  mass balance equation, 294–299
  soil zone, 284
Saturation vapor pressure, 86, 270
Scope of work, 12
S curve, 413
Seepage velocity, 290
Sensible-heat transfer through snow, 267
Separation of hydrograph components, 399
Shortwave radiation, 34
  albedo, 37
  clear sky, 35
  cloudy skies, 35
  extinction by water and snow, 39
  interception by vegetation, 36
S hydrograph, 413
Skewness, 513, 535
Snow,
  accumulation, 248
  density, 248
  development, 248
  evaporation from, 231, 266
  interception of, 249
  measurement of, 254
  retention of snow by surfaces, 249
  spatial distribution, 249, 252
  transport by wind, 250
  water equivalence, 249, 255
Snow hydrology, 247
Snowmelt, 247, 264
  air temperature as an index of, 270
  during no-rain periods, 268
  during rain periods, 268
Snowpack, 256
  albedo, 262
  change in heat content during periods of no rain, 273
  cold content, 259
  density, 257
  depth, 257
  liquid-water content, 261, 268
  porosity, 261
  routing meltwater through, 275
  thermal quality, 260
Snyder's synthetic unit hydrograph, 419
Soil Conservation Service,
  model of runoff, 366–368, 380
  unit hydrograph, 423
Soil profile, 284
Soil–rock systems,
  water-bearing potential, 287

Soil water zone, 284
Solar altitude, 23
Solar constant, 21
Solstice, 24
Souhegan River Basin, 149
Specific heat,
  constant pressure, 94
  constant volume, 93
Specific humidity, 85
Specific storativity, 295
Stability,
  atmospheric, 63, 100
Stage, 401
Stage–discharge relationships, 401
Standard deviation, 512, 535
Standard normal deviate, 524
Stefan–Boltzmann law, 33
Stomata, 219
Storage in the basin, 396
Storativity,
  specific, 295
Storm runoff, 368
Storm sewers, 406
  design of, 406
Storms,
  exterior, 128–129
  interior, 128–129
  structure, 127
Strahler's ordering system, 568
Stratosphere, 54
Stream,
  frequency, 575
  number of, 569
  order, 568
Streamflow measurements, 401
Structure of storms, 127
Sublimation, 4, 117
Subsurface runoff, 376
Subsurface storm flow, 376, 395
Suction, 350
Summer solstice, 24
Sun, temperature, 19
Superposition, 300, 305, 323, 409, 431
Superposition principle, 300, 305
  with wells, 323
Surface–groundwater interaction, 310
Synoptic scale, 128
Synthetic unit hydrographs, 419
  Gray's model, 425
  Snyder's model, 419
  Soil Conservation Service model, 423

T

Temperature,
  anomalies, 63
  distribution, 58
  horizontal distribution, 60
  measurements, 66
  statistics, 67
  temporal distribution, 58
    daily, 58
    seasonal, 60
  vertical distribution, 63
  virtual, 85
Tension pressure, 350
Terminal velocity of hydrometeors, 120
Theis equation, 317, 329
  modified, 320
Thermal conductivity, 56
Thermal inversion, 63
Thermal quality, 260
Thiessen weights, 154
Throughfall, 4
Thunderstorms, 114
Time–area curve, 448
Time base, 422
Time to peak, 421
Time–drawdown curve, 321
Tokunaga's characterization of channel networks, 572, 616
Total probability theorem, 508
Trade winds, 76
Transmissivity, 298
Transpiration, 219
  plant control of, 219
  stress, 228
Transport processes, 56
Triple point, 83
Tropic of Cancer, 24
Tropic of Capricorn, 24
Troposphere, 54
Turbulent eddy diffusivities, 57, 195
Turbulent heat transport, 57, 195
Turbulent momentum transfer, 57, 195

Turbulent transport equations, 57
Turbulent transport evaporation equation, 57, 195

U

Uncertainty, 505
  sources of, 505
Unconfined aquifers, 286, 296
  Dupuit approximation, 296
  linearization of equations, 303
Union, 506
  probability of, 507
Unit hydrograph, 2, 409
  estimation of, 418
  mathematical formulation, 417
  matrix formulation, 417
  S hydrograph, 413
  synthetic, 419
Units,
  energy, 20
  power, 20
  temperature, 20
Unsaturated porous media, 349
  conservation of mass in, 352
  Darcy's law analogy, 350
  flow in, 349
  hydraulic conductivity in, 350
  intrinsic hydraulic conductivity in, 351–352
  matrix potential in, 350
Unsteady flow in channels, 478–494
Urban hydrology, 406

V

Vadose zone, 349
Vapor,
  density, 85
  pressure, 84
  pressure gradient adjustments, 205
  saturation, 86
Variable source runoff, 375
Variance, 512, 535
Ventilation effects, 122
Vernal equinox, 24
Virtual temperature, 85, 199
Viscosity, 56
  dynamic, 56
  kinematic, 58
Volume of water above a threshold discharge, 546
Von Karman constant, 196

W

Walnut Gulch Basin, 149
Warm fronts, 110
Water,
  phases, 82
Water balance, 5
  of Earth, 4
  of Nile River, 7
  methods for evaporation, 189
Water budget equation, 5
Water equivalent, 119
Water Resources Council, 533
Water vapor. *See also* Vapor
  gas constant, 121
Watts, 20
Well fields, 323
Well function, 317
Well tests, 325
Wells, 313. *See also* Hydraulics of wells
Westerlies, 76
Wet-bulb thermometer, 87
Wetted perimeter, 401
Width function, 579
  relationship to IUH, 602
Wind speed elevation adjustments, 205
Wind shields, 133
Winds, 73
  geostrophic, 78
Winter solstice, 24

Z

Z–*R* relationship, 136